IN[illegible]

CHEMICAL ENGINEERING

GENERAL EDITOR: P. V. DANCKWERTS

VOLUME 8

UNIVERSITY OF GLAMORGAN
LEARNING RESOURCES CENTRE

Pontypridd, Mid Glamorgan, CF37 1DL
Telephone: Pontypridd (0443) 480480

Books are to be returned on or before the last date below

12. MAR
12. MAR
23
- 5 JUN 1999
- 5 MAY 2000
30.
02. JUN 93
23. JUN 1995
- 5 MAY 2000
12 FEB 1996
25. NOV
11 JAN 2002
10 JAN 1997
22. APR 94
- 4 FEB 1997
22 MAR 2004
25. OCT
13 MAR 1997
- 1 OCT 2004
22. NOV
13. JAN 95
- 4 JUN 1997
- 8 1995
26 SEP 1997
- 2 DEC 1998
15. JAN 93
12 FEB 93

OTHER TITLES IN THE SERIES IN
CHEMICAL ENGINEERING

Vol. 1. WILKINSON: Non-Newtonian Fluids
Vol. 2. JAMRACK: Rare Metal Extraction by Chemical Engineering
Vol. 3. NAGIEV: The Theory of Recycle Processes in Chemical Engineering Techniques
Vol. 4. BRADLEY: The Hydrocyclone
Vol. 5. ŠTERBÁČEK and TAUSK: Mixing in the Chemical Industry
Vol. 6. HOBLER: Mass Transfer and Absorbers
Vol. 7. ROSENBROCK and STOREY: Computational Techniques for Chemical Engineers
Vol. 9. KING: Phase Equilibrium in Mixtures
Vol. 10. BROWN and RICHARDS: Principles of Powder Mechanics
Vol. 11. BRETSZNAJDER: Prediction of Transport and other Physical Properties of Fluids
Vol. 12. GRASSMANN: Physical Principles of Chemical Engineering
Vol. 13. KEEY: Drying

THE POLYTECHNIC OF WALES
LIBRARY
TREFOREST

INDUSTRIAL GAS CLEANING

SECOND EDITION

The principles and practice of the control of gaseous and particulate emissions

W. STRAUSS
Reader in Industrial Science, University of Melbourne

PERGAMON PRESS
OXFORD · NEW YORK · TORONTO
SYDNEY · PARIS · BRAUNSCHWEIG

628.512
STR
628.53
STR

U. K.	Pergamon Press Ltd., Headington Hill Hall, Oxford OX3 0BW, England
U. S. A.	Pergamon Press Inc., Maxwell House, Fairview Park, Elmsford, New York 10523, U.S.A.
CANADA	Pergamon of Canada, Ltd., 207 Queen's Quay West, Toronto 1, Canada
AUSTRALIA	Pergamon Press (Aust.) Pty. Ltd., 19a Boundary Street, Rushcutters Bay, N.S.W. 2011, Australia
FRANCE	Pergamon Press SARL, 24 rue des Ecoles, 75240 Paris, Cedex 05, France
WEST GERMANY	Pergamon Press GmbH, 3300 Braunschweig, Postfach 2923, Burgplatz 1, West Germany

Copyright © 1975 W. Strauss

All Rights Reserved. No part of this publication may be reproduced, stored in a retrieval system or transmitted in any form or by any means: electronic, electrostatic, magnetic tape, mechanical, photocopying, recording or otherwise, without permission in writing from the publishers

First edition 1966
Second edition 1975

Library of Congress Cataloging in Publication Data

Strauss, Werner.
Industrial gas cleaning.

(International series of monographs in chemical engineering, v. 8)

Bibliography: p.

1. Gases—Cleaning. 2. Fume control. I. Title.

TP242. S75 1975 628.5'3 74-8066

ISBN 0-08-017004-8

ISBN 0-08-019933 x (f)

88530

Printed in Hungary

(3/8/94)

CONTENTS

PREFACE TO THE SECOND EDITION

As readers of the first edition are aware, this book attempted to review critically the science, technology and "state of the art" of cleaning industrial process and waste gases, with particular emphasis on the control of atmospheric pollution. During the six-year period since the completion of the manuscript of the first edition, there have been some major developments in the technology of gas cleaning, with lesser, but still significant, theoretical developments.

These technological developments include processes for removing sulphur dioxide from flue gases, improvement in the design of direct-flame incinerators, filter materials suitable for temperatures to 400°C and a reverse-flow cyclone, to name only a few. These developments have made the revision of the book an urgent one. The new edition also gives the writer an opportunity to correct a number of errors found in the first edition. He wishes to express his thanks to all those who have written to him pointing these out.

The first edition has been found suitable by many as a textbook for graduate courses and special courses in air-pollution-control technology. In view of this, a brief section has been appended giving some examination questions which were used by the writer. If teachers and practitioners in the field develop new questions, particularly numerical ones, the author would be pleased to receive these to incorporate in future editions.

A criticism levelled at the first edition was that no cost information was included. At that stage this was deliberate as, with the exception of the work of Stairmand,[804] very little had been done in this area. More recently, however, some attempt has been made to assess costs in the United States, England and Australia, and this will be published elsewhere by Mr. J. R. Alonso.[15] A new chapter, Chapter 12, has been added which will help in assessing cost of plants, although it is not intended as a guide to detailed costing.

The author wishes to acknowledge the help by discussions and friendly criticism he has received in the preparation of this new edition from his colleagues and research students at the University of Melbourne and at the Westinghouse Research Laboratories, in particular Prof. S. R. Siemon and Dr. E. V. Somers. He also wishes to acknowledge gratefully the help he has received in the careful revision of the manuscript from Miss V. Carter and Mrs. D. Muir.

University of Melbourne WERNER STRAUSS

PREFACE TO THE FIRST EDITION

THE control of air pollution at its major and most varied source, the industrial process, is a task confronting an increasing number of engineers and applied scientists. When starting his postgraduate studies some years ago the author was set the problem of finding a new method of reducing the fumes from steel-making processes, with particular emphasis on the fine orange-brown fumes from the then recently introduced oxygen lancing in the open-hearth and the pneumatic steel-making processes. The new process was to be cheaper than the conventional methods used—electrostatic precipitation or venturi scrubbing—although the efficiency requirements were not quite as rigorous. Subsequently a pebble-bed type filter was developed and found to be over 90 per cent efficient in some circumstances.[832]

At the beginning of the programme of research no details of fume emission at the various stages of open-hearth steel making could be found in the published literature, and even if the fume characteristics had been defined, the possible mechanisms of collection were not set down in a textbook or monograph, together with the relevant equations which could be used to test the applicability of a particular mechanism. The author found further that gas cleaning was treated by its practitioners largely as an art based on practical experience and rule of thumb. Although the basic theory of gas-cleaning mechanisms was published in scattered papers, no integrated account of these was available to engineers who wished to use them in "scale-up" calculations or in predicting the effect of a change in one of the process variables, such as gas velocity through a cleaning system. Even now only two aspects of gas cleaning, absorption and electrostatic precipitation, have received extensive treatment, while the gas-cleaning methods involving particle mechanics generally have not been comprehensively reviewed. The author hopes that this book will go some way towards filling this gap, and that it will prove useful to the engineer designing or specifying gas-cleaning plant as well as to the applied scientist developing new methods of gas cleaning.

The author wishes to acknowledge the assitance and encouragement he received from Prof. M. W. Thring, who first directed his interests towards this field, and from his colleagues, in particular, Mr. R. S. Yost, who is responsible for Appendix I, Mr. C. H. Johnson, who read and made many helpful suggestions with respect to Chapters 4 and 7, and Mr. J. B. Agnew, who similarly assisted with parts of Chapter 3. The author is particularly indebted to the Engineering Librarian, Mr. J. Greig, and his staff, for their help with references and diagrams, and Mrs. F. M. Beissel, for her careful typing of the manuscript.

University of Melbourne WERNER STRAUSS

LIST OF SYMBOLS

Latin Letters

a	constant
	interfacial or surface area per unit volume
	height of cyclone entrance line
	distance from node
	coefficient of diffuse reflection
A	surface area
	aggregate surface area of particles in unit volume
A_p	external surface area of a porous solid
$\mathcal{A}$	area of plates in electrostatic precipitator
	amplitude of sound vibrations
b	constant
	time fraction wind is in a 45° sector
	entrance width of cyclone
	(ab = cross-sectional area of cyclone entry)
B	breadth of settling chamber
	diameter of opening at cone apex in cyclone
c	constant
	concentration (mass or volume) of particles in gas
c_0	number of particles at zero time
	centre-line particle concentration
C	Cunningham correction factor
C_D	drag coefficient
C_{DA}	drag coefficient for accelerating particles
C_0	function of applied voltage $\mathcal{V}$ and electrode geometry ($\mathcal{V}/\ln R_2/R_1$)
C_p	specific heat, constant pressure
C_v	specific heat, constant volume
C_y	generalized eddy diffusion coefficient—cross wind (Sutton[843])
C_z	generalized eddy diffusion coefficient—vertical (Sutton[843])
$\mathcal{C}$	constant
d	diameter of spherical particle
d'	diameter of sphere of influence
d_A	area diameter (diameter of circle with same projected area as that of particle)

d_c	diameter of cloud
d_e	drag diameter
d_s	surface diameter (diameter of sphere with same surface area as particle)
d_v	volume diameter (diameter of sphere with same volume as particle)
d_{crit}	diameter of smallest particle 100 per cent collected
d_{50}	diameter of particle 50 per cent collected
D	diameter of cyclone
	diameter of collecting body (fibre, droplet, rod or sphere)
	diameter of sampling probe
D_C	diameter of core
D_e	diameter of exit pipe in cyclone
$\mathcal{D}$	diffusivity (Brownian)
e	electronic charge
	charge on an ion
e	turbine efficiency
E	strength of heat source relative to surrounding atmosphere ($J\,s^{-1}$; W)
	field strength of electric field ($V\,m^{-1}$)
E'	strength of charging field
$\bar{E}$	energy intensity (sonic) ($J\,m^{-3}$)
E_c	critical field strength for electrical breakdown of gases
f	free falling speed of particles
	Fanning friction factor
F	fluid resistance force on particle
F_h	hydrodynamic attractive force in sonic field
F_r	radiation pressure in a sonic field (N)
F_t	thermal force
F_c	fluid resistance to clouds
F_E	electrostatic force
F_{EI}	electrostatic image force (image of collector induced on particle)
F_{EM}	electrostatic image force (image of particle induced on collector)
F_{EC}	coulombic force
F_{ES}	space charge force
F_W	fluid resistance corrected for wall effects
g	gravity acceleration
g_c	gravity-acceleration constant
G	potential temperature gradient in the atmosphere ($°C\,m^{-1}$)
	force on a particle
	friction constant (Stairmand) for cyclones
$\mathcal{G}$	gravitational settling parameter
h	height of cylindrical section of cyclone
H	total height of smoke plume (effective stack height)
	height of settling chamber
	height of cyclone (or length, if cyclone body curved)

$\mathbf{H}$	field strength of magnetic field (A-turn m^{-1})
H_S	chimney stack height
H_t	height of a transfer unit (gas absorption)
H_T	buoyancy rise of plume
H_V	momentum rise of plume (velocity rise)
$\mathscr{H}$	Henry's law constant
i	ionic current per unit length of conductor
I	light intensity
	index of agglomeration (sonic)
$\mathscr{I}$	sound intensity
J	variable in Bosanquet buoyancy rise equation
k	orifice coefficient
	extinction coefficient
$\mathbf{k}$	Boltzmann's constant
k_f	mass-transfer coefficient—gas to solid surface
k_G	gas-film mass transfer coefficient
k_L	liquid-film mass transfer coefficient
K	correction factor
K_G	overall mass-transfer coefficient (pressure units)
K_L	overall mass-transfer coefficient (concentration units)
l	distance between two particles
	distance for absorption
$\boldsymbol{l}$	distance between enclosing walls
L	rate of liquid flow
	depth of filter bed
	length of settling chamber
	distance between wire and plate in electrostatic precipitator
	distance moved by particle in sonic field
L_1	ratio of liquid flow to gas flow in scrubber ($l m^{-3}$)
m	mass flow
	mass of particle
	irregularity factor—a function of wire condition
M	molecular weight
M'	weight of a molecule
$\mathbf{M}$	rate of deposition (mg m^{-2}/day)
$\mathscr{M}$	function of weights of particles and ions
n	turbulence index (Sutton)
$\mathbf{n}$	number of times greater than the force of gravity
N	molecules per unit volume
	number of points
N_1	ions per unit volume
$\mathbf{N}$	number of revolutions
$\mathscr{N}$	number of revolutions of gas stream in cyclone

N_A	rate of molecular transfer of species A
p_A	partial pressure of component
p_{AM}	logarithmic mean partial pressure of component A
p_i	partial pressure at interface
P	total gas pressure
P_c	cyclone gas pressure (average)
P_i	cyclone inlet gas pressure
Pe	Peclet number
$\mathcal{P}$	power
$\mathbf{P}$	probability of collection
Δp	pressure drop
Δp_{CF}	pressure drop with constant gas flow
q	charge on a particle (subscript I refers to particle I)
Q	gas-flow rate ($m^3 s^{-1}$, $m^3 h^{-1}$)
	charge on a collecting body
Q'	rate of emission of pollutant gas
r	radius of particle
	distance between centre of particle and centre of collector
$\mathbf{r}$	resistivity of dust layer
$\bar{r}$	average pore radius
R	universal gas constant
	radius of a circle
	radius of precipitator wire and tube
	interception parameter (d/D)
R_A	drag on accelerating particles
Re	Reynolds number ($u\varrho D/\mu$—pipes, $ud\varrho/\mu$—particles)
Re_c	Reynolds number—collecting body($v_0 D\varrho/\mu$)
R_h	half distance between sphere centres for hexagonal particle arrangement
S	cross-sectional area of absorption tower
	retentivity of charcoal (Turk's equation 3.60)
	stopping distance
	width of corona layer on discharge electrode
	influence factor (ratio of diameter of sphere of influence of particle and actual particle diameter)
	depth of cyclone exit pipe within cyclone
S_*	dimensionless stopping distance
Sc	Schmidt number ($\mu/\varrho\mathcal{D}$)
$\mathcal{S}$	collection surface in electrostatic precipitator per unit volume gas flow
t	time
T	absolute temperature
T_s	stack exit gas temperature
T_1	absolute temperature at which density of stack gases equals density of atmosphere (K)

u	gas velocity
	wind velocity
	molecular velocity
$\bar{u}$	average velocity
	average velocity of gas molecules
u_g	velocity amplitude of gas
u_H	axial velocity
u_i	inlet velocity
	ionic mobility in a field of unit strength
u_p	velocity amplitude of particle
u_R	radial velocity
u_s	superficial velocity
u_t	terminal velocity
u_T	tangential velocity
u_τ	shear velocity
U	average gas velocity
U_S	velocity of sound
v	velocity
	stack exit velocity
v_0	undisturbed upstream fluid velocity
Δv	relative velocity of two particles
V	rate of gas flow (absorption)
	volume of settling chamber
V_a	swept volume
V_A	volume of species A at normal boiling point
$\mathcal{V}$	voltage, potential difference
V_1	voltage of collecting body
$\mathcal{V}_c$	corona starting voltage
V_d	potential across deposited dust layer
w	thickness of refractory
W	rate of solids emission
	weight of adsorbing solid
	distance between successive wires in electrostatic precipitator
x	downwind distance from stack (plume dispersion)
	mole fraction
	downstream distance in precipitator
	thickness of boundary layer
	thickness of dust layer
	dust content of filter cloth
	exponent in voltage/corona current equation
x_e	equilibrium dust content of filter cloth
X	function in Bosanquet buoyancy equation
X_g	amplitude of gas vibration

X_p	amplitude of particle vibration
y	cross wind distance (plume dispersion)
	mole fraction
Z	function in Bosanquet equation
	height of absorption tower
	diffusion collection parameter
z	particles per m^2 of fibre

Greek Letters

α	constant
	blade angle
	entrance loss coefficient for cyclone
	packing density
β	constant
	loss constant
	volume concentration correction factor
γ	specific heat ratio (C_p/C_v)
$\in$	porosity or voidage of a packed bed or porous solid
$\in$	dielectric constant of an aerosol particle
$\in_0$	specific inductive capacity of space ($8{\cdot}85 \times 10^9$ A^{2}N^{-1})
ε	loss number
ζ	dimensionless pressure-loss factor
η	efficiency
η_C	interception-collection efficiency
η_D	diffusion-collection efficiency
η_I	inertial-impaction collection efficiency
η_0	overall efficiency
η_{ICD}	combined efficiency
η_z	particle collection efficiency on fibre
θ	angle of particle movement
	parameter for cylindrical coordinate
$\varkappa$	permeability coefficient
	thermal conductivity
	coagulation constant
$\varkappa_g$	thermal conductivity of gas
$\varkappa_p$	thermal conductivity of particle
$\varkappa_{gtr}$	translational part of thermal conductivity of gas
λ	mean free path of gas molecules
	wavelength of sound waves
μ	viscosity of gas
μ_d	viscosity of droplet
ν	frequency

ξ	constant for free vortex formula
ϱ	gas density
ϱ_F	fibre bed density
ϱ_L	density of liquid
ϱ_V	density of vapour
ϱ_p	density of particle
σ	molecular diameter
	surface tension
	space charge per unit volume
σ_{AB}	sum of radii of two interacting molecules or ions (distance between centres)
τ	dimensionless time parameter
	time constant
	dimensionless precipitator length ($x/Lf(\mathcal{D})$)
	period of vibration
τ_0	fluid shear stress at wall
T	parameter ($= mu_T^2/3\pi\mu dR_2$)
φ	friction factor for cyclones (Stairmand formula)
Φ	current density ($A\,m^{-2}$)
	dimensionless drift velocity parameter
χ	internal porosity of solid granules
	circularity of particles
ψ	inertial impaction parameter
Ψ	sphericity
ω	drift velocity
ω'	effective migration velocity

INTRODUCTION

I.1. WHY CLEAN GASES?

There are two fundamental reasons for the cleaning of gases in industry, particularly waste gases: profit and protection. For example, profits may result from the utilization of blast furnace gases for heating and power generation, but particulate impurities have to be removed from the gases before they can be burned satisfactorily. On the other hand, sulphur dioxide can be extracted from flue gases and economically converted into sulphur,[951] or germanium can be recovered from the fly ash of certain coals.

Protection, of the individual working in industry, the public in general, and of property, is the other reason for cleaning gases in certain cases. For example, waste gases which contain toxic constituents such as arsenic or lead fumes constitute a serious danger to the health of plant operatives and the surrounding population.[353] Other waste gases such as those containing fluorine compounds or sulphur dioxide, although not normally endangering health in the concentrations encountered, may kill plants,[612, 858] damage paintwork and buildings or merely discolour wallpaper and curtains, making an industrial town a less pleasant place to live in.[500]

The extent to which industries can clean their gases remains primarily a question of economics. In some instances provision of one type of plant may prevent profitable operation, or require major reconstruction, while another type of plant, although not quite as effective, will mean continued production. When the material to be collected constitutes the major process product, as for example the particles from a spray-drying operation, an economic balance of loss of product against cost of increased efficiency of collection will decide the optimum type of collector.

Since, in practice, the most extensive gas cleaning is undertaken to prevent atmospheric pollution, the nature of such pollutants will be discussed here.

I.2. NATURE OF AIR POLLUTANTS

Air pollution may be either natural or man-made. Natural pollution may arise from sea spray, soil erosion or volcanic explosions, the most famous of the last being the eruption of Krakatoa in Indonesia in 1883, which caused sky darkening hundreds of miles away. Ocean spray, largely sodium chloride, is a significant contributor to the water-soluble content in deposited matter collected up to about 30 km from the coastline, while further inland naturally occurring salts, such as calcium sulphate, predominate.[840] Biological decay, particularly bacterial action in the soil, gives rise to large quantities of hydrogen

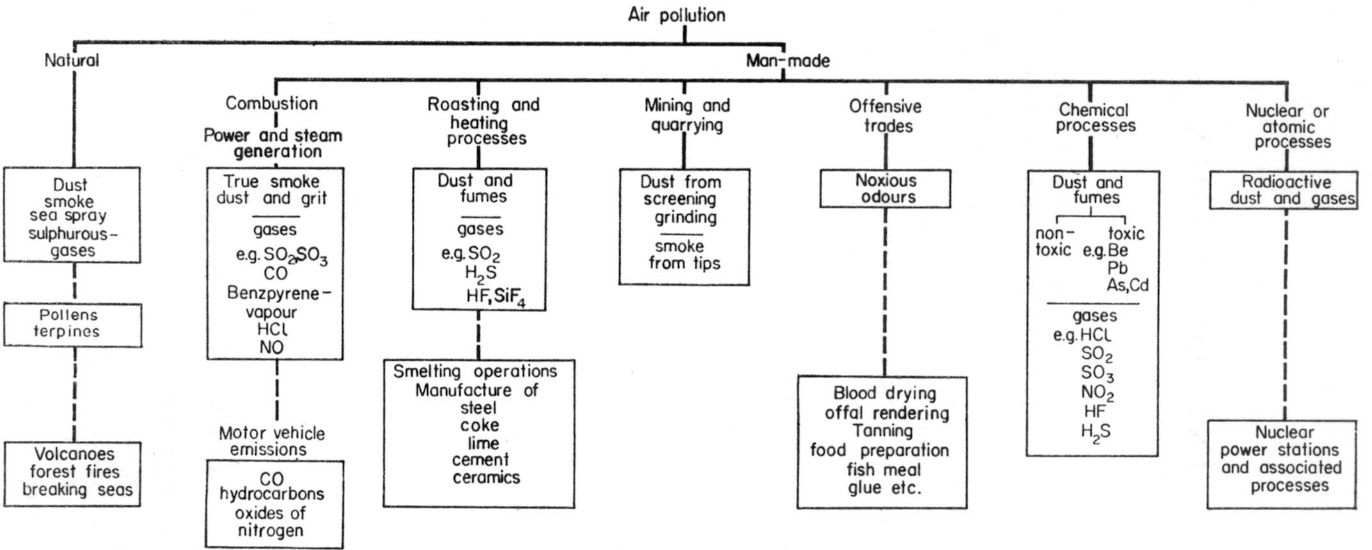

FIG. I.1. The sources of air pollution.[603]

sulphide, ammonia, hydrocarbons, oxides of nitrogen (N_2O, NO, NO_2) and the oxides of carbon (CO, CO_2). In all these cases the production by natural sources far exceeds that of man-made sources. The exception to this are the emissions of carbon monoxide, approx. 220×10^9 kg each year, which arise almost wholly from motor cars, and far exceed the amounts produced from natural sources, the major one being forest fires.[612, 688] Man-made pollution originates either from combustion of carbonaceous materials—coal and coal products, oil and wood—or from certain industries: the manufacture of chemicals and cement or the processes of metallurgy, mining and quarrying, and the incineration of refuse. Figure I.1 shows the chief sources and the main constituents of air pollution.[603]

The most important of these from the point of quantity are the products of combustion: both gaseous; such as carbon monoxide, carbon dioxide, sulphur dioxide and sulphur trioxide, and particulate matter; such as fly ash, which is largely inorganic material, and unburned carbon. Figure I.2 illustrates these for a modern highly mobile and industrialized

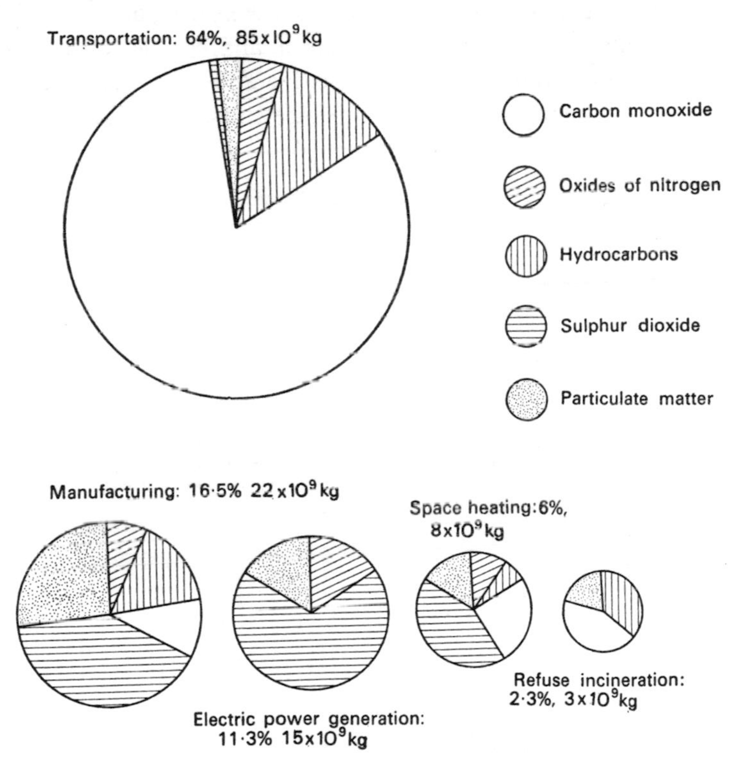

FIG. I.2. Major air pollutants in the United States (1965).[251]

society, the United States. The collection of particulate matter is today incorporated into standard power station practice, while maintaining a minimum carbon monoxide concentration in the effluent gases is the aim of efficient operation. Thus, although exposure to cold undiluted flue gases, which may contain about 0·2 per cent carbon monoxide, is very dangerous, the concentrations which occur near power stations never present a hazard, whereas the cumulative effect of motor-car exhausts in a road tunnel traffic jam may lead to harmful concentrations and long road tunnels should be monitored against this risk.

The removal of sulphur dioxide from flue gases, where it may be present in concentrations up to 0·4 per cent, depending on the sulphur in the fuel, remains one of the most difficult air-pollution problems. The maximum allowable concentration (M.A.C.),* which is the recommended maximum 8-hour exposure limit for a normal person, is five parts per million (ppm) for sulphur dioxide, but concentrations as low as 0·2 ppm will cause damage to pine trees, clovers and lucerne, although other plants, for example cabbage types, are somewhat more resistant.[958] Nonhebel[603] has pointed out that the potential sulphur dioxide–sulphuric acid emission from a modern base load 1000 MW power station is $5 \cdot 5 \times 10^5$ kg/day and so particular care has to be taken to remove the sulphur dioxide before emission, or to ensure that the emission does not cause ground-level pollution above 0·1 ppm even when atmospheric conditions are extremely unfavourable for dispersion.

Fluorides, which harm plants in concentrations as low as five parts per thousand million (5 pptm: 7–9 days continuous exposure), are found in appreciable concentrations in waste gases from fertilizer manufacture, aluminium smelting and other processes where fluoride compounds form part of the raw materials or fluxing agents. The M.A.C. value of fluorides is 3 ppm.

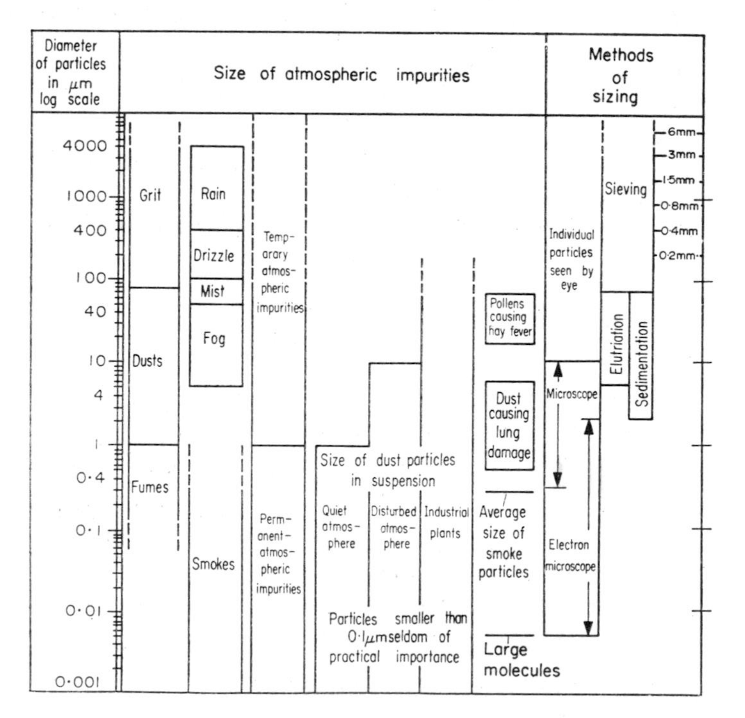

FIG. I.3. Size classification of atmospheric impurities, and the chief methods of particle sizing.[779]

* Threshold Limit Values (T.L.V.) are now commonly used instead of M.A.C. values in the U.S.A., U.K. and elsewhere. They are numerically identical with M.A.C. values.

Apart from gaseous pollutants, small particles and mist droplets are the major problem in industrial gas cleaning and air pollution control. The fumes from the processing and purification of low melting point metals such as lead, arsenic, beryllium, cadmium and zinc are extremely toxic, and a high degree of efficiency is necessary in their collection. Acid mists, for example from sulphuric or phosphoric acid manufacture, are often controlled by law, and effective collection equipment is generally installed.

Particles and droplets are variously described as grit, dust, fume, smoke, mist, aerosol or smog. The meanings generally given to these terms, and those adhered to here, are listed in Table I.1. The grouping of particles in the various categories, their approximate size ranges, the chief methods of sizing and particle visibility are shown in Fig. I.3.

TABLE I.1. CLASSIFICATION OF TERMS

Grit:	Coarse particles, greater than 76 μm which is the size of the opening in the 200 mesh sieve.
Dust:	Particles smaller than 76 μm (i.e. able to pass through a 200 mesh sieve) and larger than 1 μm.
Fume:	Solid particles smaller than 1 μm.
Mist:	Liquid particles, generally smaller than 10 μm.
Fogs:	Mists are sometimes called fogs when they are sufficiently dense to obscure vision.
Smoke:	This is the term used generally to describe the waste products from combustion, and may either be fly ash or the products of incomplete combustion, or both. The particles can be liquid or solid.
Smog:	This portmanteau word—a combination of smoke and fog—is used to describe any objectionable air pollution. There are two kinds, known as the the Los Angeles and the London type. The Los Angeles smog is photo-chemical and comes from motor-car exhausts. The London type comes from the incomplete combustion of coal and is characterized by its relatively high sulphur dioxide concentration and particle content.
Soot:	Soot is the aggregated particles of unburned carbon produced by incomplete combustion.
Aerosols:	Initially this term was used for the fine relatively stable aerial suspensions. In recent years the term has been generally applied to all air-borne suspensions.

I.3. HOW CAN GASES BE CLEANED?

All process gases, whether waste gases or those about to undergo further treatment, flow along a duct or pipe which can be interrupted by the proposed method of cleaning. The method adopted depends of course on the nature of the material to be removed. If this is a gas, two alternatives are possible. The gas can either be passed through or brought into contact with a medium which will absorb the required gas in a preferred way, or it can be changed chemically. Gas absorption in liquids is widely used on such gases as hydrochloric acid vapour, ammonia, sulphur dioxide and carbon dioxide which form an appreciable part of the gas stream, while adsorption on a solid is more common with small quantities or traces of gases such as water vapour on silica gel, carbon dioxide on lime or organic vapours on activated carbon.

Changing the chemical nature of the gases usually implies combustion or catalytic processing, particularly catalytic oxidation of organic materials, but of course also applies to such techniques as gas retention in a holding chamber to allow a process to go to completion instead of "freezing" it by immediate expulsion to the atmosphere.

The processes of removing gases are therefore chemical reaction, absorption or adsorption. In nearly all cases only one of these acts at any one time. Standard chemical engineering methods can therefore be applied to the design of plant.

The removal of small particles and droplets is however much more complex and a rigid physical classification is not possible as several mechanisms may, and often must, occur in combination. The basic mechanisms that can be employed are:

(a) Gravity separation.
(b) Centrifugal separation.
(c) Inertial impaction.
(d) Direct interception.
(e) Brownian diffusion.
(f) Eddy diffusion.
(g) Thermal precipitation.
(h) Electrostatic precipitation.
(i) Magnetic precipitation.
(j) Brownian agglomeration.
(k) Sonic agglomeration.
(l) Turbulent deposition.

In most cleaning plants more than one of the above mechanisms is usually responsible for the cleaning action, although in the majority of cases one is the controlling mechanism for the particular types of particles considered.

Thus, in filtration, inertial impaction, direct interception and Brownian diffusion are responsible, but Brownian diffusion is of major importance for particles in the sub-micron ranges, while impaction and interception are the collection mechanisms in the micron sizes. Electrostatic forces can also play a vital part here, partly because of the charges induced by charged particles on uncharged filtration media.

I.4. TACKLING A GAS-CLEANING PROBLEM

Before a suitable plant can be recommended and designed, it is first necessary to know what has to be removed from the gas stream, the extent of the gas stream and its condition. Thus an analysis of the gas stream and its contents must be undertaken. The most important items of information required are:

(1) Gas-flow rate.
(2) Gas temperature.
(3) Gas composition.
(4) Nature of material to be removed.
(5) Degree of removal required.

The degree of removal required will be discussed in Chapter 1, while the methods of obtaining items (1–4) will be outlined in Chapter 2. It may be possible, and in the case of exceptionally difficult problems may become economic necessity, to avoid gas cleaning altogether either by changing the basic process, by removing the plant to a more suitable site or simply by putting up a higher chimney.

If the cleaning plant is to be used on a new process, an estimate of the factors (1–3) must be made, while some information about the nature of the material can often be found in the literature concerning the process or from tests on similar processes operating elsewhere. In the case of gas cleaning plant over-design can prove very expensive and laboratory investigations may be warranted.

When the problem has been adequately analysed it is possible to carry out a series of calculations on the removal of gaseous constituents or particles which will indicate the type of cleaning mechanism that is possible, and the approximate dimensions and complexity of the required plant. From this an economic evaluation of the methods of gas cleaning will indicate the most suitable method and its approximate cost.

Another aim of the detailed analysis of cleaning mechanisms in this book is that it may help those who have found a new technique to analyse its principles, and so assist them in scale-up calculations.

CHAPTER 1

TO WHAT EXTENT MUST GASES BE CLEANED?

1.1. INTRODUCTION

If gases are used for chemical processing, and any constituent, either gaseous or particulate, is deleterious to the process, it must be removed. For example, carbon monoxide acts as a catalyst poison in ammonia synthesis, and has to be reduced to an acceptably low concentration before the hydrogen and nitrogen can be passed to the catalyst tower. The degree of removal required is set by the upper concentration limit that can be tolerated by the process. The cost of gas cleaning here is an integral part of the economics of the whole process.

Other gases, mainly natural, refinery or coal gases, are produced in very large quantities for sale for industrial and domestic heating and cooking, and these contain appreciable quantities of hydrogen sulphide which corrodes pipelines when the gases are cooled below their dew points. They also give off an unpleasant odour on combustion, and can affect certain processes such as steel reheating. It is therefore necessary to reduce the hydrogen sulphide to a limit acceptable by the consumer.

Waste gases from processes such as combustion, metallurgical and chemical operations are usually emitted to the atmosphere, and these may contain constituents which could be harmful, such as arsenic oxides or radioactive materials, or merely unpleasant, like clouds of smoke. Limits are imposed on these emissions by three main considerations:

(i) The concentrations which will harm plant and animal life.
(ii) Legal limitations, imposed by the country, state, county or city.
(iii) Reducing air pollution to establish civic goodwill.

These are not independent, as for example the legal limits on emissions are governed by the concentrations which damage the life of animals and plants.

Sections of this chapter will discuss what are harmful concentrations of air pollutants, and what concentrations are likely to be produced at ground level when the gases are emitted from a tall stack. Geographical factors which are likely to influence these concentrations are briefly mentioned. A further section will deal with legislative control of air pollution in some of the major industrial countries.

1.2. AMBIENT AIR-QUALITY STANDARDS

A number of industrialized countries have set up, or are in the process of setting up "ambient air-quality standards", which will determine what can be emitted from a pollution source in terms of what is present in the area or region being considered. The World Health Organization (W.H.O.) in its Inter-Regional Symposium on Criteria for Air Quality and Methods of Measurement (1963)[953] agreed on the definition of four levels of pollution, expressed in terms of pollutant concentration and exposure times, on which air-quality standards can be based. These levels, which were subsequently endorsed by the W.H.O. Expert Committee on Atmospheric Pollutants,[954] are as follows:

Level 1: Concentrations and exposure times below which, according to present knowledge, neither direct nor indirect effects (including alteration of reflexes or of adaptive or protective reactions) have been observed. This is the level of "detectable" effect.

Level 2: Concentrations and exposure times at, and above which, there is likely to be irritation of the sensory organs, harmful effects on vegetation, visibility reduction and other adverse effects on the environment.

Level 3: Concentrations and exposure times at and above which there is likely to be impairment of vital physiological functions or changes that may lead to chronic diseases or shortening of life.

Level 4: Concentrations and exposure times at and above which there is likely to be acute illness or death in susceptible groups of the population.

An extensive list of air-quality standards, which correspond to *Level 1*, has been issued for the U.S.S.R. This is given, in part, in Table 1.1.[881] It has been pointed out however that the U.S.S.R. air-quality standards are "hygienic standards" which should be aimed towards, although temporary "sanitary standards", depending on conditions, have to be used until such time that these can be achieved.[687a]

The United States Air Quality Act of 1957[880] sets up the machinery for similar standards for the United States. The first criteria recommended in 1967 under this act, which were subsequently withdrawn, are for sulphur dioxide. The recommended limits are:[30]

	Maximum (ppm)	1 percentile (ppm)
24-hr average	0·05–0·08	0·04–0·06
1-hr average	0·12–0·20	0·05–0·11
5-min average	0·10–0·50	0·05–0·14

TABLE 1.1 *(cont.)*

Mixtures. When more than one substance with toxic properties is present, the maximum permissible concentration (M.P.C.) is derived from

$$\text{M.P.C.} = A/M_1 + B/M_2 + C/M_3 + \ldots$$

where A, B, C, etc., are the concentrations of the substances, and M_1, M_2, M_3, etc., their respective M.P.C. values. If certain mixtures (SO_2 and phenol, SO_2 and HF, SO_2 and NO_2, etc.) are present, then the M.P.C. should not exceed 1·0 mg m^{-3}. In other mixtures ($CO+SO_2$, H_2S+CS_2, phthalic anhydride, maleic anhydride and α naphthaquinone) the M.P.C. values of the individual substances should not be exceeded.

Table 1.1. Air-quality Standards for the U.S.S.R. (1967)[881] Maximum Permissible Concentrations (M.P.C.)

Pollutant	Maximum at any one time		24-hr average maximum	
	mg/m^3	ppm	mg m^{-3}	ppm
Acetic anhydride	0·1	0·022	—	—
Acrolein	0·3	0·12	0·1	0·04
Ammonia	0·20	0·265	0·20	0·265
Aniline	0·05	0·012	0·03	0·007
Arsenic	—	—	0·003	—
Benzene	1·5	0·43	0·80	0·23
Carbon disulphide	0·03	0·0088	0·01	0·0029
Carbon monoxide	3·0	2·4	1·0	0·80
Chlorine	0·10	0·032	0·03	0·0095
Dimethyl sulphide	0·08	0·027	—	—
Dust (inert)	0·5	—	0·15	—
Ethylene	3·0	2·4	3·0	2·4
Ethylene oxide	0·3	0·15	0·03	0·015
Fluorine compounds (as F)				
gaseous HF, SiF_4	0·02	0·022	0·005	0·0055
soluble inorg. NaF	0·03	0·016	0·01	0·005
insol. inorg. (AlF_3, Na_3AlF_6, CaF_2)	0·2	0·10	0·03	0·015
Hydrochloric acid as HCl	0·2	0·12	—	—
as H^+	0·006	0·0037	0·006	0·0037
Hydrogen sulphide	0·008	0·0053	0·008	0·0053
Methanol	1·0	0·70	0·5	0·35
Methyl acetate	0·07	0·02	0·07	0·02
Methyl mercaptan	$9 \cdot 10^{-6}$	$4 \cdot 4 \times 10^{-6}$	—	—
Nitric acid as HNO_3	0·4	0·14	—	—
as H^+	0·006	0·002	0·006	0·002
NO_2	0·085	0·042	0·085	0·042
Pyridene	0·08	0·023	0·08	0·023
Soot	0·15	—	0·05	—
Sulphuric acid as H_2SO_4	0·3	0·07	0·1	0·023
as H^+	0·006	0·0014	0·006	0·0014
Sulphur dioxide	0·5	0·167	0·15	0·050
Vanadium	—	—	0·05*	0·0167
pentoxide	—	—	0·002	—
Mercury	—	—	0·0003	—
Lead (except tetraethyl lead)	—	—	0·0007	—

* This is specially designated as a second, lower maximum permissible concentration, and corresponds closely to the United States of America value.

TABLE 1.2. AMBIENT AIR-QUALITY OBJECTIVES—STATE OF NEW YORK[29]

(a) Regional and Subregional Objectives

		Subregions			
		1	2	3	4
Sulphur dioxide (less than these values 99% of time ppm)	recreational regional objective "A"	0·1 or 0·25 for 1 hr	0·1 or 0·25 for 1 hr	0·1 or 0·25 for 1 hr	0·15 or 0·25 for 1 hr
	industrial regional objective "D"	0·1 or 0·25 for 1 hr	0·1 or 0·25 for 1 hr	0·15 or 0·40 for 1 hr	0·15 or 0·40 for 1 hr
Suspended particulates (mg m^{-3}) (24 hr)	recreational region "A"	50% < 40 84% < 60	45 70	55 85	65 100
	industrial region "D"	50% < 65 84% < 100	80 120	100 150	135 200
Settleable particulates (mg cm^{-2}) (30 days)	recreational region "A"	50% < 0·30 84% < 0·35	30 35	60 80	60 80
	industrial region "D"	50% < 0·60 84% < 0·80	0·90 1·20	1·20 1·70	1·50 2·25

(b) All Regions

Sulphuric acid mist	0·10 mg m^{-3}	all subregions
Beryllium	0·01 mg m^{-3}	all subregions
Hydrogen sulphide	0·10 ppm for 1 hr	all subregions
Carbon monoxide	15 ppm for 8 hr	all subregions

		Subregions			
		1	2	3	4
Oxidants (inc. ozone, etc.) less than these values 99% of time		0·05	0·05	0·10	0·10
(ppm)	1 hr	0·15	0·15	0·15	0·15
	4 hr	0·10	0·10	—	—
Fluorides (as HF in air) (pptm)		1	1	3	4
Soluble fluoride (as F) in forage for livestock consumption (ppm)		35	35	—	—

More recently (February 1969) the modified criteria which will probably be adopted state that it is reasonable and prudent to conclude that concentrations of 0·1 ppm or more of sulphur oxides in the atmosphere may produce adverse health effects in a particular segment of the population. The limit set on particulates is 80 micrograms per cubic metre (i.e. 0·08 mg m^{-3}). Thus the standard for sulphur dioxide is greater than the U.S.S.R. "hygienic standard", while the one for particulates, which may be comparable to the U.S.S.R. "soot" is of much the same order.

Before the publication of these values, various states (New York, California, Colorado, Oregon and Florida) had air-quality standards or objectives. New York aimed at both regional objectives based on whether the region was commercial/industrial at the one extreme, or recreational at the other, and within these, the subregional classification of rural, residential, commercial/light industrial and heavy industrial. Table 1.2(a) lists the sulphur dioxide and particulate concentration levels, and Table 1.2(b) lists several other pollutants. The California Administrative Code lists three levels of ambient air quality: adverse, serious and emergency, which correspond to Levels 2, 3 and 4 of the W.H.O. levels. Table 1.3 lists these for a number of pollutants. The air-quality standards for other areas are more limited in scope.

TABLE 1.3. AMBIENT AIR-QUALITY STANDARDS—CALIFORNIA[29]

Pollutant	Adverse level	Serious level	Emergency level
Photochemical pollutants (hydrocarbons, nitrogen dioxide, oxidants, ozone, photochemical aerosols)	0·15 ppm for 1 hr (by potassium iodide)	not fully determined	not fully determined
Carbon monoxide	not applicable	30 ppm for 8 hr or 120 ppm for 1 hr	not determined
Ethylene	0·5 ppm for 1 hr or 0·1 ppm for 8 hr	not applicable	not applicable
Hydrogen sulphide	0·1 ppm for 1 hr	not determined	not determined
Sulphur dioxide	1 ppm for 1 hr 0·3 ppm for 8 hr	5 ppm for 1 hr	10 ppm for 1 hr

Guidelines are also given for sulphuric acid, carcinogens, hydrogen fluoride and lead.

Particulates	Sufficient to reduce visibility to 3 miles when relative humidity $<70\%$	n.a.	n.a.

1.3. HARMFUL CONCENTRATIONS OF POLLUTANTS

While the air-quality standards, particularly in the U.S.S.R., are the desirable maximum pollutant levels, which may be achieved in time, with better control methods, a relatively clear definition of what are harmful pollution levels, corresponding to Levels 3 and 4 of the W.H.O. classification is essential. The determination of harmful levels of concentrations and exposure times is very difficult, and much depends on the plants and animals exposed. In general, plants are more susceptible to pollutants than animals and in particular, man. For industrial hygiene purposes, the American Conference of Governmental and Industrial Hygienists (A.C.G.I.H.) has recommended threshold limit values (T.L.V.) for a large number of substances, and these are the maximum allowable concentrations in which a healthy person may be permitted to work for 8-hr periods daily, without adverse effects. These values, which are listed for some materials in Table 1.4, may be taken as a guide in these instances where no further information on the substance is available.

TABLE 1.4. THRESHOLD LIMIT VALUES OF IMPURITIES IN SOME COMMON INDUSTRIAL WASTE GASES[236]

Substance (gas)	A.C.G.I.H.* (ppm)	Suggested values	
		(ppm)	($mg\ m^{-3}$)
Aniline	5	5	19
Benzene	25	25	18
Bromine	0·1	0·1	0·7
Carbon dioxide	5000	5000	9000
Carbon disulphide	20	20	60
Carbon monoxide	50	50	55
Carbon tetrachloride	50	25	65
Chloroform	50	50	240
Chlorine	1	1	3
Ethanol	1000	1000	1600
Fluorine	0·1		0·2
Formic acid		10	16
Hydrogen bromide	3	3	10
Hydrogen chloride	5	5	7
Hydrogen cyanide	10	10	11
Hydrogen fluoride	3	1·5	1
Hydrogen peroxide 90%	1	1	1·4
Hydrogen sulphide	10	10	15
Nitrogen dioxide	5		9
Ozone	0·1	0·1	0·2
Pyridine	5	5	15
Sulphur dioxide	5	5	13
Toluene	200	200	750

* A.C.G.I.H.—Recommendations of the threshold limits committee of the American Conference of Governmental Industrial Hygienists (1966).

TABLE 1.4 (*cont.*)

Substance dust and fume	A.C.G.I.H. mg m^{-3}	Suggested mg m^{-3}
Aluminium (fume)		15
Antimony	0·5	0·5
Arsenic	0·5	0·25
Beryllium	0·002	0·002
D.D.T.	1	2
Diphenyl		2
Iron oxide fume	15	15
Lead tetraethyl		0·1
Lindane	0·5	0·5
Mercury	0·1	0·1
Pyrethrum	2	1
Sulphuric acid	1	1
Titanium dioxide	15	
Vanadium pentoxide	0·1	0·1
Zinc oxide	15	15

Mineral dusts	Millions of particles m^{-3} air
Aluminium oxide (alumina)	1770
Asbestos	177
Mica	708
Portland cement	1770
Silica > 50% free SiO_2	177
5% < free SiO_2 < 50%	708
< 5% free SiO_2	1770
Silicon carbide	1770

Radio-isotopes	½ life	T.L.V. in air[98] μ Ci m^{-3}
Carbon^{14}	5568 years	3·5
Cobalt^{60}	5·3 years	0·35
Tritium (H^3)	12·4 years	22
Iodine^{131}	8·1 days	0·013
Strontium^{90}	19·9 years	0·00013
Sulphur^{35}	87 days	0·27
Zinc^{65}	250 days	0·11

It should be noted, for example, that during the 1952 London smog the sulphur dioxide concentration averaged 0·7 ppm and rose to a maximum of 1·7 ppm, both values still well below the T.L.V. of 5 ppm. Thus, while T.L.V. values are applicable to healthy persons, much lower values will have to be used in framing even "sanitary standards" for the general population.

Plants are adversely affected by much lower concentrations of pollutants than animals. The gases for which extensive tests have been carried out are sulphur dioxide, hydrogen fluoride, oxides of nitrogen and ozone. The widespread damage that can result from the sulphur dioxide released in copper smelters has been recognized for many years, while more recently the large quantities of sulphur dioxide in flue gases from very large power stations are also considered a hazard. Table 1.5 lists the maximum tolerances of a number of plants to sulphur dioxide for exposure times of 150 hr.[958] Certain trees, particularly conifers, are particularly susceptible to sulphur-dioxide damage.[346, 746, 958]

TABLE 1.5. EXPERIMENTAL UPPER TOLERANCE CONCENTRATIONS OF PLANTS TO SULPHUR DIOXIDE FOR AN EXPOSURE PERIOD OF 150 HOURS.[958]

Type of plant	Maximum concentration range (ppm)
Lucerne and clovers	0·15–0·3
Summer wheat, spinach	0·2–0·3
Beans, lettuces	0·2–0·4
Strawberries, roses	0·2–0·8
Potatoes, radishes	0·3–0·8
Beet sugar, cauliflower	0·4–0·8

Thomas and Hill[859] obtained relations for leaf destruction of alfalfa, which are still being used[30] which were based on their own work and that of O'Gara.[613] These relations were for concentrations c, in ppm and time t in hours.

(a) Traces of leaf destruction $$(c-0{\cdot}24)t = 0{\cdot}94 \qquad (1.1a)$$

(b) 50 per cent leaf destruction $$(c-1{\cdot}4)t = 2{\cdot}1 \qquad (1.1b)$$

(c) 100 per cent leaf destruction $$(c-2{\cdot}6)t = 3{\cdot}2 \qquad (1.1c)$$

A relation for intermittent fumigation damage has been developed by Zahn[959] for a series of plants (lucerne, sugar beet, spinach, red currants).

Fluorides, particularly hydrogen fluoride and silicon tetrafluoride, are the second major pollutant, which not only damages plants, but accumulate in these. These compounds occur in the waste gases from superphosphate fertilizer plants, and a number of melting operations, particularly of aluminium.[346a]

If they are then used for fodder, fluorosis of ruminants, particularly cattle and sheep, can result. Gladioli have been shown to be most susceptible to hydrogen fluoride, and several other plants (e.g. apricot, sweet pea) also show signs of "burn" with low levels of hydrogen fluoride fumigation. Alfalfa and white pine, on the other hand, did not, in certain tests,[2] show these signs, but chemical analysis did give high accumulation (ppm) in the plant on completion of the tests. A survey of some of these is shown in Table 1.6. Other workers have found that even 0·1 ppm will injure the most sensitive gladiolus types in 5 weeks' exposure.[169]

Accumulations of fluorine on forage plants of 30 ppm (dry weight basis) are marginally non-toxic to cattle, but amounts above 50 ppm are positively toxic.[646] As the accumulations in Table 1.6 show, these concentrations (dry-weight basis) are easily achieved in plants exposed to gases contaminated even by 1·5 pptm of hydrogen fluoride.

TABLE 1.6. FLUORIDE FUMIGATION AND EXPOSURE TIME WHICH CAUSES INCIPIENT FOLIAR INJURY. FLUORIDE ACCUMULATIONS ARE ALSO GIVEN

	Daylight exposure						Darkness exposure			
Fumigation level (pptm)	1·5		5		10		1·5		5	
	$t \times c$	a	$t \times c$	a	$t \times c$	a	$t \times c$	a	$t \times c$	a
Plants										
Gladiolus	97	37	119	46	137	57	82	59	122	44
Alfalfa*	356	149	506	203	506	182	200	25	552	132
Apricot	225	58	213	83	163	107	336	130	368	84
Sweet pea	313	327	307	148	374	141	—	—	—	—
Carrot	284	323	398	723	425	307	—	—	—	—
White pine*	202	70	665	138	496	67	367	136	—	—

* No visible burn on completion of fumigation.

$t \times c$ = Exposure factor [t (time hours)$\times c$ (concentration pptm)],
a = accumulation: fluoride concentration in exposed leaves, ppm (dry weight basis).

Chlorine as a pollutant is not as widespread as sulphur dioxide and fluoride pollution, although hydrochloric acid gas from the Leblanc soda process was one of the first serious industrial pollutants. Chlorine gas is less toxic than fluorine or fluoride. Levels of 0·31 ppm caused no damage to tomato plants (3-hr exposure) while 0·61 ppm caused slight damage and 1·38 ppm severe damage for similar exposure times.[116] Other workers[369a] have reported that alfalfa and radish are injured by 2-hr exposures at 0·10 ppm.

Data for the oxides of nitrogen is more limited but these tend to be about one-fifth of the toxicity of sulphur dioxide. It is estimated that the plant toxicity threshold level is 0·4 ppm[346a]

Peroxyacetyl nitrate (PAN), a photochemical reaction product of oxides of nitrogen and hydrocarbons, is a very serious phytotoxicant in those areas which are polluted with effluents from combustion of liquid fuels. Plant damage can be caused by 8-hr exposure to concentrations as low as 0·05 ppm. Severe damage on bean, petunia and tomato is caused by 0·5 ppm for 1 hr and 0·1 ppm for 5 hr.[346a]

Ozone, in quite low concentrations (0·033 ppm average to 0·106 ppm maximum) has some effect on leaf crops, while 0·8 ppm (1-hr exposure) has been noted to affect tobacco plants. Of considerable importance also is the synergistic effect of sulphur dioxide and ozone.[369a, 505] In the case of tobacco plants, 2-hr exposure to either ozone (0·03 ppm) or sulphur dioxide (0·24 ppm) caused no leaf damage, while the combination of these led to 38 per cent leaf damage.

There is little general information as to the effect of dusts on plants. In most cases dusts are chemically inert and cause little damage, except when concentrations are very high and the deposits exclude sunlight from leaves and flowers until such time when they are washed clean by rain. Cement dusts, however, are not inert chemically[97, 191] and their effect on vegetation has been studied extensively. The cement dust from cement works forms a hard crystalline crust when deposited on the leaves. Calcium hydroxide solution is liberated from these crystals by atmospheric moisture and this penetrates through the leaf epidermis destroying the cells.

For radio-isotopes, a table of maximum exposures recommended in the report of the Committees on Permissible Doses for Internal Radiation (1958 revision) and the book by Blatz[92] may be used as a guide. Here, however, concentrations which are absorbed by plant may be much lower than this, but will still be a problem when eaten by cows resulting in radioactive milk. In general, emissions of radioactive materials must be kept as low as possible.

1.4. GROUND-LEVEL CONCENTRATIONS FROM ELEVATED-POINT SOURCE EMISSIONS

The concentrations to which plants and animals are exposed at ground level can be reduced by emitting the gases from a process at great heights, from a tall chimney, so that the gases are dispersed over a very large area. This is often the cheapest way of dealing with an air-pollution problem. In other cases, where it is not technically or economically possible to reduce the concentrations of certain gases to an acceptable value, this method enables a low effective ground level concentration to be maintained, even with adverse meteorological conditions such as persistent atmospheric inversions.

Normally, the temperature of the atmosphere decreases with increasing height (lapse conditions) and the warm gas emitted by a chimney will rise to great heights. Under certain conditions it is possible that a warm layer of air lies on top of a colder layer next to the ground. This is an inversion layer, and chimney plumes have difficulty in penetrating it, unless their temperature is greater than that of the layer. The waste gases from chimneys then collect in the restricted region below the inversion layer, and concentrations at ground level increase until winds and changed conditions disperse the accumulated gases. These ground level concentrations depend on the nature of the material being emitted, the gas temperature and velocity of emission, the actual stack height, wind velocity and the atmospheric temperature conditions. Observations have shown and calculations have indicated that at low wind velocities the buoyancy rise due to the gas temperature is likely to carry the waste gases to great heights, while at high wind velocities turbulence will rapidly disperse the waste gases. It is at moderate wind velocities of 6–9 $m\,s^{-1}$, that the highest ground-level concentrations are likely to occur.

The problem of estimating ground-level concentrations may be considered in two sections. First the path of the plume and the maximum height reached (called the effective stack height, H) is calculated, and then the diffusion of the gases from this point must be considered. As a *first approximation*, the effective stack height can be considered as the sum of the actual height of the stack, H_S, the rise of the plume due to the velocity of the issuing gases, H_V, and the buoyancy rise H_T, which is a function of the temperature of the gases being emitted and the atmospheric conditions:

$$H = H_S + H_V + H_T. \tag{1.2}$$

For the velocity of the issuing gases to be effective in increasing the height of the plume, "downwash" at the chimney exit must be prevented by keeping the stack exit velocity above a critical value which has been found experimentally as a function of the wind velocity[766] (Fig. 1.1). A guideline to suitable efflux velocities is given in the U.K. Memorandum on

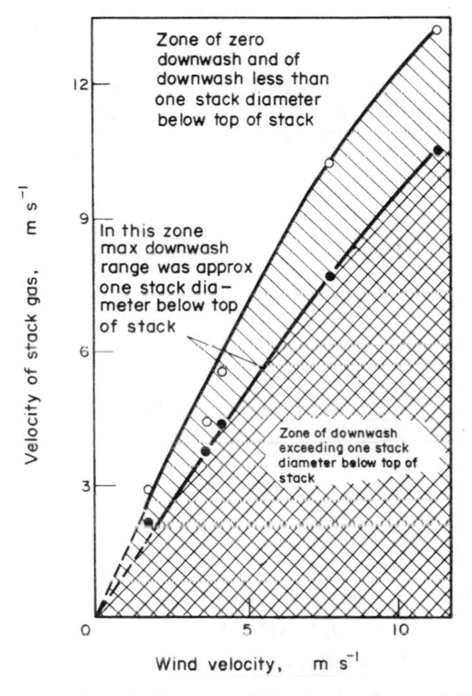

FIG. 1.1. Correlation of stack exit velocity and wind velocity for avoiding "downwash" of smoke plume.[766]

Chimney Heights (1963, 1967)[875] which recommends that small boilers (rated up to 13,600 kg h^{-1} evaporation/hr) should have velocities not less than 6 m s^{-1} at full load when equipped with forced-draught fans, and not less than 7·6 m s^{-1} when equipped with induced-draught fans. This is increased to a maximum of 15·2 m^{-1} at full load for boilers rated at $2{\cdot}04 \times 10^5$ kg h^{-1}.

To achieve a stack exit velocity above the critical value in order to avoid downwash it may be necessary to remodel the stack exit, and a venturi nozzle design has been found most effective, as well as minimizing pressure losses. A suggestion has been made that smoke puffs rather than the usual plume will penetrate inversions more easily, and would not be affected by winds to the same extent.[262]

The buoyancy rise effect H_T depends on atmospheric conditions, and in carrying out calculations of the type described here it is assumed that atmospheric conditions are stable over great heights and distances, an assumption that is only rarely justified in practice. When the temperature of the atmosphere decreases with increasing height at a rate greater than the adiabatic lapse rate (i.e. when the atmospheric temperature decreases by more than 1°C/100 m) then the hot plume of gases emitted from a stack will rise to great heights, and no ground level pollution will occur. However, under mild lapse conditions (when the atmospheric temperature gradient G is less than 1°C per 100 m) or when there is a temperature inversion, H_T tends to a minimum value.

The rise of a plume can be calculated in a large number of ways, which give predictions of various degrees of accuracy. These have been reviewed in detail by Strohm[837] and Somers[784] and several methods will be presented here. These are the method of Bosanquet, Carey and Halton[102], which gives a conservative estimate of the plume rise for H_T and H_V, a modification of the method by Priestley[660] for the buoyancy rise H_T, which gives a value which has been confirmed experimentally to within 3 per cent, and some empirical equations.

The formulae by Bosanquet, Carey and Halton are:

(1) for the stack exit velocity rise H_V

$$H_V = H_{V_{max}}\left(1 - \frac{0{\cdot}8 H_{V_{max}}}{x}\right) \quad \text{when} \quad x > 2H_{V_{max}} \tag{1.3a}$$

and

$$H_{V_{max}} = \frac{4{\cdot}77}{1 + 0{\cdot}43u/v} \frac{\sqrt{(Qv)}}{u} \tag{1.3b}$$

where $H_{V_{max}}$ = maximum momentum rise of plume (m),
x = distance downwind (m s^{-1}),
u = wind velocity (ft sec^{-1} or m s^{-1}),
v = stack exit velocity (ft sec^{-1} or m s^{-1})
Q = gas flow rate from stack (ft^3 s^{-1} or m^3 s^{-1}) measured at the temperature at which the density of the stack gases would be equal that of the ambient atmosphere.

(2) for the buoyancy rise H_T

$$H_T = \frac{6{\cdot}37 g_c Q}{u^3 T_1}(T_S - T_1)\left(\ln J^2 + \frac{2}{J} - 2\right) \tag{1.4}$$

where g_c = gravity acceleration constant (9·81 m s^{-2}),
T_1 = absolute temperature at which density of stack gases equals the density of the atmosphere (K)
T_s = stack exist gas temperature (K)

and

$$J = \frac{u^2}{(Qv)}\left(0{\cdot}43\sqrt{\left(\frac{T_1}{g_c G}\right)} - 0{\cdot}28\frac{v}{g_c}\frac{T_1}{(T_S - T_1)}\right) + 1 \tag{1.5}$$

where G = potential temperature gradient of the atmosphere (0·3°C m^{-1}).

The calculation can be simplified by replacing $(\ln J^2+2/J-2)$ by Z in equation (1.4) and finding Z as a function of X from Fig. 1.2,
where

$$X = \frac{ux}{3{\cdot}57\sqrt{(Qv)}}. \tag{1.6}$$

A nomographic solution to these equations (1.2–1.6) drawn up by Strauss and Woodhouse[835] gives an answer of sufficient accuracy for practical cases.

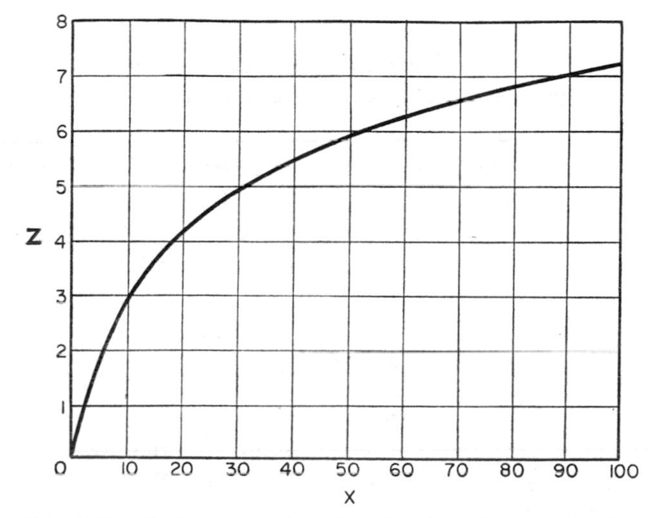

FIG. 1.2. Correlation for buoyancy rise of smoke plume between functions X and Z.[102]

The method by Priestley[660] is based on the assumption that the rising gas is in turbulent motion generated by the plume itself, and an allowance is made for the entrainment of air in the plume, the spreading of the plume and its loss of momentum and heat by lateral diffusion as well as for the buoyancy of the plume. This self-generated turbulence by the plume is additional to the atmospheric turbulence due to wind, and is generally the predominant factor in the initial plume rise.

This work also shows that the buoyancy and velocity rises are not strictly additive, the assumption made in equation (1.1), but that the combined rise is less than the sum of the two considered separately. Although Priestley's theory does not lead to a single formula covering all conditions, explicit approximate formulae can be derived for important ranges of conditions. Thus, for calm conditions, the total buoyancy rise is given by:

$$H_T = 1{\cdot}55c^{-\frac{1}{2}}\left(\frac{g_cG}{T_1}\right)^{-\frac{3}{8}}\left(\frac{Eg_c}{C_p\pi\varrho T_1}\right)^{\frac{1}{4}} \tag{1.7}$$

where E = strength of the heat source relative to the surrounding atmosphere (B.t.u. s^{-1} or $1{\cdot}055\times10^3$ W),

C_p = specific heat of the gas at constant pressure (B.t.u./lb or $2{\cdot}326\times10^3$ J kg^{-1}),

ϱ = density of the stack exit gases,

and c takes the value of 0·09. In strong inversion with light winds, which often in practice combine to provide the most serious set of conditions, the same formula can be used, with c replaced by $0{\cdot}265u^{\frac{1}{2}}$ (i.e. $1{\cdot}55^{-\frac{1}{2}}$ replaced by $0{\cdot}414u^{-\frac{1}{4}}$).[661] In wind speeds above 3 m s^{-1} and lapse rates not too far from adiabatic, the same dependence on E, but a different one on wind-speed, results in the form

$$H_T = \frac{\alpha E^{\frac{1}{4}}}{u}. \tag{1.8}$$

This has been verified experimentally by a series of measurements at two sites, which yield values of α of 4900 and 6200 respectively.[532] For smooth ground and a neutral atmosphere a value of α of 5700 is suggested.

These equations emphasize the importance of the heat in the exit gases in obtaining a large buoyancy rise.

The formula adopted by the American Society of Mechanical Engineers (A.S.M.E.)[780] is along similar lines, using the one-fourth power of the heat source; but involves the actual stack height H_S

$$H_T = (6{\cdot}81+0{\cdot}0677H_S)E^{\frac{1}{4}}/u. \tag{1.9}$$

Moses and Carson[583] have published empirical equations based on experience at the Argonne National Laboratory and other published observations, and the form of which was based on the Holland equation. Three stability classifications were used; based on the change in potential temperature with height, from the top of the stack to the top of the plume, $d\theta/dz$:

Unstable: $d\theta/dz < -0{\cdot}22$ K/100 metres,

Neutral: $-0{\cdot}22 \leqslant d\theta/dz \leqslant 0{\cdot}85$ K/100 metres,

Stable: $d\theta/dz \geqslant 0{\cdot}85$ K/100 metres.

The plume rises are

Unstable: $$H_T = (3{\cdot}47vD+6{\cdot}40\sqrt{E})/u, \tag{1.10a}$$

Neutral: $$H_T = (0{\cdot}35vD-3{\cdot}28\sqrt{E})/u, \tag{1.10b}$$

Stable: $$H_T = (-1{\cdot}04vD+2{\cdot}78\sqrt{E})/u \tag{1.10c}$$

where D is the exit diameter of the stack.

The units of H_T, D are m, and v, u are m s^{-1}.

A full-scale study of plume rises from a large coal-fired electric generation station indicated that most equations over-estimate plume rise at low wind speeds, while for moderately high wind speeds (greater than 3 m s^{-1}) the Carson and Moses equations give the best fit.[151]

The two methods of calculating ground-level concentrations arising from the release of gases from a point source, those by Bosanquet and Pearson[74, 103] and by Sutton,[843] give similar answers, and only the Sutton equation, which is the more widely used, is given here.

The calculation requires the selection of appropriate eddy-diffusion coefficients C_y and C_z which are functions of the mean wind velocity profile, the degree of turbulence and other factors. They can be selected for the appropriate atmospheric conditions and plume height from Table 1.7. The coefficients assume an aerodynamically smooth surface, a condition only rarely met in practice, and so tend generally to give over-estimates in the ground-level concentration.

TABLE 1.7. GENERALIZED EDDY DIFFUSION COEFFICIENTS[311] C_yC_z AND TURBULENCE INDEX n

Height of source above ground (m)	Large lapse C_y	Large lapse C_z	Zero or Small lapse C_y	Zero or Small lapse C_z	Moderate inversion C_y	Moderate inversion C_z	Large inversion C_y	Large inversion C_z
0	0·128	0·073	0·073	0·043	0·046	0·027	0·036	0·021
9·75	0·128	0·073	0·073	0·043	0·046	0·027	0·036	0·021
25	0·073		0·043		0·027		0·021	
30	0·070		0·040		0·026		0·020	
46	0·061		0·036		0·023		0·019	
61	0·058		0·033		0·021		0·017	
76	0·055		0·030		0·020		0·016	
92	0·049		0·027		0·017		0·014	
107	0·040		0·021		0·014		0·011	
n	0·20		0·25		0·33		0·50	

Sutton's equation for the ground-level concentrations of a gaseous pollutant, in parts per million by volume (ppm) at a distance x m downwind and y m crosswind from the source is

$$c = \frac{2\times 10^6 Q'}{\pi C_z C_y u x^{2-n}} \exp\left[-\frac{1}{x^{2-n}}\left(\frac{y^2}{C_y^2}+\frac{H^2}{C_z^2}\right)\right] \quad \text{ppm} \tag{1.11}$$

where Q' = rate of emission of pollutant gas (ft^3 s^{-1} or 0·0283 m^3 s^{-1}),
n = turbulence index (dimensionless).

This gives a maximum value at a distance downwind $x_{\max}$ of

$$C_{\max} = 2{\cdot}35\times 10^5 . \frac{Q'}{uH^2} . \frac{C_z}{C_y} \quad \text{ppm} \tag{1.12}$$

where

$$x_{\max} = \left(\frac{H}{C_z}\right)^{2/(2-n)} \quad \text{m} \tag{1.13}$$

In practice the maximum concentration occurs approximately 10 to 15 times the effective height of the plume downwind from the source, which is approximately 25–35 times the actual stack height in the case of a hot gas emission. When the emissions consist of particles which are so small that they are carried along like gas molecules (i.e. if they are in the sub-micron size ranges) then the calculation will give ground-level concentrations in g m^{-3} instead of ppm by volume if the stack emission value in equation (1.11) which is in (m^3 s^{-1}) is replaced by kg s^{-1}.

The average rate of deposition of particulate emissions can be found from[102]

$$\mathbf{M} = 2{\cdot}956\times 10^{9}\frac{Wb}{H^2}\left[\frac{\left(\dfrac{20H}{x}\right)^{20f/u+2}}{\Gamma\left(\dfrac{20f}{u}\right)}\exp\left(-\frac{20H}{x}\right)\right] \tag{1.14}$$

where $\mathbf{M}$ = deposition rate (mg m^{-2} d^{-1}),
W = rate of solid matter emission (kg s^{-1}),
b = fraction of time wind is in 45° sector under consideration,
f = free falling speed of particles (m s^{-1}).

This may be written as

$$\mathbf{M} = \frac{Wb\times 10^6}{H^2}\cdot\mathcal{F}\left(\frac{f}{u},\ \frac{x}{H}\right) \tag{1.15}$$

where $\mathcal{F}(f/u,\ x/H)$ represents the function in the diaresis in equation (1.14). This function has been plotted for various values of f/u and x/H (Fig. 1.3). In practice, particles of different sizes are emitted into a rising plume in a wind of varying velocity. It is then necessary to calculate the rate of deposition of particles in different size groups, with different wind velocities at a series of distances from the source to obtain an accurate picture. It is usually possible to assume an average wind velocity of 6 m s^{-1} for an approximate calculation.

For particles with diameters less than 20 μm (density unity), a simpler equation by Hawkins and Nonhebel[604] can be used:

$$\mathbf{M} = 3{\cdot}246\times 10^{10}\frac{Wbf}{uH^2}. \tag{1.16}$$

The estimates obtained by these calculations do not allow for some important practical phenomenon frequently connected with uneven temperature distributions in the atmosphere. These include "looping", in which a plume as a whole eddies up and down through considerable vertical distances, and "fumigation" in which a high concentration built up clear of the ground overnight is quite suddenly brought down to the surface in the early morning with the onset of convection. The above formulae do, however, give a guide to the concentrations and deposition rates that may be expected.

An allowance to be made for the complex behaviour of plumes rising from rows or groups of chimneys has been suggested. When stacks are close together, their plumes will combine;

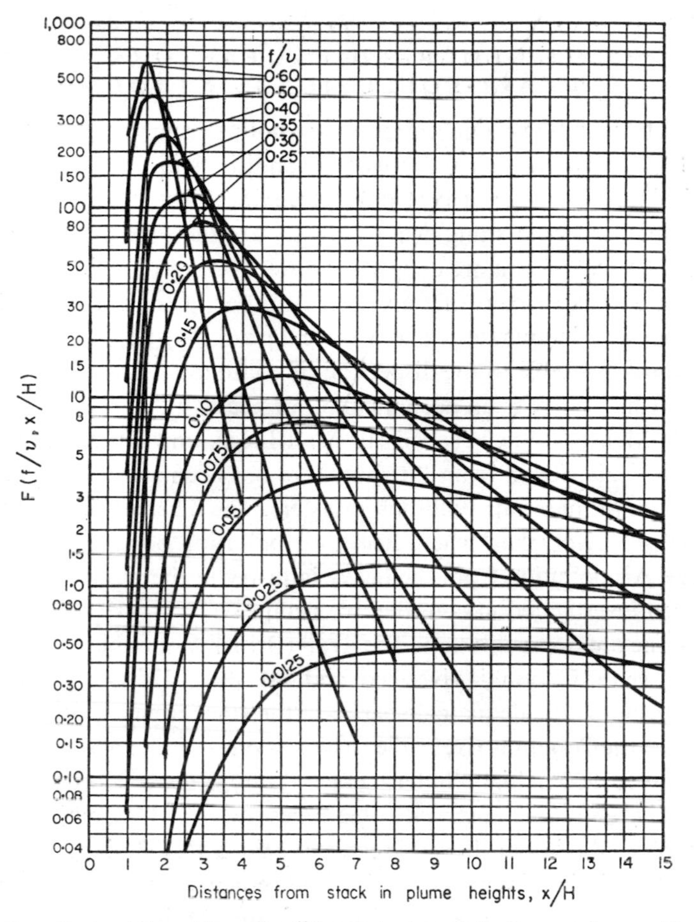

FIG. 1.3. Variation of rate of deposition of solid pollutant particles from a plume with distance from stack, based on falling speed and wind velocity.[102]

when they are further apart, and the direction of the wind is not in line with the line of the stacks, their combined effect is more complex. For equally loaded stacks, the A.S.M.E.[39] procedure is recommended. A simple rule for dealing with the emissions from two or more stacks is to add a contribution from the new stack to the effluent of the first stack, by the simple formula[542]

$$\text{Quantity of pollutant emitted } (Q' \text{ or } W) \times (1+\tan\theta) \tag{1.17}$$

where θ is the angle subtended at the base of the new stack by the first one, i.e. $\tan\theta = H_S/D'$ where D' is the distance between the stacks. This can be extended to three or more stacks by using the multiplier

$$(1+\tan\theta_1+\tan\theta_2+\ \ldots) \tag{1.18}$$

where θ_1 and θ_2, etc., are the angles subtended by the new stacks.

1.5. GEOGRAPHICAL FACTORS IN AIR-POLLUTION CONTROL

The estimates of ground-level concentrations of pollutant materials obtained by calculations (Section 1.4) are based on stable weather conditions in flat open country. However these geographical features may not exist at the place where a works is situated. For example, important economic factors in siting heavy industry are the sources of raw materials, such as the proximity of a coal or iron ore deposit for a steel works, or a river valley where barges and ships provide a cheap means of transport. Another factor is availability of land and closeness of a town for a labour supply. Therefore a plant may be situated where geographical features are not favourable to the dispersion of pollutants. The most common problems of this type arise where there is an accumulation of heavy industrial and domestic sources of pollution. The Meuse Valley between Huy and Liège, and the Monongahela Valley at Donora in Pennsylvania, 50 km from Pittsburg, are both centres of metallurgical industry, both have been sites of major air pollution disasters. The Thames river plain in London is another example of a river plain where heavy air pollution occurs, but here largely from domestic sources. Air pollution from the heavy industry in the Don river basin in Sheffield, England, has been investigated in considerable detail.[292]

High stacks alone are not a solution. The Trail (British Columbia) Smelter is located on a terrace 55 m above the Columbia river bed, and has stacks 125 m high. At certain periods this is not enough to obtain adequate dispersal of the smelter gases, because of the geographical feature of the Columbia river valley which is one-third to half a mile wide and bounded by mountains 450–900 m high. Sintering operations at the smelter are controlled by the sulphur dioxide concentration observed some miles away at Columbia Gardens near the United States border. Maximum allowable concentrations are 0·3 ppm for 40 min in summer and 0·5 ppm for 60 min in winter. At Ducktown (Tennessee) two copper smelters created considerable local damage because of sulphur dioxide emissions. High stacks, in fact, extended the damage over a distance of 50 km, and an acid plant was eventually erected to utilize the gases, after a U.S. Supreme Court investigation. The Anaconda smelter erected 90 m stacks on a spur of the Rocky Mountains, 213 m above the furnaces and 335 m above the valley floor, but arsenic poisoning due to arsenic trioxide particles, formed because of the condensation to small particles below 190°C, could be traced over a distance of 56 km along the path of the prevailing winds.

These examples illustrate how geographic features—prevailing winds, incidence of fogs and topographical features—can determine the extent to which it is necessary to clean the gases before releasing them to the atmosphere.

1.6. AIR-POLLUTION CONTROL LEGISLATION

Clean air legislation has been written into the statute books of a number of countries, while in others control is exercised under health acts, regulations regarding nuisance, and also by common law. The legislation may cover a whole country, as in the United Kingdom,

France, Germany, New Zealand, Poland and Belgium, individual states as in Australia, or it may vary from county to county within the same state, as in parts of the United States. Because of the widespread damage that can be caused by air pollution, which can even affect two countries, as in the case of the Trail smelter, the most widespread uniform legislative control is desirable. Some of the specific air pollution control legislation in the United Kingdom, United States, U.S.S.R., Germany, France, Netherlands, Belgium, Scandinavia, Italy, Ireland, Japan, Australia and New Zealand will be given here.

1.6.1. United Kingdom

The three relevant Acts of Parliament governing the control of sources of pollution in England and Wales are:

(i) The Alkali, etc., Works Regulation Act (1906).
(ii) The Public Health Act (1936).
(iii) The Clean Air Act (1956).

Other Acts, such as the Smoke Nuisance (Scotland) Acts (1857 and 1865) and the Public Health (Scotland) Act (1897–1939), are used in Scotland, and are in some ways similar.

The Public Health Act (1936) had important provisions which enabled local authorities to make by-laws, subject to the confirmation of the Minister of Housing and Local Government, to regulate the emission of smoke from industrial sources. The usual regulation prohibited the discharge of smoke for more than 2 min within a continuous period of 30 min. These regulations have, however, been more effectively covered by the Clean Air Act (1956)[291] which gives local authorities much wider powers. This Act has provisions regarding the limiting of emission of dark smoke, which is defined as being as dark as, or darker than, shade 2 on the Ringelmann Chart (Fig. 1.4), except when lighting up a furnace, or as a result of unforeseeable plant failure.

The Clean Air Act requires that new furnaces should be as far as possible smokeless and be equipped with approved grit and dust-arrestment equipment if they burn more than 1 tonne per hour. The height of chimneys is also controlled, and in all cases, to prevent down draught,

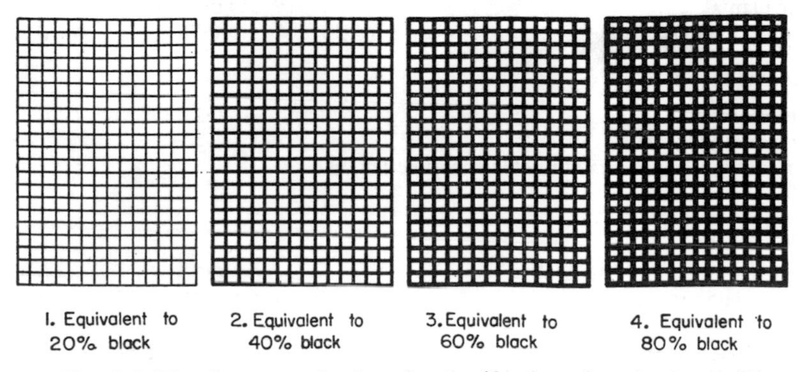

FIG. 1.4. Ringelmann smoke chart for classification of smoke density.[469]

TABLE 1.8. STATUTORY LIMITS FOR EMISSIONS UNDER THE ALKALI, ETC., WORKS REGULATION ACT (1906)

Hydrochloric acid in gases from Alkali (Salt cake) process	0·2 gr./ft^3	(0·46 g/m^3)
Sulphuric acid (chamber process) (SO_3 equivalent)	4 gr./ft^3	(9·2 g/m^3)
Sulphuric acid (concentration process SO_3 equivalent)*	1·5 gr./ft^3	(3·5 g/m^3)

Other emission limits recommended by the Alkali Inspectorate

A. *Gaseous emissions:*		
Fluorine compounds (as SO_3 equivalent)	0·1 gr./ft^3	0·23 g m^{-3}
Sulphur dioxide (as SO_3) (depending on height of discharge)	0·5 gr./ft^3 (up to 100 ft)	1·15 g m^{-3} (up to 30 m)
	1·0 gr./ft (over 100 ft)	2·3 g m^{-3} (over 30 m)
Chlorine	0·1 gr./ft^3	0·23 g m^{-3}
Hydrogen sulphide	5 ppm	
Nitrogen oxides (nitric acid by ammonia oxidation) (SO_3 equivalent)	0·75 gr./ft^3 (as SO_3)	1·7 g m^{-3} (as SO_3)
Other processes, e.g. nitrogen	1 gr./ft^3 (as SO_3)	2 g m^{-3} (as SO_3)
B. *Particulate emissions:*		
Arsenic works	0·05 gr./ft^3 as As_2O_3 for <5000 ft^3/min	0·115 g m^{-3} as As_2O_3 for <8500 m^3/hr
	0·02 gr./ft^3 as As_2O_3 for >5000 ft^3/min	0·05 g/m^3 as As_2O_3 for >8500 m^3/hr
Antimony works	As for arsenic works, but calculate as Sb_2O_3 instead of As_2O_3	
Cadmium works	0·017 gr./ft^3 (as Cd)	0·039 g m^{-3} (as Cd)
Cement works	0·1 gr./ft^3	0·23 g m^{-3}
Iron and steel works:		
(i) Fume emissions	0·05 gr./ft^3	0·115 g m^{-3}
(ii) Oxygen processes	0·05 gr./ft^3	0·115 g m^{-3}
(iii) Sinter plants	0·05 gr./ft^3	0·115 g m^{-3}
(iv) Hot blast cupolas	0·05 gr./ft	0·115 g m^{-3}
Power stations:		
(i) Built before 1958	0·2 gr./ft^3	0·5 g m^{-3}
(ii) New works (2000 MW) electrostatic precipitators with efficiencies better than 99·3%		
Sundry dust emissions:		
Dust (particulate matter $>$ 10 μm)	0·2 gr./ft^3	0·5 g m^{-3}
Fume (particulate matter $<$ 10 μm)	0·05 gr./ft^3	0·115 g m^{-3}

* Sulphuric acid (contact process)—loss of acid gases to air not to exceed 0·5 per cent of sulphur burnt and the emission to be free of persistent mist.

the recommended height is two and a half times that of adjoining buildings or 37 m. In any case, the chimneys have to be approved by the local authorities, and the topography of the surrounding buildings and land would be considered. The Act also empowers local authorities to set up smoke-control areas or "smokeless zones" in residential and commercial areas, and other suitable areas such as new industrial estates. These areas may prohibit the installation of processes which cause smoke or fume, even if dust and fume recovery plant is installed to minimize the emissions. Other provisions of the Act permit the Minister of Housing and Local Government to exempt certain industries whose processes present special problems in air-pollution control, as in the case of metallurgical fumes, and places these under the provision of the Alkali Act, which is also under his direction.

The Alkali, etc., Works Regulation Act of 1906 arose out of earlier legislation (1863) which regulated the emission of hydrochloric acid gas from the manufacture of sodium carbonate by the Leblanc or "saltcake" process. The Alkali Act today, through its list of scheduled processes, covers nearly all industrial processes which are actual, or may appear to be potential emitters of noxious gases. The operators of these scheduled processes are required to register the process, and a condition of the registration is that the Chief Alkali Inspector, who administers the Act, must be satisfied that the "best practicable means" of reducing the discharges of noxious or offensive gases are employed. The scheduled processes are subject to inspection, and provision must be made for sampling and flow measurements by the inspectors. Certain statutory limits are prescribed in the Act, examples of which are given in Table 1.8.

Limits for substances not controlled directly under the Statute are also prescribed by the Chief Alkali Inspector and are called "Presumptive Limits". These have the advantage that they can be introduced and varied at the discretion of the Chief Alkali Inspector, without the time-consuming process of requiring an Act of Parliament. The word "Presumptive" arises

TABLE 1.9. CHIMNEY HEIGHTS REQUIRED IN THE UNITED KINGDOM FOR CHEMICAL AND RELATED INDUSTRIES

A. *Contact sulphuric acid plants:*

(Based on a calculated 3 min mean ground level concentration at 0·20 ppm.) (Note: the minimum height for a contact acid plant is 120 ft (36·4 m)

Acid productions per day tonnes H_2SO_4	Basic chimney height* ft. (m) at stack exit velocities of:			
	20 ft/sec (6 m /sec)	30 ft/sec (9 m/sec)	40ft/sec (12 m/sec)	50 ft/sec (15 m/sec)
100	104 (31·5)	101 (30·4)	99 (30	96 (29
300	175 (53)	167 (50·6)	163 (49·3)	159 (48)
600	248 (75)	241 (73)	235 (71)	230 (69·5)
1000	819 (96·5)	310 (94)	303 (92)	296 (89)
1500	391 (119)	381 (116)	372 (113)	364 (112)
2000	452 (137)	439 (134)	430 (131)	419 (127)

* See foot p. 30.

TABLE 1.9 (*cont.*)

B. *Nitric acid plants:*

(Based on an efflux velocity of 24·3 m/sec, exit gas concentration of (5·3 g/m^3 (NO_2) and a g.l.c. of 0·16 ppm NO_2, wind velocity of 6 m/sec)

Acid production tons HNO_3 per day	Gas volume ft^3/min (m^3/hr) N.T.P.	Effective height ft (m)	Plume rise ft (m)	Basic chimney ht. ft (m)
175	14,000	205	27	180
	(24,000)	(62·1)	(8·2)	(54·6)
350	28,000	287	39	250
	(47,500)	(87)	(11·8)	(76)
530	42,000	353	47	300
	(71,000)	(107)	(14·2)	(90)
700	56,000	412	55	350
	(95,000)	(125)	(16·5)	(106)
1060	84,000	468	68	400
	(142,000)	(142)	(20·6)	(121)

C. *Warm sulphur dioxide emissions:*

Rate of emissions, tons SO_2/day	3·6	7·5	13	21	30	40
Basic chimney ht. ft	100	150	200	250	300	400
m	30	45	60	75	90	120

D. *Copper works where scrap, swarf and residues are melted for recovery of ore:*

Rate of melting tons/24 hr	25	50	100	150	200	300
Basic chimney ht.* ft.	72	102	144	177	204	250
m	22	31	43·5	53·5	61·6	75

Lead works	Max. conc.		Max. total loss of lead/week	
	gr/ft^3	g/m^3	lb	kg
Small works < 3000 ft^3/min 5000 m^3/hr	0·05	0·115	100	45·3
Medium works 3000–10,000 ft^3/min 5000–17,000 m^3/hr	0·05	0·115	400	182
Large works > 10,000 ft^3/min > 17,000 m^3 /hr	0·01	0·023	1000	453
Very large works > 140,000 ft^3/min > 240,000 m^3/hr	0·005	0·0115	>1000†	7453

† Higher stack will be required.

* Actual chimney height H_s can be calculated from the basic chimney height A to allow for surrounding buildings, etc., by the following formula:

in ft (or m) $H_s = 0{\cdot}625A + 0{\cdot}935B$; B is the building height; if the answer is required in metres, B (and A) must also be in m.

from the fact that if an emission standard is laid down and complied with, it is *presumptive* evidence that the "best practicable means" required under the Alkali Act are being used.

In addition to prescribing limits, the Chief Alkali Inspector also prescribes chimney stacks for sulphuric-acid manufacture, nitric-acid manufacture, copper works, and miscellaneous warm emissions of sulphur dioxide. These are given, in part, in Table 1.9.[549]

*1.6.2. United States**

Air Pollution Legislation in the United States shows the gradual entry of the Federal Government through the Department of Health Education and Welfare (D.H.E.W.) in which a Division of Air Pollution has been set up*, into an area which was previously a matter for the individual states. Initially, in 1955, Congress passed an Act which authorized D.H.E.W. to conduct research and render technical assistance to State and local governments. The next Clean Air Act (December 1963) reaffirmed State and local responsibility for control activities, but authorized D.H.E.W. to further support of local activities. A 1965 amendment set national standards for motor-car exhausts, and the most recent legislation (Air Quality Act, November 1967) sees the further entry of the U.S. Federal Government into Air Pollution Control. This Act authorizes D.H.E.W. to define regions, i.e. broad atmospheric areas, and specific air-quality control regions, and to publish air-quality criteria, on which the States must base air-quality standards. The recent legislation is very powerful in that it overrides State legislation if this is found to be inadequate.

The 1967 Act sets up a Presidential Air-quality Advisory Board, and advisory boards to assist D.H.E.W., and also requires the registration of fuel additives. Further aspects deal with research into different aspects of research into control and training of personnel.

In 1962 sixteen States had specific air-pollution-control legislation and by 1967 this had extended to all States[31] except Alabama, Maine, Nebraska, South Dakota, Vermont, and the territories of Guam and Puerto Rico. In these latter states and territories the problem was still covered by public health and nuisance laws.

The most comprehensive legislation is in California, and applies particularly to the Los Angeles County Air Pollution Control District (L.A.C.A.P.C.D.) and the Bay Area (B.A.A.P.C.D.). Emissions are divided into visible, particulate and gaseous categories, and where these overlap all regulations can be applied. In Los Angeles and the Bay Area, smoke of any colour darker than Ringelmann 2 for more than 3 min/hr is prohibited. Other regions are even more stringent, for example Utica (N.Y.) (population approx. 100,000 in 1960) permits only Ringelmann 1 (20 per cent obscuration) while in contrast to this Albany (N.Y.) and Boston (Mass.) will permit Ringelmann 3 (60 per cent obscuration).[747]

Particulate emissions permitted by L.A.C.A.P.C.D. and B.A.A.P.C.D. are limited to 0·3 gr./ft^3 (0·7 g m^{-3}), while in Pennsylvania the Allegheny County Control Agency applies standards to open-hearth furnaces, cupolas and other metallurgical operations varying between 0·2–0·5 kg/1000 kg (0·28–0·59 g m^{-3}). Other rules used relate the size of the operating unit to the amount of particulate matter that can be produced. The amount of gaseous

* In December 1970 the Division of Air Pollution of the U.S.A., D.H.E.W., became part of the Environmental Protection Agency (E.P.A.) and was renamed the Air Pollution Control Office (A.P.C.O.).

material that may be permitted depends on the type of gas. Sulphur dioxide is generally limited to 0·2 per cent by weight in most counties, although B.A.A.P.C.D. uses the resulting ground level concentration instead of the emission as a basis for control. Limits are also set for emissions of fluorides, carbon monoxide, hydrocarbons and odorous gases. The trend in legislative control is that it is tending to become a multi-county or state-wide matter, rather than of local concern only, and the regulations will tend to become more restrictive.

1.6.3. U.S.S.R.

In the U.S.S.R. air pollution is controlled through the "Office of Inspection and Technical Control of Gas Purifying Installations in Industrial Plants". This office supervises:

(i) The performance of gas purifying and dust-collection equipment.
(ii) The enforcement of maximum allowable concentration levels (M.A.C. values) for industrial plant.
(iii) The implementation of government policy with respect to clean air.

Supervisors of emissions from plants are the officers of the State Sanitary Inspectorate in the case of plants subordinate to the republics, or regions, or the local stations of the Sanitary and Epidemological Department in the case of cities. Supervision of air-pollution-control facilities in the construction of new plants is exercised at the stages when the works are being located and designed, and inspections are frequently carried out by the local authorities. An order issued in 1949 stipulated that no electric power station could be built without dust-collection plant, and similar regulations regarding collection equipment are applied to non-ferrous metallurgical works, blast furnaces in steel works, and to organic solvent recovery plants.

Very strict regulations regarding M.A.C. values, not only for 8-hour periods, but also for continuous exposure are prescribed, these being lower than those prescribed in the United States or in the United Kingdom. These values are listed in Table 1.1. As was pointed out, however, in Section 1.2, the U.S.S.R. Standards are hygienic standards, representing the optimum control situation, while realistic "sanitary standards" are not available.

1.6.4. German Federal Republic

Recent legislation (1st June 1960) in the German Federal Republic[557] is paving the way for federal control and supervision of industries, power plants, etc., which may cause air pollution (or noise). This legislation alters sections 16 and 25 of the law affecting trades and section 906 of the Civil Code. Previous legislation varied with the administrative "Länder" divisions, but the new section 16 of the law affecting trades requires the registration of trades and industries which may cause air pollution, and sets up a committee of local and federal government, technical and economic experts to advise the controlling authorities of the technical and economic limitations of pollution (and noise) control. Section 25 of the legislation limits the registration period and also empowers the authorities to supervise emissions from the industries at the industries' cost. The authorities can also require industries to establish pollution control plant as far as is economically feasible.

The alteration of section 906 of the German Civil Code which was enacted at the same time departs from the previous point of view that if an industry producing noxious gas was indigenous, no claims as to damages to the surrounding population or property could be admitted. Now, although there can be no injunction preventing the noxious process, the operator of the process may be liable for damages.

This legislation supersedes the previous regulations under the Health Acts of the administrative "Länder" (States) and municipalities. Under the new law the German "Bundesregierung" (Federal Government) issued a regulation covering fifty-two types of industrial equipment which are subject to installation permits. This covers virtually all types of plant and equipment which emit polluting substances (4th August 1960).

The next restrictions which effectively set emission standards were the Technische Anleitung zur Reinhaltung der Luft (TAL) (Technical Instructions for Air Pollution Control) which was an administrative directive addressed to the relevant administrations (8th September 1964). These contain the provisions and specific requirements for different plants. They follow the Verein Deutscher Ingenieure (VDI) Richtlinien (Directives) which have been established by the VDI Commission for Air-pollution Control. Some of these standards are shown in Table 1.10. The Standards for fly-ash emissions by boilers over 10 tonnes per hour capacity are given in VDI Standards Nos. 2091–2093 (Bituminous Coals) and 2096–

TABLE 1.10. PROCESS EMISSION STANDARDS OF THE VEREIN DEUTSCHER INGENIEURE[814]

Process	VDI Standard	Emission ($g\ m^{-3}$)
A. *Particulate Emissions*		
Ore sintering	2095	0·7
Flared blast furnace gas	2099	0·05
Coke screening, crushing and grinding	2100	0·3
Primary copper smelting	2101	0·5
Secondary copper smelting	2102	0·5

	VDI Standard	Permissible emission continuous exposure		Concentration intermittent exposure	
B. *Gaseous Emissions*		$mg\ m^{-3}$	ppm	$mg\ m^{-3}$	ppm
Nitrogen dioxide	2105	1·0	0·5	2·0	1·0
Hydrogen sulphide	2107	0·15	0·1	0·3	0·2
Sulphur dioxide	2108	0·5	0·2	0·75	0·3

Notes: (a) These values are half-hour mean values which should not be exceeded.
(b) The sulphur dioxide value is such that plants are likely to be affected, and lower values should be established when technology permits this.

2098 (Brown Coals). They consist of a series of graphs giving maximum emissions and have been discussed by Schwarz.[744]

More detailed standards qualify permissible emissions from cement plants by stack height, topography, production and works location factors.[814]

1.6.5. France

French legislation dates from the law of Morizet[44] (1932) and involves two principles.

(i) If stack emissions are toxic, or even dangerous, then precautions must be taken for minimizing these before permission is granted for the process to be carried out.

(ii) For non-toxic material, the maximum solids concentration emitted is limited to 1·5 g m^{-3} (0°C, 1 atmosphere pressure) and the maximum emission to 300 kg/hr. This last figure is most conservative and cannot be applied to modern power houses, where, even with efficient equipment, the solids output from an 360,000 kg/hr steaming capacity will be 10 times the legal limit.

The department of the Seine limits smoke to densities of Ringelmann 1 or less, except for about 5 per cent of the normal working time of the furnace. Also specified for the waste gases are maximum concentrations of sulphur dioxide (2 per cent) and carbon monoxide (1 per cent).

More recently a law has been promulgated which will enable more specific regulations to be enforced.[709] One of these (17th September 1963) provides for the creation of special protective zones, and another (19th August 1964) concerns the granting of permits to industrial installations which produce "atmospheric pollutants and smells, which constitute a public nuisance, endanger health or public security, or are harmful to agriculture, the preservation of structures, public monuments and scenic beauty".[955]

1.6.6. Holland

In the Netherlands a plant cannot be built without permission of the Inspector of Works. In the case of possible air pollution, a measure for controlling this is recommended and can be enforced by the local authority. For new medium capacity boilers the recommendations follow the "code of good practice" which was proposed by a committee in 1950. This code suggests:[517]

(i) That the dust content of the gases leaving the stacks must be less than $(H/50)^2$ g m^{-3} where H is the chimney height in metres. This means that for a 50-m stack, the maximum dust concentration may be 1 g m^{-3} while for a 100-m stack the concentration may rise as high as 4 g m^{-3} (0°C, 1 atm.).

(ii) The content of coarse particles, greater than 50 μm diameter, must be restricted to 0·025 $(H/50)^2$ g m^{-3} or one-fortieth of the total emission.

(iii) The height of the stack should be at least 1·5 times the height of the highest neighbouring building.

The rules (i–iii) are suggested only for installations in flat, open country. In built-up areas and for large power stations more stringent rules should apply.

1.6.7. Belgium

The height of chimneys that carry smoke and flue gases containing sulphur dioxide is prescribed, and based on the concentration of sulphur dioxide present.

Belgium also adopted a new law on 28th December 1964 which prevents certain kinds of pollution, regulates the use of devices and regulation for the control of pollution. The Act has to be supplemented by implementory provisions. There are also local regulations for Brussels and surrounding communes, which prohibit smoke, soot, fumes, dust, odours, vapours and toxic or corrosive gases.

1.6.8. Sweden and Denmark

No special legislation for the control of air pollution exists, but control over plants can be exercised by either local or central health authorities.

1.6.9. Italy

Italian air-pollution legislation dates from the early 1930s and is contained in the unified sanitary laws of 1934. Those factories producing unhealthy vapours must be situated in the country, while a second category deals with those industries which require some protective measures for the surrounding population. In many instances, however, population has in fact grown up around the industries of the first group and special precautions must be adhered to if the plant is to be kept in operation.

There are also specific regulations in Milan (1952) and Turin (1963).

1.6.10. Australia

Clean air legislation is a matter for the individual states, and four, New South Wales, Queensland, Western Australia and Victoria, have passed specific Clean Air Acts while most others, including the Federal Government Territories, rely on the existing legislation to prevent nuisance to control possible sources of air pollution. The exception is South Australia, which has amended its Health Act.

The Australian States legislation falls into two categories.[492] The first copies many of the provisions of the U.K. Alkali Act, with scheduled premises, which pay a licence fee (N.S.W. 1961, Queensland 1965, and Western Australia 1967), while the second resembles the U.K. Clean Air Act (1956) (Victoria 1958, South Australia 1963). The legislation in all States sets up Clean Air Councils or Advisory Committees (in Western Australia both, the latter being "technical") with different numbers of members (from 9 to 14), and the Acts also give powers to the Health Departments to make Regulations under the Acts. This latter power has been used to issue detailed Regulations similar to the "Presumptive Limits" of the U.K. Chief Alkali Inspector, and effectively give the different States very similar limits on emissions. As the Victorian and New South Wales Acts are typical of the others, some details are given below.

The New South Wales Clean Air Act (1961) empowers the Department of Health, which has an Air Pollution Control Branch to license scheduled industries, and charge a licence fee. Its officers can enter the scheduled premises and conduct tests and require the installation or alteration of existing equipment for controlling pollution. The department may also prohibit scheduled processes in certain areas, and has very wide advisory powers. To assist with this, the Act requires an advisory body of twelve members drawn from industry, trade unions, government departments and the universities to be set up.

Scheduled industries are those which present particularly difficult air-pollution-control problems, such as the iron and steel industry. Non-scheduled industries, such as small boilers, are controlled by the local authorities.

The Victorian Clean Air Act of 1958 has as its basis the Clean Air Act (England and Wales) of 1956 and specifically prohibits the emission of dark smoke from industrial chimneys. Other aspects of the legislation include the control of plans for new industrial establishments to ensure the fullest precautionary measures to prevent pollution, and the establishment of a Clean Air Committee which has powers to advise the Minister as to the abatement of pollution from power stations, locomotives, ships, aircraft and motor vehicles. An Air Pollution Control Division in the Environment Protection Authority administers the Act, and has powers of entry into premises and the investigation of pollution.

Important Regulations under the Victorian Act (1961) require that installers of plant which produce gaseous wastes are required to submit the plans to the Health Department, where engineers check these and make sure that "best practicable means" of preventing air pollution are installed. These regulations also require that plant operators are provided with means of checking the chimney exits of their plants, so that visible discharges can be detected from within the plant. Later regulations specify smoke-density indicators or closed-circuit television as alternatives to direct sighting.

An Australian Commonwealth Senate Select Committee on Air Pollution was set up in 1968, and as a result it is probable that uniform legislation will be enacted by the individual States, or overriding Commonwealth legislation could also result.

1.6.11. Canada

Here, as in Australia, clean air legislation is a state matter, and Ontario passed an Air Pollution Control Act in 1958. This legislation enables municipalities to pass by-laws, and gives them power to administer and enforce the air pollution abatement policy. In 1955 the Windsor area in this state, which is heavily industrialized, had an ordinance which did not define the prohibited smoke density.[747] Detailed information regarding air pollution control in a particular area must be obtained from the local authorities.

The municipal act of the province of Nova Scotia (1955) was amended in 1960 to give municipal councils power to regulate the emission of air pollutants (smoke, gases and odors). The regulations enable standards to be set up and enforced, to require the installation of control equipment, and furthermore, officials have the power and authority to enter premises.

A new Regulation (22nd April 1960) in Manitoba replaced an earlier Regulation (No. 91,

1945). This defined air pollutants (air contaminant, dust, fly ash, fume, etc.) and the emission limits which are legally enforceable. Smoke emissions have been defined in terms of the Ringelmann Chart: Ringelmann greater than 1 and less than 2 is permissible for no more than 4 min per half hour, or 8 min at any one time; Ringelmann greater than 2 and less than 3 for no more than 3 min per 15 min or more than 6 min at any one time. Solid emissions are limited to less than 0·4 gr/ft^3 (0·92 g/m^3) and sulphur dioxide to 0·2 per cent of the volume of gases present.

The regulation for Alberta (30th August 1961) is similar to those in Manitoba, but also requires the submission of plans of pipelines and plants to the Provincial Board of Health before their construction.

1.6.12. New Zealand

The New Zealand Health Act (1956) has many similarities to the United Kingdom Alkali, etc., Works Regulation Act (1906), and in fact was drafted after a visit from the Chief Alkali Inspector. Plants which produce gases termed "noxious and offensive" must have effective means of preventing their discharge, and the emissions from sulphuric-acid plants are restricted. "Chemical Inspectors" have been appointed with right of entry, inspection and investigation to supervise these processes.

The New Zealand Air Pollution Regulations (1957) which were established under the 1956 Act list similar limits to the Alkali Inspectors Statutory and Presumptive Limits. An important difference is the much stricter control on emissions from sulphuric-acid plants (1·5 gr/ft^3, 3·5 g m^{-3}). A specific New Zealand development is the control of "Animal Products Works" (New Zealand Chemical Works Order 1960), for controlling odours from the rendering of animal and fish offal, which is a much more significant process in New Zealand than in England. Another unique New Zealand problem is the release of hydrogen sulphide from geothermal sources in built-up areas, particularly the town of Rotorua. No specific regulations are so far available for this source, but it is significant because of the use of hydrogen-sulphide-laden steam for domestic heating and other services.

1.6.13. Ireland

The Republic of Ireland amended its Local Government (Sanitary Services) Act in 1962, introducing very powerful clauses, which are similar to those of the U.K. Alkali, etc., Works Regulation Act. This Act enables the Minister for Local Government to make regulations for the following:

(a) controlling sources of pollution of the atmosphere including the emission of smoke, dust and grit;
(b) regulating the establishment and operation of trades, chemical and other works and processes which are potential sources of atmospheric pollution;
(c) specifying maximum concentrations of pollutants in the atmosphere;
(d) measurement of emissions of pollutants into the atmosphere;

(e) investigation of, and obtaining information on emission of pollutants into the atmosphere;
(f) testing, measuring and investigating atmospheric pollution;
(g) regulating potential sources of pollution of the atmosphere from radioactive materials;
(h) specifying particular controls of atmospheric pollution for particular areas;
(i) licensing of persons engaged in specified works or processes, being works or processes discharging pollutants into the atmosphere, and prohibiting the engagement in such works or processes of persons other than licensed persons;
(j) licensing of premises from which pollutants are discharged, and prohibiting discharge from other than licensed premises;
(k) the cancellation or suspension of licences.

The Act also provides for the right of entry by authorised persons into any premises for the purpose of checking on contravention of regulations. The Act also covers ships in harbours or rivers, which, although not specifically within the district of a sanitary authority, is deemed to be within the district of the nearest one. This is different to most Acts, which exclude shipping and harbours.

1.6.14. Japan

A National Law "With Regard to Regulate, etc., the discharge of Smoke" was promulgated in June 1962 and enforced since April 1963. Before this a few prefectures including Tokyo and Osaka had local smoke-control regulations. The term "Smoke" refers to soot, dust and sulphur dioxide generated by combustion processes (including thermal power generation). The law also designates "Toxic Substances" (HF, H_2S, SeO_2, HCl, NO_2, SO_2, Cl_2, SiF_2, $COCl_2$, CO_2, HCN, NH_3).[847]

CHAPTER 2

BASIC DATA REQUIREMENTS

2.1. INTRODUCTION

Before it is possible to specify suitable gas-cleaning plant, the characteristics of the gases and the quantity of gas to be treated must be known. The temperature and chemical nature of the gases and the types of particles to be collected will determine the type and size of plant as well as the materials of construction. Consideration should also be given to the dew point of the gases. This may be surprisingly high if sulphur trioxide is present, and sets a minimum operating temperature in some cases, such as bag filters. A high dew point may be an advantage in other cases as it indicates very often a favourable operating temperature for electrostatic precipitators collecting fume with high electrical resistance. Here sulphur trioxide is sometimes added to the gases to achieve the dew-point elevation, and the quantity required has to be known.

If a constituent gas is to be scrubbed out into solution, the rate of mass transfer of the gas to the scrubbing liquor must be known, while the removal of small particles requires an understanding of the physical characteristics of the particles: size distribution, density and shape. Whether the particles are good conductors, or whether they are magnetic, may also be of importance.

This chapter will discuss the measurement of gas temperature and the measurement and calculation of gas-flow rate and dew point and the methods of sampling of gases and particles. Also the chief methods of particle size determination will be outlined, although a detailed treatment is beyond the scope of the present volume.

2.2 DETERMINATION OF GAS FLOWS

The volumes of gas to be treated can be found either by direct measurement of the gas flow in a duct, pipe or chimney, or by calculation, if the gas composition is known, from the quantity and composition of reactant materials and the chemical nature of the process which is usually combustion.

2.2.1. Measurement of gas flows

Gas flows are most frequently measured with a differential head device—orifice plates, venturi tubes or pitot tubes—although for work at ambient temperatures in non-toxic gases, particularly air conditioning, a vane anemometer is also frequently employed. For very low rates of flow, when a pitot tube response is very small and may not be stable, a thermo-anemometer, which depends on the resistance change with variable loss of heat from a hot wire placed in the gas stream, can be very useful, but the instrument is not very robust and has not yet found wide industrial application.

Orifice plates and venturi tubes are usually permanently installed in a plant. Orifice plates require a straight run of duct of at least nine duct diameters (D.D.)—six before and three after the plate—for ordinary conditions, although the introduction of very turbulent conditions may require longer calming sections. Venturi tubes require even longer straight duct sections to allow for the long diffuser section and are more complex to construct than simple orifice plates, but have the advantage of a much lower pressure loss. Details of construction and the limitations of these devices may be obtained from British Standard No. 1042 or the appropriate American Standard.

Pitot tubes are very versatile. They do not require a long calming section in a duct as they measure local velocities: they are small so that they can be inserted through a small hole in the side of the duct without causing plant shut down and do not cause any appreciable pressure loss in the gas stream. Their chief disadvantage in finding a total gas flow is that a number of velocity measurements have to be made to establish the velocity profile of

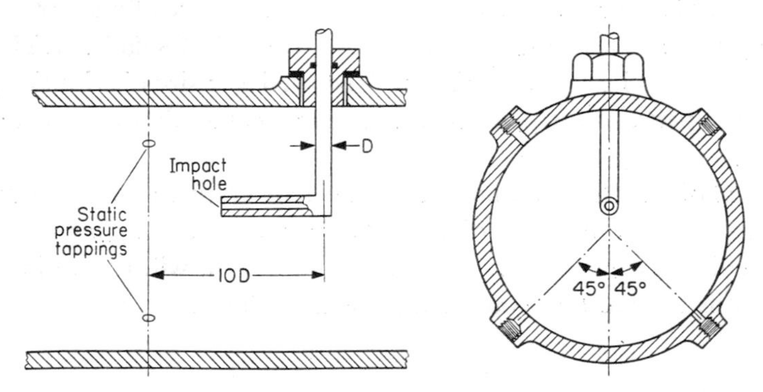

FIG. 2.1. Hook-type pitot tube for measuring impact pressure with separate static pressure tapping in pipe or duct wall.[185]

the gas stream. This is then integrated, usually graphically. Consequently if there are sudden fluctuations in the gas stream, erroneous flow rates may be deduced.

Essentially pitot tubes consist of an impact and a static pressure tube. In its simplest form (Fig. 2.1), the "hook" type, these may be separate tubes with the static pressure being measured at the duct wall. This may be used for very small ducts, where multiple wall tubes may be too large and cause flow pattern deformation or where a standard pattern tube is not available. The tube nose is best tapered with an 8° total cone angle, but it may be square ended as shown.

Usually the impact and static pressure tubes are combined into a single unit. The British Standard (British Standard No. 1042)[128] patterns are shown in Fig. 2.2. The ellipsoidal head shown in Fig. 2.2(b) is to be recommended at low-flow velocities. It is more robust than the N.P.L. sharp-ended tube shown in earlier editions of the standard.

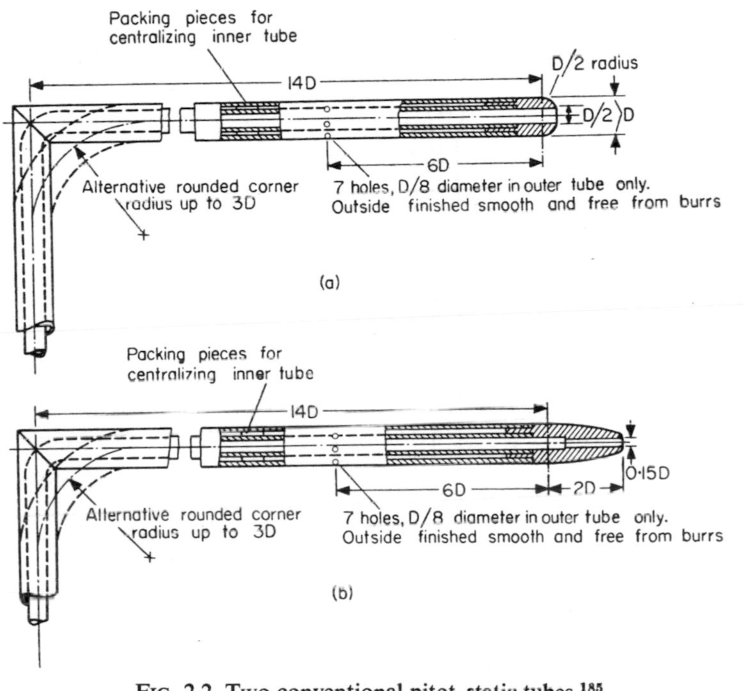

FIG. 2.2. Two conventional pitot–static tubes.[185]

(a) Hemispherical head pitot–static tube.
(b) Ellipsoidal head pitot–static tube.

In addition to the conventional pitot tube, a reverse type is also made (Fig. 2.3) which has the advantages of being more easily inserted into a duct, and the pressure differential is greater than in the types shown in Figs. 2.1 and 2.2.

The gas velocity in these can be calculated from the empirical relation

$$U = 1{\cdot}256\,\sqrt{(\Delta h/\varrho)}\ \text{m s}^{-1} \qquad (2.1)$$

where Δh = pressure difference (pascals),

ϱ = density of gas (kg m^{-3}).

For lower gas velocities hot-wire anemometers are useful. They are temperature dependent and require comparatively frequent calibration. More recently thermistors have been successfully used to temperatures up to 50°C, and the relation between air flow, temperature, humidity and pressure determined.[838] These should prove of great value in measuring gas-stream velocities continuously, at ambient temperatures.

When it is necessary to measure both speed and direction of gases, such as in jets or other swirling gases, a *yawing* pitot tube can be used. One such instrument, with a spherical head and water cooled for use in very hot conditions, has been developed by the United Steel's Research Laboratories.[630]

When using the pitot or other "point" velocity measuring instrument, velocity measurements must be made at a number of points, preferably in a straight section of the duct.

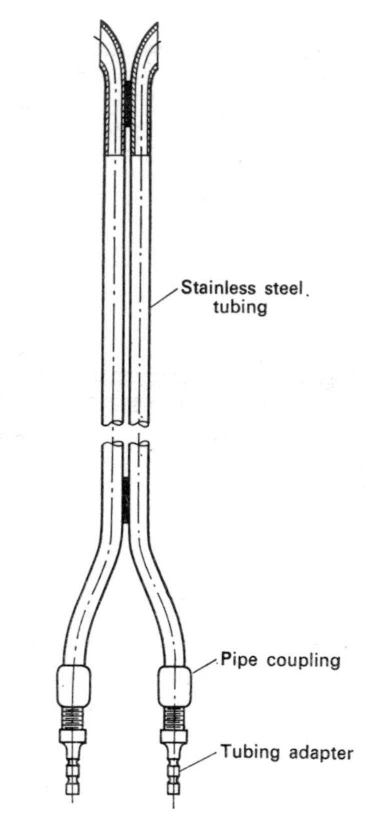

FIG. 2.3. Double or reverse type of pitot tube.[920]

Circular ducts are usually divided into a number of annuli of equal area and the velocity measured on two diameters intersecting at right angles giving four velocities for each annulus. When it is difficult to measure the flow near the wall, this may be estimated from a plot of velocity against distance from wall on a log–log scale. In rectangular ducts, or ducts of odd shapes, the duct cross-section is divided into a number of small rectangles and the velocity measured in the centre of each. This is most conveniently carried out from a number of traverse entry points either in one wall of the duct, or preferably, in two adjacent walls. With fluctuating gas flows it is best to measure the variation in gas flow at one point only. If the flow is in the streamline region, the velocity at the centre of a circular duct is twice the average velocity. Alternately it is possible to measure the velocity at a radial position of 0·762 times the pipe radius from the axis of the pipe.

For relatively small gas flows a number of other meters are used, such as displacement-type wet gas meters or the bellows type which is universal for town gas. Piston, fan and variable aperture-type meters ("Rotameters") are also used for moderate gas-flow measurement, particularly for process control. These require the whole of the gas flow to pass through the meter, and are therefore normally used to measure the volumes of gas sampled in a sampling train rather than the total flow in a duct.

Comprehensive details of flow measurement equipment may be obtained from specialized books such as *Flow Measurement and Control*[185] and from the appropriate standards and handbooks.

2.2.2. Calculation of flow rates

Very frequently a gas-cleaning plant has to be designed before the process to which it is to be attached is in operation. In the most common case the process is combustion of coal or fuel oil, the compositions of which are known. The composition of the waste gases can either be calculated, assuming complete or partial combustion and a representative fuel–air ratio or determined by gas analysis on the plant itself or a similar one. The calculation of the mass flow of the gases then becomes a problem in material balances. If the gas temperature is also known or is estimated from an energy–heat balance, then the gas volume can be calculated. A typical example is shown in Appendix I.

2.3. MEASUREMENT OF GAS TEMPERATURE

The determination of actual gas temperatures to a reasonable degree of accuracy is essential to a proper choice of cleaning mechanism and cleaning medium.

When the gases to be cleaned are in a vessel, or flowing along a duct, the walls of which are not at the same temperature as the gas, the method of measuring the temperature of the gas must allow for the thermal radiation effects inherent in the system. Thus, when the gases are colder than the enclosing walls, the walls will radiate heat to the temperature-measuring element, and the temperature recorded will be warmer than the actual gas temperature. Conversely, for hot gases flowing in a colder duct, the temperature-measuring element will radiate heat to the enclosing walls, and so will be unable to reach the actual gas temperature. This radiant heat transfer results in a temperature difference and increases very rapidly with temperature and at 1500°C it may be of the order of 200–300°C. In a well insulated vessel or duct, where the inside wall temperature is close to that of the gases, the temperature measured by the temperature-sensing element will be virtually that of the gas.

The temperature-sensing elements

Thermometers. The ordinary mercury in glass thermometer will cover the temperature range from below 0°C to about 350°C. An alcohol-in-glass thermometer extends the lower end of this range to −80°C.

Platinum resistance thermometers are very accurate, and have a range from −200 to +600°C and sometimes higher.

Thermocouples. Copper–constantan thermocouples cover the range from −200 to about +400°C, while other base metal thermocouples, the most common being the chromel–alumel couple, have an upper limit of 1200°C.

TABLE 2.1. MATERIALS FOR THERMOCOUPLES FOR USE AT TEMPERATURES UP TO 2800°C*

Material	Temperature range†	Comments
Copper vs. constantan	−200–400°C (500°C)	
Iron vs. constantan	0–800°C (1100°C)	
Chromel vs. alumel	0–1100°C (1300°C)	
Platinum vs. Pt 8·7–13% Rh	0–1600°C (1700°C)	
Pt 20% Rh vs. Pt 60–40% Rh	0–1800°C (1850°C)	
Iridium vs. Ir 40–60% Rh	1400–2000°C (2100°C)	(neutral atmosphere)
Iridium vs. tungsten	1000–2200°C (2300°C)	(vacuum, neutral atmosphere or reducing conditions)
Tungsten vs. W–Re	up to 2500°C** (2800°C)	(neutral atmosphere)

Ir = Iridium, Pt = Platinum, Re = Rhenium, Rh = Rhodium, W = Tungsten.

* Some of the data were obtained from[25]; others are quoted by courtesy of Engelhard Industries Ltd.[241]
† Figures in brackets are extreme temperatures which can be attained for "spot" readings.
** Temperature–EMF relation (non-linear) calibration is provided for each batch supplied by the makers. The calibration of these thermocouples has been described by Lachman.[471]

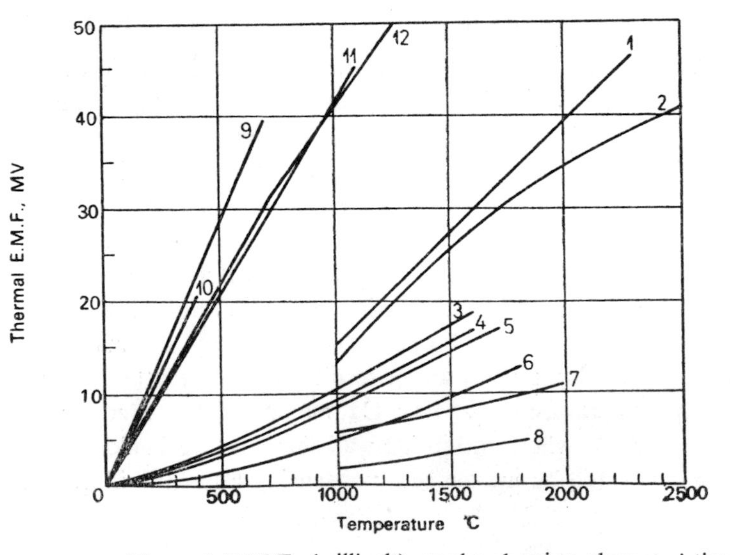

FIG. 2.4. Temperature vs. Thermal E.M.F. (millivolt) graphs showing characteristics of thermocouples.[241] 1. Iridium vs. tungsten. 2. Tungsten-rhenium vs. tungsten. 3. Platinum vs. 87% platinum-13% rhodium. 4. Platinum vs. 90% platinum-10% rhodium. 5. 99% platinum-1% rhodium vs. 87% platinum-13% rhodium. 6. 95% platinum-5% rhodium vs. 80% platinum-20% rhodium. 7. Iridium vs. 60% iridium-40% rhodium. 8. 80% platinum-20% rhodium vs. 60% platinum-40% rhodium. 9. Iron vs. constantan. 10. Copper vs. constantan. 11. Chromel vs. alumel. 12. Platinel.

Noble metal thermocouples can now be used to temperatures as high as 2400°C continuously, or 2800°C intermittently (e.g. a tungsten–rhenium couple). The most common thermocouples and their characteristics are given in Table 2.1 and Fig. 2.4, which gives their temperature–EMF curves.

For use in corrosive atmospheres, thermocouples must be sheathed for protection. Metal sheaths are satisfactory below 1000°C, while silica sheaths can be used to slightly higher temperatures, but are themselves corrosive with respect to noble metal thermocouples. Recrystallized alumina sheaths may be used to about 1850°C. Beryllium oxide (BeO) can be used to higher temperatures.

Calibration of temperature sensing elements. The primary standards of temperature are the melting point of ice (0°C) the boiling point of water (100·0°C) the boiling point of sulphur (444·60°C) and the melting points of silver (960·5°C) and gold (1063°C). The melting point of palladium, at 1552°C, is a secondary standard, while the melting point of nickel (1452°C) can also be used. Platinum melts at 1769°C. The melting points of some other metals (rhodium, 1960°C, iridium 2443°C, etc.) can be used to extend the scale upwards. Lachman[471] used a standardized micro-optical pyrometer when calibrating noble metal thermocouples to 2200°C in a high-temperature research furnace. More recently Zysk and Toenshoff[965] have given details of furnace designs used up to 2400°C with a tungsten-tube furnace (Fig. 2.5). Special precautions are taken to eliminate emissivity corrections by

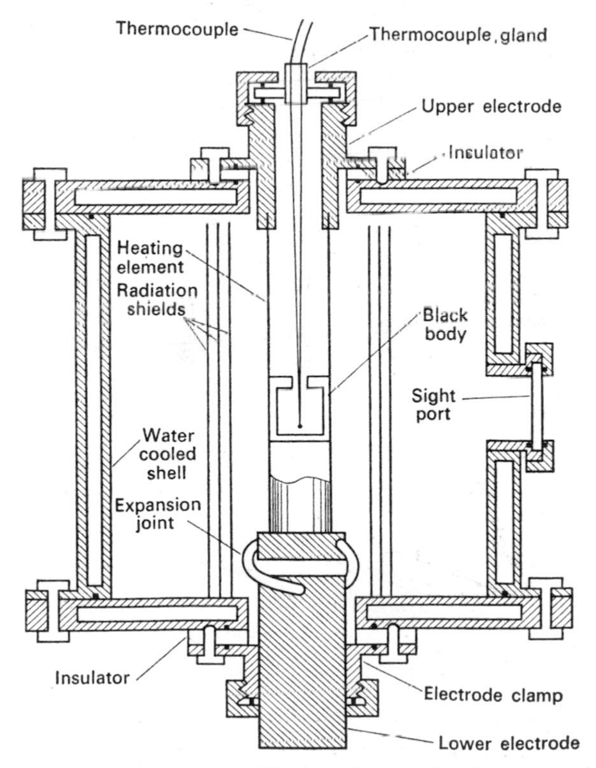

FIG. 2.5. High-temperature calibration furnace for thermocouples.[965]

containing the thermocouple in a black-body consisting of a molybdenum cylinder with a BeO cover.

The radiation losses from temperature-sensing elements can be reduced in several ways.

(a) *Thermocouples of different sizes.* The radiation heat loss from a thermocouple is proportional to the surface area of the head formed where the two wires are joined. Thus, the losses will be less with smaller thermocouples. If then a series of thermocouples of different sizes are used, and the temperatures measured plotted graphically against the surface areas of the thermocouples, the curve obtained (which is usually virtually a straight line), can be extrapolated to the temperature ordinate at zero surface area. This temperature is then a measure of the actual gas temperature.[861] The method is particularly useful in closed systems, or where withdrawal of the large quantities of gases required for suction and pneumatic pyrometers would interfere with the gases to a marked degree.

(b) *Suction pyrometers.* Another way of finding the actual gas temperature is to shield the temperature measuring element from the radiation of the walls by surrounding it with

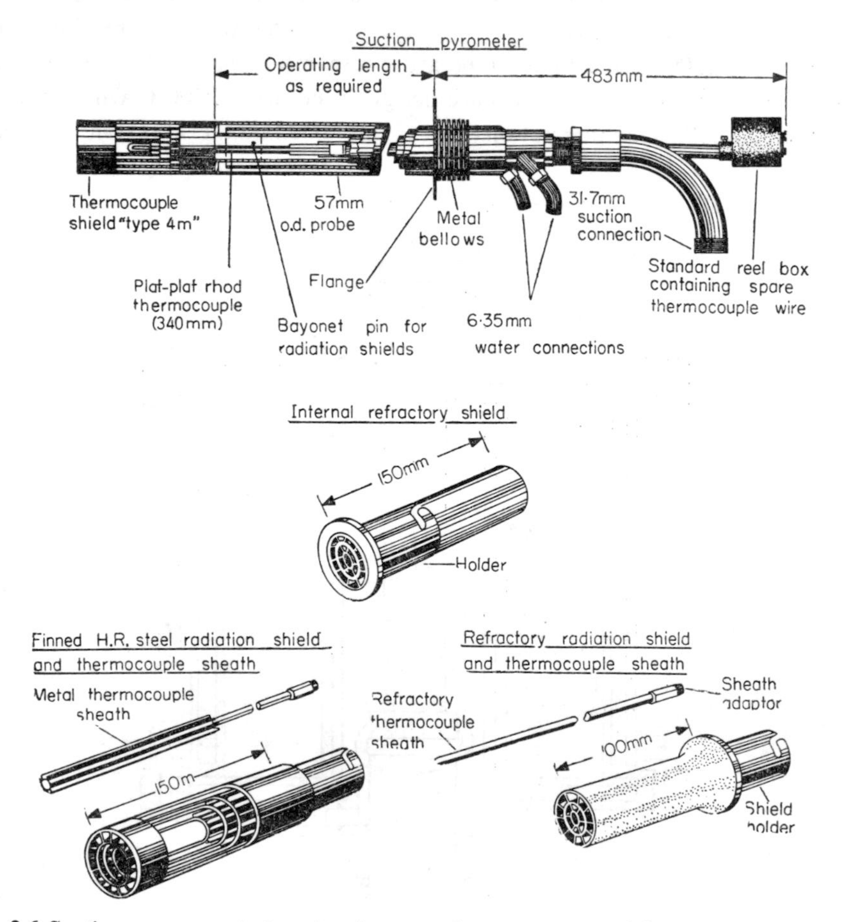

FIG. 2.6. Suction pyrometer design, showing general arrangement and three types of radiation shield.[481]

so-called "radiation shields" and aspirating the gases past the element (usually a thermocouple) at high velocities in order to increase convective heat transfer to the element. The number of shields required is a function of the gas temperature, while the construction of the shields depends on the chemical properties of the gases as well as on the temperature.

Thus, Inconel (75–80 per cent nickel, 15–20 per cent chromium, rest iron) is satisfactory to temperatures of 1100–1200°C when only a little sulphur dioxide is present in the gases. For gases with a high sulphur-dioxide content, a high chromium stainless steel is preferable and may be used to about the same temperature. At higher temperatures the sheaths are made of refractory cements (mullite, alundum).

The thermocouple itself is usually sheathed to protect it against oxidation from the gases. The assembly of thermocouple, sheath and radiation shields is mounted on the end of a stainless-steel probe which must be water cooled if it is to be used at temperatures exceeding 900°C. A typical suction pyrometer is shown in Fig. 2.6 together with three types of radiation shields. The gas flow is provided by an ejector and measured with an orifice plate.

Since the refractory multiple shields are very fragile, they have been modified for mounting within the walls of the suction pyrometer, as is shown in Fig. 2.6. The shields extend 150 mm back from the tip of the pyrometer. It has been shown experimentally[54] that the best position for the thermocouple in this type of suction pyrometer is about 35 mm from the end; further forward the couple tends to receive some radiation from the walls through the gas entry, while somewhat further back the water-cooled walls of the probe radiate cold to the couple.

TABLE 2.2. EFFICIENCY PER CENT FOR SHEATHED THERMOCOUPLES WITH BLACKENED METAL MULTIPLE RADIATION SHIELDS. GAS VELOCITY 150 m s^{-1} [479]

Temperature, °C	Number of radiation shields									
	1	2	3	4	5	6	7	8	9	10
400	98									
600	88	98								
800	71	93	98							
1000	54	81	93	97	99					
1200	39	69	85	92	95	98	99			
1400	28	55	74	83	91	95	97	99	99	
1600	20	43	62	73	83	90	94	96	97	98

The accuracy of a temperature measured by a suction pyrometer is a function of the actual gas temperature, the temperature of the surroundings and the velocity at which the gases are being drawn past the thermocouple. If no gas is sucked past the thermocouple, the reading obtained has a certain error, but when the gas velocity is increased, the amount of the error will be reduced. The fraction of this reduction in error is called the "efficiency" of the pyrometer. The efficiency of a suction pyrometer operating at 150 m s^{-1} has been found for a series of temperatures to 1600°C and with up to ten radiation shields (Table 2.2).

The effect on the efficiency of using velocities other than 150 m s^{-1} can be obtained by multiplying the actual number of radiation shields by a factor obtained from Table 2.3 which gives the equivalent number of shields at 150 m s^{-1}.

TABLE 2.3. THE EFFECT OF USING GAS VELOCITIES OTHER THAN WITH MULTIPLE SHIELD SUCTION PYROMETERS[479]

Velocity (m s^{-1})	3	6	15	30	60	90	120	150	180	210
Equivalent number of shields at 150 m s^{-1}	0·21	0·28	0·40	0·52	0·69	0·82	0·93	1·00	1·07	1·14

An allowance has also to be made for the cooling effect of aspirating the gases past the thermocouple. This is given in Table 2.4, and is of the order of 20°C at velocities of 200 m s^{-1}.

As the gases at high temperatures have lower density than at ambient temperatures, and the mass flow is measured after the gases have been cooled in the water-jacketed probe, a density correction must be applied to the flow measurement at the orifice plate when estimating the gas velocity past the thermocouple. The values for the mass flow of air given in Table 2.5 enable the gas velocity past the thermocouple to be estimated from the orifice plate gauge reading.

TABLE 2.4. COOLING OF GASES DUE TO ASPIRATING AT HIGH SPEEDS[479]

Velocity (m s^{-1})	30	60	90	120	150	180	210	240
Temperature drop (°C)	$\frac{1}{2}$	$1\frac{1}{2}$	$3\frac{1}{2}$	$6\frac{1}{2}$	10	14	19	25

If metal radiation shields are used, these soon become blackened and act as black-body emitters (emissivity = 1) but if refractory radiation shields are used, these glow at high

TABLE 2.5. MASS FLOW OF AIR, kg h^{-1} mm^{-2}

Temperature (°C)	Velocity, m s^{-1}								
	30	60	90	120	150	180	210	240	270
200	0·141	0·163	0·245	0·362	0·408	0·489	0·571	0·652	0·733
400	0·059	0·118	0·177	0·236	0·295	0·354	0·413	0·471	0·531
600	0·044	0·089	0·132	0·176	0·221	0·265	0·309	0·353	0·397
800	0·036	0·072	0·108	0·143	0·179	0·215	0·251	0·287	0·323
1000	0·030	0·060	0·091	0·121	0·151	0·181	0·212	0·242	0·272
1200	0·026	0·052	0·079	0·105	0·131	0·157	0·183	0·209	0·236
1400	0·023	0·046	0·069	0·092	0·115	0·138	0·162	0·184	0·207
1600	0·020	0·041	0·062	0·082	0·103	0·124	0·144	0·165	0·186

temperatures, and have an emissivity of less than 1. The data in Tables 2.2 and 2.3 assume the use of black shields.

The equivalent number of other types of shields can be calculated by the use of a factor "f" which is a function of the ratio of the thickness of the refractory tube w and its thermal conductivity $\varkappa$, where:

$$f = 4aT^3 \frac{w}{\varkappa} \tag{2.2}$$

w = refractory thickness—mm
$\varkappa$ = refractory thermal conductivity cal/(°C) (mm) (sec),
T = temperature—°C
a = constant.

TABLE 2.6. VALUES OF f FOR VARIOUS VALUES OF $w/\varkappa$[479] (C.G.S. UNITS)

Temperature °C	$w/\varkappa \times 10^{-2}$			
	10	20	50	100
1000	0·1	0·2	0·6	1·1
1200	0·2	0·3	0·9	1·8
1400	0·3	0·5	1·3	2·6
1600	0·4	0·7	1·8	3·6

The values of f in terms of the ratio $w/\varkappa$ can be obtained from Table 2.6. From the known emissivity of the refractory and the value of f the effective number of single metallic shields for which each refractory shield is equivalent can be read from Table 2.7.

TABLE 2.7. EFFECTIVE NUMBER OF SIMPLE METALLIC SHIELDS TO WHICH EACH REFRACTORY SHIELD IS EQUIVALENT[480]

f \ Emissivity	1	0·8	0·6	0·4	0·2
0	1·0	1·2	1·5	2·0	3·0
1	1·4	1·6	1·8	2·2	3·2
2	1·7	1·9	2·1	2·4	3·3
3	2·0	2·1	2·3	2·6	3·5
4	2·2	2·3	2·5	2·8	3·6

The refractory shields may require a fin allowance which is equal to $\sqrt{2}$, thus multiplying the equivalent number of shields by 1·41. If the refractory shields are of the packed tube pattern (Schack type) this allowance is a factor of 2·5. An example illustrating this is given in Appendix II.

Instead of using Tables 2.2 and 2.3 to calculate the suction pyrometer efficiency, a more rapid method for obtaining an estimate of the actual gas temperature is based on the fact that the efficiency of the instrument increases very rapidly (from zero) when the gas is first speeded up, and less rapidly with further increases in velocity. The shape of the velocity–temperature curve from zero to the actual flow rate can therefore be used to indicate the efficiency at the flow rate. An index suggested for specifying the curve shape is based on the temperature $T_{1/4}$ shown by the instrument at a quarter of the specified maximum flow:

$$\text{Shape factor} = \frac{T_{max} - T_0}{T_{max} - T_{1/4}} \tag{2.3}$$

where T_0 = indicated temperature with zero flow velocity,
T_{max} = indicated temperature with maximum flow velocity.

Table 2.8 gives the efficiency at the maximum measured flow rate for a number of shape factors.

TABLE 2.8. EFFICIENCY FOR VARIOUS SHAPE FACTORS[479]

Shape factor	2	$2\frac{1}{2}$	3	4	6	8	11
Efficiency %	63	80	87	93	97	98	99

The response of suction pyrometers is relatively fast and depends on the material of the head. Thus metal heads take about 2 min to reach equilibrium temperature, while zircon heads take twice as long.

When gas densities are low (i.e. sub-atmospheric) the rate of heat transfer from the gas to the thermocouple is reduced, and at 30 kPa errors with conventional suction pyrometers tend to be of the order of 5 per cent. This problem can be overcome by placing the thermocouple junction into the gas stream moving past it at sonic velocities, just behind the throat of a convergent nozzle machined into the end of the inner radiation shield.[14]

(c) *The pneumatic pyrometer.* If a gas is drawn through a restriction in a pipe, the pressure drop across the restriction is a function of the geometry of the restriction, the mass flow of the gas, and its density. The density, in turn, is a function of the absolute pressure of the gas, its absolute temperature, and its composition. In the pneumatic pyrometer a continuous sample of the hot gases are drawn through a restriction, cooled and then drawn through a second restriction, where the temperature of the gases is also measured. The hot gas temperature is calculated from the pressure losses at the two restrictions and the temperature at the second (cold) restriction. In the most fully developed of these instruments, the calculation is carried out automatically by means of a simple analogue computer.

Basically, for a non-compressible gas, the pressure loss, Δp, across a constriction in the tube is related to mass flow, m, by the equation:

$$\Delta p = km^2/\varrho \tag{2.4}$$

where k is the orifice coefficient and ϱ is the gas density.

From the perfect gas law,

$$\varrho = MP/RT \tag{2.5}$$

where M = molecular weight of the gas,
P = total gas pressure,
R = universal gas constant.

In the pneumatic pyrometer, the two constrictions are in series, so the mass flows are the same for both restrictions. If subscript 1 refers to the hot restriction and subscript 2 to the cold restriction, then

$$\frac{P_1 \Delta p_1 M_1}{k_1 R T_1} = \frac{P_2 \Delta p_2 M_2}{k_2 R T_2}. \tag{2.6}$$

If the pressure loss between the two restrictions is negligible, $P_1 = P_2$. Furthermore, since the dissociation of gas molecules, e.g. in flue gas, is negligible up to 2000°C, $M_1 = M_2$. Even at 2500°C the effect is calculated to be only about 5 per cent for a typical flue-gas composition.[392] It is also important that no condensation occurs between the restrictions and that the flow is in the turbulent range.

Equation (2.5) can then be simplified to give

$$T_1 = k \frac{\Delta p_2}{\Delta p_1} T_2 \tag{2.7}$$

where $k = k_2/k_1$.

An instrument developed largely by the B.C.U.R.A.[392] uses this principle in the venturi pneumatic pyrometer, shown in Fig. 2.7.

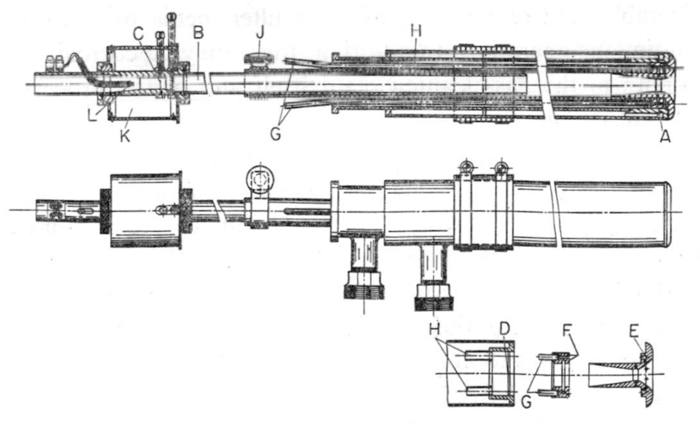

FIG. 2.7. Venturi pneumatic pyrometer.[392]

A — Hot venturi throat.
B — Gas tube.
C — Cold venturi throat.
D — Position for sealing ring.
E — Hot venturi insert.
F — Upstream pressure tapping point.
G — Pressure line.
H — Pressure lines.
L — Temperature sensing element.

The probe, of 57 mm dia, is water cooled and made either of brass or of stainless steel. Its length depends on the particular application. The design of the hot venturi presented many problems. It is made of corrosion resistant steel and is easily replaced owing to the use of rubber O–rings to seal the connection of the pressure tapping to two tubes in the water-cooled jacket. A brass liner is provided for the water-cooled tube to prevent condensation of the moisture in the gases. The cold end venturi is placed after the water-cooled section, and the gas temperature is measured with a platinum resistance thermometer because these have a resistance change proportional to the absolute temperature in the required temperature range.

The measuring equipment[482] must automatically compute equation (2.6). The pressure differentials Δp_1 and Δp_2 are therefore converted to electric currents using pressure transducers. These are Beaudouin 0–5 kPa units supplied with 1 kHz a.c. from a transistorized oscillator. The transducers each incorporate a differential transformer so that their output is proportional to the pressure differentials applied to them. The a.c. output is rectified with a silicon rectifier and then one current is passed through the resistance thermometer and the other to the slidewire of a potentiometer. The potential difference across the resistance is applied to the input of the potentiometer, which then indicates directly the absolute temperature of the hot gas.

This arrangement is much simpler than earlier methods which used a thermocouple instead of the resistance thermometer[393].

2.4. ESTIMATION OF DEW POINT

An accurate knowledge of the dew point of the gases to be cleaned is of primary importance, for collection below the dew point causes deposition of water droplets in which corrosive substances (e.g. sulphur trioxide) may be dissolved. These shorten the life of plant very considerably, and reduce that of some filter media to a few hours instead of some years of continuous operation. In addition, there may be corrosion of the materials of construction, both in the gas-cleaning plant itself and in the process plant—e.g. the economizers in a boiler plant.

In combustion gases the moisture comes from the air used, as well as from hydrogen in the fuel. Consequently, on days with high atmospheric humidity, the waste gases, after cooling to ambient temperatures, will almost certainly be supersaturated and deposit moisture on collecting media, if these are below the dew point. If the fuel contains some sulphur, sulphur dioxide and sulphur trioxide will be present in the waste gases. Sulphur trioxide in very small quantities (0·005 per cent) with about 10 per cent water vapour will raise the dew point of the gases to about 150°C.[402]

The dew point can be found by an instrument which measures the temperature of a smooth, non-conducting surface set between two electrodes. When the material is dry, its resistance is infinite, but when a film of moisture condenses on the surface, the resistance is lowered, and a current passes between the electrodes. One of the earliest models was made by Johnstone[402] and a recent, more elaborate modification has been described by Bassa and Beer.[62]

Alternatively, the dew point can be estimated if the sulphur-trioxide content of the gases is known. The sulphur-trioxide determinations are made more difficult by the fact that frequently about 10 to 100 times as much sulphur dioxide is present, and the sulphur dioxide slowly oxidizes to sulphur trioxide. This may be overcome by adding 6 per cent pure benzyl alcohol as an inhibitor to the absorbing (N/5 alkali) solution. Benzaldehyde, mannitol or *p*-amino phenol hydrochloride can also be used. The sulphur trioxide is estimated as the sulphate. Summaries of the standard methods of estimation have been given by Corbett and Crane.[175]

The estimation of dew-point elevation can be made very simply from the ratio of the partial pressure of sulphuric acid, which may be taken to be the same as that of the sulphur trioxide (SO_3), and of the water vapour, using the graph derived by Müller[586] (Fig. 2.8).

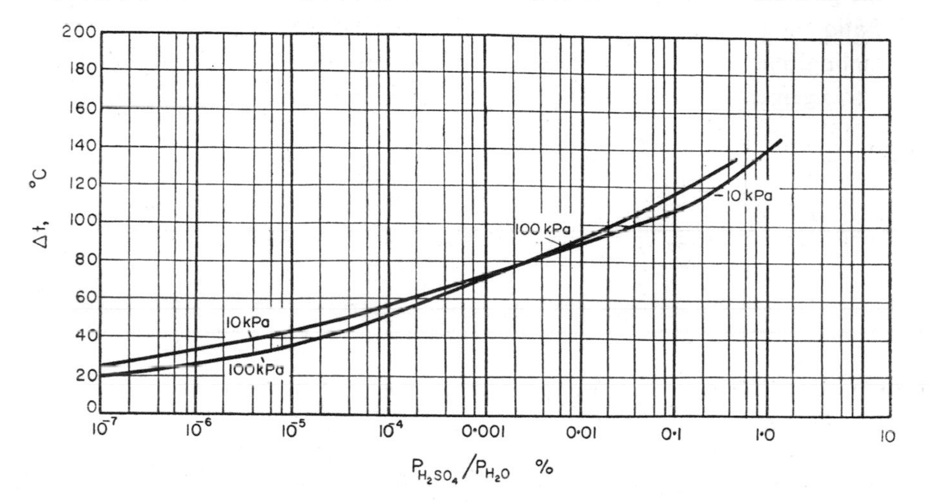

FIG. 2.8. Elevation of dew point by sulphur trioxide. Correlation of temperature elevation and partial pressure ratio $P_{H_2SO_4}/P_{H_2O}$.[586]

The curves are drawn for water-vapour pressures of 10 and 100 kPa, and the dew-point elevation can be read off directly. The dew point of air and pure water vapour only may be found in standard tables, and the dew point of the mixture obtained by addition. For example, a flue gas with a 10 volume per cent water vapour and 0·01 volume per cent sulphur trioxide will give

$$\frac{P_{H_2SO_4}}{P_{H_2O}} = \frac{0{\cdot}01}{10} \times 100 = 0{\cdot}1 \text{ per cent.} \tag{2.8}$$

From Fig. 2.8, the dew-point elevation for this ratio is given as 105°C. The dew point for $P_{H_2O} = 10$ is 45°C, and so the actual dew-point temperature is 150°C. Müller found agreement to within about 5°C between the calculated values and those found experimentally by various authors, so this method may be used with some confidence.

2.5. CLASSIFICATION OF SAMPLING

When sampling the gas stream for which gas cleaning plant is to be specified and designed it may be important to determine one or more of the following:

(i) The major gaseous constituents and their concentration.
(ii) The minor gaseous constituents, including trace quantities.
(iii) The composition and concentrations of liquid droplets and solid particles in the gas stream.

In some cases it is desirable, for control purposes, to measure a component continuously, for example oxygen in the flue gases from combustion or traces of dangerous materials, but mostly the gases are sampled as a preliminary step so that satisfactory techniques may be specified and adequate materials of construction may be selected.

In sampling, part of the gas stream is diverted through a sampling "train", of which a general arrangement is shown in Fig. 2.9. The train consists of a nozzle placed in the gas

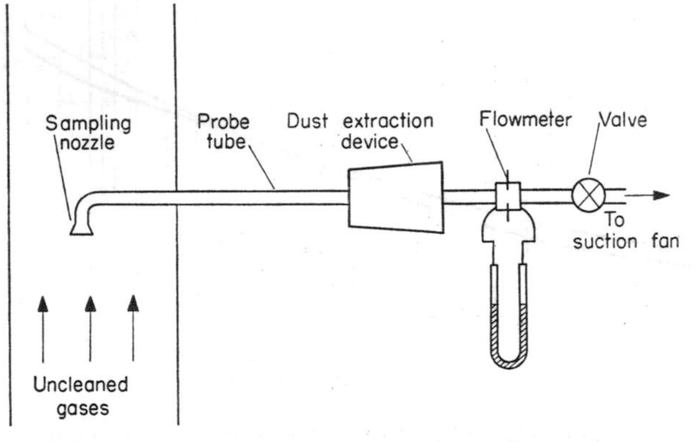

FIG. 2.9. General arrangement of sampling train.

stream, usually facing upstream; a collection device; a volume-measuring device, which may be either indicative or integrating; and some means of drawing the gases through the train, either a pump or an ejector. In detail, the methods depend on which of the three types of constituents are to be sampled.

2.6. COLLECTION AND ANALYSIS OF MAJOR GASEOUS CONSTITUENTS

When inserting a probe into a duct care must be taken that the sample is representative, and has not been collected from a "dead" zone where stagnation has taken place, or from a duct with stratified layers of gases. Leakage of air into the sample has also to be avoided. If the gases sampled are combustion or reaction gases, where the reaction is incomplete, care must be taken that the reaction does not progress in the sampling train, or the sampling tube act as a catalyst for further reaction. This can often be achieved by using a water-cooled probe or similar device for "freezing" the reaction. Since undesirable condensation may result with the condensate absorbing one of the gas constituents preferentially, collection above the dew point is generally expedient. If dust is present it may have to be filtered out before collecting the sample.

The sampling probe in the case of gases is a simple tube. Mild steel can be used to about 300°C, ceramic lined to 500°C, and stainless steel of the 18 per cent chromium, 8 per cent nickel type to about 800°C. Water-cooled probes can be used to much higher temperatures (copper to 1250°C, stainless steel to 1600°C), while ceramic tubes have a very long life in the highest temperatures encountered commercially, those in metallurgical furnaces. Blockage of the sampling lines by high dust loads and condensation are common and special precautions may have to be taken, such as heated sampling lines and frequent or continuous replacement of the dust filter. To prevent corrosion of the sampling line, possibly due in combustion gases to sulphuric acid from sulphur trioxide and condensed water, special materials may be required, particularly when the sampling line is to be left in place for continuous sampling.

Those techniques where gas is sampled continuously for reaction control* can also be used for analysis of gases for plant design. The major techniques for analysis are density, thermal conductivity, infra-red absorption, differential absorption in solvents, change in electrical resistance of solvents and specific physical properties, such as the paramagnetic property of oxygen or the radioactivity of certain gases from radioactive sources.

Density instruments can be used to measure a component of higher molecular weight than the gas without this component. The most common example is carbon dioxide in flue gases. Carbon dioxide with a molecular weight of 44, has a considerably higher density than the other main components; oxygen (32), nitrogen (28) and carbon monoxide (28). The density meter is usually a comparison-type instrument where two fast-moving impellers immersed in air and gas respectively are driven in opposite directions and face two discs which are connected to a pointer on a scale. This instrument, known commercially as the Renarex or Pyrorex indicator, could also be applied to other dense gases mixed with air.

Thermal conductivity is used in a balancing-type instrument, built on the Wheatstone bridge principle. Carbon dioxide has about 60 per cent of the thermal conductivity of air, while carbon monoxide is only slightly lower than air. Unfortunately sulphur dioxide has only one-third of the thermal conductivity of air, and so if flue gases contain several per cent sulphur dioxide (from high sulphur fuel) then the carbon dioxide reading will be much higher than if no sulphur dioxide were present.

Infra-red absorption presents a very elegant technique developed in recent years for routine industrial application[553] for gas mixtures whose molecules are dissimilar, e.g. SO_2, CO, CO_2, H_2O, N_2O, CH_4 and organic vapours. Some of these can be present together as long as the absorption bands do not overlap those of the gas being determined. Even then filter tubes containing the interfering gas can be placed in the optical paths so that the radiation emerging from these tubes will contain wavelengths which can be absorbed by the gas to be detected but none capable of being absorbed by the interfering gas in the sample.

* The British Standards Institution has prepared a standard for the continuous automatic analysis of flue gases.[131]

This type of arrangement, using either a split-beam (Fig. 2.10a) or a dual-beam arrangement (Fig. 2.10b) is shown. When the difference in absorbance at two wavelengths is to be measured, the beam splitter allows for this, compensating for interfering gases or particulate materials. On the other hand, the dual-beam arrangement measures the absorbance at a single wavelength through two beams (one measuring and one reference) giving precise differences in absorption.

Absorption in the *ultra-violet spectrum* has also been successfully used for continuous monitoring of gases in process streams (including flue gases). These include sulphur dioxide,

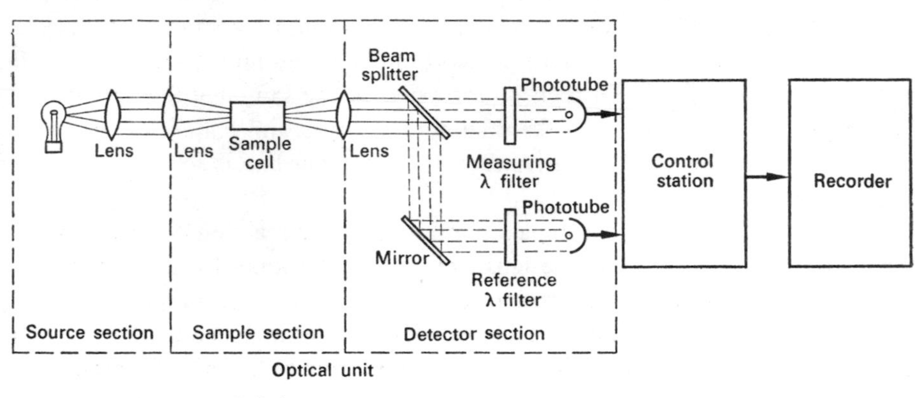

(a) Split beam arrangement

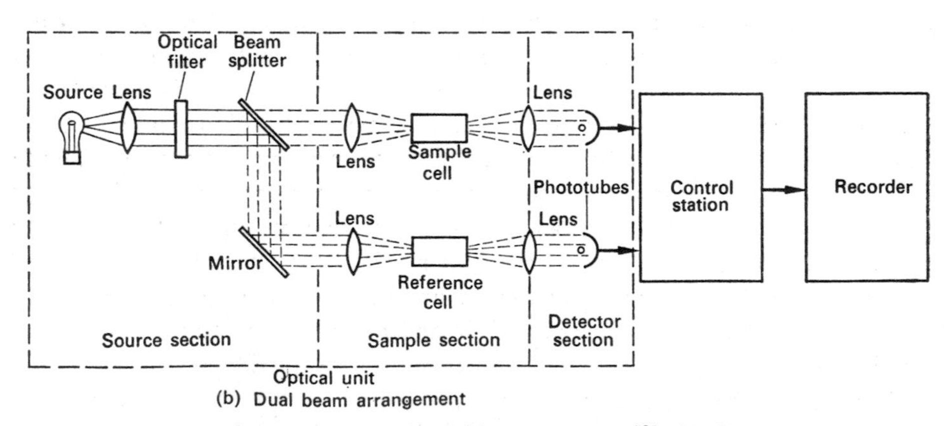

(b) Dual beam arrangement

FIG. 2.10. Schematic arrangement of precision photometer[586a] (du Pont model 400).

nitrogen dioxide, chlorine, bromine, fluorine, hydrogen sulphide, acetone, ammonia, benzene, dimethylformamide, furfural, toluene, cyclohexanone and phenol. Typical concentrations and the detection capability of a commercial instrument are given in Table 2.10. In a typical application to flue-gas analysis for both sulphur dioxide and nitrogen dioxide a split-beam arrangement is used. The 302-nanometer (nm) wavelength is strongly absorbed by SO_2, the 546-nm reference wavelength is not absorbed. Interference from any NO_2 in the gas is minimized since this gas absorbs almost equally at both these wavelengths. For

NO_2 analysis a 436-nm measuring wavelength and a 546-nm reference wavelength are used, since SO_2 has no absorption at either wavelength and so does not interfere. An optical filter for blocking out wavelengths below 436 nm is installed between the lamp and the sample cell eliminating photochemical reaction. Manual or automatically operated filter switching can be used, so that both gases can be monitored alternatively. In order to obtain a linear output from the photometer a logarithmic amplifier is used.

Preferential absorption of one of the component gases is of course the classical method of gas analysis. Carbon dioxide is easily absorbed in caustic potash, oxygen in alkaline pyrogallol, and carbon monoxide in one of a number of solvents such as ammoniacal cuprous chloride. All of these are used in series in flue gas analysis in either the Orsat apparatus or one of its many modifications, where a sample of gas is progressively quantitatively reduced by successive absorption of the components. The process has been mechanized in such units as the "Mono Duplex" (made by James Gordon and Co. Ltd.) where the initial carbon-dioxide absorption and volume reduction is followed by combustion of unburned gases (carbon monoxide and hydrogen), and a second carbon-dioxide absorption, this volume reduction representing the proportion of unburned gases in the original sample. A more elegant technique also applicable to much smaller carbon dioxide concentrations is to measure the electrical conductivity of the caustic-potash solution before and after carbon-dioxide absorption has taken place.

Oxygen is *paramagnetic*, that is it seeks the strongest part of a magnetic field, while most common gases are diamagnetic, seeking the weakest part of a magnetic field. This is utilized in the instrument based on the magnetic wind principle devised by Lehrer and shown in Fig. 2.11.[566] The gas to be measured is drawn into the cell, traverses the annulus and leaves at the opposite side. The cell contains a horizontal tube which supports two identical platinum windings which are joined in a Wheatstone bridge circuit and become heated by the application of a potential across the bridge. A magnetic field is placed across part of the tube, the poles being arranged so that the magnetic flux is concentrated around one winding. The oxygen in the gas is drawn into this field, enters the heated tube, and has its susceptibility reduced, thus allowing fresh oxygen to enter the tube, causing a continuous current of oxygen to pass along the tube from left to right. The left-hand platinum winding which heats the gas becomes cooler than the right-hand winding, and so unbalances the bridge, giving a measure of the oxygen in the gas. Detailed description of the commercial applications of this principle have been reviewed by Sterling and Ho.[813]

Gases with *radioactive components* are passed through special cells with Geiger–Müller or scintillation–photomultiplier unit tubes and connected to counting and recording equipment.

The methods described above all refer to continuous sampling and/or immediate analysis of the gas. However, in many cases this equipment is not available, nor is it possible to carry out the analysis at the point of sampling. It is then necessary to draw a sample of the gas into a gas sampling pipette, which may be of either glass or metal. Sampling pipettes are bulbs with a stopcock at both ends, and usually have a capacity between 25 and 500 ml. Before use, the pipettes are cleaned and filled with a liquid that will not absorb any of the gases to be sampled. For flue gases this is usually water with some strong acid added to

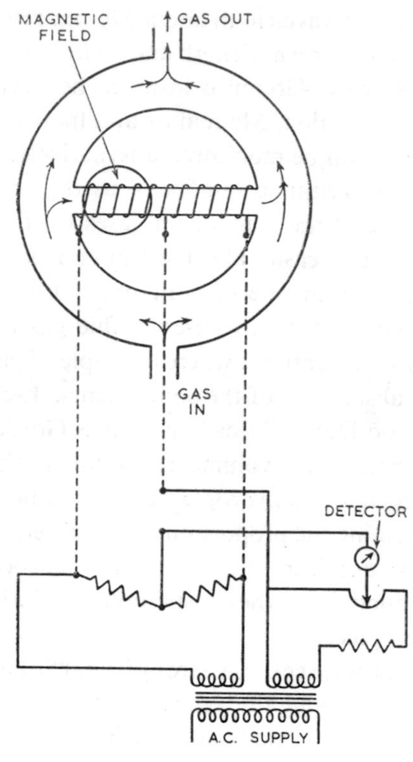

FIG. 2.11. Principle of oxygen concentration measurement using the paramagnetic property of oxygen ("magnetic wind" method).[566]

minimize carbon-dioxide absorption. In other cases mercury is often satisfactory, although heavy and more difficult to handle. During sampling the liquid is displaced by the gas, and if the pressure in the gas line is not too low, the liquid running out of the pipette can draw the gas sample in. Samples may also be collected in syringes, which is a favoured technique if gas chromatography is to be used for the analysis. If gas chromatography is unavailable or unsuitable for the gas, an absorption method,* either constant pressure (Orsat type) or constant volume (Bone and Wheeler type) apparatus, can be used. It should be noted that sulphur dioxide and carbon dioxide are usually absorbed together in alkaline solvents and differentiation of these components requires special precautions and techniques.

2.7. COLLECTION AND ANALYSIS OF MINOR GASEOUS CONSTITUENTS

Many of the techniques used for the major gaseous constituents, absorption, infra-red absorption, etc., can also be used for minor constituents but certain modifications may have to be applied. In addition to these, gas chromatography comes into its own for trace analysis, particularly of hydrocarbons.

* The methods to be used in the case of flue gases are covered by a British Standard Specification.[129]

Infra-red absorption will give full scale sensitivity for the tabled volume concentrations on a particular commercial instrument which is able to analyse continuously (Table 2.9).

When *gas chromatography* is used, spot samples of the gases are taken, and analysed subsequently. If a large number of samples are taken at close time intervals it becomes possible to find the variation in trace concentrations. The application of gas chromatography to these measurements is complex and cannot be simply described so reference should be made to one of the general textbooks and special papers such as Ettre's review of the application of gas chromatographic methods to air pollution studies.[246]

Most widely used are the *absorption methods*, where the gas stream sampled is passed through an absorbing solution (or in some cases a bed of solids) and the volume sampled is measured in a flowmeter until enough material has accumulated for analysis. The main disadvantage is that only an average concentration for the sampling time can be obtained. The infrared absorption technique avoids this difficulty, while spot sampling for chromatography overcomes it at least in part, because of the speed of taking a spot sample.

Details of methods of analysis* have been fully surveyed in recent reviews by Kay[428] and books by Jacobs,[394] Stern[815] and Leith.[503]

TABLE 2.9. FULL-SCALE DEFLECTION AT MAXIMUM SENSITIVITY

Gas	Vol.% for full scale
Carbon monoxide	0·05
Carbon dioxide	0·01
Nitrous oxide	0·01
Methane	0·1
Water vapour	0·25
Organic vapours	0·1 to 1

The air contaminant most frequently found in combustion waste gases is sulphur dioxide, and some of the most common methods of determination will be described here. In concentrations as low as 0·5 ppm sulphur dioxide can be absorbed in starch iodide solutions in a counter-current column. The light absorbed by the unchanged and partly decolorized reagent is compared by photoelectric cells coupled to a galvanometer.[189] Concentrations of the order of 5 ppm can be determined by a colorimetric comparison using *p*-rosaniline hydrochloride and formaldehyde which produces a red–violet compound.[361] A more complex method which entails hydrogen reduction of sulphur dioxide absorbed on silica gel, followed by molybdenum blue complex formation has also been suggested.[1] Other methods based on oxidation to sulphur trioxide with subsequent determination as sulphuric acid have been suggested, but EDTA methods are the most common analytical methods in use at present.

* Analysis of minor constituents in flue gases is covered by a British Standard Specification.[132]

Conductivity methods for continuously monitoring sulphur dioxide have been developed commercially and are widely used. Besides a number of instruments by manufacturers in the United States, there are also Japanese, German, French and British makes. The main problem appears to be that sulphur dioxide is not the only acidic (or ionic) gas present in the atmosphere and nitrogen dioxide and ammonia interfere, giving somewhat greater values for the conductivity method than for those based on West-Gaeke (*p*-rosaniline) or other colorimetric methods.[99, 498, 853] Furthermore, the differences between the two methods show seasonal variation, being greater in summer than in winter.

A method of determining sulphur trioxide in the presence of sulphur dioxide has recently been described by Seidman[749] but presents considerable difficulty because of the much greater proportion of sulphur dioxide and its continuous slow oxidation.

A recommended method for measuring both SO_2 and SO_3 in flue gases is to simply condense out the SO_3 as sulphuric acid, by passing the gas mixture through a cooling coil maintained at 60 to 90°C, well below the dew point, and then absorbing the SO_2 in hydrogen peroxide. This method is widely used and gives reproducible results.[306] The on-stream continuous method of SO_2/SO_3 determination has also been developed by Nacovsky[591] using the same principle; condensation of the trioxide followed by absorption of the dioxide. Conductance measurements can be used for monitoring the concentrations obtained.

Hydrogen sulphide can be determined by passing the gases through a moistened strip of lead acetate paper. The darkening of the strip can be measured and related to a hydrogen sulphide concentration.

When hydrogen sulphide and sulphur dioxide are both present in the gas stream, a feasible analytical method must separate these. This can be effected by passing the mixture

TABLE 2.10. DETECTION CAPABILITY OF AN ULTRA-VIOLET SPECTRUM PROCESS GAS MONITOR

Pollutant gas (vapour phase)	Typical stack concentration ppm	Detection capability of UV analyser ppm
Sulphur dioxide	2000	8
Nitrogen dioxide	50 to 500	2
Chlorine	2000	10
Bromine	1000	4
Fluorine	2000	100
Hydrogen sulphide	1000	3
Acetone	1000	50
Ammonia	1000	50
Benzene	2000	5
Dimethyl formamide	2000	12
Furfural	1000	0·04
Toluene	2000	5
Cyclohexanone	2000	50
Phenol	0·1 to 10	0·002

through a column of silica gel impregnated with silver sulphate (Ag_2SO_4) or quartz wool drenched in $KHSO_4/Ag_2SO_4$ solution followed by absorption of the residual sulphur dioxide in acidified dilute hydrogen peroxide.[105, 139] The hydrogen sulphide absorbed on the silica gel can be desorbed with stannous chloride ($SnCl_2$) and concentrated hydrochloric acid, with subsequent sulphur determination using molybdenum blue,[139] while the sulphur dioxide in the peroxide solution can best be determined by Persson's torin method.[634]

Traces of carbon dioxide are usually absorbed in alkali and the conductivity change measured. Carbon monoxide can be similarly determined by scrubbing the gas with alkali, oxidizing the carbon monoxide by passing it over a catalyst (heated copper oxide) and then treating as carbon dioxide.

2.8. ISOKINETIC SAMPLING AND PROBES FOR SOLID AND LIQUID PARTICLES

The method of sampling particles from a gas stream is essentially the same as that of sampling gases, but extra precautions have to be taken to ensure a representative sample. When particles are about 10 μm or larger, and the sampling velocity is lower than the velocity in the duct (Fig. 2.12b), some of the gases will be deflected around the probe, but some

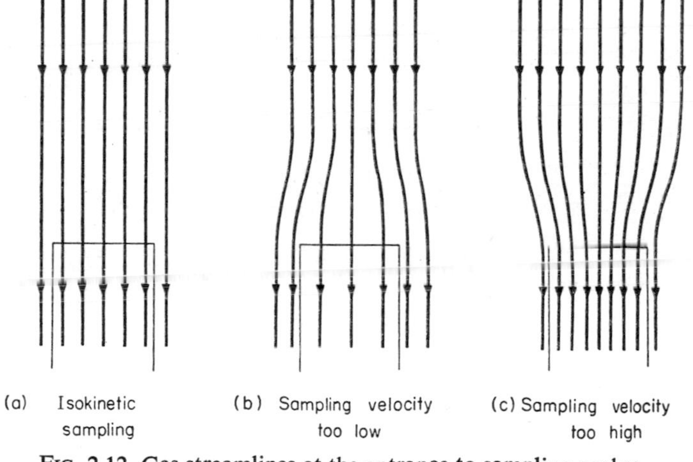

FIG. 2.12. Gas streamlines at the entrance to sampling probes.

of the particles, because of their inertia, will enter the probe and so a greater number of these larger particles will find their way into the sample. Conversely, if the sampling velocity is greater than the duct velocity (Fig. 2.12c) some of the larger particles will be deflected around the sampling nozzle. Only when the sampling and duct velocities are the same—that is, when the velocity is isokinetic—will a representative sample of the larger particles be collected by the probe (Fig. 2.12a). The actual error involved has been estimated by Badzioch[49] for a steady sampling velocity of 7·6 m s^{-1} varying duct gas velocities, and 5- and

10-μm particles (with a density of 2). Figure 2.13 shows that, even with a duct velocity twice the sampling velocity, the error for 5-μm particles is only 3 per cent, but approaches 20 per cent for 10-μm particles.

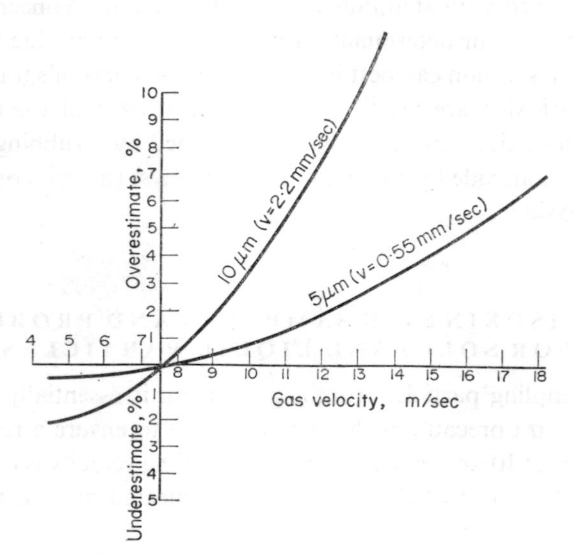

FIG. 2.13. Errors in concentrations of particles of 1 and 10 μm sampled at 7·6 m s^{-1} with a 25-mm dia. nozzle at ambient conditions.[49]

Davies[208] has suggested the following formula for estimating the sampling error for non-isokinetic conditions when using a sharp-edged tube facing upstream:

$$\frac{c_s}{c_a} = \frac{u_a}{u_s} - \frac{u_a/u_s - 1}{(4\psi+1)} \tag{2.9}$$

where c_s = concentration of particles in sample,
c_a = concentration of particles in gas,
u_a = velocity of gases in duct,
u_s = velocity in probe,
ψ = inertial impaction parameter (Section 7.2)

$$= d^2C(\varrho_P - \varrho)u_a/18\mu D \tag{7.10}$$

where D = diameter of probe,
d = diameter of particle,
ϱ = density of gas,
ϱ_P = density of particle,
μ = viscosity of gas.

As long as the velocity ratio u_s/u_a is between $\frac{1}{2}$ and 2, the concentration error is less than 20 per cent for $\psi < 0{\cdot}1$. A similar estimate is found from the equations given by Badzioch.[49]

More recently Davies[205] has considered the implications of sampling in still air and cross-wind, as well as using large sample heads with small openings, the latter being common when the probe is a water-cooled design. The many alternative possibilities lead to a number of formulae that can be used to calculate the minimum and maximum sampling velocities, depending on physical dimensions, orientation and fluid direction.

Isokinetic sampling can be achieved in one of two ways: either by using a null-type sampling probe, or by measuring the velocity of the gas stream with a pitot tube as near to the probe as possible without interfering with the gas flow, and then adjusting the sampling velocity. In a null-type probe a static balance is maintained between the inner and outer tube walls. A typical design is shown in Fig. 2.14. The static tap lines are connected to

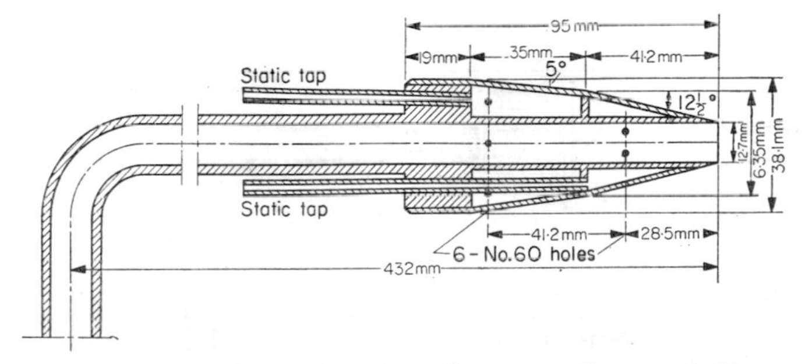

FIG. 2.14. Typical design of a null-type sampling nozzle.[920]

both arms of a U-tube manometer and the sampling rate adjusted during sampling. However, it has been shown[216] that at the point of static balance, particularly with low gas flow rates (less than 6 m s^{-1}), the sampling velocity is not really isokinetic, and at these velocities even a small error in the static balance can introduce large errors in sampling. At greater gas flows the error is smaller (less than 5 per cent above 15 m s^{-1}) with the type of probe shown. If a null-type probe is to be used, calibration over the test range would be advisable.

Several types of sampling probes where the duct velocity is measured separately have been developed, particularly by research associations such as the British Iron and Steel Research Association[312] and the British Coal Utilization Research Association[350, 351] and are now produced commercially. The B.C.U.R.A. sampling train (Fig. 2.15) is particularly interesting because the problem of minimizing condensation in the probe and duct filter has been solved by mounting the dust extraction cyclone, backed up by a filter, on the end of the probe inside the duct. Several screw-on nozzles with different entrance diameters are provided so that approximately constant volume sampling rates can be used over a wide range of velocities in the duct. The combination will withstand temperatures up to 350°C, but above this the filter must be mounted on the other end of the probe, outside the duct. The cyclone will operate up to 850°C within the duct, and also can be connected to the other probe end for work at higher temperatures. If at these higher temperatures unreacted particles (e.g. unburned coal) are collected, they may continue to burn, and so a method of "freezing" the reaction may be necessary. This is achieved in the sampling head developed by the International Flame Research Foundation.[378, 630] This is shown in Fig. 2.16. Here

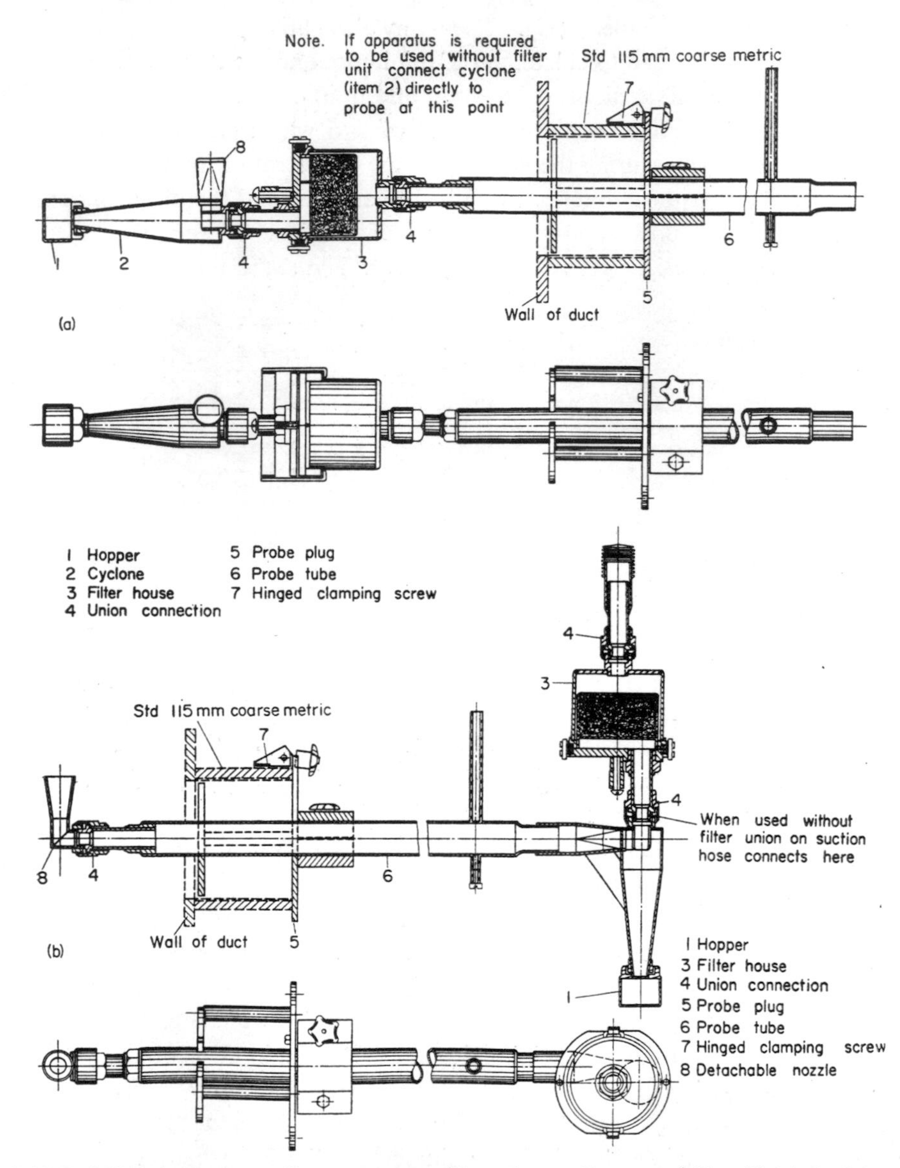

FIG. 2.15. B.C.U.R.A. Duct sampling apparatus with cyclone collector and filter.[351] (a) Cyclone and filter positioned at end of probe, inside the duct (to 300°C). (b) Cyclone and filter positioned outside duct (for use at higher temperatures).

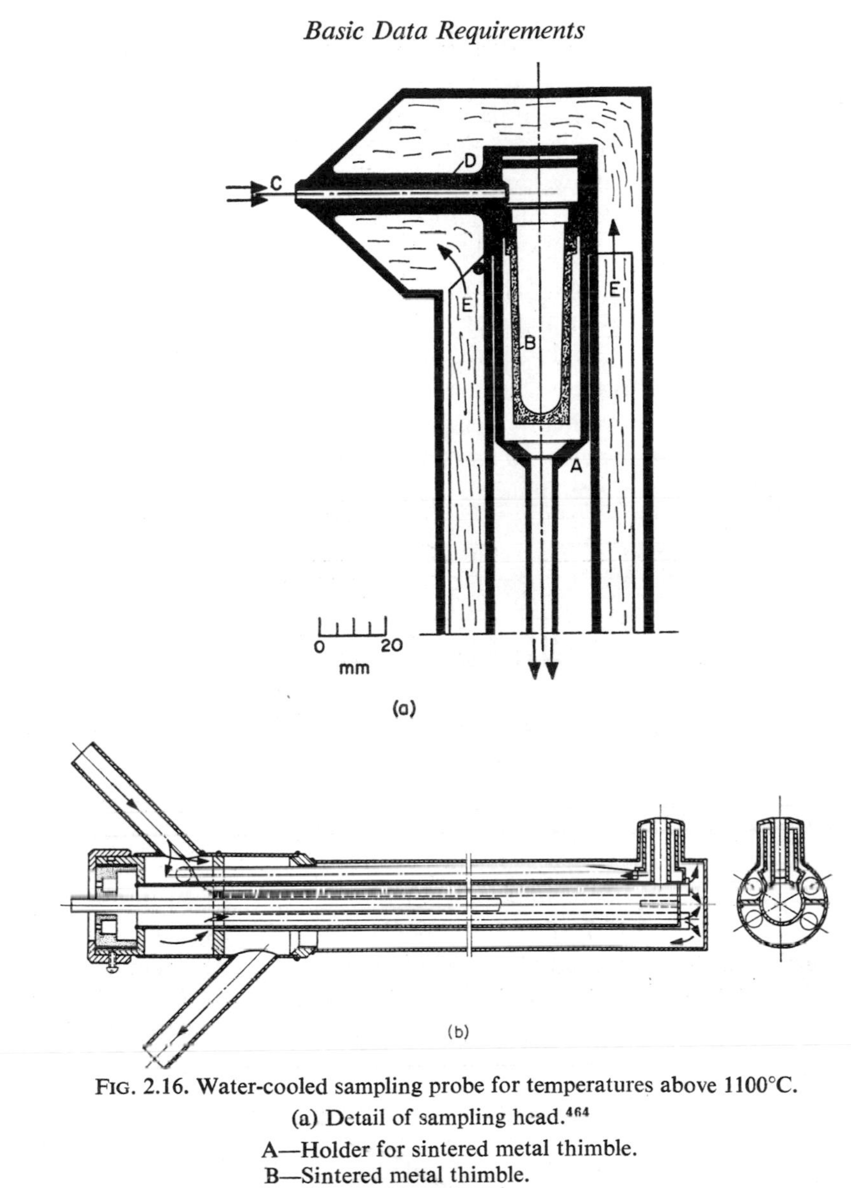

FIG. 2.16. Water-cooled sampling probe for temperatures above 1100°C.
(a) Detail of sampling head.[464]
A—Holder for sintered metal thimble.
B—Sintered metal thimble.
C—Entrance to probe.
D—Sampling tube liner.
E—Water inlet lines.
(b) General arrangement of water-cooled probe.[378]

the collection device, a sintered brass or stainless-steel thimble, is mounted in the head of the water-cooled probe. The nozzle, which is also water cooled, is fitted with a replaceable stainless-steel liner and the material in the liner as well as the filter can be collected and analysed.

The flow pattern at the sampling nozzles has been studied[903] but in practice the actual shape of the probe was found to have little influence on the accuracy of sampling.[936] So long as a chamferred end was provided then the length of the end was not significant. The probes must be constructed so as to resist corrosion and oxidation. For flue gas analysis mild steel is satisfactory to about 400°C, but oxidation may become a problem at higher temperatures. Stainless steel with welded construction is suitable to about 850°C without water cooling. With water-cooled probes, copper is a possible material (for short probes) to about 1200°C while stainless steel should be used for operation at higher temperatures. The cooling water may be introduced to the tip of the probe by running it through concentric tubes, in at one annulus and out through the other. At lower temperatures it is better to run the water in through the outer annulus, ensuring warmer water and less condensation in the probe. At higher temperatures the procedure should be reversed to allow cooler water to reach the tip. For sampling at the highest temperatures the cold water should be introduced in separate small tubes finishing close to and cooling the hottest parts of the probe tip and having the return flow in a single annulus.

2.9. SAMPLE COLLECTORS FOR SOLID AND LIQUID PARTICLES

Some of these have been briefly mentioned in the previous section. The *cyclone collector* in the B.C.U.R.A. sampling train is of the Stairmand high efficiency pattern (Section 6.7, Fig. 6.16) 38 mm dia. and will collect all particles above about 5 μm dia. A similar pattern developed by Walter[901] appears to be satisfactory for particles above 2 μm. In the cyclones a fairly high rate of flow has to be maintained for maximum collection efficiency (about 8·5–17 $m^3 h^{-1}$ in the B.C.U.R.A. Stairmand pattern) so the system of interchangeable nozzles has to be used to cover a reasonable range of duct gas flows. In practice, a flow of 8·5 $m^3 h^{-1}$ measured in the flowmeter at ambient temperatures means a flow through the cyclone of 12·3 $m^3 h^{-1}$ at 120°C and 15·7 $m^3 h^{-1}$ at 230°C. At higher sampling temperatures the measured rate also has to be reduced to stay within the cyclone performance limits.

Thimble filters of sintered brass or stainless steel were mentioned as forming part of the I.F.R.F. sampling probe head. Filter thimbles may be of alundum, which is able to withstand considerable temperatures and so these may also be incorporated either in a filtering unit inside the duct[143] or in a heated filtering unit outside the duct. Alundum and porcelain filter tubes both have a very high gas-flow resistance and block up quickly. Up to 350°C, bags of fibre-glass cloth have been successfully used inside the duct. With colder gases paper filter thimbles become practicable, and these have the advantage that they need only be used once and standard laboratory extractive methods can be applied to the recovery of the collected material. A very neat internally heated collection unit, with fibre glass or cotton bags, able to operate at exceptionally high duct and fume concentrations has been described by Guthmann.[328] This unit can also be used at low concentrations as provision has been made for substituting a filter paper as the filtering medium.

The earliest units used laboratory filters papers but these are not normally very efficient collectors of fine aerosols in the sub-micron ranges. In recent years asbestos bearing sheets have been shown to be virtually 100 per cent efficient.[798] Equally efficient are deep, loosely packed glass fibre filters. The B.C.U.R.A. filter (Fig. 2.15) is usually filled with two layers of glass wool: a superfine wool, fibre diameters 1–4 μm, about 6 mm deep, followed by a 38 mm layer of a coarser (6 μm dia.) glass fibre pad. For work at higher temperatures quartz and aluminium silicate fibres extend the operating range to the temperature limit of their metal containers. Glass containers with quartz fibres have been successfully used to 350°C[832] for the collection of open-hearth fume particles, which are almost exclusively sub-micron sized. Quartz fibres are chemically stable to nearly everything except hydrofluoric acid and this assists the subsequent leaching of collected material.

Another technique of recovering collected material is the use of a filter made of soluble material. Because of the water present in most gases it is recommended, when no organic vapours are in the gas stream, that only materials soluble in organic solvents be used. For example,[43] a tetrachlornaphthalene pad will retain two micron particles and is insoluble in water, but dissolves in benzene, and will also sublime. The pad may be prepared either by condensing a layer on paper or by dissolving tetrachlornaphthalene in ether, precipitating with ethanol and then filtering on to a copper gauze support.

Membrane filters[147] are very widely used in sampling aerosols, and there is a most extensive literature on their application. They are made by a number of firms in the United States and Europe (e.g. Millipore Filter Corpn.,[573] Gelman Instruments Co.,[295] Sartorius Membranfilter G.m.b.H.[722]). Membrane filters have holes which are very closely sized and range from 0·01 μm to 8 μm. The most common grade used for aerosol filtration has holes 0·8 μm±0·05 μm, while 0·65 μm±0·03 μm and 0·45 μm±0·02 μm are also in common use. Membrane filters, therefore, have a screening action in particle collection, rather than inertial impaction and diffusion. Furthermore, they are 100 per cent efficient in collecting particles larger than the diameter of the holes. Most membrane filters are made of cellulose materials, and collected particles are retained on the surface of the filter. A microscope count can then be carried out using incident light. Transmitted light can also be used after making the filter transparent, using optical immersion oil. The filter material dissolves in suitable organic solvents (esters: e.g. ethyl acetate; ketones: e.g. acetone, methanol, pyridine, etc.) and so the particles can be readily recovered. Membrane filters are also made from heat-resistant materials, expoxy resins for acid resistance, or polyvinyl chloride, which is resistant to some organic solvents. The filters can also be used for identifying specific materials using spot tests. Common tests are for ammonium, calcium, halide, lead, sulphate and nitrate ions. One account of aerosol filtration followed by electron microscopy is given by Spurny and Lodge[795] and a multifilter apparatus for sequential sampling has been described by Baum and Riess[63] and Friedrichs.[282]

Electrostatic precipitators are not usually employed for collection of samples from ducts, although a number of models are available for atmospheric sampling. Because of the low flow velocities normal in precipitators, a unit of impracticable size would be required to collect the relatively large volumes sampled from ducts. Moreover, the necessary high-voltage, direct-current source equipment is invariably heavy and not readily transportable.

In selecting a collection technique, the subsequent treatment of the collected material must be considered. If only the overall weight is required, nearly any of the above methods is satisfactory, as the collector can be weighed before and after sampling, the net weight of the sample being obtained by difference. If the sample is to be analysed chemically it will have to be removed from the filter, either by leaching or by the use of a soluble filter medium. Perhaps of greatest importance is the sizing of the collected particles and this presents the greatest technical difficulties. If liquid sedimentation or elutriation is to be used for sizing, leaching the filter with the sedimentation liquid is an acceptable method. However, for both liquid and air elutration or sedimentation a major problem remains in redispersing the collected sample into the particles and agglomerates which existed in the gas stream.

Connecting a cascade impactor (see Section 2.13) directly to the probe is possibly the only solution which gives a size classification of the particles as they are collected by the probe. This is particularly useful with droplets. If an optical or electron microscopic examination of the particles is to be undertaken, it is probably best to avoid the problem of

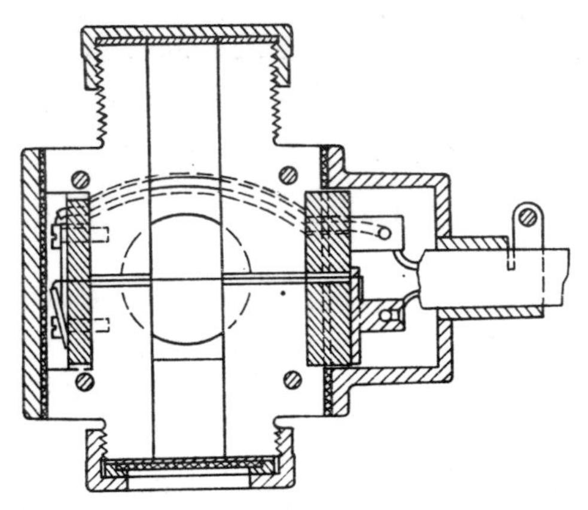

FIG. 2.17. Head of thermal precipitator for sampling very fine dusts.[153]

The sampling head is composed of two blocks of brass held together by screws to form a cube. The channel is formed between them by thin strips of insulating material which also isolate the resistance wire from the brass. Two holes in opposite faces accept the cover glasses which are kept in position by closely fitting brass plugs, for heating the wire a known current is passed through it from a battery. A shallow aspirator of 300 ml capacity fitted with a glass outlet jet, provides through the head a flow of air which is below the critical velocity.

dispersion of the bulk sample by collecting a special unagglomerated sample. This may be done by allowing the particles to settle in a chamber containing a sticky slide, or by drawing the particles in the gas through a thermal precipitator (Fig. 2.17). If the time for settling is known, an allowance for agglomeration during this period can be made.

2.10. SAMPLE VOLUMES

To find the volume of the gases sampled the collection system should be followed both by a means of measuring flow rate (either an orifice or an area flow meter) and by an integrating flow meter (frequently a bellows type gas meter). This last can be omitted if the flow rate is measured continuously and then integrated over the sampling time. It is important to cool the residual gases and to include in the train a catchpot, such as a conical flask, to remove excess water vapour, which will otherwise settle out in the flow-measuring device. A thermometer or thermocouple pocket should also be provided at or near the flow-measuring system so that the volumes can subsequently be related to a standard temperature.

The pressure at this point should also be measured, usually by a mercury-filled U-tube manometer, so that the volume at a standard pressure can be found.

The gases can be drawn through the sampling train by an ejector or a pump which may be driven electrically or pneumatically. If the gases sampled are explosive or combustible, electrical motors and other equipment must be suitably protected.*

2.11. CLASSIFICATION OF SIZE-ANALYSIS METHODS

Size-analysis methods fall into three groups:

(a) Those based on particle geometry.
(b) Those based on particle aero- and hydrodynamics.
(c) Those based on surface area.

Screen analyses and visual examination by microscope or electron microscope are the methods in the first group. The second comprises elutriation, sedimentation and impaction, while the third includes permeability, direct surface area (B.E.T.) measurements, β back scattering, etc.

In the design of gas-cleaning plant one is principally concerned with the aerodynamic behaviour of particles, so those techniques which enable this to be predicted are most valuable. For larger particles in the regions above 75 μm (200 mesh) screen analyses and microscope analysis are easily carried out, but electron micrographs are often the only method of penetrating into the regions of sub-micron sized particles where aerodynamic methods are no longer feasible because of excessive Brownian motion. Surface area methods can sometimes be used, particularly permeability measurements which are very fast, but they give only an average size.

2.12. SIZE ANALYSES BASED ON PARTICLE GEOMETRY

Sieves are commonly used for particle size analyses down to 200 mesh screens, which have sieve openings of 74 μm (Tyler (U.S.) scale), 63 μm (Institute of Mining and Metallurgy (I.M.M.) (British) scale), or 76 μm (British Standard sieves). Tyler sieves are available down to 400 mesh (38 μm) and this represents the lower limit of ordinary screen analysis. Special electro-formed precision micromesh sieves[192] enable sieve analysis to be carried down as far as 5 μm particles. It should be remembered that sieves may pass irregular particles which are longer than the sieve opening if their two shorter axes are within the limits of that opening. Small fine particles tend to agglomerate and screening by shaking may take a very long time, so an air jet can be used to blow undersize particles through the sieve. This principle has been incorporated in the "air jet" sieve[497] in which a controlled air stream blows the fine particles through, giving rapid and accurate separation.

* SPECIAL WARNING. If the gases sampled are toxic or likely to contain toxic constituents (e.g. blast furnace gases contain an appreciable fraction of carbon monoxide) great care must be taken as to the disposal of the residual gases expelled from the train during sampling.

Microscopic examination without determining the size distribution is rapid and gives an excellent idea of the appearance of the particles. Their chemical composition also may often be inferred from colour and from knowledge of their source. If only a bulk sample was collected, its redispersal for microscopic sizing will nearly always require the preparation of a suspension and subsequent evaporation of some drops thereof placed on a slide. A much better method is to avoid redispersion by collecting a sample on a sticky slide placed in the gas stream or in a settlement chamber. Coagulation during settling can be allowed for. Jet impactor slides are also suitable for microscopic examination. Here a particle count is usually all that is required, as the stages of the jet impactor (Fig. 2.20) have classified the particles. The use of a microscope for particle sizing has been explained elsewhere, for example by Fairs[248] and Green and Lane.[317] Special graticules, a selection of which is shown in Fig. 2.18, are available to simplify the counting and sizing. The spheres on the graticules are sized in a $\sqrt{2}$ progression, as in standard Tyler screens. Although particles as small as 0·14 μm can be detected under white light, 1 μm represents the practical minimum

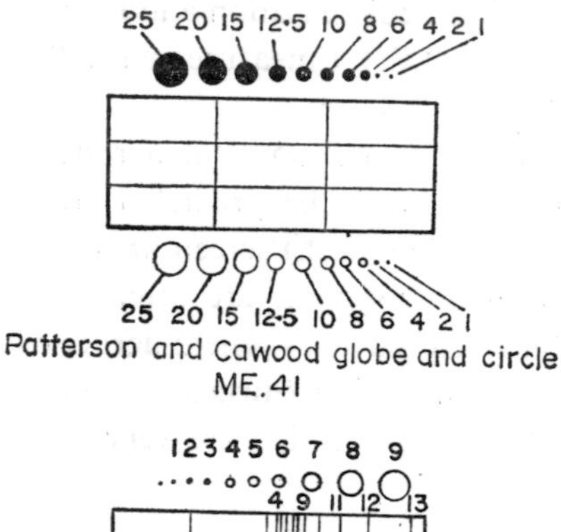

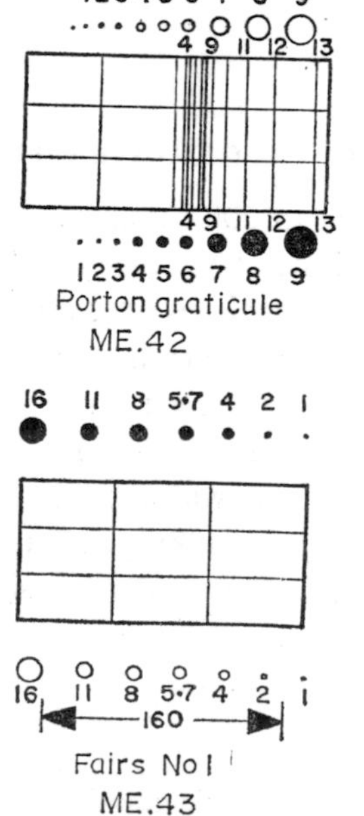

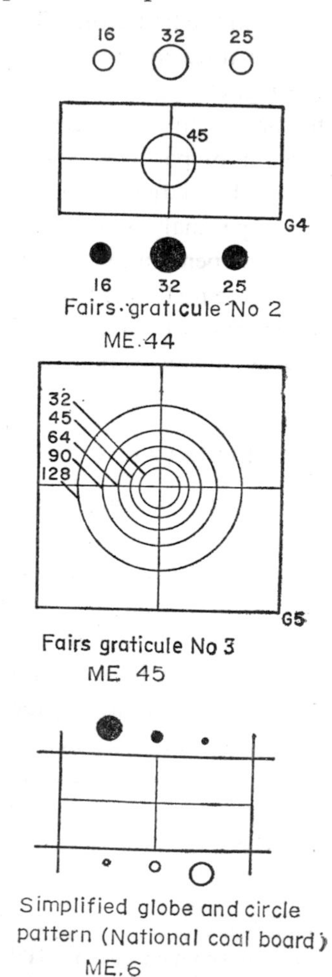

FIG. 2.18. Various graticules used for particle sizing and counting.[846]

for normal sizing for the optical microscope. It should be remembered that when particles are observed under a microscope they are normally in their position of greatest stability, so if they are thin plates they will present their largest dimensions for measurement.

Electron microscopy is a highly specialized technique. Smoke and fume samples are usually collected with a thermal precipitator in which the material is deposited on a membrane on a cold slide near a heated wire. A thermal precipitator will not retain particles greater than 20 μm[919] but will readily collect all particles in the sub-micron ranges. Membrane (Millipore) filters are also suitable for sample collection. The resolution of simple electron microscopes is about 4 nm—while high resolution models are able to improve on this by factor of 10. This is, however, well beyond the size of any particles normally encountered in industrial collection.

The tedious nature of particle sizing and counting, whether through a microscope or from electron micrographs, has been overcome by the development of two techniques for automatic particle sizing and counting. The first of these depends on mechanical scanning with photoelectric detection, coupled with high speed counting devices.[578, 579] The second involves scanning the sample with a flying spot from a cathode ray tube, picking up the scattered light pulses from the particles individually on a photocell,[287] and recording the pulses on counters.

2.13. PARTICLE ANALYSES BASED ON AERO- AND HYDRODYNAMICS

Sedimentation is the simplest method of particle size grouping utilizing the aero- or hydrodynamic behaviour of the particles. One of the earliest sedimentation pipettes, the Andreasen pipette,[22, 412] is still in use, sometimes in modified forms, such as the one developed by Stairmand,[799] a robust and easily used unit. The settling medium is usually water to which a peptizing agent may have been added. The particles are shaken up in the fluid and then allowed to settle, samples of the mixture near the bottom of the apparatus being taken at regular intervals. These are then evaporated and weighed. The size distribution therefore depends on the viscosity of the fluid, so that temperature uniformity during sedimentation is important. An adequate sample (approximately 1 gram) is required. The process has been made automatic by the development of the sedimentation balance[47, 65, 104, 281] where the increase in the load of a pan suspended in the bottom of a beaker containing the mixture is counteracted by some form of torsion arrangement, the changes in which are recorded on a time base. A very elegant technique of finding the rate of sedimentation is to measure the back-scattered β radiation from the deposited material at which a one millicurie ^{90}Sr source has been directed. Centrifugation can of course be used to speed up the gravity-settling process.[223, 421]

Sedimentation in air can be used for particle size-distribution measurement in apparatus of the settlement dust-counter type[152] in which a volume of air is enclosed in a cylinder at the base of which is an arrangement for exposing a number of microscope cover glasses in sequence. By timing the exposures and counting the number of particles collected on each slide, a particle size distribution can be obtained.

The range for sedimentation is from approximately below 325 mesh (42 μm) to about 0·2 μm.

In elutriation the fluid is passed upwards counter-current to the sedimenting particles. To obtain steady flow requires careful control. Liquid elutriators, using both gravity and centrifugal forces (in miniature cyclones),[432] are available commercially. The particles separated in them are much coarser than in simple sedimentation. Fine particles can be separated by air elutriators. A simple model has been described by Stairmand[802] while the

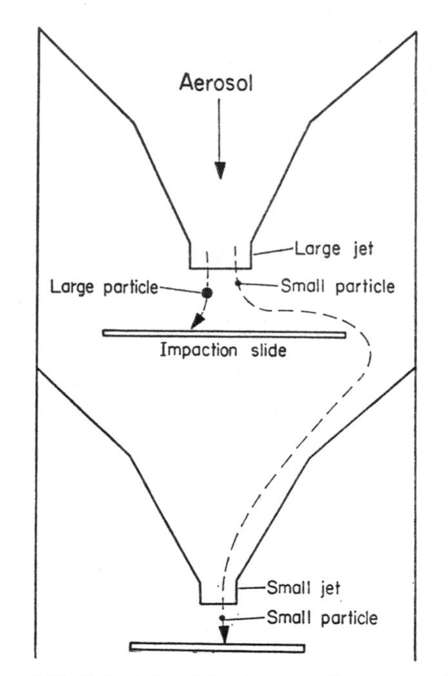

FIG. 2.19. Principle of the "cascade" impactor.[575]

multi-stage Haultain "Infrasizer"[11] has been available for many years. The centrifugal elutriator developed by Gustavsson[327] and sold commercially as the Bahco air elutriator has proved very useful for particle size analysis of the type encountered industrially.[457] Except for the Bahco elutriator, elutriation methods are normally restricted to mineral particles which are comparatively coarse and where large samples are available.

Multi-stage or cascade impaction as a development of the impingement principle was first introduced in the Greenburg–Smith impinger[320] in which a stream of dust-laden gases is passed through a nozzle and the resultant jet allowed to impact on a plate before being diverted. The multistage impactor was introduced by May[565] and the principle is shown in Fig. 2.19. The large jet with low velocity will deposit the larger particles on the slide, while the small high-velocity jet will deposit the smaller particles on a subsequent slide. The design of a six-stage model with after-filter developed by the Battelle Institute[575] is shown in Fig. 2.20. The efficiencies of collection on each stage vary considerably. As seen for May's four-stage model (Fig. 2.21), the "cut-off" in each stage is not very sharp, and with six

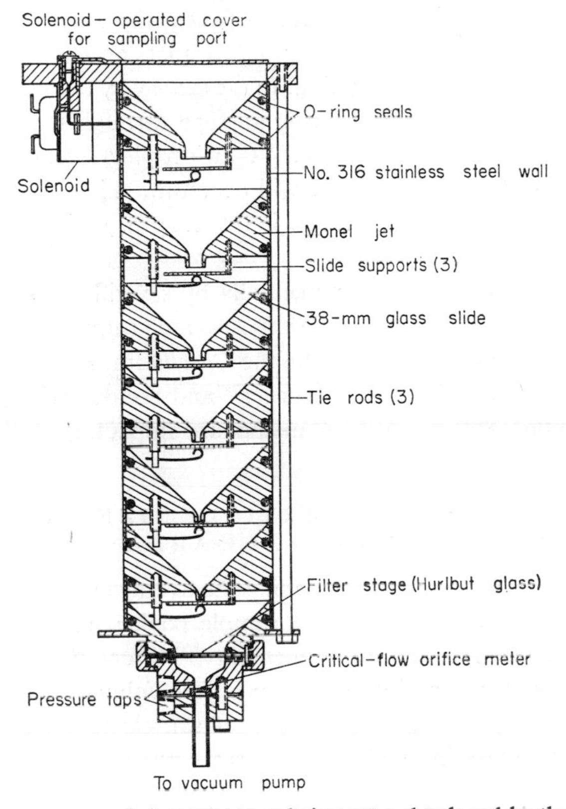

FIG. 2.20. General arrangement of six-stage cascade impactor developed by the Battelle Institute.[575]

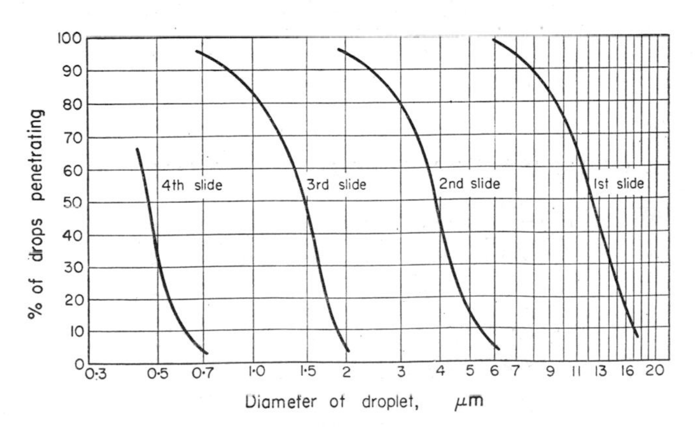

FIG. 2.21. Droplets penetrating past the various stages in a four-stage impactor.[565] (The range of the six-stage unit is similar.)

stages considerable overlapping will occur between the stages. However, the cascade impactor presents the only method of sampling droplets and obtaining an immediate size distribution with a minimum of coagulation. In the same way the impactor will size particles while sampling as the agglomeration during passage through the impactor is very small. However, particles have less sharp "cut-off" than droplets because of particle "bounce" and "shatter". Use of sticky coated slides will overcome this to some extent. Both for the four-stage and six-stage models the range is 0·5–15 μm, which probably represents the practical limits of impactors.

Cascade impactors are produced commercially by several firms. Two of these (*Casella* and *Unico*) are four-stage impactors, while the *Anderson* sampler is a six-stage unit. Friedrichs[283] has compared the three types, while the *Casella* impactor has also been calibrated independently by Berner and Preining.[73] Berner[72] and Sundelöf[842] have also discussed the particle-size distribution obtained with non-standard impactors.

2.14. AVERAGE PARTICLE SIZE BASED ON SURFACE AREA

Permeability has been widely used for finding the average sizing of particles in powders. The passage of a volume of air through a sample packed under standard conditions is timed. The specific surface A found by permeability[150, 499] can be related to average particle size by consideration of the porosity ϵ (fraction of free volume) which is found experimentally from the density of material and weight of sample tablet of known volume. The equivalent capillary theory[245] in simple form gives the particle diameter d as

$$d = \frac{3}{2}\left(\frac{1-\epsilon}{A}\right). \tag{2.10}$$

Specific surface area can also be found from absorption measurements either of gases (B.E.T. surface area determinations)[138] or of dyestuffs (particularly methylene blue), or by the heat of wetting of the surface.[321] Some of these later methods give the total specific surface area of the particles including their internal surfaces even if these are only pores a few nm diameter. Their application to particles which may themselves be porous (for example carbon particles in smoke) can lead to incorrect surface areas.

2.15. OTHER METHODS OF PARTICLE SIZING

Two other methods of particle sizing which are not easily classified under the headings in Section 2.11 are the electric gating technique and the light extinction method. The electric gating technique is incorporated in the Coulter counter in which a suspension of the particles in an electrically conducting liquid flows through a small aperture between two electrodes. If a relatively non-conducting particle passes between the electrodes, a voltage decrease occurs between the electrodes proportional to the size of the particle.[563] The apertures may be between 10 and 1000 μm diameter, and the minimum particle measured is about 0·3 μm, similar to the minimum for sedimentation.[71]

Particles in suspension in a gas or liquid absorb, reflect, or scatter light depending on their size, shape, and surface texture and on the wavelength of the incident light. This can be utilized for particle size analysis by applying the Lambert–Beer law[776] to a system where the particles are dispensed in a fluid medium:

$$I_i = I_0 \exp(-kcl/d) \tag{2.11}$$

where I_0 = intensity of the light in the absence of obscuring particles (i.e. the light transmitted by the pure fluid),
I_i = intensity of beam after passing through the suspension,
k = the extinction coefficient (a constant),
c = mass concentration of particles/unit volume (g cm^{-3}),
l = length over which absorption takes place,
d = diameter of particles (μm).

In practice, the extinction coefficient differs for particles of different sizes, so Rose[702] and Rose and Sullivan[703] give the equation for the apparent specific surface A' (in m^2 g^{-1}):

$$A' = \frac{4}{cl} \ln I_0/I_i. \tag{2.12}$$

The true specific surface A can then be found from empirical relations:

$$A = 4{\cdot}5A'^{(0{\cdot}77)}/\varrho_p^{\frac{1}{4}} \quad \text{for the range} \quad 600/\varrho_p < A' < 60{,}000/\varrho_p \tag{2.12a}$$

and

$$A = 120A'^{\left(\frac{1}{2}\right)}\Big/\varrho_p^{\frac{1}{2}} \quad \text{for the range} \quad 60{,}000/\varrho_p < A' < 180{,}000/\varrho_p \tag{2.12b}$$

where ϱ_p = density of the particle.

The development of devices using light scattering has been initially due to Gucker, O'Konski and Doyle[322–4, 614] who devised a photoelectric smoke-penetration meter or penetrometer. This was sensitive to concentrations as low as 0·001 μg l^{-1} dioctyl phthalate (approx. 0·3 μm dia.) (Fig. 2.22) More sophisticated in theory is the use of higher order Tyndall spectra and their dependence on particle size. Sinclair and La Mer[772–3] devised a simple method applicable to monodisperse aerosols in which a light beam is passed through a cylindrical tube and the colours are picked up from a window over an angle of 180°. These colours follow the spectrum in the order violet, blue, green, yellow, orange and red and in the region of 90° the colour sequence is reversed. The saturation and brightness of the colours increase with uniformity of particle size and the number of times the spectral series is repeated increases with particle size. At those angles at which the intensity of the scatter ed red is greater than green, the scattered light is red in colour. Only when the aerosol is very uniform, when the number of spectra can be counted, is it easier to identify the order by counting the number of reds. The observations are made with a telescope, mounted on a traversing carriage of the type described by Lancaster and Strauss[477] and fitted with a plane

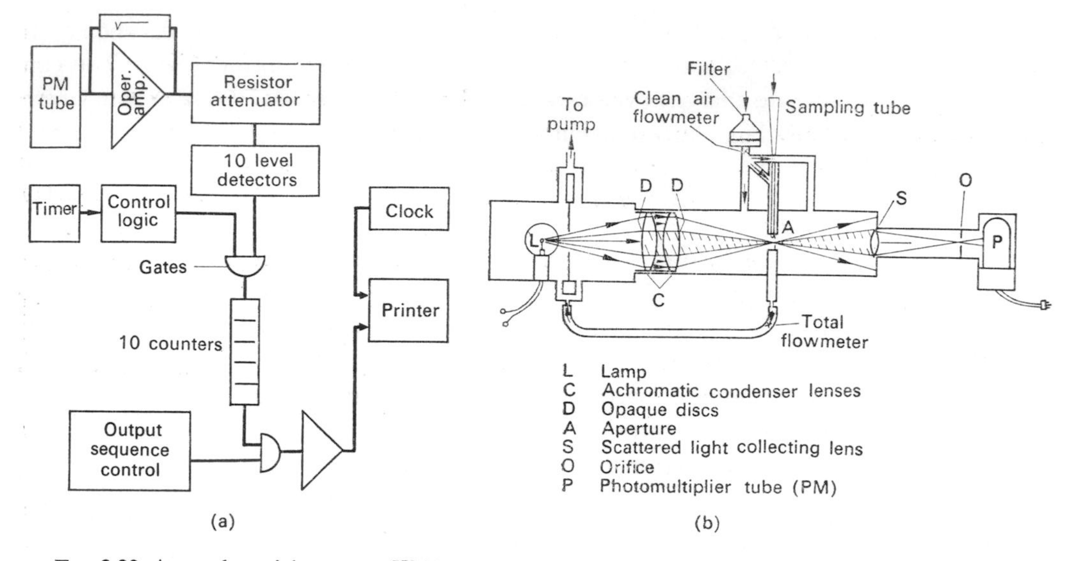

FIG. 2.22. Aerosol particle counter.[772] (a) Block diagram of electrical system. (b) Schematic diagram of optical and sampling system.

polarizer set so that it passes vibrations perpendicular to the plane of observation. The spectral colours appear at a particle diameter slightly below 0·4 μm diameter and the upper limit is about 1·5 μm. As an alternative to white light, a mono-chromatic light source may be used and the scattered light measured by photoelectric means. The particle size can then be computed from the output.[683] The problem here is that when an aerosol consists of concentric spheres of different refractive index, the comparison of scatter intensity of two-plane polarized light beams is necessary.[559–61] In practice there are considerable problems with both approaches, because of "noise" inherent in the circuits used in the detection and amplification system. This can be overcome using the high-intensity laser beams which have become available. An aerosol spectrometer using small angle ($< 7·5°$) scattering of a He–Ne laser beam has been developed at the Battelle Institute (Frankfurt).[294] Because of the narrow forward scattering (from the diffracted portion of the scattered light) the result is independent of the shape and optical properties of the particles. The method is useful for concentrations up to 10^4 particles cm^{-3} because the volume used for measurement is 0·01 mm^3. The lower particle size limit using this method is 0·17 μm diameter, and the upper limit is approximately 1·5 μm. These workers have also developed an instrument that can be used for high particle concentrations (5×10^6 cm^{-3}) in a flow system.[98]

In a similar category are smoke-density meters where a beam of light shining on a photoelectric cell across a duct is partially interrupted by particles moving along the duct. Although difficult to interpret in terms of particle size and concentration, smoke-density meters will indicate fluctuations in particle numbers.

2.16. MONITORING ATMOSPHERIC POLLUTION

The previous sections have dealt essentially with the measurement of pollutants and other contaminants in process gas streams and flues before they are released to the atmosphere and dispersed. Recent years have seen great efforts in the measuring and monitoring of pollution in the environment. Here the main problem is that concentrations are very small, and show great variation depending on weather conditions and seasonal conditions. Some of the methods described in the previous sections are suitable for measurement of ambient concentrations, either by increasing sampling times, when an absorption technique is used, or increasing path length for a light beam when an optical method is used.

In the United States, at the present time, there are a large number of monitoring stations, strategically placed in urban and rural locations. The surveillance systems currently in use measure the following:

Total suspended particulates
Sulphur dioxide
Carbon monoxide
Total oxidants
Total hydrocarbons
Nitric oxide
Nitrogen dioxide

Suspended particulates are generally collected with a high-volume sampler, using glass fibre or membrane filters. The particles are then measured by standard microchemical methods or with a mass spectrometer.

The suspended particulates are analysed for fluoride, nitrate, sulphate and ammonium ions, arsenic, beryllium, bismuth, cadmium, chromium, cobalt, copper, iron, lead, manganese, molybdenum, nickel, selenium, tin, vanadium and zinc. Asbestos, boron and silicates are also collected and analysed.

Some gases, principally *sulphur dioxide*, can be monitored on a 24 hr (or lesser period) average by passing the gas through a bubbler into a suitable solvent; dilute acidified hydrogen peroxide in the case of sulphur dioxide, and then analysing the solution by normal methods. Much more elegant instruments are currently available for rapid continuous analyses. These are based either on measurement of conductance of a solution of acidified hydrogen perioxide (0·1 ml 30 wt% H_2O_2, 1 ml N/10 H_2SO_4, 0·1 ml wetting agent, diluted to 1 litre), or by measuring the coloration of the West-Gaeke (*p*-rosaniline-formaldehyde) reagent. The West-Gaeke method suffers less from interfering materials, and tends to be more reliable. A flame luminescence method for determining sulphur is currently being developed. Another important development in sulphur-dioxide analysis is the novel method of producing stable concentrations of this gas in a flowing stream of diluent gas by using a small, sealed Teflon tube filled with liquid sulphur dioxide. The sulphur dioxide diffuses through the walls at a constant rate as long as the temperature is held constant. *Hydrocarbon gases* are detected and monitored either in a modified gas chromatograph or with a burner/flame ionization detector device. Carbon monoxide is monitored with non-dispersive infra-red analysers using long path-length cells. Nitric oxide (0–1·0 ppm) and nitrogen dioxide (0–1·0 ppm) are determined by an automated wet chemical method using a diazo-coupling reaction. The air sample is divided into two streams, and nitric oxide in the one stream is converted to the dioxide by passing through an acidified solution of potassium permanganate. Both streams are then passed into counter current absorbers where they are absorbed in solutions of sulphanilic acid, *N*-(1 napthyl) ethylene diamine dihydrochloride and acetic acid. The colours of the solutions, measured by automatic colorimeters, give the nitrogen dioxide and ($NO+NO_2$) concentrations. The efficiency of conversion is between 70 and 95 per cent, depending on bubbler design. Details of this method are given by Katz.[426]

Total oxidants (0–0·5 ppm), which are principally ozone, are determined by the neutral iodide reaction, carried out in an electric cell,

$$O_3+H_2O+2I^- \rightarrow I_2+O_2+2\,OH^-$$

$$NO_2+2I^-+2H^+ \rightarrow I_2+NO+H_2O$$

The iodine is reduced at the cathode by the reaction

$$I_2+2e \rightarrow 2I^-$$

and the corresponding oxidation reaction occurs at the anode. The resultant current flow in the circuit is served by a current amplifier, and the instrument is calibrated to give a total oxidants reading.

Ozone by itself will, in the future, be determined by an apparatus using the Rhodamine B reaction.[373, 681] However, Regener's early reagent, in which the dyestuff was absorbed on silica gel, was moisture-sensitive. More recently chromatographic sheets covered with Rhodamine B combined with a silicone resin have been produced which give a suitable sensor.

Amines in trace amounts from the primary aliphatic series from C_1 to C_6 can be collected in dilute hydrochloric acid. Amines are then extracted with amyl acetic ester, and the extract reacted with ninhydrin for a photometric amine determination.[334]

Low concentrations of hydrogen sulphide (down to 0·05 ppm) can be measured using fluorescein mercuric acetate in 0·01 N sodium hydroxide solution, and measuring the fluorescence with a photofluorometer.[338]

New types of monitoring instruments are currently being investigated, and may soon be available. These include using the fuel-cell principle for measuring carbon monoxide, sulphur dioxide and nitrogen dioxide, gas chromatography (in an automated form) for monitoring carbon monoxide, methane, total hydrocarbons, hydrogen sulphide and certain organic sulphides, atomic absorption spectroscopy for continuous measurement of lead, specific ion electrodes for measurement of fluorides, and a method for determining oxides of nitrogen in which ozone is situated in the gas phase with a reactive gas, and the reaction produces light by chemiluminescence. Lidar (light-detection and ranging) instruments for particulate concentrations are also being investigated.

The use of plants as indicators of low pollution levels is also possible. Ozone, sulphur dioxide, fluorides, nitrogen oxides, peroxyacetyl-nitrates and other phytotoxic pollutants (e.g. hydrogen sulphide, hydrogen chloride, chlorine, etc.) affect a whole range of plants, and sensitive species can be planted as pollution indicators. The systematic recognition of plant damage is made possible by the "atlas" edited by Jacobson and Hill.[395] Lichens have also long been recognized as indicators of air pollution. For example, Pyatt[666] points out that with changing pollution, the position of *Lichen thalli* on trees changes; in polluted areas most species are confined almost exclusively to the base of the tree boles. However, as atmosphere-pollution changes (e.g. with greater distance from a major steel works) the lichen species occur higher up the trees. Further systematizing and planting of test patches could help to monitor low-level concentrations on a continuous basis.

CHAPTER 3

THE REMOVAL OF A GASEOUS CONSTITUENT: ABSORPTION, ADSORPTION AND COMBUSTION

3.1. INTRODUCTION

It is often necessary to treat a gas stream for the removal of one or more of its gaseous constituents, which may be harmful, obnoxious, or commercially valuable. For example, it is important to remove carbon monoxide from the gases for ammonia synthesis before they enter the catalyst columns, as carbon monoxide is a catalyst poison.

It is essential in areas of dense or predominantly residential population, to remove the obnoxious odours, essentially organic nitrogen and sulphur compounds, produced by processes such as roasting coffee or rendering offal.

It would be very desirable to remove sulphur dioxide from flue gases, both because it contributes to atmospheric pollution, and because it can be converted to the commercial commodities, sulphuric acid and sulphur.

There are three methods of removing gaseous constituents; gases may be absorbed in a liquid, adsorbed on a solid surface, or changed chemically into a harmless gas. The last usually involves the combustion of the organic material, either directly or with the assistance of catalysts. The basic mechanism in all these is the diffusion of the particular gas either to the surface of an absorbing liquid or adsorbing solid or catalyst, or to the reaction zone of a chemical reaction. This is in contrast to the more complex collection of particles and droplets where a combination of mechanisms; inertial impaction, interception, settling, electrostatic and thermal forces, play a part, in addition to diffusion. A detailed discussion of adsorption, absorption and combustion processes is beyond the scope of the present book which is concerned with the application of these methods to gas purification and so it deals only briefly with the fundamental processes of these operations. A full treatment of these topics is given in specialized books on absorption,[582, 608, 768] adsorption[138, 550, 887] and combustion.[862] Those aspects which are of particular interest in the control of air pollution will be dealt with in some detail. These include the removal of sulphur dioxide from flue gases, the removal of hydrogen sulphide, fluorides and oxides of nitrogen from process waste gases, and the combustion, both direct and catalytic, of organic vapours and odours.

Absorption and adsorption of gases depend on the transfer of molecules from the bulk of the gas to the liquid or solid surface. In the case of the liquid, the gas molecules then diffuse to the bulk of the liquid, while at the solid surface they are held by physical (van der Waals) or chemical forces (chemisorption). When the solid or liquid surface is brought into contact with a stagnant gas, the diffusion of the gas molecules is by molecular diffusion, and the rate of this depends on the pressure and temperature of the gas and the molecular species in the gas. The rate of molecular transfer N_A, in molal units per unit area in unit time, is given by Fick's law:

$$N_A = -\mathcal{D}\frac{dc_A}{dx} \tag{3.1}$$

where dc_A/dx is the concentration gradient in the direction of diffusion (c_A being the concentration and x being the distance), and $\mathcal{D}$ is the molecular diffusivity, which has the dimensions (length)^2(time)$^{-1}$.

3.2. CALCULATION OF MOLECULAR DIFFUSIVITY

The diffusivity in the gas phase can be calculated from an equation based on the kinetic theory of gases; and for a single species, this is

$$\mathcal{D}_G = \tfrac{1}{3}\lambda\bar{u} \tag{3.1a}$$

where λ = mean free path of gas molecules,

$\bar{u}$ = average velocity of gas molecules.

From the kinetic theory, which considers the gas as a mixture of hard spheres without intermolecular forces of attraction and repulsion, the mean free path is given by

$$\lambda = 1/\sqrt{(2)}\pi N\sigma^2 \tag{3.2}$$

where N = number of molecules per unit volume,

σ = diameter of a molecule

and the average velocity is given by

$$\bar{u} = \sqrt{(8RT/\pi M)} \tag{3.3}$$

where T = absolute temperature,

R = universal gas constant,

M = molecular weight.

For the interdiffusion of two species of molecules A and B of different molecular weights, M_A and M_B, and different sizes, σ_A and σ_B, the diffusivity is

$$\mathcal{D}_{AB} = \frac{1}{3}\,\frac{N_A\lambda_A\bar{u}_A + N_B\lambda_B\bar{u}_B}{N_A + N_B} \tag{3.4}$$

where N_A and N_B are the numbers of molecules of each of the species A and B within a unit volume.

Using a mean molecular diameter, σ_{AB}, instead of individual diameters, equation (3.4) can be simplified to give[398]

$$\mathcal{D}_{AB} = \frac{1}{3\pi N(\sigma_{AB})^2}\sqrt{(\bar{u}_A^2+\bar{u}_B^2)} \tag{3.5}$$

$$= \frac{2\sqrt{(2)}\sqrt{(RT)}}{3\pi^{\frac{3}{2}}N(\sigma_{AB})^2}\sqrt{\left(\frac{1}{M_A}+\frac{1}{M_B}\right)} \tag{3.6}$$

using equation (3.3) for average velocities of the molecular species. As collision diameters are not easily obtained it is more convenient to use the molar volumes at the normal boiling points V_A and V_B, which are proportional to the cube of the collision diameter σ_{AB}:

$$\mathcal{D}_{AB} = b\frac{T^{\frac{3}{2}}}{P\left(V_A^{\frac{1}{3}}+V_B^{\frac{1}{3}}\right)^2}\sqrt{\left(\frac{1}{M_A}+\frac{1}{M_B}\right)} \quad \mathrm{m^2\,s^{-1}} \tag{3.7}$$

where b is a constant. This constant could be calculated from the kinetic theory of gases, but more realistic diffusivities are obtained when using an empirical value of $4{\cdot}3\times10^{-11}$ (SI units, with atm, °K).[301] Molecular volumes can be found in tables or by adding together atomic volumes.[609]

More sophisticated values of molecular diffusivities, allowing for attractive and repulsive forces, using the Leonard-Jones 6 : 12 potential (i.e. the attractive force varies inversely as the sixth power, the repulsive force as the inverse twelfth power) can be obtained from the equation by Hirschfelder *et al.*,[370]

$$\mathcal{D}_{AB} = \frac{bT^{\frac{3}{2}}}{P(\sigma_{AB})^2 W^1}\sqrt{\left(\frac{1}{M_A}+\frac{1}{M_B}\right)} \quad \mathrm{m^2\,s^{-1}} \tag{3.8}$$

where b is $2{\cdot}63\times10^{-11}$ in SI units
and W^1 is the collision integral which is a function of the molecular potential energy parameter characteristic of the interaction, in °K and 10^{-10} m (Angstrom units).

Details of this method of calculating the diffusivity may be obtained from reference 370.

Diffusivity in the liquid phase may be calculated from Einstein's suggestion that an osmotic force acts on molecules in the direction of decreasing solute concentration c:

$$F = -\frac{\mathbf{k}T}{c}\cdot\frac{\mathrm{d}c}{\mathrm{d}x} \tag{3.9}$$

where $\mathbf{k}$ = Boltzmann's constant.

The resistance to the motion of the molecules, considered as spheres, is given approximately by Stokes law (see Section 4.1)

$$F = 3\pi\mu\sigma u \tag{3.10}$$

where σ = diameter of molecule,

μ = viscosity of solution,

u = velocity of molecule,

$$\therefore \quad u = -\frac{\mathbf{k}T}{3\pi\mu\sigma c} \cdot \frac{dc}{dx}. \tag{3.11}$$

The rate of diffusion N_A is the product of the velocity of the molecules and the concentration

$$N_A = uc = -\frac{\mathbf{k}T}{3\pi\mu\sigma} \cdot \frac{dc}{dx}. \tag{3.12}$$

The diffusivity can now be calculated from Fick's law [equation (3.1)] and is

$$\mathscr{D}_L = \frac{\mathbf{k}T}{3\pi\mu\sigma}. \tag{3.13}$$

Values for diffusivity given by the Stokes–Einstein equation are very approximate, and in practice experimental values, such as are given by the *International Critical Tables*[387] or Landolt–Börnstein[486] should be used when these are available.

In some cases the empirical equation by Wilke and Chang[608] gives a reasonable value. This is

$$\mathscr{D}_L = \frac{7 \cdot 4 \times 10^{-9} (xM)^{\frac{1}{2}} T}{\mu V^{0 \cdot 6}} \quad \text{m}^2\ \text{s}^{-1},$$

V = molecular volume of solute (cm^3); μ = viscosity (Pa s);
x = degree of solvent association (2·6 water; 1·9 methanol, 1·5 ethanol, 1·0 for unassociated solvents such as benzene).

3.3. STEADY STATE DIFFUSION OF TWO GASES

When a gas is being continuously absorbed from a gas mixture, which is constantly renewed, an equilibrium is set up, with constant concentration gradients. The transport of the gas molecules no longer occurs by simple molecular diffusion, as is the case with a stagnant gas, but also by bulk transport of the gas, in order to replenish the concentration of the molecules being removed from the interface. If there are two species A and B, with molal concentrations c_A and c_B respectively, when the system is in equilibrium; at constant pressure

$$c_A + c_B = \text{constant}. \tag{3.14}$$

Differentiating with respect to x, the distance in the direction of the diffusion and bulk transport, perpendicular to the interface, gives a relation between the concentration gradients of the components:

$$\frac{dc_A}{dx} = \frac{-dc_B}{dx}. \tag{3.15}$$

The bulk flow in the x direction carries components A and B in the proportion of their partial pressures p_A and p_B. So for component A, the total flow N_A is the sum of the bulk flow fraction $N_A p_A/P$ (P = total pressure) and the molecular diffusion, given by equation (3.1)

$$N_A = N_A \frac{p_A}{P} - \mathcal{D}_G \frac{\mathrm{d}c_A}{\mathrm{d}x}. \tag{3.16}$$

For an ideal gas, the concentration c_B is related to the partial pressure p_B by

$$c_B = p_B/RT. \tag{3.17}$$

Substituting (3.17) in (3.16), $(P-p_B)$ for p_A, and using (3.15) this gives

$$N_A = N_A\left(1-\frac{p_B}{P}\right) + \frac{\mathcal{D}_G}{RT} \cdot \frac{\mathrm{d}p_B}{\mathrm{d}x}. \tag{3.18}$$

Rearranging:

$$N_A \frac{p_B}{P} = \frac{\mathcal{D}_G}{RT} \cdot \frac{\mathrm{d}p_B}{\mathrm{d}x} \tag{3.19a}$$

and

$$N_A \int_0^x \mathrm{d}x = \frac{P\mathcal{D}_G}{RT} \int_{p_{B_1}}^{p_{B_2}} \frac{\mathrm{d}p_B}{p_B} \tag{3.19b}$$

which on integration gives

$$N_A = \frac{\mathcal{D}_G P}{RTx} \ln (p_{B_2}/p_{B_1}). \tag{3.20}$$

If the logarithmic mean partial pressure of component B across the distance through which the transport takes place, x, is defined as

$$p_{BM} = \frac{p_{B_2}-p_{B_1}}{\ln p_{B_2}/p_{B_1}}. \tag{3.21}$$

Equation (3.20) can be written as:

$$N_A = \frac{\mathcal{D}_G}{RTx} \frac{P}{p_{BM}} (p_{B_2}-p_{B_1}) \tag{3.22a}$$

$$= \frac{\mathcal{D}_G}{RTx} \frac{P}{p_{BM}} (p_{A_1}-p_{A_2}). \tag{3.22b}$$

3.4. THEORIES OF ABSORPTION

When a gas stream is in motion across a surface, and the bulk of the gas is in turbulent motion (i.e. the Reynolds number exceeds the critical value of 2100) then turbulent mixing maintains an homogeneous composition throughout the bulk of the gas. Close to the interface the gas movement is slower, and a laminar layer occurs, while at the interface itself the gas is usually assumed to be stagnant. The most widely applied theory, which assumes steady state transport across the stable boundary layers, is called the "two film theory" and was proposed by Whitman and Lewis in 1924.[509, 937] Here the gas being absorbed diffuses by molecular diffusion across the laminar boundary layer and stagnant sublayer, and if a liquid absorbent is used, then similar layers exist on the liquid side. It is further assumed that an equilibrium exists at the liquid–gas interface. The partial pressure at the interface p_i and the concentration at the interface, c_i, are then related. When steady state conditions of transfer have been reached, the rate of transfer N_A from the gas stream to the interface, and from the interface to the bulk of the liquid stream must be equal, and

$$N_A = k_G(p-p_i) = k_L(c_i-c) \tag{3.23}$$

where p = partial pressure of the transferring component in the bulk of the gas stream,

p_i = partial pressure at the interface,

c_i = concentration at the interface,

c = concentration in the bulk of the liquid,

k_G = gas film mass transfer coefficient,

k_L = liquid film mass transfer coefficient.

Comparison with equation (3.22b) shows that the gas film mass transfer coefficient can be calculated from

$$k_G = \frac{\mathcal{D}_G}{RTx}\,\frac{P}{p_{BM}} \tag{3.24}$$

which is seen to be a function of the diffusivity, the log mean partial pressure of the non-absorbing component, and x, the distance through which diffusion is taking place, which is the film thickness. On the liquid film side the mass transfer coefficient may be calculated from

$$k_L = \mathcal{D}_L/x_L \tag{3.25}$$

where $\mathcal{D}_L$ = liquid phase diffusivity,

x_L = liquid film thickness.

Although, as has been shown, the diffusivities can be calculated, the film thickness cannot be found directly. It is, however, possible to find the equivalent film thickness representing the combination of transfer mechanisms through laminar boundary layers from an experimental mass transfer coefficient and a calculated or experimental diffusivity. It is then possible to use the value for a similar type of absorption. This has been done by Gilliland and Sherwood,[302] who correlated the results of a number of experiments using a wetted wall column:

$$\frac{k_G RTD}{\mathcal{D}_v}\frac{p_{BM}}{P} = \frac{D}{x} = 0{\cdot}023\ \mathrm{Re}^{0{\cdot}83}\ \mathrm{Sc}^{0{\cdot}44} \tag{3.26}$$

where D = diameter of the column,

Re = Reynolds number $uD\varrho/\mu$,

Sc = Schmidt number $\mu/\varrho\mathcal{D}_v$,

$\mathcal{D}_v$ = gas phase diffusivity.

Here u is the velocity of the gas relative to the column; while ϱ and μ are the density and viscosity of the gas.

This correlation is satisfactory where the rate of absorption is controlled by the rate of transfer through the gas film, but unfortunately the rippling which occurs in the liquid curtain surface makes this an unsatisfactory method of finding the equivalent film thickness for liquid-film-controlled absorption.

When the system is controlled by the liquid film resistance, the mass transfer coefficient can be measured in a disc column,[812] which gives a better correlation than the wetted wall column. The mass transfer coefficient for the liquid film k_L can be calculated from the equation

$$\frac{k_L}{\mathcal{D}_v} = 21{\cdot}0\left(\frac{4L}{\mu}\right)^{0{\cdot}7}\left(\frac{\mu}{\varrho\mathcal{D}_v}\right)^{0{\cdot}5} \tag{3.27}$$

where L = the rate of liquid flow (e.g. kg h^{-1}) per unit width of surface (kg h^{-1} m^{-2}).

However, even without detailed knowledge of equilibrium data and film coefficients, overall coefficients can be found, and these are defined by

$$N_A = K_G(p-p^*) = K_L(c^*-c) \tag{3.28}$$

where p^* = equilibrium partial pressure of solute over a solution having the same concentration c as the main liquid stream,

and c^* = concentration of a solution which would be in equilibrium with the solute partial pressure existing in the main gas stream.

These points are shown in the equilibrium diagram (Fig. 3.1).

The over-all coefficients K_G (using pressure units) and K_L (using concentration units) can be determined experimentally and used directly in the design equations.

The overall driving force is $(p-p^*)$ in pressure units and (c^*-c) in concentration units. The point B on the equilibrium curve represents the compositions of the two phases at the

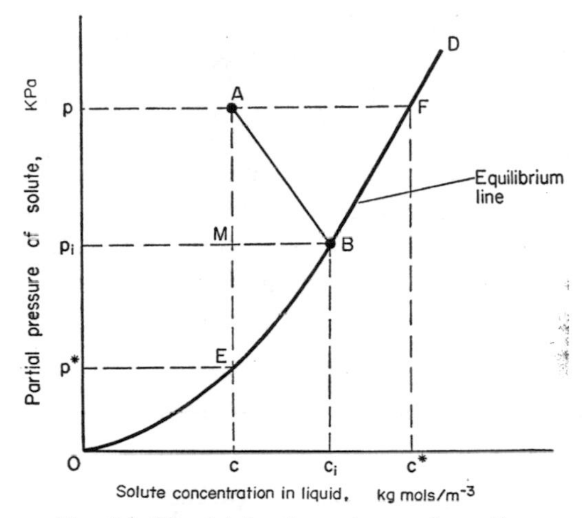

FIG. 3.1. The driving forces in gas absorption.

interface, and the driving forces $p-p_i$ and c_i-c are represented by the vertical distance *AM* and horizontal distance *MB* respectively. Thus, from (3.23)

$$\frac{p-p_i}{c_i-c} = \frac{k_L}{k_G} \tag{3.29}$$

and the ratio $-k_L/k_G$ represents the slope of the line *AB*.

Where the equilibrium relation is of the form

$$p^* = \mathcal{H}c \tag{3.30}$$

and $\mathcal{H}$ is a constant, then Henry's law applies. From equations (3.23) and (3.28) it can be deduced that

$$\frac{1}{K_G} = \frac{1}{k_G}+\frac{\mathcal{H}}{k_L} \tag{3.31a}$$

and

$$\frac{1}{K_L} = \frac{1}{k_L}+\frac{1}{\mathcal{H}k_G} \tag{3.31b}$$

If $\mathcal{H}$ is small, $K_G = k_G$ and the absorption is gas-film controlled: conversely, if $\mathcal{H}$ is large, $K_L = k_L$ and the absorption is liquid-film controlled. The equations (3.31) show also that for intermediate values of the coefficient $\mathcal{H}$ both films must be considered. Furthermore, if $\mathcal{H}$ is not a constant, i.e. if the equilibrium curve is not a straight line, then the overall coefficients will vary with the concentration dependence of $\mathcal{H}$.

While the Whitman–Lewis two-film theory, as outlined above, is useful in the design of absorption systems, it presupposes stagnant boundary layers and steady state conditions for mass transfer, which rarely exist in actual practice. Thus the gas tends to break up the stagnant layers, and the turbulent gas penetrates to the liquid surface, while the liquid in the surface film is continuously being replaced by fresh liquid from below. To overcome the problem of unsteady state diffusion, particularly when the gas–liquid interaction is short, Higbie suggested a fictitious model using Stephan's equation for molecular diffusion into a column of infinite depth. Writing equation (3.1) for unsteady state mass transfer, it was solved for a set of boundary conditions, which state that the concentration in the bulk of the liquid, at zero time, gives a positive "boundary-layer" thickness, at the interface. At some time t, there is no boundary layer, while at the same time, in the bulk of the liquid, there is an infinite layer (corresponding with Stephan's theory). The concentration of gas below the surface can then be determined, and furthermore, the instantaneous rate of mass transfer of the gas across a unit area of interface is given by

$$N_A = \frac{2}{\sqrt{\pi}}(c_i - c)\sqrt{(\mathcal{D}r)}. \tag{3.32}$$

Thus the instantaneous rate of transfer decreases with time for any period during which the liquid remains quiescent. The total amount can be obtained by integrating equation (3.32) over the time interval, while the average rate of transfer can be obtained by dividing this total amount by time t. Thus, the average rate is given by

$$N_A = 2(c_i - c)\sqrt{(\mathcal{D}/\pi t)}. \tag{3.33}$$

This equation is applied to turbulent-gas absorption by assuming that after time t, during which the liquid is quiescent, instantaneous mixing takes place, and this process is repeated indefinitely. The average rate of absorption is thus given by equation (3.31). If the average time t is replaced by a rate of renewal r, defined as $r = 1/t$, then the rate of mass transfer is

$$N_A = \frac{2}{\sqrt{\pi}}(c_i - c)\sqrt{(\mathcal{D}r)} \tag{3.34}$$

and from the definition of k_L in equation (3.23)

$$k_L = \frac{2}{\sqrt{\pi}}\sqrt{(\mathcal{D}r)}. \tag{3.35}$$

It should be noted that while this value of k_L has, by definition, the same numerical value as for the two-film theory, it has a different physical meaning.

More recently Danckwerts[196] introduced an important modification to Higbie's model, in the frequency with which particular vertical elements are mixed, designating r as this frequency. He then assumed that the mixing process, i.e. the exposure of gas to surface, and the frequency, are random processes. Similar mathematical arguments and boundary conditions can be used to determine the rate of mass transfer, which is now

$$N_A = (c_i - c)\sqrt{(\mathcal{D}r')} \tag{3.36}$$

and

$$k_L = \sqrt{(\mathcal{D}r')} \tag{3.37}$$

where r' = average exchange rate, which differs from the "rate of renewal" by the factor $2/(\pi)^{\frac{1}{2}}$. The Danckwerts' penetration model, applied to a liquid of depth H, gives

$$k_L = \sqrt{(\mathcal{D}r)}.\tanh\sqrt{(r'H^2/\mathcal{D})}. \tag{3.38}$$

As the term $\sqrt{(r'H^2/\mathcal{D})}$ approaches larger values, with greater depth, the value of k_L approaches the "infinite depth" of the elementary Danckwerts' theory.

3.5. DESIGN OF ABSORPTION SYSTEMS

A suitably designed unit for absorbing gases should be able to operate at the highest possible efficiencies, with maximum flexibility of throughput and with the lowest capital and operating charges. Absorption units can be classified into two groups. In the first, bubbles of gas are dispersed in the liquid in either a continuous or multi-stage system, while

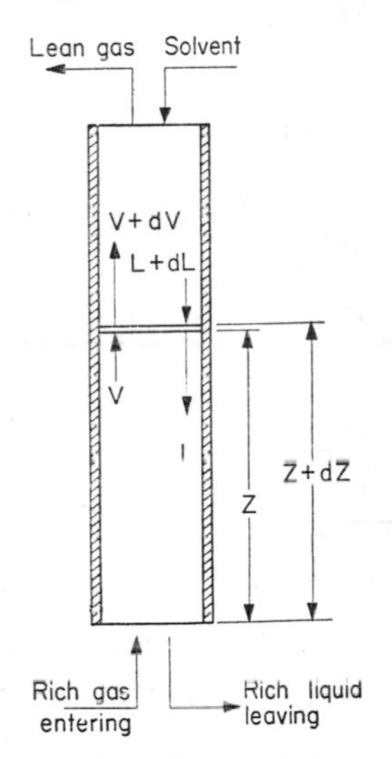

FIG. 3.2. Diagrammatic representation of a counterflow continuous absorption tower.

in the second, liquid drops are dispersed through the gas. In nearly all units except single stage absorbers, the operation of the plant is essentially a counter-current flow absorption operation of the type shown diagrammatically in Fig. 3.2, although some scrubbing plants operate with concurrent flow. Only continuous plant will be discussed here.

If

V = molar rate of flow of gas phase,
L = molar rate of flow of liquid phase,
y = mole fraction of gaseous component being absorbed,
x = mole fraction of absorbed component in liquid,

and subscript 1 = bottom of contacting tower,
subscript 2 = top of contacting tower,

then, for steady state operation, a material balance over a differential tower section is

$$dV = dL \tag{3.39}$$

and a component balance over the same section is

$$d(Vy) = d(Lx). \tag{3.40}$$

On integration, using as one limit the bottom of the tower (subscript 1) and as the other any level within the tower, equation (3.40) becomes:

$$Vy - V_1y_1 = Lx - L_1x_1 \tag{3.41a}$$

$$Vy + L_1x_1 = Lx + V_1y_1. \tag{3.41b}$$

This is the equation of the operating line, and is valid for all values of x between x_1 and x_2 and all values of y between y_1 and y_2 (Fig. 3.3a).

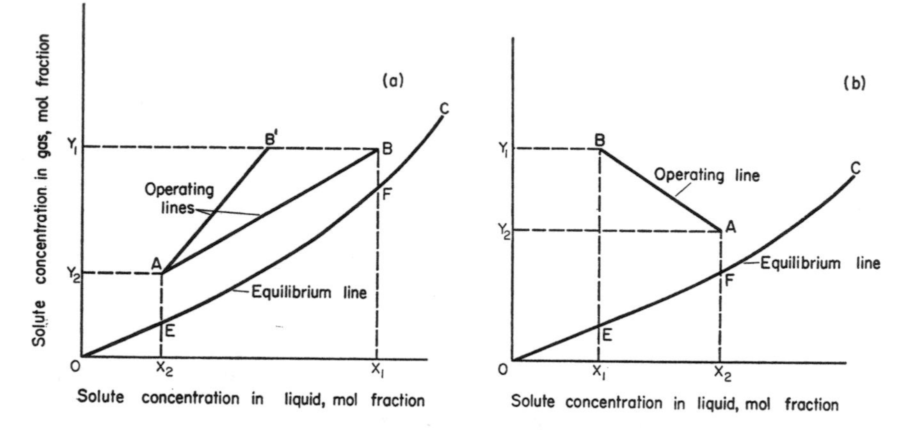

FIG. 3.3. (a) Operating and equilibrium lines in a counterflow column.
(b) Operating and equilibrium lines in a co-current contacting system.

In the general case, V and L vary at different points up the tower, and the operating line will be curved. In many gas cleaning problems, however, the component being absorbed constitutes only a small fraction of the gas and the absorbing liquid, and then the operating line is essentially linear.

In a differential section of the column, height dZ, the intersurface area is dA, and the rate of absorption is—from equations (3.23) and (3.28)

$$\begin{aligned} N_A\,dA &= K_G(p-p^*)\,dA = K_L(c^*-c)\,dA \\ &= k_G(p-p_i)\,dA \\ &= k_L(c_i-c)\,dA. \end{aligned} \tag{3.42}$$

If a is the interfacial area per unit of tower volume, and S is the tower cross-section,

$$dA = aS\,dZ. \tag{3.43}$$

The value of a is not usually known accurately in practice, in the case of packed towers, where only a fraction of the available surface may be absorbing, or in the case of spray towers, where the surface area of the droplets is not known. It is therefore usual to combine the mass transfer coefficient with the area in a composite coefficient, $k_G a$, $k_L a$, $K_G a$ or $K_L a$. Substituting (3.43) in equation (3.42):

$$\begin{aligned} \mathrm{d}(Vy) &= N_A\,\mathrm{d}A = k_G a(p-p_i)S\,\mathrm{d}Z \\ &= k_L a(c_i-c)S\,\mathrm{d}Z \\ &= K_G a(p-p^*)S\,\mathrm{d}Z \\ &= K_L a(c^*-c)S\,\mathrm{d}Z \end{aligned} \tag{3.42a}$$

The mole fraction y of thc soluble component in the gaseous phase is p/P and the mole fraction x in the liquid phase is c/ϱ_{mean}, then equation (3.42a) can be transformed to

$$\begin{aligned} \mathrm{d}(Vy) &= k_G aP(y-y_i)S\,\mathrm{d}Z = k_L a\varrho_m(x_i-x)S\,\mathrm{d}Z \\ &= K_G aP(y-y^*)S\,\mathrm{d}Z = K_L a\varrho_m(x^*-x)S\,\mathrm{d}Z \end{aligned} \tag{3.44}$$

$(y-y^*)$ is represented by the vertical lines BF at (x_1, y_1) and AE at (x_2, y_2) in Fig. 3.3a.

For gas absorption, where V and L are not constant, it can be shown that $\mathrm{d}(Vy) = V\,\mathrm{d}y/(1-y)$. Combining this with equation (3.44) gives the equation for the height of the tower Z, with gas film coefficients:

$$\begin{aligned} \int_0^z \mathrm{d}Z &= \int_{y_1}^{y_2} \frac{V}{k_G aPS} \cdot \frac{\mathrm{d}y}{(1-y)(y-y_i)} \\ &= \int_{y_1}^{y_2} \frac{V}{K_G aPS} \cdot \frac{\mathrm{d}y}{(1-y)(y-y^*)} \end{aligned} \tag{3.45a}$$

or for liquid film coefficients:

$$\begin{aligned} \int_0^z \mathrm{d}Z &= \int_{x_1}^{x_2} \frac{L}{k_L a\varrho_m S} \cdot \frac{\mathrm{d}x}{(1-x)(x_i-x)} \\ &= \int_{x_1}^{x_2} \frac{L}{K_L a\varrho_m S} \cdot \frac{\mathrm{d}x}{(1-x)(x^*-x)} \end{aligned} \tag{3.45b}$$

Simplifying these equations by multiplying both terms in the integral by the logarithmic mean driving force,

$$(1-y)_{LM} = \frac{y_2-y_1}{\ln\dfrac{1-y_1}{1-y_2}} \quad \text{and} \quad (1-x)_{LM} = \frac{x_2-x_1}{\ln\dfrac{1-x_1}{1-x_2}}$$

and assuming that the product of this and the mass transfer coefficient is a constant, then

$$
\begin{aligned}
Z &= \frac{V_{av.}}{k_G a(1-y)_{LM}PS}\int_{y_1}^{y_2}\frac{(1-y)_{LM}}{(1-y)(y-y_i)}\,dy \\
&= \frac{V_{av.}}{K_G a(1-y)_{LM}PS}\int_{y_1}^{y_2}\frac{(1-y)_{LM}}{(1-y)(y-y^*)}\,dy \\
&= \frac{L_{av.}}{k_L a \varrho_m(1-x)_{LM}S}\int_{x_1}^{x_2}\frac{(1-x)_{LM}}{(1-x)(x_i-x)}\,dx \\
&= \frac{L_{av.}}{K_L a \varrho_m(1-x)_{LM}S}\int_{x_1}^{x_2}\frac{(1-x)_{LM}}{(1-x)(x^*-x)}\,dx. \qquad (3.46)
\end{aligned}
$$

It should be noted that the two equations above using the overall coefficients K_G and K_L are only strictly correct when Henrys law applies, i.e. $\mathscr{H}$ is constant.

The group outside the integral sign is the height (H_t) of a transfer unit, while the group within the integral sign represents the number (N) of transfer units. The height of the tower is then

$$Z = H_t \times N. \qquad (3.47)$$

The height of a transfer unit can be evaluated from the mass transfer coefficient, the characteristics of the system and the initial and final concentrations of absorbing gas or solute. The number of transfer units can be evaluated by the use of the equilibrium diagram for the system and one of the following methods:

(a) Graphical integration: If no relation between the variables y and x is available except an experimental equilibrium curve, it is necessary to evaluate the number of transfer units graphically, plotting

$$(1-y)_{LM}/(1-y)(y-y^*)$$

against y, if $K_G a$ is known, or against similar functions in the other cases.

(b) Simplified graphical integration of $1/(y^*-y)$ can be used if an arithmetic average of $(1-y)$ and $(1-y^*)$ can be substituted for $(1-y)_{LM}$ without introducing any undue error. The complex integral from equation (3.47) then simplifies to

$$N = \int_{y_1}^{y_2}\frac{dy}{y-y^*}+\frac{1}{2}\ln\frac{1-y_2}{1-y_1} \qquad (3.48)$$

(c) If the concentrations are very dilute, and both operating and equilibrium curves are linear, with the equation

$$y = mx+C \quad \text{for the equilibrium line,}$$

then it can be shown that

$$N = \frac{y_2-y_1}{(y-y^*)_{LM}} \qquad (3.49)$$

where

$$(y-y^*)_{LM} = \frac{(y-y^*)_1-(y-y^*)_2}{\ln \dfrac{(y-y^*)_1}{(y-y^*)_2}}.$$

When the liquid flow rate is large compared to the gas flow rate, the operating line, slope L/V, will be steep, and the driving forces large. With low L/V ratios, the driving force will be reduced until the operating and equilibrium lines touch, when equilibrium will be attained, and the minimum liquid rate can thus be found. This minimum could only be achieved in an absorber of infinite length.

Co-current flow is not usually used in packed towers or simple spray towers. It is common, however, in various scrubbers of the venturi or cyclonic spray types which are used for lean gas mixtures. The calculation of the number of transfer units in the co-current case is similar to that in the counter-current, except that the operating line has a slope in the opposite direction as shown by the operating line in Fig. 3.3b.

An empirical equation for the number of transfer units in a cyclonic spray scrubber (Fig. 9.7) absorbing sulphur dioxide has been derived by Johnstone and Silox.[406]

$$N = \frac{1{\cdot}09 \times 10^{-5}\, nP}{V^{0{\cdot}8} S} \tag{3.50}$$

where n = number of nozzles in the scrubber,
S = cross-sectional area of gas inlet (m^2),
P = total pressure of gas (kPa),
V = gas flow rate through entry: kg moles s^{-1} m^{-2}.

A more general correlation for scrubbers, based on the power consumption of the plant has been suggested by Lunde[536] which indicates that the number of transfer units in a spray tower is proportional directly to the power introduced in the liquid $\mathcal{P}_L$ and inversely to the power introduced by the gas $\mathcal{P}_V$.

$$N \propto \mathcal{P}_L/\mathcal{P}_V^{0{\cdot}1} \tag{3.51}$$

However, constants to be used in this equation have not been determined.

3.6. CONSTRUCTION OF ABSORPTION PLANT

Counter-current gas–liquid contacting may be carried out in following ways:

(1) Bubbling the gas through a vessel containing the absorbing liquid. Vessels usually contain a pipe with holes emitting the gas, called a porous gas sparger, and the degree of gas dispersion may be assisted by agitating the liquid. This is not a usual method for industrial gas cleaning.

(2) The gas can be passed through a series of bubble cap plates or other types of plates in a plate column. The design of these columns is fully discussed in texts on the design of

distillation columns. Plate columns are more expensive to construct than packed columns, but they have other advantages, such as the ability to handle very high liquid rates, being easier to clean, more amenable to interstage cooling or heating, and having a lower total weight. Turbogrid[70] and sieveplate columns can also be used for gas absorption. These are sometimes cheaper to construct than the traditional bubble cap type columns, and may be more effective.

(3) The gas can be passed through a simple packed column. This is the most common equipment used for absorption, and will be more fully discussed below.

(4) Spray towers. These are widely used for the collection of particles and their design for this purpose is discussed in Chapter 9. Contrary to what might be expected they are sometimes used for gas absorption where there is a high liquid-film resistance,[768] because the circulation within the drops presents fresh surfaces for absorption to take place. In general, however, spray towers cannot be used for absorbing high concentrations because of the low liquid–gas ratio that can be maintained. They have the advantage that they do not easily block up because of their open construction, have a very low pressure drop and can be operated at high gas rates, particularly when a cyclonic spray eliminator is employed.

Packed columns are constructed very simply (Fig. 3.4) with the packing either randomly

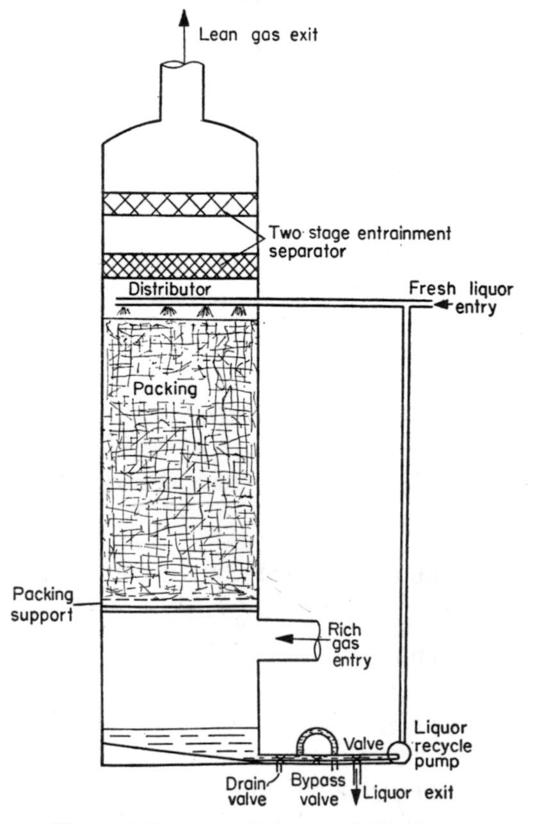

FIG. 3.4. Components in a packed tower.

distributed or stacked in an empty tower. It is supported above the gas entry and the liquid runs down over it. The packing may simply be broken rock or stone, or take the form of rings (Raschig rings, Lessing rings, Pall or Dixon rings), saddles (Berl saddles, Intalox saddles, McMahon saddles), grids (wood or carbon) or a number of other types of tower filling packing such as wooden slats and Spraypack,[547] which consists of an expanded metal mesh structure filling the column. Rings smaller than 50 mm dia. and saddles are usually randomly poured into the column, while larger rings are carefully stacked, particularly if they are fragile. The packing rings can be metal, glass, ceramic or plastic, depending on the corrosive conditions in the tower. The liquid must be carefully distributed in order to wet all the packing, particularly near the top. Spray eliminators must also be provided. The packing support must be sufficiently strong mechanically to carry both the packing and liquid if this should fill the column during flooding, and its structure must also be more open than the packing so that it does not become flooded and hold up the gas or liquid.

The diameter of an absorption column is determined by two factors, the first being a satisfactory rate of relative liquid and gas flow, and the second being the most economic tower dimensions. At low liquid and gas-flow rates, an orderly trickling of the liquid over the packing is observed. When the gas rate increases, the pressure drop increases, and a point occurs when some of the liquid is held in the packing (liquid hold up). This is the *loading point.* At a higher gas rate *flooding* occurs, and the pressure rises very quickly as is shown in Fig. 3.5. The point where this happens is the *flooding point*, which has been shown to be a characteristic of the viscosity of liquid as well as of the density and mass flow rates of the liquid and gas. The flooding rates of a large number of fluids and packing materials have been correlated (Fig. 3.6),[524] and can be determined in a particular case by using this curve. Appropriate data for some common packings are listed in Table 3.1.

It is usual to operate columns at gas flow rates less than 50–60 per cent of the flooding point rate; and this determines the minimum tower diameter. A more rapid graphical

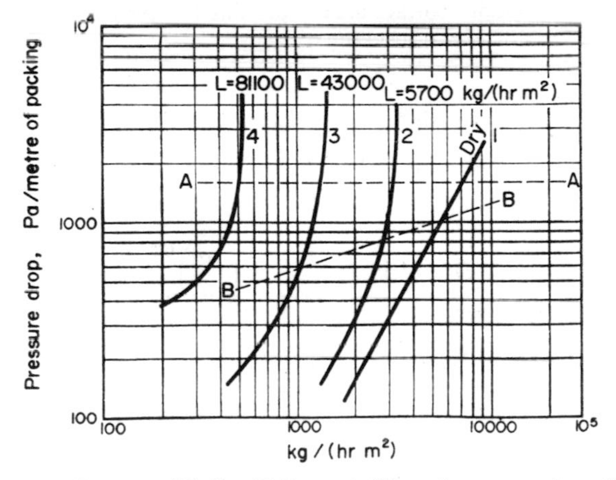

FIG. 3.5. Relation between gas flow rate (V), liquid flow rate (L), and pressure drop (Δp) in a packed column.

AA—The flooding point line,
BB—The loading point line.

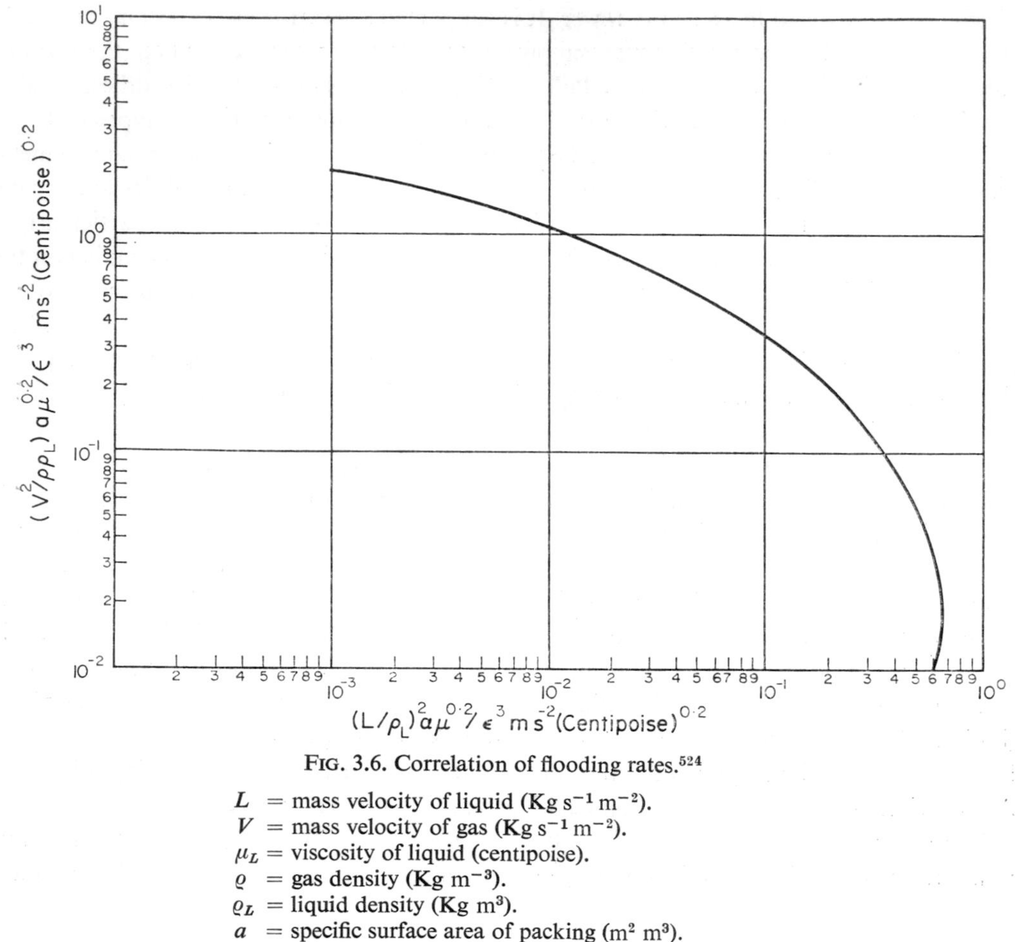

FIG. 3.6. Correlation of flooding rates.[524]

L = mass velocity of liquid (Kg s^{-1} m^{-2}).
V = mass velocity of gas (Kg s^{-1} m^{-2}).
μ_L = viscosity of liquid (centipoise).
ϱ = gas density (Kg m^{-3}).
ϱ_L = liquid density (Kg m^3).
a = specific surface area of packing (m^2 m^3).
ϵ = porosity of packing (dimensionless).

method for the minimum tower diameter is based on similar principles and has recently been published by Chen.[157] Here the tower diameter D (based on 50 per cent of flooding rate) can be found from

$$D = 16{\cdot}28\left(\frac{V}{\varphi L}\right)^{0{\cdot}5}\left(\frac{\varrho_L}{\varrho}\right)^{0{\cdot}25} \quad \text{m} \tag{3.52}$$

where V = gas flow rate (kg s^{-1}),
φ = correlation coefficient (parameter) from Fig. 3.7.

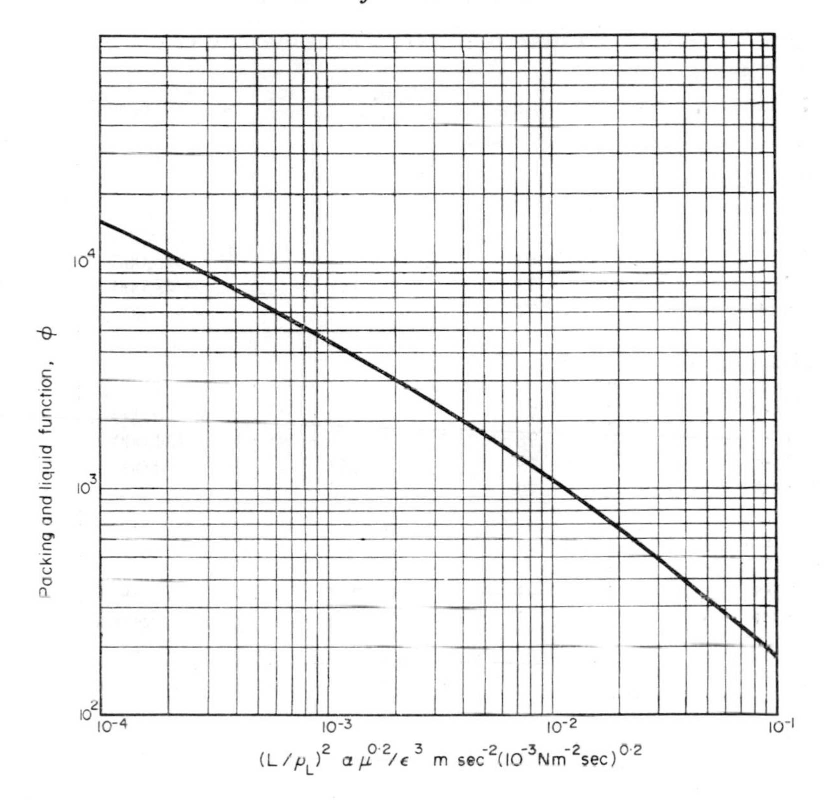

FIG. 3.7. Correlation for calculating tower diameter based on 50 per cent of flooding rate.[157]

φ = correlation parameter.

Other notation as in Fig. 3.6.

Once the minimum tower diameter has been found, the design diameter has to be based on economic considerations, so long as these indicate that the most economical diameter is not less than this minimum.

The type of packing has to be decided. This has to withstand corrosive conditions within the tower when they exist. The more complex packing materials with a larger effective surface area are more expensive, but require a smaller tower for the same duty. The cost of a tower shell is found to increase approximately with the square of the shell diameter. However, a wider tower requires a shallower packing, with a lower pressure loss for the gas flow, reducing fan power requirements. A balance of these factors will give the most economic tower diameter, and this should be chosen if it gives a gas rate of less than 50 per cent of the flooding rate. If the most economic diameter is less than this, the minimum applicable to the design requirements should be chosen.

TABLE 3.1. CHARACTERISTICS OF SOME COMMON PACKINGS

Packing	Porosity ϵ %	Specific surface (a m^2 m^{-3})	Number per m^3	Dumped weight (kg m^{-3})
Raschig rings (ceramic) mm				
6·3	63	825	3,120,000	930
12·7	64	365	373,000	837
25·4	73	191	48,000	725
50·8	74	92	5800	386
Intalox saddles (ceramic) mm				
6·3	75	990	4,150,000	837
12·7	78	626	734,000	725
25·4	78	257	100,000	547
50·8	79	118	9400	530
Pall rings (metal) mm				
25·5	93	218	53,500	530
50·8	94	120	7500	442

3.7. GAS-ABSORPTION PROCESSES

3.7.1. *Sulphur dioxide* (SO_2)

Sulphur dioxide is the most common gaseous air pollutant because of the wide occurrence of sulphur in fuels and mineral ores. The problem has two aspects. In the treatment of sulphur-bearing minerals (lead, zinc, copper, tin and other ores) sulphur constitutes a major part of the ore, and this is released during sintering and smelting as sulphur dioxide in high concentrations which causes considerable damage, even at a large distance from the point of release. This occurred in the state of Washington, the cause being the smelter gases from Trail in British Columbia, just across the United States border. Damages of over $ 420,000 were awarded against the Consolidated Mining and Smelting Co. of Canada Limited[439] up to 1937. In cases like these treatment of the smelter and sinter plant waste gases is imperative. Restrictions in many parts of the United States stipulate that the stack concentration of sulphur dioxide from a smelter should not exceed 0·75 per cent by volume, while the Los Angeles County specifies 0·20 per cent.[427]

Much more general is the problem of sulphur dioxide from combustion gases from power generation. Here the concentration of sulphur dioxide is much lower, as is shown in the values in Table 3.2.

TABLE 3.2. CONCENTRATIONS OF SULPHUR DIOXIDE IN COMBUSTION WASTE GASES (USING 15% EXCESS AIR IN COMBUSTION)

	% sulphur	SO_2 in flue gas %
Coal	1	0·06
Coal	4	0·25
Fuel oil	2	0·12
Fuel oil	5	0·31

The quantity of sulphur dioxide produced is, however, very large. A modern coal-fired power station, capacity 1000 MW, burning 9000 tonnes of coal with 2 per cent sulphur per day produces 360 tonnes of sulphur dioxide. The treatment of the enormous quantities of waste gases from power stations has been attempted in several cases, and a number of processes are on the verge of commercial application.

The absorption of sulphur dioxide from smelter gases has been satisfactorily tackled in several ways. Unfortunately sulphur dioxide is only slightly soluble in water (Fig. 3.8), but it is readily soluble in alkaline solutions. Those actually used contain ammonia, xylidine

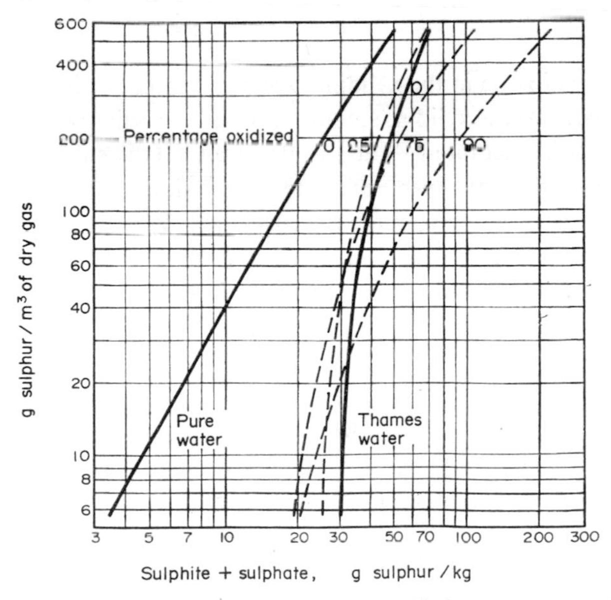

FIG. 3.8. Solubility of sulphur dioxide in water and in dilute alkaline solution such as Thames river water.[678] Solubility with various degrees of oxidation in alkaline solution are also given (broken lines). Temperature: 40°C. Gas volumes have been reduced to N.T.P.

or dimethyl aniline, although numerous other amines have been suggested.[75] The Lurgi "Sulfidine Process" uses a 1 : 1 xylidine–water mixture as the absorbent and produces pure sulphur dioxide, which can be used in other processes such as sulphuric acid manufacture. Toluidine has also been tried, but not found as successful. Xylidine and water are normally immissible, but as sulphur dioxide reacts with the xylidine, some xylidine sulphate, which is water soluble, is formed, and when the concentration of sulphur dioxide approached 100 kg m^{-3} the mixture becomes homogeneous. An equilibrium curve for the sulphur dioxide–xylidine water system is shown in Fig. 3.9. Essentially the process consists of

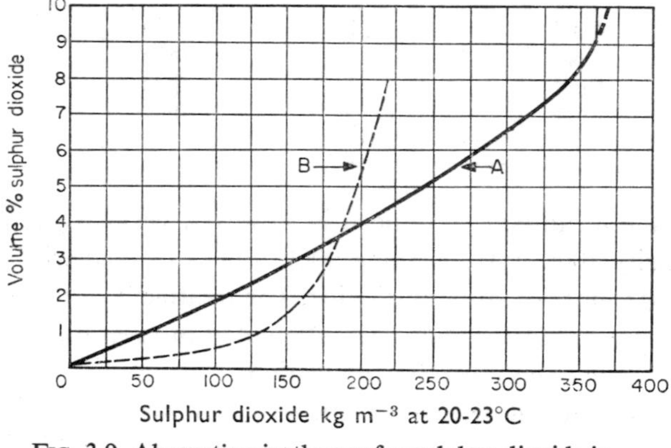

FIG. 3.9. Absorption isotherms for sulphur dioxide in
(A) Anhydrous dimethyl aniline.
(B) 1 : 1; xylidine–water mixture.[261]
(Basis of Sulfidine process.)

absorbing the sulphur dioxide in the xylidine–water mixture. Sodium carbonate (soda ash) solution is added to convert the xylidine sulphate formed to sodium sulphate, and the sulphur dioxide is stripped off in a column heated to 95–100°C. This sulphur dioxide, with some xylidine, is water washed to produce pure sulphur dioxide. Some dilute sulphuric acid is used to recover xylidine vapours from the exhaust gases. When a gas containing 8 per cent sulphur dioxide is used, the concentration after scrubbing is reduced to 0·05–0·10 per cent sulphur dioxide. At low sulphur dioxide concentrations the process ceases to be economical because of the losses of xylidine. A flow sheet is shown in Fig. 3.10.

The dimethylaniline process, developed by the American Smelting and Refining Company, and known as the ASARCO process, also produces pure sulphur dioxide.[261] Particulate impurities are first removed from the gases, using electrostatic precipitation (Chapter 10) and the gases are then scrubbed with pure dimethylaniline. The equilibrium diagram (Fig. 3.9) shows that dimethylaniline is a more efficient absorber than the xylidine–water mixture for sulphur dioxide concentrations greater than about 3·5 per cent; although at lower concentrations the xylidine-water mixtures have an economic advantage. The waste gases are scrubbed with sodium-carbonate solution to remove traces of sulphur dioxide and dimethylaniline, and also with dilute sulphuric acid which absorbs the last traces of dimethylaniline. The sulphur dioxide rich liquors are passed to a steam-distillation column which

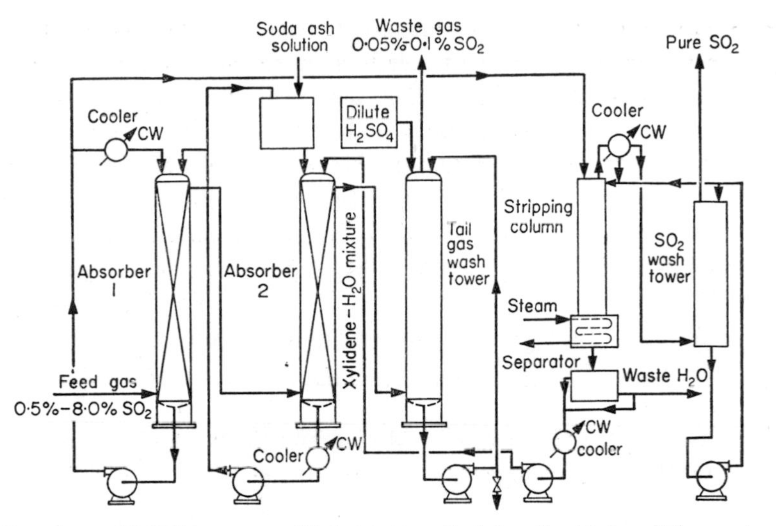

FIG. 3.10. Flow sheet of Sulfidine process:[455] absorption of sulphur dioxide in xylidine–water mixtures.

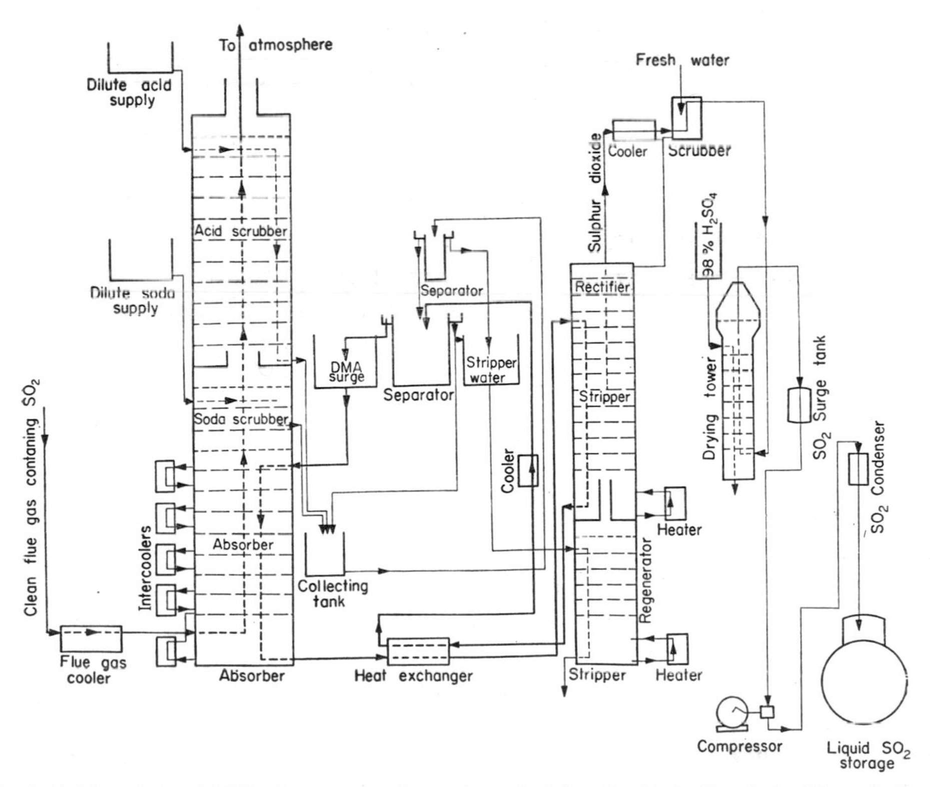

FIG. 3.11. Flow sheet of ASARCO process: absorption of sulphur dioxide in dimethyl aniline solution.[261]

strips the sulphur dioxide from the liquors, and the gases are then passed to a scrubber, to remove the dimethylaniline, a drying tower to remove moisture, and then to a sulphur-dioxide storage for further processing, such as acid manufacture.

A plant constructed at Selby, California, with a capacity of 20 tonnes per day (sulphur dioxide) recovers 99 per cent of the sulphur dioxide in the waste gases from a Dwight–Lloyd sintering machine which contains 5 per cent sulphur dioxide. It uses 0·5 g dimethylaniline, 16 g of sodium carbonate and 18 g of sulphuric acid per kg of sulphur dioxide produced, as well as 1·1 kg of steam and 0·52 MJ of power, and cooling water at the rate of 8·2 kg h^{-1} (18°C). A flow diagram of the process is shown in Fig. 3.11. Bubble-cap columns have been used for both absorber and stripping columns, and a number of operation such as absorbing soda and acid scrubbing are combined in each tower, reducing the cost of the plant.

Ammonia is used by the Consolidated Mining and Smelting Company at Trail, and is known as the COMINCO process, a flow diagram being shown in Fig. 3.12. The gases are

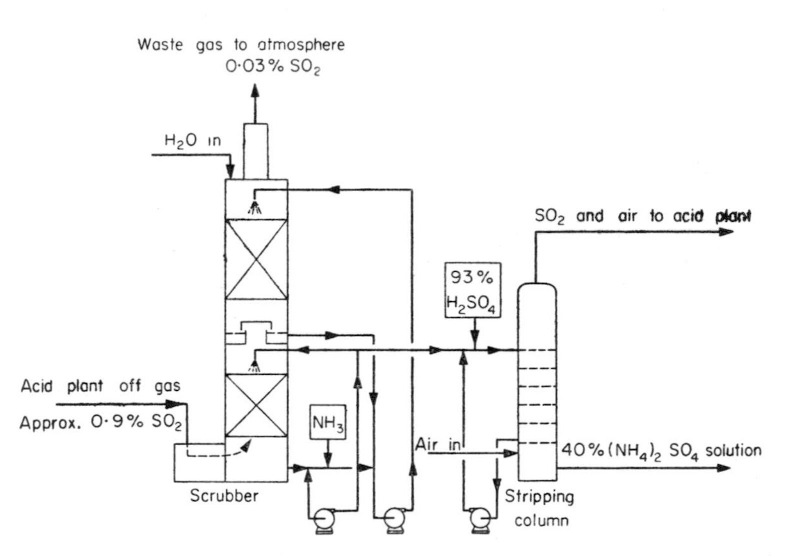

FIG. 3.12. Flow sheet of COMINCO process: absorption of sulphur dioxide in ammonia solution.[455]

passed through a two-stage column filled with wooden slat packing. The absorption is carried out in a dilute ammonium sulphate solution to which ammonia is added. The sulphur dioxide is stripped from the ammonium sulphite by adding concentrated (93 per cent) sulphuric acid, producing sulphur dioxide and ammonium sulphate. The sulphur dioxide is used in further acid manufacture.

A scrubbing process which is able to handle a wide range of sulphur dioxide concentrations (0·1–1·5 per cent), and so can be used for flue gases from combustion of fuels and dilute smelter gases, is the Lurgi Sulfacid Process[293, 540] (Fig. 3.13). The waste gases from a roaster or other combustion process are cooled and then passed to beds of activated carbon where a

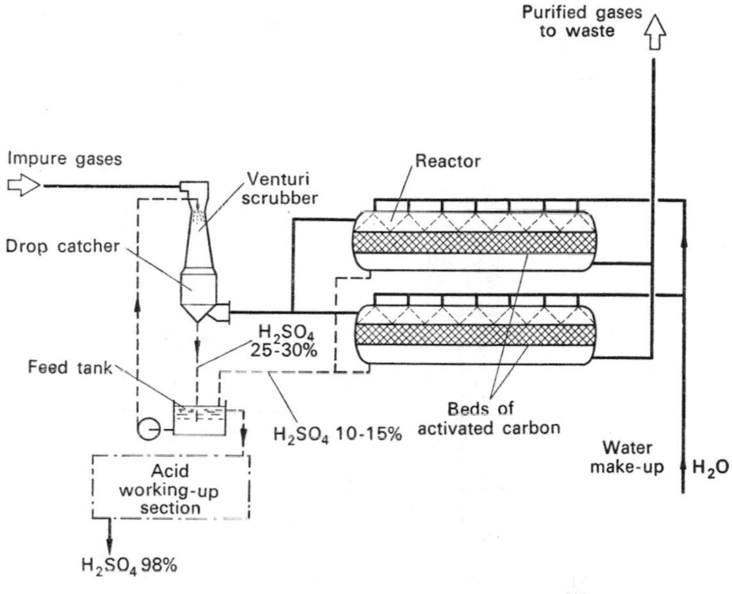

FIG. 3.13. Lurgi Sulfacid process flowsheet.[293]

catalytic oxidation takes place. These beds are contained in rubber-lined cylindrical vessels supported on hurdles. The resultant acid is washed out and then concentrated to 65–70 per cent acid using the heat of the incoming gases. Removal efficiencies are approx. 95 per cent, and the sulphur-dioxide concentration in the tail gases is below 750 ppm for 1·5 per cent sulphur dioxide in the feed gases, or 150 ppm, if 0·3 per cent sulphur dioxide is in the feed gases.

Sulphur dioxide removal from flue gases is much more difficult because of the enormous quantities of gas that have to be handled as well as the low concentrations, and the scrubbing processes considered above, with the exception of the Lurgi Sulfacid process are not very suitable. Ordinary river water is not suitable for scrubbing, but the Thames river water is slightly alkaline and has been used for scrubbing the flue gases at the Battersea and Bankside power stations. The process is limited because the discharge from the scrubbers into the river increases the calcium-sulphate concentration which will form scale in heat exchangers, as well as affecting marine life in the river. Furthermore, since the flue gases after scrubbing are cold, they will not rise and disperse the residual sulphur dioxide over a wide area, but cause heavy local pollution.

At Bankside gases from a fuel oil-fired station, using oil with 3·8 per cent sulphur, are scrubbed, and 95 per cent of the sulphur dioxide is removed. A flow sheet of the process is shown in Fig. 3.14, and equilibrium data for sulphur dioxide Thames river water is included in Fig. 3.8. A chalk slurry is added in practice to make the river water more alkaline, and therefore more effective for scrubbing, and some manganese sulphate is added, together with air, to the effluent liquors as an oxidizing agent, so that the water can be passed back to the river as the sulphate and not the sulphite. The absorption tower consists of timber-lined cast iron with grids of red cedar and teak. The aeration tanks and other parts of the plant are

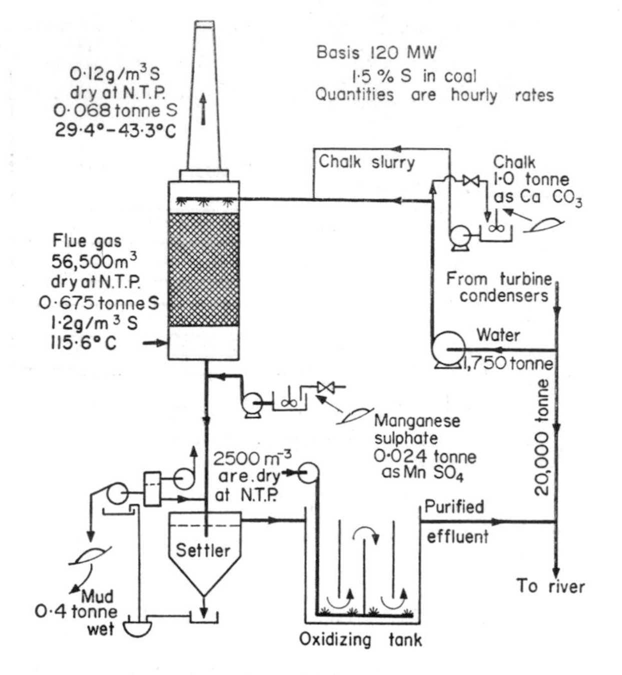

FIG. 3.14. Battersea process: absorption of sulphur dioxide in Thames river water to which an alkaline slurry has been added.[677]

lined with acid-resistant materials. There are, however, serious maintenance problems connected with the plant.[678]

Sea water, from the tidal estuary of the River Derwent, has been successfully used for treating the waste gases from a zinc-blende roasting plant by the Electrolytic Zinc Co. plant at Risdon, Australia.[433] Four circular towers, each with 12·1 m^2 of wooden packing area, 5 m dia. and 5 m high, not unlike the Battersea system, are being used to treat gases 85,000 $m^3\ h^{-1}$ at 180–200°C containing 5 per cent sulphur dioxide. An experimental scrubber, using water flows of 53 kg m^{-3} gave efficiencies better than 95 per cent. In the final unit 80,000 $m^3\ h^{-1}$ were treated with $3{\cdot}25 \times 10^6$ kg h^{-1} following an initial washing of 42,500–51,000 $m^3\ h^{-1}$ by 407,000 kg h^{-1}. The scrubbers used an electrostatic precipitator as a mist eliminator. This reduced inlet concentrations of 2·2 per cent to as low as 2 ppm with a pressure drop of 100 Pa. With lower water flows the efficiency was less, and stack gas concentrations up to 30 ppm were obtained.

Pilot-plant studies of sea water scrubbing of flue gases with sulphur dioxide concentrations of 900 ppm have been carried out by Bromley and Read.[135] The liquid gas ratios were of the order of 25 kg sea water per kg stack gas or higher, and absorption efficiencies of 90 per cent were obtained in a simple spray absorber. With a packed column absorber, 1·52 m depth of packing, 99 per cent absorption efficiencies, and even higher, were obtained. The resultant sulphate produced by oxydation of the dissolved sulphur dioxide can then be used as feed for a desalination plant in which the need for sulphuric acid addition (to prevent scale formation) can be eliminated. The amount of sea water required can be reduced by a factor of 5 if more sophisticated absorption equipment, such as multistage plates on fluidized bed absorbers, are used. The pH of the resultant solution is then between 2 and 3. If this is to be returned to the sea directly, it will require neutralization by passing over limestone.

A process similar to the Battersea one, but not relying on a slightly alkaline river to provide the bulk of the alkali, is the I.C.I.–Howden Cyclic Lime Process.[628] A lime or chalk slurry (5–10 per cent by weight) is circulated through a wooden grid absorption tower and calcium sulphite is formed, which then oxidizes to the sulphate with oxygen from the flue gases.

A flow sheet for the process is shown in Fig. 3.15. The problem of scaling which could occur with the supersaturated calcium sulphate solutions has been solved by passing the solution from the scrubber to a delay tank where a slurry of lime, adjusted for pH is added.

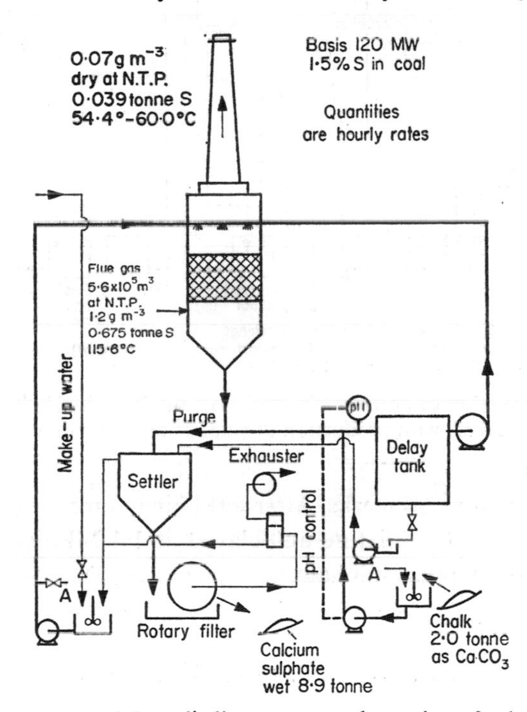

FIG. 3.15. Howden–I.C.I. cyclic lime process: absorption of sulphur dioxide in lime slurry liquor.[628, 678]

Here the excess calcium sulphate crystallizes on the existing crystals, before the solution is recycled back through the tower. Some of the crystalline slurry is passed to a settling tank and the crystals are removed. No use is made of these, because they are contaminated with fly ash, but an ammonium carbonate digestion, which would yield ammonium sulphate, has been suggested.[270]

The Japan Engineering Co. and Mitsubishi Heavy Industries have patented a process which is similar in principle to the ICI–Howden Cyclic Lime Process, but which produces high-purity gypsum ($CaSO_4$) crystals.[40, 825] In the Soviet Union a suspension of natural limestone has also been used to remove sulphur dioxide from the gases from ore agglomeration plants when the sulphur content in the ore is small.[649] The residual SO_2 in the tail gases is below 0·15 per cent. The sludge containing a mixture of calcium sulphate and sulphite is not utilized.

The other process which has been applied on a large scale (flue gas rate 95,000 $m^3 h^{-1}$ is the Fulham–Simon–Carves process.[435, 900, 951] Here (Fig. 3.16) the flue gases are scrubbed

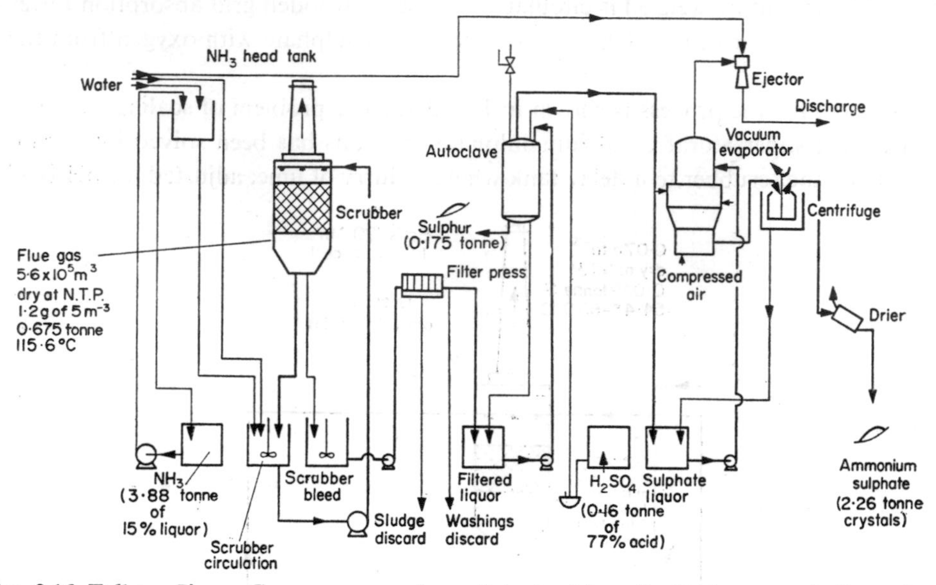

FIG. 3.16. Fulham–Simon–Carves process: absorption of sulphur dioxide in ammonia liquor from gas works or coke ovens, with production of sulphur and ammonium sulphate.[435, 678]

with ammoniacal liquors from gas works. After scrubbing, some sulphuric acid ((77 per cent) is added to the liquors and the mixture autoclaved at 1·4 MPa and 170°C for 3hr. This produces sulphur and ammonium sulphate.

The overall equation is

$$2NH_4HSO_3 + (NH_4)_2S_2O_3 = 2(NH_4)_2SO_4 + 2S + H_2O.$$

The first plant was installed at the Fulham power station in 1939, and a second large plant is at the North Wilford Power Station (Nottingham) where it was tested in 1957.

The relative economics of the three flue-gas washing processes which have been operated indicate that the Howden–I.C.I. cyclic lime process is the cheapest to operate if no suitable river or source of cheap ammonia is available. If ammonia is readily available, for example as a by-product from a gasworks, then the Fulham–Simon–Carves process is the most economic.

Pietelina[649] reports a modification of the Fulham–Simon–Carves process which has been investigated in the U.S.S.R. in which the saturated sodium disulphite solution is treated with sulphuric acid and decomposed in an autoclave at 600 kPa and 147°C. The removal efficiency claimed is 93 to 97 per cent, and both sulphur yield has been increased and autoclave treatment times reduced from the original Fulham conditions. A pilot plant treating 50,000 $m^3 h^{-1}$ was built in 1968.

Mitsubishi Heavy Industries[40] have also developed a process for scrubbing flue gases with ammonia, but unlike the Fulham–Simon–Carves Process, crystalline ammonium sulphate is produced. Sulphur-dioxide-removal efficiencies of 95 per cent are claimed with initial concentrations of 0·1–0·2 per cent SO_2, and ammonia losses are slight. The ammonium sulphate liquors are supplied to the crystallizer in a 45 per cent solution, using a waste-heat-recovery process.

Another process which also produces ammonium sulphate is the Mitsubishi Manganese Oxyhydroxide Process.[40] A 3 per cent manganese oxyhydroxide slurry is used as an absorbent. Ammonia and oxygen are supplied to the sulphide complex which is formed, ammonium sulphate is produced and the slurry is recycled. It is considered an improvement on the ammoniacal liquor process, with an absorption efficiency of 97 per cent SO_2 from a 0·1–0·2 per cent SO_2 gas stream producing tail gases with 30–60 ppm SO_2 and no losses of ammonia. As in another Mitsubishi process, 45 per cent ammonium sulphate solution can be supplied to a crystallizer without using an evaporator.

Pietelina[649] also reports a *Cyclic Magnetite* process used in the U.S.S.R. for removing sulphur dioxide from flue gases from thermal power stations or for agglomeration plants where the ores have a high sulphur content. The gases are scrubbed with a suspension of magnesium sulphite and magnesium oxide crystals in a solution of the magnesium sulphite and disulphite. The slightly soluble disulphite crystals, which are rod shaped, are removed from the solution, dried and roasted at 850–900°C. This decomposes the disulphite to the oxide and sulphur dioxide. The oxide is returned to the absorbing slurry, while the sulphur dioxide, which is now present in a concentration of 10–15 per cent, is fed to an acid plant. Pilot plant tests have been carried at a thermal power station (14,000 $m^3 h^{-1}$) and at an agglomeration plant (8000 $m^3 h^{-1}$), and removal efficiencies of 95–97 per cent have been obtained. Four full-scale plants, treating 4 million $m^3 h^{-1}$ have been installed during 1969 at the Magnitogorsk steel works, with an annual production of 150,000 tonnes of sulphuric acid.

Another process using ammonia as an absorbent has been developed by Shale, Simpson and Lewis[763] at the U.S. Bureau of Mines. In this process, which will also remove part of the nitrogen dioxide in the flue gases, the dust-free flue gases are contacted with a mixture of ammonia and water, in the vapour phase, in a reaction chamber. The gases are then cooled and the reaction products are formed and removed in the condensate, while the cleaned gas passes through a water extractor before being released at about 35°C. The products in the condensate are reacted with zinc oxide, forming zinc sulphide. This is subsequently decomposed, and liquid sulphur dioxide obtained with regeneration of the zinc oxide.

Other possible processes include contacting the flue gases with sodium sulphite and bisulphite, which increases the bisulphite content.

$$SO_2+NaHSO_3+Na_2SO_3+H_2O = 3NaHSO_3$$

The solution is then reacted with zinc oxide, giving zinc sulphite.

$$NaHSO_3 + ZnO = ZnSO_3\ \text{(ppte.)} + NaOH$$

This is then calcined to give zinc oxide and sulphur dioxide which can be used for other processes.[407]

$$ZnSO_3 = ZnO + SO_2$$

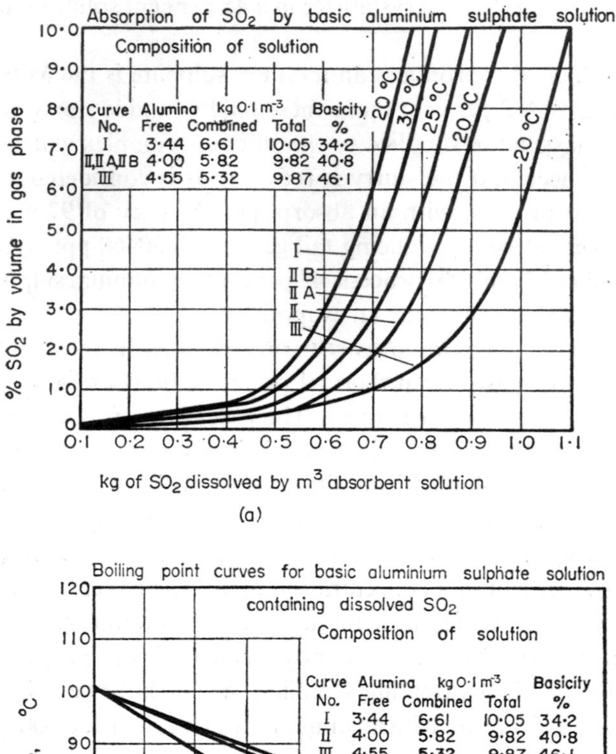

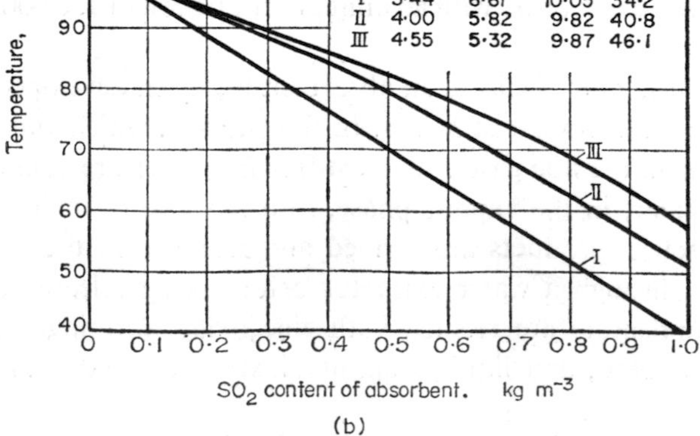

FIG. 3.17. I.C.I.—Basic aluminium sulphate process.[33]

(a) Absorption isotherms of sulphur dioxide in basic aluminium sulphate solution.
(b) Boiling point curves of basic aluminium sulphate solution containing dissolved sulphur dioxide (for steam stripping).

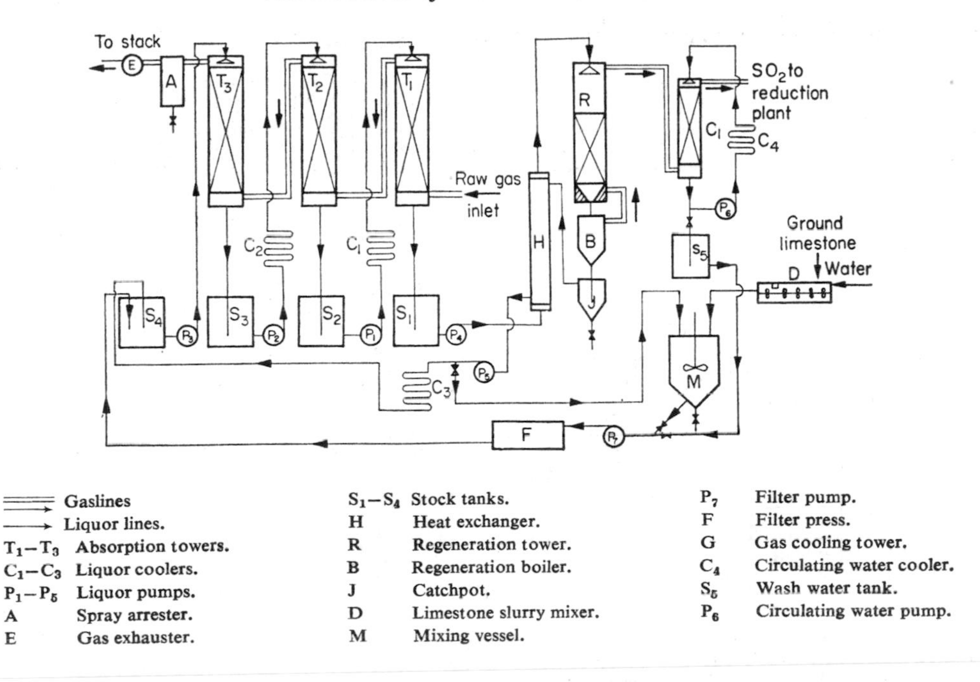

FIG. 3.17 (*continued*)

(c) Flow sheet for Outokumpu plant (only three absorption towers of the four erected are shown).

Basic aluminium sulphate has also been used as an absorbent, with recovery of the sulphur dioxide, by stripping in a steam–heated column (Fig. 3.17b).[33] The equilibrium diagram for the sulphur dioxide–basic aluminium sulphate system is shown in Fig. 3.17a. This process was developed by I.C.I. and put into operation at the Imatra Smelter of the Outokumpu Copper Company. The gases are first scrubbed with water to remove particulate impurities and then passed through four absorption towers, each with 10·6 m of wooden grid packing, the liquid moving counter-currently as shown in Fig. 3.17c (where only three towers are shown). The saturated liquors, containing 60 kg m^{-3} of sulphur dioxide then passed to a pre-heater and on to the sulphur dioxide regenerator. To resist corrosion, the plant is constructed throughout of lead-lined equipment, acid resisting brick covered with lead sheet, or wood for the grids in the absorption towers.

A recent development by Mitsubishi Heavy Industries Ltd. is the Red Mud Process[40] (Fig. 3.18) where "red mud", which is spent bauxite or "Luxmasse", is used as an absorbent in a slurry. The "Luxmasse" is the waste product after the alumina has been extracted from bauxite by the Bayer process. A typical composition is Al_2O_3 18·9 per cent; SiO_2 17·4 per cent; Na_2O 8·3 per cent; Fe_2O_3 39·3 per cent; TiO_2 2·8 per cent; ignition losses 10·5 per cent. It is the complex of sodium aluminium silicate which is the active reagent in absorbing sulphur dioxide. After the absorption stage, the slurry, which contains the absorbed SO_2 as sodium sulphate solution, can be treated to give commercially worth-while products or can be discarded. The treatment consists of removing all caustic soda and sodium sulphate,

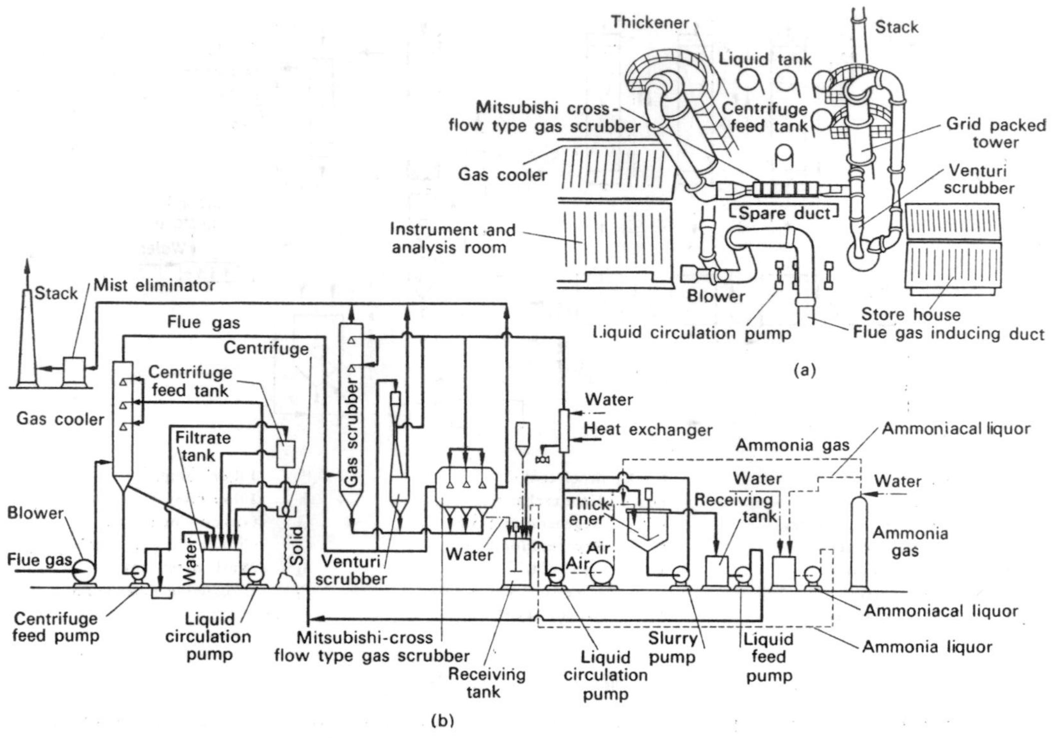

FIG. 3.18. Mitsubishi spent bauxite "Red Mud" process.

(a) Sketch showing arrangement of pilot plant.
(b) Process flow sheet for pilot plant (solid line indicates absorbent flow).

and then treating the residue with sulphur dioxide. Three types of materials are formed in the residues, which are easily separated and have the following compositions:

(1) Al_2O_3 99·2 per cent; SiO_2 0·8 per cent.
(2) SiO_2 91·9 per cent; Al_2O_3 8·1 per cent.
(3) Fe_2O_3 76·8 per cent; TiO_2 5·5 per cent; Al_2O_3 7·4 per cent; SiO_2 10·5 per cent.

The alumina and silica fractions [(1) and (2)], separated in this way, are pure white powders, free from iron oxides, and so can be used as industrial raw materials.

A high-temperature liquid scrubbing process, developed by the Atomics Division of North American Rockwell Corporation, uses a eutectic mixture of alkali carbonates.[615–16] The composition of the mixture used is 32 wt. per cent Li_2CO_3, 33 wt. per cent Na_2CO_3 and 35 wt. per cent K_2CO_3 and this melts at 397°C. It is a clear mobile liquid, which at 425°C has properties not unlike a water–ethylene glycol mixture: viscosity 0·012 Pa s, specific gravity 2, specific heat 0·40 and thermal conductivity 1·4 W m^{-2} $°C^{-1}$. The mixture absorbs more than 99 per cent of the sulphur dioxide in a mixture containing 0·3–3·0 per cent leaving 20 ppm in the tail gases. The stages in the process are a scrubbing stage, followed by reduction and regeneration of the absorbent. In the absorption, the sulphur oxides in the gas stream react with the carbonate, to form the metal (M) sulphite and sulphate:

$$SO_2+M_2CO_3 \rightarrow M_2SO_3+CO_2$$
$$SO_3+M_2CO_3 \rightarrow M_2SO_4+CO_2$$

The equilibrium constants are 320 and 10^{11} for the two reactions at 425°C, and so 40 per cent of the carbonate is converted before the sulphur dioxide partial pressure in equilibrium with the melt exceeds 200 ppm. The rates of reaction are very fast, and the absorption is limited by the mass transfer of the sulphur dioxide. A spray scrubber, using gas velocities of 7·5 m s^{-1}, gives removal efficiencies of 95 per cent. The scrubbing reactions are slightly exothermic, and this partly compensates for the heat losses from the scrubber. The reduction stage uses producer gas or reformed natural gas (75 per cent H_2, 21 per cent CO) at a temperature of 600°C, and at this temperature a disproportionation of the sulphite precedes the reduction, viz.

disproportionation: $$4M_2SO_3 \rightarrow 3M_2SO_4+M_2S$$
reduction: $$M_2SO_4+4H_2 \rightarrow M_2S+4H_2O$$
$$M_2SO_4+4CO \rightarrow M_2S+4CO_2$$

The reduction reactions are slow, and as a batch process the time is 40 to 60 min. A more rapid counter-current process is being investigated. In the regeneration stage, the sulphide reacts with a mixture of carbon dioxide and water at 425°C:

$$M_2S+CO_2+H_2O \rightarrow M_2CO_3+H_2S$$

This reaction is rapid, and is carried out in a sieve tray reactor. The resultant melt is recirculated to the scrubber. The exit gases from this regeneration stage, which contain 30 per cent hydrogen sulphide with 35 per cent each of carbon dioxide and water, can be fed directly to a Claus plant for producing sulphur. It is necessary to remove fly ash from the incoming waste gases, because it will dissolve in the melt if not removed, and turn this into a paste which cannot be circulated. Stainless steels can be used for materials of construction for those parts of the plant kept below 485°C while nickel or cobalt-based alloys are preferred for those sections where the temperatures are higher (up to 670°C).

A scrubbing process for which few details are available has been investigated on a pilot plant by Wellman–Lord Inc. at the Gannon Power Plant of the Tampa Electric Co.[32] The process is claimed to remove 90 per cent of the sulphur dioxide and sulphur trioxide, together with the residual fly ash which remains after electrostatic precipitation. Following reaction and treatment, pure sulphur dioxide is stripped off for use in an acid plant or for sulphur recovery.

The absorption of sulphur dioxide from flue gases presents two problems. The first is that the quantities of gases handled are very large; thus an 800-MW power plant produces 2·5 M m^3 h^{-1} which have to be treated, while the second is that the gases and liquors are highly corrosive. The early absorption towers constructed at Battersea and elsewhere were very large, lead lined and filled with wooden packing. Recent years have seen much effort devoted to the development of better packings and towers. A special wire packing has been tested with an ammoniacal liquor-absorption system by Klimicek, Skrivanek and Bettelheim[448] which has much better characteristics than the conventional Raschig ring packing. The height of a transfer unit with this packing is two-thirds of Raschig rings, and it permits a superficial velocity of 5 times that with conventional packing. Furthermore, the cost is only 30 per cent of the conventional packing.

A multi-stage turbulent contact absorption column has been tested with sodium and calcium carbonate solutions.[653] The turbulent contact absorber uses a non-flooding packing consisting of low-density spheres placed between retaining grids sufficiently far apart to permit turbulent and random motion of the spheres. This arrangement gives the unit a high absorption rate at high gas and liquid velocities, with low pressure drops. The equipment is non-clogging and so is useful in handling dusty gases or where a solid is formed in the reaction. In pilot plant tests with a four-stage absorber, sodium carbonate removed 88–96 per cent of the sulphur dioxide, while calcium carbonate removed 78–87 per cent.

Mitsubishi Heavy Industries Ltd. have developed a cross-flow scrubber[562] which consists of a series of horizontal V-section trays, close packed in vertical rows so that the gases are forced to penetrate the liquid surface. The efficiency of absorption of sulphur dioxide (0·11–0·24 per cent) by sodium sulphite solutions was 90–95 per cent with pressure drop in the range 0·7–1·5 kPa. This is much lower than for a venturi scrubber with similar capacity but higher than for the simple grid packing.

Potassium formate, either as a molten salt at 177°C or as an aqueous solution at 93°C,[956] has also been found a suitable absorbent, averaging an absorption efficiency with 0·39 per cent sulphur dioxide in the feed stream, of 88 per cent.

The essential sequence of reactions is as follows:

(i) Scrubbing (aqueous solution)

$$2KOOCH + 2SO_2 \rightarrow K_2S_2O_3 + 2CO_2 + H_2O \qquad (a)$$

(ii) Regeneration

$$4KOOCH + K_2S_2O_3 \rightarrow 2K_2CO_3 + 2KHS + 2CO_2 + H_2O \qquad (b)$$

$$2KHS + CO_2 + H_2O \rightarrow K_2CO_3 + 2H_2S \qquad (c)$$

$$K_2CO_3 + 2CO + H_2O \rightarrow 2KOOCH + CO_2 \qquad (d)$$

If a molten-salt system were used, K_2S rather than KHS would appear in equations (b) and (c). KHS appears in aqueous solution because of the high ionization constant. The net reaction for the overall process, in which the recycled formate and carbonate is eliminated, is

$$SO_2 + 3CO + H_2O \rightarrow H_2S + 3CO_2$$

The hydrogen sulphide is recovered as sulphur by the conventional Claus process. The stack gas scrubber has no unusual or special requirements when used with 80–85 per cent aqueous potassium formate, and a pressure drop of 1·8 kPa is anticipated in commercial equipment. The spent solution from the scrubber, containing 20–25 per cent $K_2S_2O_3$, 15–20 per cent water and the remainder KOOCH, is sent to the regeneration section. The three stages of regeneration are the reduction reaction [equation (6)] which occurs readily (and exothermically) under a pressure of 3·5 MPa and at 280°C, a stripping stage, where the H_2S is stripped out with CO_2, and formate synthesis stage. The reduction is fast, and 15 min is adequate for 96 per cent conversion.

At the present time of writing the technology of all stages had been established on bench scale and laboratory development studies, and a pilot plant was to be built at the Consolidation Coal Company's Research Division near Pittsburgh, Pa. The proposed system has the advantage that it will not require reheating the flue gases, and has no waste-disposal problems other than that for fly ash. Furthermore, only commercially available liquid/gas-handling equipment would be needed for all stages of the process.

3.7.2. *Fluorine and fluoride compounds* (HF, SiF_4)

Hydrofluoric acid gas (HF), silicon tetrafluoride (SiF_4) and the combination of these, fluosilicic acid (H_2SiF_6), are emitted with the waste gases from superphosphate fertilizer manufacture, when the phosphate rock is treated with sulphuric acid in "dens"; in the smelting of aluminium when fluorspar (CaF_2) is used as a flux; in the electrolytic manufacture of aluminium, where aluminium oxide (alumina Al_2O_3) is fused in cryolite (Na_3AlF_6) at 900–1000°C and electrolysed; and in calcium metaphosphate furnaces. An extensive review

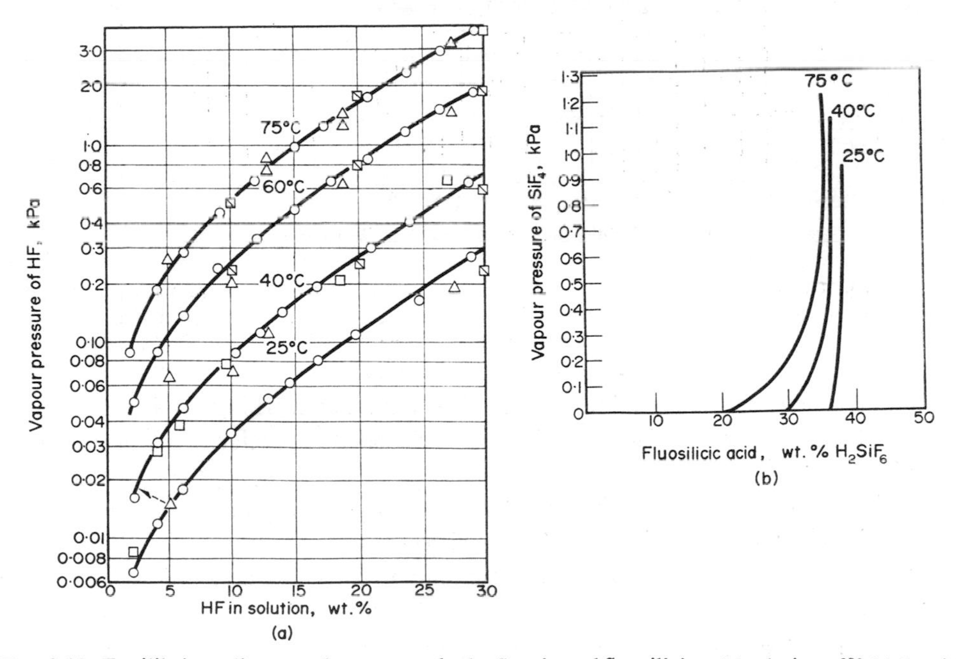

FIG. 3.19. Equilibrium diagrams for aqueous hydrofluoric and fluosilicic acid solutions.[938] (a) Partial pressure of hydrogen fluoride over hydrofluoric acid solutions.[938] (b) Partial pressure of silicon tetrafluoride over aqueous solutions of fluosilicic acid (H_2SiF_6).[938]

of the sources and concentrations of fluoride emissions has been published by Semrau.[751] As was seen in Chapter 2, very low concentrations of fluorides (of the order of one part per hundred million) may damage vegetation, and somewhat higher concentrations may lead to chronic fluoride poisoning of sheep and cattle, so fluorine emissions have to be carefully controlled. Fortunately, as is shown in the equilibrium diagrams (Figs. 3.19a and 3.19b) both hydrogen fluoride and fluosilicic acid (with which silicon tetrafluoride is in equilibrium) are very readily soluble in water, and water scrubbing is nearly always effective in reducing the fluoride concentration to an acceptable value.

Counter-current spray towers, co-current flow spray towers, venturi scrubbers, wet cell washers and packed towers have all been used in actual plants or pilot for treatment of these gases.

Lunde's correlations of power consumption with the number of transfer units, N, for spray scrubbers and venturi scrubbers is shown in Fig. 3.20. These can be converted into

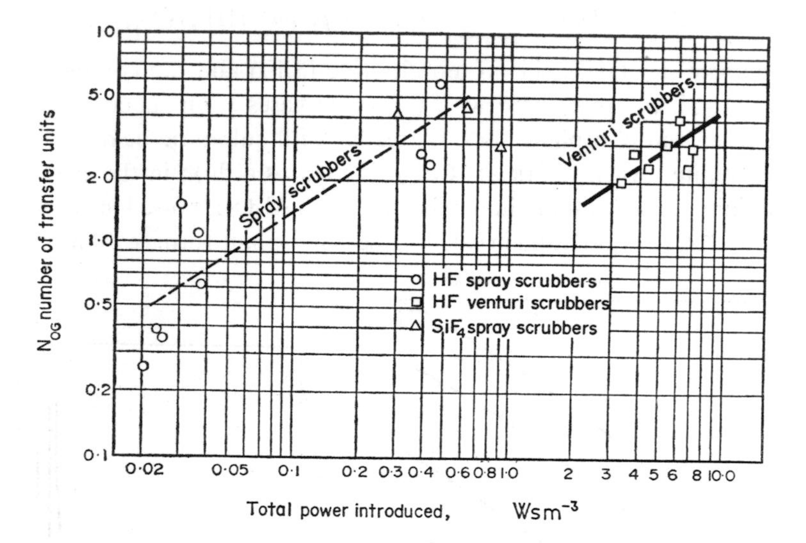

FIG. 3.20. Correlation of scrubber efficiency with respect to hydrogen fluoride with power input. Scrubber efficiency is given in transfer units.[536] Points for silicon tetrafluoride are based on data by Pettit and Sherwin.[455]

efficiencies (Appendix III). A simple counter-flow scrubber working on a nodulizing kiln was found to be 97 per cent efficient.[548] Other data on this particular plant are given in Table 3.3.

Wet cell washers, where the cells consist of coarse and fine fibres, have been tested by First *et al.*[260] One unit, shown in Fig. 3.21, consisted of three stages:

Stage 1. A 20-mm wet cell, with 255-μm Saran fibres using 2·05 $m^3\ h^{-1}$ water.

Stage 2. A fibre-glass (10 μm fibre) pad, 500×685 mm, 50 mm deep, packed to 7·35 kg m^{-3} (dry).

Stage 3. One wet cell and one dry pad in series.

The gas flow rate was 1020 $m^3\ h^{-1}$ and the pad resistance 600–690 Pa.

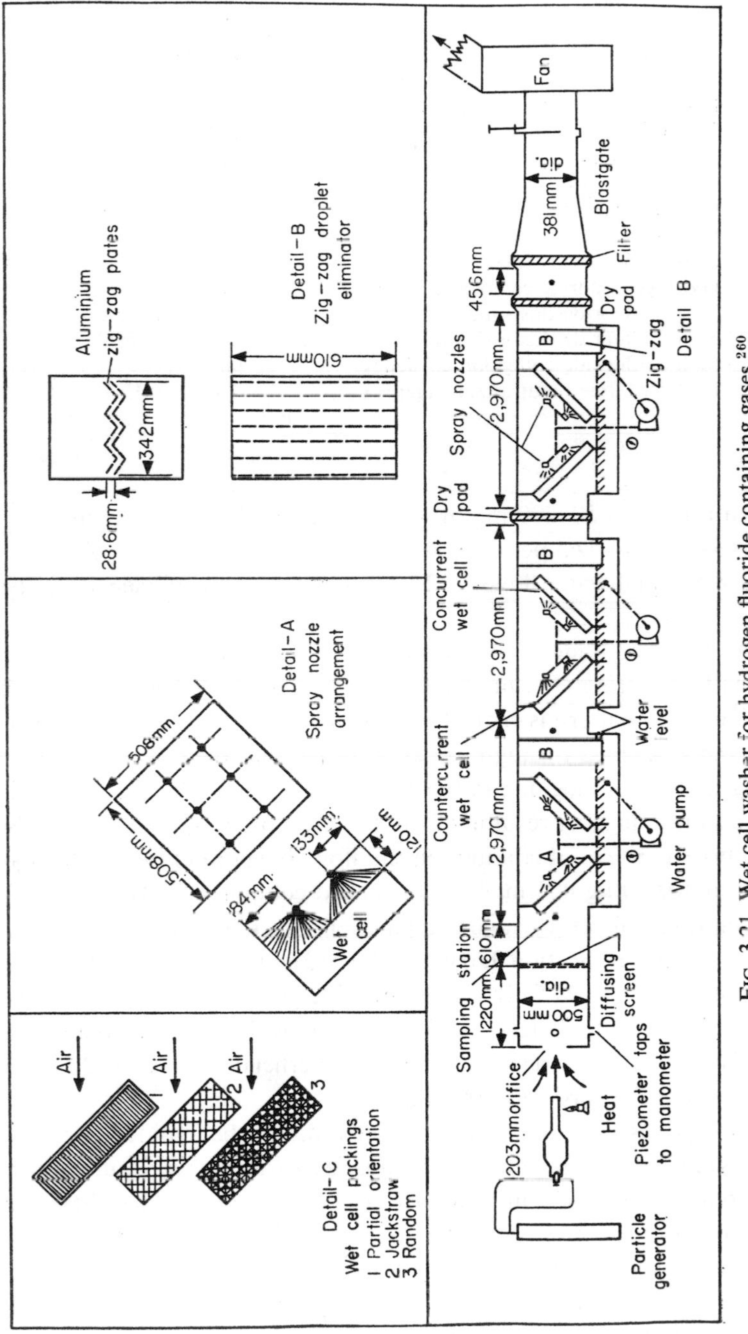

FIG. 3.21. Wet cell washer for hydrogen fluoride containing gases.[260]

TABLE 3.3. DATA FOR SIMPLE COUNTERFLOW SCRUBBER ON A NODULIZING KILN

Tower height	24·2 m
Tower diameter	457 mm
No. of spray-injection points	6
Water flow	153 $m^3 h^{-1}$
Gas-inlet temperature	300°C
Gas-exit temperature	72°C
Gas-inlet rate	88,500 $m^3 h^{-1}$
Gas-exit rate	45,000 $m^3 h^{-1}$
Hydrogen fluoride inlet rate	75 kg fluorine h^{-1}
Hydrogen fluoride removal efficiency	97·6%
Sodium fluoride dust-inlet rate	6·4 kg fluorine h^{-1}
Sodium fluoride dust-exit rate	0·25 kg fluorine h^{-1}
Sodium fluoride collection efficiency	96%
Hydrated lime (neutralizing agent) used	430 kg h^{-1}

With concentrations of 50–200 mg/m^3 of hydrofluoric acid, the cumulative efficiency of the stages was 94, 97 and 99·5 per cent. A unit using two similar wet cells, but with a lower rate of 184 $m^3 h^{-1}$ and water flows of 2·1 $m^3 h^{-1}$ (cell) gave efficiencies of better than 99 per cent.

Teller[852] has reported that spray concurrent packed scrubbers have been built by the Wellman–Lord Engineering Co., in which spray sections have been followed by a packed-bed section. The spray pattern is critical because no by-passing can be tolerated. After each set of sprays, the gases pass through irrigated baffles, and the larger particles are removed here, reducing the rate of clogging in the packed-bed section. This is especially significant because of the solids plugging resulting from the absorption of SiF_4, and is also the reason for using concurrent instead of counter-current flow. Irrigation rates are 50,000 kg $h^{-1} m^{-2}$ with gas rates of 10,000 kg $h^{-1} m^{-2}$ in the packed column with Tellerette packing. Teller also reports venturi scrubber–cyclone combinations, such as are shown in Fig. 9.25.

A floating bed scrubber (Section 9.6) where the bed consists of lightweight moulded plastic spheres has been used successfully on the fluoride-containing waste gases from alumina electrolysis:[437] 95 per cent efficiency in fluoride removal has been obtained with a 0·3 m deep bed of hollow polyethylene spheres and a superficial velocity of 2·52 m s^{-1} resulting in operating pressure drops of 870–1000 Pa.

When fluorine gas is present, the use of water should be avoided because the fluorine does not always react with water, and also because explosions have occurred in some systems. In this case sodium hydroxide in 5–10 per cent concentrations has been found to be a satisfactory absorbent. A flow sheet for a commercial fluorine and hydrogen fluoride-disposal plant is shown in Fig. 3.22.[485]

The sodium hydroxide solutions are passed down through a packed column at 38–65°C. Caustic concentrations below 2 per cent are avoided as this leads to the formation of fluorine oxide (OF_2), which is extremely poisonous. This compound is also formed when the contact time between gas and caustic is about 1 sec, and minimum contact times of 1 min

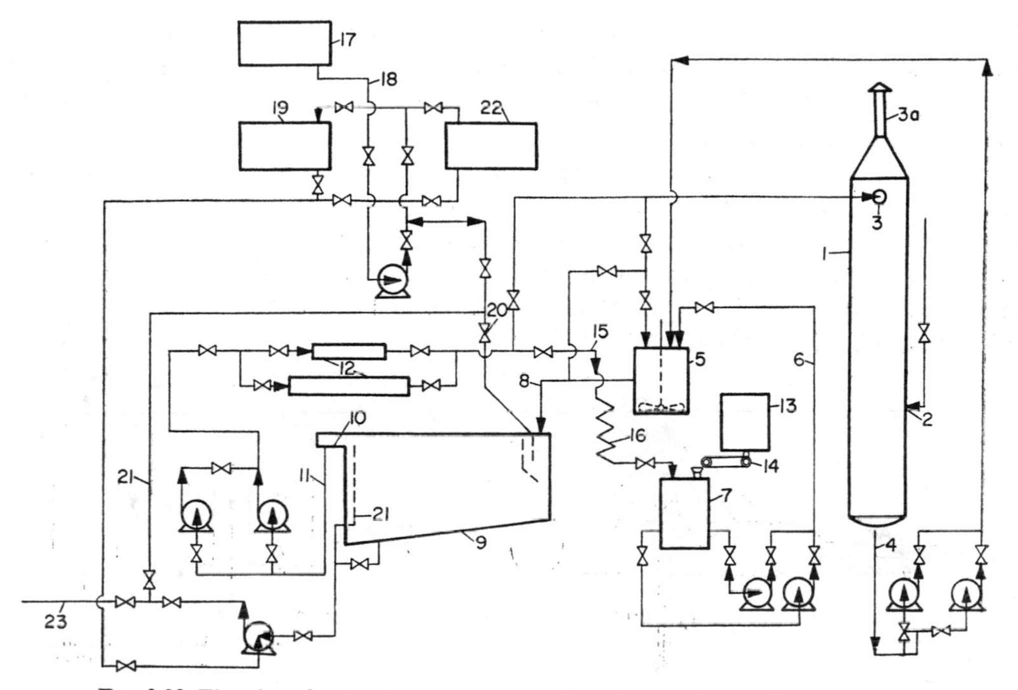

FIG. 3.22. Flowsheet for fluorine and hydrogen fluoride containing effluent gases.[485]

The fluorine-containing gases are introduced to a packed absorption tower, 1, through nozzle 2. The tower is fed with a counter-current stream of 5–10 per cent sodium hydroxide solution, which is introduced at the top of the tower through nozzle 3. Inert gas is vented through stack 3a. The effluent liquid from the tower, containing sodium fluoride in solution, is continuously withdrawn from the tower through line 4, and passed into a regeneration tank, 5, to which is supplied, through line 6, a small stream of lime slurry which enters from slaker 7. In the regeneration tank, 5, the sodium fluoride formed is converted to calcium fluoride by reaction with lime, under conditions of good agitation. The mixture flows through line 8 into a settling tank 9, wherein the calcium fluoride and excess lime are settled out. The clear, regenerated liquor overflows weir 10 and is discharged through line 11 back to the absorption tower 1.

In order to maintain temperature control on the tower, the discharge from settling tank 9 first passes through a heat exchange system, 12. This system is automatically regulated to maintain a constant temperature of 37·8°–65·6°C on the tower feed.

The lime slurry is prepared by the addition of lime (quick or hydrated) from bin 13 to tank 7, using belt feeder 14. The slaking or slurrying medium is a portion of fresh tower feed recycled to tank 7 through line 15. This solution may be cooled by passage through exchanger 16.

Incoming 50 per cent sodium hydroxide solution is pumped from tank car 17 through line 18 to storage tank 19, which is partially filled with water to make a 25 per cent solution. Make-up alkali can be withdrawn from 19 as needed and pumped through line 20 to the settling tank, 9.

When sufficient solids have accumulated in the settling tank, the clear liquor is decanted off through a swing pipe and pumped through line 21 to decantation tank 22. After the settling tank has been cleaned, the clear liquor returned thereto from the decantation tank, 22.

Waste solids in the settling tank may be removed by adding sufficient water to make a slurry, and pumping through 23 to disposal.

are recommended,[484] when the fluorine and hydrogen fluoride both react with caustic soda to form sodium fluoride (NaF):

$$F_2 + 2NaOH = \tfrac{1}{2}O_2 + 2NaF + H_2O$$

$$HF + NaOH = NaF + H_2O$$

The sodium fluoride is then treated with lime to regenerate the sodium hydroxide.

$$2NaF + CaO + H_2O = CaF_2\ (\text{ppte}) + 2NaOH$$

Other reasons for this regenerative stage are that sodium fluoride, which has limited solubility in the caustic system, is an objectionable contaminant leading to plugging and eroding of the equipment, and also because it is poisonous, and could not be discharged with the effluent water without further treatment.

This process can be used where there is residual fluorine and hydrogen fluoride after preparing fluorocarbons, before these are passed on to further treatment. The other possibility of removing fluorine, when this occurs in process waste gases, is to burn it with hydrocarbon gases, in particular hydrogen, to form hydrogen fluoride, which can then be absorbed in water. A system showing a special fluorine burner and a scrubber system incorporating this is shown in Figs. 3.23a and 3.23b.[871]

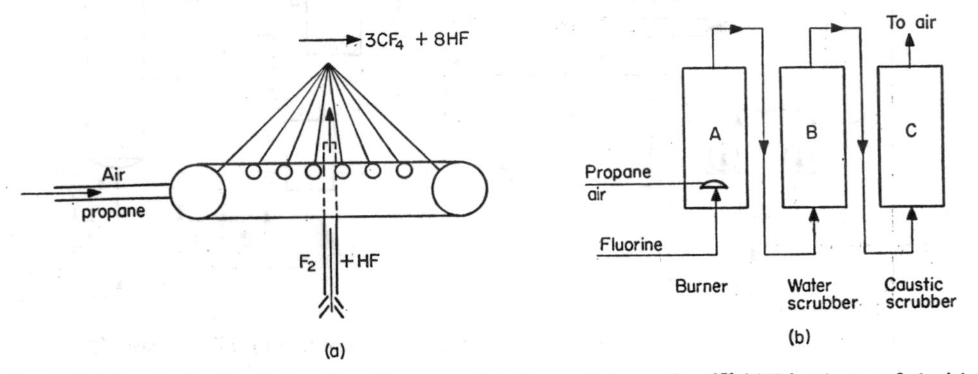

FIG. 3.23. Fluorine and hydrogen fluoride gas combustion and absorption.[871] (a) Ring burner fed with air–propane mixture. (b) Absorption process flow sheet.

Materials of construction are a problem with fluorine hydrogen fluoride gas and other fluorides. Absorption columns are either of wood with wooden grid packing or moulded plastic. Carbon bricks have also been found to be a satisfactory lining for gases containing elemental fluorine. Nickel and nickel alloys such as Monel can be used because a nickel fluoride film forms which protects the metal. In the case of steel, the iron fluoride is a powdery, non-adhesive material, and so steels are not usually used in contact with these gases, particularly at elevated temperatures.

3.7.3. Chlorine and chlorides

The absorption of hydrochloric acid gas in water, in which it is very soluble, is a standard process in the manufacture of hydrochloric acid. Chlorides are found in the waste gases of smelting aluminium; from scrap where salt (NaCl) is used as a flux, and similar processes, while the absorption of elemental chlorine is an essential stage in the purification of this gas before further processing.

The absorption of hydrochloric acid vapour generates a considerable amount of heat, and so cooling has to be provided. The process is usually a two-stage one,[184] where a cooler–absorber, frequently of carbon, is used in the first stage, and the tail gases are then passed through a packed tower with stoneware packing for the removal of the last traces of the acid gas. Equilibrium data for the system is given in Table 3.4.

Kempner, Seiler and Bowman[434] have tested a number of commercially available scrubbers treating 4250 m^{-3} h^{-1} of a gas stream containing 20 mg m^{-3} of HCl gas. Several of the units tested had excellent efficiencies, close to 100 per cent when water gas ratios in the range of 0·26 kg m^{-3} were used. One of these was an extended surface scrubber which was a modified, two-stage plate tower. This consisted of a vertical fibre-glass reinforced plastic unit

TABLE 3.4. EQUILIBRIUM DATA FOR HYDROGEN CHLORIDE GAS AND WATER

Wts. of HCl per 100 wts. of H_2O	Partial Pressure of HCl, kPa				
	10°C	30°C	50°C	80°C	110°C
78·6	114	—	—	—	—
66·7	31·0	83·4	—	—	—
56·3	7·67	25·0	71·2	—	—
47·0	1·57	5·92	18·7	82·9	—
38·9	0·30	1·32	4·75	25·0	101
31·6	0·057	0·289	1·18	7·25	33·6
25·0	0·011	0·064	0·294	2·07	11·0
19·05	0·0021	0·014	0·073	0·620	3·72
13·64	0·00041	0·0031	0·018	0·178	1·24
8·70	0·000078	0·00068	0·0046	0·052	0·41
4·17	0·0000091	0·00010	0·00085	0·013	0·12
2·04	0·0000015	0·000020	0·00019	0·0033	0·037

with a 510×790 mm cross-section having two stages, each of which was a liquid distribution plate, a 50-mm rubber-coated nylon fibre reaction pad and a 50-mm sloped mist eliminator of similar fibre placed above the reaction pad in an inverted V configuration. Under test conditions the two-stage unit gave an efficiency of 99·25 per cent with a total water flow of 0·07 kg m^{-3}. The single stage had an efficiency of 95 per cent. Another unit which had high efficiencies was a vertical tower of fibre-reinforced plastic (760 mm internal diameter) packed with 50 mm polypropylene Intalox saddles to a depth of 1170 mm. The distributor gave excellent distribution through forty small holes in a Christmas-tree-shaped feeder, below the surface, allowing the upper section to act as a spray eliminator. The manufacturer's specification of 2 kg m^{-3} gave almost 100 per cent efficiency, while 0·3 kg m^{-3} still had efficiencies of 98·8 per cent, with a pressure drop of 1·04 kPa. A horizontal scrubber consisting of sprays and packed sections using 1–2 kg m^{-3} had an efficiency of 96 per cent and a plate-tower unit with valve-type trays, using approximately 0·5 kg m^{-3}, had an efficiency of 97·8 per cent. Only a "fan spray" scrubber had poor performances, removing only 57 per cent of the HCl gas at the manufacturer's recommended water rate of 0·1 kg m^{-3}, and so this unit is not recommended for this application.

The chlorides usually encountered from smelting and other processes are not gaseous but very fine crystals, which can be collected by methods such as scrubbing, settling or filtration of the gases, depending on the circumstances.

Chlorine itself may be collected by scrubbing the gases with water. The reactions which occur are complex, and are discussed in detail elsewhere.[768]

As with the fluorides materials of construction are a major problem. Carbon steels and stainless steels are not suitable, but at low temperatures rubber-lined towers with stoneware packing may be used. Nickel, silver and tantalum metal alloys can also be used.

3.7.4. Hydrogen sulphide

Natural gas, refinery gas and coal gas, all of which are used for industrial and domestic heating as well as chemical processing, contain hydrogen sulphide as their major sulphur impurity. Depending on their source they may also contain smaller concentrations of carbon disulphide (CS_2), carbonyl sulphide (COS), thiophene (C_4H_4S) and thiols (mercaptans; RSH), pyridine bases, hydrogen cyanide, carbon dioxide and ammonia. Hydrogen sulphide also occurs in the waste gases from the evaporation of kraft pulping liquors and smelting operations. Process or fuel gases which contain hydrogen sulphide are corrosive when cooled below their dew point, have an objectionable odour, cause difficulties in steel making and heat treatment, and create other problems. It is therefore necessary to remove hydrogen sulphide and some of the other compounds from these gases. Some municipalities require hydrogen sulphide concentrations as low as 0·0115 g m^{-3} for domestic gas, although concentrations of 0·35–0·70 g m^{-3} are often allowed. For metallurgical processes higher concentrations, of the order of 1·15 g m^{-3}, are generally permissible.[310]

Because of the economic importance of hydrogen-sulphide removal a large number of processes have been developed. As these have been fully described by Kohl and Riesenfeld[455] they will only be reviewed briefly here.

The most common process for purifying (or sweetening, as this is sometimes called) natural and refinery gases, which contain small quantities of carbon disulphide and carbonyl sulphide and no ammonia, are the ethanolamine or Girbotol processes (named after the Girdler Corporation). Essentially, hydrogen sulphide and carbon dioxide, both of which are present, are absorbed in a bubble column in an aqueous solution of the ethanolamine, forming a complex at low temperatures, which is then passed to a stripping column where, on heating, the acid gases are regenerated and the ethanolamine solution returned to the absorption column (Fig. 3.24a). The first compound to be used was triethanolamine (TEA), but it was subsequently found that for the absorption of both carbon dioxide and hydrogen sulphide, a 15–20 per cent aqueous solution of monoethanolamine (MEA) was more suitable as it has a higher capacity per unit weight of solvent, a higher reactivity and is more easily regenerated. Its disadvantages are that it forms a heat-stable compound with carbonyl sulphide, diethanolurea ($CO(NHCH_2CH_2)_2$) resulting in loss of amine, and that it has a relatively high vapour pressure, requiring the scrubbing of the acid gas after stripping to remove entrained monoethanolamide vapour. Monoethanolamine is therefore commonly used for natural gas sweetening, while the diethanolamine (DEA), which does not form diethanolurea is used for refinery gases, where some carbonyl sulphide is present besides being less volatile and so reducing losses. Dipropanolamine is also used. Monoethanol amine treatment of natural gas lowers the hydrogen-sulphide concentration below 0·00575 g m^{-3}, which is a satisfactory level, even for the most stringent domestic requirements. When selective absorp-

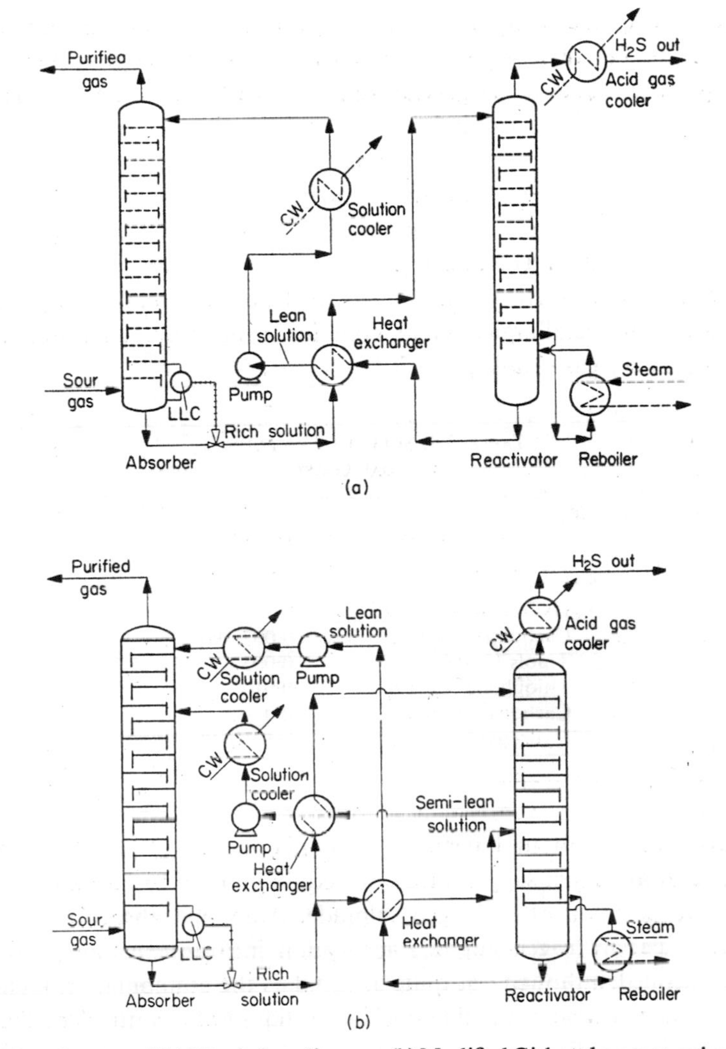

FIG. 3.24. Girbotol process.[455] (a) Basic flow diagram. (b) Modified Girbotol process, using split streams to reduce steam consumption.

tion of hydrogen sulphide in the presence of carbon dioxide is required, triethanolamine or methyldiethanolamine (MDEA) (30 per cent aqueous solutions) is used.

A process which is similar to the Girbotol processes, but more suitable for removal of higher concentrations of hydrogen sulphide, such as are found in some natural-gas deposits, such as the Lacq field, is the diethanolamine process of the Société Nationale des Pétroles d'Aquitaine (SNPA–DEA process).[906] While the operation of this process is similar to the Girbotol processes, higher amine concentrations and pressures are used. This results in a smaller plant, lower capital cost, less problems with foaming, and the simultaneous absorption of carbonyl sulphide which is subsequently released in the flash gases, without degradation of the DEA. The purified gases contain less than 2 ppm of hydrogen sulphide.

If dry hydrogen sulphide is required, scrubbing with water cannot be used to recover the volatilized amine, and diethylene glycol (or sometimes triethylene glycol) is used instead as a scrubbing liquor. Owing to the large-scale operation of these processes, it is economically advantageous to reduce steam requirements for stripping, and to improve the recovery of heat. This has been achieved by modifications such as carrying out the initial absorption in a heat exchanger and using a split-stream stripping method where only a part of the solution is stripped to a low acid gas concentration, reducing the quantity of vapour rising through the stripping column (Fig. 3.24b).

Coal gas, made either in vertical continuous retorts, used primarily for gas making, or in horizontal retorts, for metallurgical coke, contains a much greater range of impurities than natural or refinery gas (Table 3.5).

TABLE 3.5. SULPHUR COMPOUNDS IN COAL GAS[655]

Compound	Volume %
Hydrogen sulphide	0·3–3·0
Carbon disulphide	0·007–0·07
Thiols (RSH)	0·003
Thiophene (C_4H_4S)	0·010
Carbonyl sulphide	0·009
Hydrogen cyanide	0·10–0·25

The gases also contain about 1 per cent ammonia and 1·5–2 per cent carbon dioxide, and the presence of ammonia suggests that this could be used to give an alkaline solution in water for the absorption of hydrogen sulphide. Ammonia absorption is gas-film controlled and very fast, hydrogen-sulphide absorption into aqueous ammonia solutions is also gas-film controlled, although not quite as rapid as the ammonia, while carbon-dioxide absorption into water or weakly alkaline solutions is liquid-film controlled. This is therefore a process which will give the selective absorption of the two major impurities, ammonia and hydrogen sulphide, and also of some of the minor impurities such as carbonyl sulphide and hydrogen cyanide. The selective absorption, however, only results in the removal of about 90 per cent of the hydrogen sulphide, so a second scrubbing or final purification, probably with the dry iron oxide process, is necessary.

Usually the ammonia produced will not remove more than 30–50 per cent of the hydrogen sulphide, and it is necessary to strip some of the acid gas and recycle the aqueous ammonia. The removal of sulphide and ammonia, together with the recycle (in broken lines), is shown in Fig. 3.25. The end product is ammonium sulphate, and hydrogen sulphide gas, which can be turned into the acid for adding in the process for ammonium sulphate, as well as sulphur.

There are a number of variants of this process; for example, it is possible, by using concentrated ammonia solutions, to absorb the hydrogen sulphate in the first stage, and then pass the ammonia on to a second absorber (Collin Process). The acid gas is then stripped

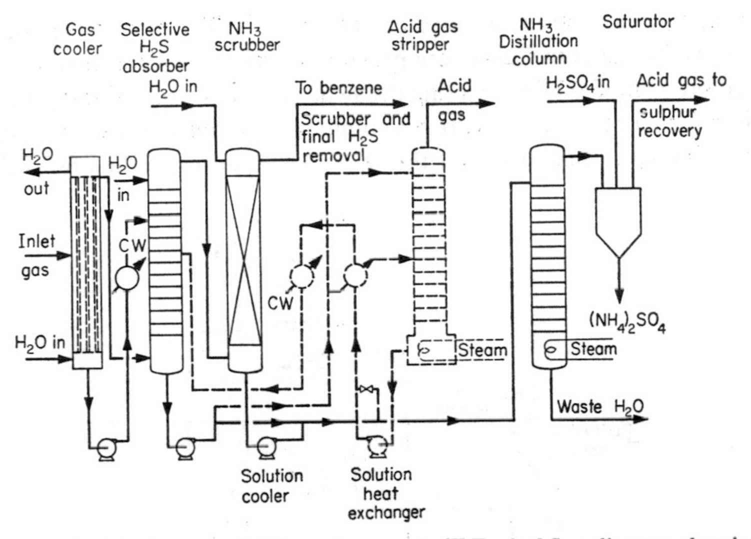

FIG. 3.25. Removal of hydrogen sulphide and ammonia.[455] Typical flow diagram showing selective hydrogen-sulphide removal process without solution recycle (solid lines) and with partial solution recycle (dashed lines); indirect ammonium sulphate recovery process.

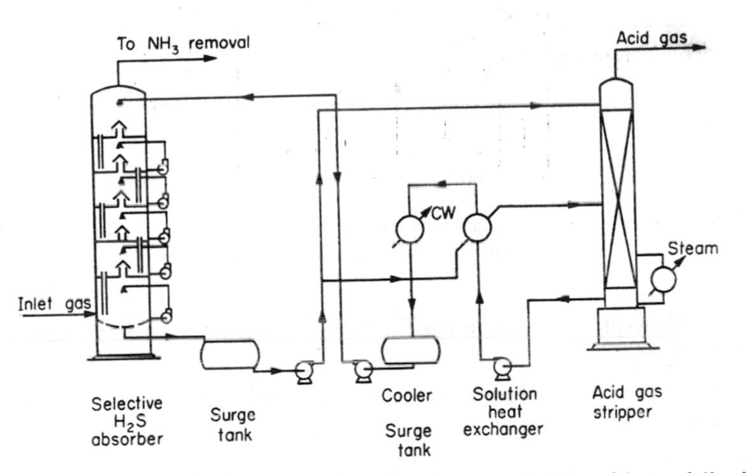

FIG. 3.26. Collin process:[455] selective absorption of hydrogen sulphide with total liquid recycle.

from the concentrated ammonia solution (Fig. 3.26). The details of these processes are discussed elsewhere.[455]

Sodium carbonate (3–$3\frac{1}{2}$ per cent aqueous solution) is used to absorb hydrogen sulphide in a number of processes developed by the Koppers Co. Inc. These are based on the reaction

$$Na_2CO_3 + H_2S = NaHCO_3 + NaHS$$

In the first of these processes, the Seaboard process, a large quantity of air is used to strip the hydrogen sulphide from the absorbent. The hydrogen sulphide cannot be recovered, and some of the sodium sulphide is oxidized to thiosulphate, leading to a weakening of the absorption liquors, which have to be replaced at intervals. The more recent development of these processes is the Vacuum Carbonate process, where the hydrogen sulphide is stripped under reduced pressure 15·6 kPa, lowering the steam required for stripping by reducing the weight required for the removal of hydrogen sulphide. This reduces the sensible heat requirement, as the stripping can be at nearly the same temperature as the absorption, and also

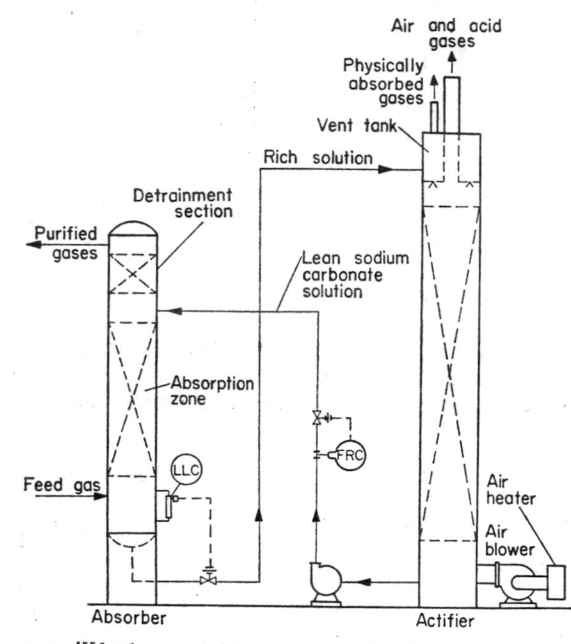

FIG. 3.27. Seabord process:[455] hydrogen-sulphide recovery by absorption in dilute sodium carbonate solution and regeneration by air.

enables the use of low-pressure steam for the reactivation stage. A flow sheet of this process is shown in Fig. 3.27.

An alternative to the Girbotol process is a process which was developed by Shell and uses a 40 per cent aqueous solution of potassium phosphate (K_3PO_4) as absorbent instead of the ethanolamines:

$$K_3PO_4 + H_2S = K_2HPO_4 + KHS$$

This process has the advantage that the phosphate is more stable than the ethanolamines, the process is more selective towards hydrogen sulphide in the presence of carbon dioxide, and in addition live steam can be used to strip the hydrogen sulphide although steam consumption is somewhat greater.

The Alkacid processes, used in Germany on a large scale since before 1939, use an inorganic alkali combined with a weak, non-volatile, organic acid. Three solutions, designated "M", "Dik" and "S", are used, depending on the application. Solution M, which contains sodium alanine, is used for either hydrogen sulphide, carbon dioxide or both; Dik, which contains the potassium salt of dimethyl glycine, is used for hydrogen sulphide in the presence of carbon dioxide; while "S", which contains sodium phenolate, can be used when the gases contain hydrogen cyanide, ammonia, carbon disulphide, thiols and tars. The former reagents M and Dik are very corrosive, while a similar phenolate process, developed by the Koppers Corporation, had severe operating difficulties.

Water itself can be used for hydrogen-sulphide absorption, but the rate of absorption is such that unless the size of the plant required for effective absorption were so large as to be uneconomic, the degree of removal would be inadequate.

In all of the above processes, the regeneration of the absorbing solution results in the production of hydrogen sulphide. However, a useful step would be the production of elemental sulphur by oxidation, particularly in the case of coal-gas purification, where sulphuric acid is required to precipitate ammonium sulphate. This would have to be supplied by an external source, if no sulphur, which can be made into sulphuric acid, were available. The most important processes in this group use atmospheric oxygen as the oxidizing agent in the presence of a catalyst such as iron oxide (Ferrox, Gluud and Manchester processes), nickel sulphate (Nickel process), sodium thioarsenate (Thylox process), iron cyanide (Fischer and Staatsmijnen–Otto processes) or organic oxidation catalysts such as hydroquinone (Perox process).

There are several iron oxide processes: the Ferrox, developed by Koppers Co. in the United States; the Gluud, in Germany; and the Manchester process, developed by the Manchester Corporation Gas Department in England. These are all essentially the same, using sodium or ammonium carbonate solution (3 per cent) to absorb the hydrogen sulphide.

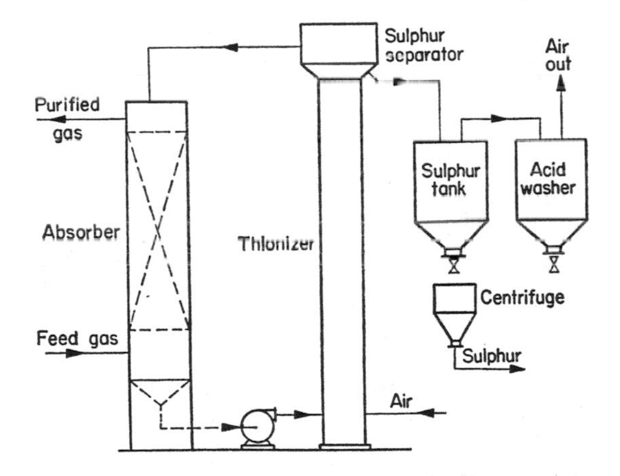

FIG. 3.28. Gluud process:[455] absorption of hydrogen sulphide in dilute ammonium carbonate solution, followed by atmospheric oxidation for sulphur recovery, using iron-oxide catalyst.

The solution also contains a small concentration ($\frac{1}{2}$ per cent) of ferric oxide, which acts as a catalyst in the oxidation of the hydrogen sulphide in the regeneration stage which takes place, in the case of the Ferrox process, in a fairly shallow tank where very fine bubbles of air are bubbled through the solution, while in the case of the Gluud and Manchester processes, very tall aeration towers (20–30 m) are used. The sulphur, on formation, clings to the rising air bubbles and is then scooped off as a froth. The essential difference between the

Ferrox, Gluud and Manchester processes is that the first two use single-stage contact, while multi-stage washing with fresh solution in each washing stage is used in the Manchester process, as well as separate delay vessels to ensure complete reaction. In these processes a considerable amount of iron oxide is collected together with the sulphur in the froth, leading to a poor sulphur product as well as reagent losses. A flow sheet of the Gluud process is shown in Fig. 3.28.

The Thylox process uses sodium thioarsenate as both absorbing and oxidizing solution. The sulphur yield from this process is larger, and the reagent losses lower than in the iron oxide catalysed processes. The main reactions are

(a) Absorption

$$H_2S + Na_4As_2S_5O_2 = Na_4As_2S_6O + H_2O$$

(b) Regeneration

$$2Na_4As_2S_6O + O_2 = 2Na_4As_2S_5O_2 + 2S$$

The process equipment requirements are similar to the iron oxide processes, as is seen from the flow sheet (Fig. 3.29).

The recent Giammarco–Vetrocoke process[296] also uses complex arsenic compounds, but the hydrogen sulphide reacts with the arsenite to form the thioarsenite, which subsequently reacts with the monothioarsenate. This compound decomposes to elemental sulphur and

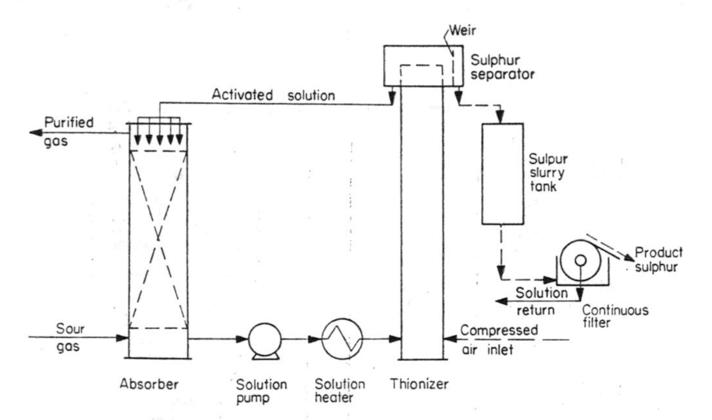

FIG. 3.29. Thylox process: absorption of hydrogen sulphide in sodium thioarsenate, with atmospheric oxidation of the sodium thioarsenate complex ($Na_4As_2S_6O$).[677]

arsenite by acidification with carbon dioxide under pressure. The sulphur is filtered off, and the arsenite is oxidized in part to arsenate by atmospheric oxygen, and then recycled. The process produces a tail gas with less than 1 ppm hydrogen sulphide and also avoids undesirable side reactions associated with the Thylox process.

The Stretford process[28, 600–1] has been widely applied since successful pilot-plant experiments in 1959. Not only does it avoid the difficulties associated with the handling of highly poisonous arsenites on a large scale, it is capable of reducing a wide range of hydrogen-sulphide concentrations (from 100 to 10,000 ppm) to a tail-gas concentration of 1 ppm. As a result plants with capacity to 60,000 m^3 h^{-1} have been constructed. In the Stretford process the hydrogen sulphide is absorbed by an alkaline solution (pH 8·5–9·5) containing, besides sodium carbonate, sodium ammonium vanadate and anthraquinone 2 : 6 and 2 : 7 disulphonate (ADA); these two reagents being present in equimolar concentrations. (In earlier plants the metavanadate was used instead of the ammonium vanadate.) In addition, Rochelle Salt (potassium sodium tartrate) is added to prevent precipitation of the vanadate. The reaction can be summarized as follows:

(a) Absorption:

$$H_2S + Na_2CO_3 \rightarrow NaHS + NaHCO_3$$

(b) Production of sulphur:

$$2NaHS + H_2S + 4NaVO_3 \rightarrow Na_2V_4O_9 + 4NaOH + 2S$$

(c) Recovery of the vanadate with ADA:

$$Na_2V_4O_9 + 2NaOH + H_2O + 2ADA \rightarrow 4NaVO_3 + 2ADA \text{ (reduced)}$$

(d) Oxydation of the ADA with atmospheric oxygen:

$$2ADA \text{ (reduced)} + O_2 \rightarrow 2ADA + H_2O$$

The process plant for the Stretford process is very simple, consisting of an absorption column with conventional packings, a delay tank, and a sulphur-extraction plant. The process has been used at pressures up to 5 MPa[717] and so would be suitable for H_2S removal after hydrodesulphurization or pressure gasification without lowering the pressure for pipeline gas. The sulphur produced in the Stretford process is of comparatively high quality, but some of the reagents tend to be occluded in the cake. This can be removed by washing with water, but considerable quantities are needed.

The processes using ferrocyanide complexes are fairly complicated chemically and because of their limited utilization will not be discussed further here. The Perox process appears to be coming into favour, particularly in Germany, for large scale application, but it is essentially the same as the other oxidation processes discussed.

In considering which process to use for removing hydrogen sulphide from process or waste gases, the initial concentration and the degree of removal required, the presence of other impurities, and whether or not carbon dioxide also has to be removed, are first considerations. Power and steam costs, and the costs of reagents, as well as the capital costs, which depend on the complexity of the plant and the materials of construction demanded by the corrosive nature of the liquids, will require careful assessment for the various possible processes to determine the most economic one.

3.7.5. Nitrogen compounds

(i) *Ammonia, amines and the pyridine bases.* Nitrogen compounds in the carbonization of coal are, in part (50–80 per cent), turned into basic nitrogen compounds in the gas. Thus, a typical distribution of these compounds in raw towns gas would be 1·1 per cent ammonia, 0·1–0·25 per cent hydrogen cyanide, 0·004 per cent pyridine bases, with traces of nitric oxide, and some (approximately 1 per cent) free nitrogen. Like the sulphur compounds (Table 3.5) which occur in gas in similar quantities, the nitrogen compounds are both corrosive and toxic. Thus, there have been a number of attempts made to develop processes for the simultaneous removal of hydrogen sulphide and ammonia, with the recovery of ammonium sulphate and elemental sulphur. Kohl and Riesenfeld[455] point out that some of these processes have been operated commercially with limited success, but no generally accepted process has been evolved.

Ammonia is very soluble in water and in strong acids, as it reacts with these to give ammonium ions. Because of the reaction, the absorption isotherms do not obey Henry's law and the experimental absorption data in the literature[768] should be used. The solubility of ammonia in water also falls off rapidly with temperature, and so two-stage recovery systems are commonly used. In the first stage the gases are cooled to 30–50°C either in a shell and tube heat exchanger (indirect cooler) or in a direct cooler, which is a packed tower, filled with wooden hurdles, into which weak ammoniacal liquors are sprayed. The liquors contain some tars, which are separated out, and the liquors are then cooled and used to wash the gases in an ammonia washer. The final washing is with water. In a typical gasworks, some 0·14 m^3 of strong ammoniacal solution are recovered for each 1000 kg of coal carbonized. The ammonia washer is usually a packed tower with wood, metal, ceramic or fibre packings.

A more modern system has been described by Nord[605] (Fig. 3.30) which is essentially a single-stage absorber and crystallizer. The gases enter a cylindrical tank with conical

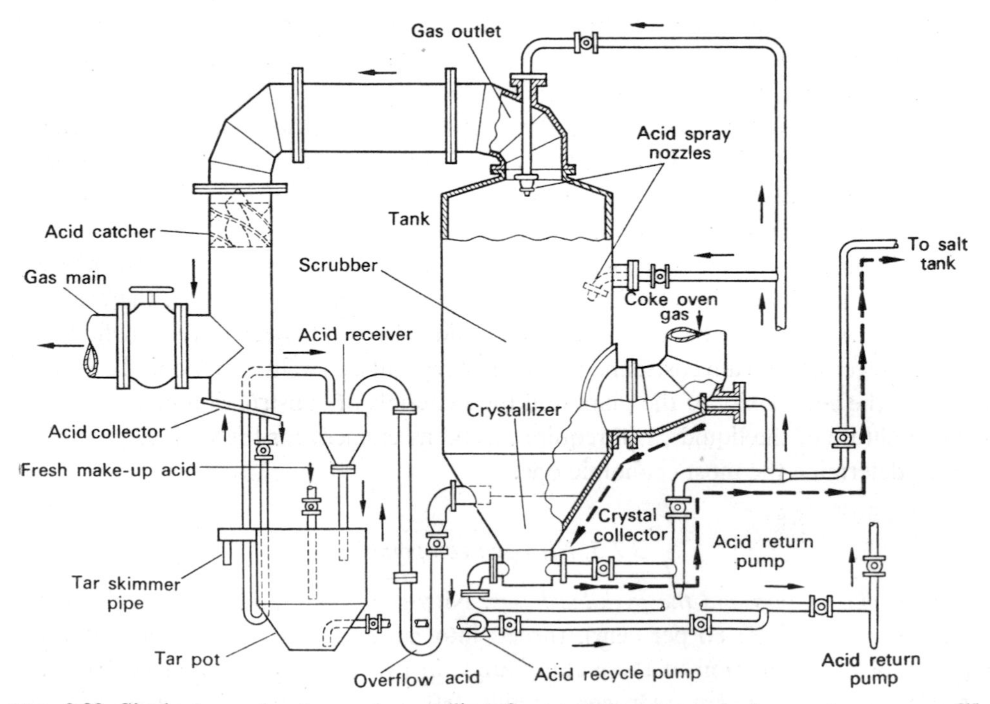

FIG. 3.30. Single-stage absorber and crystallizer for scrubbing ammonia from coke oven gas.[605]

base, into which sulphuric acid is sprayed. The ammonium sulphate crystallizes out in the conical bottom of the tank, while the carry-over is recovered in a centrifugal collector and recycled.

The more complex basic nitrogen compounds, loosely referred to as pyridine bases (pyridene, picoline, collidine, aniline and quinoline), are removed in the same scrubbing processes as ammonia.

(ii) *Nitrogen oxides*. There are at least six stable oxides of nitrogen (nitrous oxide, N_2O; nitric oxide, NO; nitrogen dioxide, NO_2; nitrogen sesquioxide, N_2O_3; nitrogen tetroxide, N_2O_4 and nitrogen pentoxide, N_2O_5) and one unstable oxide (nitrogen trioxide, NO_3). A number of these, particularly nitric oxide and nitrogen dioxide, are formed in combustion processes and in coke-oven gas, while the oxide, dioxide and nitric acid mist are formed in nitric-acid manufacture. Nitric oxide is considered a hazard in coke-oven gas as it forms resins with diolefins in the gas which can cause clogging and subsequent explosions. The formation of nitric oxide (NO) and nitrogen dioxide (NO_2) in flames is complex and is discussed elsewhere. The dioxide appears to be responsible for the brown coloration of Los Angeles-type smog. The dioxide is also formed in pickling operations in anodizing plants and a number of other industrial operations. Different scrubbing processes for industrial effluent gases containing comparatively small concentrations are used, but only with limited success.

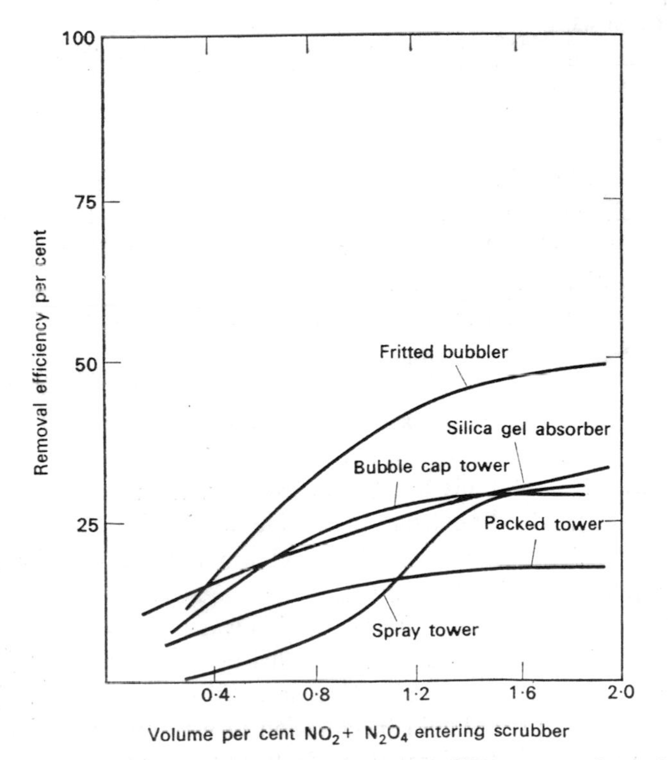

FIG. 3.31. Absorption efficiency of nitrogen oxides in water in different types of absorbers and on silica gel (Peters[635]) (NO_2 in air, at 100 kPa, 25°C).

Peters[635] studied the absorption of nitrogen dioxide (which reacts to give the tetroxide) in water in different single-stage absorbers. These were a spray tower, a bubble-cap tower, a packed tower and a fritted bubbler. The experimental results are shown in Fig. 3.31 and Table 3.6. These indicate that the best results for a single-stage absorber (50 per cent)

TABLE 3.6. CONDITIONS FOR REMOVAL OF NITROGEN DIOXIDE FROM AN AIR/NO_2 MIXTURE (25°C) FOR RESULTS SHOWN IN FIG. 3.31[635]

Equipment	Gas rate (m^3h^{-1})	Liquid rate (m^3h^{-1})	Pressure drop (kPa)	Notes
Fritted glass bubbler, one stage	0·9	0·018	5·84	Medium frits. Liquid head over frits 95 mm
Bubble cap tower, one stage	1·8	0·018	0·277	Liquid depth, 25 mm. slot velocity, 0·35 m s^{-1}.
Packed tower, 6-mm glass Raschig rings	0·9	0·009	0·645	Superficial vapour velocity, 0·6 m s^{-1}. Packed height, 120 mm. Efficiency reported per unit height of packing
Spray tower, I No. T58 1-mm nozzle	0·9	0·022	0·097	Superficial vapour velocity, 0·6 m s^{-1}. Tower height 130 mm
Silica gel absorber, No. 5 commercial gel [See *Ind. Engng. Chem.* **43** 986, (1951)]	0·9	—	—	Superficial vapour velocity, 0·6 m s^{-1}. Packed height, 300 mm. Fraction saturated, 0·90. Time per cycle, 0·5 h. Efficiencies calculated from reference 5.

was for a fritted glass bubbler having an uneconomically high pressure drop (5·8 k Pa) A much lower efficiency (30 per cent) was obtained in a bubble cap tower, but here the pressure drop was only 0·45 kPa per stage, and a multi-stage bubble cap system may give a worth-while absorption efficiency with an acceptable pressure loss.

A two-stage cocurrent–counter-current flow absorption tower, using a polyethylene waved-sheet packing, has been used by Strauss,[826] where an efficiency of 65–72 per cent was obtained in the scrubbing of nitrogen dioxide containing gases from anodizing. The pressure drop in this case was 0·5–0·8 kPa and the gas-flow rate 11,000–14,600 m^3 $m^{-2}h^{-1}$ and the initial NO_2 concentration was of the order of 0·1 per cent.

Absorption in alkaline solutions has been tried by a number of workers. They used solutions of sodium hydroxide,[826] calcium hydroxide, sodium carbonate,[289, 836] ammonium carbonate,[42, 836] ammonium bicarbonate[712, 741] and ammonium sulphite or bisulphite in a variety of gas liquid contacting equipment.

Sodium hydroxide in a 3 per cent (by wt.) solution was used in the tower with polyethylene waved-sheet packing described above. The efficiency was somewhat greater than with water, but the carbon dioxide in the air being drawn through the unit also reacted with the alkali, and maintaining the alkali concentration proved uneconomic. In addition, the alkaline adsorption resulted in the formation of a troublesome and corrosive mist, and the practice was discontinued.

Calcium hydroxide solutions have been used in a horizontal rotary disc scrubber with three turbine wheels,[289] 890 mm diameter, and operating at 180 rev/min. The gas-flow rate was 200–400 $m^3\ h^{-1}$ and efficiencies of 92–97 per cent were reported with initial concentrations of 0·5–1·0 per cent NO. The efficiency of absorption fell off with the higher gas flows and lower concentrations.

Atsukawa and others[42] used ammonium carbonate solutions in a packed tower filled with 25 mm Raschig rings to a depth of 2·6 m. The absorption efficiency of the order of 65 per cent is similar to that obtained by other workers (Table 3.7). Much better results

TABLE 3.7. ABSORPTION OF NO AND NO_2 BY AMMONIUM CARBONATE SOLUTIONS IN PACKED TOWERS AND IN CROSS-FLOW SCRUBBERS[42]

Gas rate ($kg\ m^{-2}\ h^{-1}$)	Liquid rate ($kg\ m^{-2}\ h^{-1}$	Mass transfer coefft., $K_{OG}\ a$ ($kg\ mole\ m^{-2}\ h^{-1}$ kPa)	$NO+NO_2$ conc. (ppm) Inlet NO_2	Outlet NO_2	Absorption efficiency %
A. Raschig ring 25 mm packing; tower diameter 0.4 m, packing depth 2.6 m					
2063	15,050	0·29	1970	630	68
2580	15,050	0·34	2040	700	65·7
2990	15,050	0·37	1920	710	63·0
B. PVC waved-sheet packing; two-stage absorption					
			Inlet NO NO_2	Outlet $NO+NO_2$	
1180	5250	40·5	3400 2100	270	94·6
2360	5250	77	3600 2100	590	86·7
2950	5250	87·5	3400 1800	800	84·6

(to 94·6 per cent efficiency) were obtained with PVC waved-sheet packing. This was contained in a two-stage horizontal flow absorber, 1·9 m long and 5·5 m high. The flow diagram is shown in Fig. 3.32.

Streight[836] reports pilot-plant and full-scale scrubbing of tail gases from nitric acid and nylon intermediate plants using a packed tower. The pilot-plant tower (0·3 m dia.) used 25 mm Raschig rings, and the full-scale tower (2·4 m dia.) 50 mm rings, with a packing depth of 12·6 m. In early work sodium bicarbonate solutions were used, but in the final plant a waste ammonia caustic solution (0·5 per cent free NaOH, 0·45 per cent Na_2CO_3 and 0·03–0·5 per cent NH_3), cooled to 35°C. The gas-flow rate was 56·7 $m^3\ m^{-2}\ h^{-1}$, and the liquid rate 55 $m^3\ h^{-1}$. A venturi scrubber was used to prevent an ammonium carbonate plume from issuing from the plant. The performance of the plant gave efficiencies of 74 to 84 per cent.

The use of oxidizing absorbing solutions has also been tried, largely in experiments on scrubbing of diesel exhaust gases, by workers at the U.S. Bureau of Mines.[212] Aqueous

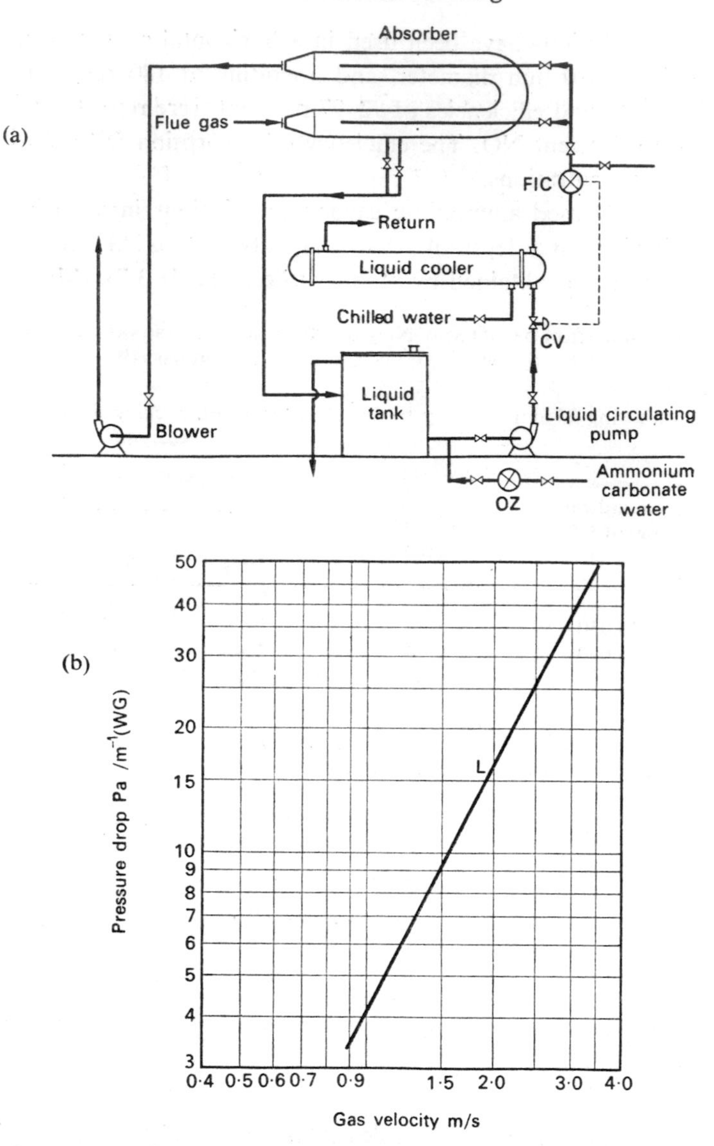

FIG. 3.32.(a) Flow diagram of two-stage experimental absorber filled with wave sheet packing.[42] (b) Pressure drop of PVC waved sheet packing[42] (L = 10,000 kg/(m^2) (hr)).

solutions and suspensions of ferrous sulphate (10 per cent, by weight), potassium permanganate (15 per cent), potassium dichromate (5 per cent), activated charcoal powder (10 per cent), and hydroquinone (0·5 per cent) were tried in different combinations. The absorption was generally low, the two best results of 49 per cent being ferrous sulphate and hydroquinone at 38–47°C. A chromic acid solution (10 per cent) followed by a sodium hydroxide solution (10 per cent) absorbed 53 per cent. Equally unsatisfactory absorption efficiencies (20 per cent) were obtained by Taigel who used potassium permanganate and sodium hydroxide.[850] Much better results (40–80 per cent approx.) were obtained using an alkaline solution (pH

9·5–12) of sodium hypochlorite in a Japanese patent,[424] but no details of the absorber have been reported.

More successful than simple absorption have been units where the gases are passed through layers of irrigated activated carbon followed by irrigated plastic packings. Ammonia gas has been added as well to give an alkaline medium. The unit, manufactured by Keram Chemie,[436] is washed out each week to remove the nitrate solution as well as dust which settles on the catalyst.

In nitric-acid manufacture dilute nitric acid is a suitable absorbent, and this is circulated in bubble-cap columns using cooled trays. Venturi scrubbers are also used, and the entrained air helps to reoxidize the nitric oxide (NO) to the dioxide. The nitric oxide comes from the reactions

$$2NO_2 + H_2O \rightarrow HNO_2 + HNO_3$$

$$2HNO_2 \rightarrow HNO_3 + H_2O + 2NO$$

In older plants the absorption towers were built of granite and filled with broken quartz. The towers were up to 25 m high. A detailed discussion of the absorption in nitric-acid manufacture can be found in Sherwood and Pigford,[768] while experimental data on the rate of nitrogen-dioxide absorption has been published by Decker, Snoek and Kramers.[215]

Catalytic processes for the oxidation and reduction of nitrogen oxides will be discussed in Section 3.11.2.

3.8. ADSORPTION OF GASES ON SOLIDS

Since the mechanism of adsorption of gas molecules on the surface of a solid is very complex, and depends on the physical and chemical nature of the gas and the solid in each particular case, it is more difficult to deduce a general approach to the design of adsorption equipment than countercurrent absorption plant. In practice most of the designs are based either on experience with other similar plants or on pilot plants. None the less an understanding of the principles is of considerable assistance in deciding whether adsorption is the best process for the removal of certain gases, in the selection of suitable adsorbing materials and in scaling up designs.

When a gas molecule is adsorbed on the surface of a solid it settles on it very much like a condensing molecule, and is then held on the surface either by physical attractive forces (London–van der Waals forces) or in certain cases, depending on the chemical nature of the molecule and the surface, by chemical forces (chemisorption). In a particular system both types of adsorption may occur as well as intermediate types. The solids best suited to adsorption are very porous, with very large effective surface areas, which are obtained with materials such as carbon, alumina or silica gel. Some surface characteristics, as for example crystal dislocations, or oxygen atoms with unshared electron pairs available for hydrogen bond formation, assist in chemisorbing specific molecules. The exact nature of these surface characteristics is imperfectly understood and more research is required to create surface characteristics which will assist in selective chemisorption of particular molecular types or groups of molecular types. Other desirable characteristics of the adsorbing solids are that the granules should be hard, so that they do not collapse under their own weight when packed into a tower, and that they should not powder or fracture easily, so that they can be transported and poured into containers without breakage. The three solids mentioned earlier all have these properties. It is frequently a requirement that the solid, after it has been saturated with the gas molecules, can be easily regenerated and re-used.

The adsorption of a gas on a solid takes place in several stages. The first is the movement of the gas molecules to the external surface of the solid, and this is the same as the diffusion of the gas molecules through a stationary layer to the liquid–gas interface in absorption (Section 3.3).

The rate of transfer N_A for this stage is expressed by an equation similar to (3.23), with appropriate allowances for the granular condition of the adsorbing material:

$$N_A = k_f \frac{A_p \epsilon}{\varrho_B}(p - p_i) \tag{3.53}$$

where k_f = mass transfer coefficient (gas to solid surface),
A_p = external surface area of the solid,
ϵ = voidage between granules,
ϱ_B = bulk density of the packing,
p, p_i = partial pressure in the bulk of the gas and at the surface, respectively.

The mass transfer coefficient can be found from an equation similar to (3.26).[942]

$$k_f = 1{\cdot}82U\left(\frac{DU\epsilon\varrho}{\mu}\right)^{-0{\cdot}51}\left(\frac{\mu}{\varrho\mathcal{D}}\right)^{-0{\cdot}67} \tag{3.54}$$

where U = mean linear velocity of gas relative to the solid,
μ = gas viscosity,
ϱ = gas density,
$\mathcal{D}$ = diffusivity of gas molecules being adsorbed,
D = equivalent spherical diameter of the granules, i.e. diameter of a sphere of the same volume as the granule.

The second stage in adsorption is the penetration of the molecules into the pores of the solid; the third stage is the actual adsorption of the molecule on the site in the pore. Sometimes a molecule then penetrates through the solid by internal diffusion, but this last stage does not affect the rate of adsorption. The actual adsorption on the site is very fast compared to the first two stages, and it is these which determine the rate of adsorption. The rate of diffusion through the pore is given by the *pore diffusivity* $\mathcal{D}_{\text{pore}}$, which can be found from the following equation,[922] for cases where the pores are smaller than the mean free path of the gas molecules λ:

$$\mathcal{D}_{\text{pore}} = \frac{\mathcal{D}\chi}{2}\{1-\exp(-2\bar{r}\bar{u}/3\mathcal{D})\} \tag{3.55}$$

where $\bar{r}$ = average pore radius,

$\bar{u}$ = average molecular velocity [eqn (3.3)],

χ = internal porosity of the solid granules.

Heat is evolved when molecules are adsorbed. When the adsorption is purely physical, then the heat evolved is the same as the latent heat in condensation. When the adsorption is a chemical reaction, the heat evolved is greater. The removal of molecules from the surface requires the heating of the surface to drive off (vaporize) the molecules. When the adsorption is chemisorption, this may mean removing some of the surface atoms with the desorbing molecule, thus changing the nature of the surface. This may decrease or increase its adsorptive capacity.

Equation (3.53) shows how the transfer of molecules from the bulk of the gas stream is a function of the size and shape of the particles, the bed voidage, and the driving force. This last function is the difference between the concentration in the bulk of the gas and at the surface, which in turn depends on the degree of saturation that has been reached. Equation (3.54) indicates how the mass transfer coefficient is a function of the gas velocity as well as of the carrier gas properties and the diffusivity of the gas being adsorbed, while the diffusivity in the pores [equation (3.55)] is mostly a function of the pore voidage (χ) and the diffusivity. To decide whether stage 1 or 2 is the rate determining step requires a comprehensive knowledge of the system, and this is only rarely available. Empirical design procedures are therefore almost invariably used. Some of the most common adsorbents and the gases for which they are used are therefore discussed, together with a description of the type of plant in which they are used.

The adsorbents can be classified into three groups:

(a) Non-polar solids, where the adsorption is mainly physical.
(b) Polar solids, where the adsorption is chemical, without changing the chemical structure of the molecules and the surface.
(c) Chemical adsorbing surfaces, which adsorb the molecules and then release them after reaction which may be either catalytic, leaving the surface unchanged or non-catalytic with the surface atoms, requiring their replacement.

3.8.1. Non-polar adsorbents—Carbon

The only important material in the group of non-polar solids is carbon, which consists almost wholly of neutral atoms of a single species, and presents a surface which is a generally homogeneous distribution of electrical charges at the molecular level, without potential gradients. Such a surface does not bind polar molecules in preference to non-polar ones at particular sites. Carbon is therefore very effective in adsorbing non-polar organic molecules, particularly near their normal boiling points. Even if there is water vapour in the gas stream, the organic molecules will be adsorbed preferentially because the polar water molecules are more strongly attracted to one another than to the non-polar carbon surface.

Thus large organic molecules are very readily adsorbed, smaller organic molecules and large inorganic ones less easily, smaller inorganic ones less still, and those which are permanent gases hardly at all. A list classifying substances into four groups is given in Table 3.8.[56]

TABLE 3.8. RELATIVE ADSORPTIVITY OF MOLECULES ON ACTIVE CARBON[56]*

2—Acetaldehyde	1—Carbon dioxide	3—Hydrogen sulphide
4—Acetic acid	1—Carbon monoxide	4—Isopropyl alcohol
3—Acetone	4—Carbon tetrachloride	4—Masking agents
3—Acrolein	3—Chlorine	4—Mercaptans
4—Alcohol	4—Chloropicrin	4—Ozone
2—Amines	4—Cigarette smoke	4—Perfumes, cosmetics
2—Ammonia	4—Cresol	4—Perspiration
3—Anaesthetics	3—Diesel fumes	4—Phenol
3—Animal odours	4—Disinfectants	2—Propane
4—Benzene	4—Ethyl acetate	4—Pyridine
4—Body odours	1—Ethylene	4—Ripening fruits
2—Butane	4—Essential oils	4—Smog
4—Butyl alcohol	2—Formaldehyde	3—Solvents
4—Butyric acid	4—Gasoline	4—Stuffiness
4—Cancer odour	4—Hospital odours	4—Toluene
4—Caprylic acid	4—Household smells	4—Turpentine

The capacity index has the following meaning:

4—High capacity for all materials in this category. One pound takes up about 20 to 50 per cent of its own weight—average about $\frac{1}{3}$ ($33\frac{1}{3}$ per cent). This category includes most of the odour-causing substances.

3—Satisfactory capacity for all items in this category. These constitute good applications but the capacity is not as high as for Category 4. Adsorbs about 10–25 per cent of its weight—average about $\frac{1}{6}$ (16·7 per cent).

2—Includes substances which are not highly adsorbed but which might be taken up sufficiently to give good service under the particular conditions of operation. These require individual checking.

1—Adsorption capacity is low for these materials. Activated charcoal cannot be satisfactorily used to remove them under ordinary circumstances.

Activated carbon (frequently called charcoal—if the source is wood) is made by the pyrogenic decomposition of suitable coals and woods in special retorts. The raw materials may be lignites and bituminous coals, woods and nuts. Some of the purest carbon is made from Eucalyptus Marginata (Australian Jarrah) which is exceptionally low in mineral content, while good quality commercial activated charcoals with very large effective surface areas come from coconut shell, which has been charred at about 1150°C and subsequently "activated" by treatment with steam at about 600°C.

* The carbon is a 50-min (chloropicrin test) activated coconut shell charcoal.

The basic parameters which determine the action of a carbon are the surface area available for adsorption and the diameters of the pores to determine whether the molecules being adsorbed can reach the micro-structure. The surface area is determined most accurately by the gas adsorption (B.E.T.) method.[237] The average pore diameter can be found from the pore volume, determined by the high pressure mercury injection technique[383] or electron microscopy.[21] Some typical data for adsorbent materials is shown in Table 3.9.[922]

TABLE 3.9. SURFACE AREA, PORE VOLUME AND MEAN PORE DIAMETER FOR ADSORBENT MATERIALS[922]

Material	Area ($m^2\,g^{-1}$)	Pore volume ($cm^3\,g^{-1}$)	Mean pore diameter (10^{-10} m)
Activated carbon	500–1500	0·6–0·8	20–40
Silica gels	200–600	approx. 0·4	30–200
Activated alumina	175	0·39	90
Kieselguhr	4·2	1·14	22,000

The practical effectiveness of an active carbon is generally measured by testing the material by passing air, which was saturated with carbon tetrachloride vapour at 0°C, at 25°C and 100 kPa pressure through a bed and measuring the amount adsorbed, expressing the amount as a percentage of the original weight. The retentivity is then tested by blowing dry air at 25°C for 6 hr through the bed.

The test used for gas mask charcoals is the length of life of the bed when chloropicrin vapours are adsorbed in a thin bed under standardized conditions. Typical values for the adsorptive capacity and the retentivity of carbon tetrachloride and other organic materials are given in Table 3.10.[550]

TABLE 3.10. ADSORPTIVE CAPACITY AND RETENTIVITY OF ACTIVE CHARCOAL FOR ORGANIC MATERIALS[550]

Substance	Adsorptive capacity (weight) %	Retention after removal (weight %)	
Carbon tetrachloride	80–110	27–30	
Gasoline (motor spirit)	10–20	2–3	
Benzene	45–55	5·9	(steam)
Methanol	50	1·2	1 hr at 150°C
Ethanol	50	1·05	1 hr at 150°C
Isopropanol	50	1·15	1 hr at 150°C
Ethyl acetate	57·5	4·87	1 hr at 150°C
Acetone	51	3·0	1 hr at 150°C
Acetic acid	70	2·5	1 hr at 150°C

Activated carbon is used for the recovery of hydrocarbon solvents, the removal and recovery of hydrocarbon gases from coal gas and natural gas, and the removal of odours and other trace impurities from gas streams. In the first two cases the recovery of the hydrocarbons is an economic project, as they can be used commercially. The plant consists of two deep beds of carbon housed in pressure vessels which are alternatively charged with the hydrocarbons and then steamed at 100–150°C to remove the adsorbed material (Fig. 3.33). The beds are 300–900 mm deep and velocities of 0·15–0·60 m s^{-1} are used. The pressure drop through such beds is given by the graph in Fig. 3.34.

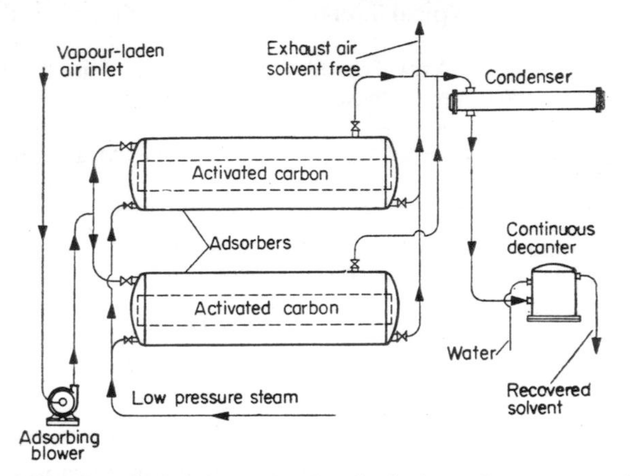

FIG. 3.33. Deep bed carbon adsorber for hydrocarbon recovery.[879]

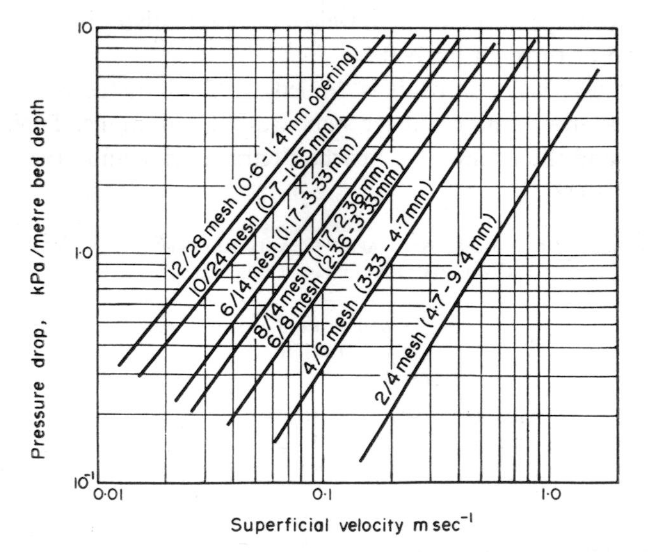

FIG. 3.34. Pressure drop through dry packed granular carbon beds.[879]
Air flowing downwards—100 kPa pressure, 21°C.

The control of odours with active charcoal is very effective. Thin beds with high flow rates and a life of 12–18 months or even longer are used continuously before replacement becomes necessary. The carbon is housed in canisters or in porous folds (Figs. 3.35a and 3.35b) and the flow rates used are similar to those in air conditioning filters, with which they are often combined. A suitable design value is 0·03 m s^{-1} for a 12 mm bed of carbon.

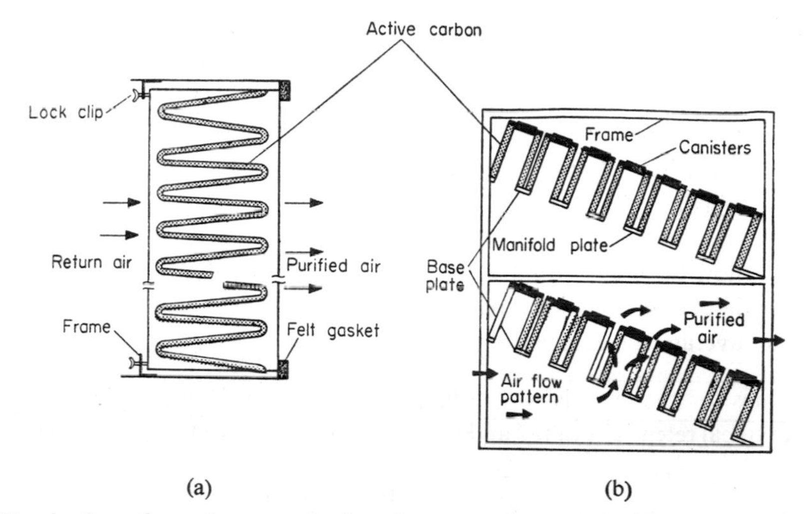

FIG. 3.35. Housing for active carbon granules for odour control, operated without cooling period.[56]

Reactivation of this carbon is more difficult than for solvent recovery systems, requiring temperatures of 600°C, and is therefore usually only carried out by the filter material manufacturer or a similar specialist organization.

TABLE 3.11. RETENTIVITY OF VAPOURS BY ACTIVE CARBONS[870] ALSO SOME ODOUR THRESHOLD CONCENTRATIONS

	Formula	Molecular weight	Odour threshold concentration, ppm by volume	Retentivity* S
Acetaldehyde	C_2H_4O	44·1	—	0·07
Acrolein (heated fat odor)	C_3H_4O	56	1·8–17	0·15
Amyl acetate	$C_7H_{14}O_2$	130·2	—	0·34
Butyric acid	$C_4H_8O_2$	88·1	0·00083 –2·4	0·35
Carbon tetrachloride	CCl_4	153·8	—	0·45
Ethylacetate	$C_4H_8O_2$	88·1	190	0·19
Ethyl mercaptan	C_2H_6S	62·1	—	0·23
Eucalyptole	$C_{10}H_{18}O$	154·2	—	0·20
Formaldehyde	CH_2O	30·0	—	0·03
Hexane	C_6H_{12}	86	—	0·16
Methyl chloride	CH_3Cl	50·5	—	0·05
Phenol (carbolic acid)	C_6H_6O	94	0·29–1	0·30
Putrescine	$C_4H_{12}N_2$	88·2	—	0·25
Skatole	C_9H_9N	131·2	$3{\cdot}34\times10^{-7}$ –0·22	0·25
Sulphur dioxide	SO_2	64·1	3·0	0·10
Toluene	C_7H_8	92·1	—	0·29
Valeric acid (body odour)	$C_5H_{10}O_2$	102	7	0·35

* Fractional retentivity S at 20°C and 100 kPa pressure.

The service life of such carbons can be estimated from an equation by Turk.[870]

$$t = 2{\cdot}41 \times 10^7\, SW/Qc\eta M \tag{3.56}$$

where t = life of unit—hours,
S = retentivity of gas (fraction) in adsorbing solid,
W = weight of adsorbing solid (kg),
η = fraction efficiency of adsorption,
Q = flow through bed $m^3\ h^{-1}$,
c = inlet concentration—parts per million,
M = average molecular weight.

Typical values for the retentivity S to be used in this equation are given in Table 3.11.[870]

Using a typical retentivity of 0·2 and molecular weight of 100, the equation (3.56) reduced to

$$t = 4{\cdot}8 \times 10^4\, W/Qc \text{ hr} \tag{3.57}$$

which should give an approximate value for the life of an active carbon filter. For odour control filters velocities of about 0·13–0·18 m s^{-1} are used through the thin layers, compared with the higher velocities in solvent recovery filters. Temperatures should not exceed 52°C, as above this temperature the gases are not so easily absorbed. Also relative humidities greater than 50 per cent should be avoided, because the retentivity is somewhat reduced at higher humidities.

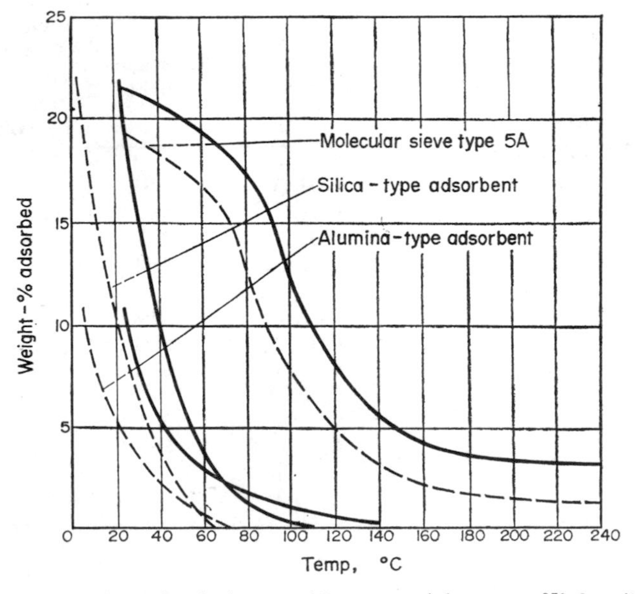

FIG. 3.36. Water vapour adsorption isobars at 10 mm partial pressure[878] for silica, alumina and artificial zeolite (molecular sieve) type adsorbents.

This isobar shows that at 120°C molecular sieves have over four times the drying power of alumina and silica type adsorbents, assuming complete regeneration. The dashed line shows the effect of 2 per cent residual water at the start of adsorption.

3.8.2. Polar adsorbents

Polar adsorbents are oxides, either silica or metallic oxides. The siliceous materials are silica gel, Fuller's earth, diatomaceous earths such as kieselguhr and synthetic zeolites. These materials have an affinity for polar as well as non-polar molecules and will adsorb polar molecules in preference to non-polar ones. Metallic oxides, aluminium oxide in the form of activated alumina or activated bauxite have an even higher attraction for polar molecules. These materials are therefore commonly used for removal of water vapour from gas streams rather than organic molecules. The synthetic zeolites, sometimes called "molecular sieves", are sodium or calcium aluminosilicates, activated by heating which drives off water of crystallization. The chief advantage of the molecular sieves is that they can be

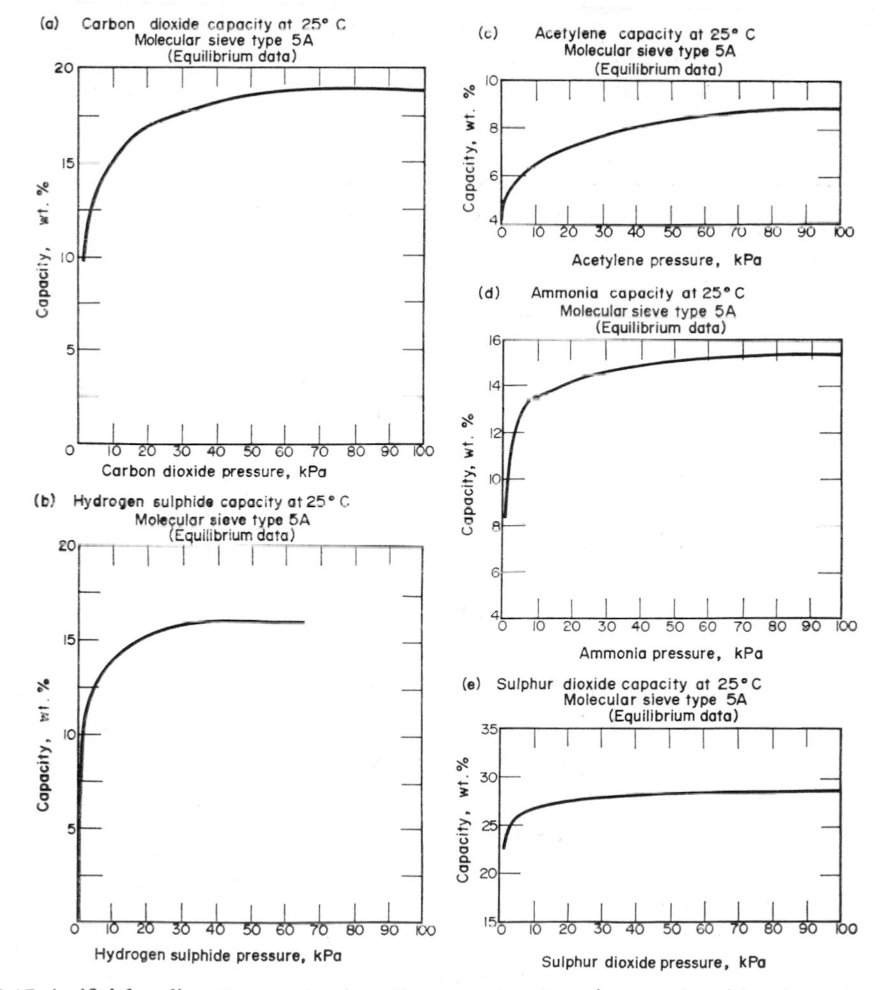

FIG. 3.37. Artificial zeolites ("molecular sieves"), Type 5.A; adsorptive capacity with polar molecules.[8]

(a) Carbon dioxide. (b) Hydrogen sulphide. (c) Acetylene.
(d) Ammonia. (e) Sulphur dioxide.

used for drying at elevated temperatures, where silica gel and alumina lose their effectiveness (Fig. 3.36). The other application is the use of these adsorbents for the selective adsorption of polar molecules such as water, carbon dioxide, ammonia, acetylene, hydrogen sulphide and sulphur dioxide (Fig. 3.37). They are, therefore, used for the purification of inert gases, the removal of carbon dioxide and water from ethylene before subjecting this to polymerization for polyethylene, and the purification of natural gas.

The plant used in these cases is nearly always similar to the one shown in Fig. 3.38 for drying natural gas and consists of two or more vessels with the dessicant. Of these one is on

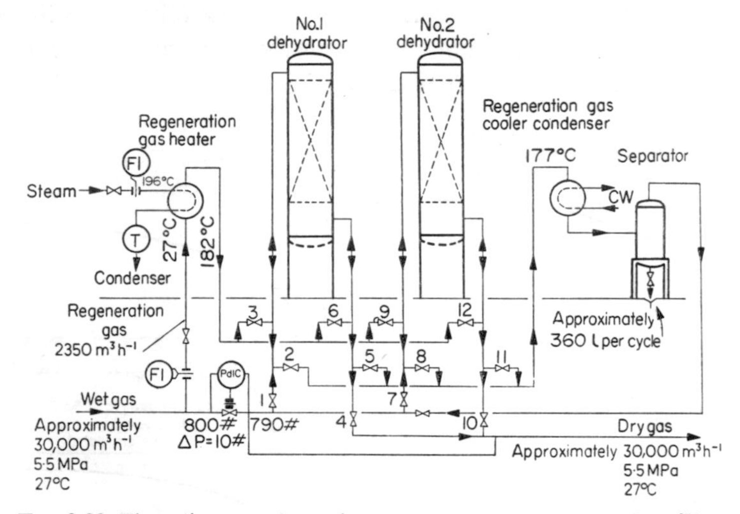

FIG. 3.38. Flow diagram of a typical natural gas dehydration plant.[455]

stream while the others are being regenerated by the passing of the regeneration gas, either air heated by steam or combustion gases, or directly heated by combustion gases at about 200°C, through the bed.

3.8.3. Adsorption with chemical reaction

Adsorbent surfaces which react chemically with the gas molecules, changing these to more useful or desirable ones before release, open the whole field of surface reactions and catalysis. Only a few examples of particular interest in air pollution control and gas purification can be given here. As in the case of adsorption without chemical reaction, it is necessary to have a large surface area of reagent accessible to the gas.

(i) *Hydrogen sulphide.* Important among these processes are the dry oxidation processes for the removal of sulphur compounds from coal gas. Chemically the hydrogen sulphide is converted to sulphur, using oxygen in carriers which react readily with it at ordinary temperatures. The most important of these is ferric oxide, and the basic chemistry of the process can be written as a reaction and regeneration stage:

$$\text{Reaction:} \quad 6H_2S + 2Fe_2O_3 = 2Fe_2S_3 + 6H_2O$$

$$\text{Regeneration:} \quad 2Fe_2S_3 + 3O_2 = 6S + 2Fe_2O_3$$

Only the α and γ forms of ferric oxide are reactive, and the reaction proceeds best at a temperature of 37·8°C and in an alkaline environment. The iron oxide is used in a finely divided form either mixed with peat or fibrous matter, containing 40 per cent water and with a porosity of about 60 per cent. Alternatively the iron oxide can be supported on a large surface area such as wood shavings or crushed slags, and the pH is carefully controlled to pH 8–8·5.

The pellets, or iron oxide on shavings, are distributed in large containers called "dry boxes" or on trays in towers (Figs. 3.39a and 3.39b). The process is a two-stage one, the

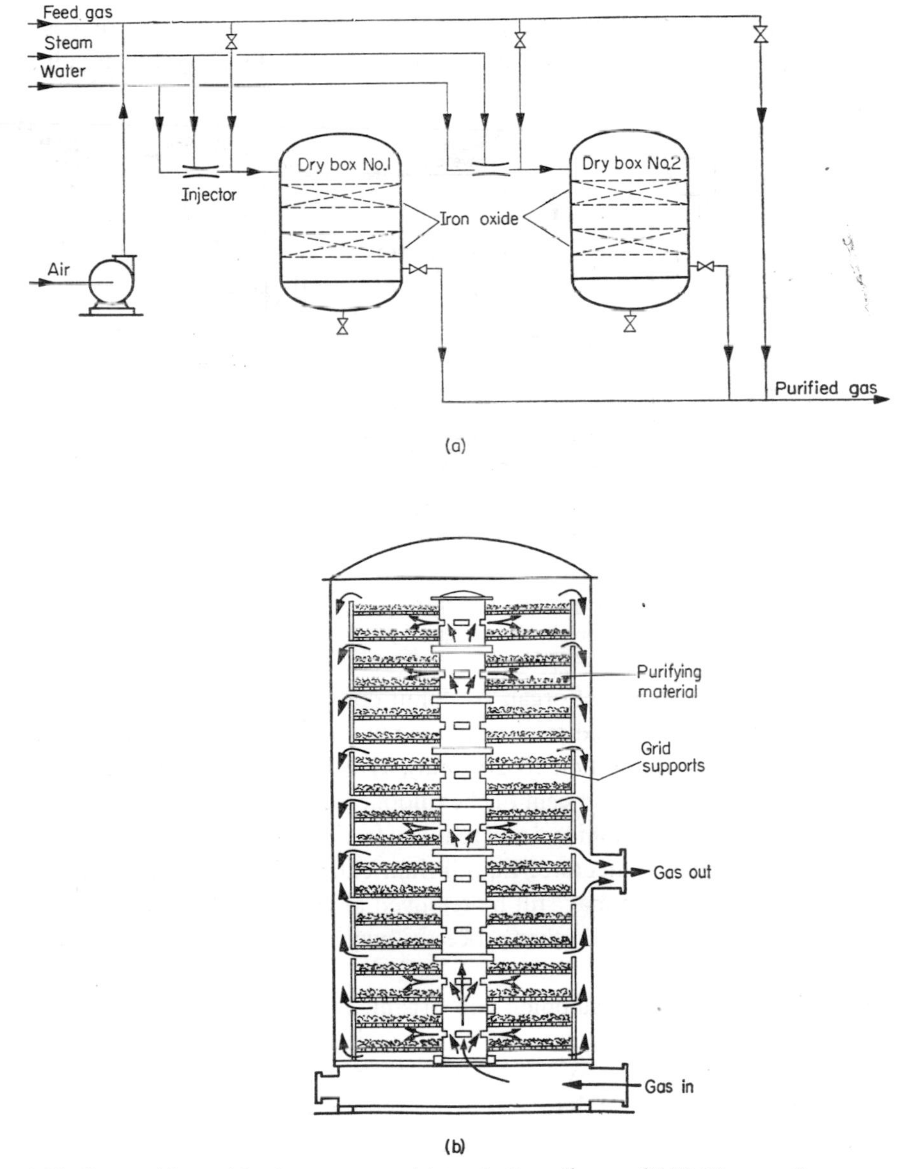

FIG. 3.39. Iron-oxide purification process. (a) Basic flow diagram.[455] (b) Thyssen–Lenze tower.[310]

first stage removing the hydrogen sulphide, while the second stage reoxidizes the iron sulphide to the oxide. In some countries the sulphur deposited on the oxide is recovered either by combustion or by solvent extraction using solvents such as perchlor-ethylene.

The soda iron process uses iron oxide mixed with sodium carbonate as a catalyst and oxidizes the hydrogen sulphide to sulphur trioxide which can be directly dissolved to give sulphuric acid.

(ii) *High-temperature processes for hydrogen sulphide removal.* Hot iron-oxide pellets in a series of fluidized beds have been successfully used to remove hydrogen sulphide and organic sulphur compounds from coal gas in the Appleby–Frodingham process at 340–360°C[141, 680] (Fig. 3.40). The coal gas used contained 14 g m^{-3} hydrogen sulphide and

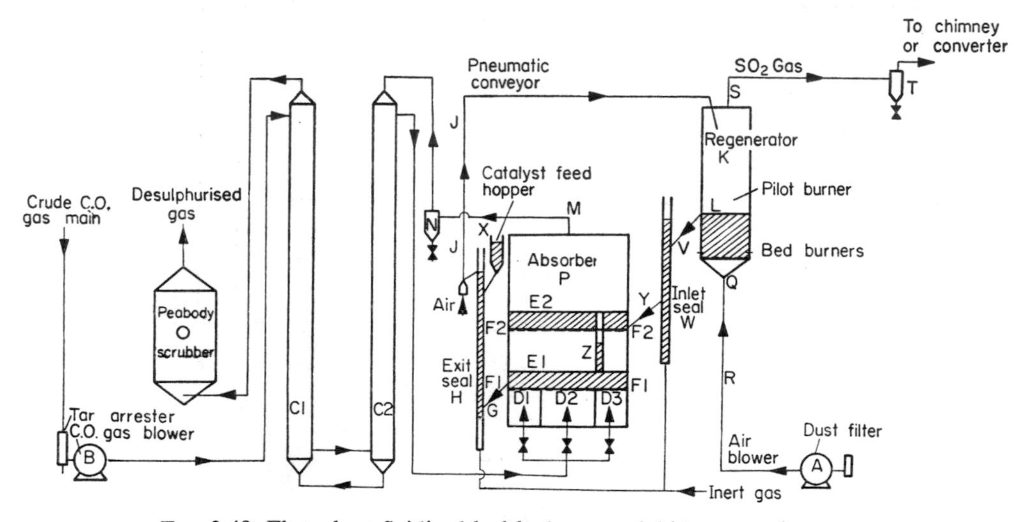

FIG. 3.40. Flow sheet fluidized bed hydrogen sulphide removal process (Appleby–Frodingham process).[680]

approximately one-tenth of this of organic sulphur compounds. Treatment of the gases in a four-stage fluidized bed adsorbed resulted in the removal of 99·7–99·9 per cent of the hydrogen sulphide (final concentration 10–20 ppm), 70–80 per cent of organic sulphur (not including thiophene) and 30–45 per cent of the thiophene. The ferric oxide pellets are recovered by roasting the partly sulphided and reduced ferric oxide in air at 800°C, producing sulphur dioxide for acid manufacture. Equilibrium calculations indicate that satisfactory amounts of hydrogen sulphide may still be removed at 600°C, although processing conditions would be more difficult, and other processes may be more attractive at this temperature.

Small quantities of hydrogen sulphide and organic sulphur compounds can also be removed at temperatures from ambient to 450°C by a series of commercial catalysts. These are of the zinc oxide, promoted iron oxide, chromic alumina, cobalt-molybdenum or activated carbon types. However, these catalysts are costly, and do not at present appear economical for removing large quantities and comparatively high concentrations present in fuel gases or natural gas.

A new process developed by the U.K. Gas Council first converts organic sulphur compounds, carbonyl sulphide and carbon disulphide to hydrogen sulphide over a catalyst bed consisting of uranium oxides (U_2O_3 and U_2O_8) supported on a refractory bed at 500°C. The hydrogen sulphide is then adsorbed on iron oxide pellets made by extrusion of spent bauxite (Luxmasse). These will retain 40 per cent of their own weight of hydrogen sulphide at temperatures up to 350°C. In contrast, zinc oxide adsorbents will only retain up to about 15 per cent of their own weight.

Activated alumina and molecular sieves (alkali metal alumina silicates) have also been shown to be a useful basis for continuous hydrogen-sulphide removal at temperatures up to 250°C. The work of Munroe and Masdin[587] showed that these will convert 70–95 per cent of hydrogen sulphide in a gas stream (0·5 per cent concentration) to elemental sulphur by catalytic reaction with sulphur dioxide

$$2H_2S + SO_2 \rightarrow 2H_2O + 3S$$

This reaction, which is basically the Claus process reaction, proceeds satisfactorily with low carbon monoxide concentrations, and less than 5 per cent water vapour. Sulphur dioxide is added in stoichiometric quantities to the feed gas, either by partial combustion of the hydrogen sulphide or by burning sulphur.

Squires[797] has proposed the use of calcined dolomite. The first stage of the process is the removal of hydrogen sulphide using the calcium oxide fraction of the dolomite [CaO+MgO]

$$H_2S + [CaO + MgO] \rightarrow [CaS + MgO] + H_2O$$

The hydrogen sulphide is recovered by reacting the [CaS+MgO] compound with steam and carbon dioxide at 1·5 MPa:

$$[CaS + MgO] + H_2O + CO_2 \rightarrow [CaCO_3 + MgO] + H_2S$$

The dolomite [$CaCO_3$+MgO] is then calcined to give the adsorbent:

$$[CaCO_3 + MgO] \rightarrow [CaO + MgO] + CO_2$$

The MgO fraction of the dolomite does not participate in the chemical reaction cycle, but it is important in maintaining the physical stability of the adsorbent.

Squires has reported laboratory data and thermodynamic calculations which indicate the feasibility of the process. At temperatures of 600–650°C, and pressures of 1·0–1·5 MPa, initial hydrogen sulphide concentrations of 1 per cent (i.e. 10,000 ppm), in a gas mixture containing hydrogen and carbon monoxide, were reduced to between 2 and 140 ppm. In the recovery cycle, the dolomite–sulphide complex reacted with a gas mixture (82 per cent CO_2, 9 per cent CO, 9 per cent H_2 and remainder steam) at 550–600°C and 1·5 MPa, to give a gas mixture with 20–24 per cent hydrogen sulphide, suitable to be a feed gas in a Claus sulphur-recovery plant. Further thermodynamic calculations for the adsorption reactions indicated that satisfactory hydrogen sulphide removal with calcined dolomite can be obtained at 850°C.

(iii) *Sulphur dioxide*. A number of processes have been developed in recent years for the adsorption, with chemical reaction, of sulphur dioxide from flue gases on solids in fixed bed, fluid bed and entrained bed reactors. In general, these processes use either a very cheap raw material as adsorbent, such as dolomite, which can be discarded, together with the sulphur, after use, or a more expensive adsorbent from which the sulphur is recovered and the adsorbent is recycled. The economic comparison of the different processes for sulphur dioxide removal from flue gases has been described elsewhere.[180, 425] The chief adsorbents which have been tested on laboratory and pilot plant scale are dolomite, alkalized alumina, activated manganese dioxide, activated carbon and activated silica gel.

(iv) *Dolomite and calcium carbonate*. In an early attempt to get a commercially viable process for adsorption of sulphur dioxide at high temperatures, beds of cheap naturally occurring dolomite (calcium magnesium carbonate), crushed to 3–6 mm, were used in a vertical reactor at 600°C and a space velocity of 1150 $m^2 h^{-1}$ by Jones[409] and Coke.[164] The efficiency of removal was 90 per cent until the surface of the dolomite had been converted to the sulphate, using about 15 per cent of the dolomite. When discussing this reaction in detail, Squires[796] indicated that the calcium sulphate is formed first, and the magnesium

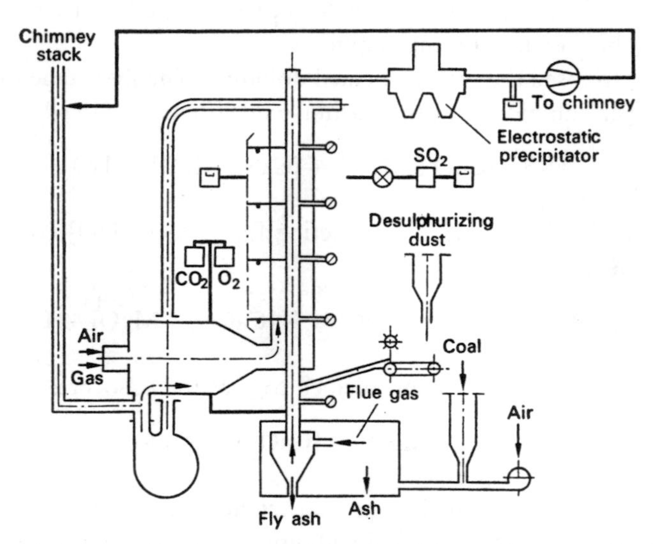

FIG. 3.41. Entrained bed reactor for absorption of SO_2 on calcium carbonate and hydrated dolomite (Jüngten and Peters).[418]

sulphate later, aided by the presence of iron oxide, which acts as a catalyst. Squires proposed the partial calcining of the dolomite, making it porous, and so permitting the gases to penetrate the adsorbent particles. Another way of obtaining greater surface area is to crush the adsorbent to smaller sizes, but these will require an entrained or fluid bed reactor rather than a fixed bed adsorber. An entrained bed reactor has been used in pilot plant experiments by Jüngten and Peters (Fig. 3.41).[418] The reactor, which was 6·35 m long, was heated externally by gases flowing through the annulus formed by an outer tube. The reactor could be controlled at temperatures between 200° and 1100°C, and dolomite, limestone, and other particles were fed to the reactor. The particles following reaction were collected by an electrostatic precipitator. Effective residence time was 1 to 4 sec. Reagents were added to the flue gases in amounts varying from 1 to 3 times the stoichiometric quantity required for 1200 ppm sulphur dioxide. Dolomite itself did not prove very efficient, but calcium carbonate ($CaCO_3$), precipitated as needles or spheres, coated with alkali gave removal efficiencies better than 90 per cent at 800°C.

A similar entrainment reactor process using calcium hydroxide has been used by Still.[963] The material is introduced into a series of inverted U-tube reactors. The gases from each inverted U-tube are passed through cyclones to remove the reagent, which is transferred

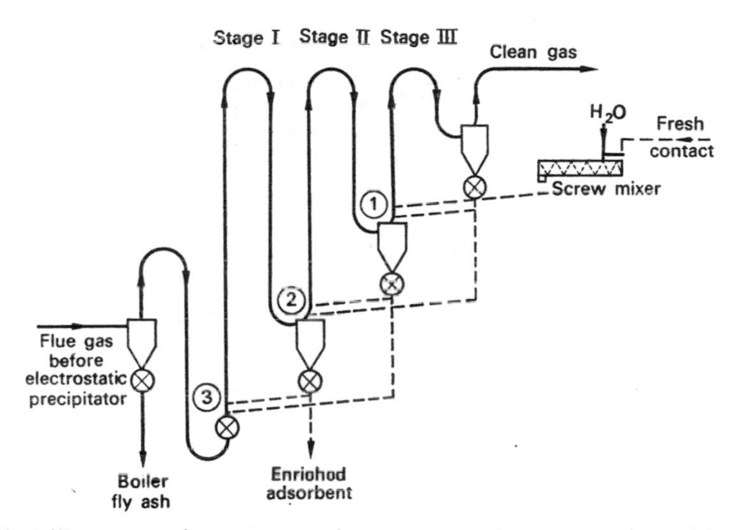

FIG. 3.42. The Still process; a three-stage entrainment reactor absorbing sulphur dioxide on hydrated calcium oxide;[963] I, II and III are the entrainment reactors; 1, 2 and 3 are the points where the material is added to the system.

back to the first of the tube reactors. Sulphur-dioxide-removal efficiencies better than 95% are expected for the Still process (Fig. 3.42).

(v) *Alkalized alumina.* As a result of an extensive study of solid adsorbents by Bienstock, Field and Myers of the U.S. Bureau of Mines, aluminium sodium oxide or *alkalized alumina* appeared to be the most promising adsorbent for sulphur dioxide as it was able to adsorb reasonable quantities of the gas and the material formed hard durable granules which could be repeatedly used in a cyclic process.[76, 77, 257]

After some preliminary studies, pilot scale plants were investigated in which the adsorbing particles, 1·6 mm diameter, fall counter-current to a rising stream of flue gases at 330°C. The technical feasibility of the process has been demonstrated in an 8-m reactor (Fig. 3.43) and adsorption efficiencies of over 90 per cent of the sulphur dioxide in the flue gas stream have been obtained. Regeneration, using H_2/CO gas mixtures at 650°, and producing H_2S has also been demonstrated, with no loss of adsorptive capacity of the alkalized alumina pellets even after 20 cycles. The difficulties with the process are associated with the losses of adsorbent pellets by attrition. If these losses can be kept below 0·1 per cent of the feed for

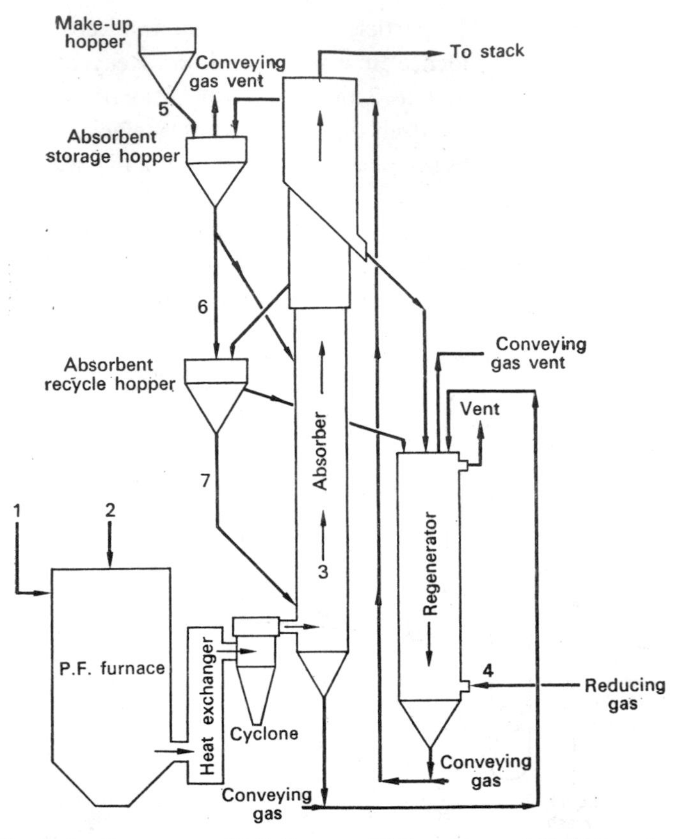

FIG. 3.43. Alkalized alumina process (U.S. Bureau of Mines Process).[76, 77] Flowsheet of pilot plant (700 $m^3 h^{-1}$ (0°C) flue gas). Gas and material flows:

1. Secondary air; 525 $m^3 h^{-1}$ (0°C) at 204·4°C.
2. Primary air; 170 $m^3 h^{-1}$ 60°C.
3. Flue gas; 700–1500 $m^3 h^{-1}$ at 330°C.
4. Reducing gas; 65 $m^3 h^{-1}$; 76·7% H_2, 14·5% CO, 7·2% CO_2, 0·3% N_2, 1·3% CH_4.
5. Absorbent makeup 0·055 kg h^{-1}.
6. Absorbent fresh feed 58·8 kg h^{-1}.
7. Absorbent recycle 588 kg h^{-1}.

each cycle, then the process will be economically feasible. It has, however, been shown that the hardness of the pellets can be increased by heating to 900°C, and further work is in progress. Different gases have been tried for regenerating the alkalized alumina, and in order of decreasing effectiveness these have been: reformed natural gas, hydrogen, producer gas and methane. Chlorine compounds (from the coal) are adsorbed by the alkalized alumina and are not removed during the usual regeneration stage, but they can be removed by treating the pellets with flue gases at 600°C. Thus, in small-scale plants, hydrogen has been used for regeneration at 650°C, while on a larger scale, reformed natural gas or producer gas has been used. This produces hydrogen sulphide, carbon dioxide and water, which is a mixture which can be fed to a Claus plant for the production of sulphur.

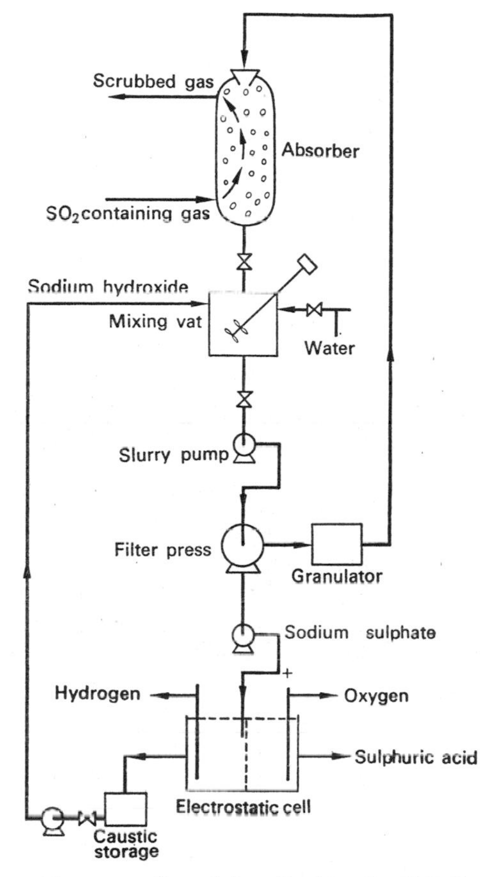

FIG. 3.44. The manganic oxide process for sulphur dioxide; the U.S. Bureau of Mines Process.[257]

The U.K. Central Electricity Generating Board has experimented with a fluidized bed adsorber as an alternative to the U.S.B.M. falling-particle system.

(vi) *Activated manganese oxide*. Manganese oxides, although found effective as adsorbents by Bienstock, Field and Myers,[76] were found difficult to regenerate. The system proposed (Fig. 3.44) by them was similar to the alkalized alumina process as regards the adsorber, but regeneration was by electrolysis. The stages suggested were:

1. the reactions of Mn_2O_3 with SO_2:

$$Mn_2O_3 + 2SO_2 + \tfrac{1}{2}O_2 \rightarrow 2MnSO_4$$

2. the manganese sulphate is slurried in water and sodium hydroxide is added:

$$2MnSO_4 + 4NaOH + \tfrac{1}{2}O_2 \rightarrow Mn_2O_3 + 2Na_2SO_4 + 2H_2O$$

The manganese oxide is recycled to the adsorber, while the sodium sulphate solution is electrolysed to give sodium hydroxide, which is used in the process, and sulphuric acid, which is recovered. The electrolytic recovery, however, requires large quantities of electrical energy and makes this process uneconomic.

Mitsubishi Heavy Industries Ltd. have developed a somewhat different activated manganese oxide process, the DAP–Mn Process[41, 533] (Fig. 3.45) where the regeneration of the adsorbent is carried out by reaction of the manganese sulphate with air and ammonia, to produce ammonium sulphate. The oxidation and subsequent regeneration take place at room temperature, the activated manganese oxide being separated by filtration. The ammonium-sulphate solution is passed through a crystallizer where ammonium sulphate is

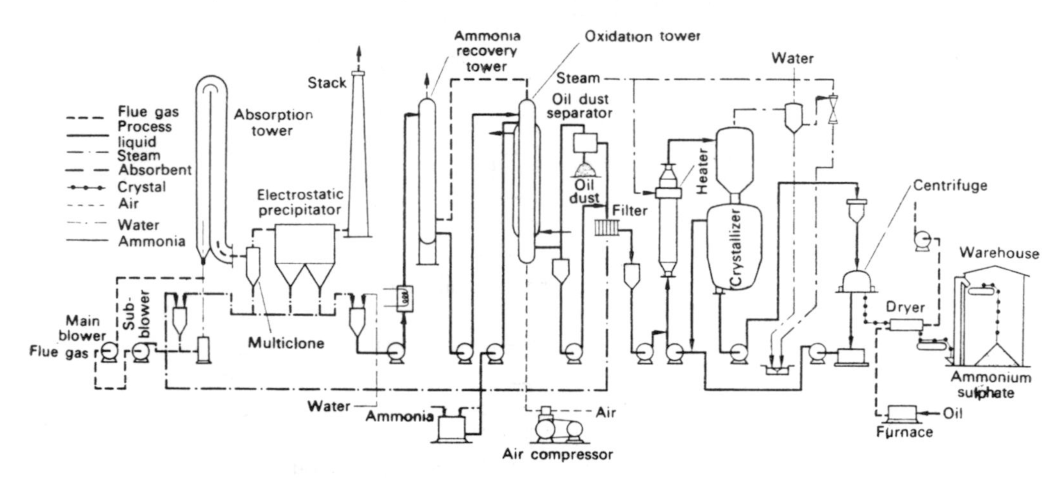

FIG. 3.45. The activated manganese dioxide process; the Mitsubishi DAP-Mn process.[562]

recovered. If the recovery of gypsum rather than ammonium sulphate is desired, lime is added to the ammonium-sulphate solution. The ammonia is then separated from solution by heating after the gypsum has been filtered off in a centrifuge. Concentrations of manganese oxide used are of the order of 150 to 250 g m^{-3} and temperatures of 135–160°C are used. The gas at entry contains 1000 ppm which is reduced by over 90 per cent before exit. There is a certain amount of oil dust (0·100–0·2 g m^{-3}) in the flue gases, but this does not affect the performance of the adsorbent. The activated manganese oxide has a composition $MnO_X.YH_2O$, where X is between 1·5 and 1·8 and Y between 0·1 and 1. This composition is claimed to be different to the manganese oxides tested by Bienstock, Field and Myers, being made by a special Mitsubishi process. The process has, so far, been tested on a 3000 m^3 h^{-1} plant but a 55-MW boiler size unit is being built at the Yokkaichi Power Station, Chubu Electric Power Company. This plant will treat 153,000 m^3 and will operate for one year before further larger plants are constructed.

(vii) *Activated carbons.* The adsorbent with greatest promise of a high degree of adsorption, because of the very high surface area, is activated carbon. The use of this material at comparatively low temperatures (below 100°C), where the sulphuric acid formed is leached out, has been discussed (the Sulfacid process). Until 1957, however, it was thought that no satisfactory adsorption on carbon at higher temperatures—above 100°C—could be effected.

New laboratory studies then showed that sulphur dioxide is adsorbed and converted to sulphuric acid and it is only possible to reverse this process above 250°C.

A process based on the use of cheap semi-coke made from peat carbonized under vacuum at 600°C has been developed which is called the Reinluft process[894a] (Fig. 3.46). The flue gases enter the lower section of the adsorber at about 150–200°C and part of the sulphur dioxide is adsorbed on the carbon and is converted to sulphur trioxide. Further up the adsorber, the gases are withdrawn, passed through a heat exchanger where they are cooled to about 110°C before being returned to the upper section of the adsorber. The gases leave the

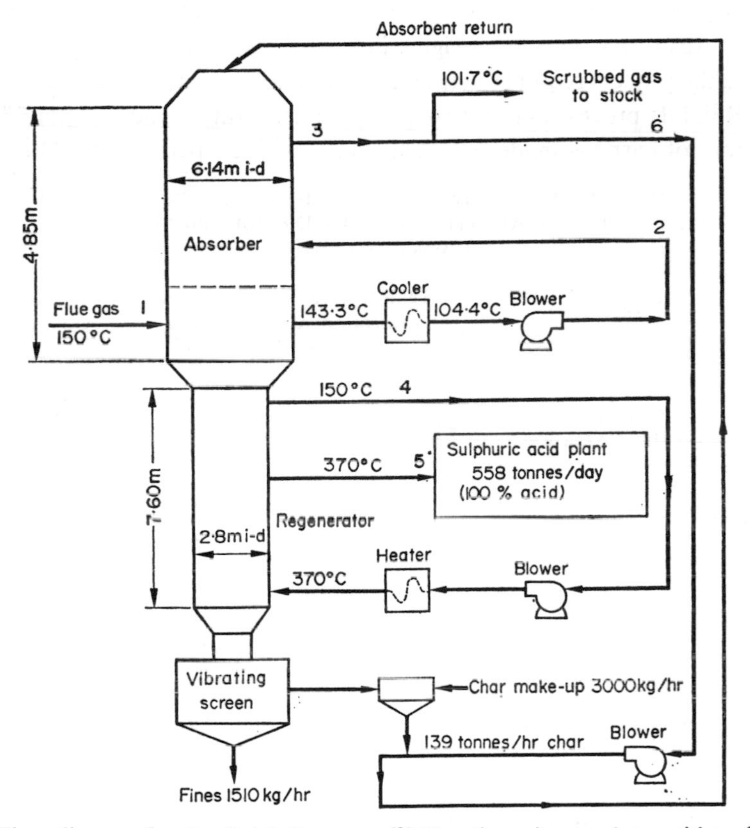

FIG. 3.46. Flow diagram for the Reinluft process.[425] The dimensions and quantities of gases and reagents are for a plant for an 800 MW power station.

top of the adsorber at just above 100°C. The spent carbon from the bottom of the adsorber, which is saturated with sulphuric acid, is dropped into the desorber section. Here the carbon is heated to 380–450°C with a stream of inert gas, the sulphuric acid is broken up, with the production of carbon dioxide, water vapour and sulphur dioxide in concentrations of 10–15 per cent, which is suitable as a feed gas for a contact acid plant. The char is recycled after the fines have been removed.

The first pilot plant, operated at the Wolfsburg Volkswagen Works,[894a] gave sulphur-dioxide removal rates varying from 45 to 96 per cent as the gas rate was decreased from

330 $m^3\ m^{-2}\ h^{-1}$ to 86·0 $m^3\ m^{-2}\ h^{-1}$. Extensive tests were also carried out at the U.K. Warren Spring Laboratory[676] where efficiencies of over 90 per cent were obtained with similar conditions. However, carbon losses due to attrition and chemical decomposition, serious corrosion problems, and the danger of combustion of the reactivated carbon, indicated that serious problems would be faced by a commercial scale plant. A more recent plant has been built to clean 34,000 $m^3\ h^{-1}$ (NTP) of gas following the electrostatic precipitator of a 150-MW boiler at the STEAG-Kraftwerk Kellermann in Lünen.[963] The flue-gas temperatures vary between 95° and 125°C with an SO_2 content of 700 to 1400 ppm. The plant was in operation for 26 weeks during 1966, and was able to remove 65–70 per cent of SO_2, which was lower than the design value of 75 per cent.

To improve the process requires a cheap carbon which is much harder than the semi-coke used in the Reinluft process and not so combustible. Dratwa and Jüngten[229] have investigated a number of carbons made from black coal with 50 per cent oxidation and of oxy-

TABLE 3.12. RETENTIVITY AND EFFECTIVENESS OF OXY-COKES AND PEATS IN ABSORBING SULPHUR DIOXIDE FROM FLUE GASES (SIMULATED)[229]

Carbon type	Temperature of absorption °C	Residence time s	% SO_2 removed
Activated carbon from black coal with 50% oxidation	80	13	60
	120	13	45
	160	13	35
	80	5·6	30
	120	5·6	20
	160	5·6	15
	80	3·8	10–5
	120	3·8	
	160	3·8	
Peat coke (three regenerations)	80	13	70
	120	13	65
	160	13	57
(two regenerations)	80	13	100
	120	13	100
	160	13	100
Oxycoke	80	5·6	97
	120	5·6	94
	160	5·6	75
	80	3·8	90
	120	3·8	73
	160	3·8	65

The results quoted are after passing 11,700 m^3 gas m^{-3} of absorbent (at N.T.P.).

cokes made by the coking of air-oxidized hard coals. These carbons, as well as peat carbons, were investigated by Jüngten for their physical properties and ability to adsorb SO_2. The oxy-coke proved to be very hard, even after repeated cycling in the adsorption system, unlike the peat coke or activated black coal carbon. The oxy-cokes also had an extensive fine-pore structure, which was a function of the pyrolysis temperature.[416] This coke was able to adsorb much greater quantities of SO_2 with lower residence time than the peat carbons (Table 3.12).

As a result of the favourable laboratory tests on the oxy-coke formed from pre-oxidized black coal, pilot-plant experiments have been undertaken in the experimental boiler plant at Mathias Stinnes A.G., Bottrop.

Jüngten, Knoblauch and Kruel[417] have reported on the operation of a continuous moving-bed absorber with cross flow of dust laden flue gases containing 0·1 per cent sulphur dioxide.

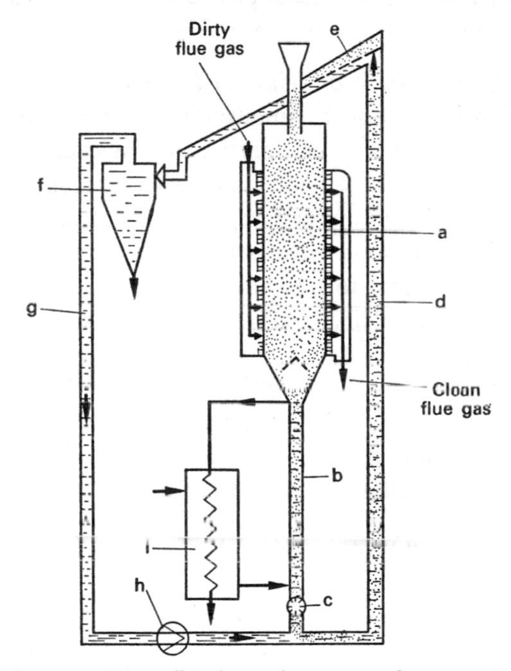

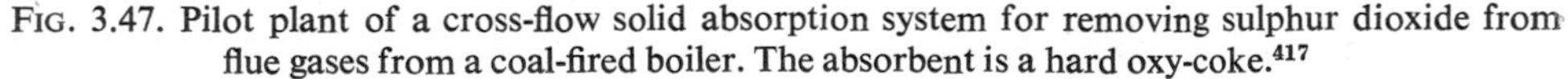

FIG. 3.47. Pilot plant of a cross-flow solid absorption system for removing sulphur dioxide from flue gases from a coal-fired boiler. The absorbent is a hard oxy-coke.[417]

a = adsorber
b = absorbent washer
c = rotary valve
d = oxy-coke transport pipe
e = screen separator
f = cyclone separator
g = water pipe
h = pump
i = acid cooler

The bed thickness was varied from 250 mm to 750 mm and a series of residence times from 5 to 40 hr was used (Fig. 3.47). These results show that almost 100 per cent removal can be obtained with a 750-mm bed and a 5-hr residence time, but thinner layers of coke, and longer residence times, resulting in lower pressure drop and less coke consumption, may still provide adequate removal of the sulphur dioxide. The work also showed that dust contents of 1·5–1·9 g m^{-3} would lead to blocking of the bed with residence times of 15 to 20 hr, but dust and fly-ash removal down to 0·05–0·10 g m^{-3} was achieved.

Another activated carbon, which may also have similar properties, has been suggested by Strauss.[824] This is made by extruding flame carbons into pellets. The flame carbons were formed from furnace oils to which activating additives had been added before burning under controlled conditions.

(viii) *Activated silica gel.* Another material which has been found suitable for the high-temperature adsorption of sulphur dioxide is silica gel which has been treated with iron salts.[360] The treatment consists of saturating the silica gel with iron-salt solutions and heating this to 600°C. The treated gel is suitable as an adsorbent at temperatures of 350° to 400°C. Regeneration of sulphur trioxide is carried out at 700°C. The proposed plant consists of two adsorption stages: in the first, the flue gases pass up through a bed of the silica gel, while above this, the granules are allowed to fall freely counter-current to the gas stream.

(ix) *Adsorption of sulphur dioxide on material introduced into the combustion chamber.* The problem of corrosion in the boiler combustion chamber, rather than sulphur-dioxide removal, prompted a number of workers, particularly Wickert,[940] to introduce materials such as dolomite into the combustion chamber. As well as preventing both high-temperature and low temperature corrosion and marked improvement in the maintenance requirements of boilers,[738, 876] the addition of these materials reduced the emissions of sulphur dioxide. There is considerable evidence that for the adsorption on the dolomite or limestone to be effective, the material has to be calcined before use, and this reaction takes place when the material is introduced into the boiler furnace near the fuel burners at temperatures above 1100°C. Calculations have shown that at this temperature it takes 0·5 sec for a 20-μm granule to be calcined.[654] Wickert calcined the material used in pilot plant work at 1250°C and he obtained results for lime (CaO) and calcined dolomite (CaO : MgO). He further found that adding 1–2 per cent Fe_2O_3 to the dolomite catalyses the reaction:

$$2MgO + 2SO_2 + O_2 \xrightarrow{(Fe_2O_3)} 2MgSO_4$$

This catalyst has no effect in the case of CaO.

Pilot plant-scale experiments have been carried out by a number of workers. Early experiments in the refinery boilers at the Mobil Bremen Refinery[940] and at Wolfsburg[360, 876] used a system which introduced the dolomite powder (optimum particle size 10–15 μm) with an air blast across the firing chamber above the burners. High degrees of sulphur removal here reported at low sulphur dioxide concentrations, for example 72 per cent removal at 250 ppm; 85 per cent removal at 136 ppm.

The most extensive large scale tests have been carried out in an ash slagging boiler with a capacity of 100,000 kg h^{-1}. Dolomite was blown in at five points, as shown in Fig. 3.48 (denoted 1 to 5). The temperature at the first point was approximately 1500°C, while at 5 it was approximately 900°C. The maximum effectiveness was obtained at point 3, where the temperature was 1150°C, which is in agreement with laboratory experiments. The dolomite quantity was 2·5 times the stoichiometric amount needed for combination with the sulphur in the coal. Here, point 3 represents a compromise between suitable temperature and re-

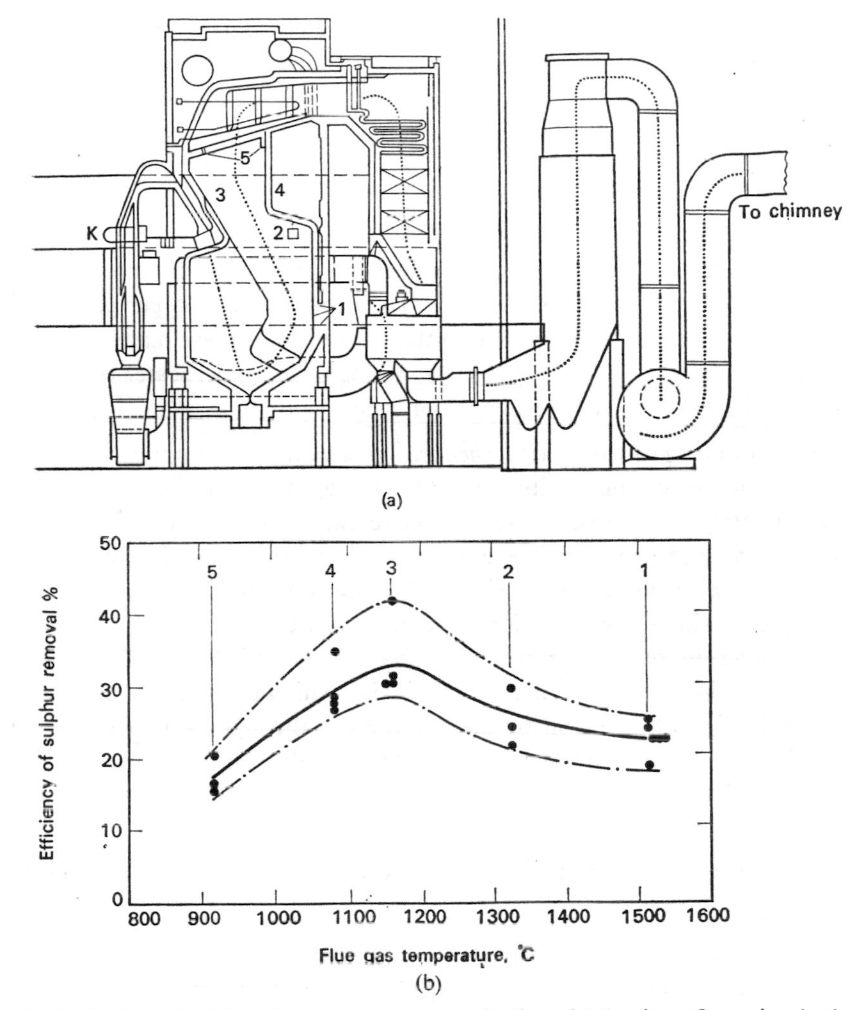

FIG. 3.48. Reduction of sulphur dioxide emissions by injection of dolomite at five points in the boiler, followed by electrostatic precipitation:[963] (a) shows the boiler and the injection points; (b) correlation of flue gas temperature at point of injection with the efficiency of sulphur dioxide removal.

sidence time. Further tests were carried out, and these are reported in some detail[963] with hydrated lime, powdered limestone, caustic potash as well as with dolomite. Hydrated lime gave the best results, which were somewhat better than those shown for dolomite.

Pollock, Tomany and Frieling[653] investigated the addition of limestone (10 per cent of the coal fired) the material being fused together with the coal. The sulphur-dioxide reduction from flue gases containing 550–890 ppm was between 40 and 80 per cent. The material here was recovered in the scrubber described earlier with the scrubber investigations by these workers. It was noted that lower concentrations of limestone had no effect.

Far more successful than the completely dry Wickert process has been an extension of the earlier work by Pollock, Tomany and Frieling, by a group from the Detroit Edison Company and Combustion Engineering Inc.[653] Dolomite is injected in the corners above the coal burners, by injectors diverging from the direction of coal injection and tilted 15 per cent upwards, the coal burners being tilted similarly downwards. The dolomite injection was at rates of approximately stoichiometric to $1\frac{1}{2}$ times stoichiometric for the SO_2 present. The flue gases were then scrubbed in a high efficiency scrubber containing a fluidized bed of 16-mm glass marbles. Removal efficiencies of 90–99·5 per cent were obtained. These high efficiencies needed slurry flow rates of 1·75 $m^3 h^{-1}$ for 1700 $m^3 h^{-1}$ of gas.

Some Japanese studies, reported by Arawaki,[34] also relate to the addition of limestone, limestone and fly ash, and limestone and magnesium hydroxide mixtures in the furnace. It was found that desulphurization was only of the order of 35–45 per cent, which is the same as reported by Zentgraf.[963] The intermittent addition of these materials, on the other hand, changed the properties of the deposits adhering to the superheater tubes which made soot blowing easier and reduced the deposition rate.

(x) *Other adsorption processes with chemical reaction.* Such compounds as peroxides, ozonides and similar compounds with —O—O— linkages are readily converted to simpler compounds on catalyst surfaces. Untreated active carbon is sometimes suitable but other compounds are only decomposed when a metal catalyst has been deposited on the carbon. Typical catalysts are metalli ccopper, silver, platinum and palladium, which are put into the catalyst by deposition from solutions of complex salts.

There are numerous other examples, such as the oxidation of carbon monoxide on copper oxide or iodine pentoxide to give the dioxide which can then be adsorbed on soda lime; the bromination of an olefin by passing the gas through carbon impregnated with bromine; iodine for mercury vapour; lead acetate for hydrogen sulphide or sodium silicate for hydrogen fluoride.

3.9. COMBUSTION PROCESSES—FLAME COMBUSTION

When atmospheric pollutants are oxidizable, such as hydrocarbon vapours from industrial solvents or paint, then their control can be achieved by combustion of the gases to give carbon dioxide and water in the case of organic material or sulphur dioxide and water from organic sulphides. If the concentration of the gas is sufficiently high to be in the flammable range, the oxidation process can be self-supporting after the mixture is ignited. The lowest vapour concentration at which this occurs is the lower flammable limit, while the maximum concentration is the upper flammable limit. Within this range controlled combustion can be achieved, but under certain circumstances explosions also occur.

The temperature above which combustion of the vapour and gases is sustained is the *autogenous combustion temperature* and this depends on the type of hydrocarbon and the *fume energy concentration* which is the available net heat of combustion per standard m^3 of gas (J m^{-3} at 21°C). The lower flammable limit is approximately 1·9 MJ m^{-3} (21°C) and fume energy concentrations higher than this are necessary for self-supporting flames, while for a flame for good combustion characteristics a fume energy concentration of about 3·7 MJ m^{-3} (21°C) is desirable.

All combustion processes require adequate oxygen for oxidizing all the combustible material. Conventional combustion methods need about 15 to 20 per cent more than the stoichiometric amount, while catalytic combustion techniques need only the stoichiometric quantity. In addition, the temperature in the combustion chamber or in the flame must be sufficiently high, there must be adequate turbulent mixing of oxygen and the combustible gases, and sufficient time for combustion has to be available. The design of the burner and combustion chamber and the degree of premixing of gases determine these factors. A pre-mixed flame tends to be shorter and hotter, and blue in colour, while a non-premixed flame tends to be luminous because of the cracking of hydrocarbon vapours which produces luminous carbon particles.

The basic chemical reaction for the oxidation of a hydrocarbon, C_mH_n, is given by

$$C_mH_n+\left(m+\frac{n}{4}\right)O_2 \rightarrow mCO_2+\frac{2}{n}H_2O \tag{3.58}$$

The heat liberated by the process is ΔH, the enthalpy or heat function at constant pressure. The second law of thermodynamics relates this to the Gibbs free energy, ΔG, by the expression

$$\Delta G = \Delta H - T\,\Delta S \tag{3.59}$$

where ΔS is the entropy change and T the absolute temperature. At 25°C and 1 atmosphere pressure, the values of these functions are called "standard" and are designated by ΔG°, ΔH° and ΔS°. For values of ΔG° less than 0, the reaction is likely to take place, while for reactions for which ΔG° is between 0 and 42,000 J kg^{-1} mole^{-1} the feasibility is doubtful, but could repay further study. When ΔG° is greater than 42,000 kJ kg^{-1} mole^{-1} the reaction is unfavourable.

The equilibrium constant for the reaction is

$$K_p = \frac{p^{m}_{CO_2}p^{n/2}_{H_2O}}{p_{C_mH_n}p^{(m+n/4)}_{O_2}} \tag{3.60}$$

and

$$\Delta G^\circ = -RT \ln K_p = -2{\cdot}303RT \log_{10} K_p \tag{3.61}$$

where p is the partial pressure of the material.

Some typical values for ΔH°, ΔG° and $\log_{10} K_p$ are given for the oxidation of formaldehyde (H.CHO), propane (C_3H_8) and ethanol (C_2H_5OH) in Table 3.13.

TABLE 3.13. FREE ENERGY AND ENTHALPY FOR THE OXIDATION OF FORMALDEHYDE, PROPANE AND ETHANOL

	298·16 K–25°C		600 K–326·84°C	
	ΔH° MJ kg^{-1} mole^{-1}	ΔG° MJ kg^{-1} mole^{-1}	ΔH° MJ kg^{-1} mole^{-1}	ΔG° MJ kg^{-1} mole^{-1}
H.CHO	–519·9	–512·8	–518·2	507·8
C_3H_8	–2044·9	–2075·0	–2041·5	–2107·7
C_2H_5OH	–2991·7	–3020·6	–3043·6	–3102·7

Direct incineration is usually carried out at temperatures of 700–800°C while catalytic combustion generally is at temperatures of 250–400°C.

Where waste gases contain hydrocarbons in a quantity to give fume energy concentrations above the minimum of 1·9 MJ m^{-3} and these hydrocarbons contain toxic gases such as hydrogen cyanide, they are usually burned in a flare stack. A problem which occurs in these cases, particularly in the petroleum industry, is the wide variation of gas flow rates. Certain hydrocarbons, particularly aromatics and others with a low carbon–hydrogen ratio, tend to burn with a smoky flame, but this can be overcome by adding water vapour in the form of steam. This sets up a water gas reaction giving hydrogen and carbon monoxide which helps to give smokeless flames. Hess and Stickel[364] have studied the theory and experimental behaviour of flares burning acetylene in considerable detail, determining the minimum levels of steam and air requirements, the limits of luminosity, flame stability and noise levels. As a result of these calculations, the volume ratio air/acetylene has been plotted against the weight of steam per unit volume of acetylene for different values of the function Φ, which is

$$\Phi = \frac{\text{Volume of air} + \text{Volume of steam}}{\text{Volume of steam}}.$$

These are plotted in Fig. 3.49. The broken line is the theoretical (stoichiometric) steam and air requirement, while the solid line marks the experimentally determined limit of lumi-

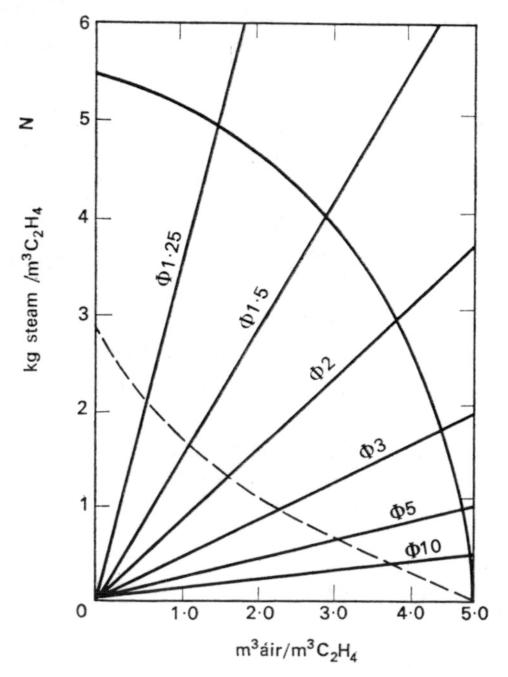

FIG. 3.49. Theoretical steam and air requirements and luminosity bands for the combustion of acetylene in air. Volumes of air and steam are at 0°C, 100 kP.[364]

— — — — — Stoichiometric limit.
———————— Luminosity limit.

nosity. This shows that to approach the most economical conditions, it is desirable to work in the range between $\Phi = 5$ and the abcissa. This also indicates that one of the main factors in maintaining a flare is to supply sufficient air. Calculations indicate that for the gas stream to entrain sufficient air it requires supersonic velocities. The usual practice is therefore to use high-velocity steam jets surrounding the gas jet, although this leads to rather loud flares on occasions. Alternative types of flares using premixing with air, on the bunsen-burner principle, or premixing with steam with subsequent entrainment of air at the jet, have not been widely used. The probable reason is that these designs do not lend themselves so readily to the very wide fluctuations in gas quantity with which the flares are expected to deal.

A typical design of a steam injection flare is shown in Fig. 3.50. The steam jets surrounding the flame induce air together with the steam to assist in rapid combustion. Steam require-

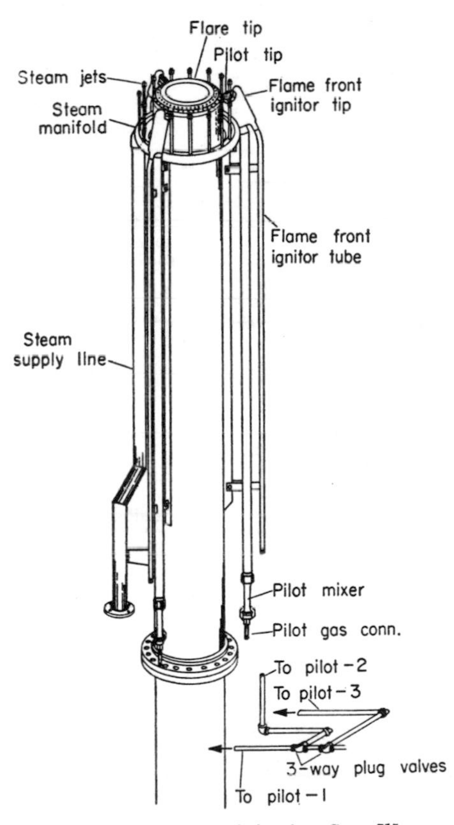

FIG. 3.50. Steam injection flare.[715]

ments vary from 0·05 to 0·33 kg steam kg^{-1} gas, depending on the quantity of aromatic constituents. A pilot flame is also provided. If the combustible gases are diluted so that the minimum fume energy concentration is not reached, then preheating of the gases may in some cases achieve an energy concentration able to support combustion. A ring burner is then used for preheating. When oil mists are found in the flare gases these must be removed at the base of the stack for safe flare operation. Flash-back precautions are also necessary when oxygen is present in the gases.

Flare stacks have been used for organic thiophosphates, which are lachrimatory, as well as for hydrocarbons.[569] In the manufacture of these compounds, the process vacuum is achieved by steam ejectors which normally discharge to hotwells, causing an in-plant pollution problem. However, the hotwell waste gases are in the combustible range, and can be vented through a flare stack.

An alternative to an open flare is to burn the gases in an enclosed combustion chamber. Typical designs use a circular flow pattern to give a high degree of turbulence and an adequate residence time (0·2–0·7 sec) in a small space.

There are a number of designs which give more or less adequate mixing or premixing, and the required residence time in the combustion chamber. Figure 3.51a shows a burner

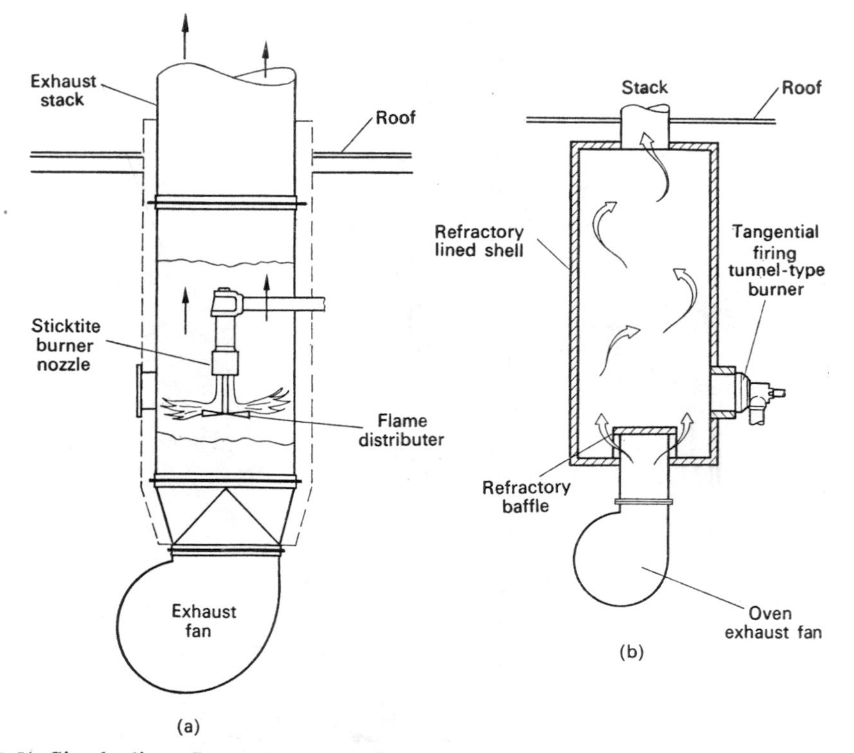

FIG. 3.51. Simple direct flame afterburners:[895] (a) afterburner with impingement flame distributor (temperature limited to 480°C); (b) tangential entry refractory lined combustion chamber (suitable for waste gases low in oxygen).

from which the gases impinge on a flame distributor, and the gases pass through the sheet of flames, at a different direction to the flame. The temperature of this type is limited to below 480°C. and so only a partial conversion of hydrocarbons is accomplished. Figure 3.51b shows a tangential entry gas burner. The exhaust gases enter the incinerator chamber against a deflecting baffle, concentrating the waste gases at the wall. This type is particularly suitable where there is little oxygen in the waste gases (below 15 per cent) and where there are widely varying gas volumes. These designs give residence times of 0·2 to 0·7 sec in small spaces. The temperature in the chamber is maintained between 480–760°C.

In an alternative design, the burner is applied directly, while the waste gases enter tangentially. This design may be arranged with an insulating refractory wall, or with the walls acting as a heat exchanger for incoming gases.

More recent designs use a premix-type burner such as the Combustifume[167] (Fig. 3.52). This has the advantage of a short flame, and thorough mixing (Fig. 3.53). The possibility of additional preheating has also been tested (Fig. 3.54) and these showed lower tempera-

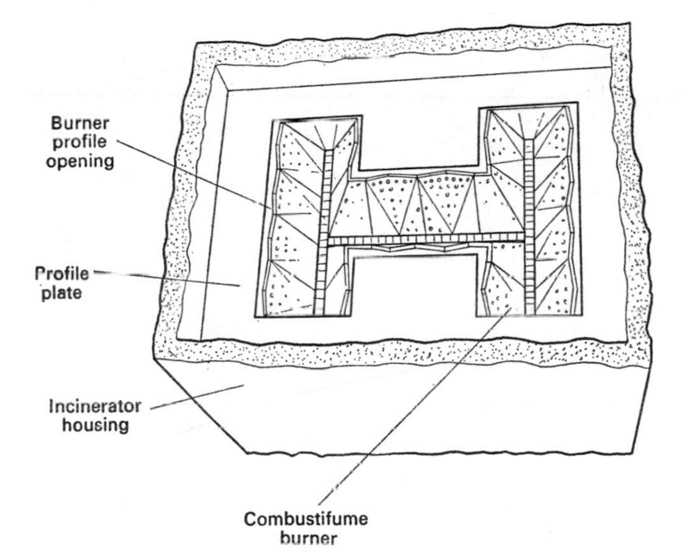

FIG. 3.52. A premix type burner (Combustifume* model) (detail). (* Registered trade name for Maxon Inc., Indiana.)[895]

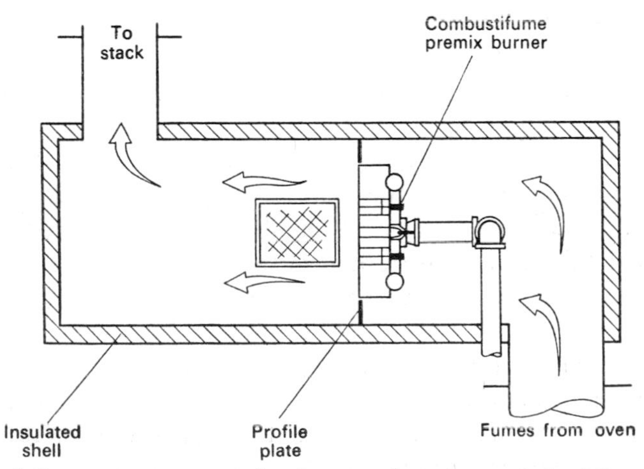

FIG. 3.53. Premix burner arrangement showing premix system and short flame length with this design.[895]

tures for the same degree of incineration with lower residence times. These burners use much higher velocities, 20–40 m s^{-1}, at the burner profile openings, compared to the 10 m s^{-1} used in the conventional combustion chambers.

Great care must be taken in the design of these combustion chambers to avoid overheating the refractory lining and to build the unit in such a way as to avoid explosions by including burner controls which protect the unit.

The waste gases can be exhausted to atmosphere or passed through a heat exchanger to recover much of the sensible heat in the gases. In these designs a supplementary gas or

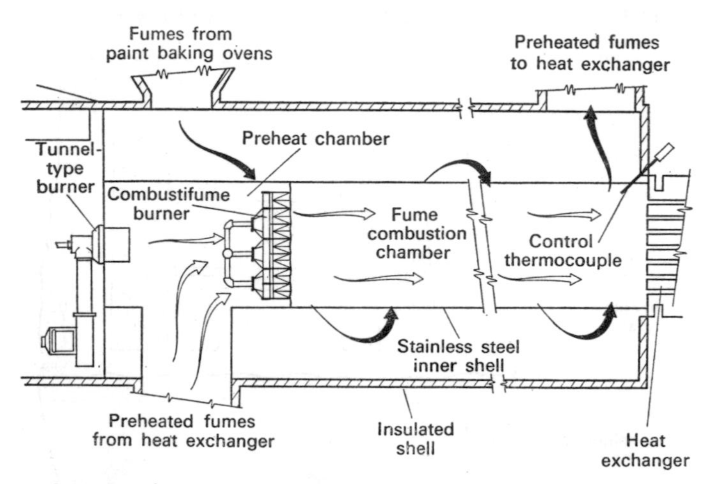

FIG. 3.54. Premix burner arranged with heat exchanger and preheating.[895]

oil fired burner is generally used to heat the gases above their autogenous combustion temperature.[713] The lower temperatures are suitable for naphtha vapour while methane and aromatic hydrocarbons require the higher temperature.

Typical applications of direct flame afterburners are in control of odours from meat and offal rendering, the organic wastes for wire enamelling plants,[69] from core ovens[419] and the oxides of nitrogen from nitration reactions.[214] In the case of the oxides of nitrogen,[214] these are premixed with natural gas before entering the burner where combustion air is added. The core oven waste gas burner described by Kafka and Ferrari[419] is a simpler design consisting of a short packed bed of refractory. Although the residence time in this was calculated to be only 0·05 sec, the relatively high temperature (800°–1000°) gave satisfactory results. Studies on a wire enamelling oven with a premixed flame burner gave over 90 per cent conversion of hydrocarbons at temperatures above 685°C while similar results were obtained for a metal coating oven.[895] The residence time in both cases was 0·3 sec.

Heat recovery from the combustion gases can aid the economic operation of direct flame (and of course catalytic) afterburners. The gas to gas heat transfer can be accomplished using either a renegerative system, such as the Ljungstrom rotary heat exchanger, or a recuperative type unit where the heat transfer is through corrugated metal surfaces. The corrugated surfaces in one design are made into flat tubes so that a cross-counterflow arrangement of great mechanical strength at high temperatures is achieved.[68] In the

Ljungstrom unit a drum-shaped heat-transfer matrix rotates continuously through adjacent but separated streams of exhaust gases and incoming gases. As the surfaces pass alternately through the two streams, heat accumulated from the hot gas is released to the incoming air or gas stream.

These combustion chambers can be applied where the variation in the gas stream is not too great, and where additional fuels, when required, are fairly cheap, because the gases, and the air required (which is in excess of the stoichiometric quantity), have to be heated to the combustion temperature. Well-designed systems can be used when there is considerable inorganic and particulate material present which precludes catalytic combustion.

3.10. CATALYTIC COMBUSTION OF ORGANIC MATERIALS

Catalytic combustion is an extension of the technique of using a combustion chamber with the advantage that oxidation on the catalyst surface will occur well below the autogenous combustion temperature, and also in gas concentrations below the fume concentration energy required for self-supporting combustion processes. An additional advantage is that only the stoichiometric quantity of oxygen (as air) has to be provided, minimizing the preheat requirements.

Normally any gas-borne organic material can be treated by catalytic combustion provided the combustion products are gaseous. Therefore organic sulphur and nitrogen compounds can generally be treated, but organic silicones and phosphates must be excluded because of the nature of the combustion products. If dusts of inorganic materials are present in an appreciable quantity they should be removed, but small quantities such as are normally found in the air will pass through the catalytic combustion unit or occasionally be caught in it. These residues can be removed by occasional (annual or semi-annual) washing of the elements.

The central feature in catalytic combustion is the nature of the oxidizing element. This may be:

(i) Active metal catalyst on a metallic carrier. The catalysts, platinum and other metals of the precious metals group, combined with promoters, are supported on a crimped nickel-alloy ribbon. One particular mixture is used for the normal range of organic compounds, while specific catalysts for selective reactions have also been developed. The finished catalyst unit takes the form of a shallow bed, although for some applications cylindrical cartridges are used. The original elements were developed by Ruff and Suter.[714] (Catalytic Combustion Division of Universal Oil Products process.)

(ii) Active metal catalyst on metal oxide carrier. A thin film of platinum metal is deposited on a carrier which is either hard-fired α-alumina or porcelain (sparkplug grade). The carrier consists of streamlined rods set in rows, offset relative to one another. The catalyst can also consist of high specific surface γ-alumina with a platinum coating.[917] (Oxycat and Degussa processes.)

The use of palladium supported on active alumina also comes into this category. This can be used to remove oxygen from gas streams containing hydrogen as well as oxygen in stoichiometric (or larger) proportions. (Deoxo process.)

(iii) Active oxide catalyst-metal oxide carrier. Active base metal oxides (such as γ-alumina) which have high surface areas can be deposited on metal oxide carrier, such as α-alumina. This combination has the advantages that it is

(a) able to withstand high temperatures;
(b) contains low-cost materials, in comparison to precious metal catalysts;
(c) can be manufactured by extrusion into slugs or pellets.[159, 890]

These were the first catalysts to be used for air purification, as early as 1927, although the method was not, at the time, followed through. In this category also are the catalysts where the active material constitutes the whole catalyst including the support, which are referred to as "unsupported" catalysts. One of these is a mixture of copper and manganese oxides (Hopcalite), which gave complete combustion of hydrocarbons at 300–400°C, with the exception of methane (30 per cent at 400°C).[890]

(iv) Active oxide catalyst-metal carrier. Houdry has patented a catalyst of an impregnated metal oxide shell covering a metal wire support, which is also a resistance heating element.[375] There are no known commercial applications of this.[890]

The temperature which is generally necessary to initiate the reaction on the catalyst depends on the hydrocarbons present. Thus hydrogen oxidizes at ambient temperatures, benzene at 227°C, while methane only oxidizes partially at 404°C. For normal paint baking

TABLE 3.14. CATALYST INLET TEMPERATURES FOR CATALYTIC COMBUSTION[618]

Industrial process	Contaminating agent in waste gases	Approximate temperature require for catalytic oxidation
Asphalt oxidizing	Aldehydes, anthracenes, oil vapours, hydrocarbons	320–370°C
Carbon black mfg.	Hydrogen, carbon monoxide, methane, carbon	*650–980°C
Catalytic cracking units	Carbon monoxide, hydrocarbons	340–450°C
Core ovens	Wax, oil vapors	320–370°C
Formaldehyde mfg.	Hydrogen, methane, carbon monoxide, formaldehyde	340°C
Nitric acid mfg.	Nitric oxide, nitrogen dioxide	†260–650°C
Metal lithography ovens	Solvents, resins	260–400°C
Octyl-phenol mfg.	Phenol	320–430°C
Phthalic anhydride mfg.	Maleic acid, phthalic acid, naphtha quinones, carbon monoxide, formaldehyde	320–340°C
Polyethylene mfg.	Hydrocarbons	260–650°C
Printing presses	Solvents	320°C
Varnish cooking	Hydrocarbon vapors	320–370°C
Wire coating and enamelling ovens	Solvents, varnish	320–370°C

* Temperatures in excess 650°C required to oxidize carbon.
† Reducing atmosphere required.

ovens, for example, it is usual to provide a catalyst inlet temperature of 330°C under starting conditions, allowing a reduction to 204°C when running and the concentrations approach a quarter of the lower flammable limit. A more detailed list of catalyst entry temperatures is given in Table 3.14.

The catalyst temperature is only just above that of the gases being discharged and the temperature rise for the gases is approximately that calculated from the heat release of the reaction.

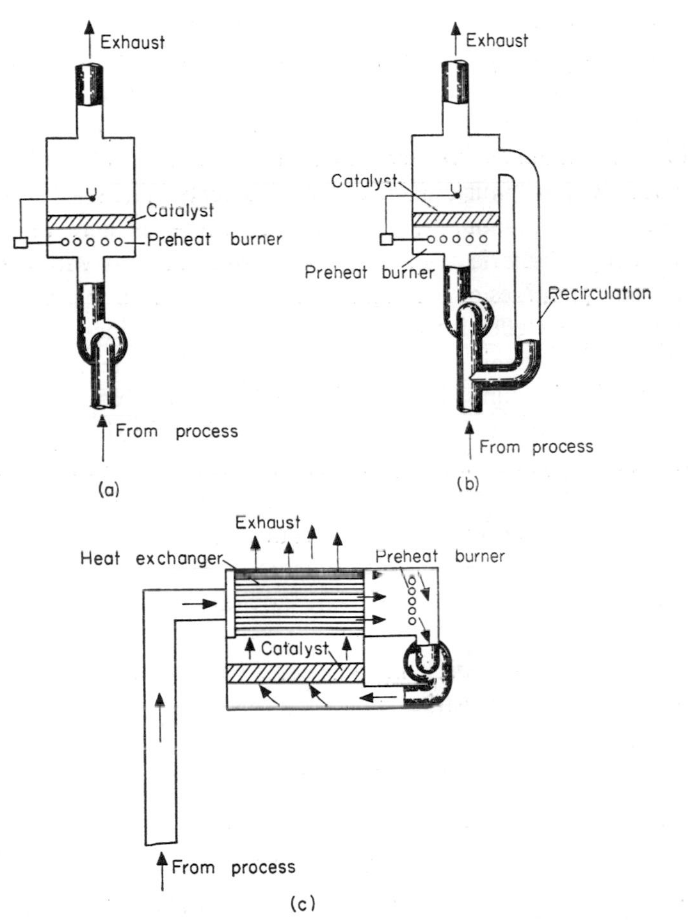

FIG. 3.55. Arrangement catalytic combustion units in gas stream.[618] (a) Basic installation, using preheat burner. (b) Installation with preheat burner, obtaining some preheat by direct recirculation. (c) Installation with preheat burner, obtaining some preheat from the use of a heat exchanger.

The arrangement of the catalyst in the gas stream can be with or without recirculation and with use of the waste gases as the preheating source or to produce steam (Fig. 3.55). For ribbon-type catalyst a face velocity of 2·5 m s^{-1} gives a statisfactory residence time with a 60 mm deep bed. The pressure drop is about 100 Pa for this design.

Catalysts which are supported on large specific surface oxides show reduced activities if they have been heated above 800°C for extended periods. In the case of the γ-alumina this is due to a combination of reduction of surface area, growth of the α-alumina modification, which has inert properties, and the extraction of the alumina with the support material.

The most difficult problem encountered in the use of catalytic combustion is the gradual deactivation or "poisoning" of the catalyst which occurs on extended use, or by the inadvertent introduction of the "poisons" into the gas stream. The deactivation is either due to the chemical interaction of the gases with the catalyst or the coating of the catalyst with a deactivating compound.

If catalysts containing copper oxide as active agent are used, chlorine or hydrogen chloride gas reacts with this forming cuprous chloride. This will sublime at temperatures above 600 °C, markedly reducing catalyst activity. If alumina catalysts are used, gases containing sulphur compounds will react giving sulphates. On the other hand, the interaction of the oxides with sulphur dioxide at 300°C is very limited.

Organic phosphorous compounds, which can be found in aerosols originating from lubricants, will form phosphoric acid, which will cover catalysts with a thin deactivating film. However, small quantities of the organic phosphides and phosphates may make little practical difference to the effective operation of a high specific-area catalyst.[917] Heavy metals, such as lead and arsenic compounds, have a similar effect to the phosphates, by producing thin deactivating films. Deactivation and fouling of catalysts can be caused by dusts. If these are refractory, such as alumina, silica and iron oxide dusts, the deactivation may be permanent although, if sintering is prevented, it may be possible to clean the filter elements and recover some lost activity. Temporary deactivation can be caused by the deposition of fine carbon particles and soot from incomplete combustion in the burner flame. This can be burned off the catalyst elements by raising the temperature to at least 350°C for short periods, but clean flames are advisable as long operating periods are essential.

The heat recovery which is possible in some cases with a catalytic oxidation unit makes this a source of heating in industries where hydrocarbon vapours are produced as an effluent.[713]

The economics of heat recovery will be considered in Chapter 12.

3.11. CATALYTIC OXIDATION AND DECOMPOSITION

Besides catalytic combustion of organic materials, catalytic oxidation in a more restricted sense can be used for the removal of sulphur dioxide from flue gases, while catalytic reduction of nitrogen oxides can be used for the treatment of tail gases from nitric acid plants, where it tends to be more effective than the scrubbing process described in Section 3.7.5.

3.11.1. Sulphur dioxide

It would appear that the simplest method of removal of sulphur dioxide from flue gases is to convert this component by oxidation, to the trioxide, and then absorb this in sulphuric acid. However, in practice there are a number of difficulties associated with the relatively high temperature at which this catalytic oxidation has to be carried out; these are concerned with removing the fly ash from the flue gases at high temperatures before these are passed to the catalyst, and the effective elimination of acid mists before emitting the cleaned gases to the atmosphere.

The oxidation of sulphur dioxide to the trioxide can be described by the simple equilibrium:

$$SO_2+\frac{1}{2}O_2 \rightleftarrows SO_3+94{\cdot}5\ \text{kJ g mole}^{-1}\ (\Delta H).$$

The equilibrium constant for this reaction is

$$K_P = \frac{p_{SO_3}}{p_{SO_2}\cdot p_{O_2}^{\frac{1}{2}}}$$

The equilibrium constant (Table 3.15) shows that below 800 K the equilibrium favours the SO_3 while at flame temperatures (above 1500 K) the equilibrium favours SO_2. The equilibrium has been extensively investigated and reviewed[944] because of the commercial importance of the process for making sulphuric acid. Rate studies and catalyst investigations are also extensive.

3.15. EQUILIBRIUM CONSTANT K_p FOR THE REACTION $SO_2+\frac{1}{2}O_2 \rightleftarrows SO_2$

Temperature		K_p	
°K	°C	Jüngten[415]	Williams[944]
600	327	4780	4015
800	527	53·0	32·3
1000	727	3·67	1·83
1200	927	0·630	0·276
1500	1227	0·114	—

It is important to note that the amount of sulphur trioxide formed in the flame tends to be greater than the equilibrium value for the molecular reaction equilibrium when excess oxygen is present. This increased concentration is probably due to the effect of atomic oxygen in the flame.[352] Homogeneous oxidation of SO_2 by the oxides of nitrogen without the presence of catalysts may also play a part at the lower temperatures (900–1050°C) which can occur in the later stages in the boiler system.[188] Jüngten[415] also reports that steel ducts in an experimental furnace at temperatures of 400–600°C and with excess air may act catalytically, increasing the SO_3 content to the order of 1 per cent of the SO_2 present.

However, to use the conversion of SO_2 to SO_3 as the basis for a method for removing sulphur oxides from flue gases, it is necessary to achieve conversions in excess of 90 per cent. Bienstock, Field and Myers[256] used a simulated flue gas with 0·35 per cent SO_2 and commercial oxidation catalysts as well as one of their own composition. The commercial catalysts, with one exception (vanadia on silica, K_2O promoted-Harshaw, V0204) converted less than 50 per cent. The exception converted up to 85 per cent at 380°C. The special alkalized vanadia catalyst (SiO_2: 39·6 per cent; K_2O: 16·5 per cent; SO_3: 27·3 per cent; V_2O: 7·1 per cent; H_2O: 9·5 per cent) was dried at 430°C and prepared in a granular form. This catalyst converted 97 per cent of the SO_2 at 365°C at flue-gas compositions and would appear to give a reasonable basis for a catalytic conversion process.

A process of this type, the Cat–Ox process (Fig. 3.56), was developed by Pennsylvania Electric Company in conjunction with the Air Preheater Company, the Monsanto Company and Research–Cottrell Incorporated, and is now being marketed by Monsanto.

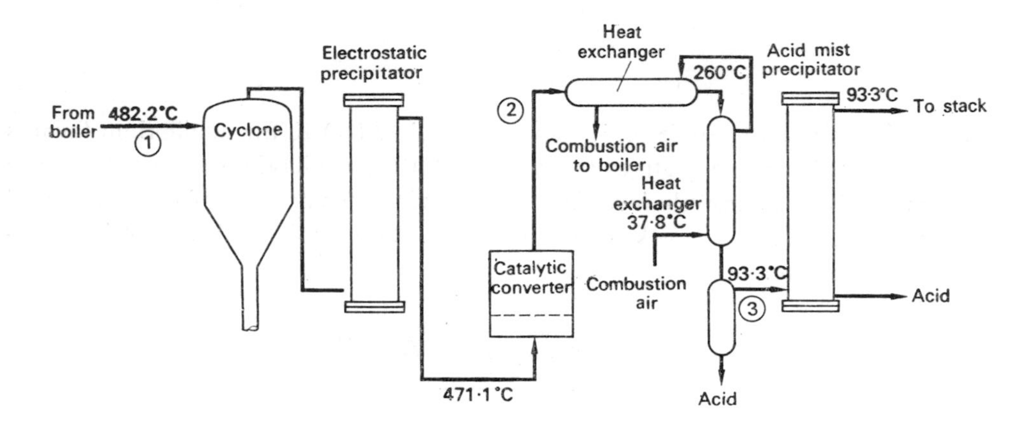

FIG. 3.56. Catalytic oxidation of sulphur dioxide; Monsanto Cat–Ox process flowsheet.[112]

Initial pilot-plant tests were carried out in 1961, with a plant having a capacity of 2550 $m^3 h^{-1}$, [112] and a prototype plant, with a capacity of 40,800 $m^3 h^{-1}$, has since been extensively tested at the 250-MW Portland No. 2 Unit of the Metropolitan Edison Company.[818] The fly ash is removed in an electrostatic precipitator operating at 485°C, which has a designed efficiency of 99·5 per cent. The gas, which contains approximately 2000 ppm sulphur dioxide and 20 ppm sulphur trioxide, is then passed through the thin bed of catalyst at 470°C, and with a conversion efficiency of 90 per cent, the gases now contain 1820 ppm sulphur trioxide and 200 ppm sulphur dioxide. Their catalytic conversion is followed by a fin-tube heat exchanger (boiler economizer) and a Ljungstrom regenerative air preheater, in which the temperature of the gases is further reduced. Following this, the sulphuric acid vapour, which has been formed from the sulphur trioxide and the water vapour in the flue gases, is brought into contact with a stream of cool sulphuric acid in a packed absorption tower, where the gases are further cooled and the acid is heated. The heated acid drains to the bottom of the tower, and is then cooled in a heat exchanger from which the heat is recovered. A substantial amount of acid mist is formed in the cooling and absorption, and these droplets are entrained in the flue gases. They are subsequently eliminated by Brink mist eliminators (see Section 8.5.2) which are fibre packed units with high efficiency at the high throughput rates.

The prototype unit has shown that virtually all fly ash, 90 per cent of the sulphur dioxide and 99·5 per cent of the sulphuric acid produced can be eliminated from the flue gases. The acid has an average concentration of 80 per cent, and this varies slightly with the stack temperature. The prototype had run for over 4000 hr by 1969, and this included continuous operation for 24 days.

An alternative catalytic oxidation process in an advanced stage of development was suggested by Kiyoura[443–444] of the Tokyo Institute of Technology (Kiyoura T.I.T. Process) (Fig. 3.57). In this process ammonia is used mixed with the oxidized SO_2 in the flue gases at

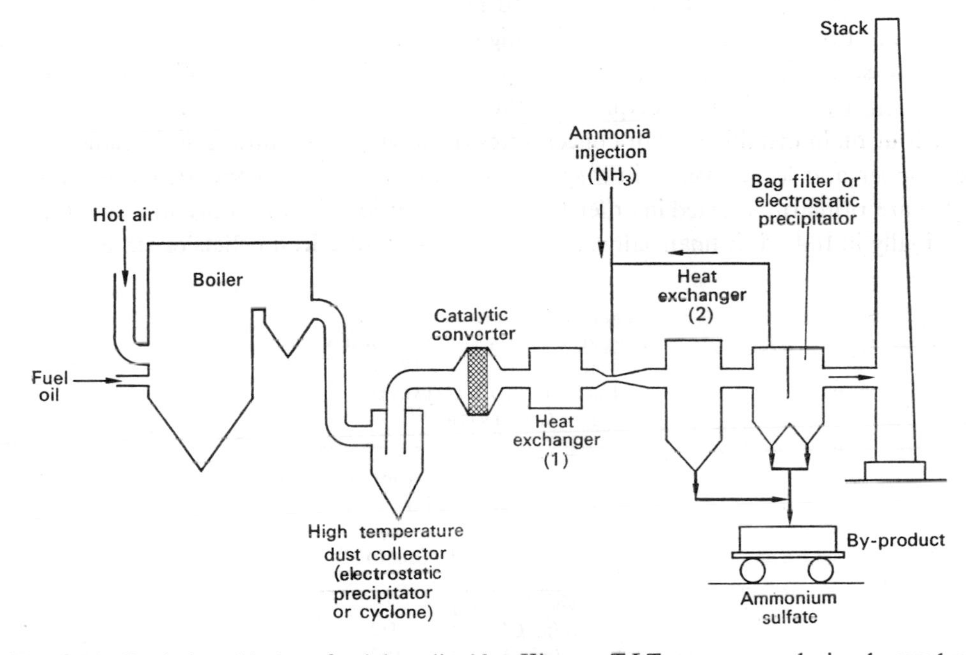

FIG. 3.57. Catalytic oxidation of sulphur dioxide: Kiyoura T.I.T. process producing by-product ammonium sulphate.[443]

220–260°C, to form ammonium sulphate. The air preheater is built as a two-stage unit, and the ammonia injection takes place in a venturi mixer between these stages. After successful bench-scale tests, a 510 $m^3 h^{-1}$ pilot plant has been tested. The plant is connected to a boiler burning a fuel oil with 3·5 per cent sulphur and the hot flue gases are cleaned with ceramic filters before being passed to a vanadium pentoxide catalyst placed in three layers at temperatures of 420–450°C. Conversion is 91–93 per cent at a space velocity of 915 $m^3 m^{-2} h^{-1}$; higher velocities (up to 2100 $m^3 m^{-2} h^{-1}$) were reported for the bench-scale plant. The hot gases are cooled in a high-temperature heat exchanger, preheated ammonia is introduced, and further cooling in a tubular cooler to 140°C takes place. The ammonium sulphate is precipitated in an electrostatic precipitator at 59–63 kV with an efficiency of 97·5 per cent. The overall recovery is, therefore, 90 per cent, and the purity of the ammonium sulphate is 99·2 per cent which approximates to reagent grade material. The pilot plant has operated continuously for several months. It could appear that one advantage of this process is that the product is non-corrosive and so the corrosion problems inherent in the sulphuric-acid product system are partially avoided. The purity of the ammonium sulphate is in contrast to the predictions made earlier by Johswich[408] who felt that the impurities, particularly tarry substances, in the flue gas would act as catalyst poisons for the catalysts which operate below 300°C, as well as contaminating the product.

The scrubbing processes for treatment of waste gases from the roasting of ores, which handle concentrations of approximately 3–8 per cent sulphur dioxide, produce tail gases with concentrations of 0·05–0·2 per cent sulphur dioxide (depending on the feed gases and process conditions) which at the upper end are the concentrations of sulphur dioxide in flue gases of thermal power stations using high sulphur fossil fuels. It is possible that in the future the scrubbing processes, or simple catalytic oxidation for making sulphuric acid, will no longer be considered adequate for the treatment of smelter gases.

The four main metallic ores or concentrates from which sulphuric acid is made are iron, zinc, copper and lead. Iron is in a special category because pyrites (FeS_2) or pyrrhotite (Fe_7S_8) are primarily roasted in order to make sulphuric acid, and only in a few locations (principally in Italy) is it financially attractive to co-produce iron pellets or sinter for metal-

TABLE 3.16. ESTIMATED UPPER IMPURITY LIMITS FOR A CLEANED 7 PER CENT SULPHUR DIOXIDE FEED GAS TO A CATALYTIC CONVERTER[224]

Substance	Approx. limits–dry basis (mg m^{-3})
Chlorides (as Cl)	1·2
Fluorides (as F)	0·25
Arsenic (as As_2O_3)	1·2
Lead (as Pb)	1·2
Mercury (as Hg)	0·25
Selenium (as Se)	50
Total/solids	1·2
Sulphuric acid mist (as 100%)	50

(Arsenic and selenium are likely to be objectionable in the product acid.)

lurgical use. In the case of the other metals, the metal oxide is the primary product, with the sulphur oxide as a by-product. If the gases are used in a conventional contact plant, the optimum sulphur dioxide concentration in the feed gases is 7–7·5 per cent (by volume); at lower concentrations (3·5–4 per cent) the plant is thermally balanced, while at still lower concentrations, external heat is required for the conversion. Metallic vapours and solids, which are carried in the gas stream from a roaster, lead to poisoning of the catalyst and plugging of the bed; so feed gases to the converter have to be carefully purified. Contamination of product acid by arsenic, mercury, lead or selenium originating in the ore also present a problem. The upper limits of impurities which are permissible for feeding the roaster or sinter plant gases to a catalytic converter are listed in Table 3.16.

The waste gases from a conventional acid plant, with a design conversion efficiency of 98 per cent, still contain 0·14 per cent or 1400 ppm sulphur dioxide, which may become unacceptable for future new roasting and sintering plant. This can be improved by using the "Double Catalysis" (Bayer) process,[576] which is sometimes called "Interpass Absorption"[225] which has a conversion efficiency of 99·8 per cent with an optimum concentration of sulphur dioxide in the inlet gases of approximately 9 per cent and a practical lower limit of 7·5 per cent. Thus the residual sulphur dioxide in the tail gases is 150 to 180 ppm, with a corresponding increase in production of sulphuric acid.

3.11.2. Oxides of nitrogen

A very attractive way of removing oxides of nitrogen is to decompose these catalytically to oxygen and nitrogen. The various aspects of this have been extensively reviewed by Bagg.[50] It was first shown by Green and Hinshelwood[319] that platinum at 100–1500°C catalysed the decomposition of nitric oxide. This decomposition was unimolecular and retarded by the presence of oxygen. Later work showed that platinum–rhodium[46] alloys and some base metal oxides, e.g. copper oxide deposited on silica gel,[271] also were effective decomposition catalysts.

Currently only the reduction of tail gases from nitric acid plants, using a platinum or palladium catalyst, together with a fuel gas, is used commercially, with efficiencies over 90 per cent. The reduction to nitric oxide, giving a colourless plume, is in some cases considered adequate. This requires stoichiometric amounts of fuel gas, for which natural gas, coke-oven gas, carbon monoxide, hydrogen and vaporized kerosene have been used. For complete reduction, further fuel gas is needed in order to react with the oxygen as well as with the nitrogen dioxide, if complete reduction is required. The temperature must be below 850°C, and if high concentrations of oxygen are present, a two-stage process must be used to prevent the temperature during reaction exceeding 850°C. Ignition temperatures for the reaction vary between 150°C (for hydrogen or carbon monoxide as a fuel) to 400°C (for methane fuel).

The most suitable catalyst is palladium, supported on the surface of a honeycomb of cells of alumina or mullite.

3.11.3. Carbon monoxide

To prevent deterioration of the vanadium pentoxide catalyst in ammonia manufacture it is necessary to remove any carbon monoxide in the ammonia synthesis gas stream at pressure of 7×10^5–14×10^5 kPa. A supported platinum catalyst is effective for this process at 160°C. Typical results are shown in Table 3.17.

TABLE 3.17. EFFECTIVENESS OF PREFERENTIAL OXIDATION OF CARBON MONOXIDE IN AMMONIA SYNTHESIS GAS OVER A PLATINUM CATALYST AT 160 °C[19]

Gas	Inlet concentration (ppm)	Outlet concentration (ppm)
CO	3000	8
O_2	4500	5
CO_2	900	900
CH_4	4600	4600

Higher CO_2 concentrations result in higher amounts of carbon monoxide in the outlet gases, especially when more oxygen is present.

CHAPTER 4

FLUID RESISTANCE TO PARTICLE MOTION

4.1. INTRODUCTION

The removal of molecules from gas streams, which was discussed in the previous chapter, depends solely on diffusion. On the other hand, the removal of particles relies only partly on their diffusion from the gas stream to a collecting surface or space, and other mechanisms such as gravitational or centrifugal separation, interception and inertial impaction, or the action of electrostatic, thermal or magnetic forces play a more important role.

Essentially, the plant for the removal of particles is a system through which the gas stream passes while the particles are acted on by forces enabling them to leave the gas stream. To be effective, these forces must be sufficiently large to take the particle out of the gas stream during its residence time in the cleaning system. The forces acting give the particles a component of velocity in a direction other than that of the gas stream, and in their movement across the streamlines the particles encounter the resistance of the gas to their movement. The motion of particles in moving fluids has not been studied extensively, although it is certain that particles spin and so have components of lift and sideways shift similar to those experienced by an airfoil. Because of the limited knowledge of fluid resistance under these conditions, it is assumed that the resistance of the moving fluid is the same as that which the particle would experience in moving through a stationary fluid.

The calculation of the fluid resistance to the cross-stream movement of the particle is essential to determining the effectiveness of a particular mechanism in removing the particle from the gas stream. For example, in a gravitational settling chamber, which is the simplest type of plant, being merely an enclosed space through which the gas stream moves, the Earth's gravitational field acting on the particles is opposed by the resistance of the gas to the falling particles. Large particles, which move more quickly, are collected, while smaller ones, which do not settle out during the residence time of the gas stream in the chamber, may escape.

This chapter discusses the calculation of the resistance of a fluid to the movement of particles when these are acted on by forces outside the fluid. Subsequent chapters are concerned with the interaction of the applied external forces with the fluid resistance, and the method of using these calculations to assess the effectiveness of gas-cleaning plant for particle removal.

The simplest calculations concerning gas–particle systems are those for a sphere in steady state motion in a continuous fluid of infinite extent. This will therefore be discussed first, because other types of particles, and other factors can be related to this system. The modifications that have to be considered are:

(i) When the fluid is not continuous.
(ii) When the particles are accelerating.
(iii) When there are walls near to the particle.
(iv) When there are numbers of particles which affect one another.
(v) When particles are not spherical in shape.
(vi) When the fluid is in turbulent motion.

A survey of many of these factors has been published by Happel and Brenner[337] and Torobin and Gauvin[865] which should be referred to for a comprehensive discussion. The following sections will, however, give a brief review of the most important working equations required.

4.2. FLUID RESISTANCE TO SPHERES IN STEADY STATE MOTION

Even for the simplified model of a sphere in steady state motion through a laminar fluid, the relation between fluid resistance and particle speed is very complex. However, the data relating these functions can be presented by a single curve[675] (Fig. 4.1). The abscissa is the

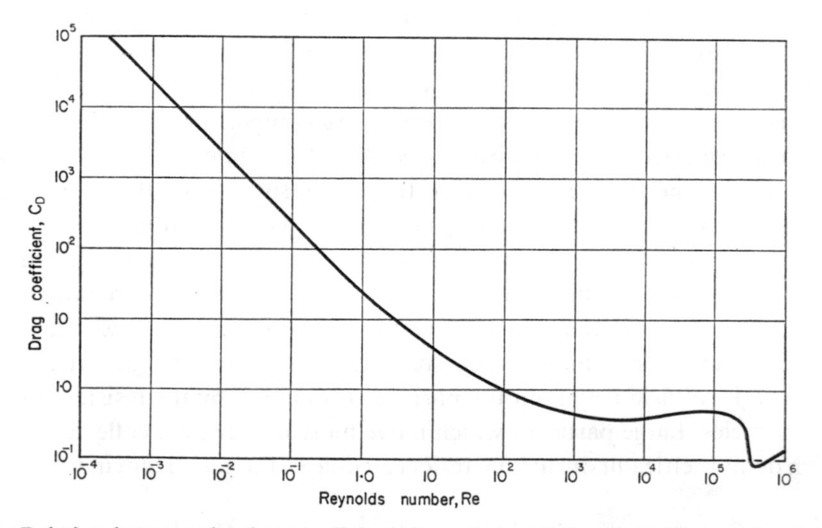

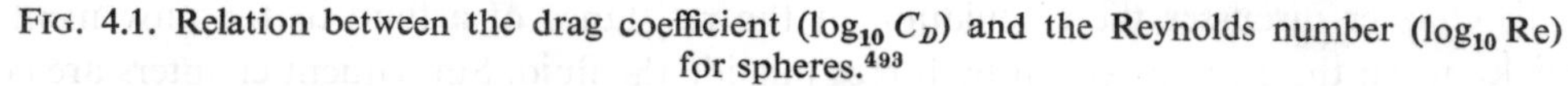
FIG. 4.1. Relation between the drag coefficient ($\log_{10} C_D$) and the Reynolds number ($\log_{10}$ Re) for spheres.[493]

logarithm of a dimensionless function of the relative velocity in the form of the particle Reynolds number

$$\mathrm{Re} = ud\varrho/\mu \tag{4.1}$$

which is the relative velocity u multiplied by a linear dimension for the particle, which is the diameter d, in the case of a sphere, the fluid density ϱ, and the reciprocal of the fluid viscosity, μ.

The ordinate is the logarithm of a function called the drag coefficient, C_D, and is given by

$$C_D = \frac{F}{A \cdot \frac{1}{2}\varrho u^2} \tag{4.2}$$

where F is a function of the fluid resistance, A is the surface area perpendicular to the direction of motion and $\frac{1}{2}\varrho u^2$ is the kinetic energy of one square unit of area of fluid moving past the particle.

In the case of a spherical particle, A is $\pi d^2/4$, and equation (4.2) becomes

$$C_D = \frac{8F}{\pi \varrho u^2 d^2} \tag{4.3}$$

The curve shown in Fig. 4.1 can be divided into four sections, each of which has certain phenomena associated with it as regards flow pattern of the fluid around the particle, and specific formulae can be used to calculate the drag coefficient in each section.

At very low velocities, associated with particle Reynolds numbers up to about 0·1, the flow around the sphere has an up and down stream symmetry (Fig. 4.2a). Elements of fluid

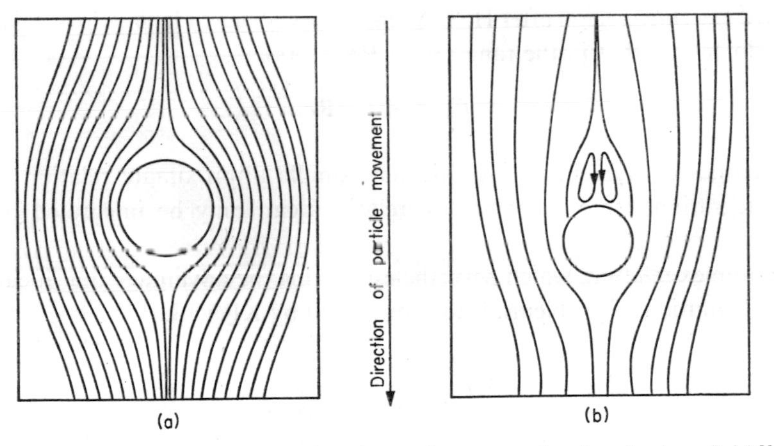

FIG. 4.2. Streamlines around a sphere in steady state motion in a laminar fluid.[204]

(a) Viscous flow region.
(b) Transition region (Re $\approx$ 2).

meeting in front of the particle are slowly accelerated sideways, and the inertia effects are too small to cause a time lag in the closing up of the flow behind it. This is the viscous, streamline or Stokes' law region. For these conditions Stokes[820] deduced that the fluid resistance can be calculated from:

$$F = 3\pi \mu d u. \tag{4.4}$$

This equation was obtained by neglecting the inertia terms in the Navier–Stokes equation for a rigid sphere in an unbounded fluid. From equation (4.4) the drag coefficient for the viscous flow region becomes

$$C_D = \frac{24}{\text{Re}}. \tag{4.5}$$

At Re = 0·05, the fluid resistance calculated from equation (4.4) agrees to within one per cent with the experimental value, but at Re = 1, the calculated fluid resistance underestimates the experimentally determined fluid resistance by about 13 per cent.

At Reynolds numbers somewhat greater than 0·1, the time lag of the elements of fluid closing up behind the particle increases, and the formation of a wake commences. To allow for this, Oseen[476] introduced a modification which partially takes into account the inertia terms in the equations of motion. This gave the equation

$$C_D = \frac{24}{\text{Re}}\left(1+\frac{3}{16}\,\text{Re}\right). \tag{4.6}$$

The drag coefficient given by this equation is 3 per cent greater than the experimental value at Re = 1.

Initially in forming a wake the fluid curls up to form stationary vortex rings (Fig. 4.2b), which grow in size as the velocity of the particle increases. This is the transition region and is often called the *Allen* region after H. S. Allen, who was one of its first investigators. Allen[13] found experimentally that in the range $30 < \text{Re} < 300$

$$C_D \approx 10/\sqrt{\text{Re}}. \tag{4.7}$$

It is sometimes convenient to use this very simple approximate relation in equations involving integrations since more precise relations can only be integrated by numerical methods.

A more complex relation, which nevertheless retains the advantage that it can be used in the analytical solution of problems, is that proposed by Klyachko[448a]

$$C_D = \frac{\text{Re}}{24} + \frac{4}{\text{Re}^{\frac{1}{3}}}. \tag{4.8}$$

This agrees to within 2 per cent with experimental values of C_D in the range $3 < \text{Re} < 400$.

An empirical equation based on a least squares fit of available experimental data has been derived by Sisk.[774] This gives an equation which is valid to within 2 per cent for the extended range of $0{\cdot}1 < \text{Re} < 3500$, and can be readily programmed on a computer. This is

$$C_D = 29{\cdot}6\ \text{Re}^{(0{\cdot}554 \ln \text{Re} - 0{\cdot}983)}. \tag{4.9}$$

More precise values of the drag coefficient for this transition range, which for practical purposes stops at Re = 1000, are based on extensive experimental data. Two of the most useful are those by Schiller and Naumann[730] for the range $0{\cdot}5 < \text{Re} < 800$ and by Langmuir and Blodgett[491] for the narrower range $1 < \text{Re} < 100$. These are:

$$C_D = \frac{24}{\text{Re}}(1+0{\cdot}150\ \text{Re}^{0{\cdot}687}) \quad \text{(Schiller and Naumann)} \tag{4.10}$$

and

$$C_D = \frac{24}{\text{Re}}(1+0{\cdot}197\ \text{Re}^{0{\cdot}63}+0{\cdot}0026\ \text{Re}^{1{\cdot}38}) \quad \text{(Langmuir and Blodgett)} \tag{4.11}$$

The latter equation contains more terms than (4.10) and can therefore be expected to give more exact drag coefficient values within the more limited range. At somewhat greater values than Re = 500, which is at the upper end of the transition region, the vortex rings break away from the body and form an extended wake. This wake is stable above Re = 1000, and the drag coefficient remains approximately constant in the range 0·38–0·5. The fluid resistance is therefore also approximately constant, according to the equation,

$$F = [C_D]_{\text{constant}} A \cdot \tfrac{1}{2}\varrho u^2 . \tag{4.12}$$

This relation was first deduced by Newton, who assumed that the drag coefficient was unity, and this region is usually referred to by his name.

At much greater velocities in the vicinity of $\text{Re} = 2\times 10^5$ the boundary layer of fluid at the front of the sphere becomes unstable, and at a still higher velocity the separation circle moves to the rear of the particle, resulting in a sharp decrease in the drag coefficient from 0·4 to 0·1.[941]

The terminal velocity reached by a particle is the velocity it attains when the fluid resistance is equal to the external force applied to the particle. If this force is G, then from equation (4.2), the terminal velocity u_t is

$$u_t = \sqrt{\left(\frac{2G}{A\varrho C_D}\right)}. \tag{4.13}$$

If the particle is a sphere moving in the viscous flow region, then equation (4.13) becomes

$$u_t = \frac{G}{3\pi d\mu}. \tag{4.14}$$

If the force on the particle is gravity, then

$$u_t = \frac{d^2(\varrho_p-\varrho)g}{18\mu} \tag{4.15}$$

where ϱ_p = density of particle.

In the general case, outside the viscous flow region, if the force on the particle is gravity, then

$$u_t = \sqrt{\left(\frac{4d(\varrho_p-\varrho)g}{3\varrho C_D}\right)}. \tag{4.16}$$

In the transition region, where C_D is a function of the Reynolds number, this equation is difficult to solve except by successive approximation. This problem has been overcome by expressing the Reynolds number in the transition region as a function of $C_D \mathrm{Re}^2$, which does not contain the velocity.[203] If the force is gravity:

$$C_D \mathrm{Re}^2 = 4\varrho(\varrho_p - \varrho)d^3 g/3\mu^2 \tag{4.17}$$

otherwise

$$C_D \mathrm{Re}^2 = 8G\varrho/\pi\mu^2. \tag{4.18}$$

Davies[203] statistically analysed reliable experimental data[36, 510, 539, 577, 736, 941] and obtained two equations:

For moderate values: $\mathrm{Re} < 4$ and $C_D \mathrm{Re}^2 < 134$

$$\mathrm{Re} = \frac{C_D \mathrm{Re}^2}{24} - 2{\cdot}3363 \times 10^{-4}(C_D \mathrm{Re}^2)^2 + 2{\cdot}0154 \times 10^{-6}(C_D \mathrm{Re}^2)^3 - 6{\cdot}9105 \times 10^{-9}(C_D \mathrm{Re}^2)^4. \tag{4.19}$$

For the range of Re: $3 < \mathrm{Re} < 10{,}000$ and $100 < C_D \mathrm{Re}^2 < 4{\cdot}5 \times 10^7$,

$$\log_{10} \mathrm{Re} = -1{\cdot}29536 + 0{\cdot}986(\log_{10} C_D \mathrm{Re}^2) - 0{\cdot}046677(\log_{10} C_D \mathrm{Re}_2)^2 + 0{\cdot}0011235(\log_{10} C_D \mathrm{Re}^2)^3. \tag{4.20}$$

Table 4.1 lists maximum sizes of spheres falling through air in the Earth's gravitational field for which the terminal velocities can be calculated using equation (4.19) and Stokes' law.

TABLE 4.1.[204] THE MAXIMUM SIZES OF SPHERES (IN MICROMETRES) FOR WHICH TERMINAL FALLING VELOCITIES CAN BE CALCULATED BY STOKES' LAW (EQUATION 4.15) AND EQUATION (4.19)[203]

Density of sphere kg m^{-3}	Maximum diameter for Stokes' law			Max. dia. with eqn. (4.15)
	within 10%	within 5%	within 1%	
	μm	μm	μm	μm
200	132	100	57	240
400	105	79	45	191
800	83	63	36	152
1000	77	59	34	141
2000	61	46	27	112
4000	48	37	21	89
6000	42	32	18	78
8000	38	29	17	70
10,000	36	27	15	65
12,000	34	25	15	62
Re	0·82	0·38	0·074	4
$C_D \mathrm{Re}^2$	21·9	9·60	1·80	133·6

The spheres are falling through air at 20°C and 100 kPa pressure ($\mu = 1{\cdot}82 \times 10^{-5}$ PaS, $\varrho = 1{\cdot}205$ kg m^{-3}).

4.3. FLUID RESISTANCE TO ACCELERATING PARTICLES

The two previous sections have discussed the resistance to a particle in steady state motion in a laminar fluid, and the terminal velocity achieved when a particle is acted on by a specific force such as gravity. However, when a particle at rest is acted on by a force it accelerates until it reaches its terminal velocity. If a particle is acted on by a constant force, it has its maximum acceleration at the start, and the acceleration decreases the more closely the particle velocity approaches its terminal value. For a particle to move with constant acceleration, the force on the particle must increase as the speed of the particle increases.

The drag force on an accelerating particle is larger than that on a particle moving with the same velocity in the steady state. Early experimenters accounted for this by assuming an increase in the mass of the particle to an effective value greater than its actual mass. However, the drag force is a function of the acceleration, and so the "added mass" concept is unsatisfactory because it assumes a constant effect. In general terms, the resistance drag, R_A, on a spherical particle accelerating through a resisting medium, is given by

$$\begin{aligned} R_A &= G - ma \\ &= 3\pi\mu\, du + \frac{\pi}{12}\varrho d^3 a + \frac{3}{4} d^3 \sqrt{(\pi\mu\varrho)} \int_0^t \frac{du}{dx} \frac{dx}{\sqrt{(t-x)}} \end{aligned} \tag{4.21a}$$

where G = external force on the particle,
a = particle acceleration relative to the medium,
m = mass of the particle,
t = time of acceleration,
x = distance travelled,
R_A = drag on the accelerating particle.

The first term in equation (4.21a) is the resistance to a sphere in steady state motion in the viscous regime [equation (4.4)], the second term represents the resistance of an ideal fluid to the accelerated motion of a sphere, which is equivalent to an increase in the mass of the particle by half the displaced medium, while the integral term expresses that part of the resistance which is due to setting the medium itself into motion.

The solution of the problem of acceleration in the viscous flow regime using all terms in equation (4.21a) appears not to have been solved, but a solution including the integral term has been carried out graphically by Fuchs[285] for the case of a particle initially at rest accelerating under a constant force. Fuchs compares this solution with the case where the integral term has been omitted, and finds that no appreciable error is introduced, regardless of the size of the particle.

In practice, the problem of accelerating particles is more likely to be important at relative velocities greater than those in the viscous flow regime, and here the method of using a modified drag coefficient, C_{DA}, proves most satisfactory. This is defined by[865]

$$R_A = C_{DA}\, \tfrac{1}{2}\varrho u^2 A \tag{4.21b}$$

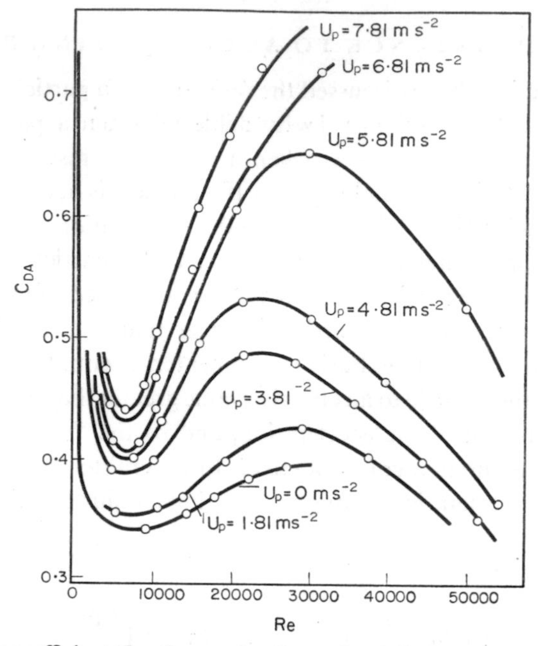

FIG. 4.3. Modified drag coefficient C_{DA} for accelerating spheres falling in air. Redrawn by Torobin and Gauvin[865] based on experiments by Lunnon.[539]

The modified drag coefficient can be found from the extensive data of Lunnon[538] which has been replotted by Torobin and Gauvin.[865] The curves are shown in Fig. 4.3. Other, more recent data, which cover other ranges, have also been reviewed by these workers,[865] but these do not cover the lower Reynolds number which are of importance in gas cleaning plant.

Integrating equation (4.21b), assuming the particle is a sphere and C_{DA} is a constant at an average value, gives

$$t = \frac{d^2(\varrho_p - \varrho)}{3C_{DA}^{\frac{1}{2}}\varrho^{\frac{1}{2}}} \sqrt{\left(\frac{\pi}{2G}\right)} \ln \frac{\frac{2}{d}\sqrt{\left(\frac{2G}{\pi\varrho C_{DA}}\right)}+u}{\frac{2}{d}\sqrt{\left(\frac{2G}{\pi\varrho C_{DA}}\right)}-u} \tag{4.22}$$

and substituting the terminal velocity (equation 4.13) to give the time taken to reach this is

$$t_t = \frac{d^2(\varrho_p - \varrho)}{3C_{DA}^{\frac{1}{2}}\varrho^{\frac{1}{2}}} \sqrt{\left(\frac{\pi}{2G}\right)} \ln \frac{\sqrt{C_D}+\sqrt{C_{DA}}}{\sqrt{C_{DA}}-\sqrt{C_D}} \tag{4.23}$$

Integrating again yields the distance travelled in time t:

$$x = \frac{4d(\varrho_p - \varrho)}{3C_{DA}\varrho} \ln \cosh \left\{ \frac{3\sqrt{(2C_{DA}\varrho G)}}{2d^2(\varrho_p - \varrho)\pi^{\frac{1}{2}}} \cdot t \right\}. \tag{4.24}$$

Since terminal velocity is an asymptotic value and cannot be reached in practice a value such as 99 per cent of the terminal velocity has to be used. Thus, in the solution of equation (4.23) the value of C_{DA} must be based on such an approximation, and an average value from experimental data is used.

The distance travelled during the time to reach 99 per cent of the terminal velocity can then be calculated by substituting the time given by equation (4.23) in (4.24).

Simpler solutions than those given in (4.23) and (4.24) have been obtained by Fuchs,[285] who neglected the second and third terms in (4.21a). Thus, it is possible to write

$$\frac{du}{dt}+\frac{u}{\tau}-g = 0 \tag{4.25}$$

where $\tau = m/3\pi\mu d$, which, for a spherical particle, is given by

$$\tau = d^2(\varrho_p-\varrho)/18\mu.$$

Integrating (4.25), and taking as one boundary condition a particle starting from rest and accelerating under gravity,

$$u = \tau g\{1-\exp(-t/\tau)\}, \tag{4.26}$$

and $$x = \tau gt+\tau^2 g\{\exp(-t/\tau)-1\}. \tag{4.27}$$

If a particle moves under a constant force, other than gravity, G, then G/m must be substituted for g, and the equation for τ without the buoyancy correction used.

In practice, the calculation is required to yield the cross-stream distance travelled by a particle during the gas stream residence time in the collection system. Assuming that the force applied to the particle is known, as well as the physical properties of the particle and the gas stream, the time and distance travelled by the particle to reach 99 per cent of the terminal velocity can be found. If the time taken to reach 99 per cent of the terminal velocity is greater than the gas stream residence time, then the cross stream distance travelled by the particle can be found by integration of equation (4.22) between the time limits set by the gas stream residence time. If this time to reach 99 per cent of the terminal velocity is less than the gas stream residence time, the distance travelled to reach this velocity must be subtracted from the total cross stream distance, and it can then be assumed that the remaining distance is travelled by the particle at its terminal velocity.

For particles smaller than 10 μm a velocity approaching the terminal value is attained in a very short distance with the forces that are used in normal gas cleaning plant (centrifugal, electrostatic, thermal, etc.) and the effects during acceleration can generally be neglected.

4.4. FLUID RESISTANCE IN A NON-CONTINUOUS MEDIUM

When particles are very small, that is of the order of the mean free path of gas molecules, or smaller, then the assumption that a gas behaves as a continuous medium with respect to the particles is no longer valid. Under these circumstances particles tend to move more quickly than is predicted by the classical theories of Stokes and others which assume a continuous medium. To allow for this "slip", Cunningham[190] calculated a correction based on the kinetic theory of gases, the form of which has been retained in the empirical equations generally used. The other important theoretical studies of the movement of particles much smaller than the mean free path were made by Epstein.[243]

In an analysis of the effect of discontinuity in the tangential velocity of molecules at the surface of particles, Epstein showed that the slip correction C could be found from

$$C = 1+\frac{2\lambda}{d}\{0{\cdot}7004(2a-1)\} \tag{4.28}$$

where C = "Cunningham" correction factor,

λ = mean free path of gas molecules based on the Chapman–Enskog equation

$$\lambda = \mu/0{\cdot}499\varrho\bar{u}, \tag{4.29}$$

$\bar{u}$ = mean molecular velocity,

$$= \sqrt{(8RT/\pi M)}, \tag{3.3}$$

a = coefficient of diffuse reflection (Millikan) or accommodation constant (Epstein),

when $a = 0$, all collisions are perfectly elastic,

when $a = 1$, all collisions are diffuse.

In an actual case of oil droplets in air, a is 0·895, while for a number of other spherical droplets and solid particles $0{\cdot}88 < a < 0{\cdot}92$.[285] Thus $a = 0{\cdot}90$ can be used in most cases.

The most precise practical calculation for the slip correction in the viscous flow regime is by using an empirical correlation of Davies[203] based on weighted averages of experimental falling speeds.

This is:

$$C = 1+\frac{2\lambda}{d}\{1{\cdot}257+0{\cdot}400\exp(-1{\cdot}10d/2\lambda)\} \tag{4.30}$$

The changes with pressure and temperature in the Cunningham Correction can be calculated as functions in changes in viscosity and molecular velocity. The latter changes as the square root of the absolute temperature [equation (3.3)], while the changes in viscosity with temperature can be calculated from the Sutherland equation and the values given in Appendix A.8.1. The effect of pressure on viscosity can be estimated from the Enskog relation for a hard sphere gas[329]

$$\frac{\mu_p}{\mu_0} = \frac{B}{V}\left(\frac{1}{y}+0{\cdot}8+0{\cdot}76ly\right) \tag{4.31}$$

where μ_p/μ_0 is the ratio of viscosities at the pressure p and 101·3 kPa,

B = second virial coefficient of the gas

$$= 2\pi N\sigma^3/3, \tag{4.32}$$

σ = molecular diameter,

$$y = \frac{B}{V} + 0{\cdot}625\left(\frac{B}{V}\right)^2 + 0{\cdot}2869\left(\frac{B}{V}\right)^3 + 0{\cdot}115\left(\frac{B}{V}\right)^4, \tag{4.33}$$

V = molar volume of the gas.

The change in dinamic viscosity with density for CO_2 is given in Table 4.2, and shows the limited influence of this parameter. The effect of temperature on very small particles, however, has a large effect on the Cunningham Correction. Calculated values of this for 0·01 to 10-μm particles, from 0° to 1600°C are shown in Table 4.3.

TABLE 4.2. VARIATION OF VISCOSITY OF CO_2 WITH GAS DENSITY

Gas density (kg m^{-3})	2·0	10·0	20·0	30·0	40·0	50·0
Viscosity ratio μ_p/μ_0	1·006	1·008	1·020	1·035	1·053	1·074

TABLE 4.3. VALUES OF CUNNINGHAM CORRECTION FACTOR FOR PARTICLES FROM 0·01 TO 10 μm AND TEMPERATURES FROM 0° TO 1600°C

Temperature (°C)	Air viscosity $\mu \times 10^6$ Pa s	Particle diameter (μm)			
		0·01	0·10	1·0	10·0
0	17·04	20·15	2·64	1·149	1·015
200	25·85	39·84	4·58	1·299	1·0297
400	32·86	59·89	6·55	1·457	1·0450
600	38·80	80·43	8·59	1·626	1·0606
800	44·05	100·99	10·64	1·801	1·0761
1000	48·76	121·5	12·69	1·982	1·0918
1200	53·10	142·4	14·77	2·171	1·1076
1400	57·13	163·3	16·85	2·363	1·1234
1600	60·90	184·2	18·94	2·557	1·1393

The modified Stokes law equation, which is usually referred to as the Stokes–Cunningham equation, is

$$F = 3\pi\mu du/C. \tag{4.34}$$

The slip correction is less than 1 per cent for 20-μm particles (density 1) in ambient air, about 5 per cent for 5-μm particles, $16\frac{2}{3}$ per cent for 1 micron particles and almost 300 per cent for 0·1-μm particles.

Outside the viscous flow region, no accurate experimental data are available, but approximate measurements by Benarie[67] indicate that Cunningham corrections calculated from equation (4.30) overestimate the slip correction by at least 0·2 per cent.

4.5. FLUID RESISTANCE TO PARTICLE MOTION IN A BOUNDED FLUID

In some types of particle collection equipment, such as settling chambers, cyclones or electrostatic precipitators, the particles are negligibly small compared with the dimensions of the plant. In other types, however, such as compressed felt filters or pebble beds with fine granules, the inter-fibre or pebble distances are sufficiently small to make the fluid passing through the filter behave as one which has one or more effective boundaries. These can increase the fluid drag on a particle moving through the filter. It should however be noted that the present theories of filtration (Chapter 7) do not include this factor. A detailed mathematical analysis of the motion of a particle in a bounded fluid is given by Happel and Brenner.[337]

A finite boundary to the fluid in which the particle is moving will have two effects. When the outward movement of fluid thrust aside by the particles is stopped by the boundary, a return flow of fluid is produced, and as the streamlines around the particle are distorted by the boundary this also acts on the particle.

The boundary effects depend on the type of boundary. Theoretical considerations or experimental work have established factors for modifying the Stokes law equation (4.4) for the following cases:

(i) Particle near a single wall.
(ii) Particle between two parallel walls.
(iii) Particle moving along the axis of an infinitely long cylinder.

The fluid resistance near a boundary, F_W, can be calculated by dividing the Stokes law fluid resistance, F, by the boundary correction factor K,

$$F_W = F/K. \tag{4.35}$$

In the three cases, the correction factor is given by the following:

(i) Sphere moving parallel to an infinite plane wall of infinite extent, at a distance $l/2$ from the wall[530]

$$K = 1 - \frac{9}{16}\frac{d}{l}. \tag{4.36}$$

(ii) Sphere moving between and parallel to, two equidistant walls, separated by a distance l[253 335]

$$K = 1 - 1{\cdot}004d/l + 0{\cdot}418(d/l)^3 + 0{\cdot}21(d/l)^4 - 0{\cdot}169(d/l)^5. \tag{4.37}$$

This equation can be used when the ratio d/l is less than $\frac{1}{20}$, but for larger d/l ratios, the correction equation (4.37) gives an underestimate of the fluid resistance. For examples when d/l is $\frac{1}{2}$, then the fluid resistance is about twice that calculated from equation, (4.35) and (4.37).

While the sphere remains fairly close to the centre line between the walls, the correction does not vary very much. When the sphere moves towards one wall, and its effect increases, the effect of the other wall is reduced by a similar amount. However, when a particle approaches fairly closely to one wall, the fluid resistance does increase to an appreciable extent. For example,[253] it has been calculated that when a sphere is only $\frac{1}{8}l$ from one of the walls, the correction factor is

$$K = 1 - 1{\cdot}305d/l \tag{4.38}$$

which gives an increase in the fluid resistance of about 30 per cent compared with the sphere moving at the centre between the walls.

(iii) Sphere moving along the axis of an infinitely long cylinder with diameter l[254]

$$K = 1 - 2{\cdot}104d/l + 2{\cdot}09(d/l)^3 - 0{\cdot}95(d/l)^5. \tag{4.39}$$

This equation has been confirmed experimentally for small values of d/l, where the correction is not very large.[510]

For d/l values up to 0·25, the correction factor[545]

$$K = [1 + 2{\cdot}25d/l + 5{\cdot}06_2(dl)^2]^{-1} \tag{4.40}$$

agrees with experimental evidence, while for larger d/l ratios, experimental curves such as those shown in Fig. 4.4 can be used.[269]

When particles are not spherical,[641] the correction factor to be used is the same as for spheres with equivalent diameters (Section 4.8).

At velocities greater than those which occur in the streamline region, Faxén,[253, 254] the author of some of the major theoretical developments (equations 4.37, 4.38 and 4.39), has

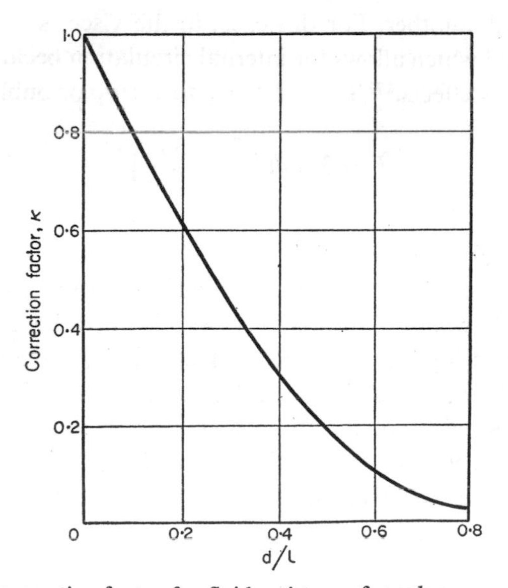

FIG. 4.4. Experimental correction factor for fluid resistance for spheres moving along the axis of a cylinder (after A. W. Francis).[269]

obtained a very complex relation, discussed by Liebster,[510] which is applicable to particles for which d/l is less than 0·05. At very high velocities (Re about 10^4) the available experimental evidence indicates that wall effects are negligible.[538]

4.6. FLUID RESISTANCE WHEN A NUMBER OF PARTICLES ARE PRESENT

Almost invariably in the removal of particles from gas streams, there are large numbers of particles, so the equations for the fluid resistance to the movement of single particles have to be modified to allow for the influence of particles on one another. Particle interactions become appreciable at quite low concentrations, so that a particle volume concentration (the ratio of particle volume to total volume) of 0·002 will increase the fluid resistance to particle movement by about 1 per cent.

The movement of a group of particles in an unbounded fluid results in movement of fluid around the group. When the particles are sufficiently close together, the fluid between the particles moves with the particles, and the group can be considered to be a cloud. When there are walls enclosing the particle group or the particles are sufficiently far apart, then the fluid will also move between the particles. In a practical case there is likely to be some movement of particles as cloud groups and other intermediate cases of temporary groups and also individual particles.

The problem is so complex that so far only partial solutions have been obtained for the prediction of cloud movement and hindered movement effects. In general, clouds tend to move more quickly than individual particles, while groups bounded by walls tend to move more slowly than single particles.

It has been suggested[142] that clouds in an unbounded fluid behave similarly to droplets of one fluid moving through another. For this case, in the viscous-flow region, a correction factor has been calculated which allows for internal circulation because of the viscous drag, but neglects surface energy effects.[475] The resistance to a drop or bubble is given by

$$F = 3\pi\mu du\left(\frac{2\mu+3\mu_d}{3\mu+3\mu_d}\right) \tag{4.41}$$

where μ_d is the viscosity of the droplet fluid.

In the case where the viscosity of the droplet fluid is the same as the surrounding fluid, the correction factor is $\frac{5}{6}$; where the droplet viscosity is much lower than the surrounding fluid (i.e. a gas bubble in a liquid) the correction factor is $\frac{2}{3}$; while for a droplet with very high viscosity compared with the fluid, the extreme case for which is a rigid sphere, the correction factor becomes unity, and equation (4.41) reduces to the simple Stokes' form.

Therefore, if it is assumed that the particle cloud is spherical and that the viscosity within the cloud is the same as in the surrounding fluid, then the fluid resistance to the cloud, F_c, is

$$F_c = \tfrac{5}{2}\pi\mu d_c u \tag{4.42}$$

where d_c is the diameter of the cloud.

The assumption of equal viscosities cannot be justified, particularly when there are ranges of sizes of particles present, and the smaller ones effectively constitute part of the fluid surrounding the larger ones in the cloud. The viscosity of a suspension μ_c is given by[126, 700]

$$\mu_c = \mu(1-c)^{-k} \tag{4.43}$$

where c = volume concentration,
volume of particles/total volume of suspension,

k = constant = 2·5 for spheres.[232]

At low volume concentrations, equation (4.43) simplifies to

$$\mu_c = \mu(1+kc). \tag{4.44}$$

A model for clouds of large and small particles together does not appear so far to have been used for the calculation of fluid resistance for particle groups.

When fluid flows between the particles in a bounded system, the resistance to movement of the particles depends on whether the particles retain their original orientation due to certain inter-particle forces or whether there is a tendency for the particles to align themselves. Considerable experimental work, particularly in connection with fluidization,[508, 684] has shown that the equation

$$F_c = F(1-c)^{-4 \cdot 65} \tag{4.45}$$

can be used to calculate the resistance to the movement of a particle group in a bounded fluid when there are no particle interactions. At low volume concentrations this simplifies to

$$F_c = F(1+4 \cdot 65c), \tag{4.46}$$

a relation which had been suggested earlier by Hawksley.[347]

In cases where there are particle interactions, as in the case of flocculated suspensions, the relation

$$F_c = F(1-c)^{-6 \cdot 875} \tag{4.47}$$

may be expected to apply, in analogy with equation (4.45). The form of this equation for low concentrations has been derived elsewhere as[142, 347]

$$F_c = F(1+6 \cdot 875c). \tag{4.48}$$

The constant 6·875 is based on the arrangement the particles take up relative to one another and assumes that surrounding particles may take all positions around a particle with equal probability.

The most satisfactory theoretical approach to the calculation of hindered motion velocities in a suspension is that of Richardson and Zaki[685] who suggest two models for the settling of equal-sized spheres. In both, the particles are arranged as centres of hexagons of fluid (Fig. 4.5a) and in one[545] the vertical distances between the particles are the same as the

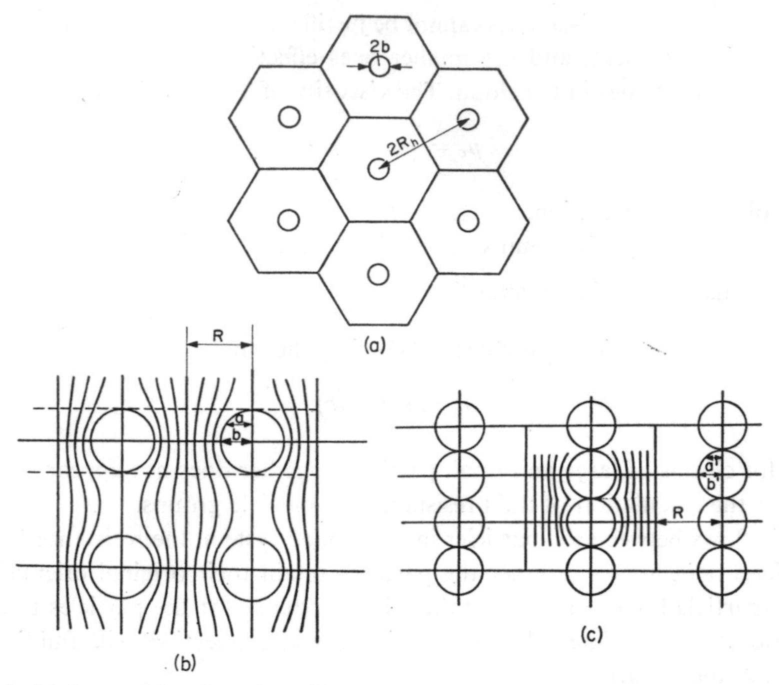

FIG. 4.5. Models for particle orientation with respect to one another for hindered settling.[685] (a) Hexagonal pattern (horizontal section). (b) Particle configuration I at 10 per cent volumetric concentration. (c) Particle configuration II at 10 per cent volumetric concentration.

horizontal distances (Fig. 4.5b) while the other[347] arranges the particles in adjacent horizontal layers (Fig. 4.5c) so as to offer the minimum resistance to the fluid.

The volumetric concentration for both configurations can be calculated from the geometry of the systems. In configuration I

$$c = \frac{\pi}{3\sqrt{(3)}}\left(\frac{b}{R_h}\right)^3 \tag{4.49}$$

and in configuration II

$$c = \pi b^2/3\sqrt{(3)}R_h^2, \tag{4.50}$$

where b is the radius of the spheres and R_h is half the distance between centres. Boundary conditions were simplified by assuming that each sphere was surrounded by a cylinder and not a hexagonal prism of fluid, and the resultant shearing force at the surface of the sphere was calculated. The resultant equation gave a correction factor, β, for the drag on a particle surrounded by others, compared with a single sphere, in terms of the sphere radius, the radius of the cylinder of equal cross-sectional area to the hexagonal section, and an elementary ring on the sphere surface. The correction factor equation was evaluated for both configurations I and II, and the solution, together with experimental data and equation (4.48) is plotted in Fig. 4.6.

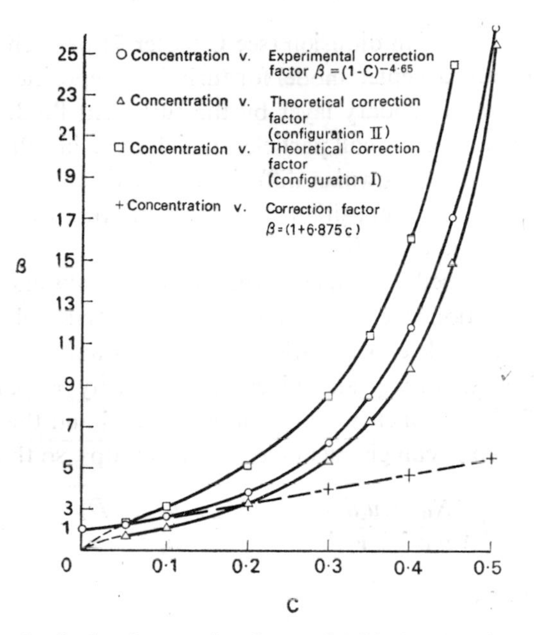

FIG. 4.6. Experimental and calculated correction factors for hindered settling, "β". The calculated corrections are based on configurations I and II. (Fig. 4.5.)[684]

The curves indicate that configuration II, which implies alignment of the particles in rows, agrees fairly well with experimental work at high concentrations (greater than $c = 0{\cdot}2$). At very low concentrations the assumption of an average pressure gradient made in the derivation of the correction factor is not entirely valid, as it tends to zero, and gives the trend to $\beta = 0$ rather than $\beta = 1$. At higher concentrations this pressure gradient is finite and the result follows closely on the experimental curve.

4.7. DEPOSITION OF PARTICLES FROM TURBULENT FLUIDS[449]

In the preceding sections the tacit assumption has been made that the flow in a duct, pipe, or gas cleaning system is laminar, although, in practice, it is frequently turbulent. The turbulence structure in pipes has been studied by hot wire anemometry, and can be divided into three regions:

(1) The turbulent central region, in which energy diffusion is the important phenomenon.
(2) The region near the wall, where there is viscous flow, and the production and the diffusion of turbulence eddies and viscous action play equally important roles.
(3) The intermediate region in which the energy obtained by diffusion is dominated by the local rate of change of turbulent energy production (buffer region).

Various workers[213, 274, 309, 617] have shown that the rate of deposition of particles from a turbulent fluid is far greater than can be explained by consideration of gravitational, thermal or electrostatic forces, Brownian diffusion (see Chapter 7) or such fluid dynamic forces as particle spin. The generally accepted model for turbulent deposition is that the particles are carried to the edge of the boundary layer by the turbulent fluid, and are then projected through the laminar layer. Very small particles, which have insufficient inertia to reach the wall, may be carried there by Brownian diffusion, but this latter mechanism makes only a very small contribution to the deposition rate, for a mixture of particles of which only a small fraction is in the sub-micron-size range.

Dimensional analysis is used to correlate experimental results and obtain useful equations for calculating the deposition rate,[449] which is called the particle flux, N_0. The other appropriate variables are the particle concentration (number per unit volume) C, particle diameter d and density ϱ_p, fluid density ϱ and kinematic viscosity ν, particle terminal velocity u_t, the Brownian diffusion coefficient $\mathcal{D}_B$ [equation (7.22)] and the fluid shear stress at the wall τ_0. These nine variables can give six independent groups, so that

$$f\left(\frac{N_0}{u_\tau c}, \frac{u_\tau d}{\nu}, \frac{\varrho_p}{\varrho}; cd^3, \frac{u_t}{u_\tau}, \frac{\mathcal{D}_B}{\nu}\right) = 0 \tag{4.51}$$

where u_τ = shear velocity

$$= \sqrt{\frac{\tau_0}{\varrho}}. \tag{4.52}$$

τ_0 can be calculated either from the Blasius formula[309] for Re $< 10^5$

$$\tau_0 = 0{\cdot}0225\varrho u^2\left(\frac{\nu}{UD}\right)^{\frac{1}{4}} \tag{4.53}$$

where D = pipe or duct diameter,
U = average gas velocity,

or approximately from the Fanning friction factor, f, so that

$$\tau_0 = U^2 f\varrho/8; \tag{4.54}$$

f can be obtained from Fig. A.5.4, p. 550, or can approximately be obtained from

$$f = 0{\cdot}288\,\text{Re} - 0{\cdot}242. \tag{4.55}$$

The first group, $N_0/u_\tau c$, is the ratio of the rate of transport of particles towards and along the wall, $u_\tau d/\nu$ is a particle Reynolds number, ϱ_p/ϱ is a relative particle/fluid density, cd^3 is a measure of particle volume, u_t/u_τ is a measure of the terminal velocity to shear velocity, and describes the effect of an external force, and $\mathcal{D}/\nu$, which is the inverse of the Schmidt number (equation 7.26), is the ratio of the molecular transport coefficients of mass and momentum. If gravity or external forces and concentration effects are neglected, then equation (4.51) simplifies to

$$f\left(\frac{N_0}{u_\tau c}, \frac{u_\tau d}{\nu}, \frac{\varrho_p}{\varrho}, \frac{\mathcal{D}_B}{\nu}\right) = 0. \tag{4.56}$$

If, furthermore, Brownian diffusion is negligible

$$f\left(\frac{N_0}{u_\tau c}, \frac{u_\tau d}{\nu}, \frac{\varrho_p}{\varrho}\right) = 0 \quad (4.57)$$

While, under other circumstances, the grouping involving Brownian diffusion, $\mathcal{D}_B$, may be more important than the relative density

$$f\left(\frac{N_0}{u_\tau c}, \frac{u_\tau d}{\nu}, \frac{\mathcal{D}_B}{\nu}\right) = 0 \quad (4.58)$$

Friedlander,[274] using the experimental work of Lin, Moulton and Putnam[513] together with the assumption that the projection velocity is $0{\cdot}9u_\tau$, showed that, for $S_* < 5$,

$$\frac{N_0}{u_\tau c} \doteqdot \frac{S_*^2}{1525} \quad (4.59)$$

where $S_* = Su_\tau/\nu$, and the stopping distance, S, is similar in sense to ψ [equation (7.10)].

For a spherical particle in the Stokes regime

$$S_* = 0{\cdot}05\left(\frac{u_\tau d}{\nu}\right)^2 \frac{\varrho_p}{\varrho}. \quad (4.60)$$

Substituting in equation (4.59) gives

$$\frac{N_0}{u_\tau c} = 1{\cdot}64 \times 10^{-6}\left(\frac{\varrho_p}{\varrho}\right)^2\left(\frac{u_\tau d}{\nu}\right)^4 \quad (4.61)$$

which is in agreement with equation (4.57) obtained from the dimensionless analysis.

Instead of the simple dimensionless stopping distance S_* Davies[209] has plotted experimental points using (S_*+r_*) as a parameter, where r_* is the dimensionless particle radius, $r_* = ru_\tau/\nu$,

$$\frac{N_0}{u_\tau c_0} = \frac{(S_*+r_*)^2}{600} \quad (4.62)$$

where c_0 = centre line particle concentration.

These equations apply to particles in the sizes ranging from 0·5 μm to 50 μm.

Davies[210] has also predicted that a minimum deposition rate occurs at a particle size where the turbulent and Brownian deposition rates are approximately the same and this has been observed experimentally[916] to occur at particle diameters between 0·5 and 2 μm somewhat below Davies' predicted minimum of 3 μm.[210]

The previous equations (4.59, 4.61, 4.62) can be used to calculate deposition rates, but they are not applicable unless the following conditions are strictly adhered to:

(1) the flow is fully developed;
(2) significant external forces, such as gravity or electrostatic fields are not present;

(3) there is no effective re-entrainment from the surfaces on which the particles have been deposited;
(4) the surfaces are smooth, and concentrations are low.

The use of the equations is best illustrated by an example.

EXAMPLE. An aerosol consisting of 2·5-μm rad. particles, density $2{\cdot}6 \times 10^{+3}$ kg m^{-3} (representing fine quartz) is flowing at 6 m s^{-1} along a 250-mm diameter smooth tube in a current of air at 20°C (density 1·2 kg m^{-3}; kinematic viscosity $1{\cdot}5 \times 10^{-5}$ m^2 s^{-1}).

The Reynolds number $R_e = UD/\nu = 10{,}000$.

The Fanning friction factor, f, is 0·0315, and, using equations (4.52) and (4.54),

$$u_\tau = U\sqrt{\left(\frac{f}{8}\right)} = 0{\cdot}376 \text{ m s}^{-1}.$$

Then substituting in equation (4.60), $S_* = 1{\cdot}7$. Thus, using equation (4.59), $N_0/u_\tau c = 1{\cdot}86 \times 10^{-3}$, which could also have been obtained by substituting directly in equation (4.61). So $N_0/c = 0{\cdot}00104$ m s^{-1}. If $c = 10^8$ particles m^{-3}, then $N_0 = 1{\cdot}04 \times 10^5$ particles m^{-2} s^{-1}.

4.8. FLUID RESISTANCE TO NON-SPHERICAL PARTICLES

While small droplets and fume particles formed from condensing vapours tend to be spherical, those particles formed by crystallization or by milling tend to have other shapes. The equations in the previous sections apply only to spheres and have to be modified when they are to be used for non-spherical particles. Furthermore, not only do shape factors have to be considered, but also the orientation of the particle and whether this orientation changes during particle translation.

Because gas-cleaning-plant calculations are concerned with the aerodynamic behaviour of the particles, the most useful particle-size data is obtained in the same flow region as is used in the plant by methods based on aerodynamics, such as sedimentation or elutriation. The particle "size" is based on the fluid-resistance equations and is expressed as the diameter of a sphere with the same fluid-resistance characteristics as the particle and the same density. This is the *drag diameter*, and can be substituted for the diameter of the sphere in the fluid-resistance equations for spheres in the earlier sections.

If the drag diameter is not known for the flow region required by the calculations, or if particle sizes have not been measured by some other means, such as those depending on particle geometry (screening, microscopic sizing), then the calculation of the fluid resistance becomes difficult, and requires a detailed knowledge of the behaviour of non-spherical particles.

Particle orientation depends on the flow region. For viscous flow it has been theoretically predicted[288] that particles with three mutually perpendicular planes of symmetry will retain their initial orientation, while those with only two planes of symmetry will take up a preferred orientation with the lines of the intersection of the planes in the direction of flow. In agreement with these predictions it has been observed that isometric particles (cubes, tetrahedra, octahedra) and some non-isometric particles (cylinders, parallel pipeds) do retain their initial orientation,[359, 641] while round discs[736] and triangular laminae[904] take up the preferred orientation. It has also been observed, however, that *Bacilli subtilis* spores, which are prolate spheroids 1·38 μm long and 0·74 μm diameter, tend to move with their longest axis in the direction of motion.[318]

Outside the viscous flow range (Re $>$ 0·05) there is a tendency for particles to take up a preferred orientation with the largest projected area perpendicular to the direction of movement. The preferred orientation has been definitely established for tetrahedra and cubes by Re = 10, and for other shapes by Re = 20. At higher Re values, between 70 and 300, particle instability sets in. Some particles wobble from side to side while others spin; rods spin about their axis while cubes and discs sideslip; tetrahedra tend to follow a spiral path.[348] The side to side oscillation of discs such as falling leaves is a common observation.

Besides particle orientation two other factors must be introduced in the fluid resistance equations for spheres if equations of the same form are to be used for non-spherical particles. These are a linear dimension equivalent to the diameter of the sphere, and a correction factor based on the surface area of the particle to adjust the surface area term in equation (4.2). Incorporating both these factors and dependent on the aerodynamic behaviour of the particles is the *drag diameter* which was introduced previously. When fluid resistance is to be deduced from the geometry of the particles, then the two factors must be considered separately. The equivalent diameter is defined in terms of either the particle surface area, volume or projected area, while the area-correction terms are dimensionless ratios called "shape" factors. The most useful of these were introduced by Wadell[894] and are the sphericity Ψ, which is excellent as a correlation factor for non-spherical particles, and the circularity χ.

The definitions of the equivalent diameters and shape factors are summarized in Table 4.4.

In the *viscous flow region*, the drag diameter can be deduced theoretically[204] for ellipsoidal particles. The drag diameter in each case depends on the orientation, which may be the

TABLE 4.4. EQUIVALENT DIAMETERS AND SHAPE FACTORS

Name	Symbol	Definition
Surface diameter	d_s	Diameter of sphere with the same external surface area as the particle
Volume diameter	d_v	Diameter of sphere with same volume as the volume of the particle
Area diameter	d_A	Diameter of circle with same area as projected area of particle
Drag diameter	d_e	Diameter of sphere with same resistance to motion as the particle in a fluid of same viscosity and at the same velocity
Sphericity	Ψ	Ratio of surface area of sphere with same volume as the particle to the actual surface area of the particle
Circularity	χ	Ratio of the circumference of a circle having same cross-sectional area as the irregular particle to the actual perimeter of the irregular particles cross-section

ellipsoidal particle moving in line, or at right angles to its axis of revolution, as shown in Fig. 4.7. In the first case ellipsoidal particles have a major axis, a, and two equal minor axes, b and c, which can be expressed as fractions, the major axis by using a multiplier, n. The other cases are variants of this. The results are graphed for four cases, these being shown in Fig. 4.8.

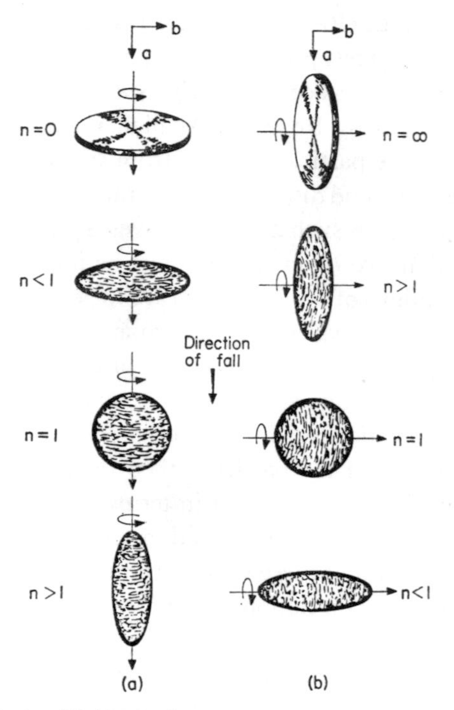

FIG. 4.7. Ellipsoids of revolution.[204] (a) Moving along axis of revolution. (b) Moving in a direction perpendicular to axis of revolution.

Case (i). Ellipsoid of revolution moving along its axis of revolution

$$a = \text{axis of revolution},$$
$$a = nb, \quad b = c.$$

Case (ii). Ellipsoid of revolution moving in a direction at right angles to its axis of revolution

$$b = \text{axis of revolution},$$
$$a = nb,$$
$$a = c.$$

Case (iii). Elliptical plate moving edgeways

$$a = nb,$$
$$c = 0.$$

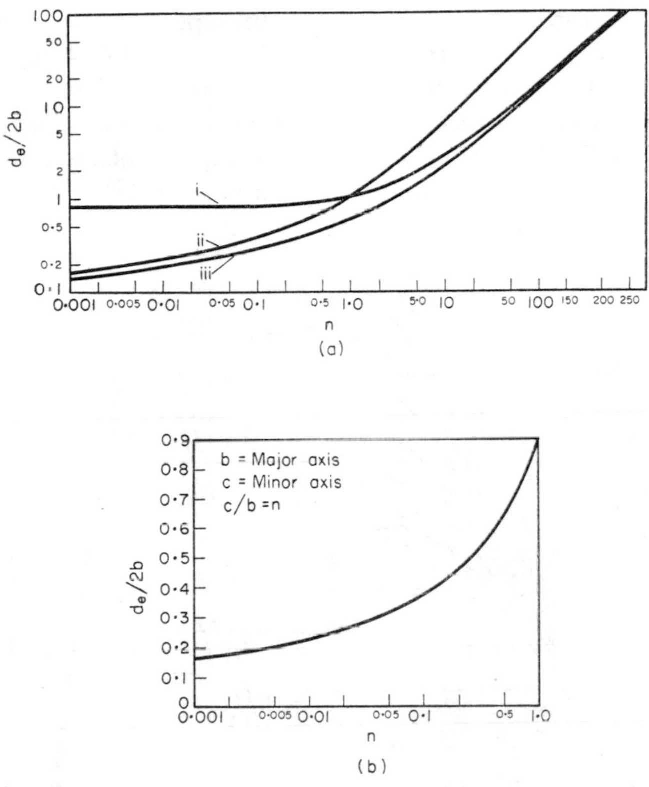

FIG. 4.8. Equivalent diameters d_e for ellipsoids of revolution.[204] (a) Moving in direction of a axis: (i) a = axis of rotation. $a = nb$, $b = c$. (ii) b = axis of rotation. $a = c = nb$. (iii) Elliptical plate moving sideways. $an = b$, $c = 0$. (b) Elliptical plate moving in direction perpendicular to its own plane.

Case (iv). Elliptical plate moving in a direction perpendicular to its own plane

$$a = 0,$$
$$b = \text{major axis},$$
$$c = nb.$$

The equivalent diameter required, d_e, can be found by equating $d_e/2b$ to the value read off the curves in Fig. 4.8 for the value of n for the appropriate case. The drag on oblate and prolate spheroids in steady state motion has also been considered in detail by Luiz[534–535] using an "effective mass" approach. The formulae derived by this method are, unfortunately, somewhat difficult to use, but they do illustrate that, as would be expected, the drag on an oblate drop is greater than that of a prolate (torpedo-shaped) drop of equal volume.

For isometric particles[158, 641]—cubes, octahedra, cube octahedra and tetrahedra—the particle velocity can be obtained by multiplying the velocity for a sphere with equivalent volume diameter, d_v, by an empirical correction factor K, given by:

$$K = 0{\cdot}843 \log_{10} \Psi/0{\cdot}065 \tag{4.63}$$

where Ψ = sphericity $= d_v^2/d_s^2$.

For non-isometric particles[359]—cylinders, parallel pipeds and spheroids—the particle velocity can be found by using a series of correlation curves (Fig. 4.9), which also yield a correction factor K. This is a function of the ratio of volume diameter to area diameter, d_v/d_A, and the particle sphericity as a parameter. These curves are probably also applicable to irregular particles. Outside the viscous flow region, experimental results are more limited, but a number of empirical correlations have been suggested.[55, 66, 546, 641, 659] The simplest method is to employ the empirical correlations using the sphericity of the particles as graphed

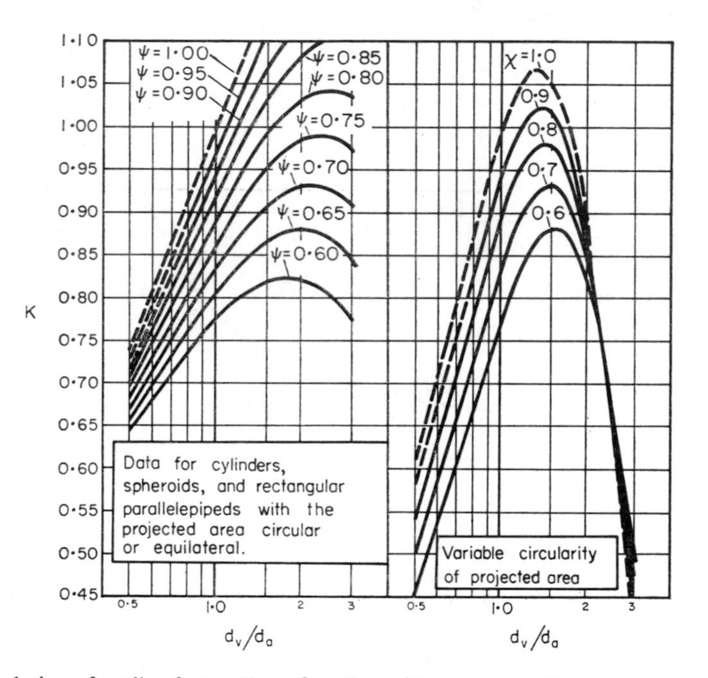

FIG. 4.9. Correlation of settling factor K as a function of the sphericity Ψ and circularity χ for various ratios d_v/d_A for non-isometric particles in the viscous flow region.[359]

in Fig. 4.10 for isometric particles. For more irregular particles it has been suggested that the drag coefficient can be calculated from[348]

$$C_D = \frac{4(\varrho_p - \varrho)gd_v^3}{3\varrho u^2 d_s^2} = \frac{\Psi 4(\varrho_p - \varrho)gd_v}{3\varrho u^2} \tag{4.64}$$

for

$$\text{Re} = \frac{\varrho u d_s}{\mu} = \frac{1}{\sqrt{(\Psi)}} \cdot \frac{\varrho u d_v}{\mu} \tag{4.65}$$

In the fully turbulent region (Re < 2000) the relation

$$C_D = 5{\cdot}31 - 4{\cdot}88\Psi \tag{4.66}$$

has been fitted to experimental values.[641]

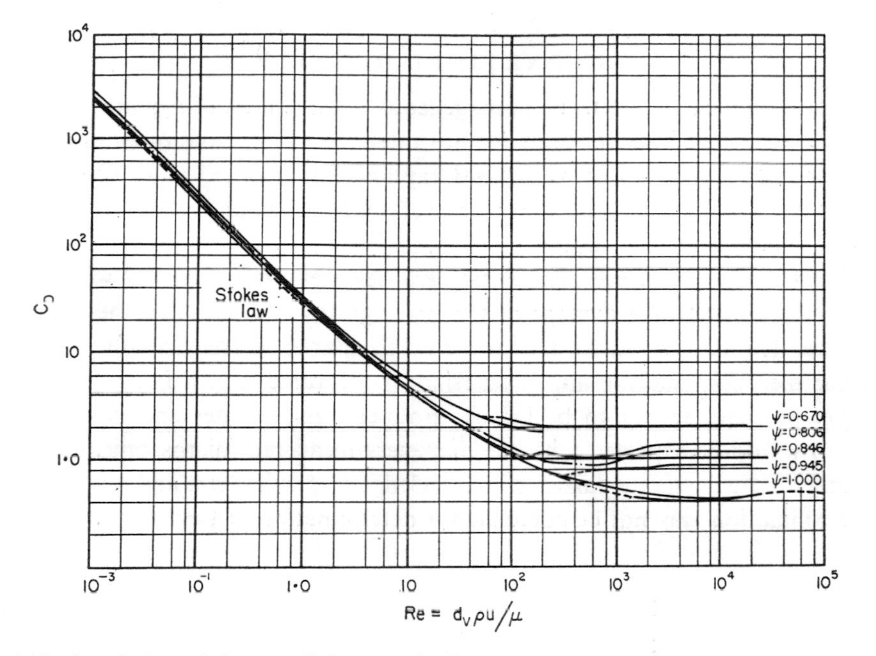

FIG. 4.10. Correlation of drag coefficient C_D for isometric particles with different sphericity Ψ.[641]

4.9. PARTICLE ROUGHNESS

At present no measure of roughness of particles exists similar to that which has been introduced for pipe flows, although the effect of roughness on particle behaviour has been observed. For example, while smooth spheres will not spin in fluids at high Reynolds numbers, rough ones have been observed to do so.

It is suggested that at low speeds rough particles carry an envelope of fluid in the indentations and the particle dimension to be considered is the external diameter of the particle.[349] At higher velocities, the critical value of C_D will occur much lower with rough particles than with smooth ones.

TABLE 4.5. OBSERVED DENSITIES OF PARTICLES (WHYTLAW-GRAY AND PATTERSON)[939]

Material	Normal density (kg m^{-3})	Maximum value		Average of low values	
		density (kg m^{-3})	diameter (μm)	density	diameter (μm)
Cadmium oxide	$6\cdot5\times10^3$	$2\cdot70\times10^3$	2·42	$0\cdot51\times10^3$	5·96
Silver	$10\cdot5\times10^3$	$4\cdot22\times10^3$	1·79	$0\cdot94\times10^3$	4·30
Gold	$19\cdot3\times10^3$	$8\cdot00\times10^3$	2·35	$1\cdot24\times10^3$	5·54
Mercuric chloride	$5\cdot4\times10^3$	$4\cdot32\times10^3$	4·53	$1\cdot27\times10^3$	3·63
Mercury	$13\cdot6\times10^3$	$10\cdot8\times10^3$	2·05	$1\cdot7\times10^3$	3·08
Magnesium oxide	$3\cdot6\times10^3$	$3\cdot48\times10^3$	3·26	$0\cdot35\times10^3$	7·29

4.10. PARTICLE DENSITY

Even particles which appear to be homogeneous often consist of agglomerates and non-homogeneous mixtures. This has been demonstrated for smokes and other particle systems from vapour phase processes by Whytlaw-Gray and Patterson,[939] who measured particle density by a method identical with that employed by Millikan in his "oil drop" experiment for measuring the electronic charge.

Table 4.5 lists some of the experimental densities measured, together with the normal densities of the material, the average of the lower group of values and the corresponding particle sizes. The data indicates that some of the particles are virtually homogeneous, for example mercuric chloride and magnesium oxide fume, consisting of tightly packed units. Other particles appear to be loosely grouped agglomerates and have much lower masses than would be expected with the diameters measured microscopically. Therefore, in many cases of agglomerate particles, a knowledge of particle size and shape factor from physical examination may not be adequate for determination of fluid resistance.

CHAPTER 5

GRAVITY AND MOMENTUM SEPARATION EQUIPMENT

5.1. INTRODUCTION

The simplest method of removing particles from a moving gas stream is to allow them to settle out under the force of gravity. Large particles will often do so on the floor of a horizontal duct, which acts in this instance as a simple settling chamber, while specially designed chambers will act as efficient collectors of coarse particles. Coarse particles, frequently designated as grit, are usually defined as those unable to pass through a 200 mesh screen, or larger than 76 μm. For these particles, particularly if they are abrasive, simple settling chambers are a preferred means of collection because of the low pressure loss through the plant as well as the long maintenance-free periods obtained with this type of plant.[136]

Momentum separators rely essentially on producing a sudden change of direction in the gas stream. The particles, because of their inertia, will continue to move in the same direction as the initial gas flow, and move into a collecting hopper, while the gas stream, freed of its larger particles, leaves the collector. Momentum separators are slightly more complex in construction than settling chambers, but have the advantage that they take up less room, and in their more sophisticated form, are able to collect particles down to about 20 μm with reasonable efficiency.

5.2. THEORY OF SETTLING-CHAMBER DESIGN

In settling chambers the gas stream is slowed down sufficiently to allow particles to settle out. In theory, a very large settling chamber would give sufficient time for even very small particles to be collected, but practical size limitations restrict the applicability of these chambers to the collection of coarse particles.[37] In the horizontal-type chamber (Fig. 5.1), an average gas velocity U (m s^{-1}) can be assumed to represent piston flow through the chamber. This can be simply derived from the gas flow rate Q (m^3 s^{-1}) and the height (H m) and breadth (B m) of the chamber:

$$U = Q/B.H \text{ m s}^{-1}. \tag{5.1}$$

From the length (L m) of the chamber, the *residence time* (t sec) of the gas can be calculated:

$$t = \frac{L}{U} = \frac{L.B.H}{Q} = \frac{V}{Q} \quad \text{sec,} \tag{5.2}$$

where V = volume of the chamber.

If a particle of size d of a particular material will settle a distance h m in t secs, then h/H represents the fraction of particles of this size that will be collected. If h is equal to or greater than H, all particles of that size (or larger) will be collected in the settling chamber. A curve

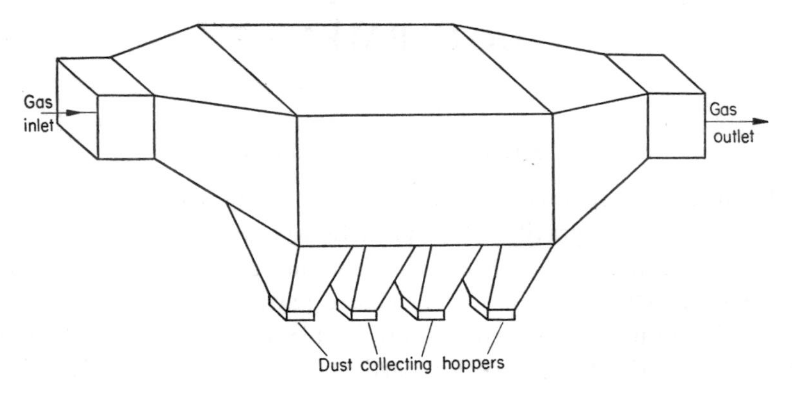

FIG. 5.1. Horizontal flow settling chamber with square cross-section.

of h/H ratios for the different sizes of a material is the *fractional* or *grade* efficiency curve for the settling chamber.

The value of h can be found from the product of the mean falling speed u_m of the particle, and the residence time t

$$h = u_m t. \tag{5.3}$$

If the particles are sufficiently small to reach their terminal velocity u_t in a negligibly small fraction of the residence time, this can be taken as equal to the mean velocity u_m. If the particles are of such a size that this assumption cannot be made, then the distance settled by the particles must be calculated in two stages. First the distance fallen, and time taken to reach 99 per cent of the terminal velocity are found (equations 4.23 and 4.24) and then the distance fallen during the remaining residence time is calculated. If the terminal velocity is not reached during the gas-stream residence time, equation (4.21a) has to be used in the integrated form with the residence time (0 to t) as boundary conditions. When the particles are smaller than 76 μm Stokes' law gives a reasonable approximation to the terminal falling speed. For larger particles, the general equation (4.16) must be used, with the correct drag coefficient C_D, calculated from equations (4.7 to 11). Settling velocities for spherical particles of unit density, both experimental and calculated from Stokes' law, are listed in Table 5.1.

TABLE 5.1. SETTLING VELOCITY OF SPHERICAL PARTICLES (DENSITY: 1000 kg m^{-3}) IN AIR

Temperature: 20°C. Pressure: 100 kPa

Particle diameter (μm)	Experimental (m s^{-1})	Calculated from Stokes' law (m s^{-1})
0·1	$8{\cdot}7 \times 10^{-7}$	$8{\cdot}71 \times 10^{-7}$
0·2	$2{\cdot}3 \times 10^{-6}$	$2{\cdot}27 \times 10^{-6}$
0·4	$6{\cdot}8 \times 10^{-6}$	$6{\cdot}85 \times 10^{-6}$
1·0	$3{\cdot}5 \times 10^{-5}$	$3{\cdot}49 \times 10^{-5}$
2	$1{\cdot}19 \times 10^{-4}$	$1{\cdot}19 \times 10^{-4}$
4	$5{\cdot}0 \times 10^{-4}$	$5{\cdot}00 \times 10^{-4}$
10	$3{\cdot}06 \times 10^{-3}$	$3{\cdot}06 \times 10^{-3}$
20	$1{\cdot}2 \times 10^{-2}$	$1{\cdot}2 \times 10^{-2}$
40	$4{\cdot}8 \times 10^{-2}$	$5{\cdot}0 \times 10^{-2}$
100	$2{\cdot}46 \times 10^{-1}$	$2{\cdot}5 \times 10^{-1}$
400	1·57	4·83
1000 = 1 mm	3·82	30·50

When Stokes' law can be used for the terminal velocity, and this is a reasonable approximation for the average settling velocity, the particle size that will be wholly retained within the chamber can be found from:

$$d_{\min} = \sqrt{\left(\frac{18HU\mu}{(\varrho_p-\varrho)gL}\right)} \tag{5.4a}$$

$$= \sqrt{\left(\frac{18Q\mu}{(\varrho_p-\varrho)gBL}\right)} \tag{5.4b}$$

These simple equations should only be used as a guide to the collection efficiency of the chamber, as the complicating factors of falling speeds outside the Stokes' law range are nearly always of some importance for the particles collected by settling chambers. Other factors which must be taken into consideration are hindered settling effects which will occur at high particle concentrations (Section 4.6). All these tend to reduce the collection efficiency.

The other major design factor in specifying settling-chamber dimensions is that gas-flow velocities must be kept below the re-entrainment or "pick-up" velocity of the deposited dust. As a general rule[493] velocities below 3·0 m s^{-1} are satisfactory for most materials. Some materials, such as starch powder and carbon black, however, are re-entrained at lower velocities, while for heavy particles which have agglomerated into larger lumps, much higher velocities could be used. Some experimental re-entrainment velocities are given in Table 5.2. Because of the factors, and the problem of obtaining an even gas flow through the chamber, the actual efficiency of the chamber is not likely to be as predicted by the simple relations given above.

TABLE 5.2. PICK-UP VELOCITIES OF VARIOUS MATERIALS[53]

Material	Density kg m^{-3} × 10^3	Median size (μm)	Pick up velocity (m s^{-1})
Aluminium chips	2·72	335	4·3
Asbestos	2·20	261	5·2
Non-ferrous foundry dust	3·02	117	5·7
Lead oxide	8·26	14·7	7·6
Limestone	2·78	71	6·4
Starch	1·27	64	1·8
Steel shot	6·85	96	4·6
Wood chips	1·18	1370	3·9
Wood sawdust	—	1400	6·8

Improved settling chamber efficiencies in the horizontal flow type chamber can be achieved by decreasing the height a particle has to fall before being collected. This has been applied in the Howard settling chamber (Fig. 5.2) where a number of collecting trays

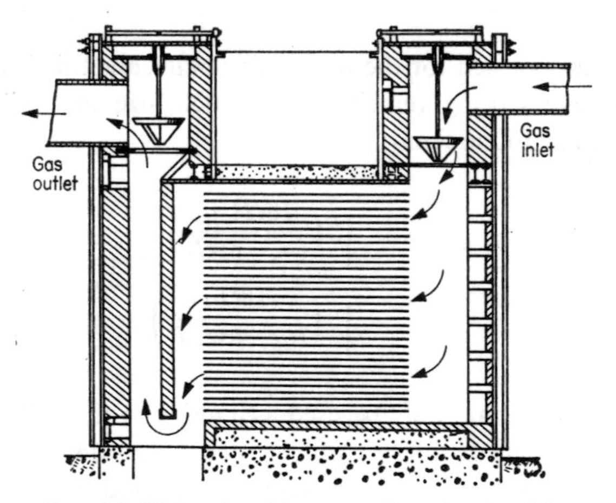

FIG. 5.2. Howard multi-tray settling chamber.[493]

have been inserted in the chamber. The main objection to the wide use of the Howard chamber is that the spaces between the trays are difficult to clean, although this could be overcome by installing a self-cleaning system, such as a spray washer.

If a chamber, height H, contains N trays, then the height in each section is approximately $H/(N+1)$, and the fractional efficiency $h(N+1)/H$.

The vertical flow settling chamber is essentially an elutriator, removing those particles whose settling speed is greater than the gas velocity. Units of this type are used as grit arrestors on small cupolas and in small boiler plant. In its simplest form as in the penthouse type for several cupolas (Fig. 5.3a) the dust moves to heaps at the side of the arrestor entry, and some material falls back into the cupola. The penthouse is cleaned manually when

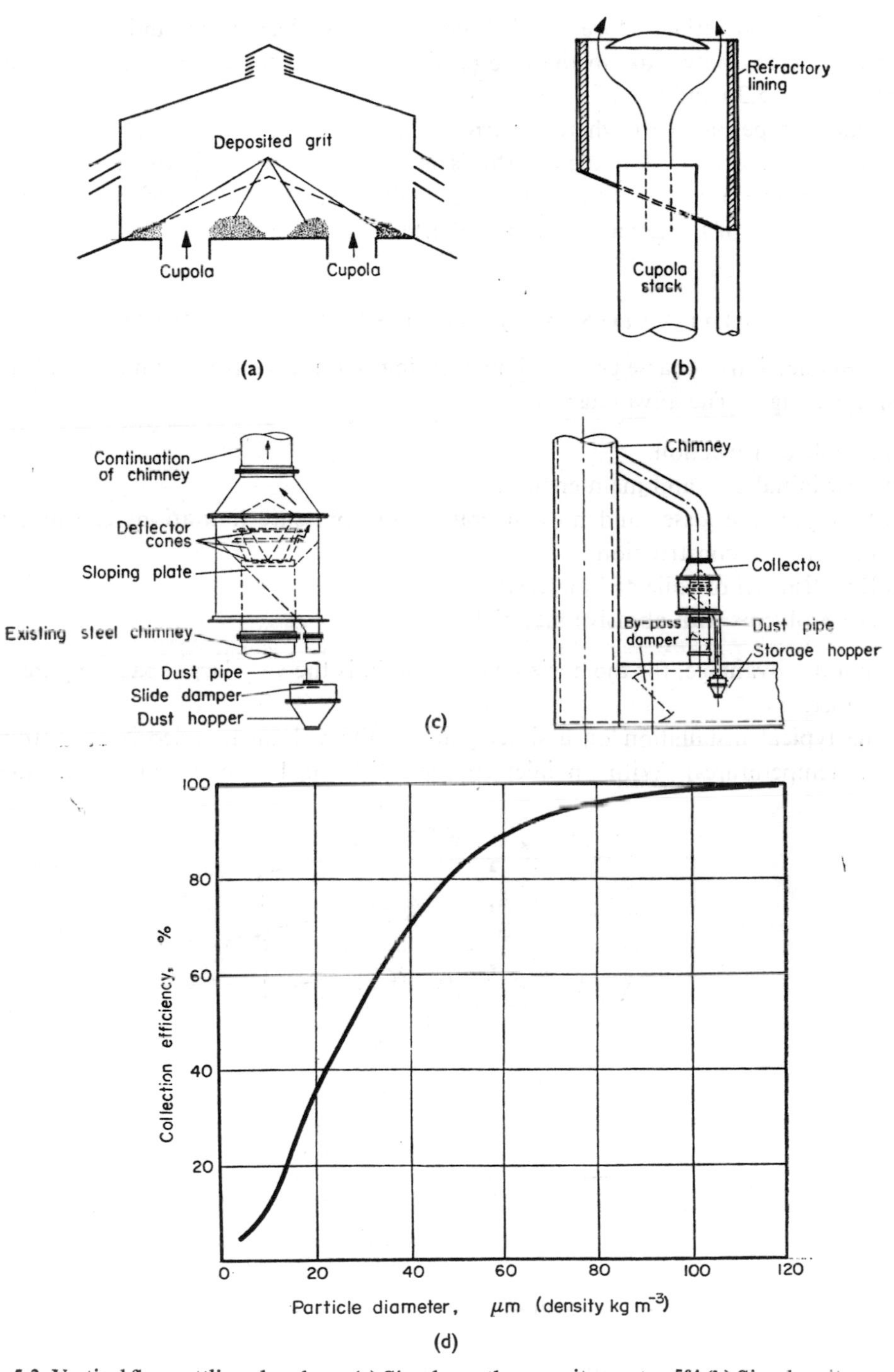

FIG. 5.3. Vertical flow settling chambers. (a) Simple penthouse grit arrestor.[764] (b) Simple grit arrestor for cupolas.[764] (c) Deflector-type grit arrestor.[657] (d) Fractional efficiency curve for arrestor shown in Fig. 5.3(c).[657]

the plant is not operating. More sophisticated are the deflector type units, where the gas stream is deflected outwards, so that the particles fall into the collecting annulus around the stack (Figs. 5.3b and 5.3c).

In a unit of type Fig. 5.3b, where the arrestor diameter is about two and a half times the stack diameter, the velocity in the arrestor is only $1/(2\frac{1}{2})^2$, i.e. $1/6\frac{1}{4}$, of the stack velocity, and so very coarse particles of the order of 200–400 μm, which are carried away in a stack with a gas velocity of 1·5–2·0 ms^{-1} will be collected in an arrestor of this type.

5.3. APPLICATION OF SIMPLE SETTLING CHAMBERS

For the collection of coarse grit the simple settling chambers have a number of advantages and disadvantages. The advantages are:

(i) Simple construction.
(ii) Low initial cost and maintenance.
(iii) Low pressure losses and no temperature and pressure limitations except those of materials of construction.
(iv) Dry disposal of collected materials.
(v) No problems with abrasive materials.

The major disadvantage, for the horizontal flow unit, is the very large space requirement of the chamber.

In one typical installation on a sinter plant,[136] the volume handled is $2{\cdot}3 \times 10^6$ m^3 h^{-1} (ambient temperatures). With an inlet concentration of 1 g m^{-3}, this is equivalent to

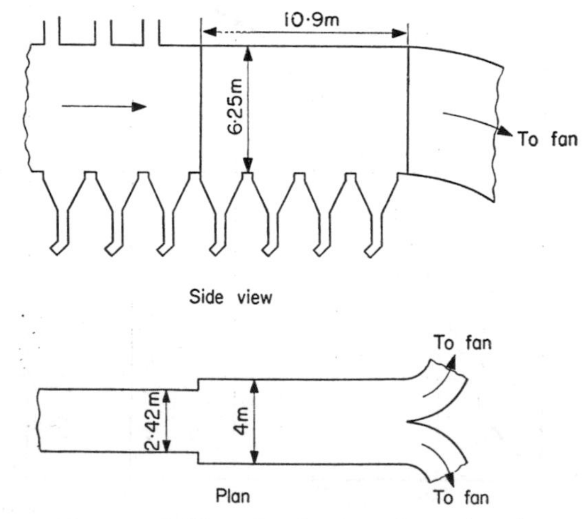

FIG. 5.4. Settling chamber in a sinter plant.[136]

2300 kg h^{-1} of solids. The settling chamber (Fig. 5.4) reduces this to 850 kg h^{-1} or 0·37 g m^{-3}. With such large volumes, the low power consumption with the low pressure loss in the unit is of great economic importance.

Jennings[399] has calculated fractional efficiency curves for the two most important dust constituents from a sinter plant, quartz (density 2600 kg m^{-3}) and iron oxide (density 4500 kg m^{-3}). The chamber in this example is 3·0 m high and wide and 6·0 m long. The volume flow is 158 $m^3 h^{-1}$. The curves (Fig. 5.5) show that all particles with a density of

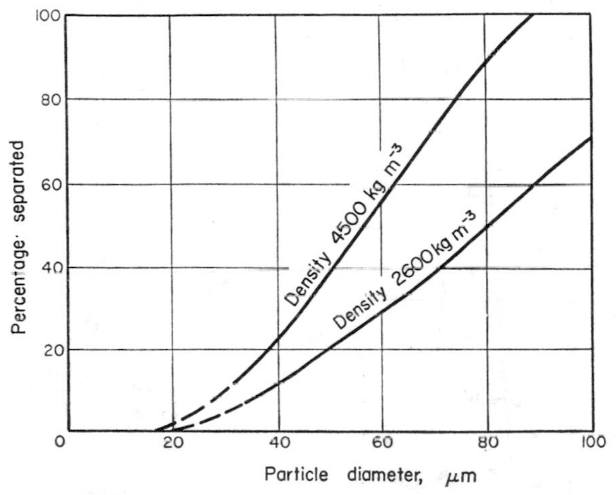

FIG. 5.5. Predicted fractional efficiency curves for dusts from a sinter plant.[399]

4500 kg m^{-3} larger than 80 μm are likely to be collected but only 70 per cent of the lighter particles when these are 100 μm in diameter.

The efficiency for a vertical flow stack collector is not very high. For a typical shell boiler plant, 75 per cent overall collection efficiency was obtained with a stack arrestor (type Fig. 5.3c).[390] For small plants this may, however, be satisfactory.

5.4. MOMENTUM SEPARATORS

The efficiency of a simple settling chamber can be improved, and space requirements reduced, by giving particles a downward momentum in addition to the gravity settling effect. The number of possible designs incorporating this principle is very great, varying from a simple bafle in the chamber to specially designed jets which give accelerated settling.

Simple baffle chambers (Fig. 5.6a) are sometimes placed in series. Although their efficiency is of the same order as the horizontal chamber, the pressure loss is greater. This can be lowered by using a pattern which has rounded baffles (Fig. 5.6b).

Other momentum collectors use a downwards facing tube to give additional downwards movement to the particles, which is of the order of $g/3$ in addition to the normal gravitational force g. One worker (Fig. 5.6c)[399] has suggested that a cone with gradually increasing diameter gives a slower moving gas at the base of the chamber and reduces re-entrainment. Similar ideas were probably responsible for features of Prockat's accelerated settling chamber such as the hopper following a deep cylindrical section (Fig. 5.6d)[663] Test results for both coal and fly ash samples are available for this model, and are shown in Table 5.3, together with the dust gradings used. The data shows that increased gas velocity reduces the chamber

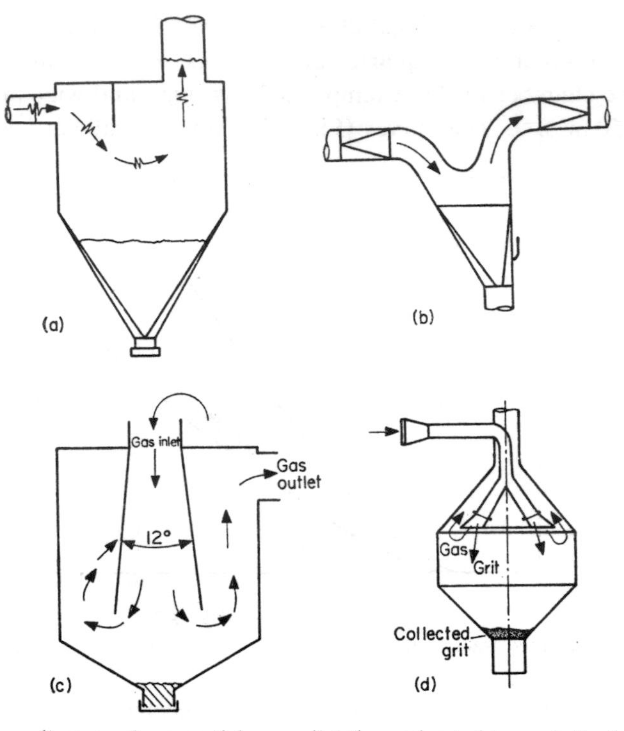

FIG. 5.6. Baffle-type collectors where particles are slightly accelerated towards the hopper. (a) Simple baffle type.[7] (b) Rounded inertial trap.[7] (c) Downward facing entry.[399] (d) Accelerated settling chamber with deepened hopper.[663]

efficiency. The depth of the turbulent zone in the cylindrical portion of the chamber must be critical in determining the re-entrainment of particles.

A modern development of a simple momentum separator with a change in direction as the separating mechanism is described by Muschelknautz.[589] The gases flow down a cylindrical duct and then change direction by 180° to pass up an annular duct surrounding the inlet, while the dust is deposited in a hopper. A very carefully designed collar is built at the end of the inlet duct. In one case additional (rotating) air is blown in at an annulus before the collar into the boundary layer, with a velocity 3 times that of the axial velocity in the main gas stream. This annular jet assists in turning the main gas stream by clinging to the boundary. The exhaust duct serves to draw off the cleaned gas and convert some of the kinetic energy to pressure energy. In a second, less effective, design, the exhaust gases are partly drawn into slits in the collar, without the addition of secondary air. Diagrams illustrating these units with typical fractional efficiency curves are shown in Fig. 5.7.

More elaborate, and also more efficient, is the collector where the gas is allowed to impinge on a surface which is shaped in such a way as to retain the particles while allowing the gas to escape. In one type, the venturi momentum collector (Fig.5.8),[848] the gas passes horizontally through a series of venturi-shaped passages (a) formed by the diamond-shaped ducts, (b) extending from within a short distance of the top to the bottom of the main duct.

TABLE 5.3. SIZE GRADING OF TEST DUST[663]

	Pulverized coal (%)	Boiler fly ash (%)
Greater than 200 μm	0	2·08
120 μm–200 μm	1·84	7·08
90 μm–120 μm	2·92	14·88
75 μm– 90 μm	6·48	5·28
60 μm– 75 μm	13·52	10·08
Smaller than 60 μm	75·24	60·60

Efficiency of the Prockat model momentum collector

Gas velocity at inlet ($m\ s^{-1}$)	Dust content ($g\ m^{-3}$)	Efficiency (%)
Pulverized coal		
7·3	34·9	74·3
9·8	91·0	63·0
12·0	23·4	47·0
Boiler fly ash		
5·3	19·8	79·7
8·6	21·1	70·5
13·4	10·3	55·5

The velocity of the gas increases as it approaches the throat of the venturi passages, and the momentum of the particles causes them to concentrate along the conveying walls. The concentrate passes through the slots (f) in the vertical ducts and is trapped in the V of the vertical duct. The dust drops into the hopper, while the gas leaves the top of the ducts. The traps are arranged in series of 6, 9 or 12 rows, and dampers (h) control the flow and act as by-pass valves.

Another device, which also has a very low pressure drop, is the D.E.P. curtain type collector[544] (Fig. 5.9). The basic element is a U profile where the jets of dust-laden gases, formed by the spaces in a row of these profiles, impinge on the base of the U, and the gas rebounds or moves in a circular motion in the curved portion. The impingement and circular movement throw the dust out of the gas stream, and it falls into the hopper below. Rapping or vibration is applied in some cases to assist the dust to fall off the U channels. Liquid sprays can also be used for this, and these prevent re-entrainment.

The system is very robust and can be used at high temperatures and in corrosive conditions. For example, in acid conditions the collector cells can be made of acid-resistant

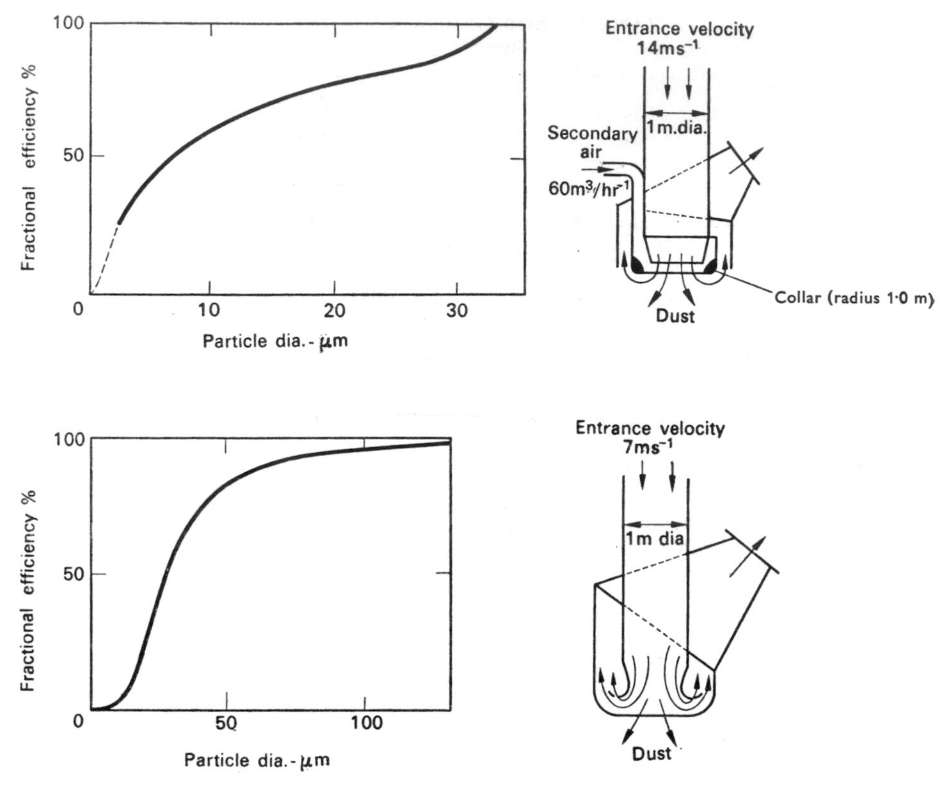

FIG. 5.7. Modern momentum separators[589] (designs and fractional efficiency curves).

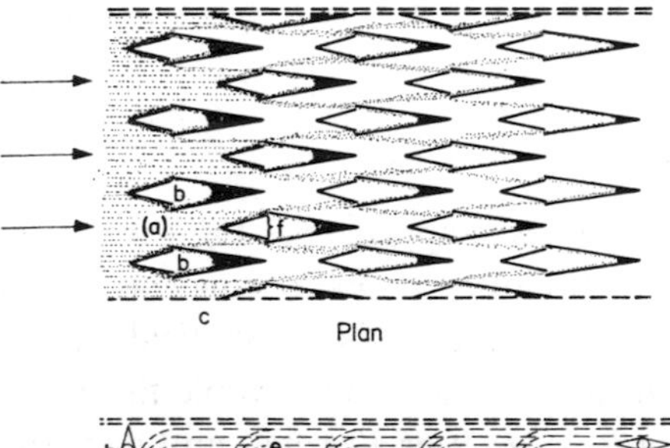

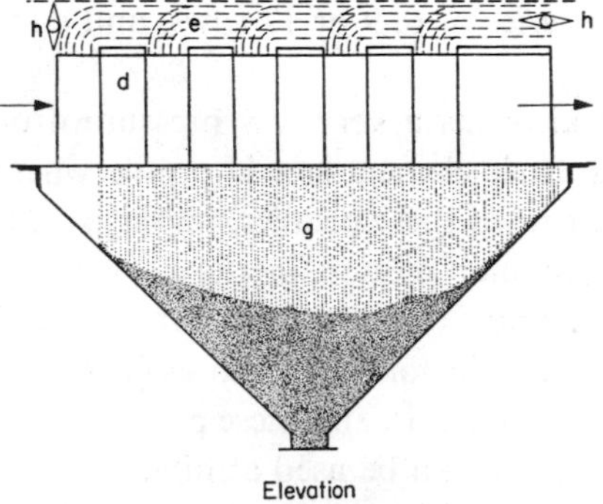

FIG. 5.8. Venturi baffle collector.[848]

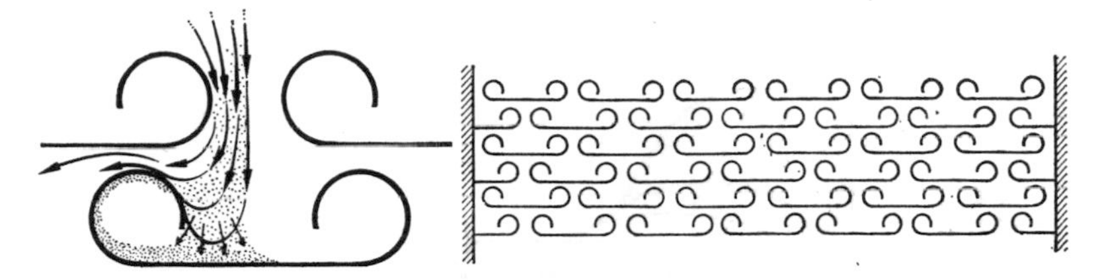

FIG. 5.9. D.E.P. curtain collector.[544]

stainless steels, and the chamber walls lined with acid resistant tiles. The pressure drop through installations varies from 25–100 Pa and may be lower in certain plants. Data on actual plants show the following results:

(a) Cement plant, limestone rotary dryer.
  Inlet concentration: 20–70 g m^{-3}. 38 per cent less than 10 μm.
  Temperature: 127°C.
  Flow rate: 800 m^3 s^{-1}.
  Pressure loss: 16 Pa.
  Efficiency: 80–91 per cent.

(b) Steam generator fed with pulverized coal.
  Pressure loss: 33 Pa.
  Efficiency (12 rows): 80 per cent.

Incomplete data on a number of other plants show generally the same results.

A collector which combines some features of both the ones just described is the reverse nozzle impingement separator (Fig. 5.10) where the dust laden gases impinge on a curved slotted surface which reverses the gas flow but allows the dust particles to pass through the slots to an enclosed channel where they can fall into the hopper.

Also in the category of momentum collectors is the Calder–Fox scrubber,[249] which is used for the recovery of acid mists (Fig. 5.11). The gas carrying the acid mist droplets is forced through orifices where agglomeration takes place and then impinges on baffles, where the agglomerated droplets are deposited. The construction of the plant uses either lead sheets with holes or alternatively, for operation at higher temperatures, strips of glass which are more fragile, but better able to withstand high temperature. In the case of lead sheets, 3 mm thick sheets are used, with 3 mm diameter holes machined on a 9 or 12 mm square pitch, followed by the impact plate, with holes on the same centres, but 6 mm diameter. This is followed by a collector plate with 2 mm diameter holes 3 mm apart. In the glass unit the orifice slots are 1·5 mm wide and the impact slots 3 mm wide, spaced on 8 mm centre lines.

The published literature shows how difficult it is to get straight-edged glass without curvature for the glass strips. Velocities in the "scrubber" are about 30–35 m s^{-1}, with a pressure drop of 870–1370 Pa. Higher velocities lead to re-entrainment, while at lower velocities, less efficient collection takes place. The efficiency of the Calder–Fox type collector is 90–97 per cent, and droplets as small as 2–2·5 μm are collected.

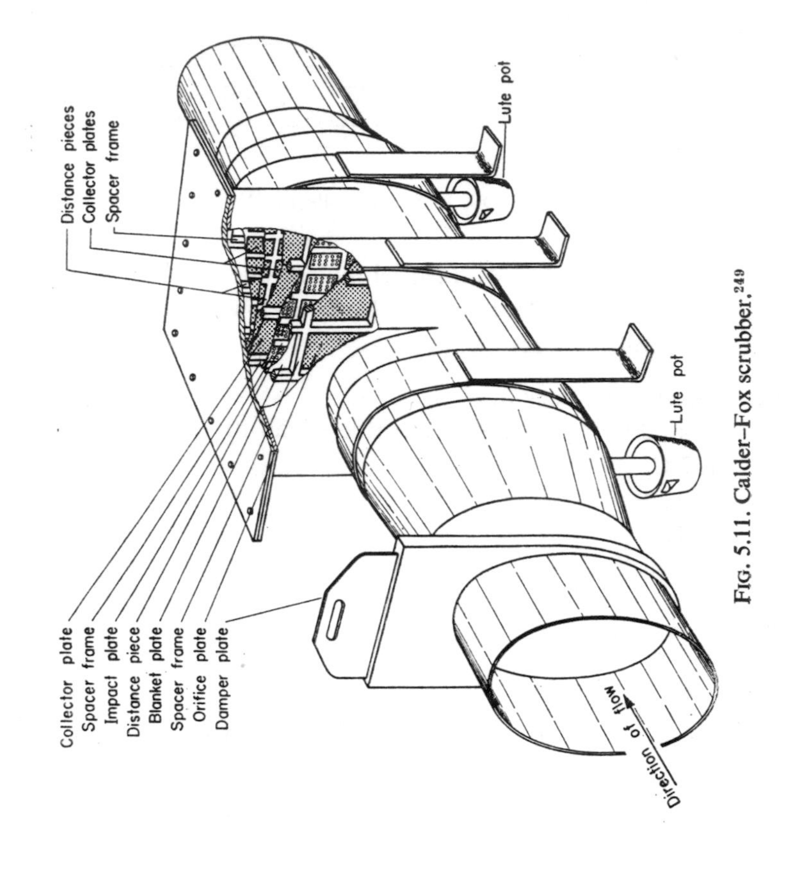

FIG. 5.11. Calder–Fox scrubber.[249]

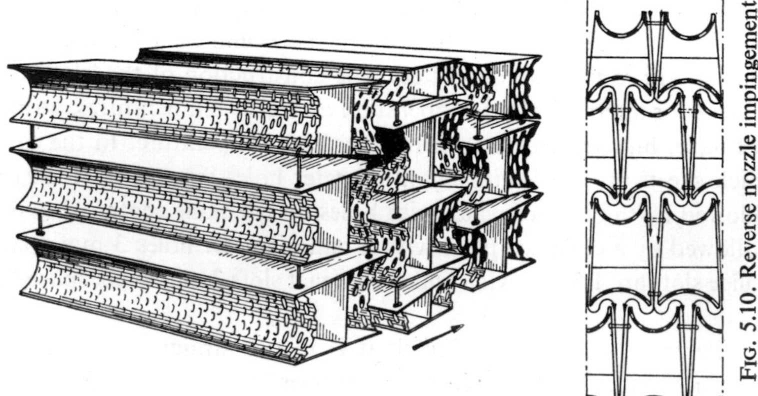

FIG. 5.10. Reverse nozzle impingement collector.[493]

An elegant recent development, which is similar in principle to the Calder–Fox scrubber, is the Petersen "Drucksprung" (pressed spring) collector,[637] which has been widely applied to acid mists, oil mists and fine particles such as dyestuffs and pigments. The collector consists essentially of a coiled spring with one end closed off, in which the spring elements are shaped so that successive layers mesh together and achieve a change of direction of the gas stream passing through. The shape of these elements, which is shown in Fig. 5.12, gives effective coagulation of the mists, and the droplets run along the spiral to fall into a sump. The advantages of this collector are first, the simplicity with which the shape of the spring

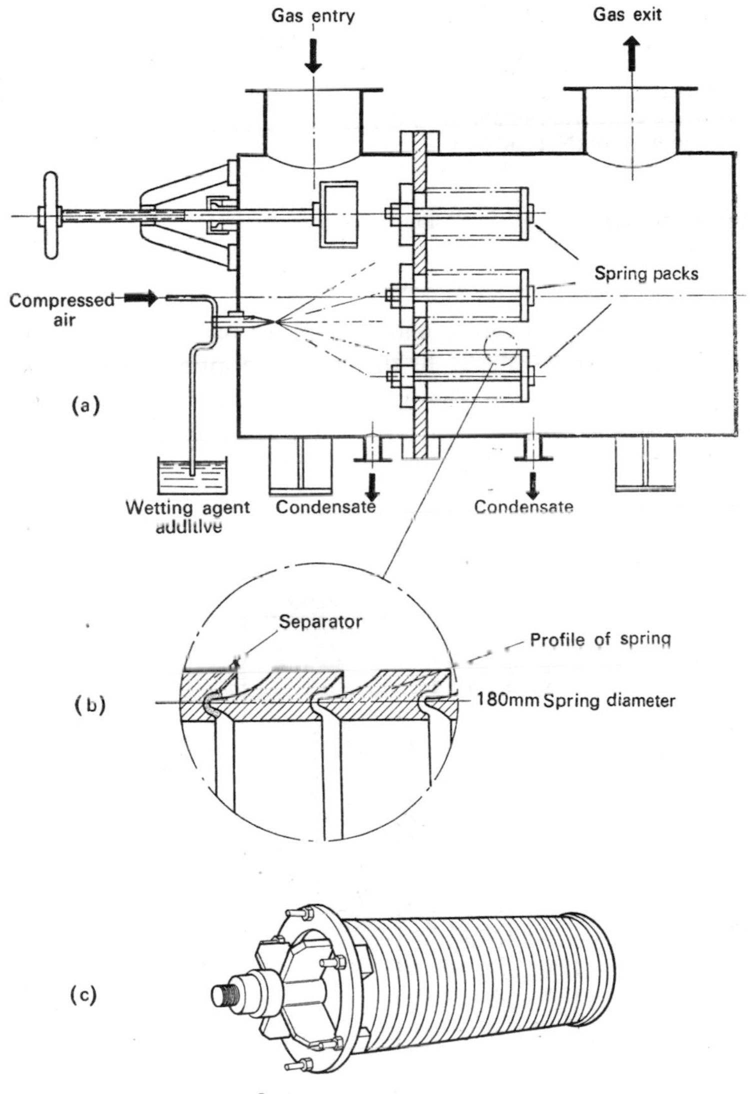

FIG. 5.12. Petersen compressed spring mist eliminator.[637] (a) Spring packet arrangement. (b) Spring detail. (c) Spring pack.

TABLE 5.4. EFFICIENCY OF PETERSON "PRESSED-SPRING" COLLECTOR

Type of gas and aerosol	Inlet concentration (g m^{-3})	Exit concentration (g m^{-3})
(1) Waste gases from phosphoric acid manufacture by burning elemental phosphorus. Phosphoric acid mist as P_2O_5	52	0·06
(2) Waste gases from sulphuric acid wet catalysts. Acid mist as SO_3	18	0·15
(3) Waste gases from $CaCl_2$ production. HCl content	17·7	0·0004
(4) HF in waste gases from superphosphate manufacture	1·8	< 0·005
(5) Waste gases from sulphuric acid manufacture (NH_3 added)		
H_2SO_4	0·8	0·05
SO_2	6	0·2
(6) Waste gases from HCl pickling (HCl)	0·08	0·0002

element can be economically extruded, and second, the relative ease with which the gaps between successive coils can be altered, by tightening the spring, changing the effectiveness and pressure drop of the collector. The pressure loss tends to vary between 1·5–4·5 kPa,

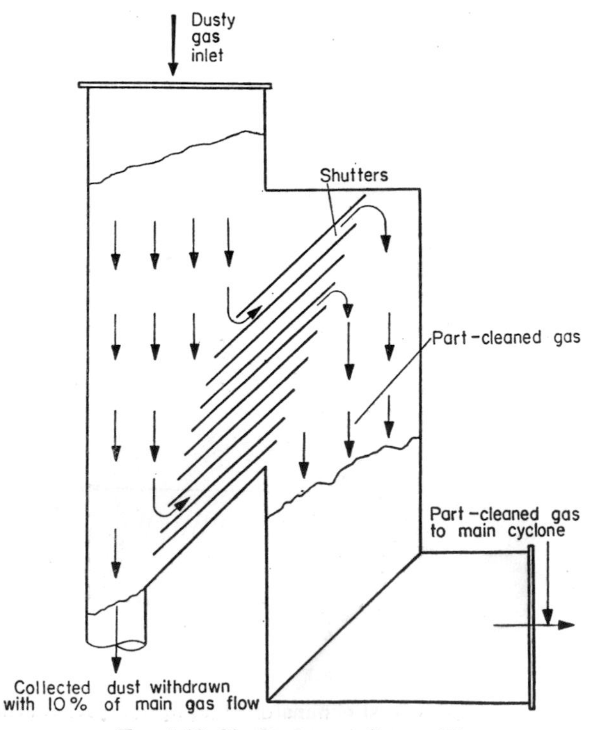

FIG. 5.13. Shutter-type collector.[803]

and the energy consumption is approximately 0·7 W m^{-3} h^{-1}. Typical efficiencies are shown in Table 5.4.

The principle that a sudden change in direction of the gas stream will leave the coarser particles to continue to move directly ahead is also applied in the shutter type of collector (Fig. 5.13), which is sometimes used as a low resistance precleaner before cyclones or bag houses. Here about 80 per cent of the gas stream, partially cleaned, is drawn through the shutters while the rest of the dusty gas is passed to a cyclone.

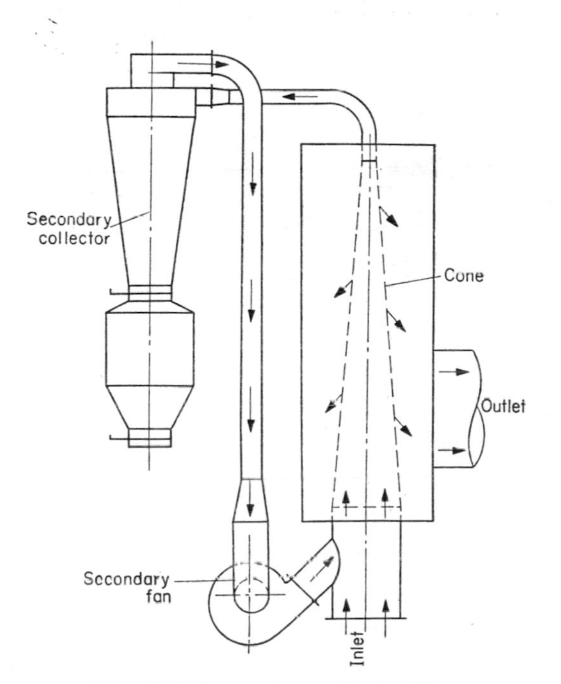

FIG. 5.14. Louvre type collector.[401]

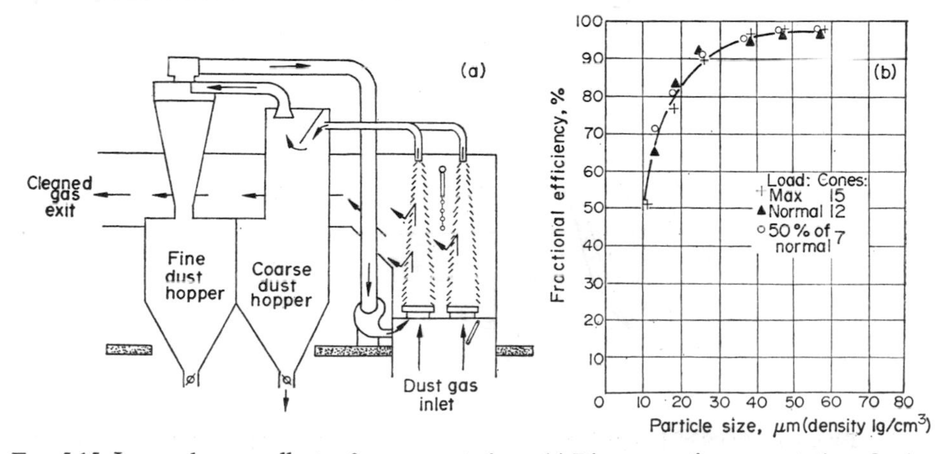

FIG. 5.15. Louvred cone collector for power stations. (a) Diagrammatic representations for louvre-type collector followed by baffled chamber and cyclone in series.[330] (b) Fractional efficiency curves.[330]

Much more efficient are the louvred collectors where the dust-laden gas stream enters the larger end of a truncated cone (Fig. 5.14), having almost its entire surface perforated with louvred slots. The gas stream changes direction to pass through the cone, while the dust is not deflected but passes on to the end of the cone, together with a small fraction of the gas stream, where it is drawn off to a secondary collector. The developers of this equipment also claim that the flow pattern near the louvres tends to send the particles towards the central gas stream in the cone. This type of equipment is very useful for the collection of fly ash in small installations with a varying load, as the efficiency of the collector seems to be relatively constant under these conditions.[330]

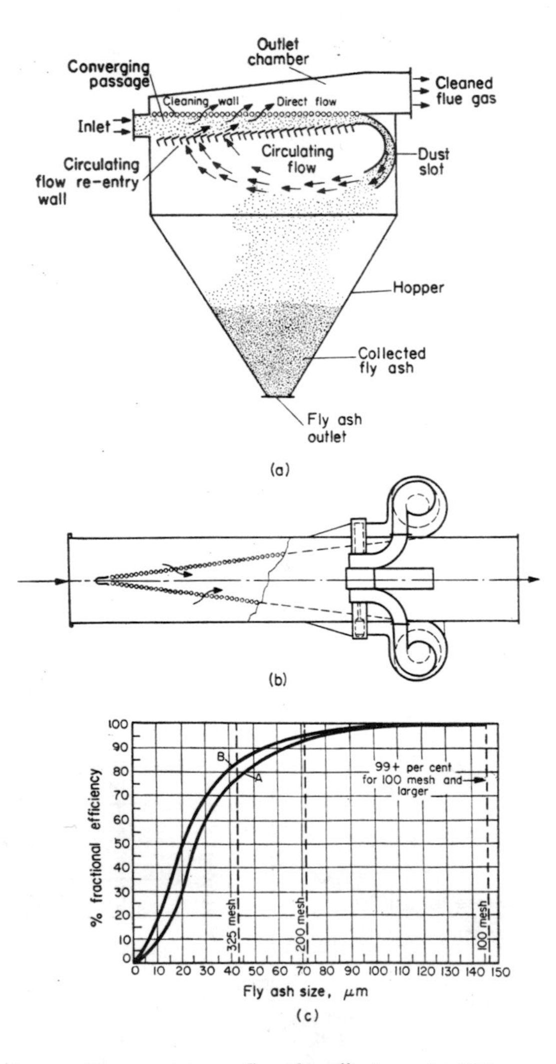

FIG. 5.16. Two other types of low-resistance fly ash collectors. (a) Using gravity settling for dust separation.[391] (b) Using cyclones for dust separation.[391] (c) Fractional efficiency graphs for these two types.[391]

For larger throughput of gases, several of these collectors can be arranged in parallel. A typical arrangement is shown in Fig. 5.15a, where the louvred collector is followed by a baffled settling chamber and cyclone in series. The efficiency of this combination has been measured[330] and the fractional efficiency curve is shown in Fig. 5.15b for fly-ash collection. It is seen that the efficiency is similar to a high throughput medium efficiency cyclone, but the pressure drop is of the order of 250–375 Pa which is somewhat lower than that of the cyclone. Other examples of louvred collectors are shown in Figs. 5.16a and 5.16b. The resistance in these collectors is lower, and so is the efficiency, as may be seen from the fractional efficiency curves (Fig. 5.16c). These units are widely used as grit arrestors in small boiler plants.

CHAPTER 6

CENTRIFUGAL SEPARATORS

6.1. INTRODUCTION

Cyclones, the generic name for collection systems where particles are removed from spinning gases by centrifugal forces, are probably the most common dust-removal systems used in industry. They are simple to construct and the conventional types have no moving parts. Because they can be made from a variety of materials, including those with refractory and corrosion-resistant properties for special applications, there are no temperature limits to the use of cyclones and maintenance can be reduced to a minimum.

The centrifugal force on particles in a spinning gas stream is much greater than gravity, therefore cyclones are effective in the removal of much smaller particles than gravitational settling chambers, and require much less space to handle the same gas volumes. On the other hand, the pressure drop in a cyclone is greater, and power consumption is much higher. Cyclones are distinguished from momentum separators, which were discussed in Chapter 5, in that in the latter there is simply a diversion of the gas stream from its original path, while in cyclones there are a number of revolutions of the gas stream. Gravity settling chambers and momentum separators are used, with few special exceptions, for the collection of coarse grit—particles greater than 76 μm—while commercial cyclones are effective in collecting particles down to 10 μm (when their density is unity) and can be used for smaller particles in certain designs.

When large grit particles are present in appreciable quantities, particularly if these are very hard, the cyclone walls may suffer from erosion and it is often desirable to introduce a collector of the momentum separator or gravity settling chamber type as a precursor to the cyclone.

The spinning motion can be applied to the gas stream in several ways, and cyclone types can be classified accordingly. The gases can be drawn through curved vanes in a duct, in a unit called the *straight-through cyclone*[515] or *vortex air cleaner*,[197] or they can be spun in a special turbine.[947] In the conventional or *reverse-flow cyclone* the gases are admitted tangentially to a cylindrical upper section which contains a centrally placed exhaust pipe penetrating below the tangential inlet, while a conical lower section is connected to the dust hopper. The gases in this case spiral down towards the apex of the cone and then are reversed up again through the exit.

In a combination of the straight-through and the reverse-flow cyclone, the dusty gases are admitted axialliy as in the staight-through cyclone, while secondary air is admitted through jets tangentially at the walls, giving a similar flow pattern to that of the "reverse-flow" type, but with improved efficiency.[445–6, 599, 731] Because of the similarity of vortex form, this is called the *tornado* collector in England, and the Drehströmungsentstauber (rotating flow deduster) in Germany, where it was developed.

This chapter discusses the theory of particle separation in a spinning gas, and the application of this to the design of the various types of cyclone. Commercial types are described, and, as far as possible, experimental fractional efficiency data is presented for these. Methods of calculating cyclone efficiency and pressure drop are reviewed in detail.

6.2. PARTICLE SEPARATION IN A SPINNING GAS

The simplest system that can be studied in the case of cyclone separators is the motion of a particle in a spiralling gas stream. If a gas stream enclosing a particle moves round the arc of a circle (Fig. 6.1), and it is assumed that the particle has the same tangential velocity as the gas stream, then the centrifugal force on the particle, F, that is the force normal to the tangent of the arc, is given by

$$F = m\frac{u_T^2}{R} \tag{6.1}$$

where m = mass of the particle,
R = radius of the circle,
u_T = tangential component of the gas velocity.

When the path of the gas is a spiral along the walls of a cylinder (the system used commercially in the straight-through cyclone), the particles will move outwards as they are carried along by the gas stream, and their path will be an expanding helix. The particle velocity can be resolved into three velocity components: the tangential velocity u_T, tangential to the gas spiral and normal to the axis; the radial drift velocity u_R, normal to the tangential component and normal to the axis; and the axial velocity u_H, along the axis of the gas spiral. The magnitude of the centrifugal force is frequently described in terms of the number of times $\mathbf{n}$ this exceeds the force of gravity, mg. This number is therefore found by dividing the centrifugal force (equation 6.1) by mg:

$$\mathbf{n} = \frac{u_T^2}{Rg} \tag{6.2}$$

Free-vortex theory. (a)

It is often assumed that the free vortex formula can be applied to the system.[58, 197, 207, 255, 822, 877] Then the tangential velocity u_T and the radius are related by the equation:

$$u_T R^n = \xi \tag{6.3}$$

where ξ = constant
and $n = 1$.

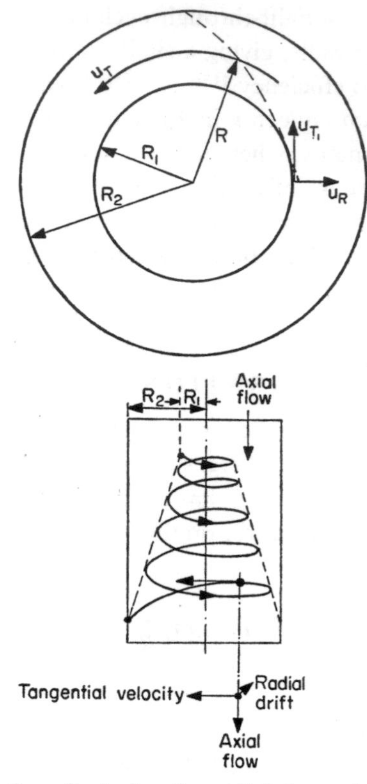

FIG. 6.1. Resolution of velocity of a particle in a spiralling gas stream.

It should be noted, however, that experimentally for gas cyclones the value of the exponent n is found to be 0·5–0·7 and not unity. The following analysis, however, has not yet been completed[822] with the experimental exponent.

If at a time t, the position of the particle in cylindrical coordinates is given by (R, θ), then the radial drift velocity component of the particle is

$$u_R = \frac{dR}{dt} \tag{6.4}$$

and the tangential velocity component is

$$u_T = R\frac{d\theta}{dt} \tag{6.5}$$

(assuming that the tangential particle and gas velocities are the same).

The radial acceleration is then equal to

$$\frac{du_R}{dt} = \frac{d^2R}{dt^2} - R\left(\frac{d\theta}{dt}\right)^2 \tag{6.6}$$

and the tangential acceleration is given by

$$\frac{du_T}{dt} = R\frac{d^2\theta}{dt^2}+2\frac{d\theta}{dt}\cdot\frac{dR}{dt}. \tag{6.7}$$

Multiplying the accelerations on the particles by the particles mass gives the forces on the particles (Newton's second law), and these forces will be opposed by the resistance of the gas to any relative movement of the particles. If it is assumed that the particles encounter the same resistance as in steady state motion through a laminar gas, and their relative velocity is in the viscous flow region, then Stokes' law [equation (4.4)] may be applied, and the equations of motion of the particle (assumed to be a sphere) relative to the gas stream are

(a) Radially:

$$\frac{m}{3\pi d\mu}\left\{\frac{d^2R}{dt^2}-R\left(\frac{d\theta}{dt}\right)^2\right\} = -u_R = -\frac{dR}{dt}. \tag{6.8}$$

(b) Tangentially:

$$\frac{m}{3\pi\mu d}\left\{R\frac{d^2\theta}{dt^2}+2\frac{dR}{dt}\cdot\frac{d\theta}{dt}\right\} = 0. \tag{6.9}$$

(The tangential component is zero because it is assumed that the particle and gas tangential velocities are the same.)

To integrate these equations (6.8) and (6.9) they are put into dimensionless form by:

(i) Expressing the radii in terms of the external radius R_2:

$$R = rR_2. \tag{6.10}$$

(ii) Expressing the velocities in terms of the velocity at the external radius u_{T_2}:

$$u_R = u.u_{T_2}. \tag{6.11}$$

(iii) Expressing the times in terms of the ratio R_2/u_{T_2}:

$$t = \tau R_2/u_{T_2} \tag{6.12}$$

where r, u and τ are dimensionless variables.

Equations (6.8) and (6.9) now reduce to

(a) Radially:

$$T\left\{\frac{d^2r}{d\tau^2}-r\left(\frac{d\theta}{d\tau}\right)^2\right\} = -\frac{dr}{d\tau}. \tag{6.13}$$

(b) Tangentially:

$$T\left\{r\frac{d^2\theta}{d\tau^2}+2\cdot\frac{dr}{d\tau}\cdot\frac{d\theta}{d\tau}\right\} = 0 \tag{6.14}$$

where

$$T = \frac{mu_{T_2}}{3\pi\mu dR_2}$$

Substituting $\pi d^3(\varrho_p - \varrho)/6$ for m, assuming that the particles are spherical, with diameter d, then

$$T = \frac{d^2(\varrho_p - \varrho)u_{T_2}}{18\mu R_2} \tag{6.15}$$

Equation (6.14) reduces to

$$\frac{T}{r}\frac{\mathrm{d}}{\mathrm{d}\tau}\left(r^2\frac{\mathrm{d}\theta}{\mathrm{d}\tau}\right) = 0 \tag{6.16}$$

which gives:

$$r^2\frac{\mathrm{d}\theta}{\mathrm{d}\tau} = \text{constant.} \tag{6.17}$$

By multiplying equation (6.5) by R on both sides

$$R^2\frac{\mathrm{d}\theta}{\mathrm{d}t} = u_T R = u_{T_2}R_2 \tag{6.18}$$

and substituting for u_{T_2} and R_2 it can be shown that the constant in equation (6.17) is unity.

Substituting (6.17) in equation (6.13) gives

$$\frac{\mathrm{d}^2 r}{\mathrm{d}\tau^2} + \frac{1}{T}\frac{\mathrm{d}r}{\mathrm{d}\tau} - \frac{1}{r^3} = 0. \tag{6.19}$$

TABLE 6.1. SEPARATING DISTANCES FOR SPHERICAL PARTICLES IN A SPIRALLING GAS STREAM[197]

Data: Air absolute viscosity = $1{\cdot}78 \times 10^{-5}$ Pa s
density = $1{\cdot}23$ kg m^{-3}
volume flow = $0{\cdot}19$ m^3 s^{-1}
mean axial velocity = $12{\cdot}20$ m s^{-1}

Dimensions: R_1 = 11·1 mm
R_2 = 25·4 mm
R_2/R_1 = 2·3
uR = constant = $0{\cdot}127$ m s^{-1}

Density of dust = $2{\cdot}7$ g cm^{-3}

Particle diameter (μm)	Optimum separating length (mm) [eqn. (6.19)]	Optimum separating length (mm) [eqn. (6.21)]
3	900	680
5	267	224
10	89	61

Equation (6.19) is a non-linear differential equation which cannot be solved directly. A similar equation has, however, been solved with a differential analyser[877] and the results of this calculation for a specific case[197] are given in Table 6.1.

If the second order differential is neglected, equation (6.19) reduces to

$$\frac{dr}{d\tau} = \frac{T}{r^3} \tag{6.20}$$

and this, on integration, gives

$$\tau_2 - \tau_1 = \frac{1}{4T}(r_2^4 - r_1^4). \tag{6.21}$$

Substituting the original dimensions, the time taken for a spherical particle to drift from radius R_1 to radius R_2, using this solution, is

$$t = \frac{9}{2}\left(\frac{\mu}{\varrho_p - \varrho}\right)\left(\frac{R_2}{u_{T_2} d}\right)^2 \left\{1 - \left(\frac{R_1}{R_2}\right)^4\right\}. \tag{6.22}$$

This equation has been used for a solution in the same case as used in the numerical solution of equation (6.19). The results are also given in Table 6.1 and it is seen that this simplification gives an estimated separating distance about 26 per cent less than the numerical solution of the complete equation. Cyclone efficiency calculations based on this simplified equation[207] will therefore tend to over-estimate the effectiveness of the unit.

Vortex theory. (b) *Singular perturbation problem solved by method of matched asymptotic expansions.* A much more sophisticated approach which does not make the "free-vortex" assumptions, and which leads to realistic values of the separation distance for small particles in the range to 10 μm diameter, has been developed by Thompson.[860]

As before, at time t, the particle will occupy a position (R, θ, Z) within a fixed frame of cylindrical coordinates, Z representing the position along the axis. The particle enters the system at $Z = 0$. It is assumed that the gas rotates in an essentially uniform way, with local velocity $\mathbf{u}$ having radial, azimuthal and axial components O, r, ω and u_H, where $\omega\ (= u_T/R)$ is a constant angular speed, and u_H is a constant vertical speed.

The effect of the boundary layer at the wall of the cyclone is neglected, in the expectation that, if u_H is sufficiently large, any new vorticity generated at the wall cannot diffuse inwards before it is blown out at the exit. Particle interactions are also neglected.

The components of the velocity of the particle $\mathbf{U}_p$ at (R, θ, Z) may be written

$$\mathbf{U}_p = (\dot{R},\ R\dot{\theta},\ \dot{Z}) \tag{6.23}$$

where the dot denotes differentiation with respect to t. The relative velocity $\mathbf{U}_{\text{rel}}$ is given by

$$\mathbf{U}_{\text{rel}} = \mathbf{U}_p - \mathbf{U}_G = (\dot{R},\ R\dot{\theta} - R\omega,\ \dot{Z} - u_H). \tag{6.24}$$

The viscous force on the particles, as they are very small, can be obtained from equation (4.4), written as

$$F = -mK\mathbf{U}_{\text{rel}} \tag{4.4}$$

where m is the mass of the particle and K is a linear function of viscosity μ:

$$K = 18\mu/(\varrho_p - \varrho)d^2. \tag{6.25}$$

Applying Newton's second law, subject only to the viscous force on the particle, and of gravity in the axial direction (i.e. $-Z$-direction), gives us the equations of motion of the particle as

$$\ddot{R} - R\dot{\theta}^2 = -K\dot{R}, \tag{6.26a}$$

$$R\ddot{\theta} + 2\dot{R}\dot{\theta} = -KR(\dot{\theta} - \omega), \tag{6.26b}$$

$$\ddot{Z} = -K(\dot{Z} - u) - g. \tag{6.26c}$$

At the entry level ($Z = 0$), $\mathbf{U}_{rel} = \mathbf{O}$, and if the initial distance of the particle from the axis is R_1, then the equations (6.26) must satisfy the initial conditions

$$R = R_1, \quad \dot{R} = 0, \quad \dot{\theta} = \omega, \quad Z = 0, \quad \dot{Z} = u_H \quad \text{at} \quad t = 0. \tag{6.27}$$

The problem is to integrate equations (6.26) and (6.27) in order to find the time t at which the particle will reach the wall at $R = R_2$. Since these equations are linear in R, it will be the same as the time from $R = R_1/R_2$ to 1. Equation (6.26c) is uncoupled from the others, and its solution with the initial conditions is

$$Z = (u - g/K)t$$

showing that the particles have constant axial speed. There is here a restriction below the axial gas speed sufficient to give rise to a vertical drag component that just balances the weight of the particle. Since, in practice, $1000 < K < 10{,}000$ this defect is small, and can be neglected, except for large particles.

The assumption that $\dot{\theta} = \omega$, also implies that $\ddot{\theta} = 0$. Equation (6.26b) then shows that $\dot{R} = \ddot{R} = 0$, and so by (6.26a) and (6.27) $R_1\omega^2 = 0$, which is possible only if the gas does not spin or the particle starts and stays on the axis. Since K is a large parameter, other terms in equations (6.26 a, b) may be compared with K, and in particular, the term $2\dot{R}\dot{\theta}$ in (6.26b) cannot be larger than $0(1)$ for small t, but the term on the right may be either $0(K)$ or $0(1)$. In the first case $\ddot{\theta}$ would also have to be $0(K)$, and so for small t there should be

$$\ddot{\theta} = -K(\dot{\theta} - \omega) + 0(1),$$

giving

$$\dot{\theta} = \omega + Ae^{-Kt} + 0(t).$$

The condition $\dot{\theta} = \omega$ at $t = 0$ requires that $A = 0$, and thus $\ddot{\theta} = 0$ instead of $0(K)$, so that the first assumption is untenable. It follows that $\dot{\theta} - \omega = 0(K^{-1})$ when t is small, which is physically reasonable since, in the limit $K \to \infty$ (i.e. infinite viscosity), in which case the particles would be tied to the surrounding gas and $\mathbf{U}_{rel} = 0$. Writing $\dot{\theta} = \omega(1 - S)$ in (6.26a), where $S < 0(K^{-1})$ compares with K, then

$$\ddot{R} + K\dot{R} - \omega^2 R = 0(K^{-1}). \tag{6.28}$$

The leading approximation to $R(t)$ must therefore take the form

$$R = Ae^{-\lambda_1 t}+Be^{\lambda_2 t} \tag{6.29}$$

with a maximum error of $0(K^{-1})$ when t is small, and where λ_1 and λ_2 are the solution of the quadratic equation $\lambda^2+K\lambda-\omega^2 = 0$, and A and B are constants. This latter equation has the solution

$$\lambda = -\tfrac{1}{2}K\left[1\pm(1-4\omega^2/K^2)^{\frac{1}{2}}\right].$$

Since ω/K is small, it is possible to specify

$$\lambda_1 = \omega^2/K-\omega^4/K^3+0(K^{-5}), \quad \text{and}$$
$$\lambda_2 = -K-\omega^2/K+0(K^{-3}).$$

Using the initial conditions (6.27) it is found that

$$R = R_1\{e^{\omega^2 t/K}+(\omega^2/K^2)e^{-Kt}\}+0(K^{-1}) \tag{6.30a}$$

or, more precisely, putting $\omega/K = \varepsilon$,

$$R = R_1(1-\varepsilon^2)e^{K(\varepsilon^2-\varepsilon^4)t}+\varepsilon^2 e^{-K(1+\varepsilon^2)t}\}+0(K^{-3}). \tag{6.30b}$$

Although the estimate $0(K^{-3})$ for the error in the last equation refers only to a solution of (6.28) when the right side is zero.

The equations (6.30a) and (6.30b) indicate that there are two distinct time scales associated with the motion of the dust. Initially there is a period in which the non-zero relative velocity is set up, i.e. $|\lambda_2|\, t = 0(1)$, which corresponds to a boundary layer at the entrance of the cyclone. When t is $0(1)$, then the boundary-layer terms of the form $e^{\lambda_2 t}$ disappear, and the terms of the $e^{\lambda_1 t}$ begin to predominate, and these now determine the drift to the walls. It is clear from (6.30a) that there is never a time when $\dot{R} < 0$.

The utility of (6.30b) depends on the actual order of magnitude of the term $\omega^2(2S-S^2)$, suppressed in (6.28). In terms of S, equation (6.26b) is

$$\dot{S}+S\left(K+\frac{2\dot{R}}{R}\right) = \frac{2\dot{R}}{R}. \tag{6.31}$$

From (6.30a) it is seen that $\dot{R}/R = 0(\varepsilon)$ for small t, so that in the boundary layer S is at most $0(\varepsilon^2) = 0(K^{-2})$, compared with K.

The primary aim is to determine the time t that a particle takes to move from a radius R_1 to R_2, or to go to $R = 1$ from given values of R_1/R_2 and a viscosity function K, but the solution of (6.26 a and b) by standard algorithms will fail because of the presence of the boundary layer near $Z = 0$. Thus, if the time is scaled to the boundary layer, the particles can be traced through the layer, but rounding errors in a direct solution for $t = 0(1)$ compared with K will introduce spurious artificial layers at each step, whose effect will be to vitiate the numerical solution in the shape of numerical instabilities. The set of equations (6.26 a and b) and (6.27) constitute together a singular perturbation problem[885] which can be solved by the method of matched asymptotic expansions. Two expansions will be found, corresponding to $t = 0(K^{-1})$ and $t = 0(K/\omega^2)$, which are referred to as the *inner* and the *outer* expansions, respectively. It is expected that the value of $R = 1$ will be attained only by the outer expansion in the range of K values that are of direct interest in the cyclone problem.

Inner expansion. In order to trace the initial motion of the particles, a scaled time is adopted, viz.

$$T = K(1+\varepsilon^2-\varepsilon^4)t$$

In full, this would be $T = |\lambda_2|\, t$, but the terms beyond $0(\varepsilon^4)$ will be neglected. Introducing this term, and writing as before $\dot{\theta} = \omega(1-S)$, it is found that (6.26 a and b) are converted to

$$(1+2\varepsilon^2-\varepsilon^4)R''+(1+\varepsilon^2-\varepsilon^4)R'-\varepsilon^2R(1-S)^2 = 0, \tag{6.32}$$

$$(1+\varepsilon^2-\varepsilon^4)\{RS'-2R'(1-S)\}+SR = 0 \tag{6.33}$$

where the prime denotes differentiation with respect to T, while the limits, from (6.27), are

$$R = R_1/R = a, \quad R' = S = 0 \quad \text{at} \quad T = 0. \tag{6.34}$$

From (6.30a) it is clear that R, R' and R'' are all 0(1) at most, and from (6.31) it follows that S and S' are both $0(\varepsilon^2)$. Hence the inner expansion takes the form

$$R = a(\mathbf{R}_0+\varepsilon^2\mathbf{R}_1+\varepsilon^4\mathbf{R}_2+\ \ldots), \tag{6.35a}$$

$$S = (\varepsilon^2\mathbf{S}_1+\varepsilon^4\mathbf{S}_2+\ \ldots) \tag{6.35b}$$

the $\mathbf{R}_i$ and $\mathbf{S}_j$ being functions of T only.

From the boundary conditions (6.34) it follows further, that

$$\mathbf{R}_0 = 1, \quad \mathbf{R}_0' = \mathbf{R}_1' = \mathbf{R}_2 = \mathbf{R}_2' = \ \ldots = \mathbf{S}_1 = \mathbf{S}_2 \ldots = 0 \quad \text{at} \quad T = 0. \tag{6.36}$$

On substituting (6.35 a and b) into (6.32) and (6.33) and equating the coefficients of each power of ε^2 to zero, it is found that

$$\mathbf{R}_0''+\mathbf{R}_0' = 0, \quad \mathbf{R}_1''+\mathbf{R}_1'+2\mathbf{R}_0''+\mathbf{R}_0'-\mathbf{R}_0 = 0,$$
$$\mathbf{R}_2''+\mathbf{R}_2'+2\mathbf{R}_1''+\mathbf{R}_1'-\mathbf{R}_1-\mathbf{R}_0''-\mathbf{R}_0'+2\mathbf{R}_0\mathbf{S}_1 = 0; \tag{6.32'}$$

$$-2\mathbf{R}_0' = 0, \quad \mathbf{R}_0(S_1'+S_1)-2\mathbf{R}_0'(1-S_1)-2\mathbf{R}_1' = 0,$$
$$2\mathbf{R}_0'\mathbf{S}_2+\mathbf{R}_0(S_2'+S_2)+S_1'(R_0+R_1)+S_1(2\mathbf{R}_0'+2\mathbf{R}_1'+S_1)-2\mathbf{R}_2'-2\mathbf{R}_1+2\mathbf{R}_0' = 0 \tag{6.33'}$$

The solution of these equations, subject to (6.36), is routine, and yields

$$R = a\{1+\varepsilon^2(T-1+\mathrm{e}^{-T})+\varepsilon^4[(\tfrac{1}{2}T^2-7T+15-2T^2+8T+15)\mathrm{e}^{-T}]\}, \tag{6.37a}$$

$$S = 2\varepsilon^2(1-\mathrm{e}^{-T}-T\mathrm{e}^{-T})+\varepsilon^4[(-14+\tfrac{4}{3}T^3+6T^2+16T+8T)\mathrm{e}^{-T}$$
$$+(44+6)\mathrm{e}^{-T}]+\ \ldots \tag{6.37b}$$

The first two terms in the inner expansion for R are identical with those obtained directly from (6.30b) by substituting $T = Kt(1+\varepsilon^2-\varepsilon^4)$ and expanding the slowly varying exponential as a power series in T and ε^2. From (6.37 a and b) it is also seen that the particles leave the boundary layer with approximately constant angular speed $\omega(1-2\omega^2/K^2)$ and radial speed ω^2/K.

Outer expansion. When $T = Kt$ becomes comparable with $\varepsilon^2 = \omega^2/K^2$, the inner expansion no longer converges. Simultaneously the elements in the solution which corresponds to the terms $e^{K(\varepsilon^2-\varepsilon^4)t}$ in (6.30b) become important, and are no longer represented by a power series in T. To examine the continuation of the solution over this period of time, the second scaled time variable

$$\tau = K(\varepsilon^2-\varepsilon^4)t = (\omega/K^2-\omega^4/K^3)t$$

is adopted. As before, $\dot{\theta} = \omega(1-S)$ is substituted in (6.26 a and b), and the following are obtained:

$$\varepsilon^2(1-2\varepsilon^2)R''+(1-\varepsilon^2)R'-R(1-S)^2 = 0, \tag{6.38a}$$

$$\varepsilon^2(1-\varepsilon^2)\,[S'R-2R'(1-S)]+RS = 0. \tag{6.39a}$$

The prime now denotes differentiation with respect to τ. The outer expansion will therefore have the appearance

$$R = a(\rho_0+\varepsilon^2\rho_1+\varepsilon^4\rho_2+\ \ldots), \tag{6.40a}$$

$$S = \varepsilon^2\sigma_1+\varepsilon^4\sigma_2+\ \ldots \tag{6.40b}$$

where ρ_i and σ_j are functions of τ only.

In accordance with the principle of matched asymptotic expansions, arbitrary constants appearing in the terms of (6.40 a and b) will be evaluated assuming the equality of the expansions in the limit $K \to \infty$, which produces the dual limit $T \to \infty$ and $\tau = 0$.

When (6.40 a and b) are substituted into (6.38a) and (6.39a) and coefficients of like powers of ε^2 are equated, the following set of differential equations is obtained:

$$\rho_0'-\rho_0 = 0, \quad \rho_1'-\rho_1+\rho_0''-\rho_0'+2\rho_0\sigma_1 = 0,$$

$$\rho_2'-\rho_2-2\rho_0''+\rho_1'-\rho_2+2\rho_1\sigma_1-\rho_0(\sigma^2-2\sigma_2) = 0, \tag{6.38b}$$

$$\rho_0\sigma_1-2\rho_0' = 0, \quad \rho_0\sigma_2+\rho_0\sigma_1'+\sigma_1(\varrho_1+2\varrho_0')-2\varrho_1'+2\varrho_0' = 0. \tag{6.39b}$$

The first equation of (6.38b) has the solution $\rho_0 = A_0e^{\tau}$ where A_0 is a constant. Since $\tau = \varepsilon^2(1-2\varepsilon^2)\,T$, $\varrho_0 = A_0+0(\varepsilon^2)$ when T is large and τ is small. To procure a match at 0(1) between (6.37a) and (6.40a), it is thus necessary that $A_0 = 1$, and so $\rho_0 = e_1^{\tau}$ corresponding exactly to the first term in (6.30). Similarly the first equation in (6.39b) yields $\sigma_1 = 2$ which agrees with the limit $T \to \infty$ in (6.37b) to $0(\varepsilon^2)$. The second equation in (6.38b) has solution $\rho_1 = -(4\tau+A_1)e^{\tau}$ where A_1 is constant. Putting $\tau = \varepsilon^2(1-2\varepsilon^2)T$ in (6.40a)

$$R = a\{e^{\tau}-\varepsilon^2(A_1+4\tau)e^{\tau}+0(\varepsilon^4)\}$$

$$= a\{1+\varepsilon^2(T-A_1)+0(\varepsilon^4)\}$$

when expanding e^{τ} as $1+\varepsilon^2T+0(\varepsilon^4)$.

The match with (6.37a) is thus extended to $0(\varepsilon^2)$ if $A_1 = 1$, so that $\rho_1 = -(4\tau+1)e^{\tau}$. This yields $\sigma_2 = -14$, which again agrees with limit $T \to \infty$ in (6.37b) to $0(\varepsilon^2)$.

Finally, the last equation of (6.38b) has the solution $\rho_2 = (4\tau^2+46\tau+A_2)e^{\tau}$, where A_2 is constant. Putting $\tau = \varepsilon^2(1-2\varepsilon^2)\,T$ in (6.40a) yields

$$R = ae^{\tau}\{1-\varepsilon^2(4\tau+1)+\varepsilon^4(4\tau^2+42\tau+A_2)+\ \ldots\},$$
$$= a\{1+\varepsilon^2(T-1)+\varepsilon^4(\tfrac{1}{2}T^2-7T+A_2)+\ \ldots\}.$$

Comparison with (6.37a) shows $A_2 = 15$. The final solution for R in the outer region is therefore

$$R = ae^{\tau}\{1-\varepsilon^2(4\tau+1)+\varepsilon^4(4\tau^2+42\tau+15)+0(\varepsilon^6)\} \tag{6.40}$$

where $\varepsilon = \omega/K$ and $\tau = \varepsilon^2K(1-\varepsilon^2)t$.

Extraction time for particle. The extraction time τ_e for any particle is the value of τ in (6.40) at which $R = 1$ (for $a = R_1/R_2$). This gives

$$\frac{1}{a}e^{-\tau} = 1-\varepsilon^2(4t+1)+\varepsilon^4(4\tau^2+42\tau+15). \tag{6.41}$$

If ε is small, then $\tau_\varepsilon = \tau_0 = \ln R_2/R_1$. At the same degree of accuracy as is presented by this the exact solution may be written in the form

$$\tau_e = \tau_0+\varepsilon^2\tau_1+\varepsilon^4\tau_2$$

where τ_1 and τ_2 are $0(1)$ compared with ε.
Substituting gives

$$1-\varepsilon^2\tau_1+\varepsilon^4(\tfrac{1}{2}\tau_1^2-T_2)+\ \ldots$$
$$= 1-\varepsilon^2(4\tau_0+1)+\varepsilon^4(4\tau_0^2+38\tau_0+15)$$

whence $\tau_1 = 4\tau_0+1$ and $\tau_2 = 4\tau_0^2-34\tau_0-\frac{29}{2}$. Thus the extraction time from $R = a = R_1/R_2$ is

$$\tau_e = \tau_0+\varepsilon^2(4\tau_0+1)+\varepsilon^4(4\tau_0^2-34\tau_0-\tfrac{29}{2})+0(\varepsilon^6) \tag{6.42}$$

where $\tau_0 = \ln R_2/R_1$. In terms of unscaled time values, $\tau_e = K(\varepsilon^2-\varepsilon^4)t_e$, where t_e is the unscaled extraction time. So, the extraction time, t_e, is given by

$$t_e = \frac{K}{\omega^2}\left\{\tau_0+\varepsilon^2(5\tau_0+1)+\varepsilon^4\left(4\tau_0^2-29\tau_0-\frac{27}{2}\right)\right\}. \tag{6.43}$$

This can be readily solved by substituting the following for K, ω, τ_0 and ε:

$$K = 18\mu/(\varrho_p-\varrho)d^2, \quad \omega = u_T/R_1$$
$$\tau_0 = \ln R_2/R_1 = 2{\cdot}303 \log_{10} R_2/R_1$$
$$\varepsilon = \omega/K = u_T(\varrho_p-\varrho)d^2/18R\mu.$$

Using the same data as in Table 6.1, the following separating distances were obtained:

3 μm diameter particle, 202 mm, 5 μm diameter, 82 mm, and 10 μm diameter, 2·5 mm.

This equation can also give an extraction time of 0·20 sec for a 5 μm diameter particle, 0·054 sec for a 10 μm diameter particle for a 50 cm diameter, 2·5-m long cyclone, with an entrance velocity of 15 m s^{-1} and a gas throughput of 6 m s^{-1}, similar in pattern to Fig. 6.16a.

6.3. STRAIGHT-THROUGH CYCLONES WITH FIXED IMPELLERS

Straight-through cyclones or "vortex" gas cleaners are able to handle very large quantities of gases in a small plant. However, because of the high gas velocities used, a large amount of re-entrainment occurs: eddies formed at the walls help to "bounce" particles back into

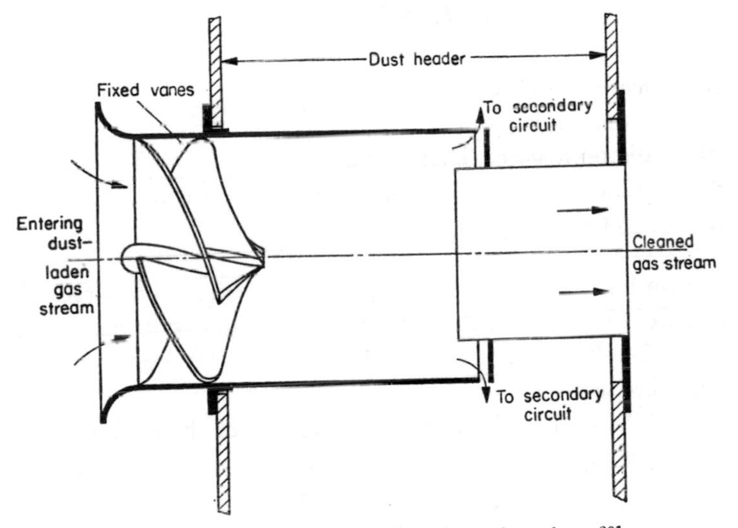

FIG. 6.2. Fixed impeller straight-through cyclone.[201]

the main gas stream. In practice this is reduced by introducing a water spray at the dirty gas inlet.

In a fixed impeller straight-through cyclone (Fig. 6.2) it is usual to have the suction fan on the clean air side, to reduce erosion damage to the fan blades. The separation chamber, following the fixed impeller where most of the pressure drop occurs, is therefore at a lower pressure than the gases entering the impeller. It is therefore necessary to provide additional suction on the dirty gas (concentrated dust) exit line to prevent sucking back of the collected particles.

Because of the simple flow pattern in the straight-through cyclone it is possible to calculate theoretically[822] the smallest particle which should be completely removed from the gas stream. Four assumptions have to be made:

(i) There are no heat gains to the cyclone from the surroundings due to the cooling effect of the adiabatic expansion through the impeller.
(ii) The pressure losses occur in the impeller only.
(iii) The free vortex equation holds for the motion of the particle in the separation chamber.
(iv) The particles leave the blades of the impeller at the blade angle.

Assuming adiabatic expansion through the impeller blades, the pressure p, temperature T and volume Q of the gas in the separation chamber (denoted by subscript c) can be calculated from the initial conditions (denoted by subscript i) before the impeller:

$$Q_c = Q_i\left(\frac{p_i}{p_c}\right)^{1/\gamma} \tag{6.44}$$

and

$$T_c = T_i\left(\frac{p_i}{p_c}\right)^{\gamma-1} \tag{6.45}$$

where γ = specific heat ratio C_p/C_v
= 1·67 for monatomic gases,
1·40 for diatomic gases (including air),
1·30 for triatomic gases (including superheated steam),
1·135 for wet steam.

The pressure drop for a straight-through cyclone, which occurs largely in the impeller, can be calculated from the surface area of the impeller and walls exposed to the gases.[854] Experimentally it is found to be about 1·25 kPa, and this may be taken as an approximate value over the impeller section.

If the diameter of the cyclone is D and the diameter of the core D_c, then the *average* velocity of the gases leaving the curved impeller blades u_c in terms of the initial velocity u_i is given by

$$u_c = u_i\left(\frac{D^2}{D^2-D_c^2}\right)\frac{Q_c}{Q_i} \tag{6.46}$$

$$= u_i\left(\frac{D^2}{D^2-D_c^2}\right)\left(\frac{p_i}{p_c}\right)^{1/\gamma}. \tag{6.47}$$

The velocity u_c can be resolved into three components—tangential, axial and radial (Fig. 6.3). If the angle at which the gases leave the blades of the impeller is the same as the blade angle, α, and the central core is extended through the separation chamber, the *average* velocity tangentially u_{CT} is:

$$u_{CT} = u_c \cos\alpha$$

$$= u_i\left(\frac{D^2}{D^2-D_c^2}\right)\left(\frac{p_i}{p_c}\right)^{1/\gamma}\cos\alpha \tag{6.48a}$$

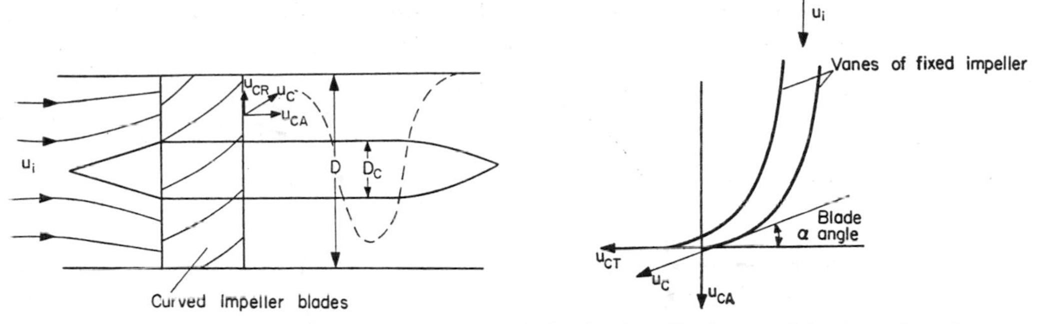

FIG. 6.3. Resolution of velocity and path of particle leaving impeller in a straight-through cyclone. (a) Path of particle. (b) Velocity resolution.

while the average velocity in the axial direction, u_{CA}, is

$$\begin{aligned} u_{CA} &= u_c \sin \alpha \\ &= u_i\left(\frac{D^2}{D^2 - D_c^2}\right)\left(\frac{p_i}{p_c}\right)^{1/\gamma} \sin \alpha. \end{aligned} \tag{6.48b}$$

The residence time in the separation chamber can now be found from the length of the chamber L and the axial velocity:

$$t = \frac{L}{u_{CA}} = \frac{L}{u_i \sin \alpha}\left\{1 - \left(\frac{D_c}{D}\right)^2\right\}\left(\frac{p_c}{p_i}\right)^{1/\gamma}. \tag{6.49}$$

In the general theory of cyclone separators (Section 6.2) an expression [equation (6.22)] was obtained for the time taken for a particle of diameter d to drift from an inner to an outer radius. Then, for given chamber dimensions, it is possible to calculate the smallest particle (diameter $d_{\min}$) that could theoretically be collected in the straight through cyclone:

$$d_{\min} = \frac{3}{4}\frac{D}{\cos \alpha}\left\{1 - \left(\frac{D_c}{D}\right)^2\right\}\sqrt{\left(\frac{\mu \sin \alpha}{(\varrho_p - \varrho)u_i}\left\{\frac{p_c}{p_i}\right\}^{1/\gamma}\right)}. \tag{6.50}$$

If the further assumption is made that the particles in the gas stream are evenly distributed on leaving the impeller, then the fractions of particles smaller than $d_{\min}$ can also be calculated. Such a calculation would tend to be conservative as separation will have started in the impeller blades.

If the tangential gas velocity in a horizontal straight-through cyclone is low (less than 15 m s^{-1} for a 0·6 m diameter cyclone) then the net normal force on the particle varies appreciably according to the particle position. Thus in the bottom position (Fig. 6.4) the gravity acceleration may be added on to the centrifugal acceleration:

$$\frac{u_{CT}^2}{\sqrt{\left(\frac{D_c^2 + D^2}{8}\right)}} + g. \tag{6.51}$$

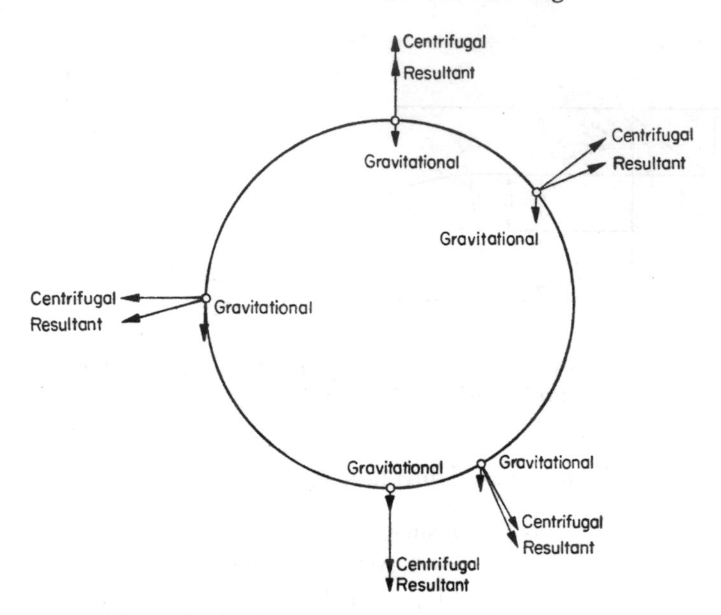

FIG. 6.4. Forces on particle moving in circular motion in a vertical plane. The force parallelograms are based approximately on particles with a tangential velocity of 0·6 m s^{-1} moving on a 0·3 m radius.

In the top position the gravity acceleration must be subtracted from the centrifugal acceleration:

$$\frac{u_{CT}^2}{\sqrt{\left(\frac{D_c^2+D^2}{8}\right)}}-g \tag{6.52}$$

while, generally, in the intermediate positions the net resolved force can be obtained by adding $g \cos \theta$:

$$\frac{u_{CT}^2}{\sqrt{\left(\frac{D_c^2+D^2}{8}\right)}}+g \cos \theta \tag{6.53}$$

where θ is the angle which the radius makes with the vertical.

For tangential velocities greater than 15 m s^{-1} when D is not too large (less than 0·6 m) the maximum correction becomes less than 0·1 per cent and can be neglected.

6.3.1 Straight-through cyclones with reverse flow

An important modification of the straight-through cyclone with fixed impellers is the type using the *reverse-flow* principle. This was patented by Zenneck and Schaufler in 1953, and gives a major improvement in cyclone efficiency by reducing entrainment by "bouncing" small particles off the wall. The principle of the system is shown in Fig. 6.5. The dusty gas enters the main cyclone through a carefully designed spiral entry, similar to Fig. 6.3(a), and leaves at the far end, dust particles having been thrown from this spiralling gas. The reverse gas stream enters through tangential jets in the wall, and spirals in the same tangential direction, but in the opposite axial direction to the dusty gas stream, entraining the dust particles thrown from the central spiral. The outer gas spiral moves into the hopper past the inlet for the dusty gases, reverses in direction, and deposits the particles. It then joins

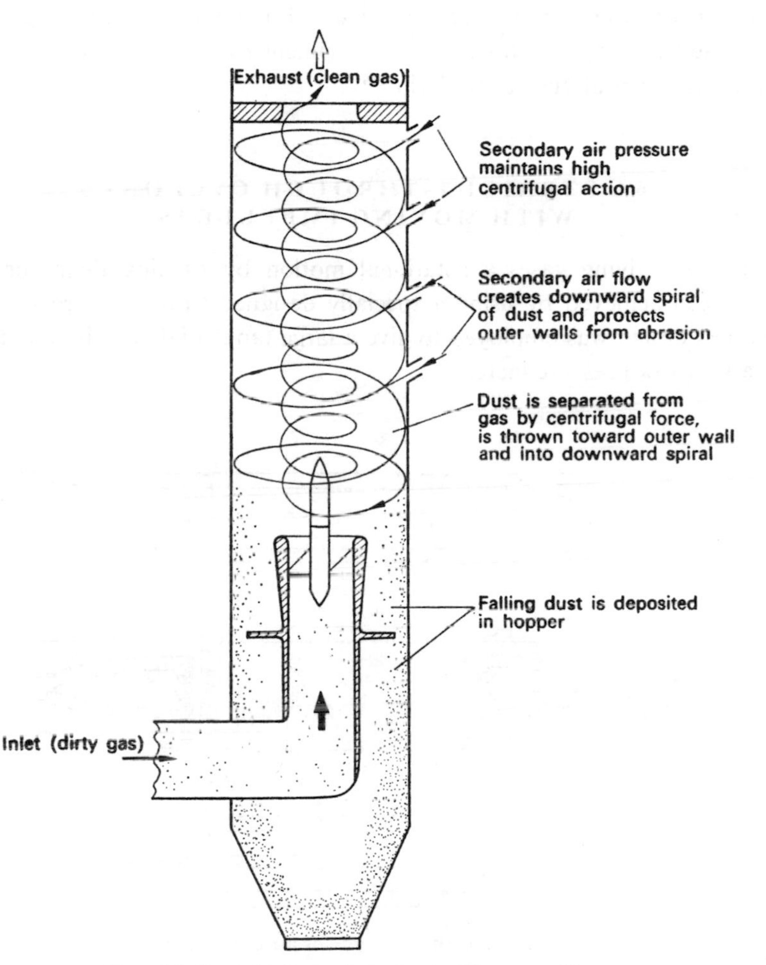

FIG. 6.5. Straight-through cyclone with reverse flow.

the dusty gas stream. The action of this double spiral is like that surrounding the vortex section of a tornado, and has been sold in some countries with this as a trade name.

The theory of this double-spiral system has been discussed in detail by Schmidt,[731] who has given a potential flow pattern $u_T = \xi/R$ to the outer spiral, a rotational flow pattern ($u_T = \xi R$) to the inner spiral, with a mixed-flow zone separating these two regions. In this mixed-flow region in which the potential flow changes to the rotational flow, a reversal of direction takes place in the secondary (axial) flow. Schmidt has set up the equations of motion for a particle in a similar way to Section 6.2, but with the complex interaction pattern, is unable to solve these for this cyclone type.

The original design of inlet had a complicated, aerodynamically well-designed annular section and central core section, but a much simpler and less costly section is found adequate in commercial designs.[446] Other design features which were found to apply were that the tangential reverse-flow gas inlets were inclined at 30° to the horizontal, and were placed at 180° intervals around the periphery of the cylinder. Furthermore, increasing the number of nozzles decreased the pressure loss. The development of optimum dimensions and other design factors has been discussed by Klein.[445]

6.4. STRAIGHT-THROUGH CYCLONES WITH MOVING IMPELLERS

An alternative to giving gases a rotational motion by sucking them through a fixed impeller has been achieved by using a specially designed turbo-compressor (Fig. 6.6).[947] A very complex design was employed to give a large tangential velocity component to the gases with a very low pressure increase.

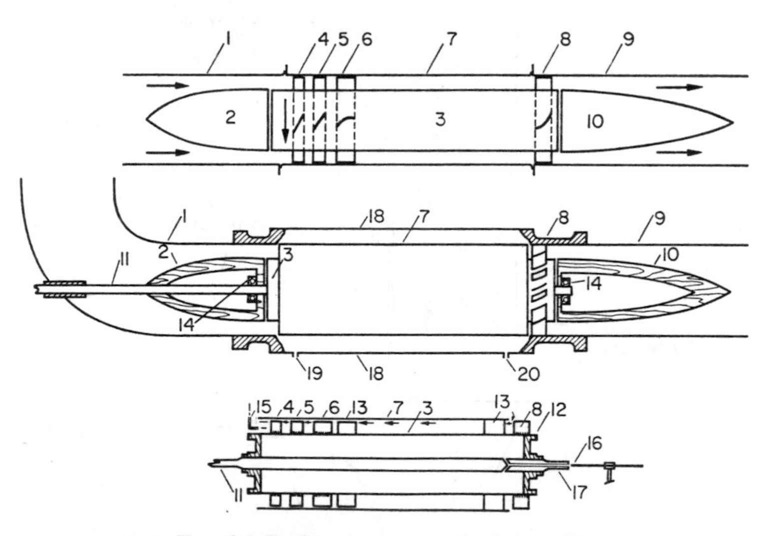

FIG. 6.6. Turbo-compressor air cleaner.[947]

1 is a fixed cylindrical inlet duct, 2 a streamlined fairing, 3 the rotating cylindrical inner sleeve of the centrifuge, and 7 the rotating outer sleeve. The compressor turbine blades 4, 5 and 6 are attached to the inner casing 3 and are separated from the outer casing 7 by small clearances. The reaction turbine blades 8 are also attached to the inner sleeve but are not included in the outer casing 7. 9 represents a cylindrical outlet duct and 10 a streamlined fairing fixed in this duct. Two clearances, left between the outer casing of the centrifuge 7 and the fixed inlet and outlet ducts, permit free rotation of the centrifuge and discharge of the liquid film.

The inlet and outlet ducts are connected by a fixed outer sleeve 18, which seals the centrifuge for operation on gases above or below atmospheric pressure. Two small outlets 19 and 20 are provided. The inlet duct 1 has been connected to a bend to permit the mounting of an electric motor for driving the centrifuge. Power is transmitted by shaft 11, which passes through fairing 2 and is supported on two sets of self-aligning ball races 14. For a large installation the motor would be mounted inside one of the fairings.

The mean velocity u_C of the gases (and, as a first approximation, of the particles) can be calculated from:

$$u_C = N\pi\left(\frac{D+D_c}{2}\right)e \tag{6.54}$$

where N = number of turbine revolutions/unit time,
e = turbine efficiency (usually about 85 per cent).

From the mean turbine velocity and each blade angle, the tangential and axial velocity components can be calculated. A procedure similar to that used with the fixed impeller unit can then be adopted for calculating fractional efficiencies for each row of turbine blades.

Three pairs of rows of turbine blades in series were used, the blade angles were 75° and 60° in the first pair, 63° and $38\frac{1}{2}$° in the second, and 43° and $-8\frac{1}{2}$° in the final pair. Rotor speeds of 83·3 s^{-1} were used and the chamber was 150 mm diameter, with a 100 mm hub, giving a tangential gas velocity of 33 m s^{-1} with a radial velocity of 9 m s^{-1}. A detailed theoretical analysis of a unit of this type is extremely difficult, and will not be attempted here.

6.5. CENTRIFUGAL-FAN CYCLONE SEPARATORS

Cyclone separators based on conventional fan design are similar to the axial flow turbo-blower (Section 6.4). These will act as induced draft fans in addition to acting as dust-cleaning devices. Their aerodynamic characteristics are those of a forward-curved centrifugal fan, for which the relations between volume flow, pressure difference and power consumption follow the standard pattern.

The dirty gas entering the fan is turned through 180° into the volute chamber, while the dust particles are accelerated by the impeller fan blades through an angle of 90°. The maximum velocity given to the particles can be calculated from the fan speed (number of revolutions per unit time N and diameter D), and the angle of the blades (α):

$$u_C = N\pi D \cos\alpha. \tag{6.55}$$

The gas-flow pattern, and the particle trajectory in these units, is complex and no detailed study is available at the present time. It is therefore not possible to calculate the size of particle for which 100 per cent collection is to be expected. In general, it can, however, be stated that while the centrifugal forces developed in these units is high, the gas-stream residence time is comparatively small.

6.6. CONVENTIONAL (REVERSE-FLOW) CYCLONES

The conventional cyclone (Fig. 6.7) has a much more complex flow pattern than the fixed impeller straight-through cyclone. Essentially the flow consists of a double spiral, the outer spiral moving down towards the hopper while the inner one moves up towards the exit pipe. Superimposed on this is a secondary gas flow from the outer towards the inner spiral.

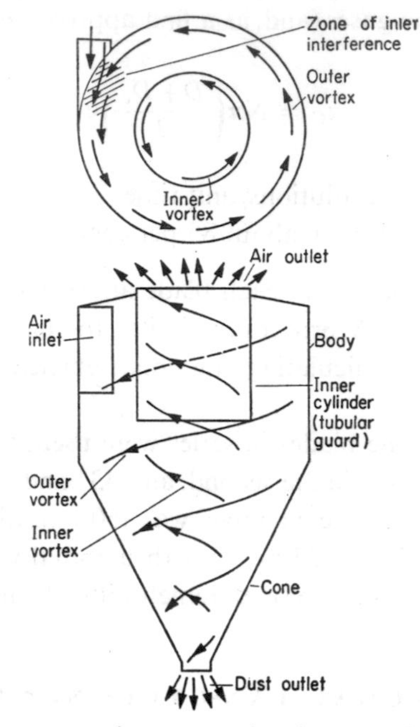

FIG. 6.7. Flow pattern in a conventional, reverse-flow cyclone.[7]

These patterns have been extensively investigated experimentally by a number of workers[258, 431, 515, 585, 765, 803] and have recently been the subject of a detailed review.[391] The methods used have included extensive pitot-tube measurements with a globe pitot tube,[585] smoke visualization[765] in gas cyclones, dye tracers,[113, 803] and aluminium powder combined with an elegant optical technique in liquid cyclones.[431]

The most detailed measurements in gas cyclones have been carried out by ter Linden at Delft,[515] in a cyclone of normal dimensions with a scroll entry. Ter Linden's diagrams, showing tangential, radial and axial velocities as well as total and static pressures, are given in Fig. 6.8. Similar patterns, but more sharply defined, were also obtained by Kelsall[431] for a hydraulic cyclone (Fig. 6.9). From the tangential and axial velocity curves, it can be seen that there is a zone near the walls where the gases spiral downwards with increasing tangential velocity, while towards the centre the gases move towards the exit with greater tangential velocity than at the same height near the wall. The tangential velocities rise to a maximum at a ring about one-half to two-thirds of the exit tube diameter. Within this ring is a central core with decreasing tangential velocity although the axial velocity tends to a maximum. The radial velocity which is very much lower than the tangential velocity tends to be almost constant over much of the cyclone cross-section. The positive value indicates drift towards the axis, except at the central core where the drift is outwards. Near the core ring the radial velocity tends to have a zero value, giving some experimental confirmation to the assumptions made in calculating the cyclone "cut" particle diameter.

The pressures at various radii show positive pressure near the wall, not very different from the pressure at the gas entry and a zone of negative pressure at the core, which is found to

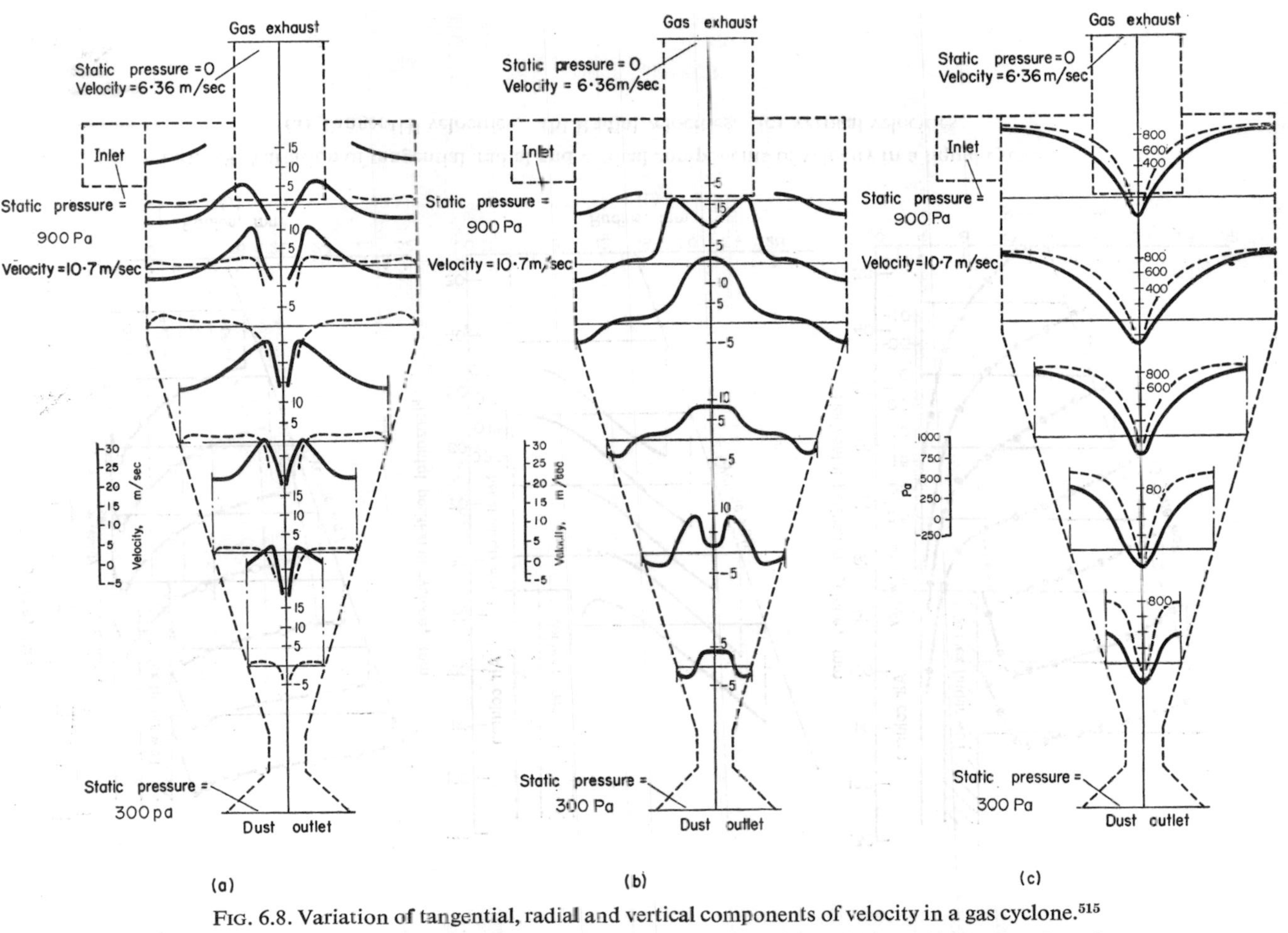

FIG. 6.8. Variation of tangential, radial and vertical components of velocity in a gas cyclone.[515]
(a) Variation of tangential velocity and radial velocity. (b) Variation of vertical velocity. (c) Total and static pressures.

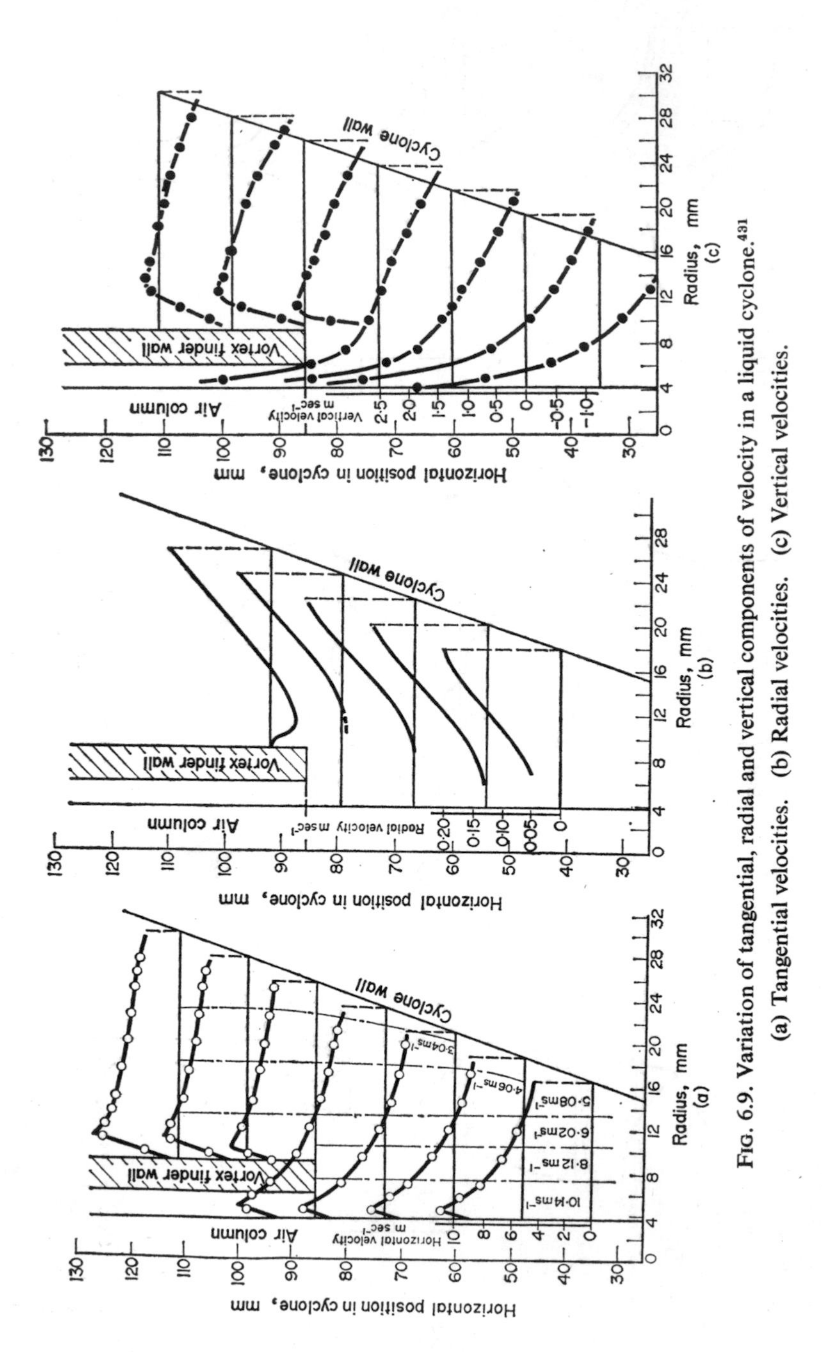

FIG. 6.9. Variation of tangential, radial and vertical components of velocity in a liquid cyclone.[431]
(a) Tangential velocities. (b) Radial velocities. (c) Vertical velocities.

extend right into the hopper. Any air leaks into the hopper tend to cause an upward current right through the cyclone and out the exit pipe, re-entraining the collected dust. Conversely when the hopper is placed under suction a marked improvement in efficiency can be obtained.[230]

There is some disagreement whether the secondary flow pattern due to the axial and radial velocity components consist of a single vortex over the whole length of the cyclone (Fig. 6.10a) or a double vortex (Fig. 6.10b), one near the top next to the exit pipe and one in the

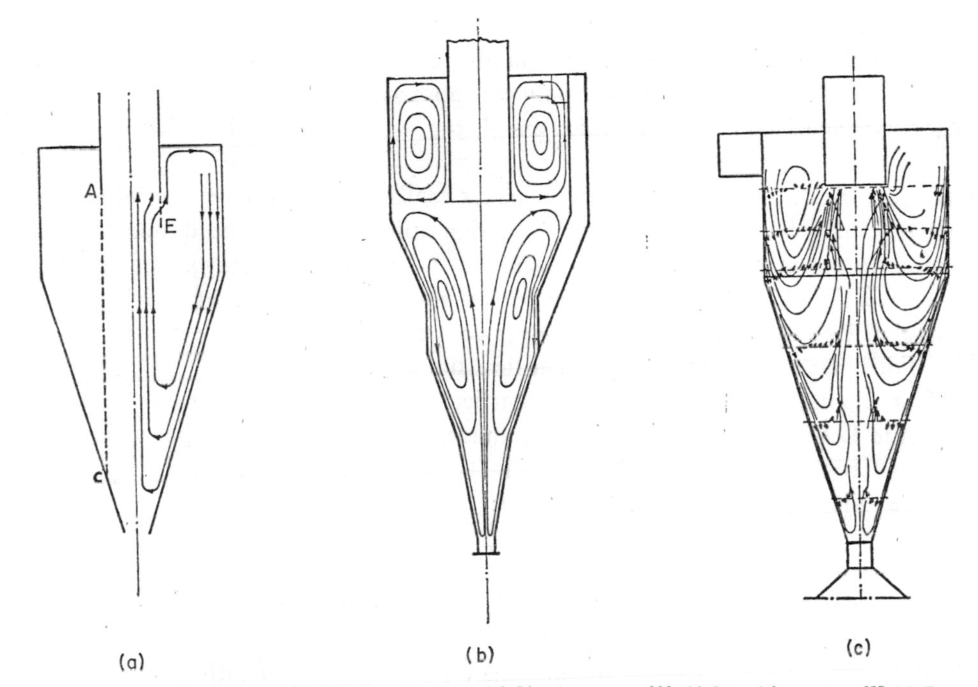

FIG. 6.10. Secondary flow patterns in cyclones. (a) Single vortex.[266] (b) Double vortex.[915] (c) Intermediate pattern.[516]

conical section. The concentration of dust near the top, and the marked improvement in cyclone performance when this is removed by a special duct (van Tongeren or Ambuco types), tends to favour the double vortex, as do some ink tracer studies in hydraulic cyclones.[264] The actual secondary flow pattern is probably an intermediate pattern of the type plotted by ter Linden[516] (Fig. 6.10c).

The variation of tangential velocity with radius outside the central core of a conventional gas cyclone follows the vortex equation (6.3) with the exponent n equal to 0·5 in all cases measured experimentally with the exception of the early work by Prockat,[663] and the more recent studies by First[258] and Alexander[9] which give 0·88 and 0·7, respectively, for the exponent n. For liquid cyclones the exponent tends to vary between about 0·7 and 1,[391] the unit values being typical of work with cylindrical vessels. Present experimental evidence therefore tends to favour 0·5 as the value most applicable to the relation between tangential velocity and radius for gas cyclones, and for the calculation of tangential velocity the equation

$$u_T R^{0.5} = \text{constant} \qquad (6.56)$$

will be used in the next sections.

6.7. PREDICTION OF CYCLONE PERFORMANCE

A number of attempts have been made to predict the effectiveness of conventional reverse-flow cyclones. Theoretical models make a number of assumptions which are not confirmed experimentally, and the predictions are only very approximate. Other methods of predicting efficiency use experimental factors which enable the whole fractional efficiency curve to be forecast with considerable precision.

A satisfactory theoretical model which makes realistic assumptions regarding the path of a particle in a cyclone of conventional design still remains to be developed, and no decision as to the most promising approach can be made at this stage. The assumptions used for several attempts to calculate the *critical particle diameter*, the particle size, which it is calculated will be collected with 100 per cent efficiency, are presented together with modifications that may lead to a more realistic estimate.

The first attempt to estimate the critical particle size was made by Rosin, Rammler and Intelmann in 1932.[706] The major assumption made was that the particle has to reach a wall by moving across a gas stream, which retains its shape after leaving the cyclone entry, in order to be collected. Other assumptions were:

(i) Particles do not influence one another.
(ii) When particles reach the cyclone wall there is negligible chance of their being re-entrained.
(iii) Stokes' law can be applied to the movement of the particles relative to the gas stream.
(iv) Buoyancy effects can be neglected.
(v) Cyclones are assumed to be cylindrical in shape, with diameter D, with entrance cross-section a by b.
(vi) The tangential velocity of the particles is constant and independent of position.

The critical particle size d_{crit} can be found by the following calculation: If the gas stream, with an entrance velocity u_i retains this velocity as it spirals $\mathcal{N}$ times around the cyclone, of diameter D, the length of the effective gas path will be $\pi D\mathcal{N}$. The critical particle size is the particle which will move a distance b, the width of the entrance pipe, during the residence time of the gas stream in the cyclone, which is found from the ratio of the path length and the steam velocity $(\pi D\mathcal{N}/u_i)$. The force moving the particle towards the wall is the centrifugal force [equation (6.1)], and the resistance force is Stokes' law [equation (4.4)]:

$$F = 3\pi\mu d_{crit} u_i b/\pi D\mathcal{N} \qquad \text{(Stokes' law resistance)},$$

$$F' = \pi d_{crit}^3 \varrho_p u_i^2/6R \qquad \text{(centrifugal force)}$$

It is assumed that $u_i = u_T$ at a distance R from the centre of the cyclone, where

$$R = D/2 - b/2 \qquad \text{(average distance of particle in inlet from cyclone axis)}$$

If these are equated, then

$$d_{\text{crit}} = 3\left\{\frac{2\mu}{\pi\varrho_p u_i}\cdot\frac{R}{N}\left(1-\frac{2R}{D}\right)\right\}^{\frac{1}{2}}. \tag{6.57}$$

The number of revolutions, $\mathcal{N}$, are to be found from:

$$\mathcal{N} = \frac{tu_i}{\pi D} \tag{6.58}$$

where t = residence time of the gas stream,
= V/Q,
Q = gas throughput per unit time,
V = volume of cyclone.

The effective volume of a cyclone with dimensions as shown in Fig. 6.11 is

$$V = \frac{\pi}{4}\left\{\left(\frac{H-h}{D-B}\right)\left(\frac{D^3-B^3}{3}\right)+D^2h-D_e^2S\right\}. \tag{6.59}$$

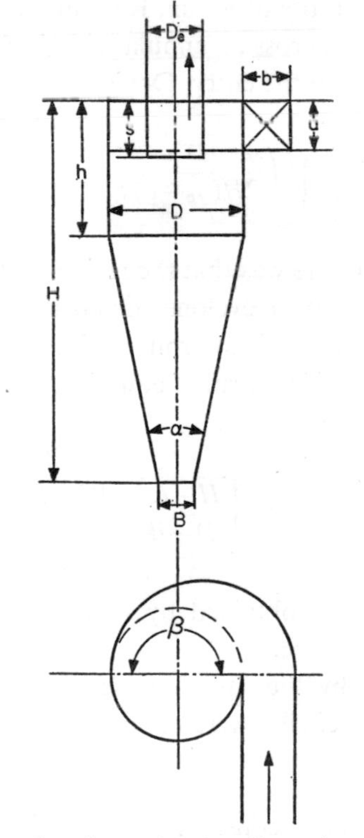

FIG. 6.11. Conventional cyclone, showing dimensions used in calculations.

Dalla Valla,[193] however, has suggested that an experimental value for $\mathcal{N}$ varying between 0·5 and 3 should be used.

The importance of the gas-stream configuration as it enters the cyclone has also been stressed by Rietema.[686] He suggests that all collection takes place in the layer of gas around the cyclone wall which has the thickness of the width of the cyclone entrance. The critical particle diameter is found from:

$$d_{\text{crit}} = \sqrt{\left\{\frac{9u_H}{u_i} \cdot \frac{D}{aH} \frac{Q\mu}{\Delta p} \left(\frac{\varrho}{\varrho_p - \varrho}\right)\right\}} \tag{6.60}$$

where u_H = axial velocity.

Rietema states that above an unspecified Reynolds number, the ratio u_H/u_i can be considered approximately constant, and then for a cyclone with known dimensions D, a and H, the critical diameter becomes a function of the volume throughput and pressure drop.

These calculations of critical diameter which depend on the width of the wall layer can yield a complete fractional efficiency curve by considering the distance through which a particular particle will move in the allotted time. Thus, the d_{50} particle size (50 per cent collected) can be obtained by using the half distance across the entry. Those particles which can move this distance will be collected, while the other half will escape.

Using assumptions similar to those of Rosin, Rammler and Intelmann, except for assuming that the particle has to move across an annular ring in free vortex motion instead of the gas stream of the same shape as in the entry, Davies[207] calculated that

$$d_{\text{crit}} = \frac{3}{2} \sqrt{\left(\frac{D^2\mu}{2H(\varrho_p - \varrho)u_i} \left\{1 - \left(\frac{D_e}{D}\right)^4\right\}\right)}. \tag{6.61}$$

The other major assumption here was that the residence time was given by H/u_i, which is very approximate, ignoring all other cyclone dimensions except overall height. A more practical residence time can be obtained from cyclone volume (equation 6.59) and gas throughput Q, which gives a modified form of equation (6.61):

$$d_{\text{crit}} = 3 \sqrt{\left(\frac{Q\mu}{2\pi u_i^2(\varrho_p - \varrho)} \left\{\frac{1 - (2D_e/3D)^4}{\frac{H-h}{D-B} \cdot \frac{D^3 - B^3}{D^2} + h - S\left(\frac{D_e}{D}\right)^2}\right\}\right)}. \tag{6.62}$$

Results similar to Davies were obtained by Feifel[255] who also assumed a free-vortex particle path in a cylindrical cyclone.

The values of d_{crit} obtained by the equations deduced by Rosin *et al.*,[706] Davies,[207] ter Linden,[515] Dalla Valla[193] and Feifel[255] were compared by Svanda[845] for a typical cyclone, and the results expressed as

$$d_{\text{crit}} = \text{“constant”} \times 10^{-6} \sqrt{\frac{1}{u_i}}. \tag{6.63}$$

TABLE 6.2. COMPARISON OF CRITICAL PARTICLE DIAMETER CALCULATIONS FOR CYCLONES USING DIFFERENT EQUATIONS (J. Svanda)[845]

Cyclone characteristics: $D = 50$ mm, $D_e = 20$ mm, $H = 220$ mm, $a = 24$ mm, $b = 10$ mm

Particle density: 2·65 g cm^{-3}, gas viscosity $1{\cdot}68 \times 10^{-5}$ kg m^{-1} s^{-1}.
Diameter of dangerous zone (ter Linden)/D = 2·3
$\mathcal{N}$ (Dalla Valla) = 3

Author of equation	"Constant" in eqn. (6.63)
Rosin, Rammler and Intelman[706]	7·32
Davies[207]	4·67
Ter Linden[515]	5·02
Dalla Valla[193]	8·92
Feifel[255]	8·95

The calculated "constant" is shown in Table 6.2 and indicates the similarity in the results obtained by the different workers, in spite of the crude assumptions made.

The assumptions made in the theoretical derivations for critical particle diameter are not confirmed in practice, and methods for calculating cyclone efficiency based on experiment tend to be more reliable. These usually calculate the *cut* of a cyclone, which is the particle size for which 50 per cent collection efficiency is achieved.

Particles spinning in a ring at the point of maximum tangential velocity have a 50 per cent chance of being collected or passing into the exit pipe. To continue spinning in this ring the outward movement of the particles towards the wall must be balanced by the inward (radial) drift of the gases.

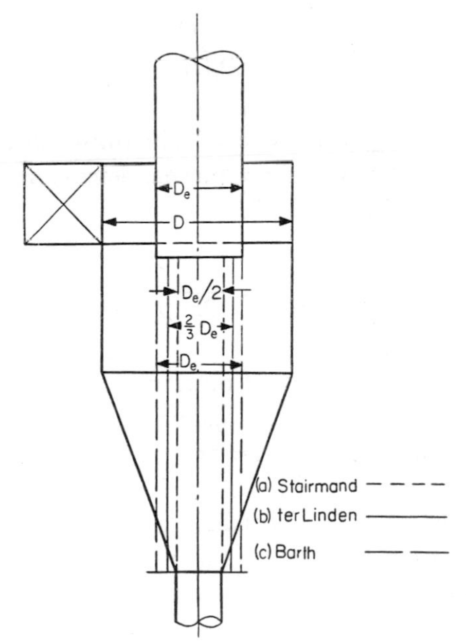

FIG. 6.12. Central cores assumed by various workers.

Stairmand[803] has suggested that the maximum tangential velocity $u_{T\,max}$ is reached at a radius one-half of that of the exit pipe (diameter D_e). The average radial drift in to the central core is given by dividing the gas flow by the surface area of the central core (Fig. 6.12a):

$$u_{R\,av.} = \frac{2Q}{\pi D_e(H-S)}. \tag{6.64}$$

If the fluid resistance encountered by the particle is given by Stokes' law (equation 4.4), using $u_{R\,av.}$ as the velocity, and this is balanced by the centrifugal force on the particle, then

$$d_{50} = \frac{3}{u_i\varphi}\sqrt{\left(\frac{Q\mu D_e}{2\pi(\varrho_p-\varrho)(H-S)D}\right)} \tag{6.65}$$

where d_{50} = *cut* diameter for the cyclone,
φ = friction factor [Section 6.8, equation (6.74)].

An additional modification used in deriving (6.65) has been that

$$u_{T\,max}^2 D_e/4 = u_i^2 D/2 = \text{constant} \tag{6.56a}$$

which was the experimental relation found by Stairmand.

Similar ideas to those introduced by Stairmand have been used by Barth,[58] except that the maximum tangential velocity is reached in a ring in line with the exit tube, and not within this (Fig. 6.12c). The drift velocity through the cylinder in line with the exit tube is

$$u_{R\,av.} = \frac{Q}{\pi D_e(H-S)}. \tag{6.66}$$

A particle with cut diameter d_{50} will have a velocity outwards equal to this, and in terms of a gravitational terminal settling velocity u_t this diameter can be found by using equation (4.15) assuming viscous-flow resistance, while

$$u_t^* = \frac{u_{R\,av.}}{\mathbf{n}} = \frac{Qg}{2\pi(H-S)u_T^2} \tag{6.67}$$

The efficiency of collection of particles of other diameters can be found by calculating their terminal gravitational settling velocities u_t, and then referring to Barth's experimental determinations (Fig. 6.13, curve a) where the ratio u_t/u_t^* is graphed against efficiency of collection.

Similar curves for calculating collection efficiency have been determined by Muschelknautz and Brunner[590] for a hydrocyclone (72 mm diameter) (curve b), two cyclones with tangential entry 400 and 300 mm diameter respectively (curves c and d) and an axial entry cyclone, 200 mm diameter. These are also shown in Fig. 6.13. The other dimensions are shown in Table 6.6.

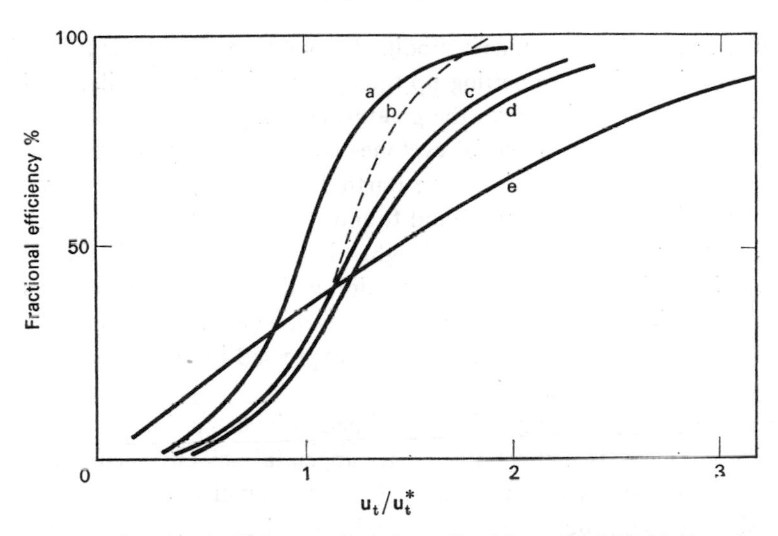

FIG. 6.13. Curves for fractional efficiency calculations (Barth's method).[589] (a) Experimental curve (Barth). (b) Hydrocyclone (D = 72 mm). (c) Cyclone with tangential entry (D = 400 mm). (d) Cyclone with tangential entry (D = 300 mm). (c) Axial entry cyclone (D = 200 mm).

TABLE 6.3. CYCLONE DIMENSIONS FOR THE FRACTIONAL EFFICIENCIES SHOWN IN FIG. 6.13. THE CYCLONE TYPES ARE THOSE SHOWN IN FIG. 6.19

Curve	Cyclone type	$\frac{a \times b}{\frac{\pi}{4} D_e^2}$	$\frac{D}{D_e}$	$\frac{2h}{D}$	$\frac{2H}{D_e}$	$\frac{u_{T\,max}}{u_i}$
b	D	0·96	3	0·3	10	2·3
c	E	0·44	4	0·17	7·5	4·3
d	D	0·90	3	0·3	10	2·8
e	A	0·75	2	(0·36)	10	2·0

A second parameter for characterizing the collection efficiency of specific cyclone types was introduced by Barth and Leinweber.[61] This they called the *separation parameter*, **B**, and defined it as

$$\mathbf{B} = \frac{4u_t^* Q}{\pi D^3}. \tag{6.68}$$

It can be seen that this combines a characteristic parameter for the particles (u_t^*), a flow rate, and hence a velocity for a particular cyclone (Q) and a cyclone dimension (D). Barth and Leinweber claim that model cyclone experiments using the **B** parameter give good agreement with full-scale operation. Leinweber, in subsequent papers,[501-2] has determined the characteristic **B** for a large number of cyclone types, together with pressure drop characteristics, explained in the next section.

Barth and Leinweber[61] also define a modified separation parameter $\mathbf{B}^*$, which should be used if space considerations for setting up the cyclone limit the available diameter to D^*, rather than the value D, which gives the greatest cyclone effectiveness.

In later studies on four commercial cyclones (with flow rates up to 220 $m^3 s^{-1}$) using brown coal dust, Petroll *et al.*[640] tested the Barth–Leinweber parameters over a wide range of dust concentrations up to 5 g m^{-3}, and found them less satisfactory as universal parameters. Their explanation is the lack of a height/diameter ratio function in the parameter **B**, as well as the difficulties of particle agglomeration which occurs in some dusts.

Quite recently, using a computer programme, cyclone efficiency and pressure drop have been optimized for given cyclone characteristics expressed in terms of the separation parameter **B** by Rumpf, Borho and Reichert.[716]

Ter Linden, on the basis of extensive experimental investigations has suggested that the critical particle diameter is a function of the tangential velocity u_T and cyclone exit pipe diameter only, while the general shape of the fractional efficiency curve is dependent on the other dimensions as well, such as H, S, D, the included cone angle α and the inlet angle β, and this shape remains unchanged as long as the relative dimensions are kept constant. The critical particle size is given by

$$d_{\text{crit}} = 3\sqrt{\left(\frac{2D_e g\mu u_R}{3u_T^2(\varrho_p-\varrho)}\right)}. \tag{6.69}$$

This equation assumes that the maximum tangential velocity is reached at two-thirds of the radius of the exit pipe (Fig. 6.12b) and that those particles which have a drift velocity outwards because of the centrifugal force equal to the radial inward drift of the gases u_R will be collected. The method of finding the fractional efficiency curve for the cyclone is then identical with the gravitational settling speed ratio method by Barth, except that only a graph for cyclones with one set of the same relative dimensions has been published[516] (Fig. 6.14).

The effect of changing cyclone dimensions on overall efficiency, using standard test dusts, has been investigated by both ter Linden[515] and Stairmand.[803] The specifications of the standard dusts used are given in Table 6.4.

TABLE 6.4. STANDARD TEST DUSTS

Size range (μm)	% by weight		
	Pulverized clay (ter Linden)	Fine sand (Stairmand)	Fly ash (U.S. Standard)
0–5	30	20	35·8
5–10	12	10	22·3
10–20	18	15	21·4
20–50	28	27	14·1
Over 50	12	28	6·5

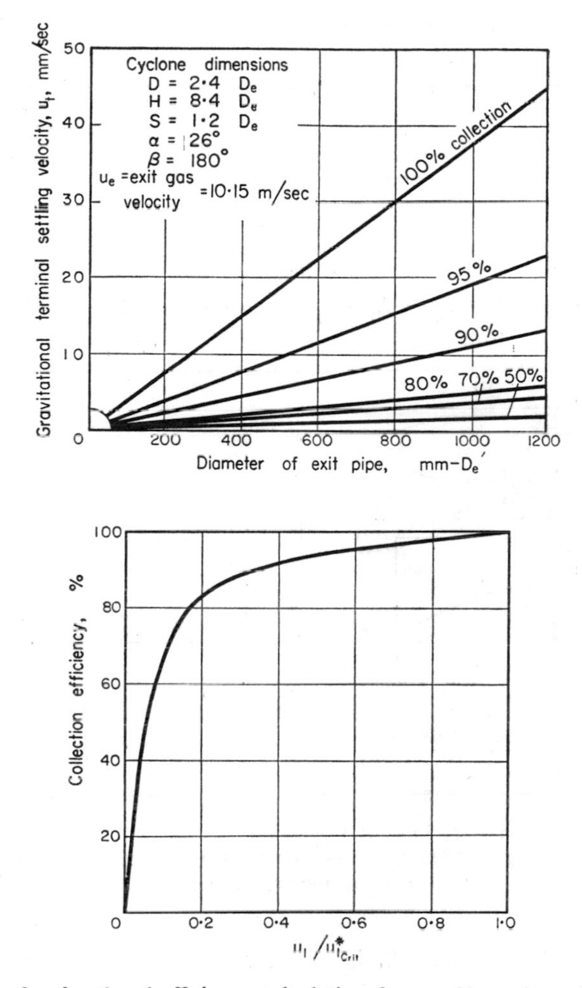

FIG. 6.14. Curves for fractional efficiency calculation for specific cyclone (ter Linden).[516]

Ter Linden studied the efficiency (and also pressure drop characteristics) of a cyclone where he was able to vary one dimension at a time, such as the cyclone diameter, *D*, length *H*, or distance of penetration of exit pipe *S*, as well as the inlet velocity u_i. The results obtained are shown in Fig. 6.15. He was able to show that increasing the relative dimension of diameter *D* to the exit pipe diameter D_e gave an increase in efficiency until a 3 : 1 ratio was achieved, after which further diameter increases have little effect on efficiency (Fig. 6.15a). Increased length *H* also tends to improve efficiency (Fig. 6.15b), while for the particular cyclone investigated, when the exit pipe penetrated one pipe diameter into the cyclone, and ended just below the entrance, maximum collection efficiency was attained (Fig. 6.15c). Increased inlet velocity was found to improve efficiency and the ratio of the cross-sectional area of outlet and inlet pipes was also found to play a part, smaller ratios giving higher efficiency (Fig. 6.15e).

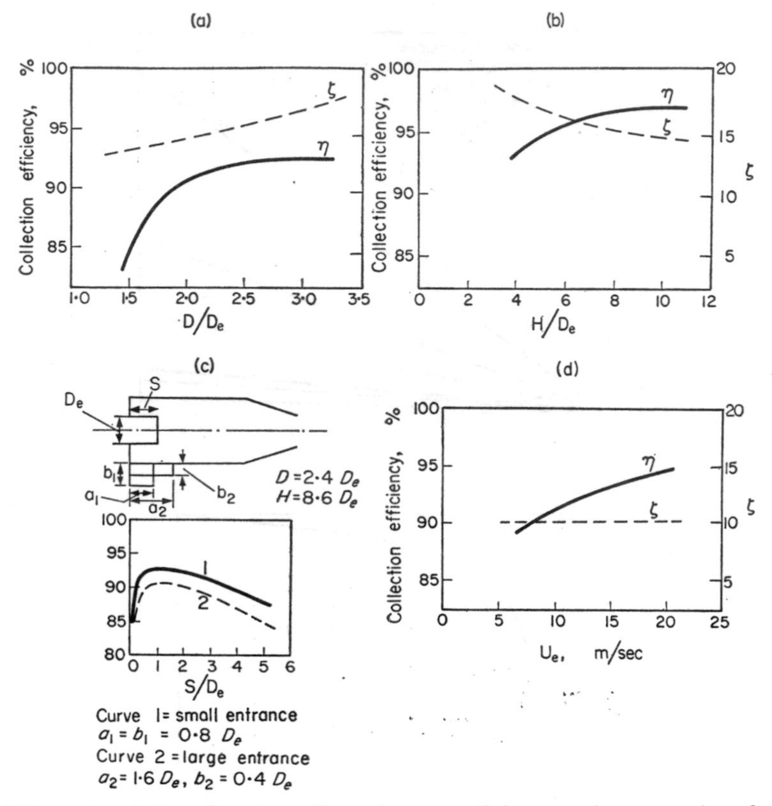

FIG. 6.15. Effect of variation of cyclone dimensions on efficiency and pressure loss factor ζ.[516] [Cyclone dimensions as in Fig. 6.14.] (a) Cyclone diameter. (b) Cyclone height. (c) Depth of exit pipe. (d) Entrance velocity u_e.

Stairmand[806] obtained fractional efficiency curves for two cyclone types, one being a high efficiency, low throughput type (Fig. 6.16a) and the other a medium efficiency, high throughput model (Fig. 6.16b). The fractional efficiency curves are shown in Fig. 6.17. These curves were obtained for 200 mm diameter cyclones, with dust of a density of 2000 kg m^{-3} and air at 20°C, operating with an entrance velocity of 15·2 m s^{-1}. These curves tend to confirm ter Linden's investigations, as the relatively longer cyclone with small diameter exit pipe is more efficient than the cyclone with the larger exit pipe and relatively shorter trunk. The performance of a cyclone for conditions other than those for which experimental fractional efficiency data are available can be predicted from these by transposition of the experimental curves as follows.[806]

(i) If there is a change in the density of the dust then the particle size of the new dust, which will be collected with the same efficiency as a particular size of the test dust, can be obtained by multiplying the test dust size by:

$$\sqrt{\left(\frac{\text{density of test dust}}{\text{density of new dust}}\right)}.$$

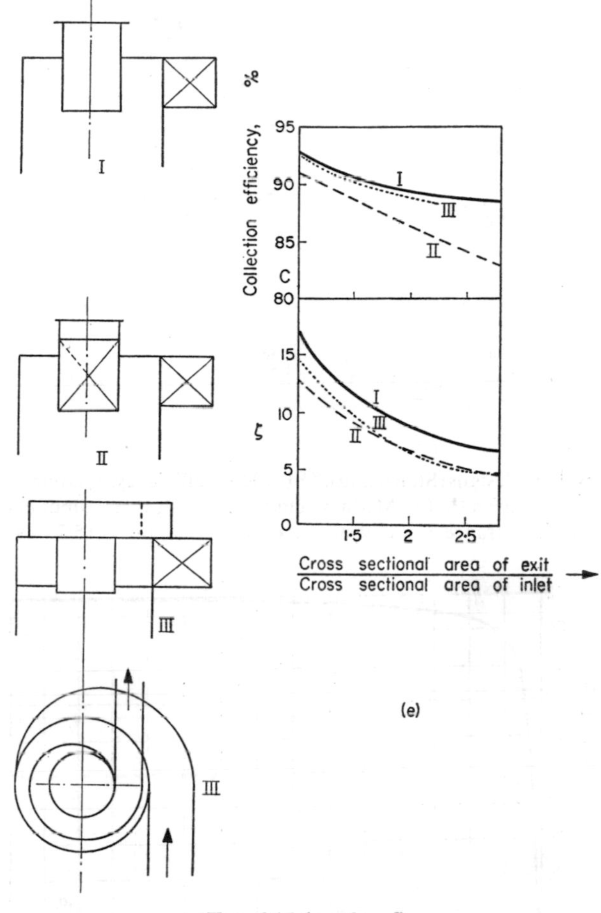

FIG. 6.15 (*continued*)
(e) Exit area as ratio of inlet area.

(ii) If there is a change in the volume flow through the cyclone, the particle size at the new flow, which will be collected with the same efficiency as a particle at the test flow rate, can be calculated by multiplying the test dust size by:

$$\sqrt{\left(\frac{\text{test flow}}{\text{new flow}}\right)}.$$

(iii) If there is a change in gas velocity, such as would accompany a temperature change of the gases, the particle size for equal efficiency can be calculated by multiplying the test dust size by:

$$\sqrt{\left(\frac{\text{new viscosity}}{\text{test viscosity}}\right)}.$$

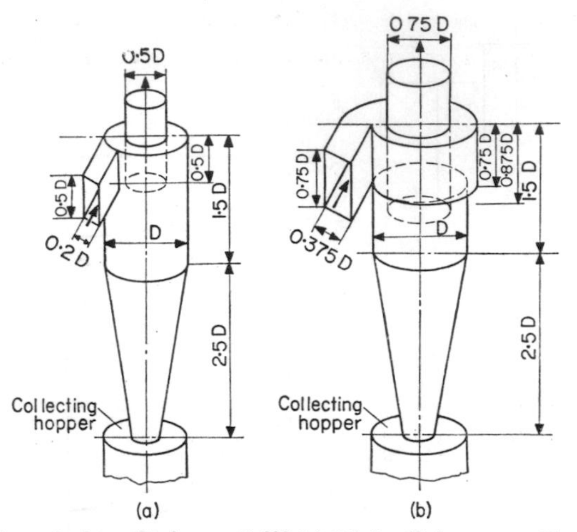

FIG. 6.16. Standard cyclone designs (Stairmand).[806] (a) High efficiency, medium throughput pattern. Normal flow rate = $500D^2$ m^3 h^{-1}. (b) Medium efficiency, high throughput pattern. Normal flow rate = $1500D^2$ m^3 h^{-1}. Entrance velocity at these flows is approx. 15·2 m s^{-1} in both types.

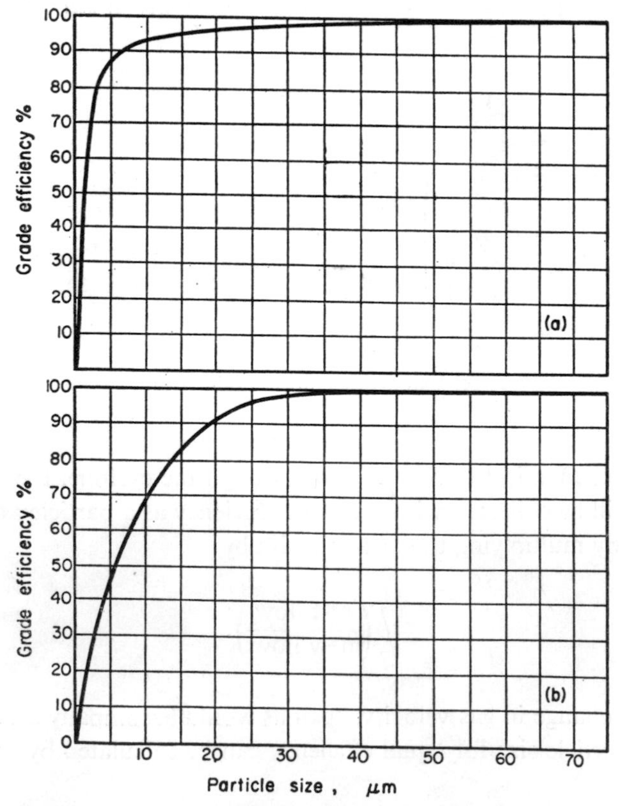

FIG. 6.17. Fractional efficiency curves for 200 mm diameter cyclones, inlet velocity 15·2 m s^{-1}, solid density 2000 kg m^{-3} in air at 20°C. (a) High efficiency cyclone, Fig. 6.16a. (b) High throughput cyclone, Fig. 6.16b.

(iv) If there is a change in the diameter of the cyclone, but retaining geometrical similarity to the test cyclone, the particle size for equal efficiency can be calculated by multiplying the original cyclone particle size by:

$$\sqrt{\left(\frac{\text{diameter of new model}}{\text{diameter of test model}}\right)}.$$

6.8. PREDICTION OF CYCLONE PRESSURE DROP

A knowledge of cyclone pressure drop and the factors which affect it is required so that power consumption can be predicted, and if possible minimized by better choice of cyclone parameters, and also so that correct fans can be selected.

Pressure losses or gains occur at the following places for the reasons given:

(i) Friction losses in the entrance pipe.
(ii) Losses due to gas expansion or compression at the entry.
(iii) Losses due to wall friction in the cyclone.
(iv) Kinetic energy losses in the cyclone.
(v) Losses at the entrance to the exit pipe.
(vi) Static head losses between inlet and exit pipe.
(vii) Recovery of energy in the exit pipe.

While Stairmand[800] and Barth[58] present detailed theories which include a number of the sources of pressure losses, Shepherd and Lapple[765] and ter Linden[516] both consider that the kinetic energy loss by the gases in the cyclone is so large compared to the other sources of pressure loss, that this is the only one that has to be considered. Ter Linden gives the pressure loss in terms of the entrance velocity and a dimensionless pressure loss factor ζ:

$$\Delta p = \zeta \frac{u_i^2(\varrho + \varrho_p')}{2g} \tag{6.70}$$

where $\varrho_p' = c(\varrho_p - \varrho)$ allows for the particle concentration c.†

For the cyclones investigated by ter Linden the experimental pressure loss factor has been plotted as broken lines in Fig. 6.15.

Barth[58] presents an elaborate method of calculating the pressure loss factor from two components which are:

(i) Pressure loss at entry and friction loss at the walls (denoted by subscript i).
(ii) Pressure loss at the central core and exit pipe entry (denoted by subscript e).

† While, in theory, the density of the medium increases because of the presence of the particles, and this should increase the pressure loss, according to equation (6.70), in practice, for other reasons, the presence of particles decreases the pressure drop through the cyclone (see Section 6.9).

The pressure loss factor ζ is expressed in terms of the inlet gas velocity, which equals $u_{T\,max}$, and is given by Barth as a function of a *loss number* ε, defined by

$$\varepsilon = \zeta \bigg/ \left(\frac{ab}{\pi D_e^2/4}\right)^2 \left(\frac{u_{T\,max}}{u_e}\right)^2 \tag{6.71}$$

where ab = cross-sectional area of entrance pipe,
u_c = velocity at entry to exit pipe,
= $4Q/\pi D_e^2$.

The loss number consists of the *sum* of two components ε_i and ε_e

$$\varepsilon = \varepsilon_i + \varepsilon_e \tag{6.72}$$

and these components can be found from

$$\varepsilon_i = \frac{\Delta p_i}{u_{T\,max}^2(\varrho+\varrho_p')/2g} = \frac{D_e}{D}\left\{\frac{1}{\left[1-\dfrac{2u_{T\,max}(H-S)\mu'}{u_e D_e}\right]^2}-1\right\} \tag{6.72a}$$

where μ' = coefficient of friction (gas and wall)
(frequently assumed to be 0·02)

$$\varepsilon_e = \frac{\Delta p_e}{u_{T\,max}^2(\varrho+\varrho_p')/2g} = \frac{K}{(u_{T\,max}/u_e)^{\frac{2}{3}}}+1 \tag{6.72b}$$

The velocity ratio $u_{T\,max}/u_e$ has been expressed by Barth in terms of the dimensions of the cyclone, the coefficient of gas–wall friction μ' and an entrance design loss factor α:

$$\frac{u_{T\,max}}{u_e} = \frac{\pi D_e(D-b)}{2ab\alpha+\pi(H-S)(D-b)\mu'}. \tag{6.73}$$

The entrance design loss factor depends on the type of entry used. Three types are shown in Fig. 6.18a. For the wrap around inlet, α is unity, while for the other types shown α is either less than or greater than one. The third inlet type is not recommended practice, while the entrance correction factor for the second inlet type can be obtained from Fig. 6.18b.

TABLE 6.5. VALUES OF K AND ζ FOR BARTH PRESSURE LOSS CALCULATIONS

		Sharp-edged exit pipe	Round-edged exit pipe
K		4·40	3·41
$\zeta\Big/\left(\dfrac{ab}{\pi D_e^2/4}\right)^2$	$\dfrac{u_{T\,max}}{u_e} > 1$	Curve *a* Fig. 6.17c	Curve *b* Fig. 6.17c
	$\dfrac{u_{T\,max}}{u_e} < 1$	2·0	1·1

Weidner[914] measured $\zeta/\{ab/(\pi D_e^2/4)\}^2$ for a series of values of the velocity ratio $u_{T\,\max}/u_e$, and these are plotted in Fig. 6.18c. Curve "a" applies to sharp edged exit pipes: curve "b" to round edged ones. Values of K and the pressure loss factor are listed in Table 6.5.

To calculate the pressure-loss factor ζ it is therefore necessary to find ε_i and ε_e from equations (6.2a) and (6.2b), and the velocity ratio $u_{T\,\max}/u_e$ from equation (6.3), and then apply equation (6.71).

The pressure-loss factor ζ and the loss number ε have also been determined empirically for a number of different cyclone types (Fig. 6.19) by Muschelknautz and Brunner.[590]

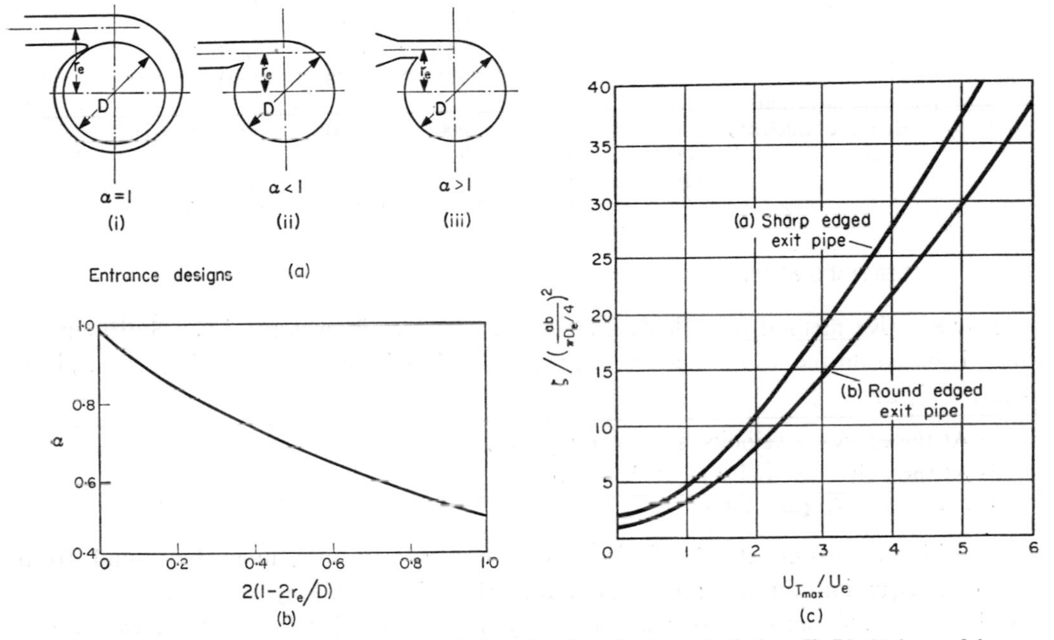

FIG. 6.18. (a) Alternative cyclone entry designs for loss factor calculations.[58] (b) Values of loss factor α for pattern (ii) when α < 1.[58] (c) Weidner's experimental curves for pressure loss function.[914]

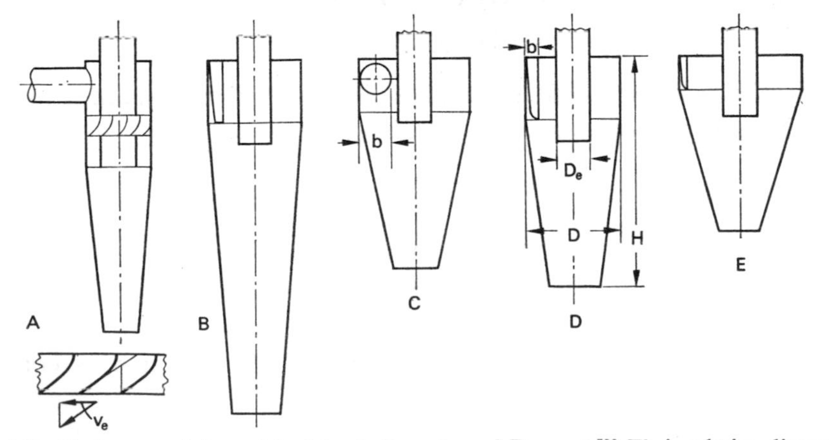

FIG. 6.19. Cyclone models used by Muschelknautz and Brunner.[590] Their relative dimensions are given in Table 6.6.

Their proportions and their characteristic parameters are listed in Table 6.6.

TABLE 6.6. CYCLONE PROPORTIONS FOR DIFFERENT CYCLONE TYPES (Fig. 6.19.) AND THEIR EXPERIMENTAL PRESSURE-LOSS FACTORS ζ AND LOSS NUMBERS[590]

Cyclone type	A	B	C	D	E
D/D_e	2	3	3·5	3	4
$(a\times b)/(\pi D_e^2/4)$	2·7	0·9	1*	0·9	0·44
$2H/D_e$	12·5	23·5	14	14·5	11
$2b/D$	(0·4)	0·27	0·57	0·27	0·17
α	0·76	0·76	0·71	0·75	0·93
Dimensionless wall friction coefficient	0·005	0·005	0·005	0·005	0·010(0·007)
$u_{T\,max}/u_i$	0·74	2·6	2·9	3	4·3
ε	12·5	2·73	2·5	2·45	2·0
ζ	6·9	18·5	21	22	37

* Here it is b^2/D_e^2.

An alternative method for calculating the pressure loss in cyclones is by Stairmand[800] which is based on the losses measured at various points in terms of velocity heads, given by $u^2(\varrho+\varrho_p')/2g$.

(i) At the entry—1 velocity head $u_i^2(\varrho+\varrho_p')/2g$.
(ii) At the exit—2 velocity heads $u_e^2(\varrho+\varrho_p')/g$.
(iii) Losses within the cyclone.

In addition the losses actually within the entrance and exit ducts must be found from the normal equations for pressure losses in ducts (Fanning equation).

Stairmand finds that the losses within the cyclone are friction losses at the wall and a kinetic energy loss. The latter was found to be twice the difference between the velocity head at the entrance and at the periphery of the inner core, i.e.

$$(\varrho+\varrho_p')\,(u_{T\,max}^2-u_i^2)/g$$

while the wall friction factor φ is found as the ratio of the spinning speed u_i' at the inlet radius $(D/2-b/2)$ and the linear velocity in the inlet duct u_i, i.e.

$$\varphi = u_i'/u_i \tag{6.74}$$

where

$$\varphi = \frac{-\sqrt{\left(\frac{D_e}{2(D-b)}\right)}+\sqrt{\left(\frac{D_e}{2(D-b)}+\frac{4GA}{ab}\right)}}{2GA/ab} \tag{6.75}$$

where G = friction constant (Stanton and Pannell's dimensionless friction loss constant)[808]
= 0·005 for gas cyclones,
A = surface area or cyclone exposed to gases,
ab = cross-sectional area of inlet duct.

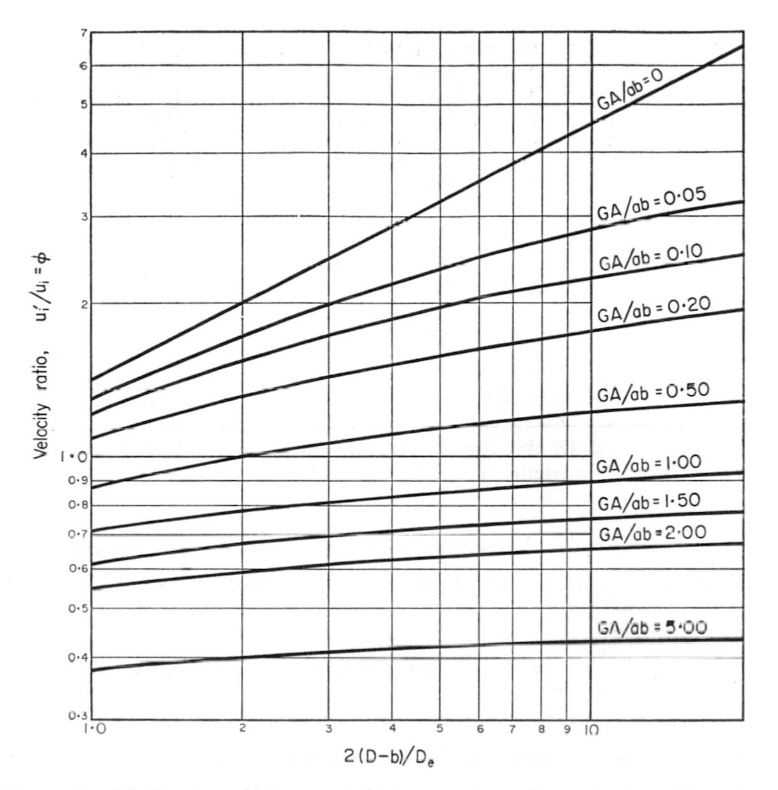

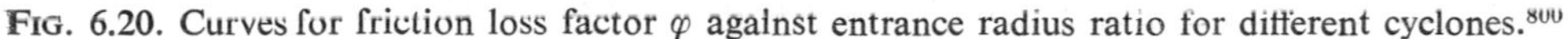

FIG. 6.20. Curves for friction loss factor φ against entrance radius ratio for different cyclones.[800]

a = depth of entrance pipe.
b = width of entrance pipe.
D = cyclone diameter.
D_e = cyclone exit pipe diameter.
A = internal surface area of cyclone.
G = friction factor − 0·005 (assumed constant).
φ = friction loss factor.

The value of the friction factor φ can be obtained from the curves of φ vs. $2(D-b)/D_e$ for various GA/ab (Fig. 6.20). The total pressure loss can then be calculated from combining the loss factors:

$$\Delta p = \frac{\varrho+\varrho_p'}{2g}\left[u_i^2\left\{1+2\varphi^2\left(\frac{2(D-b)}{D_e}\right)-1\right\}+2u_e^2\right]. \tag{6.76}$$

Stairmand found that the pressure loss calculated from this equation agreed to within 10 per cent of the values obtained experimentally.

There is some disagreement as to whether some of the rotational energy given to the gases can be recovered and so reduce the pressure loss.

Ter Linden has been able to reduce the pressure loss by 20–25 per cent by introducing a spiral at the entrance to the exit pipe, while Schiele,[729] using a solid central core shaped like a venturi, with vanes at the exit pipe entrance (Fig. 6.21), reduced the friction loss

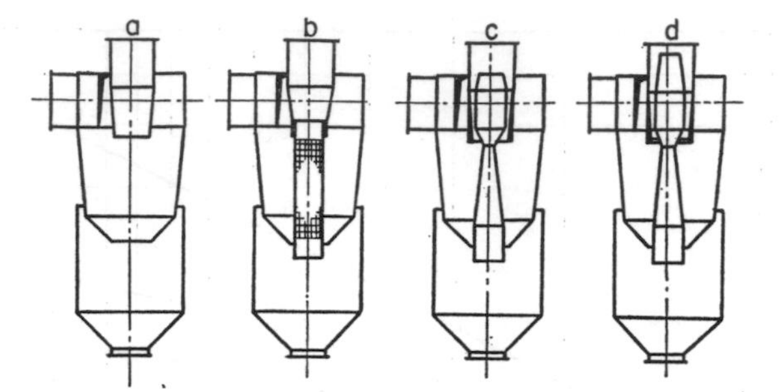

FIG. 6.21. Methods of reducing cyclone pressure drop by internal modification.[729]

(a) Plain cyclone $\zeta = 17{\cdot}4$.
(b) Sieve tube insert $\zeta = 13{\cdot}9$.
(c) Double cone insert $\zeta = 16{\cdot}4$.
(d) Double cone insert and vanes $\zeta = 10{\cdot}0$.
ζ = friction-loss factor (Barth and ter Linden).

factor ζ from 17·4 to 10, an effective reduction of 42 per cent in the pressure loss. Stairmand[800] has, however, not been able to find a reduction in pressure loss by introducing internal modifications.

6.9. EFFECT OF HIGH DUST LOADINGS

In general, the dust loadings encountered in industry are below about 12 g m^{-3}, and it is this order of concentration which was used by workers such as Stairmand and ter Linden in arriving at their fractional efficiency correlations. On some occasions, particularly in those cases where the gases are carrying a product rather than a waste product (such as boiler fly ash), the dust loadings can be much higher, even by as much as a factor of 1000. However, even such high concentrations represent comparatively low volume concentrations; thus 50 kg m^{-3} of quartz dust is only a volume concentration of 1·8 per cent. These high dust concentrations have been investigated in recent years by Kolk,[456] Sproull[793] and very extensively by Muschelknautz and Brunner.[590]

These workers have shown that the tangential velocity in the cyclone u_T is reduced by dust loading, even small concentrations having some effect. This reduces the tangential velocity in the exit pipe, which is responsible for much of the pressure loss, and so gives reduced pressure drop with increasing dust concentration, until a minimum is reached between 1 and 10 kg m^{-3}. The overall efficiency of collection increases with dust concentration, although the absolute amount which escapes may, of course, increase. The quantitative results of a series of experiments are shown for a typical cyclone in Fig. 6.22.

point

It is observed in practice that with high dust concentrations the dust concentrates into cycling strands which run down the wall, while the gas within the cyclone tends to carry only a limited amount of dust, comparable to the quantities in pneumatic conveying. Muschelknautz and Brunner have extended Barth's theory for the case of high dust loading by consideration of the forces on the cycling stands of dust.

The resistance F' to the movement of the strand is counterbalanced by the weight of the dust, m, in the strand (which is collected) and the frictional resistance drag R. The movement

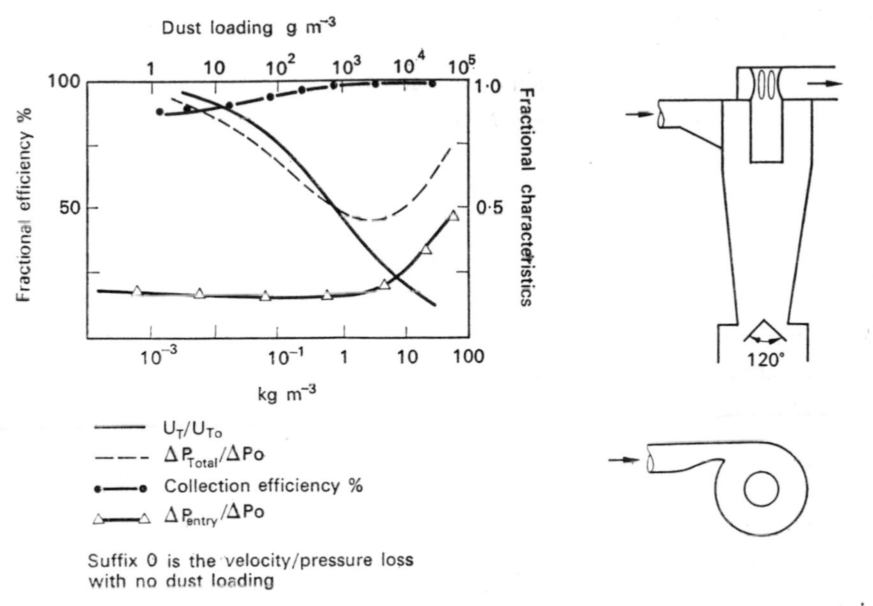

FIG. 6.22. Behaviour of cyclone characteristics with high dust loadings.[590]

Cyclone dimensions: See Fig. 6.11 for symbols.

$D = 300$ mm dia. $\quad H = 720$ mm
$D_e = 100$ mm dia. $\quad h = 220$ mm
$B = 200$ mm dia. $\quad b = 45$ mm

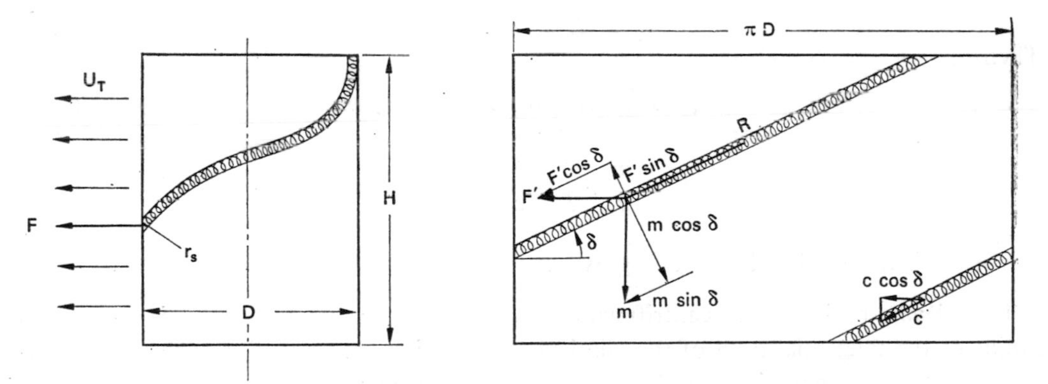

FIG. 6.23. Forces and velocities on dust strands at high dust loadings.[590]

F' resistance to strand $\quad F$ centrifugal force
m wt. of strand $\quad c$ velocity of strand
R frictional resistance drag $\quad r_s$ radius of strand

of the strand (c) was also observed experimentally to be approximately constant at 1 m s^{-1}. These forces and velocities are shown in Fig. 6.23 on a simplified cylindrical cyclone, in elevation and in development.

The force balance is given by:

$$F' \sin \delta = m \cos \delta, \tag{6.77}$$

$$F' \cos \delta + m \sin \delta = R. \tag{6.78}$$

From a balance based on particle throughput

$$m = \eta \frac{W}{c} \cdot l \tag{6.79}$$

where W = mass throughput,
l = length of the strand, and
η = efficiency of collection.

Furthermore,

$$m = A_s l(1-\epsilon)\varrho_p \tag{6.80}$$

where A_s = cross-sectional area of unit length of dust strand,
ϵ = voidage in dust strand,
ϱ_p = particle density.

Thus $(1-\epsilon)\,\varrho_p$ = bulk density of strand.

It is also necessary to define the dimensionless pressure loss factor for a dust strand ζ_s in terms of equation (6.10), where the dust-concentration system for the cyclone with strand is substituted, and a friction factor φ_s, which is defined by

$$R = \varphi_s F. \tag{6.81}$$

Then

$$F' = \zeta_s r_s H \frac{\varrho}{2}(u_T - c \cos \delta)^2, \tag{6.82}$$

$$F = \frac{m}{g} \cdot \frac{c^2}{D/2 \cos^2 \delta} \tag{6.83}$$

where g is the acceleration due to gravity.

The relevant radius for the centrifugal force on the dust strand, $D/2 \cos^2 \delta$, is the vertical radius of the tangential path of the eliptical cross-section. The velocity of the strand can be calculated from equations (6.11), (6.18), (6.81) and (6.83):

$$c = \sqrt{\left(\frac{1}{\varphi_s} \cdot \frac{gD/2}{\sin \delta \cos^2 \delta}\right)}. \tag{6.84}$$

This velocity is negligible in comparison with the mean tangential velocity.

Using

$$r_s = \sqrt{(2A_s/\pi)} \tag{6.85}$$

and the equation for the dust-carrying capacity in pneumatic conveying, for the exit pipe, one obtains

$$W = (\in \varrho_p)\left(\frac{\pi^2}{4} De_e^3 S\right) u_e \varrho \tag{6.86}$$

where u_e = velocity in exit pipe.

Further calculations show that the moment of the strand, M_s is

$$M_s = F'(D/2) = M_s \sqrt{(\eta \in \varrho_p)} \left(\frac{AD_e\varrho}{2\in \varrho_s}\right)^{\frac{1}{2}} \left(\frac{D_e}{D}\right)^{\frac{5}{8}} \times \varrho u_i u_e \pi D_e D \tag{6.87}$$

where ϱ_s = specific density of the strand,
μ_s' = new friction function given by

$$\mu_s' = \frac{\zeta_s}{2\pi} (8\varphi_s \sin\delta \cos^2\delta)^{\frac{1}{4}} \tag{6.88}$$

which depends only on the angle δ. This could be calculated from the above equations, but λ_s varies little as the values of δ are between 30° and 60°. With high particle loadings, the impulse moment on the cylinder surface in the exit pipe (M_e) is obtained by subtracting the sum of the frictional moments of the air (M_A) and the strand (M_s) from the impulse moment at the cyclone entry (M_a)

$$M_e = M_a - (M_A + M_s). \tag{6.89}$$

Using these relations, it is possible to substitute in Barth's relation (6.73), and obtain an equivalent equation for high dust loadings. This gives

$$\frac{u_{T\max}}{u_e} = \left[\frac{4ab}{\pi DD_e} . \alpha + \frac{2H}{D_e}\left\{\mu' + \mu_s'\left(\frac{\eta\pi\varrho DD_e HW}{2\in_s \varrho_p}\right)^{\frac{1}{2}}\left(\frac{D_e}{D}\right)^{\frac{5}{8}}\right\}\right]^{-1}. \tag{6.90}$$

The equivalent values of $\in_i$ and ζ_i of equations (6.71) and (6.72) are obtained by setting the function in curly brackets equal to the friction coefficient μ. The functions $\in_e$ and ζ_e depend only on the ratio $\mu_{T\max}/u_e$ and the latter can be obtained from Fig. 6.18c. A graph of u_T/u_{T_0} against

$$\left(\frac{\eta\pi\varrho DD_e HW}{2\in_s \varrho_p}\right)^{\frac{1}{2}}\left(\frac{D_e}{D}\right)^{\frac{5}{8}}$$

shows the same shape and characteristics as the equivalent plot in Fig. 6.22, and experimental points cluster around these.[590]

6.10. COMMERCIAL CYCLONES

Many commercial cyclone types are available. These range from comparatively crude designs for the collection of wood chips and sawdust to carefully designed and tested units able to collect fine particles as small as 5 μm. These designs have been arrived at after extensive developmental work by the manufacturer. This has shown the smallest particle sizes effectively collected, and the optimum gas-flow rate with the minimum pressure loss for the particular system.

In many cases the manufacturer publishes a fractional efficiency curve based on tests carried out with standard dusts as a guide for those selecting cyclones. These may be somewhat misleading with certain types of dust and so most manufacturers will only give guarantees based on tests carried out with some of the material to be collected. In this section typical examples of commercial cyclones will be presented together with fractional efficiency curves where these are available.

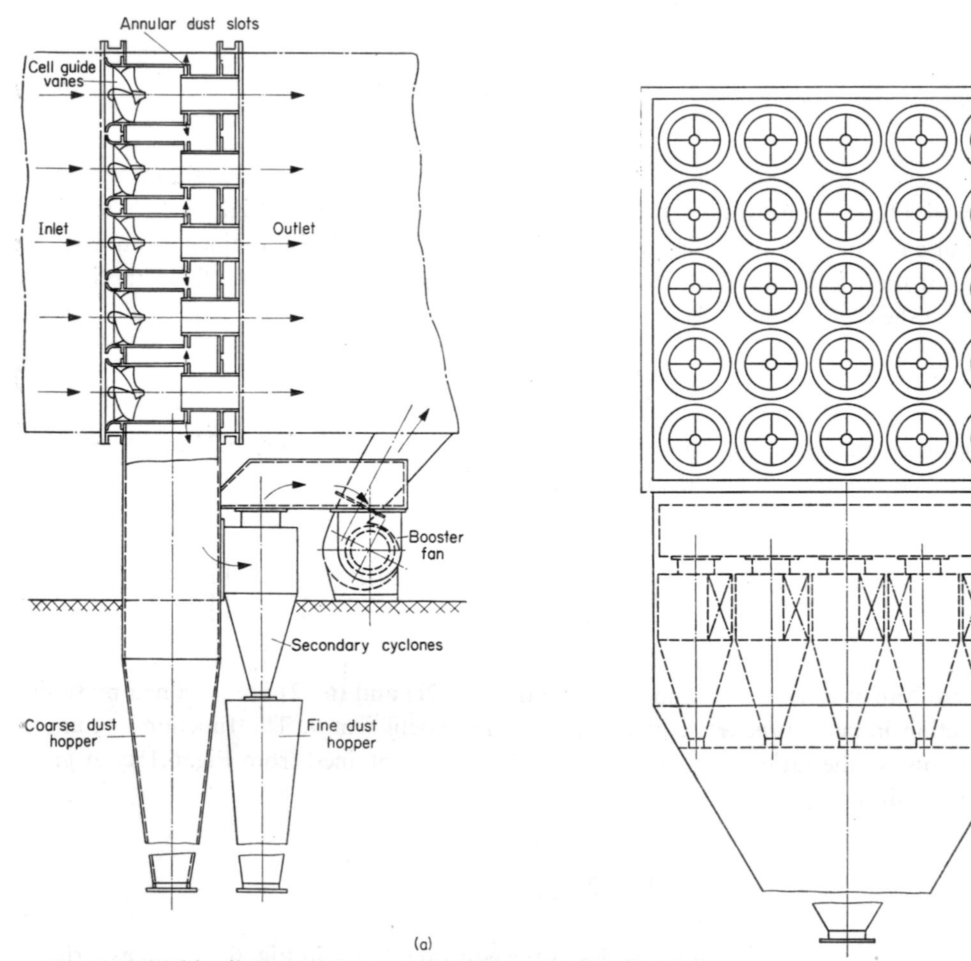

FIG. 6.24. Multi-cellular straight-through cyclones, drawing down concentrated dust through settling chamber and cyclones in series. (a) Diagram of installation.[201]

6.10.1. Straight-through cyclones with fixed impellers

Straight-through cyclones act as dust concentrators, where the concentrated dust, together with some gas, is drawn off at the periphery, and led to a secondary collector, while the clean gas is passed out axially. The secondary collector may be another cyclone of conventional pattern or a settling chamber. Multi-cellular straight-through cyclones are frequently applied as preliminary cleaners for flue gases with very high fly-ash concentrations, such as would come from burning high ash content pulverized coal, before passing the partially cleaned gases to electrostatic precipitators. A typical design for a multicell unit is shown in Fig. 6.24a. The concentrated dust is drawn to a settling chamber, then through cyclones, after which the cleaned gas joins the gas from the straight-through cyclone cells. A single straight-through cyclone cell is shown in Fig. 6.2, and a fractional efficiency curve is shown in Fig. 6.24b. A slightly different arrangement (Fig. 6.25a) allows the heavy dust concentrate particles to settle into a gravitational settling chamber, while the lighter ones are drawn off overhead to cyclones, the clean exhaust from which is also passed into the clean gas stream from the cells. The fractional efficiency curve for this arrangement is shown in Fig. 6.25b.

Large single cell units are used for applications on comparatively small boiler plant (gas volumes to 85,000 $m^3 h^{-1}$ where low pressure losses are essential, and the degree of particle removal required is not very critical, as in marine boilers. The upflow collectors (Figs. 6.26a and 6.26b) are particularly useful for installation in a chimney using natural draught. In one type (a) a settling chamber is provided, while a cyclone is shown in the second type (b). Where the unit cannot be installed in a chimney, a downflow collector with tangential entry (Fig. 6.26c) may be installed. Typical efficiency data is shown in Fig. 6.26d. As would be expected, the unit with cyclone secondary collector is more effective than the type with only a settling chamber, while the downflow unit is better than the upflow type (Chapter 5).

Very elegant designs using extended impellers have been described elsewhere[197] but are not manufactured commercially. These designs have demonstrated that the smallest particle,

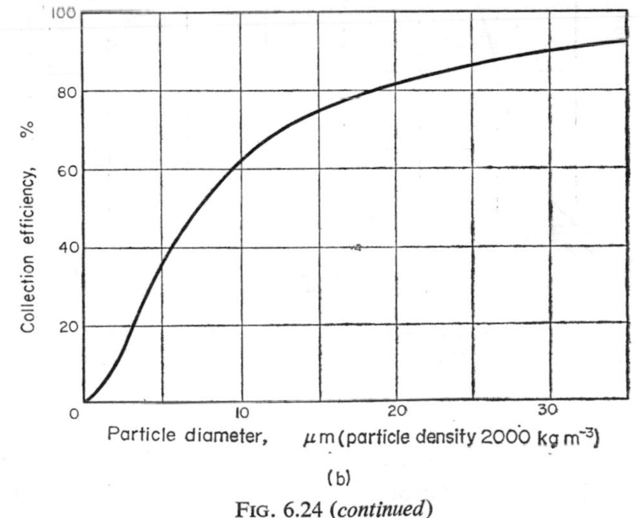

FIG. 6.24 (*continued*)
(b) Fractional efficiency curve for installation.[201]

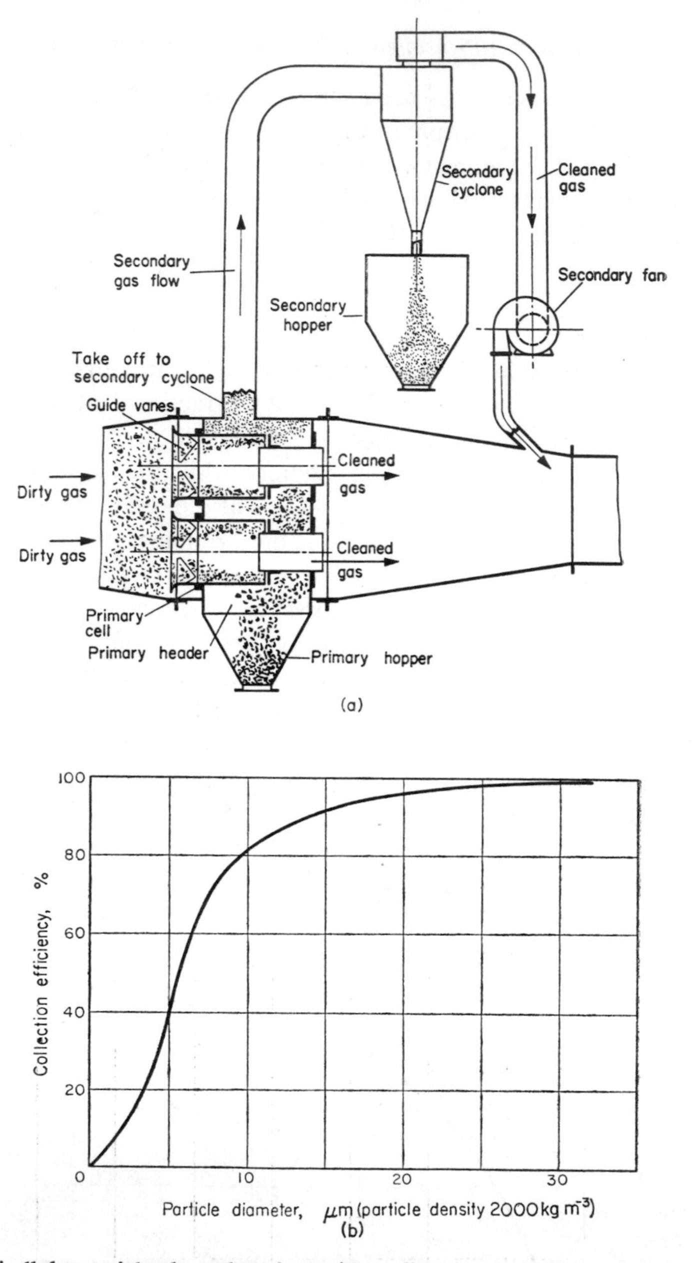

FIG. 6.25. Multicellular straight-through cyclone with settling chamber (primary hopper) and cyclones in parallel. (a) Arrangement of installations.[839] (b) Fractional efficiency curve.[839]

with a density 1000 kg m^{-3}, that can be removed is about 3–5 μm diameter, using entrance velocities as high as 210 m s^{-1}. Turbulence seems to re-entrain smaller particles, while larger ones tend to bounce off the walls and back into the main gas stream.

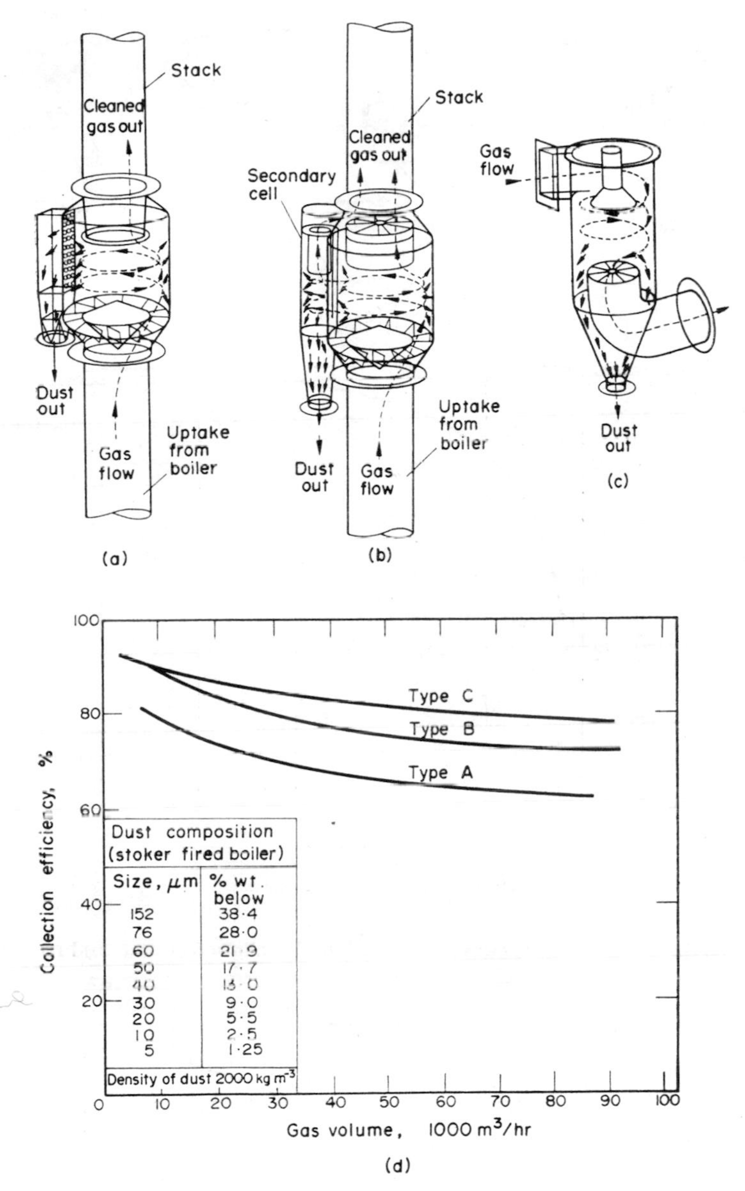

FIG. 6.26. Upflow and downflow centrifugal collector.[377] (a) Upflow collector with settling chamber separator. (b) Upflow collector with cyclone separator. (c) Vortex downflow collector. (d) Efficiency vs. volume flow correlation for types shown.

This can be overcome by irrigating the cyclone by introducing a water spray at the entry. A commercial design of this type is shown in Fig. 27a, together with a fractional efficiency curve (Fig. 6.27b). This curve, which is for coal dust, shows that efficiencies of 80 per cent for 1-μm particles are achieved with this unit.

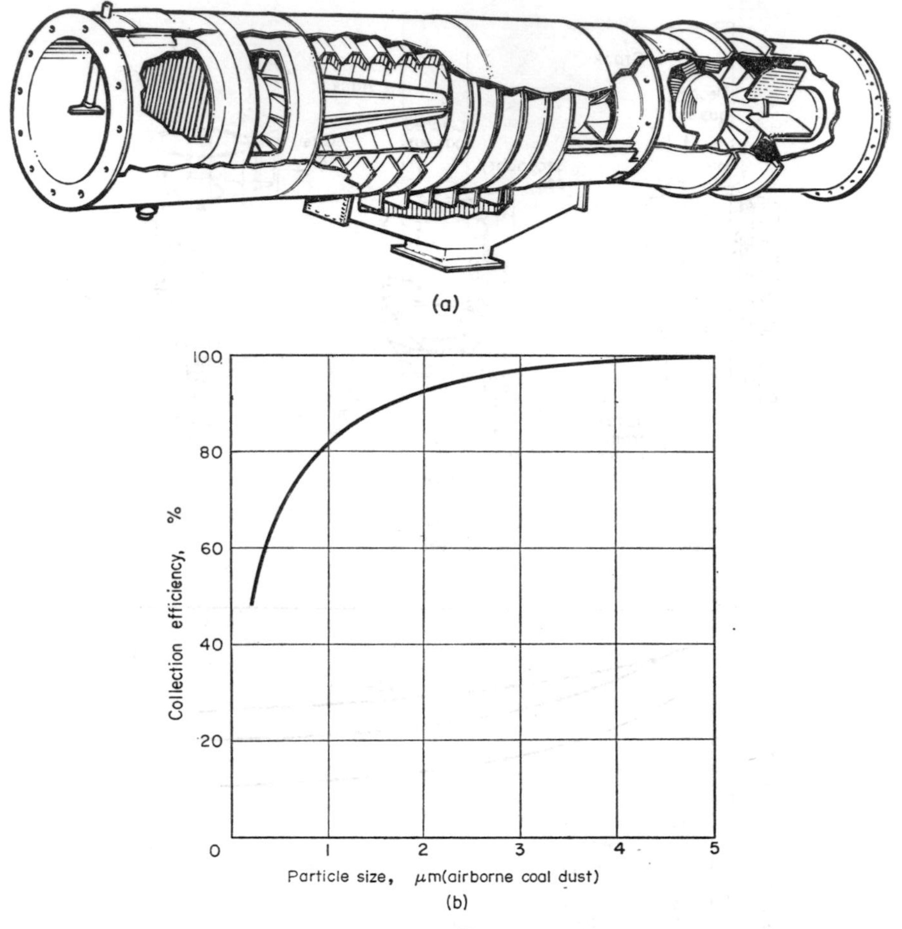

FIG. 6.27. Straight-through irrigated dust collector.[413] (a) Installation of unit. (b) Fractional efficiency curve.

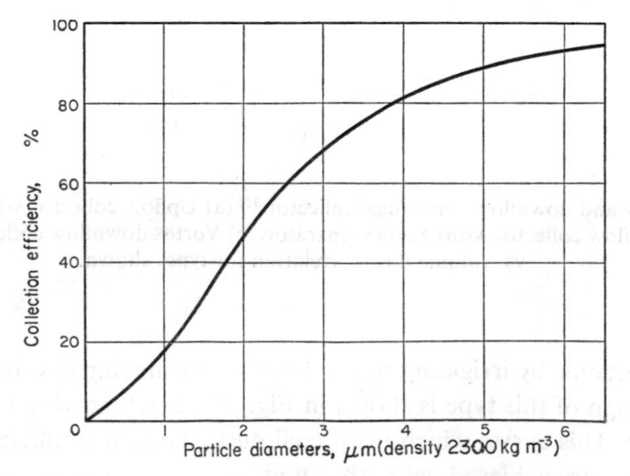

FIG. 6.28. Fractional efficiency curve for straight-through cyclone with moving impeller.[947]

6.10.2. Straight-through cyclones with moving impellers

A diagram of this unit has been shown in Fig. 6.6, and a fractional efficiency curve is given in Fig. 6.28.[947] This equipment may be expected to give efficiencies of about 90 per cent with 5-μm-sized particles (density 2600 kg m^{-3}) which is somewhat better than for conventional cyclones of the high-efficiency type.

6.10.3. Straight-through cyclones with fixed impeller and reverse flow

The principle of this design was outlined in Section 6.3.1, and in practice the secondary gas is passed into the inner chamber from an outer annular chamber allowing 75–125 mm clearance, depending on the size of the cyclone. A general arrangement is shown in Fig. 6.29. Flow rates for commercial designs range from 34 to 6800 m^3 h, and the unit can cope with dust loadings up to 230 g m^{-3}. Other advantages are the limited abrasion of the walls, and the ability to cope with hot gases, which can be cooled by the secondary air, and so do not require special heat resistant materials for the cyclone body. Multi-cellular units have also been built.

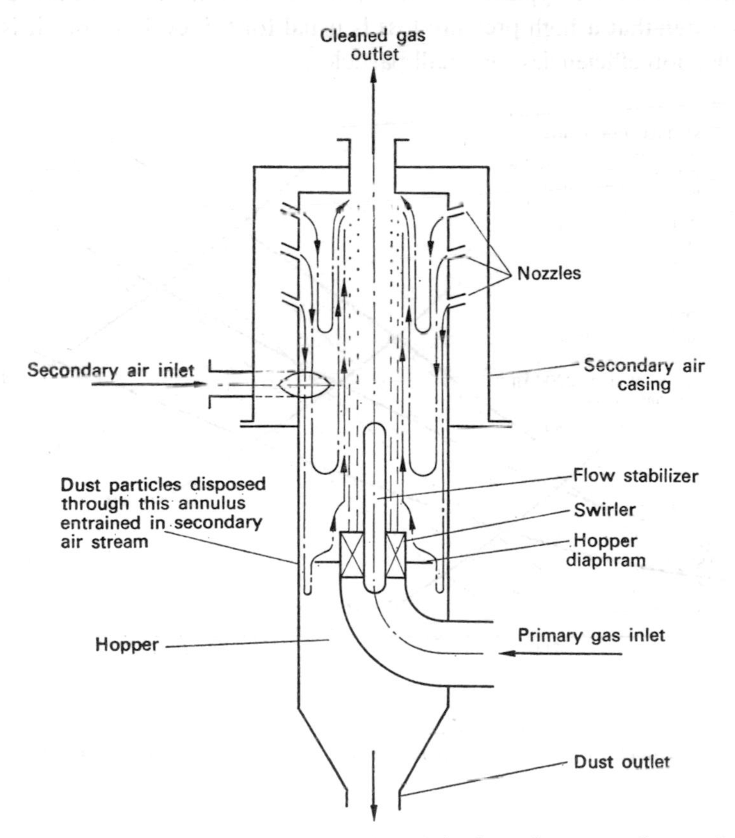

FIG. 6.29. Diagrammatic arrangement of straight-through reverse flow cyclone.

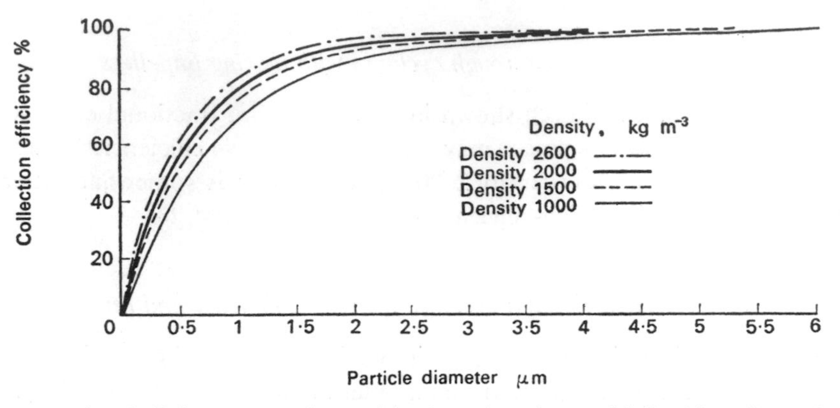

FIG. 6.30. Fractional efficiency curves for straight-through cyclones with fixed impeller and reverse flow (Aerodyne—U.S.; Tornado—U.K.).

Fractional efficiency curves for dusts ranging in density from 1000 to 2600 kg m^{-3} are shown in Fig. 6.30. The pressure loss, primary and secondary air flows, power input and collection efficiencies for a typical (200 mm diameter nominal bore) unit are shown in Fig. 6.31. It will be seen that a high pressure loss is usual for this cyclone, but it is coupled with very high collection efficiencies for small particles.

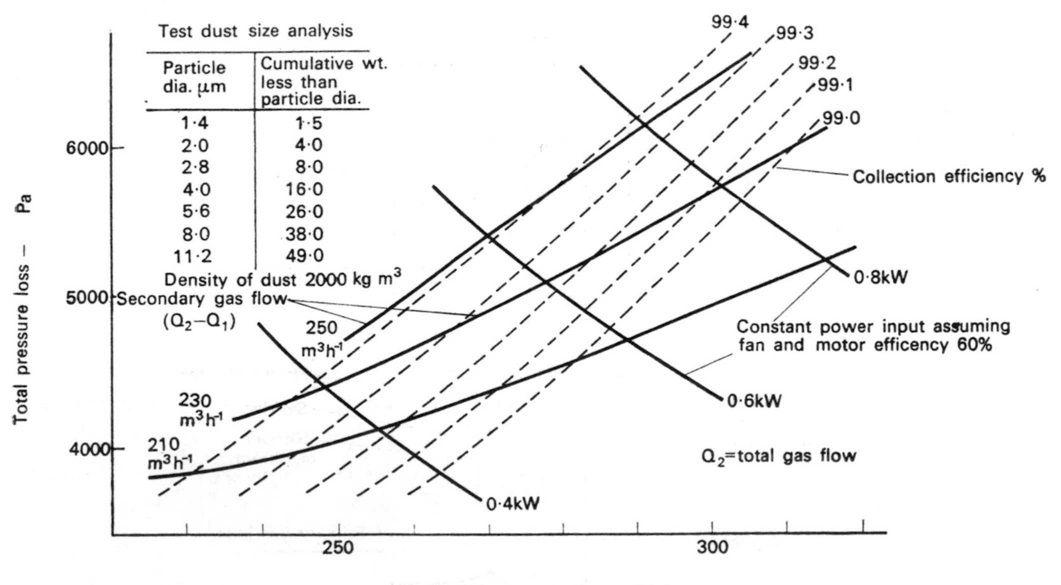

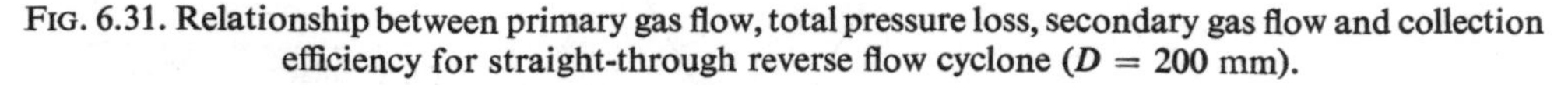

FIG. 6.31. Relationship between primary gas flow, total pressure loss, secondary gas flow and collection efficiency for straight-through reverse flow cyclone (D = 200 mm).

6.10.4. Scroll collectors

Several fixed scroll type collectors are manufactured. In these units the scroll imparts a centrifugal motion to the gas stream, concentrating the dust in the peripheral layer, which is passed to a secondary collector, while the clean gas is passed on to the exhaust. The secondary collector is almost invariably a conventional cyclone. Figure 6.32 a shows one

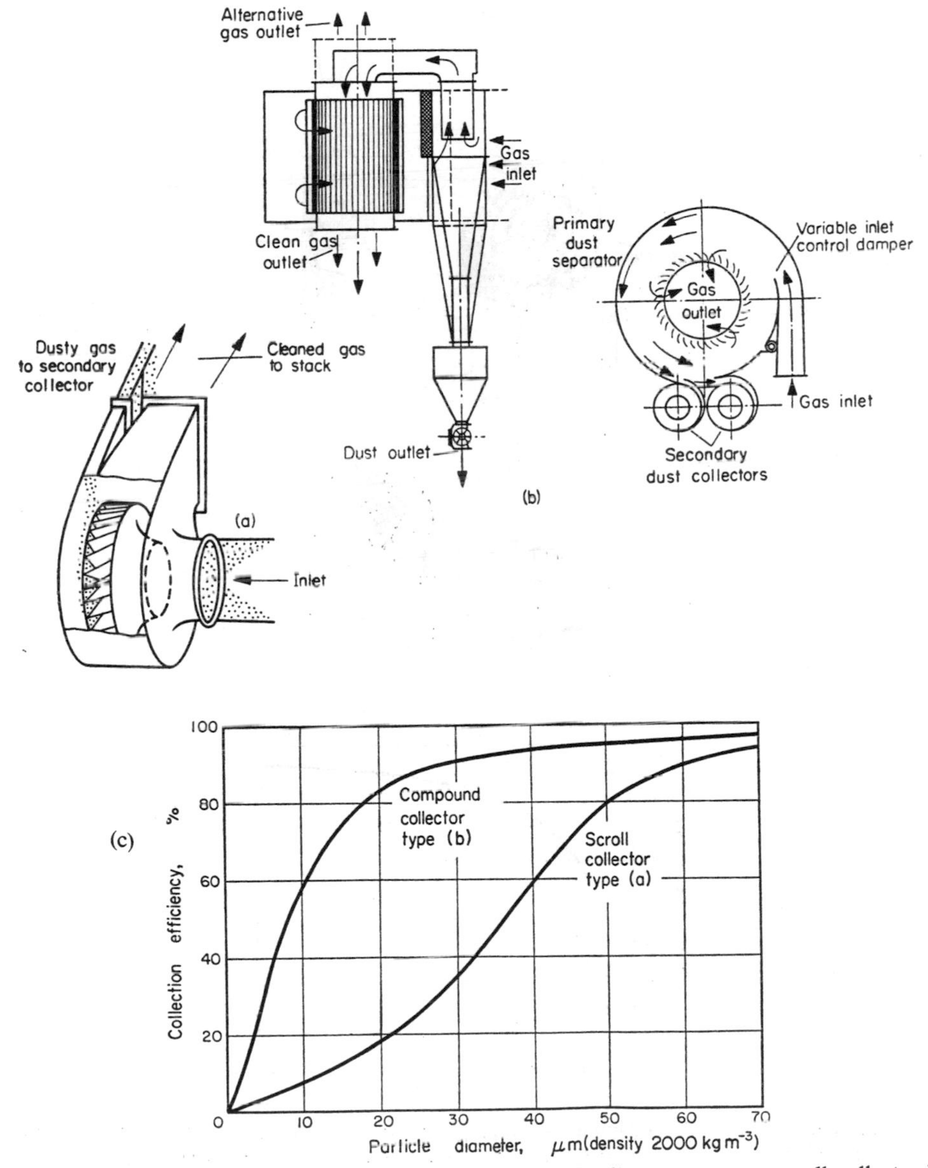

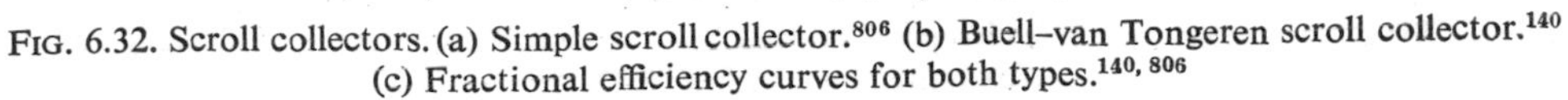

FIG. 6.32. Scroll collectors. (a) Simple scroll collector.[806] (b) Buell–van Tongeren scroll collector.[140] (c) Fractional efficiency curves for both types.[140, 806]

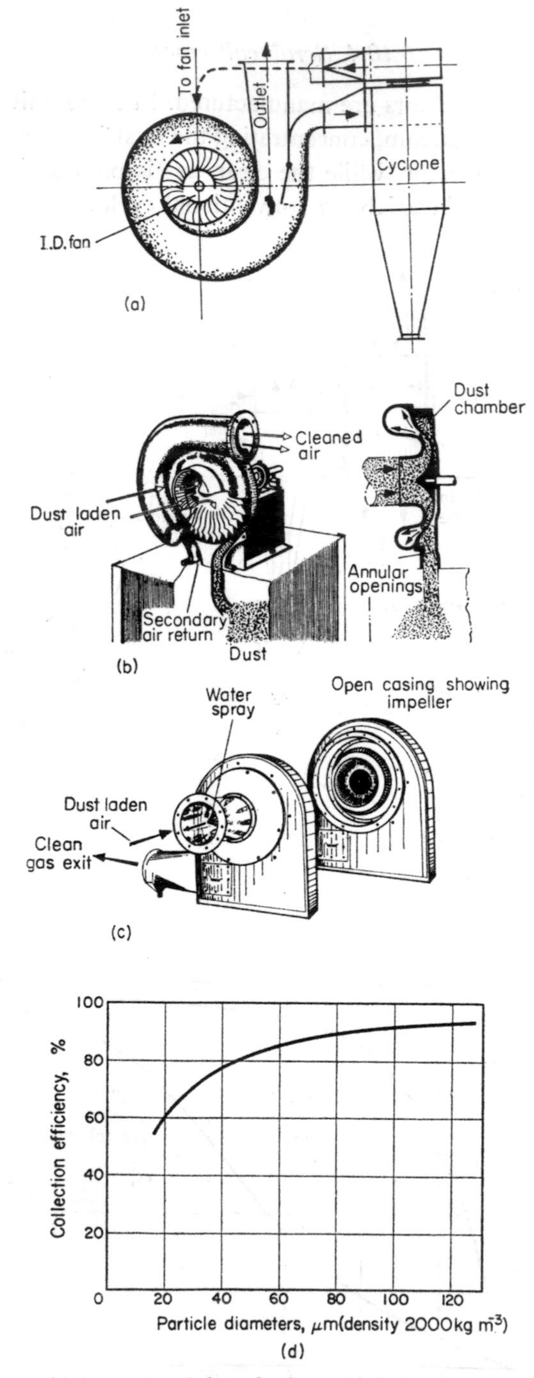

FIG. 6.33. Collectors combining induced-draught fans. (a) Induced-draught fan followed by conventional cyclone for concentrated dust.[657] (b) Induced-draught fan followed by gravity settling hopper for concentrated dust.[18] (c) Irrigated induced fan-type unit.[18] (d) Fractional efficiency curve for unit shown in Fig. 6.33a.[657]

pattern for which a fractional efficiency curve has been obtained by Stairmand[806] using 2 m diameter unit passing 6·6 $m^3 s^{-1}$. Another collector, similar in principle but with tangential entry and scroll exit is shown in Fig. 6.32b. Fractional efficiency curves for both types plotted on the same scale indicate that the second pattern has somewhat greater efficiencies at particle sizes about 30 μm (Fig. 6.32c).

6.10.5. Collectors combined with induced-draught fans

Instead of using a fixed scroll to impart rotation to the gas stream, this can be accomplished by an induced-draught fan. Figure 6.33 shows two typical arrangements, a fractional efficiency curve being shown for the first type (Fig. 6.33d). A modification of type b, shown in Fig. 6.33c uses a water spray to improve performance.

6.10.6. Conventional (reverse-flow) cyclones

The design of these cyclones has been discussed in some detail in the earlier sections on cyclone theory, performance and pressure loss. The various types produced commercially

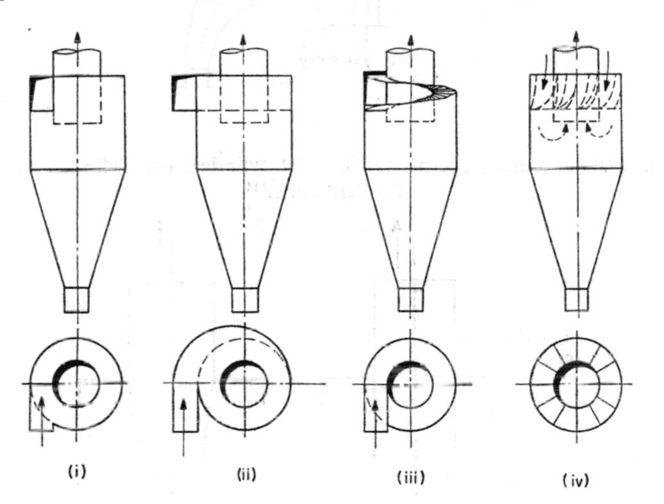

FIG. 6.34. Basic types of cyclone entry. (i) Tangential entry. (ii) Wrap-around entry. (iii) Curved entry. (iv) Axial entry.

differ as to means of dust entry, exit, and in relative proportions. Four types of entry are common (Fig. 6.34).

(i) Tangential entry, without spiral.
(ii) Wrap around inlet without spiral.
(iii) Wrap around inlet with spiral.
(iv) Axial gas inlet with vanes.

The trunk may be cylindrical or cone shaped, and in some cases, where head room is limited, can be bent through a 90° angle (Fig. 6.35). The exit pipe can be either straight,

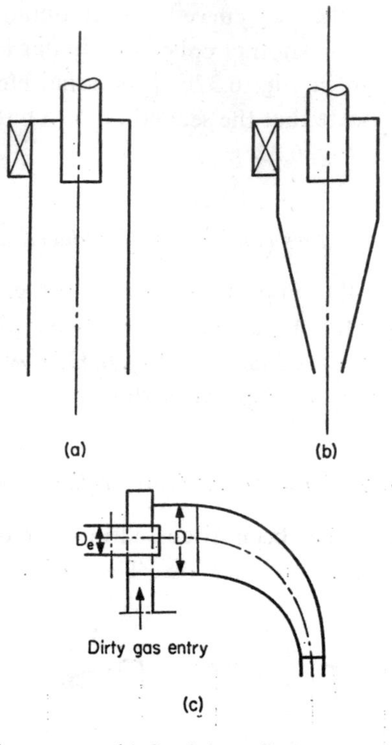

FIG. 6.35. Basic cyclone trunk patterns. (a) Straight cylinder. (b) Cylinder and cone (straight). (c) Bent cone.[514]

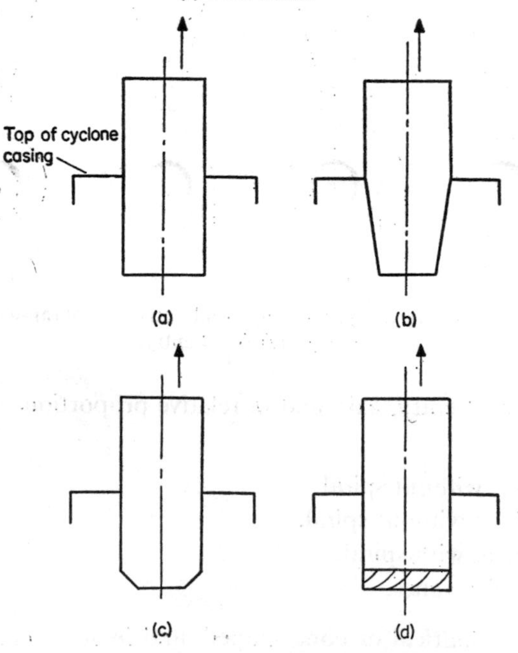

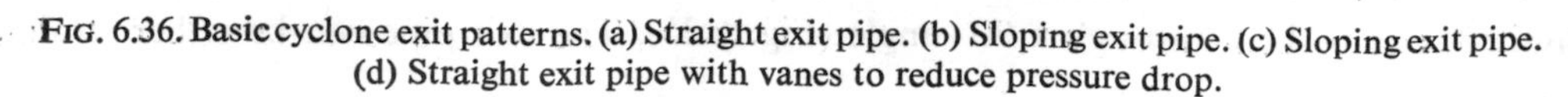

FIG. 6.36. Basic cyclone exit patterns. (a) Straight exit pipe. (b) Sloping exit pipe. (c) Sloping exit pipe. (d) Straight exit pipe with vanes to reduce pressure drop.

inclined, with vanes (Fig. 6.36) or fitted with an internal structure of the type discussed previously as reducing the pressure loss (Fig. 6.21).

The efficiency of these types in various combinations has been extensively studied by a number of workers.

The fractional efficiency curves of two typical designs (Fig. 6.16) have been given (Fig. 6.17). The efficiency of other models can be obtained by using the methods outlined in Section 6.7.

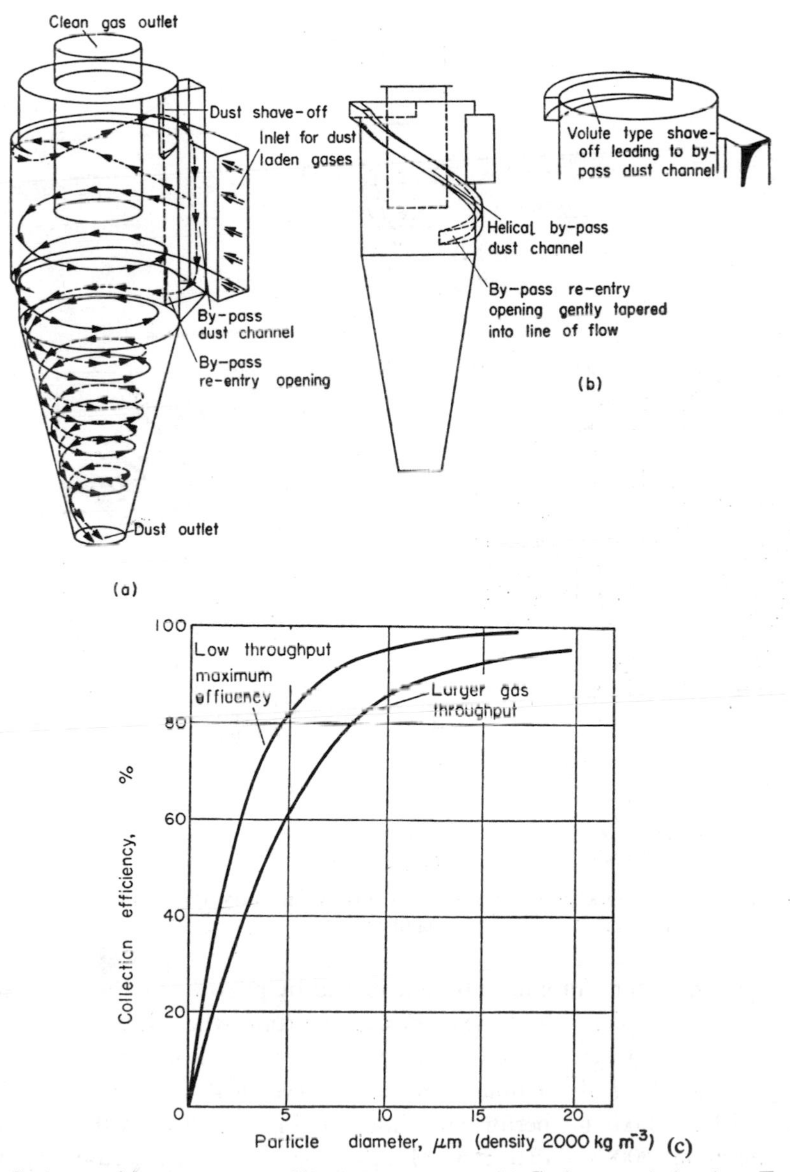

FIG. 6.37. Cyclones with dust shave-off.[140] (a) Flow pattern. (b) C—by-pass design (van Tongeren) (c) Fractional efficiency curves for cyclones with shave-off.

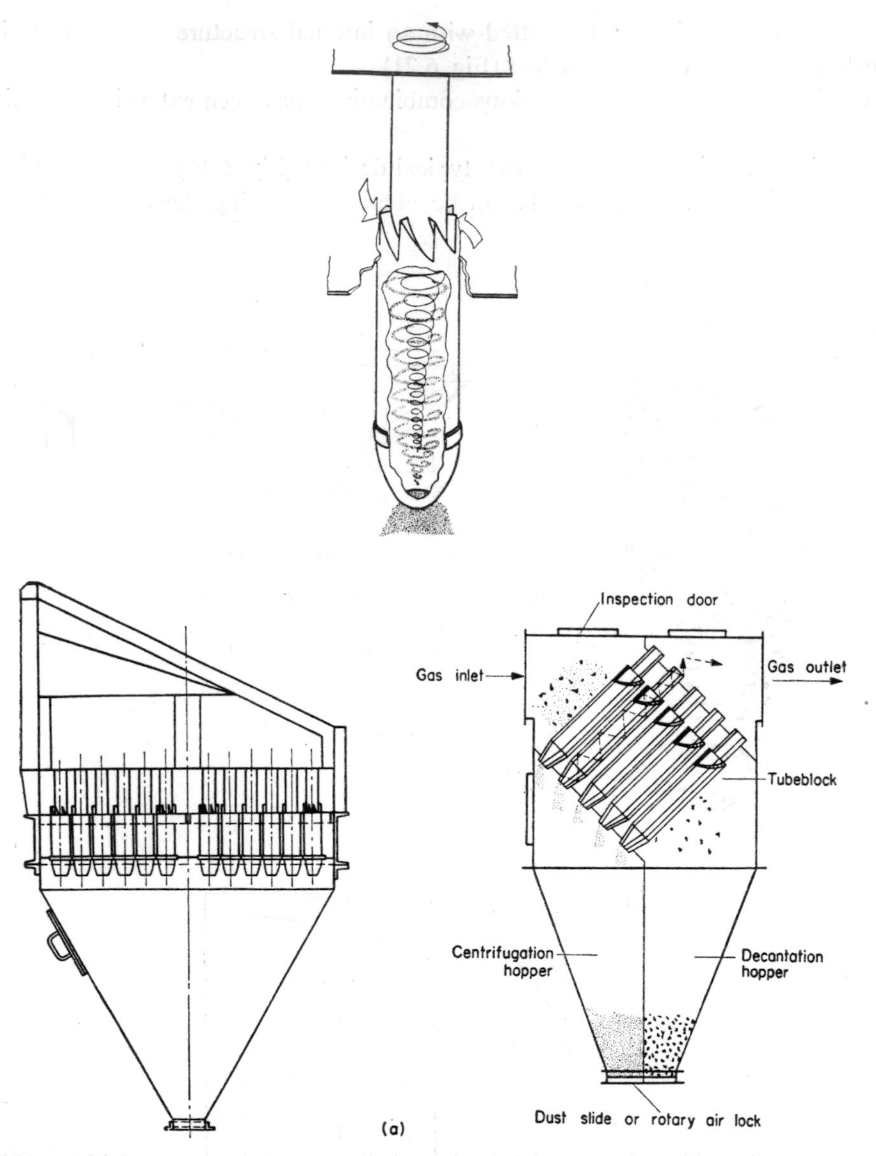

FIG. 6.38. Multiple conventional cyclone designs. (a) Axial entry type with preliminary settling chamber.[657]

Improvements in performance have been achieved by providing a dust shave-off channel which moves the dust which collects near the cyclone roof, where it is likely to find its way into the exit tube by short circuiting into the exit pipe if it is not removed into the conical section (Fig. 6.37a). Typical fractional efficiency curves for commercial installations are shown in Fig. 6.37b. Maximum performance was obtained when a low throughput was used with a cyclone about 900 mm diameter. Modified designs with increased gas throughput had somewhat lower efficiencies.

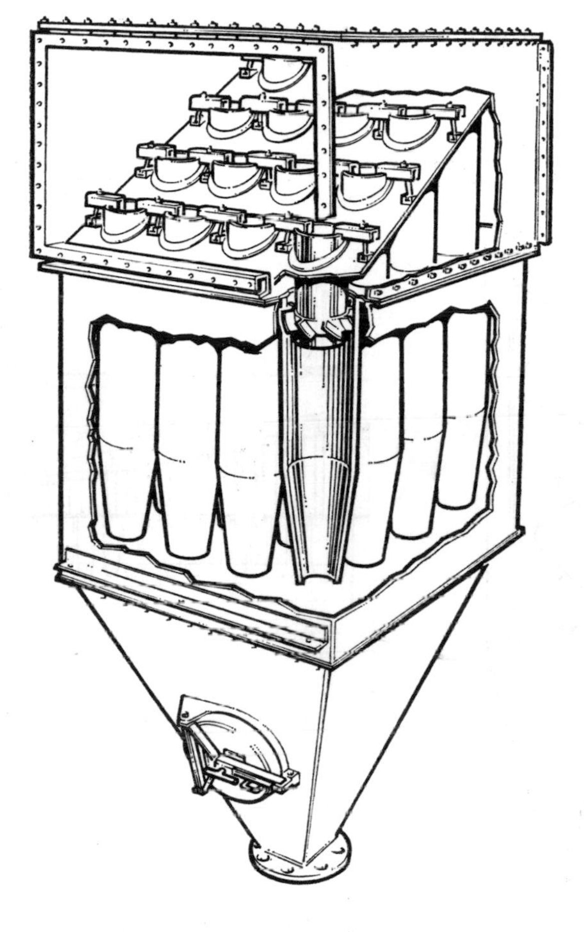

(b)

FIG. 6.38 (*continued*)

(b) Axial entry with mechanism for rapid dismantling of cyclones.[920]

6.10.7 Multiple cyclones

The formulae for critical particle size show that, in theory, smaller cyclones, using the same gas velocities, are more effective collectors of small particles than larger cyclones. In practice this improved efficiency is reduced because of the increased short circuiting that takes place in small cyclones. One worker, in comparing high efficiency cyclones with multiple small cyclone designs, has shown that the multiple cyclone arrangement is only marginally better than the large cyclone, although much more expensive to manufacture.[806] Other problems, such as blocking up of the cone and hopper, occur with certain dust types and high dust loads, and special modifications for easy cleaning have to be included. One important advantage of sets of multiple cyclones is that a single high-efficiency cyclone

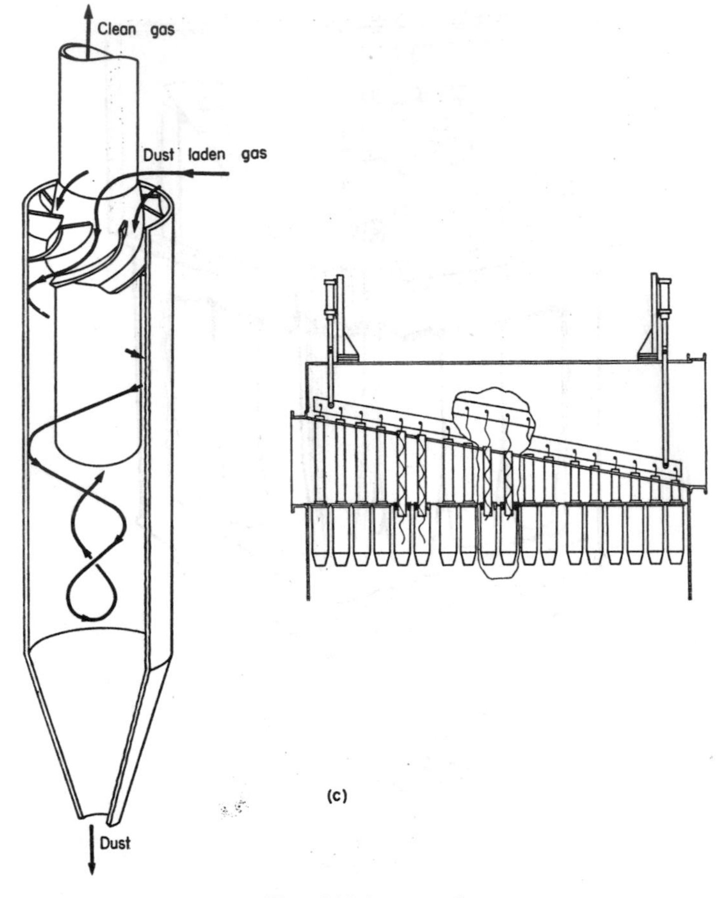

FIG. 6.38 (*continued*)
(c) Axial entry cyclones fitted with dust scraper.[377]

with a capacity of 4600 $m^3\ h^{-1}$, 900 mm diameter, requires 7·6 m headroom (cyclone, hopper and exhaust), while a multiple-cyclone set of similar capacity only requires 2·4 m headroom.

Typical multiple cyclone types with axial and tangential entry are shown in Fig. 6.38. The first type (a) is made of heavy duty cast iron for erosion resistance, type (b) shows a simplified rapid dismantling design, while type (c) is a design incorporating a wire scraper which is actuated externally and enables the cyclone to be cleaned intermittently, either manually or automatically, without dismantling.

Typical fractional efficiency curves are shown in Fig. 6.39a for standard dusts, while Fig. 6.39b shows fractional efficiency curves obtained for a type (c) unit for fly ash collection under difficult operating conditions.[330]

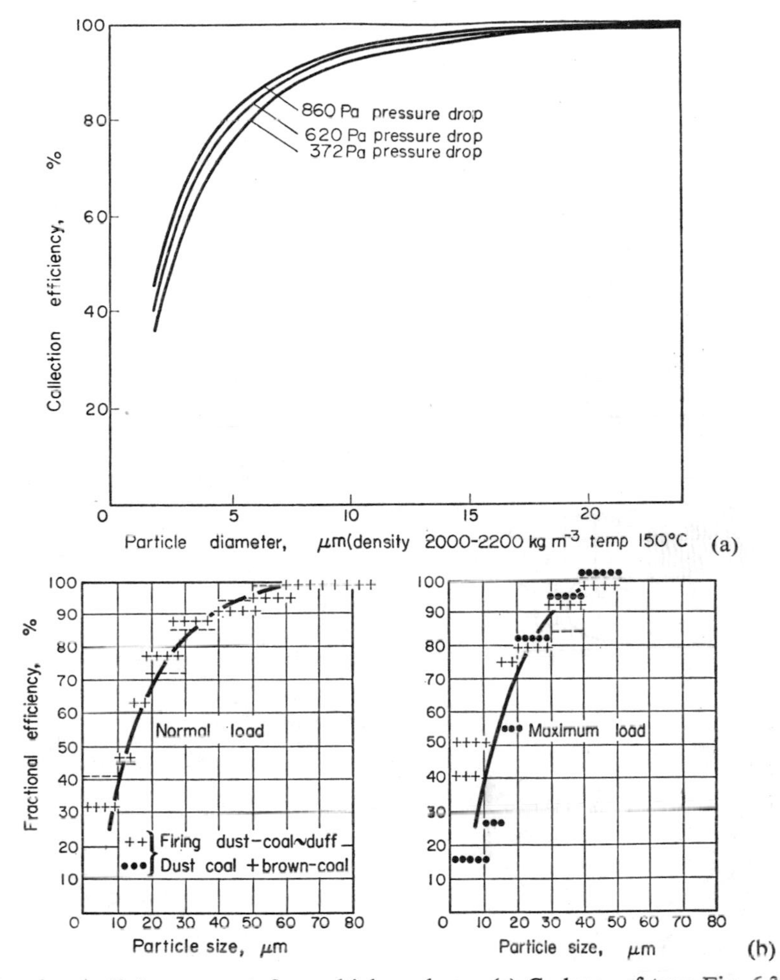

FIG. 6.39. Fractional efficiency curves for multiple cyclones. (a) Cyclones of type Fig. 6.38a (test dust).[657] (b) Cyclones of type Fig. 6.38c (power station operating on high ash coal).[330]

6.10.8 *Cyclones for liquid droplets*

When droplets, formed from condensing vapours or entrained in gases, have to be collected, the design of an appropriate cyclone has to make provision for the additional problem of creeping layers of liquid. These layers, formed from agglomerated droplets, are whipped into the exit pipe by the fast-moving gas stream. The layers creep to the edge of the exit pipe where they are easily re-entrained. The extent of the creeping depends on the liquid properties—surface tension and viscosity—and an oil tends to be more readily entrained than water. Entrainment will also occur in the low-pressure zone at the axis of the cyclone, and liquid droplets moving towards the apex in a conical cyclone may be sucked into the exit gases. It has therefore been suggested[230] that a cylindrical cyclone design is more suitable for droplet collection than the more usual conical shape. One design of this type[518] (Fig. 6.40a) has a flat plate as a false base, *b*, surrounded by a slit *a*, which permits the liquid to

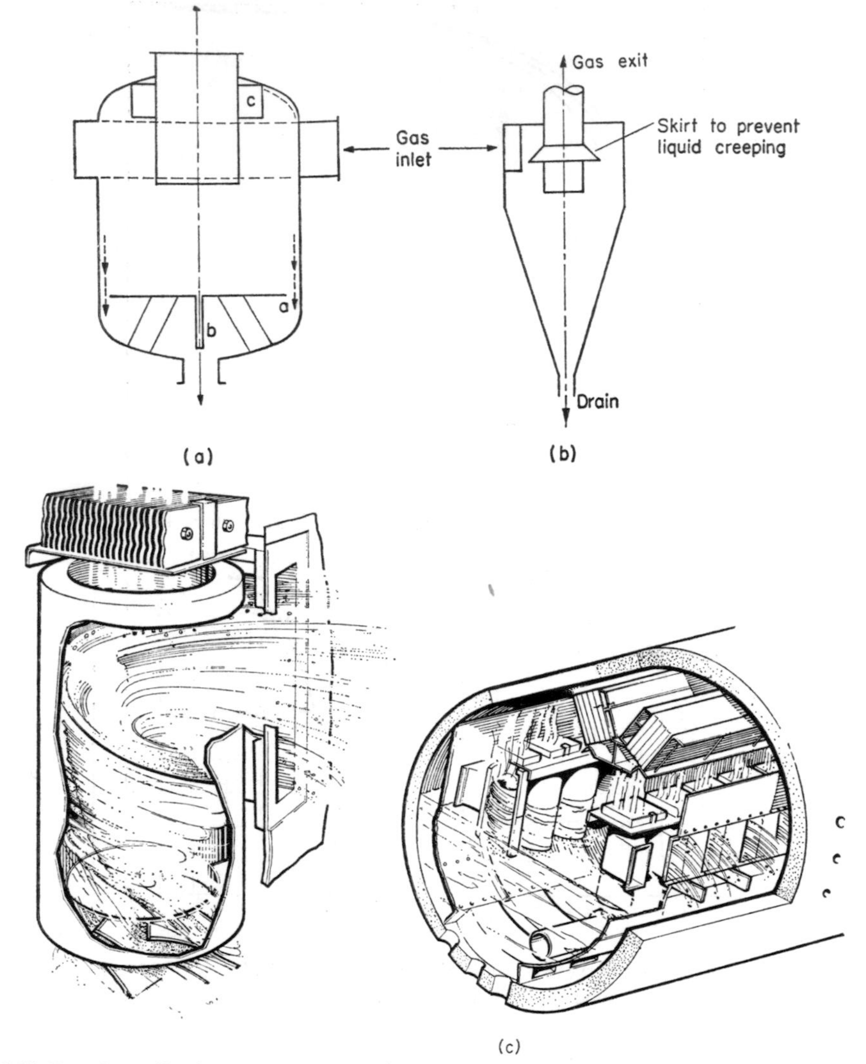

FIG. 6.40. Droplet collecting cyclones. (a) Cylindrical model with baffle (*b*) over exit drain (*c* = skirt to prevent liquid creep).[518] (b) Tangential entry conventional cyclone fitted with skirt to prevent liquid creep.[803] (c) Mist eliminator fitted into boiler drums.[45]

run off to the drain chamber. A skirt *c*, on the cyclone roof, prevents creep of liquid into the exit pipe, as the liquid is blown off the skirt and forced back against the wall. A conventional cyclone, with a skirt to prevent creep of liquid, is shown in Fig. 6.40b.

Another design, for steam–water droplet separation in a boiler drum, also has wrap around entry for the steam–water droplet mixture, parallel walls and inclined vanes at the periphery of the base for the water, as well as corrugated plates in the vapour exit to prevent spray entrainment (Fig. 6.40c). An alternative solution in a cyclone droplet separator is used in the cyclonic spray scrubber (Fig. 9.7), where the creeping liquid path is lengthened by having the gas–liquid entry at the base of the cyclone.

6.11. CYCLONES IN SERIES

When the cyclone only acts as a dust concentrator the concentrated dust must be passed to a secondary collector, which will still have to collect the same quantity of dust, but deal with a far smaller quantity of carrier gas, usually about 10 per cent of the initial gas volume. The efficiency of the combined system of course depends on the efficiency of the primary collector and cannot exceed this.

If the gas discharge from the secondary collector is returned to the gases entering the primary collector, the overall efficiency must be that of the primary collector. If, on the other hand, the secondary collector gas discharge is added to the clean gases from the primary collector, then the overall efficiency of the combined units is lower than that of the primary collector alone.

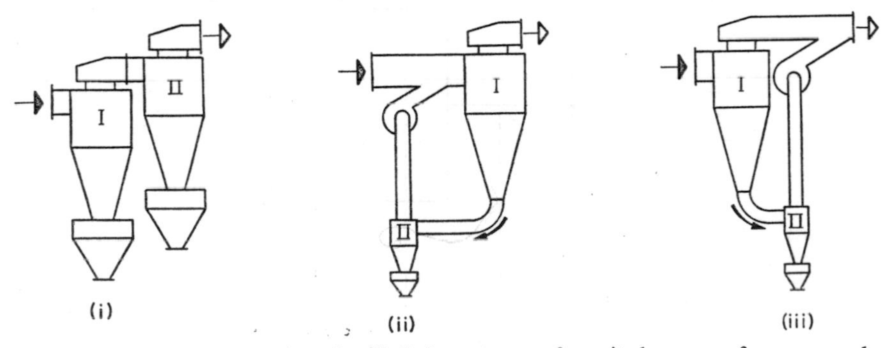

FIG. 6.41. Cyclone arrangements in series.[458] (i) Arrangement for exit clean gases from one cyclone to be fed to second cyclone. (ii) Concentrated dust and some gases bled off to secondary cyclone from base of primary cyclone, feeding clean gases to primary cyclone entry. (iii) As for (b), but passing clean gases from secondary cyclone into clean gases from primary cyclone.

To amplify this, van der Kolk[458] has considered the overall performance of two cyclones in three possible combinations (Fig. 6.41):

(i) A second cyclone operates on the clean-gas discharge of the first cyclone.
(ii) A second cyclone operates on the discharge of dust and some air at the cone tip of the first cyclone, and discharges into the entrance pipe of the primary cyclone.
(iii) A second cyclone operates, as in (ii) on the dust discharge, but in turn discharges into the clean-gas discharge of the primary cyclone.

In the first combination (i) the effect of the second cyclone is marginal, as it has to collect much finer particles than the primary cyclone. It should be emphasized that the pressure loss for this combination is twice that of a single cyclone, and only if the degree of cleaning achieved in this way is worth while can the power costs for the arrangement be justified. If two low-efficiency cyclones are to be used in series, then the result will be no better than a single high-efficiency cyclone, which accomplishes the same task with much lower pressure loss than the two cyclones, and the arrangement will not be the most economic. Van Ebbenhorst Tengbergen[231] points out that while the arrangement shown in Fig. 6.41 (i) can lead to halving the dust quantity in the exit gases, it doubles capital and operating costs. He also recommends the insertion of a single high-efficiency cyclone to obtain the same effect.

It has been found experimentally[230] that the pressure in the hopper is a critical factor in cyclone performance. If air is permitted to leak into the hopper, efficiency is reduced, while bleeding air from the hopper improves efficiency. Continuous dust removal from the cyclone apex and hopper to a secondary collector (combinations (ii) and (iii)) will enhance performance in the same way, because the spiral of gas in the cyclone core penetrates into the hopper and stirs up the collected dust re-entraining the fine particles. This has been utilized in the liquid cyclone for rapid elutriation.[432] In gas cyclones a small disc is often placed in the centre of the hopper opening to eliminate the re-entrainment (Fig. 6.42).

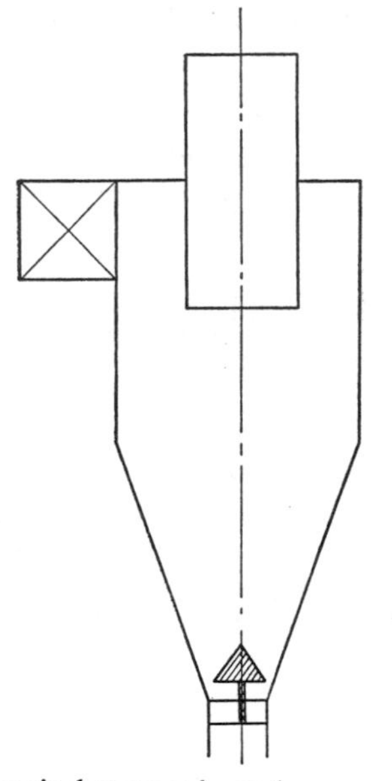

FIG. 6.42. Cyclone with adjustable conical stopper in cyclone cone to prevent vortex from reaching hopper.

The secondary cyclone in combinations (ii) and (iii) is only a very small unit compared to the primary cyclone, and its performance may be improved by the high dust concentrations present. Arrangement (ii) should give considerably better efficiencies than a single cyclone because of the continuous gas bleeding and particle removal, although some possibility of recirculation of fine particle exists. Even when (combination (iii)) the secondary cyclone discharge passes into the clean gas discharge of the primary cyclone, improved performance will result. In both these combinations (ii) and (iii) the pressure loss is only a little greater than for a single unit.

Other systems in series with cyclones are sometimes suggested. One is the possibility of improving cyclone performance by establishing an electrostatic field which will tend to assist the particles to move towards the wall. Calculations[822] show that the effect is only of the order of 2 per cent using a standard small earthed cyclone, a discharge electrode at 60 kV and normal operating conditions with charged particles entering the cyclones. Experiments with such an arrangement did not achieve any measurable improvement in cyclone performance.[639]

CHAPTER 7

THE AERODYNAMIC CAPTURE OF PARTICLES

7.1. INTRODUCTION

The collection of particles by fibrous filters and liquid scrubbers is essentially the capture of particles by collecting bodies. The gas stream passing through the filter or scrubber takes the particles close to these collecting bodies and then a number of short-range mechanisms accomplish the actual collection.

In any particular case the relative importance of these short-range mechanisms varies with the relative size and velocity of the particles, with the collecting body, and with the presence of electrostatic, gravitational or thermal attractive (or repulsive) forces.

The basic short range mechanisms are:

(i) Inertial impaction.
(ii) Interception.
(iii) Diffusion.

Mathematical models have been developed for each individual mechanism, and solutions are generally obtained by numerical methods. The combination of two or more of the short-range mechanisms has not yet been attempted on a comprehensive basis, but numerical solutions have been found for some special cases. Fortunately, in most cases one mechanism predominates, enabling simplifying assumptions to be made. Thus, for particles in the micron-size range and larger, inertial impaction and interception predominate, while diffusion is of much greater importance for sub-micron sized particles.

Electrostatic, thermal and gravitational forces modify the collection efficiencies of filters and scrubbers in special circumstances. For example, when the spray nozzles in a scrubber are insulated from the chamber and charged to a potential of 5 kV, greatly increased collection efficiencies are predicted[463] although the small charges acquired by droplets during normal atomization have been found to have a negligible effect.[404]

Under normal circumstances, thermal forces play a negligible part in filtration or scrubbing, because, to be effective, very large temperature differences between the particles and the collecting body are necessary. In practice, because of the small dimensions of the collecting fibre or droplet, these have a low heat capacity and rapidly reach the temperature of the gas stream and particles flowing through the filter or scrubber. Because of the short duration of any temperature difference, thermal forces generally do not have to be considered, and will be omitted from this chapter.

Two applications of filters have to be distinguished. In one, a relatively clean gas, such as atmospheric air, is filtered to give specially cleaned air for air conditioning, while in the other, industrial gases with a high dust concentration have to be cleaned. The collection of particles in the first case and initially in the second proceeds only rarely by a screening action because the particles are much smaller than the gaps between the fibres. Furthermore, collecting particles in the inter-fibre gaps rapidly block the filter and cause a rapid rise in pressure drop. Air-conditioning filters are replaced when the particles penetrate through them, and the pressure drop exceeds a certain, low, value, and are not cleaned *in situ*. Filters for industrial dusts, on the other hand, after capturing a thin layer of particles on the fibres, build up a cake, which is removed at frequent intervals. The effect of cake formation on pressure drop, and the timing of cleaning cycles for industrial filters will be dealt with in the next chapter.

This chapter discusses the three basic mechanisms in aerodynamic capture: inertial impaction, interception and diffusion, first individually and then in combination. The effect of temperature, external forces (gravitational and electrostatic forces) and of series of collectors are also considered in detail.

7.2. VELOCITY DISTRIBUTION AROUND A CYLINDER

When a moving gas stream approaches an infinitely long cylinder placed normal to the gas stream or a sphere, the fluid streamlines spread around the body. The streamline configuration depends on the fluid velocity. At high velocities the streamlines diverge suddenly close to the body, while at low velocities the divergence commences a considerable distance upstream.

A Reynolds number can be defined as a function of the collecting body dimensions and the relative fluid velocity:

$$\mathrm{Re}_c = \frac{v_0 \varrho D}{\mu} \tag{7.1}$$

where v_0 = undisturbed upstream fluid velocity,
D = diameter of collecting body,
ϱ = density of fluid,
μ = viscosity of fluid.

At Reynolds numbers of 0·2 a 3 per cent disturbance occurs at a distance of $100D$ upstream, while when $\mathrm{Re}_c = 2000$, there is practically no fluid disturbance at a distance of $2D$ upstream.[211]

At the high Reynolds numbers, the flow upstream of a horizontal cylinder can be approximately described with the equations for two-dimensional, friction-free, incompressible flow. The numerous equations for this system and the modifications allowing for boundary layers are given in detail elsewhere,[529, 643] and will only be briefly reviewed here.

The stream function for potential flow, Φ, is given by

$$\Phi = V_0 r \sin \theta (1 - \mathbf{R}^2/r^2) \tag{7.2}$$

where r and θ are polar coordinates, **R** is the fibre radius, and the velocity components V_r and V_θ follow from the stream function

$$V_r = \frac{1}{r}\frac{\partial \theta}{\partial \varphi}; \qquad V_\theta = -\frac{\partial \varphi}{\partial \theta}. \tag{7.3}$$

Because of flow separation and wake effects, these equations are not applicable downstream of the cylinder.

For very low Reynolds numbers ($\ll 1$), Oseen's approximation of the Navier–Stokes equation as given by Lamb is used. The stream function is φ, in the viscous flow range:

$$\varphi = \frac{\mathbf{R}V_0 \sin \theta}{2(2 - \ln \mathrm{Re}_c)}\left[\frac{\mathbf{R}}{r} - \frac{r}{\mathbf{R}} + \frac{2r}{\mathbf{R}} \ln \frac{r}{\mathbf{R}}\right]. \tag{7.4}$$

This equation is only valid at the cylinder surface. In contrast to the potential flow system, at low Reynolds numbers, as was pointed out earlier, the stream lines are deflected at much greater distances upstream, and more gradually further sideways. A more complex relation for small Knudsen numbers (λ/D) related to the cylinder (i.e. ratio of mean free path of gas and cylinder diameter) ($< 0{\cdot}25$) has been derived by Natanson,[596] which reduces to 7·4 for $\lambda/D \to 0$, while in the transition region, the velocity field has been investigated by Velichko and Radushkevich.[886] Approximate descriptions of the flow-around cylinders have been derived empirically by Sell,[750] and by a method of successive approximations by Thom[855] and Davies.[206]

Real filters consist of fibres, which can be considered as a system of randomly distributed cylinders, but the problem of mathematically describing the complex flow field is extremely difficult. For a highly porous filter, this has been successfully solved independently by Kuwabara[470] and Happel[336] who obtained what is essentially the same equation, the difference being in the numerical constant **c**. This equation is given by the stream function Φ as:

$$\Phi = \frac{\mathbf{R}v_0 \sin \theta}{2[-\frac{1}{2}\ln(1-\epsilon) - \mathbf{c}]}\left[\frac{\mathbf{R}}{r} - \frac{r}{\mathbf{R}} + 2\frac{r}{\mathbf{R}} \ln \frac{r}{\mathbf{R}}\right] \tag{7.5}$$

where ϵ is porosity and $(1-\epsilon)$ the relative fibre volume.

The constant $\mathbf{c} = 0{\cdot}75$ according to Kuwabara and 0·5 according to Happel. It should also be noted that equation (7.5) no longer is a function of the flow past the fibres, as expressed by Re_c, and furthermore is only valid for the conditions of high filter porosity (i.e. $(1-\epsilon) \ll 1$) and close to the fibres [i.e. $(r - \mathbf{R})/(\mathbf{R} \ll 1)$]. Kirsh and Fuchs[441–2] found that in experimental studies of flow distribution and pressure drop to Reynolds numbers Re_c to 0·1 and a range of fibre filter volume ratios $(1-\epsilon)$ between 0·0034 and 0·27, the Kuwabara relation (i.e. $\mathbf{c} = 0{\cdot}75$) proved more satisfactory.

Thus the Kuwabara–Happel relation is generally satisfactory, while the flow is continuous and there is no "slip" at the fibres, which is realistic for fibres greater than approximately 5 μm.

For finer collectors, with Knudsen numbers below 0·25, Pich[642-3] has modified the Kuwabara–Happel relation for the case of gas slip at the surface of the cylinders. The velocity discontinuity which exists in the layer immediately next to the surface must result in a reduction of fluid resistance, and if the tangential forces acting are proportional to this velocity discontinuity, then the proportionality factor, called the coefficient of external friction (Fuchs[285]), is μ_e, and the *slip coefficient* ζ is μ/μ_e where μ is the normal viscosity. When μ_e is very large, bodies experience a Stokes' law resistance. The modified relation is

$$\Phi = \frac{Rv_0 \sin\theta\left[\frac{\mathbf{R}}{r} - \frac{r}{\mathbf{R}} + 2\left(1 + 2\frac{\zeta}{\mathbf{R}}\right)\frac{r}{\mathbf{R}} \ln \frac{r}{\mathbf{R}}\right]}{2\frac{\zeta}{r}[-\ln(1-\epsilon) - 2\mathbf{c} + 1] - \ln(1-\epsilon) - 2\mathbf{c}}. \qquad (7.6)$$

In the case of negligible slip correction, i.e. $\zeta \to 0$ and $Kn \to 0$, then equation (7.6) reduces to (7.5).

7.3. INERTIAL IMPACTION

If an aerosol is introduced into a gas stream flowing past a collecting body, the aerosol particles will follow the gas streamlines until they diverge around the collector. The particles because of their mass will have sufficient momentum to continue to move towards the collector and break through the gas streamlines (Fig. 7.1). External forces such as gravity could assist in this.

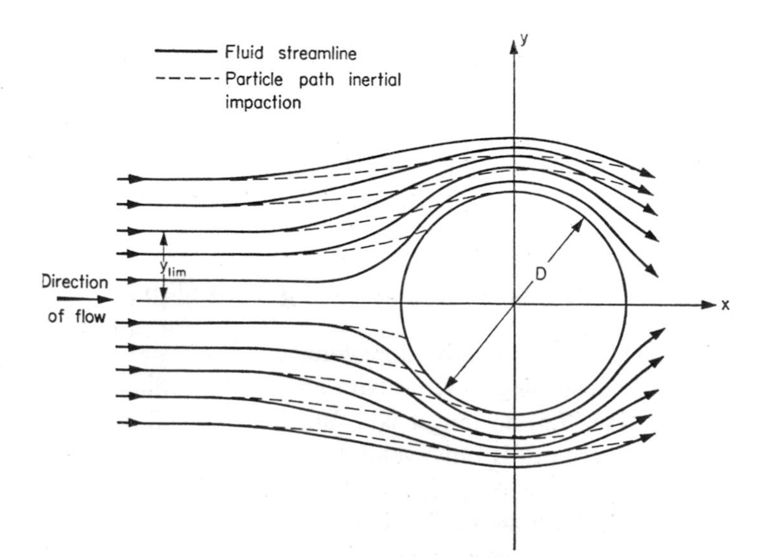

FIG. 7.1. Flow pattern around cylinder: the broken lines indicate path taken by particles being collected by inertial impaction.

In vector notation, the motion of the particle is described by the equation

$$m\frac{d\vec{u}}{dt} = \vec{F}_e - \vec{F} \tag{7.7}$$

where $\vec{u}$ = velocity of the particle,
$\vec{F}$ = resistance of the fluid [Chapter 4, equation (4.35)],
$\vec{F}_e$ = vector sum of external forces.

If the fluid resistance is considered to be in the viscous flow region, then Stokes' law (including the Cunningham correction factor C) can be applied to the fluid resistance $\vec{F}$:

$$\vec{F} = \frac{3\pi\mu d}{C}(\vec{u} - \vec{v}) \tag{7.8}$$

where $\vec{u} - \vec{v}$ is the velocity of the particle relative to the fluid.

Neglecting the external forces $\vec{F}_e$ and assuming the particle to be spherical, then equation (7.8) can be written as

$$\frac{C(\varrho_p)d^2}{18\mu} \cdot \frac{d\vec{u}}{dt} = -(u - \vec{v}.) \tag{7.9}$$

It is convenient to convert this equation into a dimensionless form by writing the up-stream and cross stream distances (x, y) in terms of the collector diameter D:

$$\tilde{x} = 2x/D, \quad \tilde{y} = 2y/D$$

and the velocities in terms of the reference velocity v_0

$$\tilde{v}_x = v_x/v_0, \quad \tilde{v}_y = v_y/v_0$$

the time in terms of both collector diameter and reference velocity:

$$\tilde{t} = 2v_0t/D$$

while the Stokes' law term can also be expressed using these terms as the *inertial impaction parameter* which is sometimes also called the *Stokes number*, when multiplied by 2.

$$\psi = \frac{C\varrho_p d^2 v_0}{18\mu D}. \tag{7.10}$$

Then, in terms of rectangular coordinates $O\tilde{x}$ and $O\tilde{y}$, the equation (7.9) is

$$2\psi\frac{d^2\tilde{x}}{d\tilde{t}^2} + \frac{d\tilde{x}}{d\tilde{t}} - \tilde{v}_x = 0 \tag{7.11a}$$

and

$$2\psi \frac{d^2\tilde{y}}{d\tilde{t}^2}+\frac{d\tilde{y}}{d\tilde{t}}-\tilde{v}_y = 0. \qquad (7.11b)$$

Outside the Stokes' law region, when the drag coefficient C_D is not given by 24/Re (where Re $= d\varrho(\bar{u}-\bar{v})/\mu$) these equations must be rewritten with a term including the drag coefficient:

$$\frac{48\psi}{C_D \mathrm{Re}} \frac{d^2\tilde{x}}{d\tilde{t}^2}+\frac{d\tilde{x}}{d\tilde{t}}-\tilde{v}_x = 0 \qquad (7.12a)$$

and

$$\frac{48\psi}{C_D \mathrm{Re}} \frac{d^2\tilde{y}}{d\tilde{t}^2}+\frac{d\tilde{y}}{d\tilde{t}}-\tilde{v}_y = 0. \qquad (7.12b)$$

Physically, the inertial impaction parameter ψ is the stopping distance, in still fluid, of a particle with an initial velocity of $2v_0/D$, assuming that the fluid resistance is in the viscous range. Many workers, particularly in the German literature, refer to the inertial impaction parameter multiplied by D (i.e. $C\varrho_p v_0 d^2/18\mu$) as the stopping distance (Bremsstrecke).

The efficiency of capture by inertial impaction is defined by the fraction of particles (considered to be evenly distributed in the gas stream) which can be collected by the rod or sphere from a normal cross-sectional area of the gas stream equal to the frontal area of the collector. To find this it is therefore necessary to determine the trajectory of the particles in this section of the gas stream, and, particularly, the trajectory of the particle which will just touch the collecting body. In the two-dimensional case it is necessary to know the distance from the x coordinate at $x = -\infty$ at which a particle starts that will just touch the surface of the collector, i.e. the efficiency of collection by inertial impaction is:

$$\eta_I = \frac{y_{\mathrm{lim}}}{D/2} = \tilde{y}_{\mathrm{lim}}. \qquad (7.13a)$$

This limiting trajectory can be shown to be a function of Re_c and ψ only.

As equations (7.11) or (7.12) cannot be solved directly, the limiting trajectory has been obtained by numerical methods involving step by step calculations. The various solutions obtained have been adequately reviewed elsewhere,[362, 950] and will therefore only be outlined here.

The first important investigation of inertial impaction was by W. Sell,[750] who determined the velocity profiles experimentally by studying streamlines in water moving past bodies of various shapes (sphere, cylinder and flat plate, all 100 mm diameter). Using the experimental streamlines, Sell calculated particle trajectories, assuming that the particles had mass, but no size, in finding the particle accelerations. Sell found that the efficiency of collection could be characterized by a dimensionless group mv_o^2/FD, which is identical with the inertial impaction parameter.

Using the potential flow equations for an ideal fluid, Albrecht[6] calculated the trajectory of the particle that would just touch the collector. Langmuir and Blodgett[490] and Bosanquet[101] similarly used the potential flow theory to determine particle trajectories. The dimensionless group derived by Bosanquet can be shown to be the reciprocal of the inertial impaction parameter. The potential flow theory gives the fluid at the surface of the collecting body a maximum velocity twice that of the upstream velocity v_0, while in fact, the boundary fluid layers make the surface velocity zero. The differences in the calculated trajectories for the various authors derive from the different starting points for the calculations, and the number of steps taken. For example, Albrecht[6] starts at $\tilde{x} = -3$ while Langmuir and Blodgett[490] start at $\tilde{x} = -4$, and use a differential analyser to compute a larger number of stages.

Measured inertial impaction efficiencies obtained by Landahl and Herrmann[483] for droplets on cylinders did not agree with Sell's predicted values; an efficiency curve was therefore derived from calculated and experimentally confirmed fluid velocities at Re_c values of 10 obtained by Thom.[855] Their results can be expressed in the empirical relation[643]

$$\eta_I = \frac{\psi^3}{\psi^3 + 0{\cdot}77\psi^2 + 0{\cdot}22}\,. \tag{7.13b}$$

Davies[207] based his efficiencies for large Re_c ($>$ 1000) on Sell,[703] Albrecht[6] and Glauert,[304] for $\mathrm{Re}_c = 10$ on Thom's velocities,[855] while for Re_c values more applicable to flow past fibres in filters, which are of the order of $\mathrm{Re}_c = 0{\cdot}2$, he used his own viscous-flow relations.[206]

The collection efficiency of droplets has also been calculated by Pearcey and Hill[627] and several other authors.[200, 263, 371, 372, 629, 645, 905]

Albrecht[6] and later workers have shown that the calculations predict a value of the inertial impaction parameter ψ_{crit} below which the inertial impaction collection efficiency is zero. For cylinders Albrecht gave ψ_{crit} as 0·09, without allowing for a viscous boundary layer. With this allowance, Langmuir[489] obtained ψ_{crit} as 0·27. Subsequent calculations by Langmuir and Blodgett[490] and by Bosanquet[101] gave the critical value of ψ, ψ_{crit} as 0·0625 for cylinders.

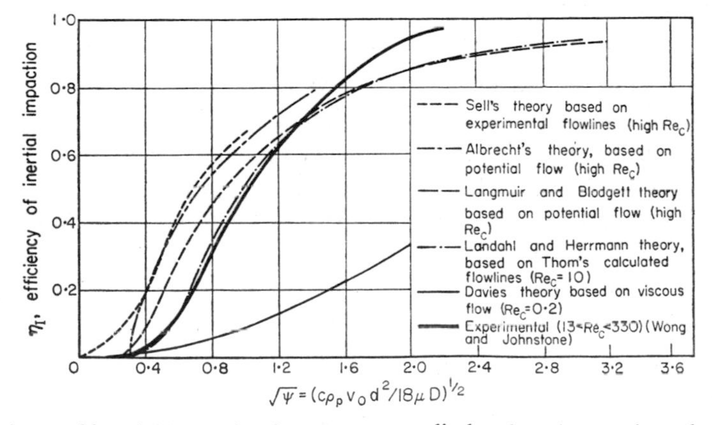

FIG. 7.2. Efficiency of inertial impaction by spheres on cylinders based on various theories and the experiments by Ranz, Wong and Johnstone.[672]

Davies and Peetz,[211] however, point out that a limiting value ψ_{crit}, below which collision is impossible, is only possible while it is assumed that the particle diameter is negligible compared with that of the cylinder. If the particle has finite size, then collision will take place when the centre of the particle is one radius upstream of the stagnation point. The velocity of impact is therefore finite, however small ψ may be, as long as the particle diameter is finite. These workers also point out that while their collection efficiencies agree with those of Langmuir and Blodgett,[490] Sell, Albrecht and Glauert all predict a higher efficiency over

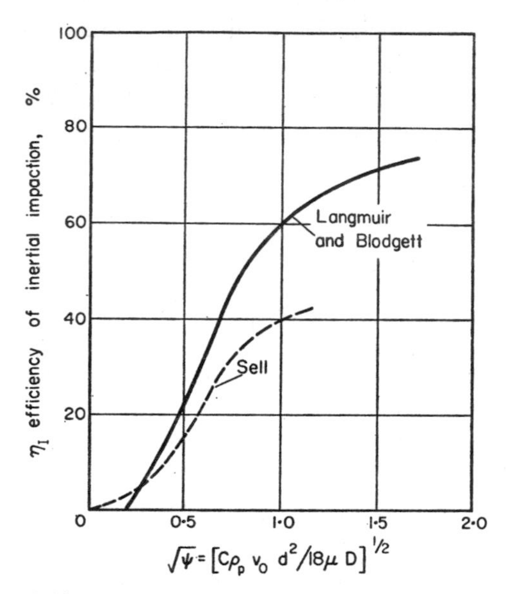

FIG. 7.3. Efficiency of inertial impaction by spheres on spheres, based on the theories of Sell[750] and Langmuir and Blodgett.[490]

large ranges of ψ, because the starting point for the calculation was too close to the cylinder without allowing for the initial velocity.

The curves for efficiency of inertial impaction on cylinders and spheres, using $\sqrt{\psi}$ as the abscissa, are shown in Figs. 7.2 and 7.3 respectively.

The most extensive measurements of collection efficiencies on cylinders have been made by Wong and Johnstone[950] (Fig. 7.4). The curve drawn through Wong and Johnstone's results, which is included in Fig. 7.2, lies close to the predicted curve by Landahl and Hermann.[483] Also it should be noted, that while practical values for Re_c in fibre filtration are about 0·2, the Re_c values for which the inertial impaction efficiency has been found have been of the order of $13 < Re_c < 330$. Until more realistic determinations of inertial impaction efficiency at low Re_c numbers have been carried out, an efficiency intermediate between the Landahl and Hermann and the Langmuir and Blodgett curves will probably give the best estimate.

Experimental work of collection of particles by spheres is frequently associated with collection of particles by raindrops or artificial water sprays. Experimental points obtained in

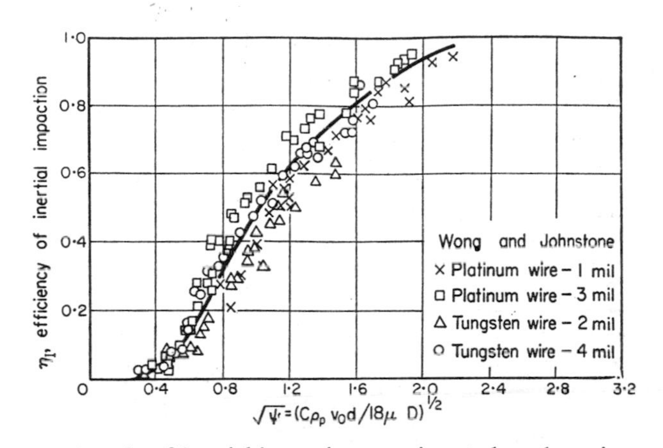

FIG. 7.4. Experimential results of inertial impaction experiments by spheres impacting on wires.[950]

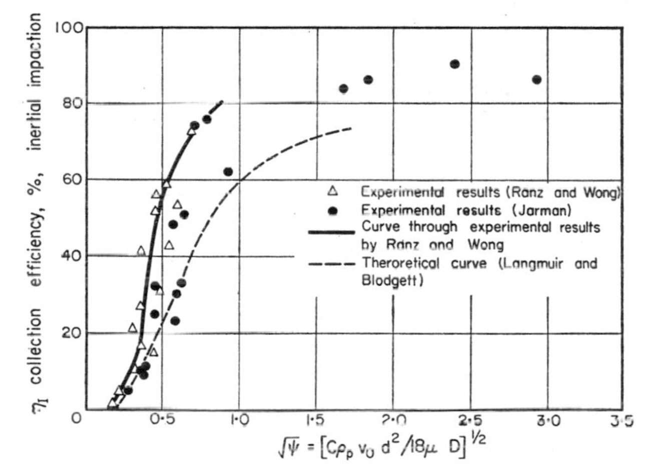

FIG. 7.5. Experimental results for inertial impaction by spheres on spheres by Ranz and Wong,[397] and Jarman,[397] together with the theoretical curve by Langmuir and Blodgett.

recent measurements by Jarman[397] together with earlier work by Ranz and Wong[672] are plotted in Fig. 7.5.

7.4. INTERCEPTION

The model considered for inertial impaction assumed particles had mass, and hence inertia, but no size, except when calculating the resistance of the fluid for cross-stream movement. In allowing for actual particle size, an interception mechanism is considered where the particle has size, but no mass, and so follows the streamlines of the gas around the collector. If the streamline on which the particle centre lies approaches to closer than $d/2$ to the collector, the particle will touch the collector and be intercepted (Fig. 7.6).

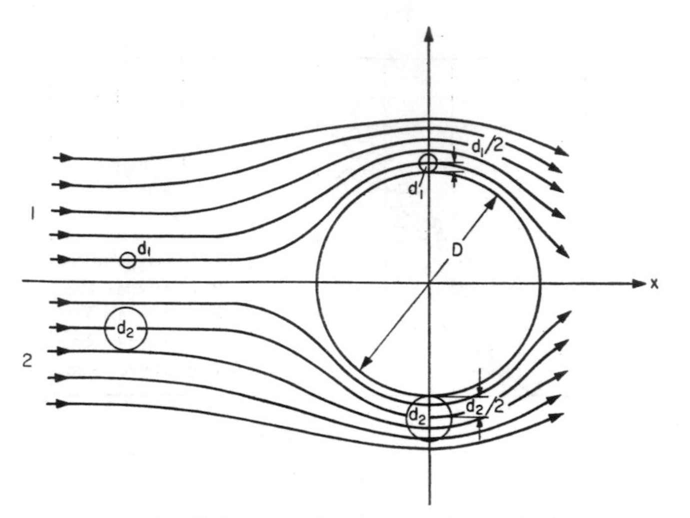

FIG. 7.6. Flow pattern around cylinder, showing interception collection mechanism for spheres of differing sizes.

Interception is characterized by a parameter R, which is the ratio of the diameters of the particle d and of the intercepting body D. ($R = d/D$). If potential flow is assumed, the efficiency of collection by interception, η_c, can be calculated from the relations[672]

$$\eta_C = 1+R-1/(1+R) \qquad (7.14)$$
(cylindrical collector)

and:

$$\eta_C = (1+R)^2-1/(1+R) \qquad (7.15)$$
(spherical collector).

Alternatively using Langmuir's viscous flow equation[489] Ranz[670] obtained the interception collection efficiency of a cylindrical target:

$$\eta_C = \frac{1}{2{\cdot}002-\ln \mathrm{Re}_c}\left[(1+R)\ln(1+R)-\frac{R(2+R)}{2(1+R)}\right]. \qquad (7.16)$$

This last equation should be used whenever possible, as it allows for the changing flow pattern with different stream velocities.

An approximate value for the collection efficiency by interception can be obtained from an equation by Friedlander[275]

$$\eta_C = 2\,\mathrm{Re}_c^{\frac{1}{2}}R^2 \qquad (7.17)$$

while for small values of $R(\ll 1)$ Natanson[596] obtained the relation:

$$\eta_C = \frac{1}{2{\cdot}002-\ln \mathrm{Re}_c}R^2. \qquad (7.18)$$

It should be noted that for the equations (7.14) and (7.15) the interception efficiency can be greater than 1. For Knudsen numbers in the range $10^{-3} < \mathrm{Kn} < 0{\cdot}25$, Pich[643] calculated the interception collection efficiency using his modified Kuwabara–Happel equation for the range where "slippage" occurs. He found that η_C increased with increasing R, increasing $(1-\epsilon)$ and λ (mean free path). This last implies improved efficiencies at reduced pressures.

The two mechanisms of inertial impaction and interception are of course not independent of one another, as has been assumed above. A much better estimate of combined interception and impaction efficiency can be obtained when an allowance is made for those

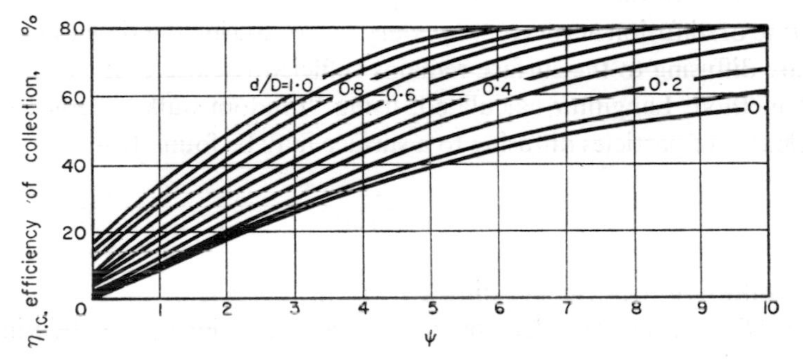

FIG. 7.7. Efficiency of collection by combining inertial impaction and interception for a Re_c value of 0·2.[207] The parameter is $d/D(=R)$.

particles whose centres lie on trajectories closer than the particle radii to the collecting body. This, however, requires the stepwise calculation of particle trajectories for different values of R and Re_c. Davies[207] has carried out this calculation for $\mathrm{Re}_c = 0{\cdot}2$, a typical value for fibrous filters. The curves are plotted in Fig. 7.7 and an equation for the combined efficiency has been fitted to these curves.

$$\eta_{IC} = 0{\cdot}16[R+(0{\cdot}50+0{\cdot}8R)\psi-0{\cdot}1052R\psi^2)]. \tag{7.19}$$

No further equations, for other values of Re_c have been calculated, so approximate combined efficiencies must be used. As a first approximation, particularly for low individual efficiencies of impaction and interception, the sum of the two can be used. However, as a particle caught by one mechanism cannot be caught again, a better estimate, which allows for this, is

$$\eta_{IC} = 1-(1-\eta_C)(1-\eta_I). \tag{7.20}$$

7.5. DIFFUSION

Very small particles, in the sub-micron size range, are rarely collected by inertial impaction or interception, because they not only follow the gas streamlines surrounding the collecting body, but also move across them in an irregular way. This erratic, zigzag movement of small particles, caused by their continued, irregular bombardment by molecules of the gas, is called Brownian motion. In a still gas small particles move freely and distribute themselves evenly throughout the gas, and if an object were placed in the gas, some of the particles would settle on it, being thus removed from the gas. In a moving gas, only limited time is available for this process of removal by diffusion, that is while the gas in the streamlines from which diffusion takes place remains sufficiently close to the collector.

The estimation of the number of particles that are removed while the gas flows past the collector then follows one of two methods. One, introduced by Langmuir,[489] calculates the average distance moved by the diffusing particle during this time, and finds the efficiency of collection from the volume cleared by diffusion as compared to the total volume flowing past. The other deduces the efficiency of collection by using theories of mass transfer to calculate the rate of diffusion across a "boundary layer" during the time the gas from which the particles are diffusing to the surface remains sufficiently close to it.

In the first method, Langmuir calculated, using "random-walk" theory, that the layer of a still gas cleared of particles diffusing to a surface may be found from

$$x = (4\mathcal{D}t/\pi)^{\frac{1}{2}} \tag{7.21}$$

where $\mathcal{D}$ is the diffusion coefficient or diffusivity of the particles.

The *particle diffusivity* may be calculated in two ways. One, introduced by Einstein, applies to particles of a size order, the same as or greater than, the mean free path of the gas molecules. The other, introduced by Langmuir, applies to particles smaller than the mean free path. Einstein,[233] from a consideration of osmotic forces, deduced that the particle diffusivity may be found from

$$\mathcal{D} = \mathbf{k}T/F' \tag{7.22}$$

where $\mathbf{k}$ = Boltzmann's constant,
T = absolute temperature,
F' = "fluid resistance to particle" term: $3\pi\mu d$.

In the particle size range considered for collection by diffusion, Stokes' law, corrected for *slip* by the Cunningham correction factor C, can be applied for the fluid resistance, and so equation (7.22) becomes

$$\mathcal{D} = \frac{C\mathbf{k}T}{3\pi\mu d}. \tag{7.23}$$

Langmuir[489] used the Stephan–Maxwell diffusion theory, which assumes that the gas molecules are not influenced by the particles. This restricts the diffusivity calculated in this way to particles which are much smaller than the mean free path of gas molecules, but still very large compared to the molecules themselves. Langmuir found that the diffusivity can be found from

$$\mathcal{D} = \frac{\bar{u}}{3N(\pi d^2/4)} \tag{7.24}$$

where N = number of gas molecules per unit volume,
$\bar{u}$ = average molecular velocity [equation (3.3)].

The number of gas molecules per unit volume is a function of the gas pressure, which may be expressed, at ordinary pressures where the ideal gas law can be assumed, as $N = P/\mathbf{k}T$. Substituting for N and $\bar{u}$ in equation (7.24) gives the diffusivity as:

$$\mathcal{D} = \frac{4\mathbf{k}T}{3\pi d^2 P}\left(\frac{8RT}{\pi M}\right)^{\frac{1}{2}}. \tag{7.25}$$

Some diffusivities calculated from both equations (7.23) and (7.25) are given in Table 7.1.

TABLE 7.1. PARTICLE DIFFUSIVITIES AND SCHMIDT NUMBERS[670]
(air at 20°C, 100 kPa (1 atm) pressure)

Particle diameter (μm)	Diffusivity ($m^2\ s^{-1}$)		Schmidt numbers using $\mathcal{D}$ calculated from	
	eqn. (7.23)	eqn. (7.25)	eqn. (7.23)	eqn. (7.25)
10	$2{\cdot}4\times10^{-12}$	—	$6{\cdot}4\times10^{6}$	—
1	$2{\cdot}7\times10^{-11}$	—	$5{\cdot}6\times10^{5}$	—
0·1	$6{\cdot}1\times10^{-10}$	$7{\cdot}8\times10^{-10}$	$2{\cdot}5\times10^{4}$	$1{\cdot}9\times10^{4}$
0·01	$4{\cdot}0\times10^{-8}$	$7{\cdot}8\times10^{-8}$	$3{\cdot}8\times10^{2}$	$1{\cdot}9\times10^{2}$
0·001	$3{\cdot}8\times10^{-6}$	$7{\cdot}8\times10^{-6}$	4·0	1·9

Diffusivity has dimensions of (area)/(time), expressed as m s^{-1} if the gas pressure is in Pa$\times10^{-1}$. The dimensionless group which includes diffusivity is the Schmidt number, Sc:

$$\mathrm{Sc} = \frac{\mu}{\varrho\mathcal{D}} = \frac{\nu}{\mathcal{D}} \tag{7.26}$$

where ν = kinematic viscosity (μ/ϱ).

Schmidt numbers corresponding to the particle diffusivities have also been listed in Table 7.1.

The other dimensionless group used in these calculations is the Peclet number (Pe), which is a measure of the transport by convective forces compared to the transport by molecular diffusion. For a system involving a gas stream, velocity v, moving past a body, diameter D, this is

$$\mathrm{Pe} = \mathrm{Re}_c\,\mathrm{Sc} = \frac{v\varrho D}{\mu}\cdot\frac{\mu}{\varrho\mathcal{D}} = \frac{vD}{\mathcal{D}}. \tag{7.27}$$

Billings[78] points out the solutions proposed for collection efficiency by diffusion have the approximate form

$$\eta_D \propto \mathrm{Pe}^{-n}$$

where $\frac{1}{2} \leq n \leq 1$, while it is generally accepted that the correct theoretical dependence is probably $n = \frac{2}{3}$. The solutions also depend on the logarithm of the cylinder Reynolds number Re_c, although its influence is slight in the range $10^{-4} \leq \mathrm{Re}_c \leq 10^{-1}$.

Langmuir assumed that collection by diffusion will take place from a surface layer, effective width x_e, during the time t that an element of the fluid moves from a point 1 to 2 (Fig. 7.8) taken as a 60° intercept upstream to a 60° intercept downstream on the collecting body, which is assumed to be a cylinder. At the point O′ the fluid element is at a distance x_0

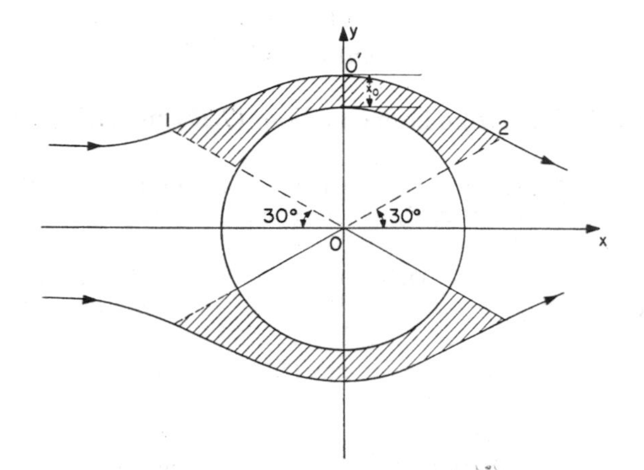

FIG. 7.8. The space near the collecting body from which diffusion is assumed to take place.[489]

from the collecting body. On the basis of the stepwise integration for viscous flow around a cylinder the effective distance x_e was found to be, as a first approximation,

$$x_e = 1{\cdot}120 x_0 . \tag{7.28}$$

While the time for the fluid to move from 1 to 2 was

$$t = \frac{0{\cdot}278 D^2 (2{\cdot}002 - \ln \mathrm{Re}_c)}{v_0 x_0} . \tag{7.29}$$

These values for t and x_e can be substituted in equation (7.21), and on rearrangement a diffusion parameter Z (similar to the interception parameter) can be found from

$$Z = \frac{x_0}{D/2} = [2{\cdot}24(2{\cdot}002 - \ln \mathrm{Re}_c)\mathcal{D}/v_0 D]^{\frac{1}{3}} \tag{7.30}$$

$$= [2{\cdot}24(2{\cdot}002 - \ln \mathrm{Re}_c)/\mathrm{Pe}]^{\frac{1}{3}} . \tag{7.31}$$

Langmuir stated that a more rigorous treatment of the diffusion problem could change the coefficient 2·24, but it would be unlikely to alter very much.[489] Natanson has subsequently suggested that it should be approximately twice Langmuir's value.[593]

The efficiency of collection by diffusion can now be calculated by an equation analogous to Langmuir's equation for interception collection efficiency (7.16):

$$\eta_D = \frac{1}{2{\cdot}002 - \ln \mathrm{Re}_c} \left[(1+Z) \ln (1+Z) - \frac{Z(2+Z)}{2(1+Z)} \right] . \tag{7.32}$$

Efficiencies of collection calculated for particles moving past a 1 μm fibre at a stream velocity of 0·1 m s^{-1} are in Table 7.2.

TABLE 7.2. PECLET NUMBERS AND DIFFUSION COLLECTION EFFICIENCIES FOR FIBRES 1 μm DIAMETER, VELOCITY 0·1 m s^{-1} AND AIR 20°C (1 atm) PRESSURE

Particle size μm	Peclet numbers Pe	Collection efficiency η_D					
		experi-mental	from eqn. (7.32)	from eqn. (7.35)	from eqn. (7.41)	$\frac{1}{Pe}$	$\frac{\pi}{Pe}$
10	$4{\cdot}2 \times 10^4$		$8{\cdot}8\times10^{-4}$	$1{\cdot}4\times10^{-2}$	$6{\cdot}6\times10^{-4}$	$2{\cdot}36\times10^{-5}$	$7{\cdot}5\times10^{-5}$
1	$3{\cdot}7 \times 10^3$		$4{\cdot}3\times10^{-3}$	$4{\cdot}6\times10^{-4}$	$3{\cdot}5\times10^{-3}$	$2{\cdot}68\times10^{-4}$	$8{\cdot}5\times10^{-4}$
0·1	$1{\cdot}63\times10$	0·18	$3{\cdot}1\times10^{-2}$	0·22	$3{\cdot}1\times10^{-2}$	$6{\cdot}0\times10^{-2}$	$1{\cdot}91\times10^{-1}$
0·01	2·5		$3{\cdot}4\times10^{-1}$	1·8	$8{\cdot}2\times10^{-1}$	0·4	1·26
0·001	0·026		3·4	17	47	37·8	120

The following equation was derived by Natanson[593] for viscous flow and low Peclet numbers (Pe ≪ 1):

$$\eta_D = \frac{2{\cdot}92}{(2{\cdot}002 - \ln \mathrm{Re}_c)^{\frac{1}{3}}} \cdot \mathrm{Pe}^{-\frac{2}{3}}. \tag{7.33}$$

Fuchs and Stechkina[286] derived an equation for the deposition efficiency based on the Kuwabara–Happel flow relation:

$$\eta_D = \frac{2{\cdot}9}{[-\frac{1}{2}\ln(1-\epsilon)-c]^{\frac{1}{3}}} \cdot \mathrm{Pe}^{-\frac{2}{3}} \tag{7.34}$$

where the terms have the same meaning as in equation (7.5).

Bosanquet,[101] using a similar approach, but assuming the contact distance to be half the perimeter ($\pi D/2$) obtained the collection efficiency for a cylinder:

$$\eta_D = [8\mathcal{D}/v_0 D]^{\frac{1}{2}} = 2\sqrt{(2)}/\mathrm{Pe}^{\frac{1}{2}}. \tag{7.35}$$

Efficiencies of collection from equation (7.23) are also given in Table 7.2.

In the same way[823] the collection efficiencies for a sphere and for one side of a strip (width W) oriented at right angles to the gas stream can be calculated.

For a sphere:

$$\eta_D = \frac{8}{3\pi}\left[\frac{2}{\mathrm{Pe}}\right]^{\frac{1}{2}} \tag{7.36}$$

For a strip:

$$\eta_D = \left[\frac{8\mathcal{D}}{\pi v_0 W}\right]^{\frac{1}{2}}. \tag{7.37}$$

Because of the mathematical difficulties involved, the calculation of collection efficiencies from mass transfer relations has not been fully developed. The general equation for unsteady state mass transfer, usually referred to as Fick's law, the limited case of which was given in equation (3.1), is

$$\frac{\partial c}{\partial t} = \mathcal{D}\frac{\partial^2 c}{\partial x^2} \tag{7.38}$$

where c refers to the particle concentration and x the thickness of the zone immediately around the collecting body. If no particle accumulation occurs in this zone equation (7.38) can be integrated to give the rate of particle diffusion per unit area of collector surface:

$$\frac{dc}{dt} = \mathcal{D}\frac{(c_0 - 0)}{x_f} \tag{7.39}$$

where x_f is the width of the zone around the collector in which the particle concentration gradient exists, c_0 is the particle concentration in the bulk of the gas, while at the surface of the collector the concentration is 0.

Johnstone and Roberts[405] suggested that a correlation based on the heat transfer analogy can be used for calculating the collection efficiency by diffusion for a spherical collector:

$$\eta_D = \frac{4}{\text{Pe}}\left(2 + 0{\cdot}557\,\text{Re}_c^{\frac{1}{2}}\,\text{Sc}^{\frac{3}{8}}\right). \tag{7.40}$$

Ranz[670] later gave a similar formula for cylinders:

$$\eta_D = \frac{\pi}{\text{Pe}}\left(\frac{1}{\pi} + 0{\cdot}55\,\text{Re}_c^{\frac{1}{3}}\,\text{Sc}^{\frac{1}{3}}\right) \tag{7.41}$$

which was used for $0{\cdot}1 < \text{Re}_c < 10^4$, and Sc less than 100. Values for the efficiency of collection calculated from equation (7.34) are also presented in Table 7.2.

Landt[487] suggested that the collection efficiency by diffusion could be found from π/Pe, while Davies[207] considered that the reciprocal of the Peclet Number (1/Pe) would have the same efficiency as corresponding values of the inertial impaction parameter ψ. However, the latter does not give realistic estimates of collection by diffusion (Table 7.2).

Goren[788] quotes Stechkina[286, 809] as giving the following for diffusion collection:

$$\eta_D = \frac{1}{\text{Pe}}\left[\frac{2{\cdot}9}{(2 - \ln \text{Re}_c)^{\frac{1}{3}}} + 0{\cdot}625\right] \tag{7.42}$$

which is based on equation (7.4).

Stern *et al.*[816] have carried out experiments on the collection efficiency of sub-micron-sized polystyrene particles on a filter at reduced pressures, where diffusion is the predominant mechanism. These workers used an equation by Torgeson.[864]

$$\eta_D = 0{\cdot}775\,\text{Pe}^{-0{\cdot}6}\,(C_{DC}\,\text{Re}_c/2)^{0{\cdot}4} \tag{7.43}$$

where C_{DC} is the fibre drag coefficient, and Re_c the fibre Reynolds number. They are characteristic of the filter and are found experimentally from the pressure drop, Δp, the fibre volume fraction α and the filter thickness h. From Chen's formula[156]

$$(C_{DC}\ \mathrm{Re}_c/2) = \frac{\pi}{4}\frac{\Delta p}{v_0}\left(\frac{1-\alpha}{\alpha}\right)\frac{D^2}{\mu h}. \tag{7.44}$$

The estimates of diffusion collection efficiency from Langmuir's equation were approximately half of the Torgeson equation estimates. These however were still conservative. For example, at $\mathrm{Pe} = 36$, experimentally η_D was 50 per cent, while Torgeson's estimate was 35 per cent and Langmuir's equation predicted 18 per cent collection. Approximately interpolating Stern's data, for $\mathrm{Pe} = 163$, experimental efficiency is 18 per cent, Langmuir's equation predicts 3 per cent (using the Einstein diffusivity equation) or 6 per cent (using Langmuir's diffusivity equation), Bosanquet's equation predicts 22 per cent, while Torgeson's equation predicts about 15 per cent. Until more extensive experimental data become available, if little is known about the filtering medium, either the Langmuir [equation (7.32)] or heat-transfer analogy equation (7.41) can be used. If, for a filter, pressure drop, filter density, fibre size and thickness of the bed can be found, Torgeson's equation will give the most reliable estimate.

Langmuir[489] modified his equations (7.28–7.32) for the combination of interception and diffusion. The effective width of the strip which is cleared of particles is actually $d/2$ wider than x_0 in equation (7.31). Because of the other assumptions made this correction is not often warranted except when the surface area for collection becomes very large.

7.6. COMBINATION OF INERTIAL IMPACTION, INTERCEPTION AND DIFFUSION

As aerodynamic capture does not proceed by the isolated mechanisms discussed in the previous sections but by two or more of these together, combined efficiencies must be considered. The fusion of interception with inertial impaction or with diffusion has been discussed in the earlier sections, but for practical calculations all three mechanisms must be combined.

In early attempts to combine the mechanisms, their separate collection efficiencies were simply added together,[801] but this makes it possible, in theory, for a particle to be collected more than once, which is inconsistent. A better approach therefore is to allow only the particles not collected by one mechanism to be collected by the others. This leads to the combined efficiency of collection η_{ICD}

$$\eta_{ICD} = 1-(1-\eta_I)(1-\eta_C)(1-\eta_D). \tag{7.45}$$

Another method suggested by Davies[207] has been to combine the inertial impaction parameter (ψ), with the diffusion collection parameter ($1/\mathrm{Pe}$), and to substitute the new parameter in the appropriate equation, e.g. in (7.19),

$$\eta_{ICD} = 0{\cdot}16[R+(0{\cdot}50+0{\cdot}8R)(\psi+1/\mathrm{Pe})-0{\cdot}1052(\psi+1/\mathrm{Pe})^2]. \tag{7.46}$$

A more general approach has been considered by Friedlander[275, 279] who uses the Smoluchowski equation. This describes the rate of collection as the sum of a diffusion term [Fick's law, equation (7.38)], and an impaction term. The equation was too difficult to solve completely, but partial solutions, when either the diffusion term or the impaction term were predominant, have been obtained. These were:

(a) *Diffusion term.* The steady flow diffusion equation was written using cylindrical coordinates, and then solved using the velocity distributions calculated by Langmuir[489] for viscous flow. This gave a proportionality

$$\eta_D \, \mathrm{Pe} \propto (B' \, \mathrm{Pe})^{\frac{1}{3}} \tag{7.47}$$

where B' is a function of the Reynolds number and can be written as $B\mathrm{Re}^{\frac{1}{2}}$, while interception is allowed for by multiplying both sides by R; the interception parameter

$$\eta_{DC} R \mathrm{Pe} \propto (BR^3 \, \mathrm{Re}^{\frac{1}{2}} \, \mathrm{Pe})^{\frac{1}{3}}. \tag{7.48}$$

(b) *Interception term.* Friedlander shows that the efficiency by interception can be found from

$$\eta_C = 2B'R^2 = 2B \, \mathrm{Re}^{\frac{1}{2}} \, R^2 \tag{7.49}$$

Multiplying both sides by $\mathrm{Pe}\, R$:

$$\eta_{DC} R \, \mathrm{Pe} = 2B \, \mathrm{Pe} \, \mathrm{Re}^{\frac{1}{2}} \, R^3. \tag{7.50}$$

The combined efficiency can be found by joining the two separate efficiencies; and finding a numerical value for the proportionality from experimental data. This was obtained by plotting $\eta_{DC} R \, \mathrm{Pe}$ against $R \, \mathrm{Pe}^{\frac{1}{3}} \, \mathrm{Re}^{\frac{1}{6}}$ on log–log scales together with data from Johnstone and Wong[950] Chen[156] and Thomas and Yoder[857] (Fig. 7.9). The experiments were approximately represented by the expression

$$\eta_{DC} R \mathrm{Pe} = 6R\mathrm{Pe}^{\frac{1}{3}} \, \mathrm{Re}^{\frac{1}{6}} + 3R^3 \, \mathrm{Pe} \, \mathrm{Re}^{\frac{1}{2}}. \tag{7.51}$$

A more useful form of this combined equation is

$$\eta_{DC} = 6 \, \mathrm{Sc}^{-\frac{2}{3}} \, \mathrm{Re}^{-\frac{1}{2}} + 3R^2 \, \mathrm{Re}^{\frac{1}{2}}. \tag{7.52}$$

Friedlander and Pasceri[279] have replotted the data according to equation (7.52) (Fig. 7.10) and show excellent agreement with available experiments.

More recently Friedlander[278] has suggested that simultaneous collection by diffusion and interception can be represented by a single function equation:

$$\eta_{DC} R \mathrm{Pe} = f\left\{R\left[\frac{\mathrm{Pe}}{2(2-\ln \mathrm{Re}_c)}\right]^{\frac{1}{3}}\right\}. \tag{7.53}$$

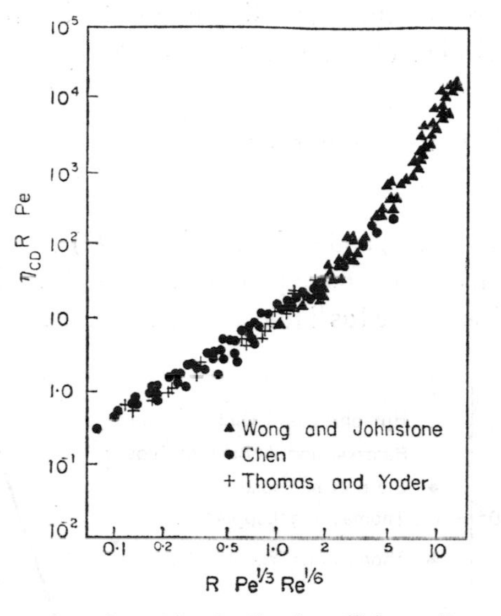

FIG. 7.9. Experimental correlation of combined collection efficiency, Reynolds number, Peclet number and interception parameter $\eta_{CD} R$ Pe *vs.* $R\,\mathrm{Pe}^{\frac{1}{3}}\,\mathrm{Re}^{\frac{1}{6}}$.

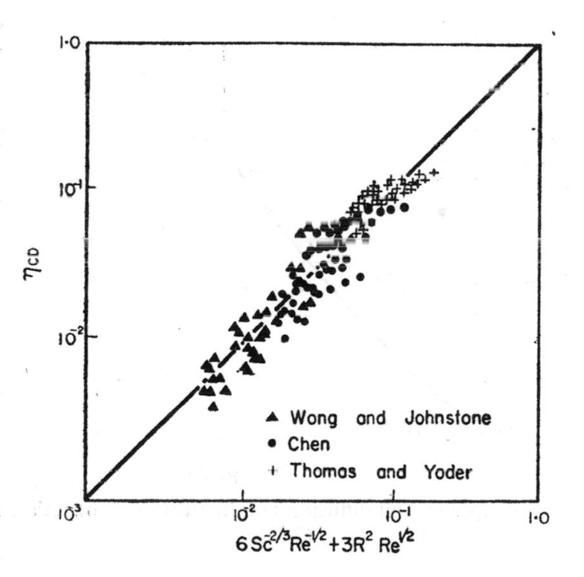

FIG. 7.10. Combined collection efficiency based on experimental results η_{CD} *vs.* $\left\{6\,\mathrm{Sc}^{-\frac{2}{3}}\,\mathrm{Re}^{-\frac{1}{2}}+3\,\mathrm{Re}^{\frac{1}{2}}\,R^2\right\}$.

For the limit $R \rightarrow 0$, which represents the collection of particles by Brownian diffusion this equation becomes equation (7.33) while for the collection of large particles, when interception is predominant, Pe $\rightarrow \infty$, and the equation reduces to equation (7.18).

Stechkina and Fuchs[286] solved the convection and diffusion boundary-layer equations, and found that their numerical results fitted the following:

$$\eta_{DC} = \eta_D + \eta_C + \frac{1 \cdot 24}{(2 - \ln \mathrm{Re}_c)^{\frac{1}{2}}} \frac{R^{\frac{2}{3}}}{\mathrm{Pe}^{\frac{1}{2}}} \tag{7.54}$$

where η_D is given by equation (7.16) and η_C is equation (7.42).

More recently Billings[78] has compared equation (7.51) with a wide range of experimental results in which solid particles as well as liquid droplets were used (Fig. 7.11).[379, 669, 816, 856, 857]

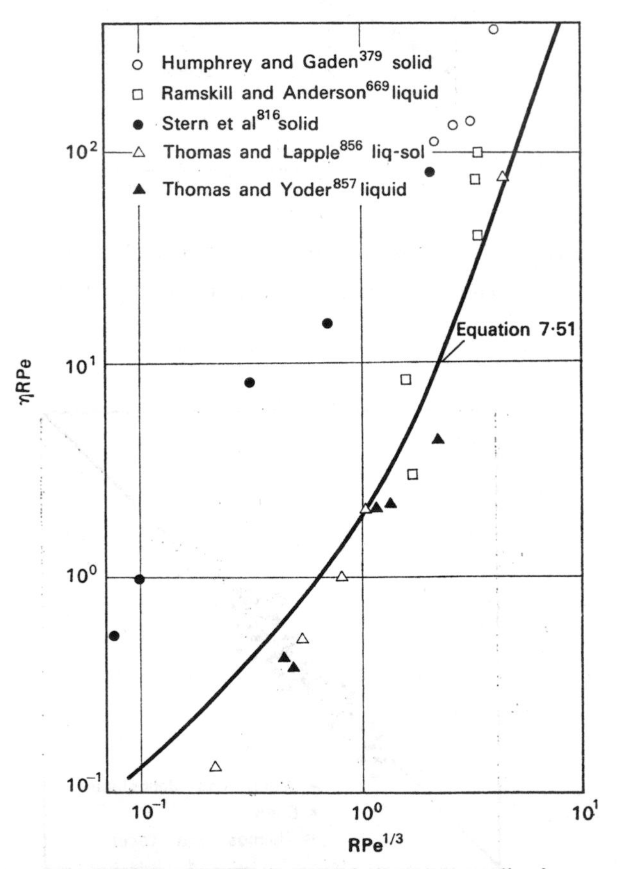

FIG. 7.11. Comparison of experimental diffusion and interception collection results with equation (7.51).

He used $R\,\mathrm{Pe}^{\frac{1}{3}}$ as the abscissa and found that the liquid droplet-collection efficiencies agreed well with the Friedlander and Pasceri equation (7.51), while the solid particles were being collected with much higher efficiencies than predicted by this equation. The reason for this difference is not known, but may be due to the way in which the aerosols were generated, the possibility of some electrostatic charging in the case of the solid, or particle accumulation effects.[78]

Hasenclever[342] also found that the Friedlander and Pasceri equation gave good agreement with his experimental collection efficiencies of 0·01–0·5-μm oil droplets and naturally occurring aerosols on 1 μm synthetic (polymeric) fibres.

Whitby,[923] however, finds that Torgeson's equation (7.43) gave the best fit to his single fibre-collection efficiencies.

Another calculation, applying the differential equation of convective diffusion to the deposition process has given similar results[667] and will not be presented here.

If particles of decreasing size moving with constant velocity approach a collector, the inertial impaction and interception efficiencies decrease with size, while collection by diffusion improves. Thus, under specific operating conditions, a particular particle size can be predicted for which the collection efficiency is a minimum. This minimum was indicated in the theories of filtration of Langmuir,[489] Davies,[207] Stairmand[801] and Friedlander,[275] and can readily be shown by differentiating equation (7.51), which has a positive second derivative.[423]

Stairmand,[801] using a simplified model, predicted that the minimum efficiency would occur for a 0·9 μm particle (density 2000 kg m^{-3}) collected on 10-μm fibres from a gas stream moving a 30 mm s^{-1}. Similar ideas were used by Landt,[488] while Davies predicted a range of minima for a range of stream velocities (Fig. 7.12).

The existence of the minimum has been confirmed by a number of workers using radioactive particles of known size,[343, 816, 856, 857] but is found to occur at a much lower value than was first predicted by Stairmand. Experimental minima are also shown in Fig. 7.12.

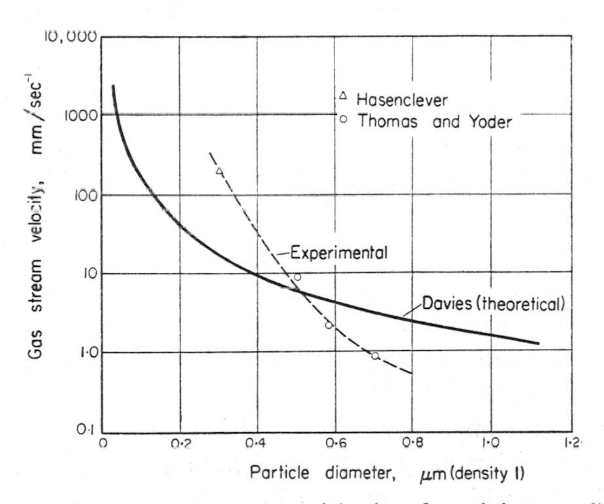

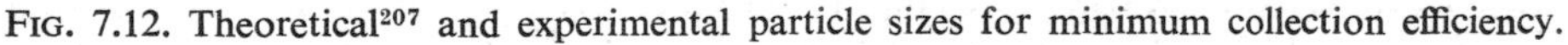

FIG. 7.12. Theoretical[207] and experimental particle sizes for minimum collection efficiency.

7.7. EFFECT OF TEMPERATURE AND PRESSURE ON THE BASIC MECHANISMS

Although the effect of temperature on inertial impaction, interception or diffusion collection efficiencies has not been investigated, it can be predicted from the temperature-dependent terms in the inertial impaction parameter [equation (7.10)], the interception efficiency [equation (7.16)] and the diffusivity and diffusion collection efficiency [equations (7.25) and (7.41)]. With increased temperatures, the efficiency of collection by both interception and inertial impaction is reduced, while collection by diffusion increases.

Results of calculations using the equations listed are plotted in Fig. 7.13.[834]

Because both the inertial impaction parameter ψ [equation (7.10)] and the particle diffusivity D [equation (7.23)] are directly proportional to the Cunningham correction C [equation

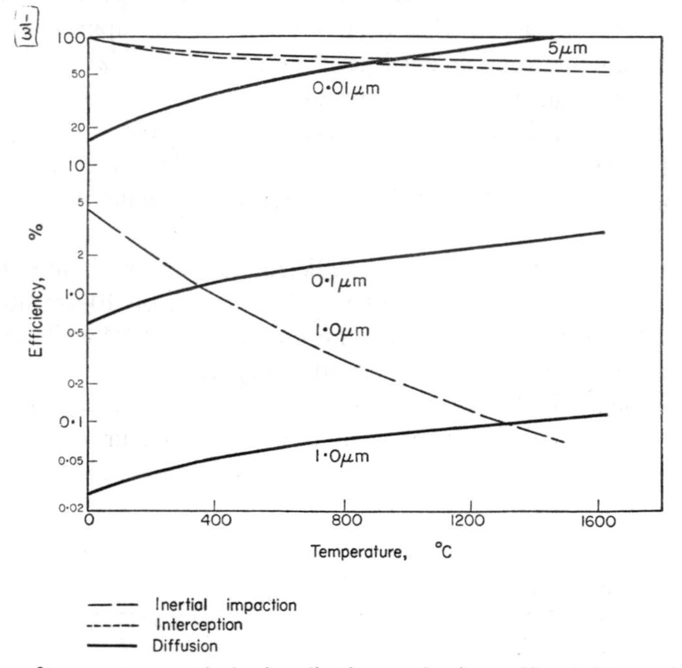

FIG. 7.13. Effect of temperature on the basic collection mechanisms of inertial impaction, interception and diffusion.[834]

(4.31)], and in turn, proportional to the mean free path of the gas molecules λ, it follows that the effect of pressure will be similar on these collection mechanisms. As increasing gas pressure reduces the mean free path, the collection efficiency by these mechanisms will also decrease with pressure.[829]

The converse of this—the improvement of collection efficiency with reduced pressures—has also been found experimentally by Stern, Zeller and Scheckman[816] and Schuster.[742] In Schuster's investigation, which was reported by Löffler, for a PVC filter (fibre diameter 19·5 μm, fibre-packing volume $(1-\epsilon)\,0{\cdot}301$) tested with polystyrene spheres (0·1–1·3 μm diameter) and paraffin oil droplets (1·08 μm diameter), an efficiency of 55 per cent was found at 86·5 kPa pressure, 82 per cent at 6·65 kPa pressure, and only 98 per cent at 0·25 kPa pressure, which was in general agreement with Stern *et al.*[816]

A quantitative prediction for the combined effect of pressure and temperature on inertial impaction has been carried out by Strauss and Lancaster.[829] The effect in the case of a 1 μm beryllium oxide aerosol in carbon dioxide, which is relevant in gas-cooled nuclear

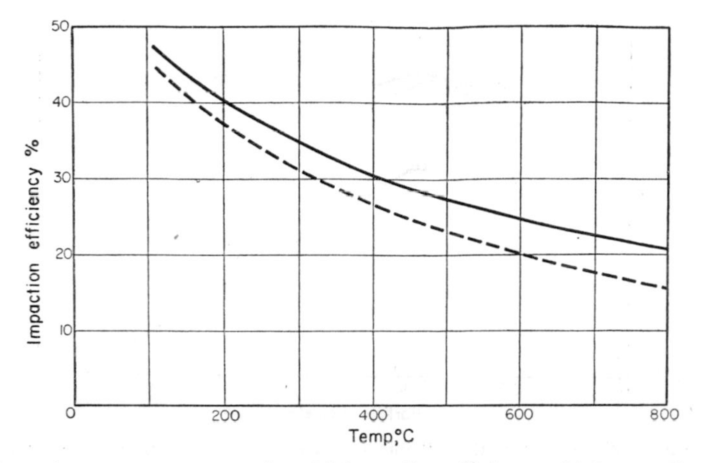

FIG. 7.14. Effect of gas temperature on inertial impaction efficiency of 1·0 μm radius BeO particles in CO_2 being carried past a 10 μm diameter fibre at 250 mm s^{-1}.[829] Continuous rule, gas density 2 kg m^{-3}; broken rule, gas density 30 kg m^{-3}.

reactors, is shown in Fig. 7.14. Although the collection efficiency by diffusion improves as the temperature increases, pressure effects tend to outweigh this and so diffusion collection efficiencies will also tend to decrease with high temperatures and pressures.

7.8. DEPOSITION BY GRAVITATIONAL SETTLING

When a slow-moving gas stream passes through a filter, the momentum of the larger particles may not be sufficient for collection by inertial impaction in all cases. Gravity settling may account for an appreciable fraction of the particles collected in this case because of the comparatively long gas-stream residence time. For example, settling is a discernible collection mechanism for one micron particles passing through a 10-μm fibre filter bed when gas-stream velocities are less than 0·5 mm s^{-1}.

The efficiency of separation by settling can be calculated from the gravity settling parameter $\mathcal{G}$ suggested by Ranz and Wong:[672]

$$\eta_G = \mathcal{G}\frac{gD}{v^2}\psi = \frac{Cd^2\varrho_p g}{18\mu v}. \tag{7.55}$$

This equation shows that settling becomes significant when gD/v^2 is greater than ψ.

Sedimentation has been found to play an important part in the deposition that occurs in the passage of gases through packed beds at low velocities. Filtration efficiency for large particles was greater for downward-flowing gases than for upflowing gases.[857] Typical curves for the penetration of dioctyl phthalate droplets into a lead shot column are shown in Fig. 7.15.

In the case of spray towers and scrubbers, the relative velocities of particles and droplets are almost always too large for gravitational settling to be important.

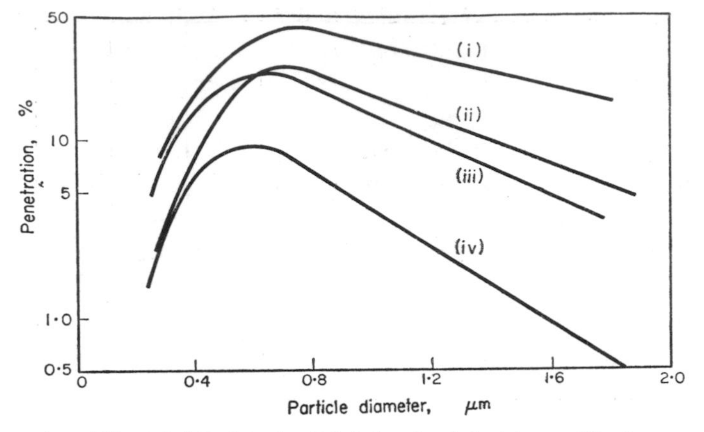

FIG. 7.15. Penetration of dioctyl phthalate droplets into a lead shot tower. The increased penetration when the flow is downwards indicates that gravitational settling may influence collection.[857]

(i) Flow up column 14·9 mm s^{-1}. (ii) Flow up column 7·45 mm s^{-1}.

(iii) Flow down column 14·9 mm s^{-1}. (iv) Flow down column 7·45 mm s^{-1}.

7.9. ELECTROSTATIC FORCES

The theories of aerodynamic capture of particles given in the earlier sections have avoided the effect of electric forces which may be present either on the particles, the collectors, or both. The fact that electrostatic forces can assist filtration has been recognized since about 1930,[331] and has led to the development of resin impregnated filters with enhanced efficiency. More recently, attention has been focused on the selection of filter fabrics with the best electrical properties for collection of specific dusts,[273] and mechanical electrostatic charging has also been used.[770]

Electrostatic charges in filter fabrics can be induced by friction; for example, wiping a lucite rod over the fabric,[770] or by the simple passage of the gas stream laden with particles through the fabric. Charges of the order of 1·2 kV are produced when air with a velocity of 1·7–2·0 m s^{-1} is passed through a chemical fibre fabric.[239]

The analysis of the effect of electrostatic forces, and the combination of this with the basic capture mechanisms is exceptionally difficult. Two attempts have met with some success. Gillespie[297] employed Langmuir's approach to particle capture[489] with the addition of static charges present during interception and diffusion, and the use of potential flow to estimate the flow velocities past the collector. The resulting equations were very complicated and will not be given here.

The graphical solutions presented by Kraemer and Johnstone[463] are much easier to apply. The calculations are based on whether the aerosol particle, the collector, or both are charged. The collection efficiencies have been found by solving the potential and viscous flow equations in these cases with the aid of a digital computer.

There are four aspects of electrical forces acting in a system of particles approaching a collector which have to be considered.

(i) When both the particles and the collector are charged, coulombic forces of attraction or repulsion act, depending on whether particles and collector have unlike or like charge. These are considered as point charges. The coulombic force has magnitude F_{EC}.

(ii) A charged collector induces an image charge on the particle surface, opposite in sign to the charge on the collector. This force has magnitude F_{EI}, and is an additional force on the particle.

(iii) If a particle is charged, it, in turn, induces an image charge opposite in sign, on the collector. This results in a force F_{EM}. This also is an additional force between particle and collector.

(iv) The particles charged in the same sense, produce a repulsion force among themselves, F_{ES}. This is called the space charge effect.

The force F_{E1} between a charged aerosol particle and a charged spherical collector (with *constant charge*) is given by

$$F_{E1} = F_{EC}+F_{EI}+F_{EM}+F_{ES} \tag{7.56}$$

where

$$F_{EC} = \frac{Qq_1}{4\pi\epsilon_0 r^2}, \tag{7.57}$$

$$F_{EI} = -\left(\frac{\epsilon-1}{\epsilon+2}\right)\frac{d^3Q^2}{16\pi\epsilon_0 r^5}, \tag{7.58}$$

$$F_{EM} = \frac{q_1^2 D}{8\pi\epsilon_0 r^3} - \frac{2q_1^2 Dr}{\pi\epsilon_0(4r^2-D^2)^2}, \tag{7.59}$$

$$F_{ES} = -\frac{q_1^2 D^3 N}{24\epsilon_0 r^2}, \tag{7.60}$$

where q_1 = charge on particle,
Q = charge on collecting body,
r = distance between particle and collector,
N = particle concentration per unit volume,
ϵ = dielectric constant or aerosol particle,
ϵ_0 = specific inductive capacity of space
$= 8{\cdot}85\times 10^9\ \mathrm{A^2\ N^{-1}\ m^{-2}}$.

When there is a constant voltage V_1 on the spherical collector, then an additional term F_{ET} is required for the force F_{E2} between the particle (charge q_1) and the total charge induced on the collector by all the surrounding particles within a radius $\mathcal{R}$,

$$F_{E2} = F_{EC}+F_{EI}+F_{EM}+F_{ES}+F_{ET} \tag{7.61}$$

where the constituent forces, in terms of the voltage V_1, are

$$F_{EC} = \frac{V_1 q_1 D}{2r^2}, \tag{7.62}$$

$$F_{EI} = -\left(\frac{\epsilon-1}{\epsilon+2}\right)\frac{V_1^2 D^2 \pi \epsilon_0 d^3}{4r^5}, \tag{7.63}$$

$$F_{EM} = \frac{q_1^2 D}{8\pi\epsilon_0 r^3} - \frac{2q_1^2 Dr}{\pi\epsilon_0(4r^2 - D^2)}, \tag{7.64}$$

$$F_{ES} = -\frac{q_1^2 D^3 N}{24\epsilon_0 r^2}, \tag{7.65}$$

$$F_{ET} = -\frac{q_1^2 D^2 \pi N R^2}{8\pi\epsilon_0 r^2}. \tag{7.66}$$

These forces are expressed in terms of force parameters, K, which are obtained by dividing the various forces by the Stokes–Cunningham equation, force $F = 3\pi\mu d v_0/C$ [equation (4.34)]. The force constants and their definition are listed in Table 7.3.

TABLE 7.3. ELECTROSTATIC FORCE PARAMETERS (KRAEMER AND JOHNSTONE)[463]

Collector shape	Type of force	Charge type	Equation no.	Parameter symbol	Parameter origin	Definition
Sphere	Coulombic attraction between charged particle and charged collector	Constant charge	7·44	K_E	F_{EC}/F	$\frac{q_1 Q_{ac} C}{3\pi\mu d v_o \epsilon_o}$
		Constant voltage	7·49	K_E	F_{EC}/F	$\frac{q_1 V_1 2C}{3\pi\mu d v_o D}$
Sphere	Force caused by image of electrical charge on collector in the particle	Constant charge	7·45	K_I	F_{EI}/F	$\left(\frac{\epsilon-1}{\epsilon+2}\right)\frac{d^2 Q_{ac}}{3\mu v_o D\epsilon_0}$
		Constant voltage	7·50	K_I	F_{EI}/F	$\left(\frac{\epsilon-1}{\epsilon+2}\right)\frac{8CD^2 V_1^2 \epsilon_0}{3\mu v_0 D^3}$
Sphere	Force caused by image of electrical charge on particle in the collector	—	7·46, 51	K_M	F_{EM}/F	$\frac{Cq_1^2}{3\pi^2\mu d v_o \epsilon_0}$
Sphere	Force caused by space charge	—	7·47, 52	K_S	F_{ES}/F	$\frac{Cq_1^2 DN}{18\pi\mu v_o d\epsilon_0}$
Sphere	Force between particle and charge induced on collector by other particles	Constant voltage	7·53	K_G	F_{ET}/F	$\frac{Cq_1^2 N R^2}{3\pi\mu d v_0 \epsilon_0 D}$
Cylindrical dipole	Coulombic attraction	—	—	K_c	—	$\frac{2q_1 mC}{3\pi^2\mu d D v_o \epsilon_0}$

F = Stokes–Cunningham resistance force to motion of particle = $3\pi\mu d v_o/C$,
Q_{ac} = charge per unit area on collector surface,
m = dipole moment of collector.

Approximate equations for the collection efficiency can be obtained if only one term of the electrostatic force relations (7·56 or 7·61) are considered and if the interception parameter R is assumed to be zero. The approximate solutions are listed in Table 7.4. These solutions are, however, very limited. They do not consider the joint action of two or more forms of electrostatic force; cannot be applied to the case of an uncharged collector; and omit the interception effect. Moreover, the collection efficiencies are based on reasonable

TABLE 7.4. APPROXIMATE COLLECTION EFFICIENCIES[463]

Collector shape	Electrostatic force	Electrostatic equation	Collector charged	Aerosol charged	Collection efficiency
Sphere	Coulombic	7·44	Yes	Yes	$-4K_E$
Sphere	Image	7·50	Yes	No	$\left(\frac{15\pi}{8} K_I\right)^{0\cdot4}$
Cylinder	Coulombic	—	Yes	Yes	$-\pi K_E$
Cylinder	Image	—	Yes	No	$\left(\frac{3\pi}{4} K_I\right)^{\frac{1}{3}}$
Plane	Coulombic	7·44	Yes	Yes	$\left(\frac{-K_E}{1-K_E}\right)^{\frac{2}{3}}$
Plane	Image	7·45	Yes	No	$\left(\frac{K_I}{1+K_I}\right)^{\frac{1}{3}}$
Dipole cylinder	Coulombic	—	Yes	Yes	K_C

assumptions only when the efficiencies are much greater than one[463] (efficiency being y_{lim}/frontal area of collector).

Natanson[594] has also considered the cases of coulombic and image forces between particles and cylinders, and derived equations for efficiency of capture, which are similar to those by Kraemer and Johnstone. They are reviewed by Pich[643] and will not be presented here.

Much more realistic solutions were obtained by using the complete force balances for the particles, and either potential or viscous flow relations for the streamlines around the collector. These were solved numerically with a digital computer and graphs of the results for three possible cases are shown:

(i) Both particles and collector charged (Fig. 7.16).
(ii) Charged collector with uncharged particles (Fig. 7.17).
(iii) Uncharged collector with charged particles (Fig. 7.18).

These calculations give lower efficiencies than those calculated by the summation of approximate efficiencies for the different mechanisms. The difference varies between 1 and 25 per cent, averaging about 5 per cent. Experimental results agreed well with the theory (Fig. 7.19)

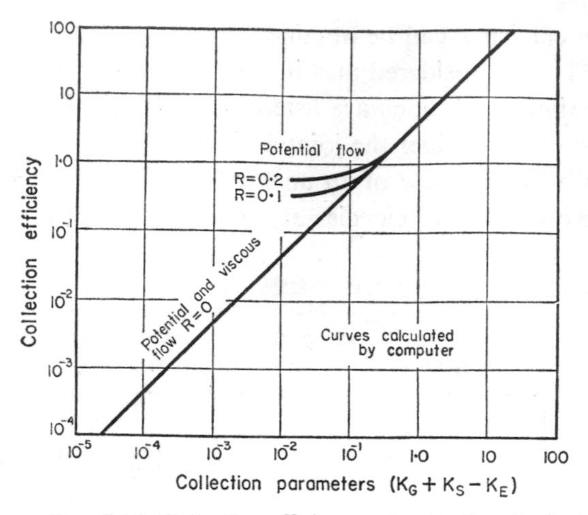

FIG. 7.16. Collection efficiency when both spherical particle and spherical collector are charged.[463]

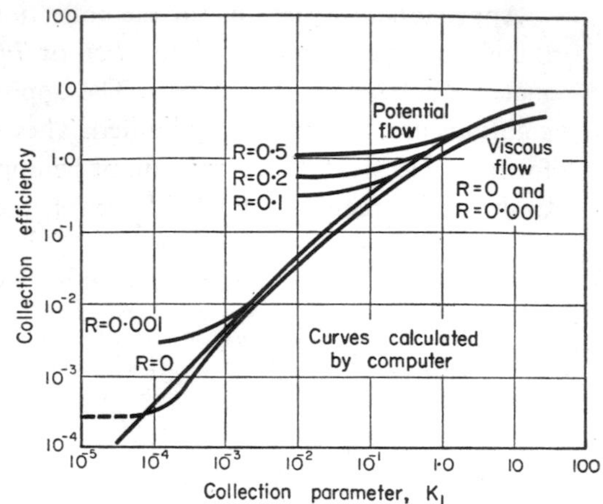

FIG. 7.17. Collection efficiency when uncharged particles are collected by a charged collector.[463]

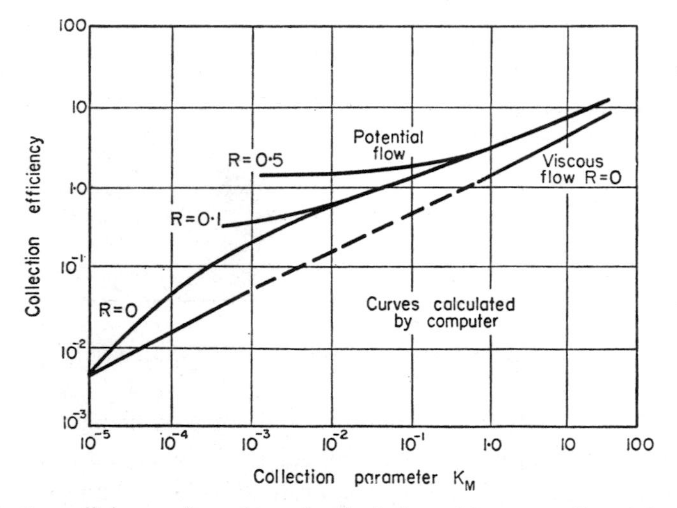

FIG. 7.18. Collection efficiency when charged spherical particles are collected by an uncharged spherical collector.[463]

although the theory tended to underestimate the efficiencies when these were very low. Calculations and experiments show that much improved collection efficiencies are obtained when both the aerosol and the collecting body are charged, while charging of even one of the elements leads to improved collection.

Gillespie,[297] was able to show that the presence of electrostatic charges increased the particle diameter for maximum penetration into a bed (Fig. 7.20). This may account in part for anomalous results such as those of Humphrey and Gaden[379] who found that the size for maximum penetration for *B. subtilis* spores which carried some electrostatic charge was 1·15 μm, this being greater than would be expected from the values in Fig. 7.12 for both theoretical and experimental maxima.

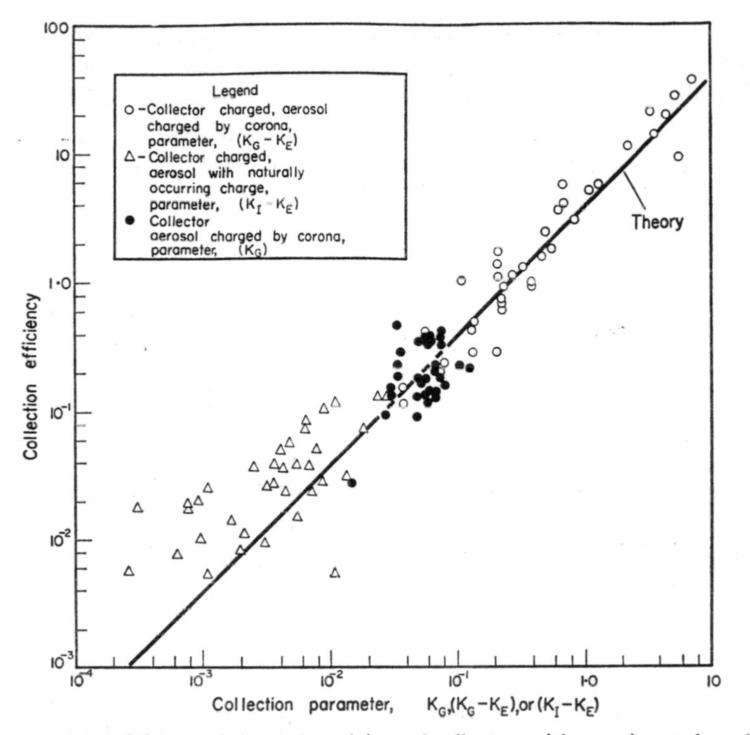

FIG. 7.19. Comparison of theory of charged particles and collectors with experimental results: the collection of dioctyl phthalate aerosol on a spherical collector.[463]

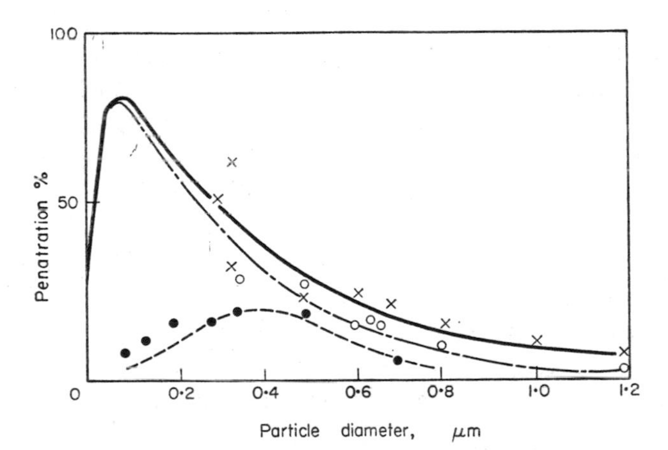

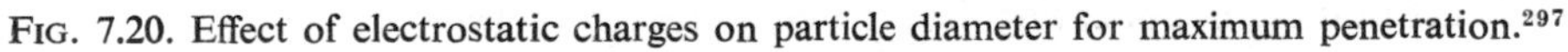

FIG. 7.20. Effect of electrostatic charges on particle diameter for maximum penetration.[297]

Experimental
- ● Charged aerosol–charged filter (polystyrene particles–irradiated resin wool).
- ○ Uncharged aerosol–charged filter (Stearic acid–irradiated resin wool).
- × Uncharged aerosol–uncharged filter (Stearic acid–resin wool).

Calculated
- – – – – heterogeneous charged aerosol–charged filter 420 V m^{-1}.
- –·–·– homogeneous aerosol–charged filter 420 V m^{-1}.
- ——— homogeneous aerosol–uncharged filter.

Superficial gas velocity – 113 mm s^{-1}.

Empirical equations for the collection efficiency with highly electrostatically charged particles have been obtained by Lundgren and Whitby[537] based on their experiments with 1-μm methylene blue particles. These workers found that a wool felt which gave a 16 per cent collection efficiency with uncharged particles improved to better than 99 per cent when 320 elementary charges were added to the particles. On this basis Lundgren and Whitby suggested the following equation for calculating the collection efficiency between charged particles and an uncharged filter at low Re_c (< 1):

$$\eta_E = 1{\cdot}5(K_m)^{\frac{1}{2}} \tag{7.67}$$

where

$$K_m = \left(\frac{\epsilon_f - 1}{\epsilon_f + 1}\right) \cdot \frac{q_1^2}{12\pi\epsilon_0 D^2 \mu\, dv_0} \tag{7.68}$$

and ϵ_f = dielectric constant of fibre.

The other terms have the meanings ascribed to them in equation (7.56). In his review of these, Löffler[529] points out that these workers and Gillespie concluded that while electrostatic effects are of little importance for larger particles, where inertial forces are predominant, for sub-micron particle collection these become a major factor.

7.10. MULTIPLE COLLECTORS

There has been only limited discussion of the collection of particles on system of collecting bodies, i.e. a randomly distributed set of cylinders, because until the Kuwabara–Happel relation became available, it was general practice to consider first single collectors and then extend this to multi-collector systems. The empirical extension of single body-collection efficiencies to realistic multi-body systems is therefore considered here.

In gas-cleaning plants such as fibre filters or spray towers, particles encounter a number of collectors during their passage through the plant. Two separate models will be considered. In one the collecting bodies are all identical and act independently of one another, while in the other, size, orientation and interference effects will be considered.

An example of the first type is a *spray tower* (Chapter 9) where a large number of almost identical spray droplets are produced. These droplets are essentially spheres falling through a slow upward-moving gas stream. When the combined collection efficiency of a single spherical droplet η_{ICD} has been found, the overall efficiency, η_0, can be found from

$$\eta_0 = 1 - (1 - \eta_{ICD})^n \tag{7.69}$$

where n is the number of collecting droplets encountered by the particles.

In most practical cases, n is large (greater than about 25), and equation (7.69) can be modified:

$$\eta_0 = 1 - \mathrm{e}^{-n\eta_{ICD}}. \tag{7.70}$$

For a spray tower the value of n can be estimated if the gas-flow rate, scrubbing liquid rate and average droplet size is known, and if it is assumed that for effective collision the droplets must cover the whole of the tower cross section:

$$n = \frac{Q_L H}{QD} . 3{\cdot}93 \times 10^3 \tag{7.71}$$

where Q_L = liquid flow rate ($m^3 s^{-1}$),
Q = gas flow rate ($m^3 s^{-1}$),
H = tower height (m),
D = average droplet diameter μm.

Equation (7.71) shows that a smaller droplet or a larger liquid flow will increase the value of n.

If a fibre filter is likewise assumed to consist of a number of identical cylindrical collectors, evenly spaced, with no inter-fibre interference, at right angles to the gas flow, equation (7.69) can also be used.

Actually *fibrous filters* consist of randomly oriented fibres with a range of diameters. It may be assumed, for fabric filters of the type used in bag filters, and other industrial filters with low resistance to the gas flow, that the fibres are relatively far apart and staggered in relation to one another. Consider an area of filter dA at right angles to the gas flow, and depth dh. If the free space velocity of the gas stream is v_S, and the packing density of the fibres is α, then the average velocity v of the gases within the filter pad is

$$v = v_S/(1-\alpha). \tag{7.72}$$

If the average diameter of a fibre is D, the total length of fibres L, and, if the overall collection efficiency of a fibre within the bed (allowing for streamline interaction) is η_α then the number of particles removed by the fibres in unit time is

$$vN \, dA \, dh \eta_\alpha DL \tag{7.73}$$

where N = number of particles entering the element of filter per unit volume.

The change in particle concentration is also given by

$$-v_S \, dA \, dN. \tag{7.74}$$

Substituting for v_S in (7.74) and combining with (7.73) gives

$$-\frac{dN}{N} = \frac{dh}{1-\alpha} \eta_\alpha DL, \tag{7.75}$$

which, on integration, gives

$$\ln N/N_0 = \frac{H}{1-\alpha} \eta_\alpha DL \tag{7.76}$$

where H = bed depth,
N_0 = initial concentration,
N = final concentration.

Now the actual volume of fibre in a unit volume of filter, which is the packing density, is given by

$$\alpha = \pi D^2 L/4. \tag{7.77}$$

Substituting for L from (7.77) in (7.76) gives

$$\ln \frac{N}{N_0} = \frac{4H}{\pi D}\left(\frac{\alpha}{1-\alpha}\right)\eta_\alpha . \tag{7.78}$$

Chen[156] points out that the average fibre diameter D to be used in equation (7.78) should be based on the ratio (D_S^2/D) where D_S is the *surface* average fibre diameter (considering total surface area) and D is the arithmetic mean diameter.

Fibre interference effect. When fibres are close together in a filter bed, there will be increases in velocity with greater packing density of the fibres. Also a change in flow pattern around a fibre will result because of neighbouring fibres. Both of these effects increase the collection efficiency from interception and inertial impaction. A higher velocity, however, will decrease diffusion collection, although the flattened streamlines may help to reduce this effect.

To obtain a measure of the interference effect, Davies assumed that the interference effect would be the same for all mechanisms. Solving the equations of motion to find the lateral displacement of the streamlines by the collectors, Davies calculated that the efficiency for interception by a single fibre in a bed of packing density α will be

$$\eta_\alpha = R(0{\cdot}16 + 10{\cdot}9\alpha - 17\alpha^2). \tag{7.79}$$

Davies then proposed that equation (7.79) should be combined with equation (7.46) to give the combined collection efficiency for a single fibre embedded in the filter:

$$\eta_\alpha = (0{\cdot}16 + 10{\cdot}9\alpha - 17\alpha^2)\,[R + (0{\cdot}5 + 0{\cdot}8R)(\psi + 1/\mathrm{Pe}) - 0{\cdot}1052R(\psi + 1/\mathrm{Pe})]. \tag{7.80}$$

However, because of the difficulty of obtaining a satisfactory average fibre diameter by direct measurement, it was suggested that this could be better deduced from a pressure loss relation based on d'Arcy's equations (8.7 and 8.10).

Chen[156] determined the fibre interference effect experimentally for values of α less than 0·10. Here

$$\eta_\alpha = \eta_0(1 + 4{\cdot}5\alpha). \tag{7.81}$$

This gives values for the fibre interference effect less than that calculated from Davies' equation (7.80), but because of its experimental basis should be used for low packing density fibre filters.

7.11. RETENTION OF PARTICLES ON COLLECTORS

In the previous sections it has been postulated that when a particle hits a collecting body it will stick and not be removed. When the particle is a mist droplet, and the collecting body a liquid, either the same as or miscible with the droplet, then the two will coalesce, forming a single body. If the collecting body is a solid or a liquid with which a particle or a mist droplet is not miscible, then the particle or droplet will sit on the surface of the collector. It may then stay where deposited, slide along to a point where it will fix itself, such as the crossing over of two fibres, or be torn off again by the gas stream moving past the collector.

Extensive observations of polystyrene latex particles approximately 1·3 μm collecting on approximately 10-μm single glass fibres were reported by Billings[78] over a Re_c range of approximately 0·1 to 0·4. After initial capture of the particles on the fibres, further particles collect in chains and Y structures. They suggest that electrostatic effects probably play some part in the collection and growth process, as similar chain formation is typical of thermal smokes where particle charging by flame ionization is effective. These deposits, which project from the fibre, act as additional collecting surface and promote further capture. Billings has expressed this actual efficiency of collection η_z as a function of the initial efficiency η_0 and a product of the local accumulation $\mathbf{z}$ (particles/mm² of fibre):

$$\eta_z = \eta_0 + \mathbf{Sz}$$

where $\mathbf{S}$ is the *particle accumulation coefficient* which for the polystyrene latex particle/glass-fibre system was $1{\cdot}36 \times 10^{-13}$ m²/particle. It follows that the more rapidly the particles accumulate, the more efficient does the filter become, although, in time, this increases the flow resistance of the filter bed. In practice, when this resistance becomes too high, filters are replaced, and discarded or cleaned for reuse.

Other photographic studies of impacting droplets indicate that particle and droplet adhesion is relatively independent of gas velocity.[300] At low velocities the drag is insufficient to detach particles even lightly adhering to the surface, while at high velocities a much greater area of contact is produced on impact to ensure adhesion.

Gillespie[297] has suggested that the critical parameter for particle retention is the angle at which a particle strikes the collecting body. When the angle of strike is greater than a certain value, the particle will not be retained by the collecting body. This concept has been incorporated in a "slippage" coefficient, i.e. the fraction of particles which do not stick on contact. The "slippage" coefficient theory has proved useful for correlating experimental data by a number of workers although it is not necessarily founded on realistic physical interpretation of the filtration process.[316]

The mechanisms of particle retention in filters has been the subject of extensive investigation in recent years by Krupp,[468] Corn,[177] Löffler[529] and Billings[78] and their work has been reviewed in the reference cited.[529] The forces retaining particles on filter media are a combination[461] of van der Waals forces, electrostatic attractive forces and, when humidity conditions permit, surface tension capillary forces. It has been found that when the humidity is high, and the capillary forces start to be significant, the electrostatic charges leak away.

7.11.1. Van der Waals forces

Earlier attempts were made to calculate the particle–solid surface adhesion by van der Waals forces on a microscopic (10 μm) scale. Lifshitz,[512] followed by Krupp,[468] developed a comprehensive macroscopic theory for the van der Waals adhesive force, denoted by $F^0_{vd\omega}$. This force for a sphere on a half space (representing a particle on a cylindrical fibre) is given by

$$F^0_{vd\omega} = \frac{\hbar\omega}{8\pi z_0^2}\mathbf{R} \tag{7.82}$$

where $\hbar\omega$ = Lifshitz–van der Waals constant,
z_0 = distance between the bodies,
$\mathbf{R}$ = radius of the contact point.

Krupp[468] estimated z_0 at 0·4 μm, which corresponds to the lattice constants for crystals with van der Waals bonds, and this has been confirmed by experiment. Krupp[467] also points out that, in practice, **R** is no longer the simple "microscopic radius", but a function of the surface roughness, where the two bodies contact elevated sections of the rough surface. Sperling[787] gives a statistical model for these based on electron micrographs.

The Lifshitz–van der Waals constant between two materials, designated by suffix 1 and 2, was estimated by Kottler *et al.*[462] to be the geometric mean of the cohesion energies of both:

$$\hbar\bar{\omega}_{12} \doteqdot \sqrt{(\omega_{11}.\hbar\omega_{22})}. \tag{7.83}$$

TABLE 7.5. COMPARISON OF VAN DER WAALS AND ELECTROSTATIC ADHESIVE FORCES FOR PARTICLES ON SURFACES (SCHNABEL[737])

z_0 (Å)	$F^0_{vd\omega}$ (a) Pa×10⁻⁴	F^0_{el} (b) Pa×10⁻⁴	$F^0_{el}/F^0_{vd\omega}$ (%)
4	1600	70	4·4
10	100	11	11·0
20	12·5	2·8	22·0
40	1·6	0·7	44·0

Notes: (a) calculated using equation (7.82) assuming $\mathbf{R} = \psi/\pi$,
(b) calculated using equation (7.85a), $\hbar\bar{\omega}$ was assumed to be 5 eV.

In general $0{\cdot}6 \leqslant \hbar\omega \leqslant 11$, where 0·6 represents synthetic fibres. For a typical case,[529] in the calculation of $F^0_{vd\omega}$, the Lifshitz–van der Waals constant, which is 0·6 eV for polymer fibres and 3·5–6·5 eV for quartz or limestone, gives a resultant $\hbar\bar{\omega}_{12}$ of 1·2–2 eV by this equation. Schnabel[737] has used a mean value of 5 eV in his calculations, and his comparisons of electrostatic and van der Waals forces are shown in Table 7.5.

7.11.2. Electrostatic forces

The case of excess electrostatic forces only is considered here. For a metallic sphere, diameter d, separated by a distance z_0 from a conducting half space, the attractive force due to a charge q is given by

$$F^0_{el} = \frac{q^2}{8\pi\epsilon_0[\gamma + \frac{1}{2}\ln(d/z_0)]^2\, dz_0} \tag{7.84}$$

where γ = Euler's constant = 0·5772.

Generally the filtration process involves non-conducting particles and fibres, where the charge distribution is limited in surface coverage and depth. For this case, Krupp found[462, 468]

$$F^0_{el} = \frac{q'^2}{8\pi\epsilon_0 d\delta}\,\frac{\ln(1+\delta/z_0)}{\{\gamma + \frac{1}{2}\ln(d/z_0)\}\{\gamma+\frac{1}{2}\ln d/(z_0+\delta)\}} \tag{7.85}$$

where q' = effective space charge (approx. in the range $0·1 < q < 0·3$),
δ = depth of charge penetration
= $(1/e)\times$surface charge.

Schnabel uses the simpler equation

$$F^0_{el} = \sigma'^2/2\epsilon_0. \tag{7.85a}$$

where σ' is the surface charge density, which, in terms of the charge difference ΔE, is given by

$$\sigma' = \epsilon_0\, \Delta E/ez_0. \tag{7.86}$$

This is approx. 7×10^8 em^{-2} if ΔE is 0·5 eV and $z_0 = 0·4\times10^{-10}$ m.

The comparison made by Schnabel of the calculated adhesive van der Waals and electrostatic forces indicate that for small particles the former is more important than the latter. Similar conclusions were arrived at by Löffler,[526] and these are at some variance with the observations by Billings.[78] At this stage therefore no general conclusions as to the relative importance of specific forces in a particular case can be predicted.

7.11.3. Capillary attraction

In a moist atmosphere a liquid bridge can be built up between a particle and the collecting body. Between a sphere and a plane, the adhesive force can be calculated from

$$F^0_{cap} = 2\pi\sigma d \tag{7.87}$$

where σ = surface tension (approximately 0·072 N m^{-1} for water at ambient temperatures),
d = diameter for a perfect sphere or contact diameter equivalent for rough surfaces.

If d = 0·2 μm, then $F^0_{cap} = 9\times10^{-8}$ N while for d = 1 μm, $F^0_{cap} = 4·5\times10^{-7}$ N.

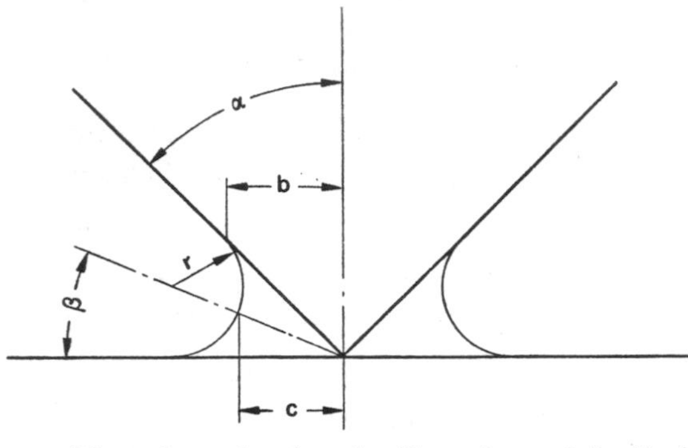

FIG. 7.21. Point contact with a plane, showing the dimensions of the liquid bridge (Löffler[526]).

Löffler[526] has modified this equation (7.87) for the important case of point contact (see Fig. 7.21)

$$F^0_{cap} = \pi\sigma b\left\{\frac{\sin\alpha}{\tan\beta} - \frac{\sin\alpha}{1-\sin\beta}\right\} \tag{7.88}$$

where α, β, b and c have the dimensions as shown in Fig. 7.21. This can be simplified for the case of $b \doteqdot c$ to

$$F^0_{cap} = \pi\sigma b\left\{\frac{\sin\alpha}{\tan\left(45^\circ - \frac{\alpha}{2}\right)} - 1\right\}. \tag{7.89}$$

F^0_{cap} becomes zero when α is approximately 33°, when the first term in parenthesis $\rightarrow$ 1. Thus, only conical particles with $\alpha > 33°$ can be held by capillary attraction. In a typical case if $b \doteqdot c = 50$ nm, for $45° < \alpha < 60°$ and water as the liquid, F^0_{cap} is of the order of $0{\cdot}8\times10^{-8}$ to $2{\cdot}5\times10^{-8}$ N. Thus, the capillary adhesive force is of the same order as the van der Waals and electrostatic adhesive forces in favourable conditions.

7.11.4. Experimental investigation of particle adhesion

The photographic studies by Billings have been described. The other two methods used in the study of particle adhesion are the centrifuge and the aerodynamic or "blow-off" method. The centrifuge has been used by Löffler,[526] Larsen,[495] Boehme *et al.*,[94] Kordecki and Orr[461] and Corn and Stein,[179] amongst others, while the aerodynamic technique has been used by Gillespie,[297] Larsen,[495] Corn and Silverman[178] and Löffler.[527] The latter method leads to the determination of entrainment velocities, although the actual determination of adhesive forces is not explicit.

The actual adhesive force retaining the particle on a fibre has been found to be of the order of 10^3 to 10^5 times the gravitational attraction.[529] Boehme,[94] Corn,[178] Larsen[495] and Löffler[529] all found that the adhesive force increases with particle size. The filtration velocity also has the effect that while the particles have a greater tendency to bounce off or slip by the collector, those which are collected tend to adhere more strongly. Similarly, if the collecting body is soft, better adhesion is found. This is borne out by the work of Boehme *et al.*[93] who find that the material which deforms on impact with the sphere (e.g. gold spheres on polymide fibres) gives 7 to 10 times the adhesion ($70–100 \times 10^{-8}$ N) compared with quartz on glass (10×10^{-8} N).

Löffler's study[528] of particle accumulation showed that when a layer of particles was stuck to a fibre, the fissured surface of this gave more adhesion partners, and so gave better adhesion. This improved collection slightly with increasing velocity, as well as much greater particle deposition. Löffler concludes that the deposited particles have a better damping action than the clean surface.

The effect of increased humidity is more marked in increasing the adhesion of larger than smaller particles and higher relative humidity was found by a number of workers[176, 495, 528] to favour increased adhesion. The widest range of relative humidities (0–95 per cent) was used by Löffler[528] who found the greatest effect with glass fibres, and smaller effects with polyester and polyamide fibres (quartz particles). The large effect with the glass fibres is probably due to the formation of condensate bridges, while the effect with the polymer fibres is due to a softening of the fibre surface at high humidities, which was confirmed by Vickers hardness measurements.[93] Löffler[529] has also reviewed the re-entrainment studies by different authors, which indicate that this is in general about 3 times the deposition velocity. This enables one to conclude that filters made of soft fibres or fibres with coarse surfaces can probably be operated at much higher velocities than is present practice. On the other hand, particles in the gas stream are likely to bombard the collected particles, which may loosen and re-entrain them.

CHAPTER 8

FILTRATION BY FIBROUS FILTERS

8.1 INTRODUCTION

The process of filtration of gases to remove particulate matter has been shown to be a combination of the mechanisms of inertial impaction, interception and diffusion. Additional factors, such as gravitational, electrostatic or thermal forces, when present, also have great influence on collection efficiency. In general, fine fibres are more effective collectors than coarse ones because of their larger inertial impaction and interception parameters, as well as their provision of a larger aggregate surface area per unit volume for diffusion to take place. Other factors, such as fibre surface roughness and hardness, may also play a part. Closer packing of fibres also tends to improve collection because of favourable fibre interference effects. Close-packed fibres will, however, increase the pressure losses, and this may not be economically desirable.

The presence of electrostatic charges on either particles or fibres will enhance collection efficiency, while if both carry charges of opposite sign, even greater efficiencies can be obtained. Conversely, if the charges on both are of the same sign, and the resultant coulombic repulsion is greater than the image force attraction, the collection efficiency will be reduced.

The fibres can be used loosely packed, pressed together in felt form or spun and woven into cloth. They can be metal, natural or chemical fibres, cellulose or glass fibres. The selection of a particular filter medium depends on the application. Thus, if the particle concentration is high, as in the waste gases from smelting, the filters have to be cleaned continuously or at frequent intervals, and a strong fabric has to be provided, able to withstand frequent cleaning. If, on the other hand, particle concentrations are low, for example in filtering air, the filter may not be loaded for many months and will not be re-used. Loosely packed fibre may then be more suitable.

Mist collecting filters are made of fibres from which the collected drops will run to a collecting point. The fibres must be sufficiently rigid to withstand the droplet load and gas pressure without matting, which reduces efficiency and increases pressure drop. High temperatures and corrosive conditions restrict the range of fibres that can be used. Glass fibres do not deteriorate at temperatures up to 316°C and some chemical fibres and wool stand up to mildly acid conditions while other chemical fibres have a long life in alkaline conditions.

This chapter discusses the construction of filter plants and filter media application to industrial waste gas cleaning. Fibre filters for mist collection and air cleaning are also dealt with briefly as they present a major field of filter application.

8.2. CONSTRUCTION OF INDUSTRIAL FILTER PLANT

The first gas filters used commercially consisted of long filter sleeves hung in rows, tied together at the bottom, while the dirty gases were ducted into the top. At intervals the sleeves were shaken manually and emptied. The sleeves were called bags, and the building containing them "bag houses", which is the name still applied to installations of this type. Equipment of the type described is still found in some industries where the duty is light and the

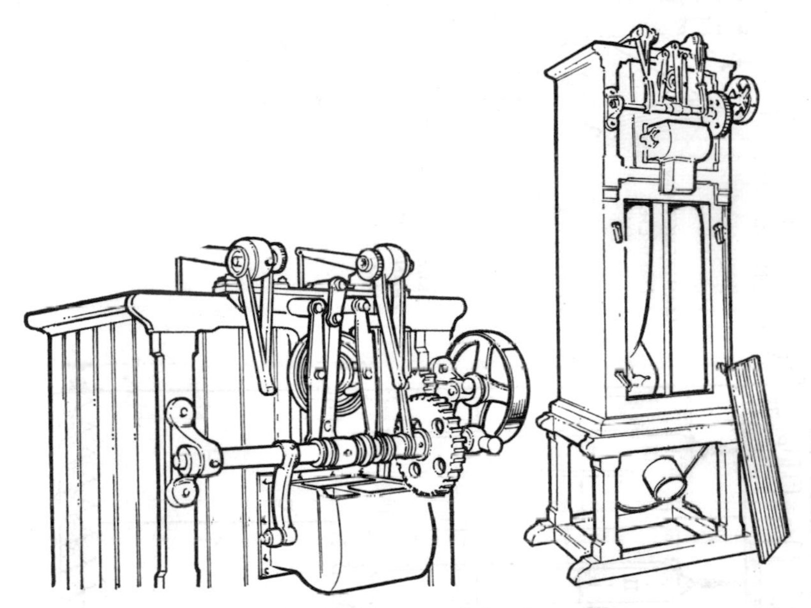

FIG. 8.1. Early automatic bag house.[381]

scale of application small, as for example in the collection of sawdust from a motorized saw-bench.

An obvious development of the rudimentary bag house described was to motorize the shaking of the bags, and install the plant in a small, separate and easily transportable housing. This was introduced in a German patent in 1881, where a mechanized shaking device, which slackened and tautened the bags, was described, and the bags were enclosed in a wooden box (Fig. 8.1). This equipment was able to handle much larger gas volumes than the traditional plant, in a smaller space, because of the frequent dust removal. Unfortunately the wooden housing of this early model tended to dry out and the cracks permitted excess air to be drawn in, reducing the amount of gas filtered. Since this early patent a large number of improvements, particularly with regard to dust removal, have been introduced

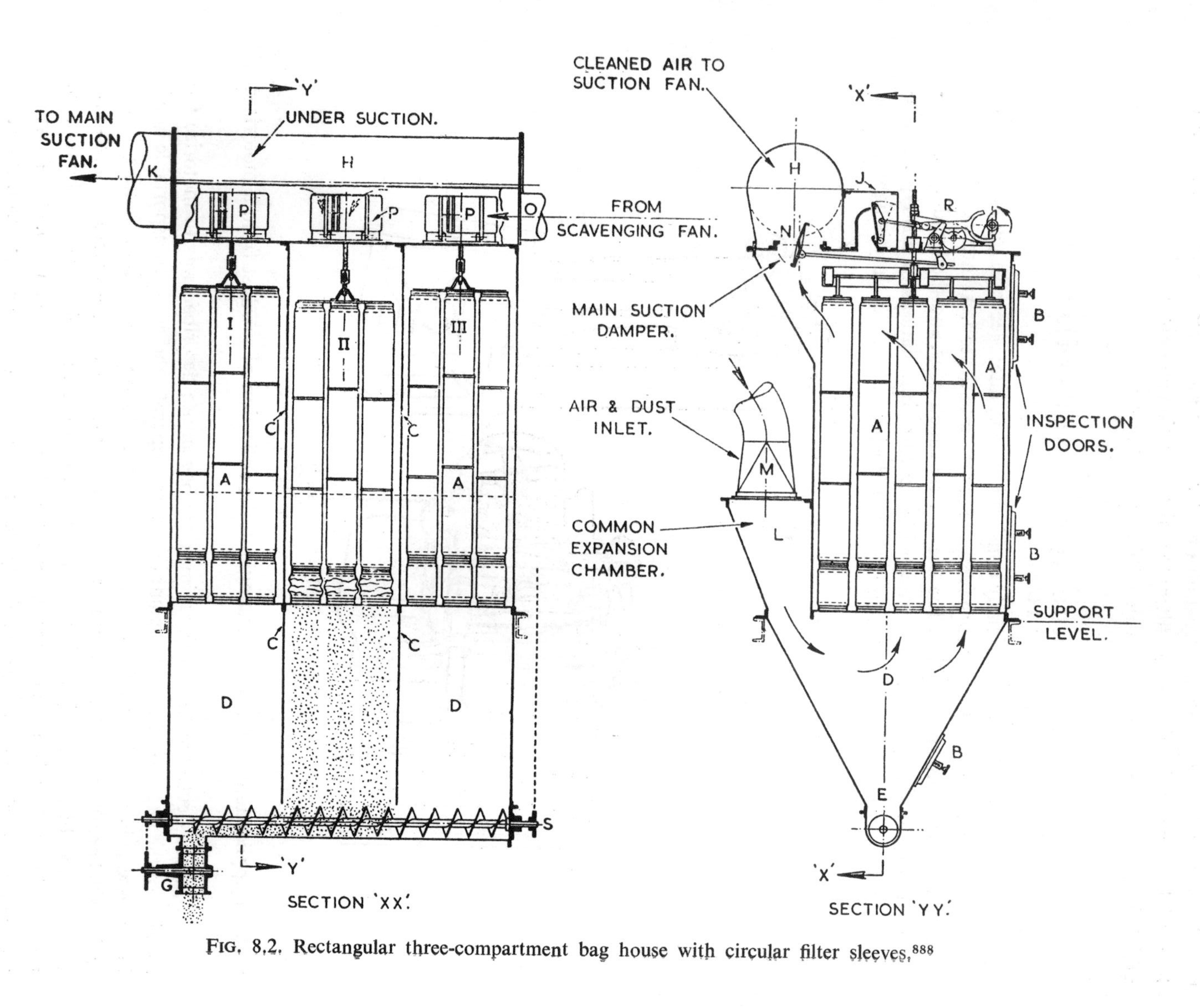

FIG. 8.2. Rectangular three-compartment bag house with circular filter sleeves.[888]

by the manufacturers of "bag houses", and a modern plant will cope with high dust concentrations, large volumes, and comparatively high temperatures up to 350°C.

The filter sleeves used can be divided into two categories. The one consists of long, circular sleeves, which are directly descended from the early "bags", while the other contains flat filter sleeves, which have the advantage that a greater filter area can be packed into a housing of smaller volume than could be used for the round sleeves.

Circular filter sleeves, where the gas flow is usually from inside to outside, do not, under these conditions, require internal support, although some rings to prevent total bag collapse during shaking are often provided. The bags are mostly suspended in a regular pattern in a rectangular steel casing, although cylindrical containers are sometimes used. In the case of the rectangular containers (Fig. 8.2) the complete "bag house" is made up of a number of

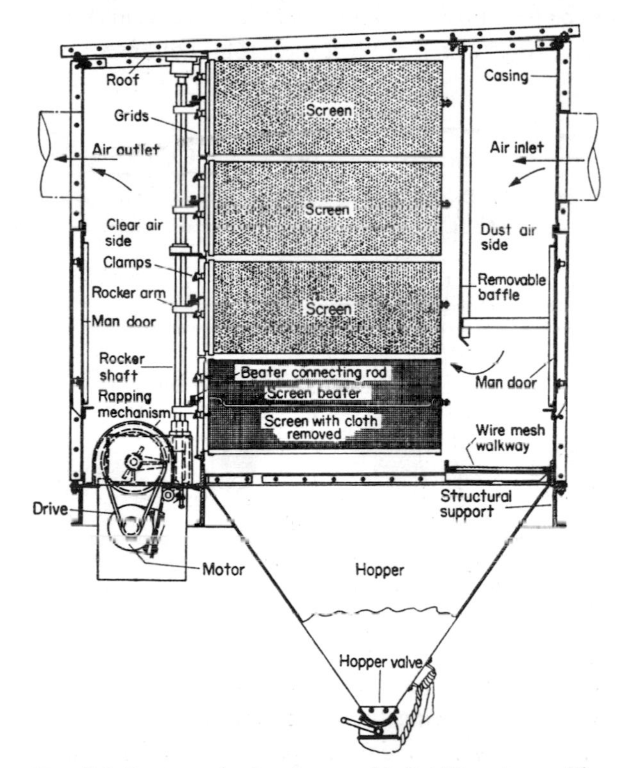

FIG. 8.3. Rectangular bag house with flat filter sleeves.[621]

similar standardized sections, each with the same number of bags and each able to be shaken separately. In the case of cylindrical containers there are usually fewer bags, but greater filter areas are obtained by using several cylindrical sections in the casing and longer bags.

When flat sleeves are used, these are supported on wire framework and the gas flow is from outside the sleeve inwards. The housing is then invariably of rectangular design, and a number of housing units make up a complete installation (Fig. 8.3).

8.2.1. Dust-cake removal methods

When the dirty gases pass through the filter cloth, the particles being removed from the gases first collect on the fibres, by the mechanisms discussed in the previous chapter. Then a cake builds up on the collected particles, and the pressure drop through the fabric increases. When the cake has built up to an optimum thickness for removal, it must be loosened from the filter cloth by some means and allowed to drop into the collection hopper.

In the first automatic unit (Fig. 8.1) the bag tension was released, the bag collapsed, and the dust cake fell through the wide sleeve into the hopper. Because the filter surface area that can be provided by wide bags was limited, the tendency has been to use narrower sleeves, which enable a much greater surface area to be fitted into the same size bag house. However, the cake tends now to be held up by the collapsed filter sleeve, and so the sleeves are stiffened at intervals by retaining rings which prevent their total collapse, and allow the cake to fall.

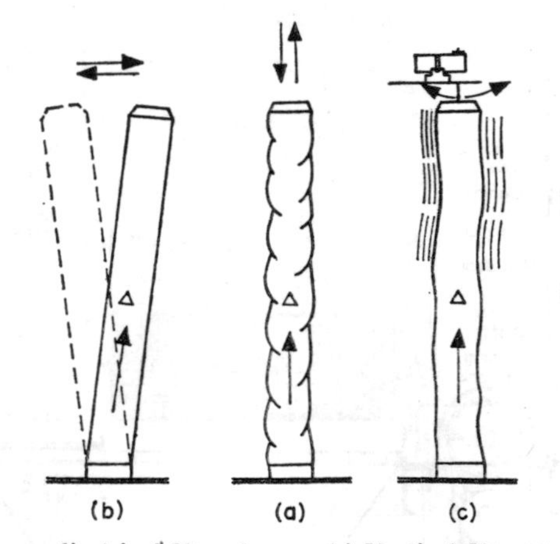

FIG. 8.4. Methods of shaking cylindrical filter sleeves. (a) Vertical filter sleeve movement. (b) Gentle sideways movement. (c) Sideways vibration of filter sleeve.

Besides vertical bag motion (Fig. 8.4a), sideways movement is used (Figs. 8.4b and 8.4c), the most effective form of this being the rapid sideways shaking which loosens the cake while vibrating the filter sleeve. At high temperatures, where the fibres are weaker, the mechanical vibration may reduce the strength of the filter cloth, and gentle sideways shaking may be preferred. Sonic vibrations and low pressure shock waves have been used in recent years, but few details of performance are available. For high-temperature operation or in corrosive atmospheres, where mechanical linkages could seize and corrode rapidly, these methods which do not require mechanical movement of the bags appear to be very suitable.

In many installations the bag shaking is assisted by reversal of the gas flow in the section being cleaned (Fig. 8.5). The bags are either kept circular by rings sewn on the bags (Fig. 8.6a) or by rings held in position by rods hanging down the axis of the bags. These systems are both used for high-temperature filters (Fig. 8.6b).

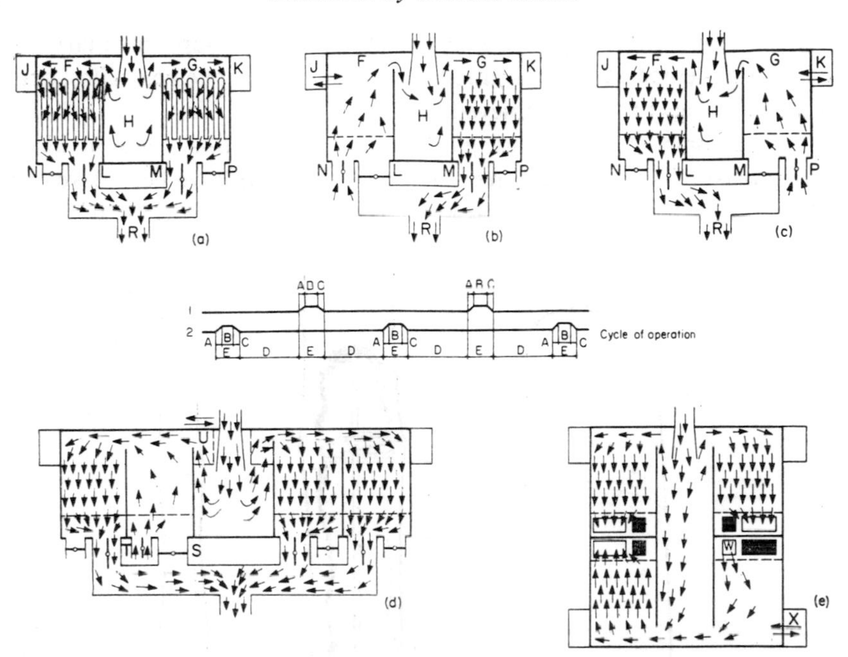

FIG. 8.5. Arrangements of gas flows to obtain reverse flow in one compartment of multicompartment filters.[778]

Two-section type:
This filter is divided into two sections F and G (Drawing (a)–(c)). Each section has its own bag shaking device with motor drive (J and K). A centre classifier (H) is highly desirable for (1) better distribution of dust-laden air, (2) more convenient dust pipe location, (3) considerable settling of dust, especially the larger particles, thereby relieving the filters of a portion of the load. The clean air chambers of both filter sections have main dampers (L and M) and small reverse air-flow dampers (N and P). Each main damper and reverse air-flow damper are connected by linkage to a small gearhead motor. A timer controls opening and closing of the dampers; whenever a main damper closes, the corresponding reverse air-flow damper opens. The two main dampers lead to a common fan duct (R) for connection to the exhaust fan.

(a) Both filter sections F and G handle the dust-laden air. This is period D in the Operation Cycle described below.
(b) Main damper L is closed, small damper N is open for reverse air-flow through filter F. This air assists shaker drive J, which is in operation removing dust from bags.
(c) Main damper M is closed, small damper P is open for reverse air-flow through filter G. This air assists shaker drive K, which is now in operation removing dust from bags.

Cycle of operation 1 and 2 represent the two sections (F and G above).
A – About 1 minute allowed for closing damper, shutting of one section.
B – Two of minutes operation of bag shaking device.
C – About 1 minute allowed for opening damper.
D – From 15 minutes to several hours when both filter compartments are in use.
E – From 4 to 6 minutes when one or the other compartment is shut off for bag shaking.

Four or more sections:
Filters of larger capacities may be subdivided into four or more sections with a common classifier. Each section has its own shaker drive, main damper and reverse air-flow damper. With a one-hour cycle, one section will be shaken every 15 minutes so that each compartment gets its turn every hour. As illustrated, filters may be arranged in single or double file, as dictated by available space.

Single file (d):
Four filters sections with second section from left closed. Main damper S is closed, small damper T is open for reverse air-flow to assist shaker drive U in removing dust from bags.

Double file (e):
Another four-section filter arrangement. Control dampers are bottom mounted. Main damper V is closed, small damper W open for reverse air-flow to assist shaker X which is in operation removing dust from bags in lower right section.

The "bag-collapse" method, of reverse gas flow without supplementary shaking, is widely used on high-temperature applications (to 280°C) such as cement kilns, and long bag life with fibre-glass bags has been achieved.

Besides shaking combined with gentle gas-flow reversal, the filter cake can also be blown off by a strong blast of reverse air. Three systems of this type have been developed. In the first (Fig. 8.7a), an external ring, slightly smaller than the filter sleeve, with a slit facing

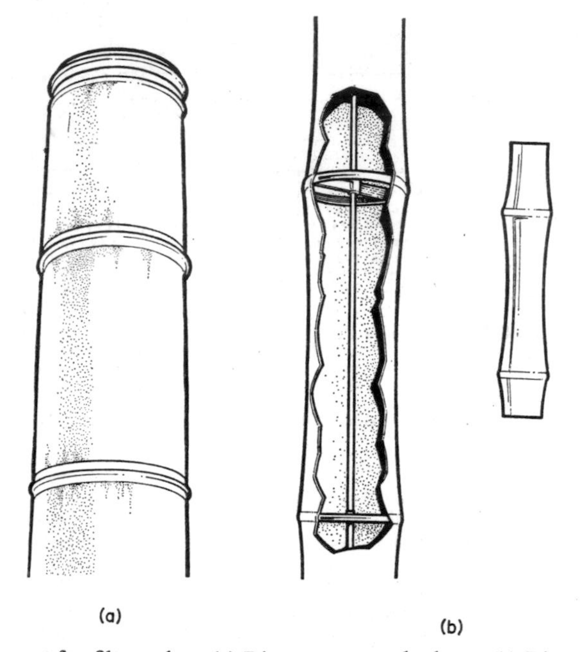

FIG. 8.6. Internal support for filter tubes. (a) Rings sewn on the bags. (b) Rings held separately on rods.[570]

inwards, is slowly and continuously moved up and down the filter sleeve. This is billowed out with the normal gas stream, so that the ring, being smaller than the sleeve, squeezes it, loosening the cake, and then the strong blast of air from the slit in the ring blows the cake off.[362a] Instead of individual rings, a framework, shown in Fig. 8.7b, can be used. In the second system, the normal gas-flow direction is reversed, being from outside inwards into a bag, supported on a framework and closed at the bottom (Fig. 8.8), while the cleaned gas escapes at the top. Intermittently a high-velocity jet, blown through a nozzle and a converging–diverging section at the top of the filter sleeve temporarily reverses the gas flow, and dislodges the collected cake, which falls into the hopper. The third system has been developed for flat filter sleeves, and as in the second system a reverse air flow is used, but in this case it is directed between two sealing off rollers which move from one sleeve to the next (Fig. 8.9).

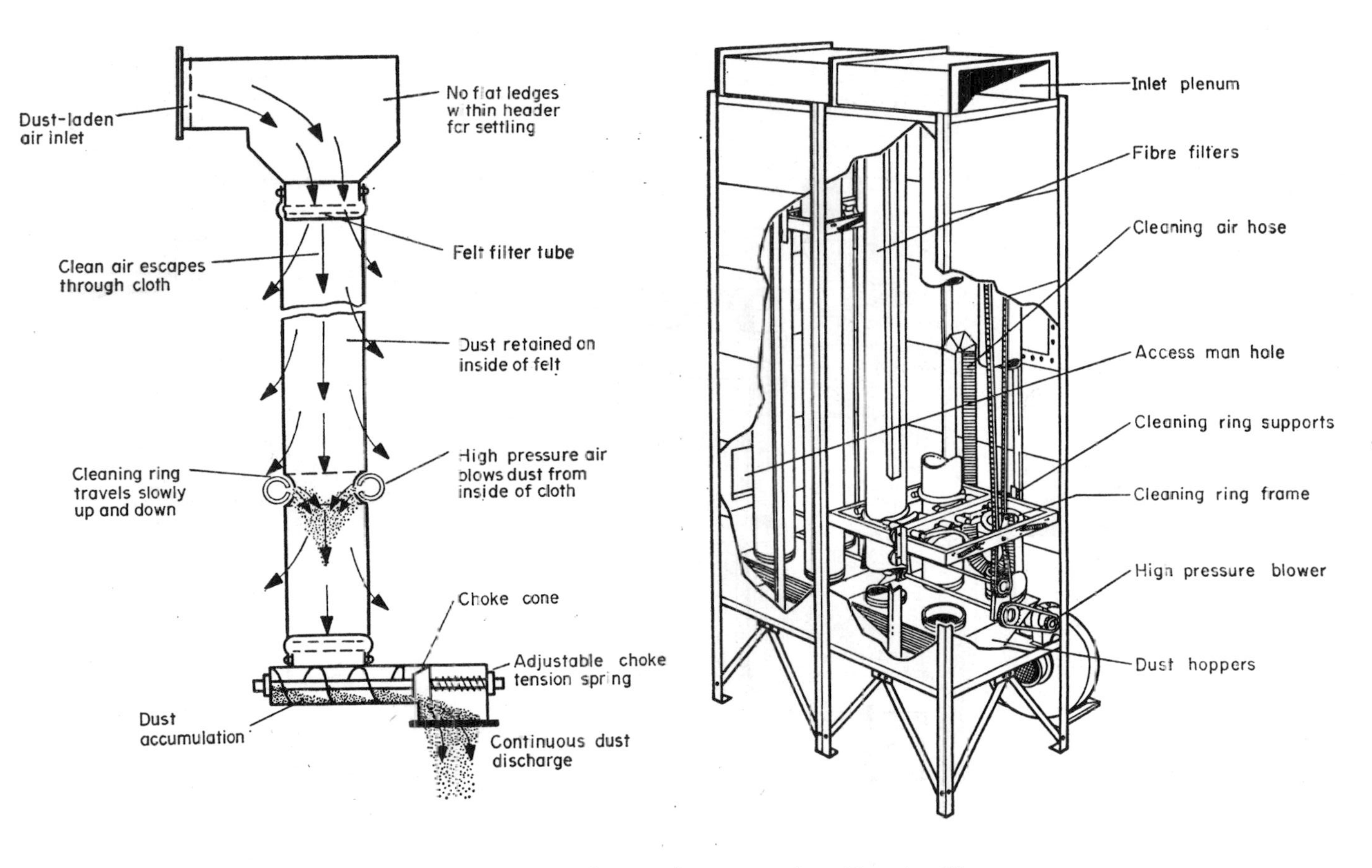

FIG. 8,7a. "Hersey" reverse jet system using a blast ring.[920]

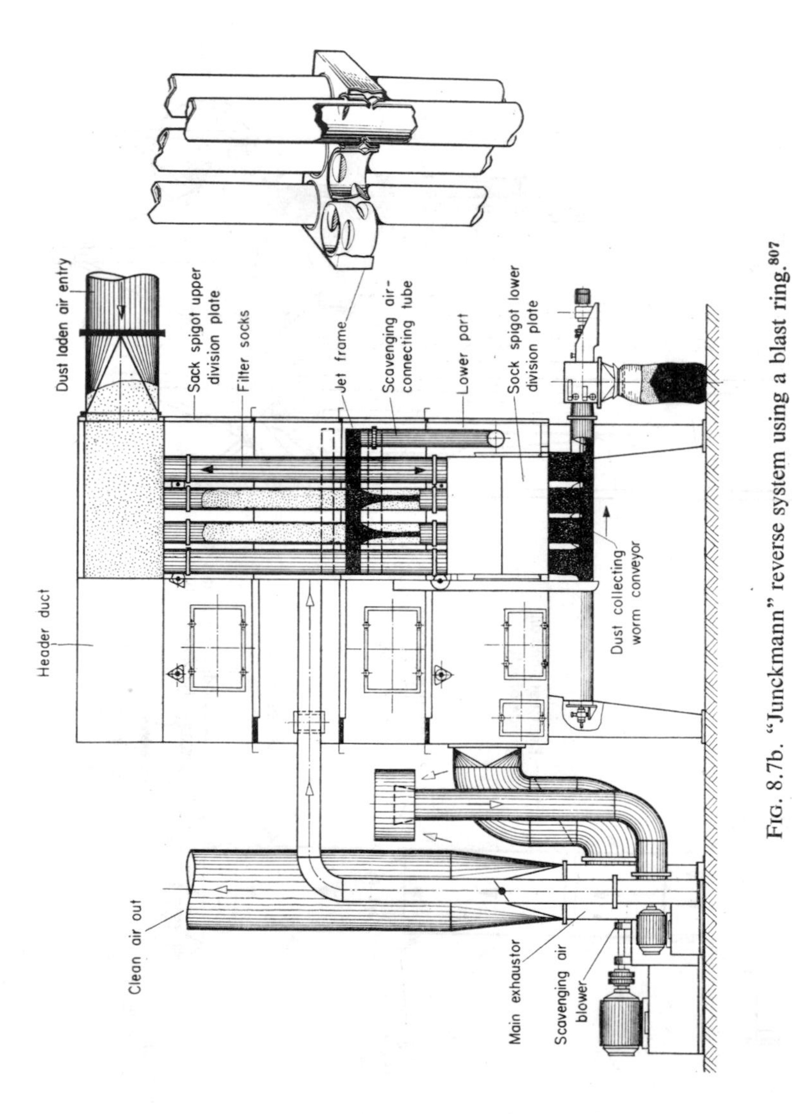

FIG. 8.7b. "Junckmann" reverse system using a blast ring.[807]

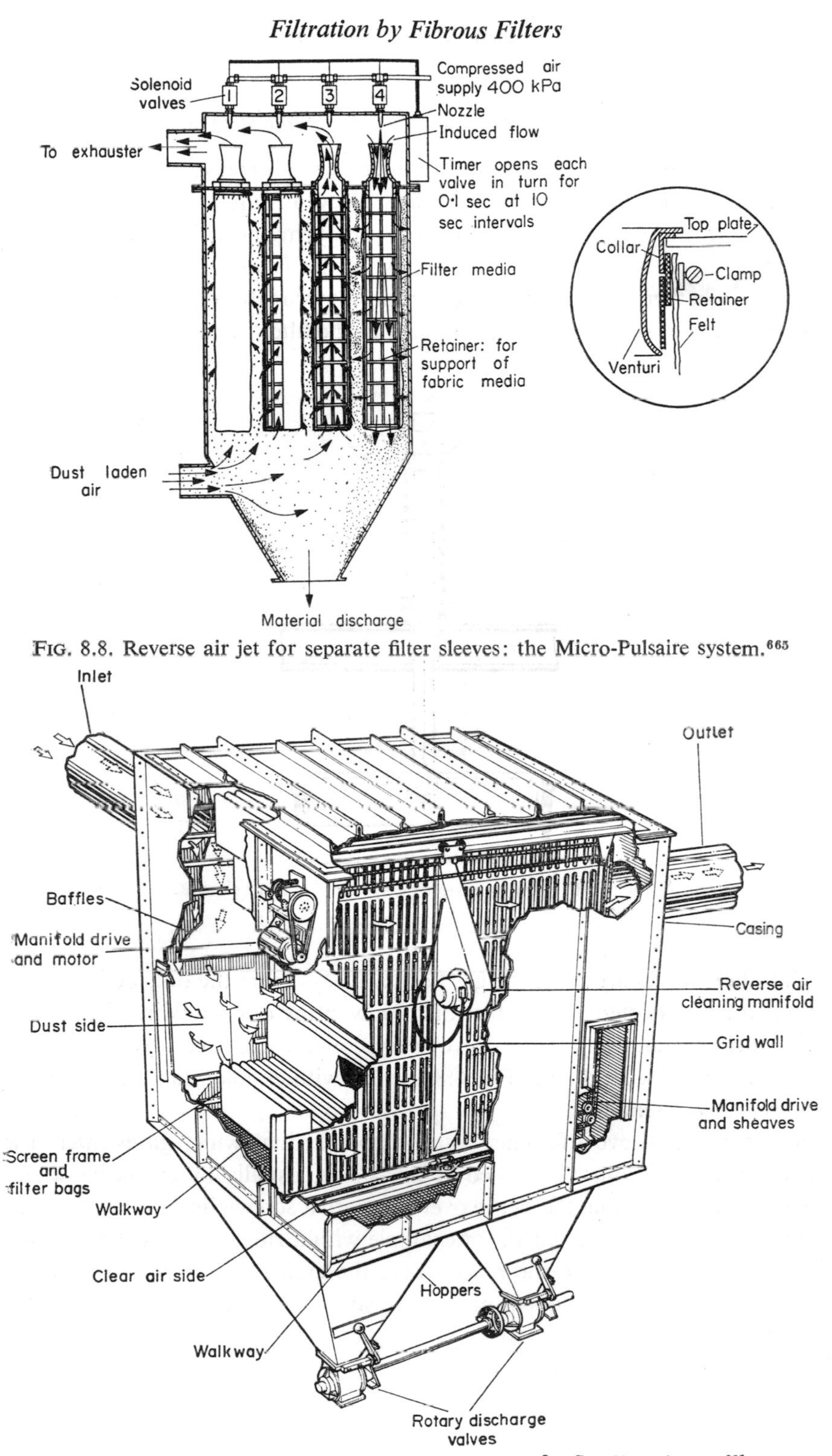

FIG. 8.8. Reverse air jet for separate filter sleeves: the Micro-Pulsaire system.[665]

FIG. 8.9. Continuous reverse air-cleaning system for flat filter sleeves.[621]

8.2.2. Suspension and support of filter sleeves

The system of suspending and supporting filter sleeves depends on the direction of gas flow and the method of dust removal. If the filter sleeve is tubular, and the gas flow is from inside the tube outwards, the tube will not collapse during normal operation. The method of shaking, the velocity of gas during reverse flow (if this is used) and the length of the tubes then determine whether rings are required for bag support during removal of the collected dust. The provision of supports adds to the cost of the filter sleeve.

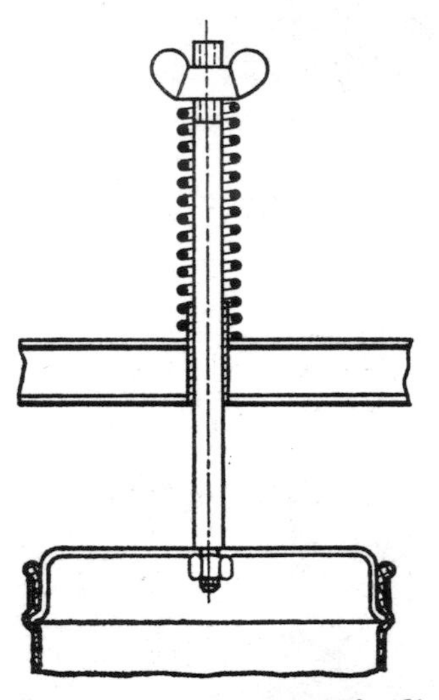

FIG. 8.10. Spring-tensioning system for filter sleeves.

When flat filter sleeves are used, the gas flow is always from the outside inwards, and the sleeves have to be supported on a framework. Flow in the opposite direction tends to blow the sleeves outwards and adjoining sleeves may contact and wear. When a reverse-flow cleaning system of the second type (Fig. 8.8) is used with circular bags, a supporting frame for the tubes is also essential.

One end of the filter sleeve is fastened round a hole through which gases enter or leave the filter sleeves. The fabric must be sealed at this point, usually by extending a short tube inside the sleeve and fastening the sleeve over this with a screw-clip or patented connection. This tube must be carefully designed, for example with a ridge, so that the bag cannot slip off during shaking. This is also the place where the greatest stresses and maximum wear occur on the sleeve, and it is frequent practice to strengthen this section. Long filter sleeves can be individually tensioned by a screw–spring system of the type shown in Fig. 8.10. Short filter sleeves may not require this.

8.2.3. Housing

The casing used for a bag filter system depends on the type of gas, gas temperature and direction of gas flow, and the type of dust cake removal system used. If the gases are of no further value, non-corrosive, non-toxic, and are brought to the inside of the filter bags, which are simply shaken, without gas flow reversal, for cake removal, then the housing for the bags serves merely as weather protection for the sleeves and shaking mechanism. Simple corrugated steel sheets, probably zinc coated against atmospheric corrosion, can then be

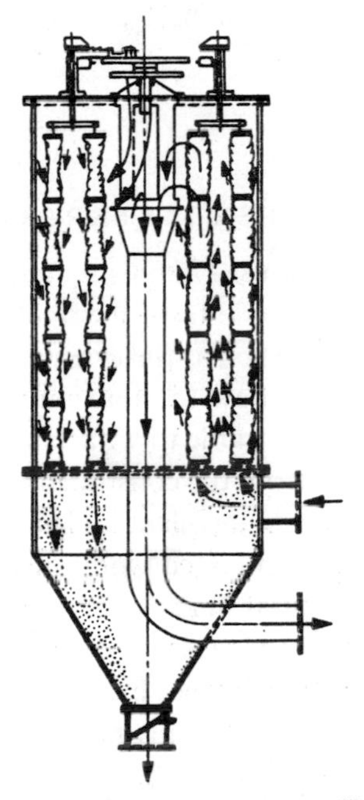

FIG. 8.11. Cylindrical housing for filter sleeves, where centrifugal separation is used for initial coarse particle separation.[807]

used for weather protection, and the cleaned gases can be allowed to escape through openings in the building. If reverse gas flow is provided by suction within the bags, then a similar simple structure will also often be found adequate.

When the gases are of some value and are required for further processing, or when they are toxic, then the "bag house" must be carefully sealed, and if the pressure in the system is inadequate, an extraction fan must be used to exhaust the gases. This system has an advantage over the open bag house in that the fan used will handle the cleaned gases, with low fan-blade erosion and a long fan life, while the fan used to push the gases through the filter sleeves of the open bag house operates on the dirty gases with resultant blade erosion and other problems associated with fans in dirty gases.

When the fan exhausts the housing, the housing must be strengthened to resist the external pressure. A cylindrical housing (Fig. 8.11) may be more economical in this case, particularly for small installations. Tangential entry into cylindrical casings also removes coarse particles by centrifugal forces. Careful sealing of the housing is necessary to reduce leaking in of excess air. In filter houses where the dirty gas is on the outside of the filter sleeves (flat sleeves or reverse-jet type, Fig. 8.8), the housing has to withstand internal pressures of the order of several inches water gauge. In this case leaks in the housing will distribute dirty gases to the surroundings.

The housing is usually made of steel, which may be treated for corrosive or humid conditions by galvanizing, painting or lining with rubber or polyvinylchloride. Some of these materials are unsuitable at high temperatures, and stainless or special steels may be required. It is frequently necessary to insulate the installation to reduce heat losses and keep the gases above their dew point.

8.3. FILTER MATERIALS

It has been emphasized that the fibres used in filter materials should be fine and that if they do possess an electrostatic charge, it should be of the right type for the dust to be collected. Other factors must also be considered, such as the orientation of the fibres, which should be perpendicular to the direction of gas flow for maximum efficiency, and the mechanical strength of the filter cloth which must be able to withstand shaking or vibration. The fibres must also withstand chemical attack from the constituents of the gases, which can be acidic or alkaline, oxidizing or reducing. When the plant is closed down mould growths can occur and in some cases (particularly with untreated wool) insects and bacteria may attack the cloth. The filters used may be of natural fibres (cotton, flax, wool, silk, asbestos) or glass or chemical fibres. There are a number of these (listed in Table 8.1), with different chemical composition and a large number of trade names by which they are normally referred to.

All natural fibres are short fibres, called staple fibres, which are either pressed into felts or first spun into yarn and then woven into cloth. The chemical fibres are made either in the form of staple fibre or as long filament yarn, the latter being much stronger mechanically. For example, polyester fibre filament yarn has approximately twice the strength of the staple fibre yarn of the same diameter.

Filter fabrics are either woven cloth or felts. The felted materials are commonly used for reverse jet air or gas-flow equipment, where the deposited cake is blown off the fabric, while woven fabrics are used for the filter bags which are shaken or vibrated, as these require greater strength. The weaves used are either by the "cotton system" which uses short staple fibre, and produces fine yarns, or "woollen system" yarns, which are made from longer staple fibres, and produce a coarser cloth, more suitable for mercerizing. If the fabric is made from a smooth filament yarn, it will have a smooth surface, which is excellent for cake release, but the knit in this case must be tighter, with lower cloth permeability. The filter cake is generally denser and cleaning by shaking will be more complete.

Essentially two types of woven cloths are made for gas filtration. These are simple, unraised fabrics or raised (mercerized) fabrics, where the raised fabric faces the dirty gas stream. The pores in woven cloths are those between the threads and between the fibres, the latter

TABLE 8.1. TRADE NAMES OF FIBRES USED AS FILTER MEDIA

Chemical name	Fibre type	Trade names
Polyacrylonitrile fibre	Staple fibre and filament yarn	Microtain
Polyamide fibre (straight chain aliphatic segments)	Staple fibre and filament yarn	Nylon Perlon Phrilon
(recurrent amide links)	Filament yarn	Nomex
Polyacrylonitrile	Filament yarn and staple fibre	Orlon: Type 42, staple fibre Type 81, filament yarn PAN Dralon-T Redon
Polyester fibre (Polyethylene tetraphthalate)	Filament yarn and staple fibre	Terylene Dacron Diolen Trevira
Polyethylene fibre	Filament yarn	Polythene Alkathene
Polypropylene fibre	Filament yarn and staple fibre	Polytain
Polytetrafluorethylene fibre	Filament yarn	Teflon PTFE Hostaflon
Polyvinylchloride	Filament yarn	Vinyon Rhovylfibro
Vinylidene chloride (90%) Vinyl chloride (10%) Copolymer fibre	Filament yarn	Saran
Vinyl chloride (60%) Acrylonitrile (40%) Copolymer fibre	Filament yarn	Dynel

making up 30–50 per cent of the voids in the cloth. When a gas is drawn through the cloth most of the flow initially will be through the holes between the threads, while only a small fraction of the gas stream passes through the interstices between the fibres, where effective particle collection takes place. The more tightly twisted the yarn the less gas in fact can pass through the fibre interstices.

After a while, with a plain cloth, coarse particles will be caught in the inter-thread holes, and the gas stream is then forced through the interstices between the fibres. The cloth then acts as a most effective filter for both fine and coarse particles, and a cake builds up. Plain, unraised fabrics tend to release the dust cake more easily on shaking the cloth, while the cake tends to be interlocked with the fibres on a raised fabric. After cake removal, the efficiency of filtration and the pressure drop with a plain cloth are again low until the first filtration stage of closing the gaps between the threads has been achieved.

When a raised or "napped" fabric is used, the particles are collected predominantly by the nap—coarse particles by inertial impaction and fine particles by diffusion—the weave acting largely as a mechanical support. Bergmann[71a] recommends light napping of the cloth particularly for use at moderate temperatures. However, if the cloth is made of a synthetic fibre at temperatures close to the softening point of the fibre, then an unnapped cloth is preferable. The particles which penetrate the nap are either lodged in the weave or are lost. Those lodged in the weave are unlikely to be dislodged during shaking, and an increasing pressure drop, with some increase in efficiency, is observed in the early stages of filter cloth life. Some of the nap fibres are torn away when the cake is shaken off, and in time the nap will be removed completely. The cloth will then act as an unraised cloth, although sometimes the nap can be reformed.

The nap is formed by passing the cloth over "nappers", which tear the surface fibres from the threads in the weave. Conventional wire nappers have proved the most satisfactory, while sanders or sueders cause "pilling" of the fibres, which is not very desirable. If cloth is to be raised, a yarn with a twist that does not exceed 200 turns/m should be selected, or napping becomes difficult. When dust loads are low, a heavier cloth (4 ply, 0·58 kg m^{-2}) has a high initial efficiency with an acceptable pressure drop, while with high dust loads a lighter cloth (0·44 kg m^{-2}) has better pressure drop characteristics.[884]

Chemical fibres tend to elongate under load and shrink at high temperatures. Thus it is essential that the cloths used are stabilized by heat setting, and furthermore, that the dust burden be kept to a reasonable minimum. It is therefore important for this reason as well as others that as much as possible of the collected material is removed during the cleaning cycle. Some fumes are very tenacious, e.g. arc furnace fumes, zinc oxide and aluminium smelting, and for these it may be desirable to singe the surface to remove the nap. A silicone resin finish has also been used to improve cake release, and this proves very satisfactory, notably if moisture is present.[71a]

Non-woven fabrics or "felts" are different from woven fabrics because they present a uniform array of fibres right through the cloth, and get their mechanical strength from the interlocking of fibres. Strong felts can only be made from crimped staple fibre. The fibres are first carded, forming a web, which is then cross lapped to build up a batt, with depth and a random fibre distribution. The fibres are mechanically locked together by passing the batt through a needle loom, and the felt is then treated with heat and chemicals to shrink the material as well as to proof it against mould and insects if necessary.

At present felts are made of wool, polyester fibres or mixtures of these. The filtration efficiency tends to be uniformly high, although the pressure drop increases with use as more particles are lodged in the cloth and not shaken off or removed with reverse gas flow. Even if reverse flow does dislodge a particle within the felt, it is likely to be recollected before reaching the felt surface.

It is generally desirable to use filter materials that present the lowest resistance to gas flow, and still give the required collection efficiency. This flow resistance is measured in terms of the cloth's *permeability*, which is defined empirically as the volume of air (m^3) which will pass through m^2 of the cloth in 1 minute with a pressure drop of 125 Pa. Typical values for permeability of chemical fibre cloths are of the order of 1 to 2 m^3/min, while wool, with a plain weave has, in one style, a permeability of 3 m^3/min.[163] (See also Section 8.4.3.)

The permeability test is specified in the A.S.T.M. Standards (No. D737, *1971 Annual Book of Standards*, Part 24, p. 111), and requires that the air flow is passed through a sample (255×255 mm) held between plates in which there is an orifice 70 mm diameter and across which a pressure drop of 125 Pa is maintained.

8.3.1. Properties and application of fibres

Cotton. Cotton is the cheapest fabric available commercially. It is strong at ambient temperatures and may be used up to 82°C. At higher temperatures cotton filter cloths rapidly lose strength, and tend to degrade in superheated steam. Cotton cloths have little chemical resistance in the case of acid or oxidizing conditions, or the presence of formic acid, organic solvents and hydrogen peroxide. Resistance to alkalis is good. Cotton fibres are relatively coarse and are not recommended for the collection of particles in the sizes below about 10 μm.

Wool. Wool fibres are much finer than cotton fibres and woollen cloths and felts have been widely used for gas filtration for many years. The cost of wool cloth is approximately twice that of cotton in most countries. Like cotton, wool is not suitable for use at elevated temperatures and continuous use above 95°C is not recommended. Wool degrades in steam and alkaline conditions, but is stable in mild acid conditions. Wool can be mixed with polyester fibre to give the cloth greater strength and the filter sleeve a longer life.

Flax and *Silk* cloths are not commonly used as filter media and no data are available.

Asbestos. Asbestos fibres are exceptionally fine and are very suitable for fine particle collection (Section 8.5.3), as well as being stable at high temperatures. Unfortunately the fibres cannot be satisfactorily spun and woven or felted into a cloth strong enough for making into filter sleeves. Mixing the asbestos fibre with 5–10 per cent cotton gives a cloth, which, although weaker than cotton at ambient temperatures, does retain some strength up to 400°C. This strength, however, is much lower than glass fibre cloths, whose use is to be preferred for high temperature applications.[521]

Glass. For temperatures between 150 and 300°C, which is the range of temperatures where nearly all natural and chemical fibres are degraded, fine glass fibres make a suitable filter fabric. The strength and flexibility of the fibres, compared to other high-temperature resistant fibres such as aluminium silicate fibres, is such that conventional bag houses with gentle shaking mechanisms can be used. The glass fibres are made from alumina-boro-silicate glass (Pyrex) or very occasionally from soda glass.[555] The basic glass constituents are melted together in a glass furnace and then cast into marbles. These are remelted in a small furnace, and the liquid glass then flows through orifices to form continuous filaments which are attenuated, sized and wound on to high-speed cylinders. Glass staple fibres can be formed by playing a jet of air on the filaments flowing from the remelting tank. The staple is collected on a drum and spun into yarn which is woven into cloth (grey goods). The grey goods are frequently treated again to remove the size by passing through an oven at 250–325°C for 48 hr. The size burns away, the glass is subsequently coronized and a permanent crimp sets into the fibres. The fibres are finally covered with a silicone resin finish, which is cured, and the finished cloth is then sewn into sleeves. The silicone resin is derived from phenyl-methyl or dimethyl silane.

The chemical resistance of the glass cloth is excellent, as it is not attacked either by acids or alkalis. Even fluoride gases with hydrogen fluoride, silicon tetrafluoride or boron trifluoride require moisture to attack the glass, and this is not available at the temperatures used.

Because filter-cake release characteristics of fibre glass cloths are excellent, gentle shaking and bag collapse or sonic vibration is found adequate and tends to lengthen the life of the filter cloths. Failure of the bags is caused by abrasion between the fibres. This is verified

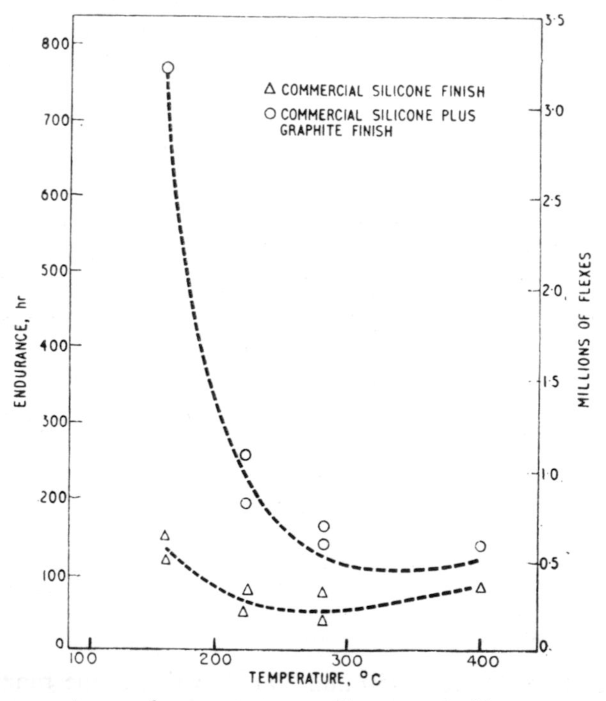

FIG. 8.12. High temperature endurance tests on silicone and silicone + graphite finished glass fabric filter bags.[570]

by the greatly increased bag life at 300°C when the bags had been treated with colloidal graphite as a lubricant[521] as shown in Fig 8.12.

Some workers consider that the burning off of the size reduces the mechanical strength of the cloth, and good results have been obtained with the untreated grey goods.[785] Some case histories for installations with glass bags are given here.[785]

TABLE 8.2. SUMMARY OF RELATIVE PROPERTIES OF CHEMICAL FIBRES (ARRANGED IN ORDER OF PERFORMANCE)

(a) Polyamide fibres (b) polyacrylonitrile fibres
(c) Polyester fibre (d) polytetrafluorethylene fibre

Resistance to dry heat	Resistance to moist heat	Tensile strength	Resistance to acids	Resistance to alkalis	Resistance to oxidizing and reducing agents	Resistance to rot and mildew	Flex and abrasion resistance	Cost
(d)	(d)	(a)	(b)	(d)	(d)	(d)	(a)	(d)
(c)	(b)	(c)	(d)	(a)	(c)	(c)	(c)	(c)
(b)	(a)	(b)	(c)	(c)	(b)	(b)	(b)	(b)
(a)	(c)	(d)	(a)	(b)	(a)	(a)	(d)	(a)

(i) *Non-ferrous metals.* (a) Lead oxide fumes at 205°C were filtered at 6 mm s^{-1}. (b) Zinc oxide fumes, filtered at 230°C, 8 mm s^{-1} velocity; using 3-m bags, 120 mm diameter. Aggregate area 20,500 m^2 of fibre-glass bags. Bag cleaning by collapse every $\frac{1}{2}$ hr, supplemented by mechanical shaking every 8 hr. Bag life was over 2 years.

(ii) *Carbon Black.* Filter velocities 2·5 mm s^{-1} at gas temperatures 205–260°C, 120-mm diameter bags 3·5 m long commonly used. Cleaned by collapse with supplementary shaking.

(iii) *Cement.* Filter velocities 11 mm s^{-1} at 260°C, 300 mm diameter woven glass tubes 7·5 m long are used. Volumes of 150,000 $m^3 h^{-1}$ are handled by a twelve-compartment filter with forty-eight tubes per compartment. Pressure drop is 0·75 kPa. Filters are cleaned by bag collapse at 60-min intervals.

(iv) *Electric arc steel furnace.* Filter velocity 5·5 mm s^{-1} at 260°C: 300 mm diameter filter tubes 7·5 m long; volume treated 180,000 $m^3 h^{-1}$. Cleaned by bag collapse.

(v) *Oxygen lanced open-hearth furnace.*[363] Design capacity is 250,000 $m^3 h^{-1}$ at 260°C. Filter velcity of 10 mm s^{-1} when nine of the ten compartments are filtering. Each compartment contains eighty bags, 300 mm diameter, 11 m long (750 m^2/compartment). Five sonic generators (or horns) are located in each compartment to aid the cleaning of dust. The sonic generators are operated with compressed air. Overall dimensions are 24 m long, 12 m wide and 20·7 m high. Outlet dust concentrations are measured to be 0·31 mg m^{-3}. After 10 months' service less than 2 per cent of of the bags needed replacement. The capital cost (1959) was $U.S. 750,000, compared with $970,000 for an electrostatic precipitator. Maintenance and operating costs were found to be approx. half of that for electrostatic precipitators.

(vi) *Pulverized-coal-fired power plant.*[100] Pilot plant tests using a four-compartment, top feed, 50,000 $m^3 h^{-1}$ bag house are reported, for boilers burning coal with 9·6 per cent ash. The inlet concentration was 0·8 g m^{-3}. Extensive tests on flow rates and filter resistance show that with current glass-fibre materials the flow rate is limited to 18 mm s^{-1}.

TABLE 8.3. PROPERTIES OF FIBRES AND CONDITIONS FOR PRE-SHRINKING FOR USE AT HIGH TEMPERATRES

Fibre	Commercial name	Type	Cost*	Melting or sublimation temperature (°C)	Useful working temperature (°C)	Wt. ($kg\ m^{-2}$)	Chemical resistance to	Conditions for preshrinking (°C)	
								(i) overfeed pin tenter	(ii) gas-heated oven
Cotton	—		1		80°				
Wool	—		2		93°	0·24			
Olefin (polypropylene)	Polytain	Spun filament needled	—		93°	24–0·29	Mineral and organic acids, dil. alkalis		
Polyester	Dacron	Spun filament combination spun and filament needled	2·7	260°	130° (150°)	0·12–0·43	Min. and organic acids, but not conc. nitric, sulphuric and carbolic	220° for 1 min	220° for 0·5 min
Polyamide (long chain)	Nylon	Spun	2·1	250°	120° (dry) 107° (moist)	0·21–0·24	Sol. in formic acid. Degrades in mineral acids. Good in alkalis.	205–220° for 0·5 min	205–220° for 0·5 min
(aromatic structure)	Nomex	Spun filament needled	6·9	371° (degrades)	230°	0·11–0·33	Min. and organic acids; alkalis at ambient temps.	—	—
Polyacrylonitrile	Microtain	Spun filament needled	2·1	Sticking temperature 225°	130°	0·095–0·36	Min. and organic acids, weak alkalis	260° for 1 min	240° for 1 min
Polytetrafluorethylene	Teflon FEP	Filament	approx. 40		288° with peaks of 316°	0·43–0·52	Min., organic acids, alkalis, oxidizing agents	—	316° for 1 min
Glass	—	Filament	2·3		300°	0·21–0·38	Dilute acids	—	—

* Relative Cost in the United States (Cotton = 1).

Chemical fibres. The chemical fibres (Table 8.1) are generally stronger and more resistant to heat and chemical attack than the natural fibres, wool and cotton. A comparison of their various properties is given in Table 8.2 (qualitative) and Table 8.3 (quantitative). Polytetrafluorethylene monofilament fluorocarbon fibre felts are chemically inert and allow continuous operation at 288°C and peak temperatures up to 316°C. The fibre is hydrophobic and comes in fine deniers, which allows the filter materials to catch fine acid-mist droplets as well as particles. The hydrophobic nature of the fibre also allows the liquid droplets to run down the surface to a collector. Another application is in the removal of solids from hot chlorine gas.

TABLE 8.4. TYPICAL COSTS OF FILTER BAGS, $U.S. (1969) COST PER BAG, BASED ON QUANTITIES OF 100 BAGS

Fibre	Size								
	150 mm diameter						280 mm diameter		
	Length (m)						Length (m)		
	1·81	2·14	2·50	2·71	3·05	3·65	6·1	7·3	10
Cotton	—	2·03	2·15	—	—	—	—	—	—
Acrylic	3·95	4·19	4·62	—	—	—	—	—	—
Polyester	4·74	5·23	5·84	—	—	—	—	—	—
Nylon	—	4·24	4·60	—	—	—	—	—	—
Glass	—	—	—	5·35	5·81	6·67	13·50	15·97	21·59

Polyester and polyacrylonitrile fabrics are resistant to acids, organic solvents, oxidizing and reducing agents, but are attacked by alkalis, while polyamide fibres are resistant to alkalis but degrade rapidly in acidic conditions. More specifically, a 100 per cent polyacrylonitrile fibre such as Microtain is better than Orlon and other acrylics as regards dimensional stability and heat resistance, being used to 126°C. While resistance to mineral and organic acids is high, it is degraded by hot, concentrated alkalis, although it is satisfactory with weak alkalis. It has been used for reverse jet as well as shaker-bag houses. The vinylidene chloride copolymer fibres have not yet been used for gas cleaning applications.

Simple olefin fibres, in which the fibre-forming substance is a long-chain synthetic polymer composed of at least 85 per cent by weight of ethylene, propylene or other olefin units, have very low heat resistance, and lose tenacity in direct proportion to increases in temperature. One of these fibres, commercially known as Polytain, a polypropylene fibre, has a maximum temperature of 93°C, above which it tends to be dimensionally unstable. On the other hand, below this temperature, the fibre has excellent resistance to mineral and organic acids coupled with good resistance to alkalis.

Experiments with polyamide fibre cloth[902] show that the material will not break up at 120°C but the filtration efficiency is reduced after use at this temperature and yellow spots are observed. Similarly, a cloth consisting of a mixture of polyester fibre with 30 per cent cotton, although retaining mechanical strength when heated above 100°C will break down locally in the fibre structure and show reduced collection performance.

A modified polyamide, having an aromatic structure instead of the conventional straight-chain aliphatic segments of Nylon 66, is a high-temperature stable fibre called Nomex (manufactured by Du Pont). Nomex will withstand both mineral and organic acids better than Nylon 66 or Nylon 6, but not as well as polyesters or acrylics. The alkali resistance of Nomex at ambient temperatures is good (better than polyesters and acrylics) but it is degraded by strong alkalis at high temperatures. The material also has good resistance to most hydrocarbons, but is degraded by oxidizing agents. The fibre is dimensionally stable and will not support combustion. The continuous use of Nomex at 220°C for filtration of metallurgical fumes has been very successful.

Filter sleeves made of chemical fibres shrink at elevated temperatures. This alters fabric dimensions, and porosity and causes cloth stiffness. When the material is to be used at elevated temperatures it must be carefully finished and preshrunk. The cloth should be boiled off under relaxed conditions using rope, Hinneken or book scours, and then dried under relaxed conditions at 150–160°C. If the fabric consists of filament yarn it can then be napped and is then preshrunk in an overfeed pin tenter or gas-heated oven. The temperatures and shrinking times are listed in Table 8.3. The fabric should be supported on rolls so that the only tension in the fabric is warpwise as it runs through the oven. Approximately 15 per cent shrinkage occurs and must be allowed for. In all cases the preshrinking temperature is higher than the recommended in-service temperatures, and preshrinking eliminates any further shrinkage of the cloth.

Chemical fibre cloths are sometimes lighter in weight than wool cloths (only 11–16 g m^{-2}). It is important that the filter sleeves are sewn with a thread of the same material or one with the same shrinkage, heat and chemical resistance.

Some typical case histories of chemical fibre applications are outlined.

(i) Filtration of 30,000 $m^3 h^{-1}$ at 135°C waste gases from grey iron cupola. A five-compartment conventional bag house was installed using gas to cloth ratios of 11·7 mm s^{-1} (five compartments) or 14·7 mm s^{-1} (four compartments); 15-min intervals between shaking in each compartment. Shaking for 1 min. The filter material was a polyacrylonitrile fibre, and the average inlet dust concentration was 15°C during charging. The filter system was virtually 100 per cent effective.[662]

(ii) Filtration of zinc oxide fumes containing some sulphur dioxide at 135°C, using an automatic-type shaker-bag house. Polyacrylonitrile filter cloth had a life of 12 months, compared to cotton (2 weeks) and wool (3 months).

(iii) Filtration of non-ferrous oxide fumes from furnace gases with some sulphur dioxide at 140°C. Polyester fibre cloth was used, without replacement for a period exceeding 12 months.

(iv) Abrasive dust with high moisture content. Polyacrylonitrile cloth lasted for $4\frac{1}{2}$ years.

(v) Polyamide (Nomex) filter material has been successfully used for metallurgical fumes from electric furnaces containing some fluorine compounds (mainly HF and SiF_4). Advantages are that this filament fabric operates with a filter velocity of

15 mm s^{-1}, with a pressure drop of 1·25 kPa, which is 50 per cent greater than for fibre glass. The filter cloths recommended for this application have weights of 0·105 kg m^{-2} (30 mm s^{-1} permeability) or 0·165 kg m^{-2} (80 mm s^{-1} permeability).

(vi) Polytetrafluorethylene (Teflon, FEP) monofilament fibre has been used for high-temperature sulphuric acid mists (200°C) and for chlorine gases at 200°C. The Frazier permeability is 130–330 mm s^{-1} with 120 Pa pressure drop.

Other examples of case histories have been published by Walter.[902] These include comparison of cotton and wool (both raised) as well as felts of synthetic fibres strengthened by woven fabrics.

Metals. Porous metals have been successfully used for filtration of very fine particles, such as catalysts fines.[619] More recently felts made of stainless-steel fibres with glass filament reinforcing and impregnated with polytetrafluorethylene have been used for some gasfiltration applications to temperatures of 270°C where high corrosion resistance has been combined with high porosity. The resistance of these materials with a gas flow of 300 mm s^{-1} varies from 40 Pa (0·34 kg m^{-2}) to 80 Pa (0·41 kg m^{-2}). Cloths made of stainless-steel fibres have also been produced commercially, but so far do not appear to have been used for filter bags.

8.4. FILTER-PLANT OPERATION

8.4.1. Gas-flow rates: gas to cloth ratios

The size of a filter plant is primarily determined by the area of filter cloth required to filter the gases. The theory of filtration requires a low gas velocity if diffusion is the predominant mechanism and a high velocity if inertial impaction and interception are predominant. The choice of a filtration velocity must also consider other factors. High velocities give greater particle penetration of the cloth, and make cake removal more difficult. This also tends to increase the pressure drop through the filter. Higher filtration rates, however, reduce the filter area required, and consequently smaller plants are required to handle the same volume.

Practical experience has led to the use of a series of *gas to cloth ratios* for various materials collected and types of equipment. Smaller plants tend to use higher filtration velocities than large ones, but this is probably due to the difficulty of getting even gas distributions in large bag houses, and within a very long filter sleeve.

The *gas to cloth ratio* is commonly taken as the cubic millimetre per second of gas filtered per square metre of filter area, and is expressed as millimetres per second (mm s^{-1}). The ratios used vary from about 5–125 mm s^{-1}. For fine dusts in a conventional tube or flat sleeve installation 10–15 mm s^{-1} are usual. Typical rates for a number of materials are listed in Table 8.5. For coarser materials rates up to 30 mm s^{-1} can be used, while with coarse materials which are easy to handle, such as dust from wood sanding machines, 50 mm s^{-1} can be employed in conventional plant.

TABLE 8.5. AIR TO CLOTH RATIOS IN SIMPLE BAG HOUSES

Type of dust	Air to cloth ratio (mm s^{-1})
Abrasives	10–12½
Asbestos†	12½–15
Blast cleaning	15–17½
Carbon	10–12½*
Cement—mills	7½–10
Cement—conveying and packing	10–12½
Clay	10–12½
Coal	10–12½
Feed	12½–15
Graphite	7½–10
Grinders	15–17½
Gypsum	10–12½
Lamp black	7½–10
Limestone	10–12½
Rubber	10–12½
Salt	12½–15
Sand	15–17½
Silica flour	10–12½
Soap	10–12½
Soapstone	10–12½
Talc	10–12½
Wood flour	10–12½

* For glass fibre installations lower air to cloth ratios 6·1–9·2 mm s^{-1} are used.

† In a recent large installation[710] it was recommended that 12·5 mm s^{-1} should not be exceeded.

In the reverse jet type of plant, where the period between cake removal blows is brief, higher velocities have been found practicable. Thus 30–35 mm s^{-1} are used for fine fumes, 45–90 mm s^{-1} for fine dusts, and velocities up to 110 mm s^{-1} for coarse dusts. Although the construction of these units is more complicated than the simple bag house, the high velocities have reduced the filter area required, and so this type of filter has become commercially feasible.[340]

8.4.2. Timing intermittent dust-cake removal

When a new filter cloth is first put into operation, it does not contain any dust, but after dirty gases have passed through the cloth and it has been shaken or had reverse air blown through it, some of the dust collected remains in the filter. After some time the amount of dust retained remains approximately constant, and this is called the equilibrium dust content of the cloth. This depends on the type of filter material, dust sizes, and the timing and type of filter cake-removal system being used.

If, in a single compartment filter, a constant gas velocity is maintained, the pressure drop increases with increasing build up of the filter cake. More commonly, a fan can maintain a constant pressure drop across a cake, and then the velocity decreases as the cake builds up.

It has been found experimentally, that with constant gas velocity, the pressure drop Δp_{CF}, measured in Pa, is a linear function of the total dust content of the cloth, x, measured in g m^{-2} of cloth:

$$\Delta p_{CF} = a+bx \tag{8.1}$$

where a and b are empirical constants, and x is greater than x_e, the equilibrium dust content. When the gas velocity, u_0 mm s^{-1}, varies, the pressure loss Δp is assumed to be over a limited range:*

$$\Delta p = u_0 \, \Delta p_{CF}. \tag{8.2}$$

The amount of dust collected when filtering V m^3 of gas per m^2 of filter cloth, is a function of the dust concentration, c g m^{-3}.

Therefore, from equation (8.1)

$$\Delta p_{CF} = a+b(x_e+cV). \tag{8.3}$$

Combining equations (8.2) and (8.3), and noting that the face velocity with constant pressure drop is a time-dependent function of the total gas flow V, $\mathrm{d}V/\mathrm{d}t$, the face velocity u_0 is

$$u_0 = \frac{\mathrm{d}V}{\mathrm{d}t} = \frac{\Delta p}{\Delta p_{CF}} = \frac{\Delta p}{a+b(x_e+cV)} \quad \text{mm s}^{-1}. \tag{8.4}$$

Integrating (8.4) for constant pressure drop gives a relation between time and total flow:

$$t = \frac{V}{\Delta p}(a+bx_e+bcV/2). \tag{8.5}$$

In an actual example[303] it was found that the equilibrium dust content of a cloth was 178 g m^{-2} and the relation for constant velocity was

$$\Delta p_{CF} = 0{\cdot}487\,x - 100. \tag{8.6}$$

Instantaneous and average face velocities were calculated for a filter with a dust burden of 4·5 g m^{-3} in the gases and a constant pressure drop of 0·25 kPa, and these are given in Table 8.6. The table shows the high initial velocity obtained of 64 mm s^{-1}, which decreases to 25 mm s^{-1} after $4\frac{1}{2}$ minutes and 11 mm s^{-1} after 27 min.

The results of the calculation show that if a filter cloth is used, such as a felt, which gives a high initial filtration efficiency, more frequent dust removal results in higher effective filtration velocities. Thus, if 27·5-min shaking intervals are used, the average velocity is 18 mm s^{-1}, but this could be increased to 30 mm s^{-1} with 6-min intervals or even 50 mm s^{-1} $2\frac{1}{2}$-min shaking intervals. This is one of the main reasons why the reverse jet filters, with their frequent cleaning cycles, are able to operate at much higher face velocities.

* This is valid in the laminar flow range.

TABLE 8.6. INSTANTANEOUS AND AVERAGE FILTERING VELOCITIES FOR A TYPICAL BAG FILTER INSTALLATION OPERATING AT CONSTANT PRESSURE DROP

Volume of gas filtered per m^2 of cloth (m^3)	Time required (min)	Face velocity at time t min (mm s^{-1})	Average face velocity after time t min (mm s^{-1})
0	0	64	—
3	1	43	64
6	2·4	32	51
8	4·1	26	43
11	5·2	23	34
15	8·9	18	28
18	11·8	16	26
30	27·5	11	18

Shorter filtration cycles, followed by brief shaking periods, can also be successfully applied to conventional filters. Experiments have shown that 85 per cent of the total dust in the cloth is removed in the first 5 sec of shaking, while 30 sec remove a further 3 per cent, and 2 min shaking only another 1 per cent.[811] These experiments were also able to show that the amount of dust removed varies with the position in the filter tube. More (95 per cent) is removed from near the top than from the centre down (80 per cent). This leads to uneven filtration rates and excess wear in certain parts of the filter tube, particularly near the top.

For minimum filter area and long cloth life a combination of short filtration cycles coupled with brief shaking times are therefore advantageous. If, however, filtration immediately after shaking is not very efficient, and a cake has to build up for effective performance then the large intial flow has the advantage that the cake first builds up rapidly, and an extended effective cleaning period follows before the flow rate is reduced to an uneconomically low rate. This method of operation is often used for very high temperature glass fibre filters where low filtering velocities and long cycling times are used.

8.4.3. Pressure losses in filter media

For the complete design of a filter plant, including the selection of the best fan, it is necessary to know the pressure drop through the filter medium. In many cases this is known from experiments on similar installations, or even from simple tests with a piece of filter material.

When it is not possible to obtain the pressure losses which occur experimentally, or if it is necessary to extrapolate existing results to higher temperatures, then theoretical or semi-empirical methods for calculating the pressure drop must be used. These methods are based on two approaches.

(i) The pressure drop is due to the friction of the gas moving through channels in the filter.
(ii) The pressure drop is due to the friction drag of the fibres placed in the gas stream.

The channel model is better for tightly packed beds while the friction drag on the fibres model is more applicable to loose fibre systems.

Channel theory.—The fibre beds are essentially slabs through which interconnected pores pass, and the pressure drop, Δp, is found from D'Arcy's formula:

$$\Delta p = \varkappa L \mu u_s \tag{8.7}$$

where L = depth of bed (bed thickness),
μ = gas viscosity,
u_s = superficial gas velocity,
$\varkappa$ = permeability coefficient.

This coefficient has been related to the specific surface area of the packing A and the bed porosity ϵ by Carman[149] in cases where orientation is not important, by the equation

$$\varkappa = k_1 A^2 (1-\epsilon)^2/\epsilon^3. \tag{8.8}$$

Here k_1 is a constant which was found to be 4·5 (using SI units) for spheres, and which will vary around this value for fibrous beds[268, 480, 841, 942] where an orientation factor must also be included. Thus, Sullivan and Hertel[841] have suggested that for porosities (ϵ) less than 0·88, equation (8.8) can be modified to include a ratio of a shape factor parameter k_2, which increases with porosity and an orientation factor k_3, which is unity for flow parallel to the fibres and 0·5 for flow normal to the fibres:

$$\varkappa = \frac{k_2}{k_3} A^2 \frac{(1-\epsilon)^2}{\epsilon^3}. \tag{8.9}$$

The ratio k_2/k_3 is as follows:

k_2/k_3 = 6·04 when fibres are normal to the direction of flow,
k_2/k_3 = 3·07 for fibres which are parallel to the direction of flow,
k_2/k_3 = 5·5 for the usual case of a fibre blanket where most of the fibres are normal to the direction of flow.

Fuchs and Stechkina,[286] on the basis of the Kuwabara–Happel velocity field [equation (7.5)], deduce the following permeability coefficient:

$$\varkappa = \frac{D^2\{1-\frac{1}{2}\ln(1-\epsilon)-\mathbf{c}\}}{16(1-\epsilon)} \tag{8.10}$$

where D is the fibre diameter and $\mathbf{c}$ is the Kuwabara or Happel constant.

Billings[78] also quotes the equation by Happel and Brenner[337] which is based on the solution in approximate form of the Navier–Stokes equation. In the present terminology this is

$$\varkappa = \frac{2(\epsilon-1)}{D^2}\left[\ln(1-\epsilon)+\frac{\epsilon(\epsilon-2)}{\epsilon^2+2(1-\epsilon)}\right]^{-1}. \tag{8.11}$$

Comparing experimental results with calculated values by equations (8.10) and (8.11) tend to give predicted values 20 to 50 per cent greater than found in practice.

A more recent model by Goren[788] is in better agreement with experiment, and is coincident with Davies empirical relation (8.13) for the range $0{\cdot}02 < (1-\epsilon) < 0{\cdot}06$, representing loosely packed filters, although for $(1-\epsilon)=0{\cdot}13$, the predictions tend to be high. Goren's[788] approach, whilst the most sophisticated, involves the generation of complex functions involving computer calculations, and so in this case the original paper should be referred to.

For beds of very fine fibres Pich[644] has deduced an equation based on the Kuwabara–Happel model. For Knudsen numbers ($\mathrm{Kn} = 2\lambda/D$) greater than 10 which represents fibres less than about 0·1 μm diameter,

$$\Delta P = 4{\cdot}58\mu(1-\epsilon)u_s L/D\lambda. \tag{8.12}$$

From which it can be deduced that

$$\varkappa = 4{\cdot}58(1-\epsilon)/D\lambda. \tag{8.13}$$

An empirical correlation which applies in the streamline flow region ($\mathrm{Re}_c < 1$) has been suggested by Davies[207] for the permeability coefficients.

$$\varkappa = \frac{64}{D^1}(1-\epsilon)^{\frac{3}{2}}\{1+56(1-\epsilon)^3\}. \tag{8.14}$$

This relation is based on a large number of materials, as shown in Fig. 8.13, for which the porosity is less than 0·98.

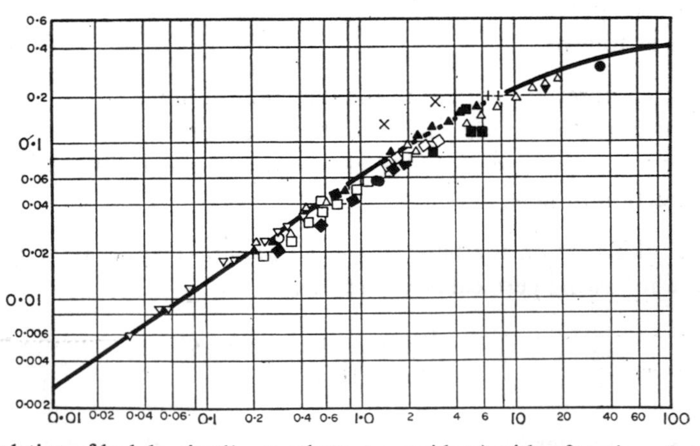

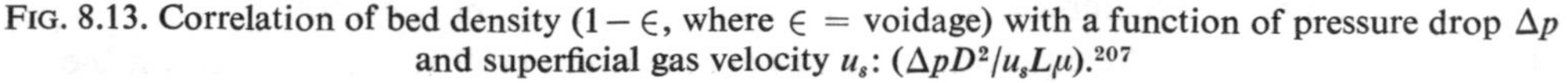
FIG. 8.13. Correlation of bed density ($1-\epsilon$, where ϵ = voidage) with a function of pressure drop Δp and superficial gas velocity u_s: $(\Delta p D^2/u_s L\mu)$.[207]

Fluid flow through fibrous material.

○ Glass wool.
● Glass wool and copper wire.
+ Glass (fibres perpendicular to flow).
× Glass (fibres parallel to flow).
△ Kapok.
▲ Merino wool.
◇ Cotton wool.
◆ Rayon.
□ Cotton wool.
■ Camel hair.
▽ Down.
▼ Glass wool.

Davies has also suggested a correlation for fibre mats with a porosity greater than 0·98, based on measurements of wool, cotton, rayon, glass, wool and steel wool pads, where the fibre size varied from 0·8 to 40 μm.

$$\varkappa = \frac{70}{D^2}(1-\epsilon)^{\frac{3}{2}}\left\{1+52(1-\epsilon)^{\frac{3}{2}}\right\}. \qquad (8.15)$$

The fibre diameter D in equations (8.14 and 15) is the effective fibre diameter. This can be found by measuring the pressure drop when gases flow through a filter pad, and it is the diameter required in the filter efficiency equations in Chapter 7. At high porosities the effective and actual mean fibre diameter is almost the same, but at low porosities, where fibres clump together, the effective fibre diameter is likely to be greater than the mean diameter.

A relation based on a regular array of fibres which was derived by Langmuir[489] is not given here as it is more difficult to apply, without gaining greater precision than is found by using Davies' equations.

The equations of Carman and Sullivan and Hertel can only be applied to very den beds, where the porosity is less than 0·95, while the empirical equations by Davies (and a Langmuir's equation) will give realistic pressure drops at higher porosities, which occ practical filters.

Fibre drag theory. Because the channel theory cannot be applied to practical gas f several formulae have been derived which consider the drag on individual fibres ir to the surrounding fibres.[156, 380] The most satisfactory of these is Chen's equatio incorporates a modification for the gas-stream interaction of a fibre in a tank of other fibres.[924] This gives

$$\Delta p = \frac{4k_4}{\pi \ln\left\{k_5/(1-\epsilon)^{\frac{1}{2}}\right\}} \cdot \frac{1-\epsilon}{\epsilon} \cdot \frac{\mu u_S L}{D_S}$$

where k_4 and k_5 are constants, which were found experimentally to be 6· tively, and D_S is the mean fibre diameter based on the surface area of th

The effect of absolute pressure on the pressure drop in a filter h experimentally by Stern *et al.*[816] at pressures between 1·7 and 100 kPa ical relation held to a high degree of accuracy in this range and can p to pressures somewhat greater than atmospheric:

$$\Delta p = \frac{u_S\, 0{\cdot}312}{1+1{\cdot}375/P}$$

where Δp = pressure drop in kPa,
P = absolute pressure in kPa

and the velocities, u_S, were less than 1 m s^{-1}.

An *empirical permeability* is quoted by many manufacturers as a characteristic of filter materials. This, as was stated on p. 329, is the number of cubic metres per minute of air able to pass through a square metre of cloth with a pressure drop of 125 Pa. It should be noted that while this is an important factor in filtration, it is not the essential measure of the suitability of the cloth. Cock and Ferris[163] point out that, for example, a 40×30 (warp $\times$ weft) woollen system fabric may have the same permeability as an 85×67 cotton system fabric, i.e. 175–200 mm s^{-1}. The lower count fabric has only $1{\cdot}9 \times 10^6$ openings per square metre compared with $8{\cdot}8 \times 10^6$ openings per square metre in the second fabric, and so the latter is much better in the collection of small non-agglomerating particles.

8.4.4. Electrostatic charges in fibre filtration

The theory of the effect of coulombic and image charges was discussed in Chapter 7. The practical effects of these charges in actual gas filters has not been investigated in detail until recently.[273]

In the case of air-conditioning filters and other filters where the collected material is retained in the filter fabric, the problem of particle release on shaking or blowing an air jet does not arise, and the fact that the charged particles are firmly held to the fibres adds to the usefulness of the filter. In the case of an industrial filter, where the collected particles have to be frequently removed, the strong electrostatic attractive forces between particles and fibre prevent particle release, and hinder filter operation by assisting the rate of plugging of filter pores.

The type and quantity of charge acquired by a filter medium is a function of the filter type and the method of charging. The rate at which a fabric loses its charge is also an important consideration. This not only depends on the conductivity of the fibres but also on the humidity of the gases passing through the filter. Thus fabrics which are poor conduc-ors retain a charge much longer than good conductors, while in humid conditions a fabric cquires a surface film of moisture which also acts as a conductor.

Charges are induced in fabrics by friction, and the type and extent of charging that a par-ular material acquires relative to others can be measured by charging a series of materials he same way. The usual technique consists of placing a strip of the material on an insu-d ring and rubbing it with a strip of the reference fabric which is mounted on an insulating ing disc.[273] The charge on the test strip is measured after a standarized number of turns e charging disc, and again after a period (frequently 2 min) to find the rate of charge ge. The maximum charge, measured immediately after charging enables the materials placed in relation to one another in a *triboelectric series* (Table 8.7).

t particles can be arranged into a similar series of positions relative to one another filter fabrics, and this can assist in the selection of filter fabrics with the most favour-arge characteristics for both particle collection and particle release, if the particles e removed by shaking, vibration or blowing.

es can be allotted to one of three categories.[5] Those which acquire a charge and do merate (Class I); those that acquire a charge and agglomerate (Class II), these active classes; and those which are not affected by the charge on the filter (Class

TABLE 8.7. TRIBOELECTRIC SERIES FOR FABRICS[273]

Volts	Fabric
Positive	
+25	
	Wool felt
+20	
+15	Glass filament, heat cleaned and silicone treated
	Glass spun, heat cleaned and silicone treated
	Wool, woven felt
+10	Nylon 66, spun
	Nylon 66, spun, heat set
	Nylon 6, spun
	Cotton sateen
+ 5	Orlon 81, filament
	Orlon 42, needled fabrics
	Arnel, filament
	Dacron, filament
	Dacron, filament, silicone treated
0	Dacron, filament M.31
	Dacron, combination, filament and spun
	Creslan, spun: Azoton spun
	Verel, regular, spun: Orlon 81 spun (55,200)
	Dynel, spun
−5	Orlon 81 spun
	Orlon 42 spun
	Dacron, needled
−10	Dacron, spun: Orlon 81 spun (79475)
	Dacron, spun and heat set
	Polypropylene 01, filament
	Orlon 39B, spun
−15	Fibraryl, spun
	Darvan, needled
	Kodel
−20	Polyethylene B filament and spun
Negative	

Polystyrene, Saran and Vinyon are at the far negative end of the series[771].

III), being inactive. The active groups are divided into finite and coarse particles. Coarse particles do not present a problem in filtration as they are easily collected on the surface layers of the fabric, usually form a loose cake, and are easily shaken off. Fine particles are much more difficult to collect because they tend to penetrate the filter medium and often leak through. By selecting a highly charged filter medium, fine particles in Class II will be agglomerated, their collection improved, and they should form a loosely agglomerated cake on the fibre filter surface. If the filter has a high rate of charge loss under these conditions, cake release will also be assisted.

Several dusts in the various categories have been experimentally investigated,[273] and their performance is listed in Table 8.8.

When filters are used for air conditioning, it is usually important that the fibres retain their charge for a long time, as the filters are only replaced at long intervals. This led to the early development of resin-impregnated wool filters which gave improved performance with only a small rise in pressure drop.[708] Other methods of obtaining a persistent charge in a filter medium were the impregnation of fibres with polystyrol, the covering of glass fibres with polystyrol or polyethylene, or the use of shredded polyethylene.[239, 913]

TABLE 8.8. RELATIONS OF FABRIC REQUIREMENTS TO DUST PROPERTIES AND DUSTS IN THE CATEGORIES LISTED[273]

Dust classification:	IA	IB	IIA	IIB	III
Relative particle size	Fine	Coarse	Fine	Coarse	Fine and Coarse
Electrostatic properties	Active	Active	Active	Active	Inactive
Agglomerating tendencies	Little or none	Little or none	Positive	Positive	—
Criteria for filtration: Leakage	P_x	Const[1]	P_x to P_c[2]	P_x to P_c[2]	Fabric Construction dictates performance
High Flow low Δp	Const[1]	P_m	P_x	Px	
Criteria for cleaning: Leakage control	D_e	Const[1]	Px to P_c[2]	P_x to P_c[2]	
Ease of cake removal	D_h	D_h	D_h	D_h	
Material	Calcined Calcium Silicate	Flux Calcined Diatomacious Earth; Commercial Finished Cement; Ball Clay	Processed Natural Diatomacious Earth; Wheat Starch; Taconite; Zinc oxide fume; Nickel furnace fume; Magnesite; Cellulose Acetate; Molybdic oxide; Sugar	Carbon SRF	Kaolin

1. Fabric construction determines property.
2. Requires low density, rapidly agglomerating dust forming large aggregates.

P_x = maximum P.D. dust and fabric. P_c = controlled P.D. dust and fabric. P_m = minimum P.D. dust and fabric
D_h = high rate of charge dissipation. D_e = low rate of charge dissipation.
[P.D. = Potential Difference]

It was assumed by Endres and van Orman[238] that these impregnated materials were self-charging. Subsequent experiments by Silverman *et al.*[771] have shown that generally mechanical friction is required, because a clean gas stream does not charge a filter, while the friction from small particles passing through the filter is not adequate. Filter media which are in themselves non-conducting can become charged when charged particles are deposited on them. Electrical fields are formed all over the surface of the filter, with concentrations at places where, due to filter weaknesses, a larger proportion of the gas stream passes. This may have been the basis of the "self-charging" filter.

Mechanical charging of filters (Fig. 8.14) has been tried by Silverman *et al.*[771] These workers achieved some increase in filtration efficiency with no increase in pressure drop. However, the improved efficiency obtained was not comparable with the efficiencies of the "absolute" filters required for radioactive waste collection. It was also found that when air

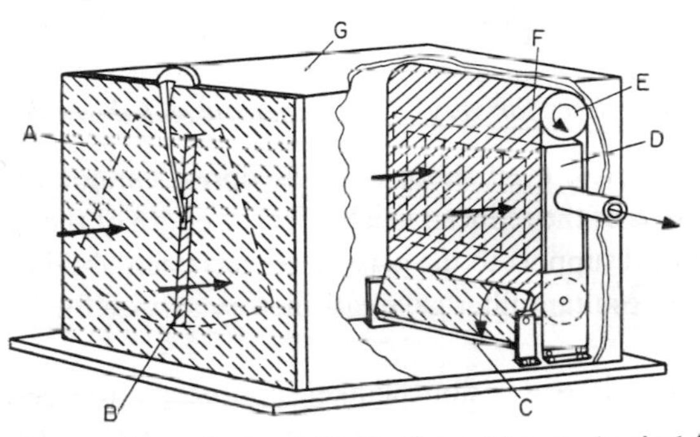

FIG. 8.14. Two-stage filter electrostatically charged by mechanical friction.[771]

A: Fabric A screen.
B: Fabric B covered windshield wiper blade.
C: Fabric A covered paddle.
D: Lucite box.
E: Lucite roller.
F: Fabric B belt.
G: Masonite box.

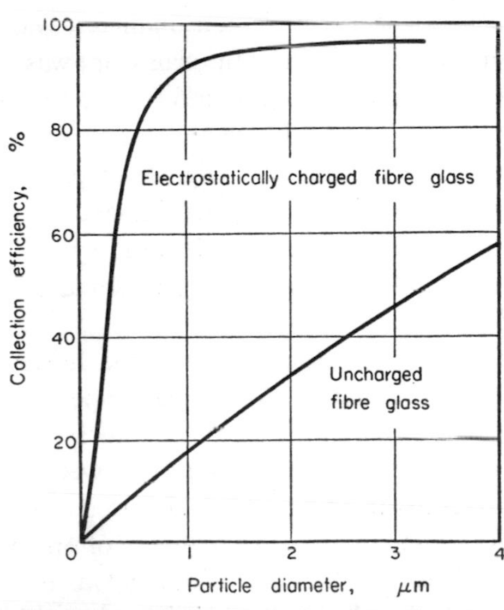

FIG. 8.15. Comparative fractional efficiency curves for charged and uncharged fibre glass filters.

with an absolute humidity exceeding 17 g water per kilogramme of dry air was passing through the filter, the filter would not retain a charge sufficiently long to be effective.

Electrostatic charging has been introduced in some bonded fibre glass filter media in commercial production, recommended for air-conditioning installations. The efficiencies for uncharged and charged media with the same type of fibre are shown in Fig. 8.15.

8.5. FIBRE FILTERS FOR SPECIAL APPLICATIONS

Fibre filters of special designs are used for a number of industrial and commercial applications which have not been covered in the discussion of bag houses. These are for use at very high temperatures, for decontamination of radioactive waste gases, for collection of mist droplets and for air conditioning installations.

8.5.1. *Filters for very high temperatures (above 400°C)*

Conventional filter sleeves, even those of glass fibres, cannot be used at temperatures above 350–400°C because of the reduced mechanical strength of the cloth at these and higher temperatures. Mechanical support to the glass fibres has been obtained by winding the fibres on a metal former. The resultant filter gave satisfactory service for long periods at temperatures of 400°C.[687]

An alternative approach is to use a fibre blanket supported on a steel mesh, which is continuously replaced when an optimum amount of dust and fume has been collected. This type of filter has been extensively investigated by Silverman *et al.*[80, 81, 82, 83, 84] at the Harvard University Air Cleaning Laboratory for open hearth furnace fumes, fly ash, acid gases and mists. Spun blast furnace slag, where 50 per cent of the fibres were less than 5 μm, 90 per cent less than 10 μm and 99 per cent less than 30 μm diameter, was used in filter beds from 10–50 mm thick. The chemical composition of the slag wool was: SiO_2, 40 per cent; Al_2O_3, 10 per cent; CaO, 39 per cent; MgO, 8 per cent and Fe_2O_3, 1 per cent.

Laboratory studies were carried out with temperatures ranging from 320° to 650°C with gas velocities between 500–1100 mm s^{-1}. High fibre packing densities and high gas velocities favoured collection, indicating that inertial impaction is the controlling collection mechanism. With favourable conditions in the laboratory collection efficiencies better than 90 per cent were consistently obtained while in some cases efficiencies as high as 97–98 per cent were noted.

Extension of this work to the continuous pilot plant scale, however, met with considerable difficulties and collection efficiencies tended to average 60 per cent (by weight). In the first design used (Fig. 8.16) the slag wool, in the form of a slurry, was fed on to a continuous steel chain conveyer belt, where it was drained, dried to a web by the hot clean gases from the cleaning section of the filter in one section, and then used as the filtering medium for the dirty gases. A later model was based on the same sequence of operations but used a rotating disc instead of the conveyer belt (Fig. 8.17). This model was even less efficient, averaging 44 per cent. This was improved, by placing the unit in series with a screw conveyer agglo-

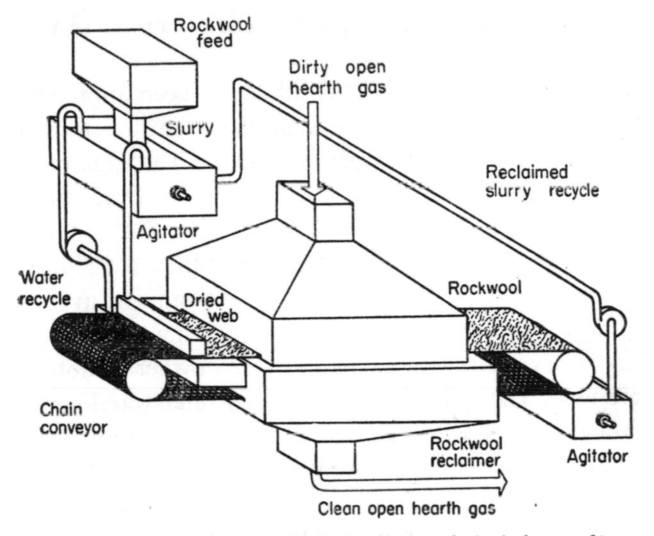

FIG. 8.16. Continuous slag wool filter—chain belt type.[84]

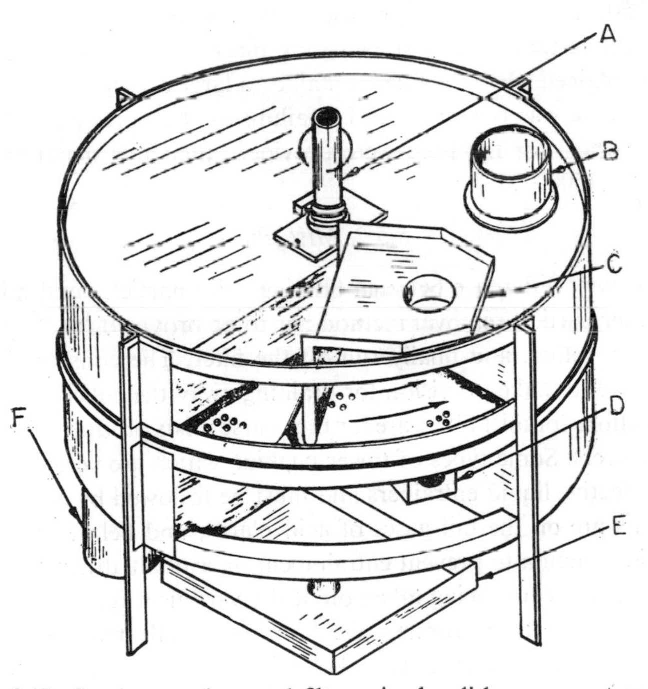

FIG. 8.17. Continuous slag wool filter—circular disk conveyor type.[82]

A. Shaft on which disc of 10 gauge perforated steel sheet is mounted.
B. Point of entry for furnace gases.
C. Tray from which slurry is introduced to disc.
D. De-watering suction box.
E. Sump for water.
F. Exit for clean gases.

merator, to 60 per cent. Stationary beds gave efficiencies similar to those which had been obtained in the laboratory.

The difficulty of cleaning the stationary bed filter has been approached by using a low-pressure shock wave operating at sonic velocities to shake the dust from the fibres. The shock wave is generated by the explosion of a paper diaphragm. The agglomerated material, after being shaken off the fibres is re-entrained and collected by a mechanical collector, for example a cyclone.

Billings and Silverman[79] tested a stainless-steel wool filter on a pilot-plant scale attached to a 400-tonne openhearth furnace. The bed depth was 50 mm and average filtering temperature was 65°C, with efficiencies as high as 96 per cent, using a filtering velocity of 500 and 750 mm s^{-1}. In laboratory tests with carbonyl iron powder combustion fume temperatures to 320°C were used. The resistance of the clean filter was low (below 0·5 kPa) but although final resistances of 8·8 kPa were usual, the construction of the filter made these possible. The filters were cleaned by shock waves created by bursting a diaphragm, and the dislodged fume was collected in a cyclone with efficiencies at 80–95 per cent.

Other studies in fly ash collection at temperatures to 980°C have used an aluminium silicate fibre which melts at 1750°C.[422] Collection efficiencies up to 90 per cent were achieved. Finer fibres and higher fibre bulk densities tended to give better collection. Very high velocities, between 750–3500 mm s^{-1}, were used, and the rising gas velocities gave reduced efficiencies, suggesting that re-entrainment took place at these velocities. Similarly, lower efficiencies were obtained when the filter contained a large amount of dust. The temperature limitation in this case was not imposed by failure of the fibre, but of the fibre support. If a refractory support for the fibre were provided, then this material could be used to temperatures up to 1500°C.

8.5.2. Mist filters

The most important difference between filtering solid particles and mist droplets is that no shaking or other particle removal method has to be provided for the latter because they agglomerate, run together, and finally run off the filter. The construction of a mist filter does, however, have to make provision for draining away the collected liquid.

Typical applications for mist filters are for the coarse mists produced in packed absorption and distillation towers. Some types of tower packing which are very effective liquid distributors are also effective liquid entrainers and must be followed by efficient mist collectors. Other applications are on the tail gases of acid plants, and below the bottom plate of a vacuum distillation column to prevent entrainment of asphalt droplets which would contaminate the product streams being taken off at the various stages of the column.

Mist eliminators for coarse droplet collection are usually made of woven metal wire, or more recently, polytetrafluorethylene. The wire is first woven and then crimped to a standard width strip of about 10–15 cm. The strip is then coiled into circular sections or other suitable shapes. For installation within distillation columns the sections are made so that they can be fitted together in the column, and readily removed for cleaning (Fig. 8.18). Various wire winding patterns are used to give mesh densities suitable for different operating conditions. Typical specifications are given in Table 8.9.

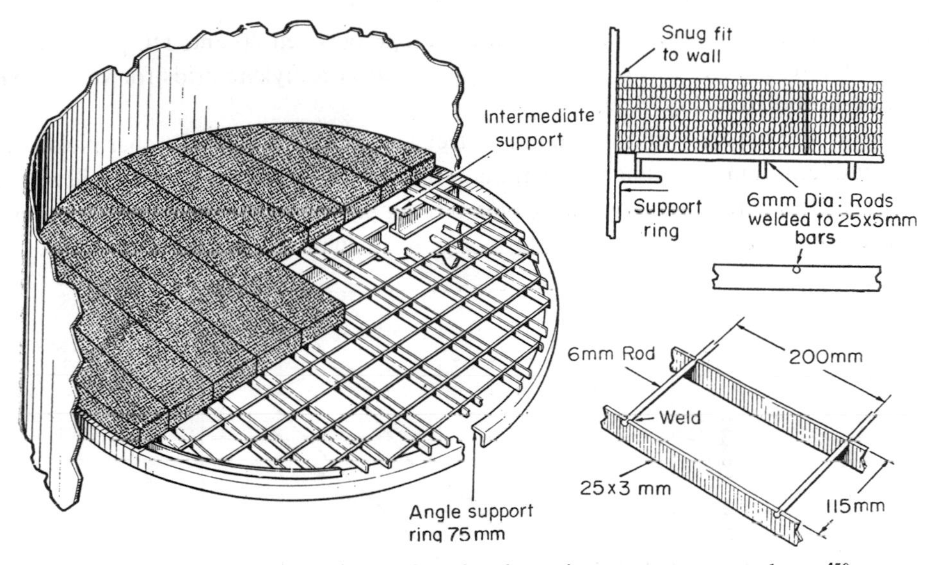

FIG. 8.18. Knitted metal wire demister for absorption tower or spray column.[450]

TABLE 8.9. WIRE MESH MIST ELIMINATOR SPECIFICATIONS

Application	Free volume per cent	Surface area (m^2 m^{-3})	Density (kg m^{-3}) (steel)
General purpose	98	330	160
Moderate velocities and clean liquids	97·5	390	190
High velocity, dirty liquids	99	195	95
High velocities, dirty liquids and particles	98·5	230	110

When droplets are to be collected at the top of a column a two-stage mist eliminator is recommended practice. The lower stage is a high density mesh (190 kg m^{-3}) which acts as droplet agglomerator while the upper stage is a low-density mesh (95–110 kg m^{-3}), which collects the enlarged mist droplets. For effective agglomeration, flooding conditions should exist in the lower mesh. This helps to scrub the gases and increase the droplet velocity, assisting collection by inertial impaction on the upper mesh. The two stages are separated by a distance of about three-quarters of the tower diameter.

The optimum design superficial vapour velocity to be used can be calculated from the liquid and vapour densities using an equation based on the Souders–Brown equation.

$$u_S = 0{\cdot}11 \sqrt{\left(\frac{\varrho_L - \varrho_V}{\varrho_V}\right)} \tag{8.18}$$

where u_S = superficial vapour velocity (m s^{-1}),

ϱ_L = liquid density kg m^{-3},

ϱ_V = vapour density kg m^{-3}.

Operating vapour velocities should be in the range between 30 and 110 per cent of the optimum. Pressure losses through steel and polytetrafluorethylene grids for various velocities can be found from Fig. 8.19.

The meshes are made of mild and stainless-steel wire as well as Monel, other nickel alloys, titanium and tantalum alloys and polytetrafluorethylene-monofilament, which will withstand 200°C at very low rates of corrosion. The mesh is supported on stainless steel or other steel bars, but these can be made of other alloys or polyvinylchloride-coated metals.

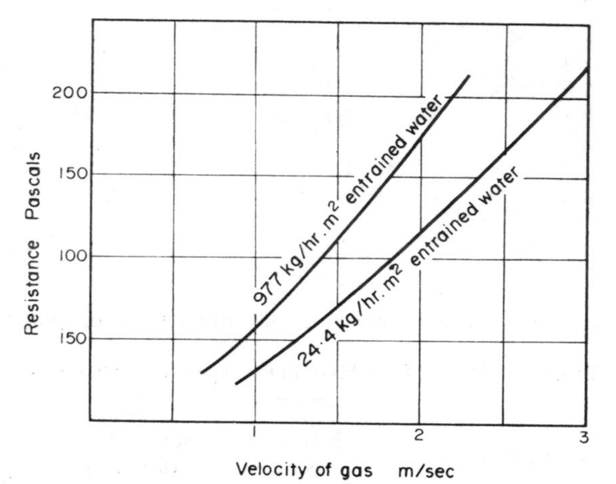

FIG. 8.19. Resistance–gas flow characteristic for 100 mm thick wire mesh demister (190 kg m^{-3}) at standard conditions.[450]

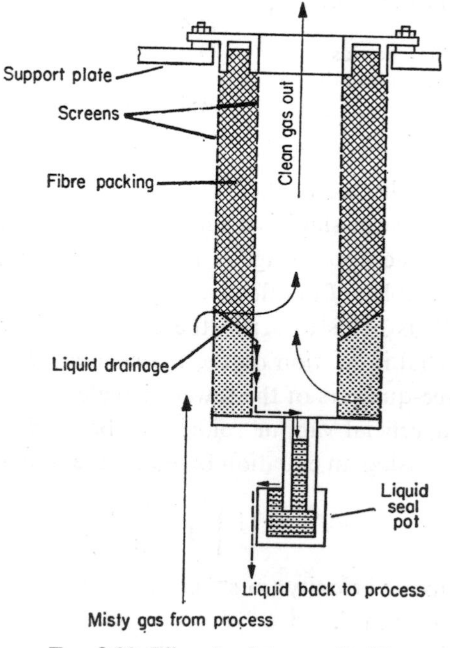

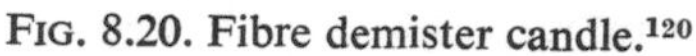

FIG. 8.20. Fibre demister candle.[120]

A typical application for a wire mesh mist eliminator made of Hastelloy C (composition per cent: 54 nickel, 15·5 chromium, 16 molybdenum, 4 tungsten, 2·5 cobalt, 5 iron) was on the tail gases of a contact sulphuric acid plant. With vapour velocities of 4·5–5·5 ms^{-1}, the acid concentration was reduced to a satisfactory 0·03–0·06 g m^{-3} at a pressure drop of 370–500 Pa.[556]

When the droplets are much finer than in the above applications, a more efficient mist collector is required which uses finer fibres than wire mesh. Special filters packed with silicone-treated glass fibre or polyester fibre have been developed[120, 250] (Fig. 8.20). The filter

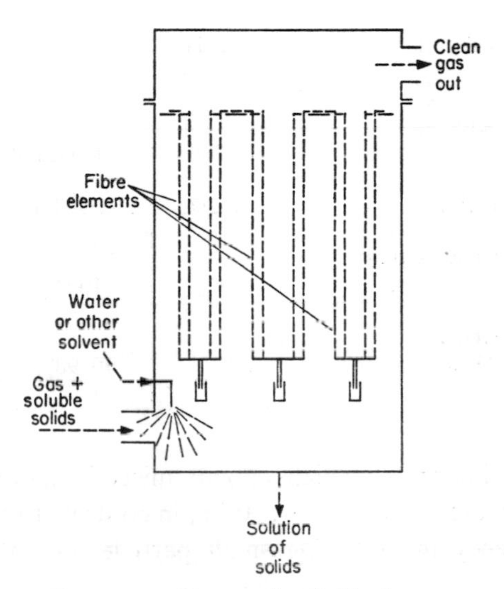

FIG. 8.21. Fibre demister candles arranged in vessel suitable for pressure or vacuum operation.[120]

"candles" can be made by wrapping the fibre on a frame or packing in a double-walled framework, which can be made of steel, PVC, or steel coated with PVC. Figure 8.21 shows the arrangement of these units in a tank for service under pressure.

The filter candle-type mist eliminator has proved satisfactory for service at elevated and high pressures, with corrosive gases. Thus, the eliminator has been used for removing organic and inorganic mist droplets from "dry" hydrochloric acid gas at 170 kPa before this was compressed in a liquid ring compressor in a recycle system. Great savings in compressor maintenance resulted.[27] In another high-pressure (1500 kPa) application, oil mist droplets have been removed from air following the third stage of a reciprocating compressor before an air heater. The air, leaving the compressor at 120°C, presents an explosion risk if the oil droplets are allowed to remain during heating to 315°C.[519] This system has also proved satisfactory in sulphonation and chloronation plants and in processes involving nitric acid process gases and methanol synthesis gas.[121]

A high-velocity fibre-pad mist eliminator has also been developed by Brink *et al.*[122, 519] The packing is supported by wire-mesh frames and these are set horizontally around a polygon. In contrast to the filter candles, where superficial velocities of 75–200 mm s^{-1} are used, the velocities in this collector range from 1500–2500 mm s^{-1} with a pressure drop of about 2 kPa. Higher pressure drops (to 3 kPa) are experienced with phosphoric acid mists in contrast to sulphuric acid mists. The operating characteristics of these units are summarized in Table 8.10.[122]

TABLE 8.10. OPERATING CHARACTERISTICS OF FIBRE MIST ELIMINATION OF SULPHURIC ACID PLANTS[122]

	High efficiency	High velocity	Spray catcher
Controlling collection mechanism	Brownian diffusion	Impaction	Impaction
Superficial velocity mm s^{-1}	75–200	2000–2500	2000–2500
Pressure drop (kPa)	1·25–3·75	1·5–2·0	0·12–0·25
Efficiency on particles greater than 3 μm	100%	100%	100%
Efficiency on particles smaller than 3 μm	95–99%	90–98%	15–30%

The spray catcher[122] is similar to the high-velocity mist eliminator, but with different packing density achieves pressure drops of 120–250 Pa, in contrast to the high-velocity eliminator. However, its efficiency in collecting small particles (less than 3 μm) is only 15–30 per cent.

If the collected material contains solid particles as well as mist droplets, the candles can be irrigated with individual water sprays directed into each unit to wash the particles off the fibre.

The efficiency of a filter pad was found to be better than 99·5 per cent for the tail gases from a sulphuric acid plant,[250] and the exit concentrations obtained were about 1·0 mg m^{-3}. Other industrial plants gave the following performances:[607]

(i) *Sulphuric acid mist from a Kachkaroff acid plant.* Inlet concentration of 0·25 g m^{-3} was reduced to 0·004 g m^{-3} sulphur trioxide, being an efficiency of 98·6 per cent with 1·75 kPa pressure drop. The plant needed no maintenance because of filter deterioration for over $1\frac{1}{2}$ years.

(ii) *Sulphuric acid mist from a contact acid plant.* Inlet concentration of 0·175 g m^{-3} sulphur trioxide was reduced to 0·0025 g m^{-3}, an efficiency of 98·5 per cent. Pressure drop built up from 2·25–2·75 kPa over 9 months due to deposition of insoluble particles on the fibre.

(iii) *Sulphuric acid mist from calciners.* Inlet concentration of 0·090 g m^{-3} sulphur trioxide reduced to 0·007 g m^{-3}, an efficiency of 92·1 per cent. The mist was exceptionally fine, and a pressure drop of 1·25 kPa was used in the pilot plant.

Note. Ceramic filters which have been normally used for this application tend to block up after about 3 months and have to be replaced.

(iv) *Phosphoric acid mist.*[119] Fibre mist eliminators reduce inlet concentration of 39 g m^{-3} (15°C) to 0·69 mg m^{-3} (15°C) on a plant producing 34,000 m^3 h^{-1} waste gases. In general, the mist eliminators remove 99 per cent of the fumes below 3 μm and virtually 100 per cent above 3 μm.

Other applications

(a) *Sulphuric acid plants: absorbing towers*

For new plants, the selection of mist eliminators may be along the following guidelines:[122]

(1) High velocity (Brink) type: straight sulphur-burning plants and ore-roasting or smelting plants, without oleum manufacture.
(2) High efficiency (Brink candle) type: any kind of plant with a by-pass oleum system or a full oleum system producing 25 per cent oleum. Any kind of a plant which includes an SO_3 boiler system heated by converter gases. Spent acid regeneration and wet gas plants.

(b) *Sulphuric acid plants: drying towers*

The entrainment from drying towers is not quite so difficult to collect, and with sulphur-burning plants, the spray catcher (low-pressure drop type) is adequate. Except in special applications such as spent-acid regeneration, the high-velocity eliminator performs satisfactorily, e.g. ore roasting.

(c) *Phosphoric acid mist*

Plants burning elemental phosphorus making orthophosphoric acid. A mist eliminator consisting of a two-stage design, similar to the wire-mesh types, but packed with materials of low mechanical stiffness (tetrafluorethylene, Dacron and polypropylene fibres) has been tested on pilot plant scale. This gave a collection efficiency better than 99·96 per cent and an outlet concentration of 0·11 g m^{-3} of 100 per cent H_3PO_4 at a pressure drop of 10 kPa. Full-scale tests showed that with gas velocities through the agglomerator/collector of 6·7–8·5 m s^{-1}, outlet concentrations tended to be below 0·03 g m^{-3}.[186]

8.5.3. *Filters for decontamination of radioactive gases*

Radioactive particles are produced at all stages of treatment of the radioactive ores; mining, milling. refining, fuel element fabrication, and in atomic reactors. While the radioactive particle concentrations are comparatively dilute during the mining and milling stages, and conventional gas-cleaning equipment, carefully applied, is adequate, when the radioactive materials are concentrated in fuel elements, and in reactors, very great care has to be taken as not only are the small particles from the fuel elements radioactive, but dust particles from the atmosphere also become contaminated. It therefore becomes vital to clean carefully not only the hot gases from gas-cooled reactors, but also the ordinary air which is used to condition reactor buildings. These protective filters also have to protect the reactor building surroundings in case of accidents. The air for radiochemical laboratories, where high concentrations of radioisotopes are being treated, also has to be carefully cleaned for the same reasons.

These topics are dealt with in detail by White and Smith[935] and Keilholtz and Battle.[430] The filters for radioisotopes must have the following specifications:

(a) The collection efficiency must be of the order of 99·99 per cent for sub-micron-sized particles, as it has been found that the radioactive particles are predominantly in the range 0·2–0·7 μm.
(b) Low initial flow resistance.
(c) Minimum maintenance requirements during operation.
(d) A life expectancy in years.
(e) High resistance to fire.
(f) Containment of materials which will not cause a secondary disposal problem.

It is usual to refer to equipment performance for efficiency of capture of radioactive particles in terms of the *decontamination factor* (D.F.) rather than efficiency, where D.F. = 1/(1—efficiency) A D.F. of 10 is equivalent to a particle capture efficiency of 90 per cent, while a D.F. of 1000 is equal to an efficiency of 99·9 per cent. D.F. values of 10^6 are frequently required for applications in cleaning waste gases from reactors.

For cleaning the air entering reactor buildings, and similar applications, the Esparto grass–asbestos fibre filters have been developed, and are discussed in section 8.5.4 in connection with air-conditioning filters. For reactor gas cleaning deep beds of asbestos and glass fibres are in use. These give innumerable surfaces for inertial impaction and diffusion.

The initial practice of the United Kingdom Atomic Energy Authority was to use canisters 330 mm diameter, 1·4 m long, filled with a mixture of asbestos and wool fibres, each having a throughput of 60 $m^3 h^{-1}$ with a pressure drop of 250 Pa. These had a methylene blue test efficiency of 99·99 per cent. However, the small volume flows led to the development of two-stage filters (Fig. 8.22) filled with cotton and asbestos, with a fibre-glass prefilter, with a throughput of 350 $m^3 h^{-1}$ and subsequently to a three-stage unit, consisting of a pre-filter and two absolute fibre glass filters, with a throughput of 1700 $m^3 h^{-1}$, at a total pressure drop of 675 Pa. This type has a methylene blue efficiency of 99·997 per cent, and floor space requirements of only 0·6 m square.[584]

United States practice has led to the use of deep bed fibre glass filters, for which the decontamination factor (D.F.) can be calculated from the empirical equation:[90]

$$\text{D.F.} = \mathscr{C} L^a \varrho_F^b u_s^c \tag{8.19}$$

where $\mathscr{C}$ = constant,
L = depth of bed (mm),
ϱ_F = fibre bed density (kg m^{-3}),
u_s = superficial velocity (m s^{-1}),
a, b and c = constants (Table 8.11).

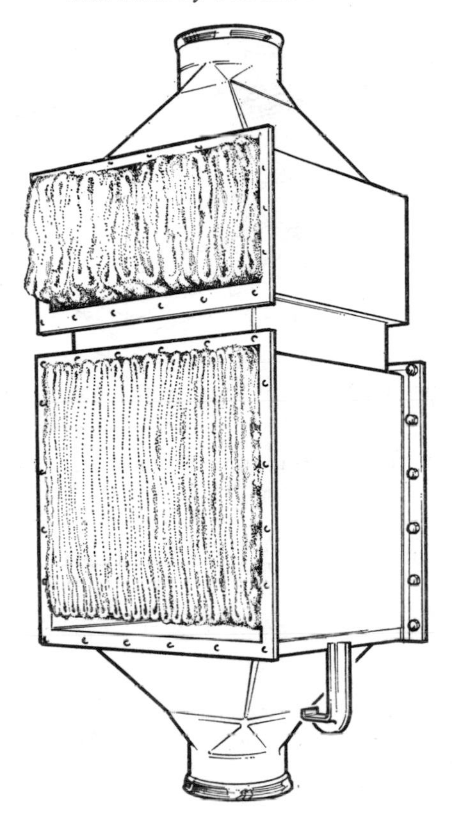

FIG. 8.22. Two-stage glass fibre, cotton asbestos filters.[584]

The empirical constants have been determined experimentally and are given in Table 8.11.[91]

Two types of multilayer deep-bed filters have been in use, one for venting a process vessel, and the other for decontamination of effluent ventilation. The process vessel venting filter (Fig. 8.23) was designed for a flow volume of 400 $m^3 h^{-1}$, and for an efficiency of 99·99 per cent with a pressure drop of 1·0 kPa.

TABLE 8.11. EMPIRICAL CONSTANTS FOR CALCULATION OF DECONTAMINATION FACTORS FOR FIBRE-GLASS FILTERS[91]

Fibre classification	Fibre size (μm)	e	a	b	c
AA	1·3	2·69	0·9	1·0	−0·2
B	2·5	—	—	—	−0·25
55	15	0·0131	0·9	1·1	−0·4
115K	30	0·0145	0·9	0·9	−0·4
450	115	—	—	—	−0·5

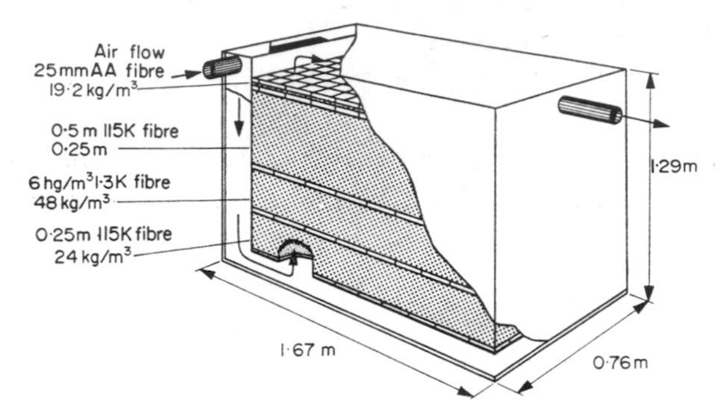

FIG. 8.23. Vessel vent filter for use on vessels with radioactive process materials.[91]

TABLE 8.12. PREDICTION OF COLLECTION EFFICIENCY OF FIBRE GLASS FILTER FOR RADIOACTIVE VENT GASES[91]

Layer	Fibre class	Fibre size (μm)	Packing density (kg m^{-3})	Depth (mm)	Pressure drop (Pa)	Predicted efficiency (%)
Bottom	115K	30	24	304	25	39
Second	115K	30	48	254	60	53
Third	115K	30	96	508	335	93
Top	AA	1·3	19·2	25	550	99·9
				1091	970	99·9

The design calculation is shown in Table 8.12.

Evaluation of this filter in service indicated that efficiencies were better than 99·9 per cent probably near the predicted value of 99·99 per cent.[91]

The filter for ventilation air consisted of 2·1 m of freely packed 115 K fibre, followed by two 12-mm sections of B and AA fibres in series (Fig. 8.24). Efficiencies of 99·9 per cent,

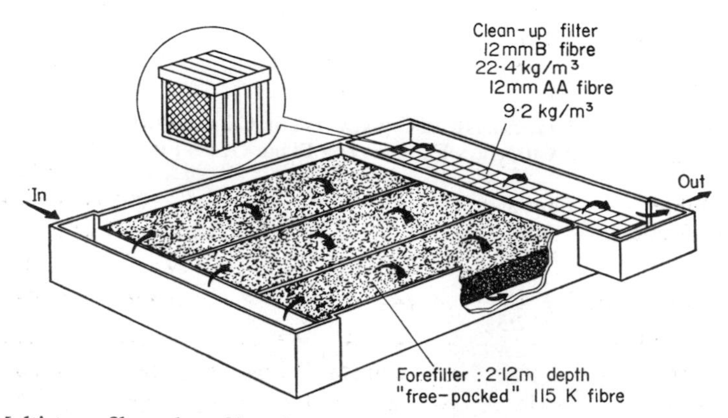

FIG. 8.24. Multistage fibre glass filter for ventilation air for radioactive process operations.[91]

with a 1·0 kPa pressure drop were predicted for this unit, and experimentally the radioactivity counts gave efficiencies of 99·84 per cent.

The late Dr. Silverman at the Harvard Air Cleaning Laboratory developed the Harvard Diffusion Board.[769] This consisted of two glass paper filters supported on an expanded metal grid and separated by a honeycomb of a flameproof material. The spaces of this honeycomb were filled with an adsorbent such as silvered silica gel or activated carbon. This combination was very successful in particulate removal, achieving efficiencies of better than 99·999 per cent, representing penetrations of 10^{-4}. In the removal of radioactive iodine vapour, silica gel gave efficiencies of 99·82 per cent, activated carbons, 95 per cent, while unfilled filters were as low as 22 per cent.

8.5.4. Filters in air conditioning

The removal of atmospheric dust from air is an essential part of air conditioning. The degree of cleanliness required varies with the application. Ordinary office and commercial air conditioning require less rigid standards than the air used in rooms for certain precision engineering operations or the manufacture of pharmaceuticals and photographic film. The amount of dust encountered in the atmosphere varies with the place, and some typical dust concentrations are given in Table 8.13.

TABLE 8.13. ATMOSPHERIC DUST CONCENTRATIONS

Locality	Concentration mg m^{-3}
Rural or outer suburban	0·4 0·8
Commercial	0·8–1·5
Light industrial	1·0–1·8
Heavy industrial	1·5–3·0
Dusty workplaces	Over 3

The atmospheric dust concentrations which are removed by the filters are much lower (about one ten-thousandth part) of the dust and fume concentrations encountered in industrial waste gases, and so it takes a very long time for an appreciable dust load to collect in the filter.

It is therefore one practice to build a filter installation of a number of small units which can be readily replaced at intervals which vary from 1 to 12 or more months, depending on filter type and dust conditions. The used filters may be of the type that can be cleaned for further use or may be discardable. In other installations, particularly useful for heavier dust burdens, provision is made for more frequent automatic replacement of the filter surface, using either a continuous cleaning bath, or a new section of cloth unwound from a roll. A low-pressure drop is essential for air-cleaning filters because very large volumes are handled, and the power consumption costs have to be kept to a minimum.

Instead of fibre filters, positive corona electrostatic precipitators are often used (Chapter 10). These units are greater in initial cost than conventional filters, but do not require the elements to be replaced as they can be washed down at intervals to remove the accumulated dust on the plates.

Air-cleaning filters are made of a number of materials: metal wire or turnings, glass or chemical fibres, asbestos or paper. The fineness of the fibre determines the filter performance. Thus, metal turnings collect only comparatively coarse particles while asbestos fibre filters are effective collectors of particles in the sub-micron size range.

Metal wire is woven into gauze, chevron crimped, and layered, so that the direction of crimping is reversed in successive layers. The wire is retained in a 5 cm wide metal surround by an expanded wire grid. When metal turnings are used these are simply supported between the wire gauze walls. The most common size of the cells is 0·5 m square (Fig. 8.25),

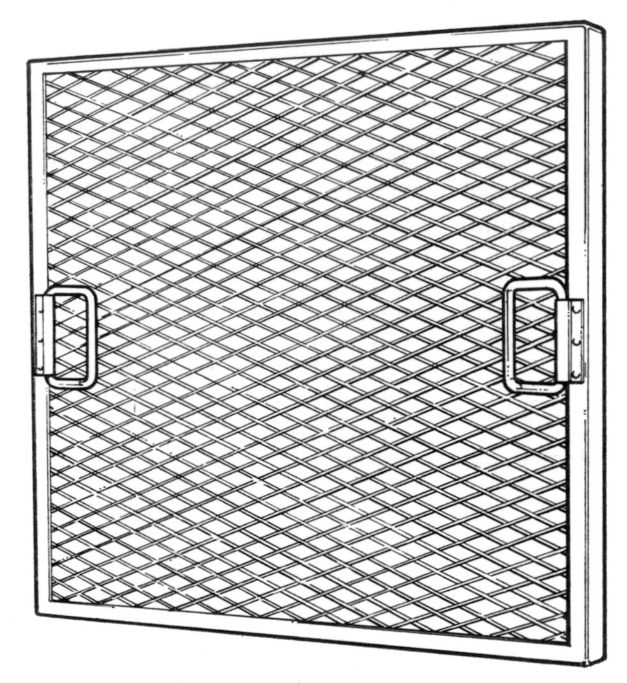

FIG. 8.25. Viscous filter cell.

although other sizes varying from $0{\cdot}3 \times 0{\cdot}3$ m to $0{\cdot}5 \times 0{\cdot}8$ m are made. A 0·5 m square filter cell weighs about 11 kg, and the cells are clipped into sectioned frameworks.

Face velocities between 1500 and 2800 mm s^{-1} are used, and a velocity of 2200 mm s^{-1} giving a throughput of 0·6 $m^3 s^{-1}$ for a 0·5 m square cell is recommended by most manufacturers. Resistance to gas flow varies with flow rate and the dust load on the filter. A resistance–flow rate characteristic for a clean filter is shown in Fig. 8.26a, while the variation of resistance with dust load is shown in Fig. 8.26b. The dust used in this experiment was a mixture of fly ash and lamp black, which is recommended for filter testing by the American Society of Heating and Ventilating Engineers.

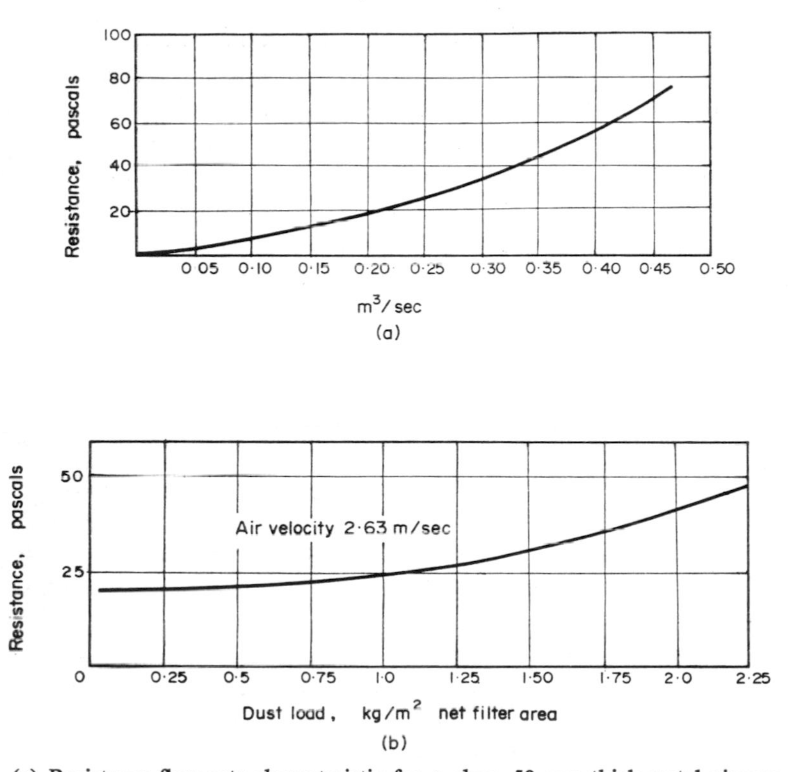

FIG. 8.26. (a) Resistance-flow rate characteristic for a clean 50 mm thick metal viscous filter.[888] (b) Resistance–dust load characteristic of a 50 mm thick metal viscous filter at 2·63 m s^{-1} face velocity. (A.S.H. and A.E. Test method: Flow rate: 2·63 m s^{-1}. Test dust: 80% Pocahontas fly ash. 20% Lamp Black. Feed rate: 12·5 mg m^{-3}).

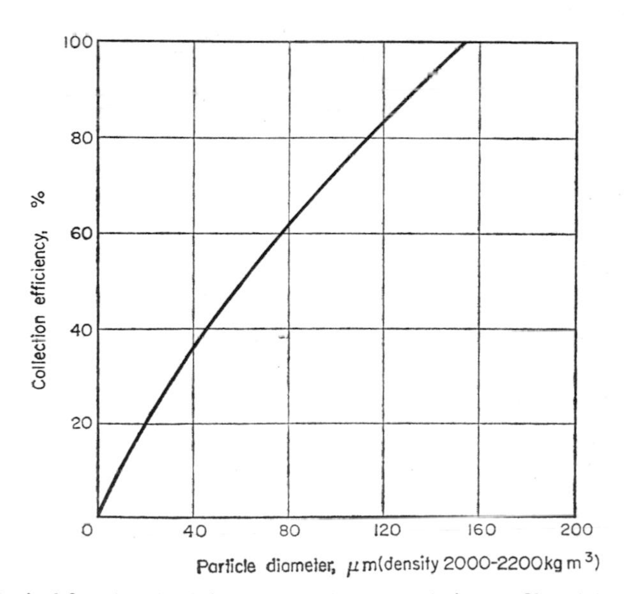

FIG. 8.27. Typical fractional efficiency curve for a metal viscous filter (50 mm thick).[888]

The metal wire in these filters is usually covered with a film of a soluble oil, which serves two purposes. It holds the collected particles on the wire and it prevents corrosion. These filters are often referred to as viscous filters. One manufacturer interleaves the metal wire with cotton gauze which helps to keep the oil in the filter as well as acting as a filter on its own. When the filters are dirty they are exchanged, steam cleaned and recoated with oil. The simple filters have a comparatively low efficiency; an overall gravimetric efficiency of about 90 per cent, and a typical fractional efficiency curve is shown in Fig. 8.27. The low efficiencies for the small particles indicates that inertial impaction is the predominant collec-

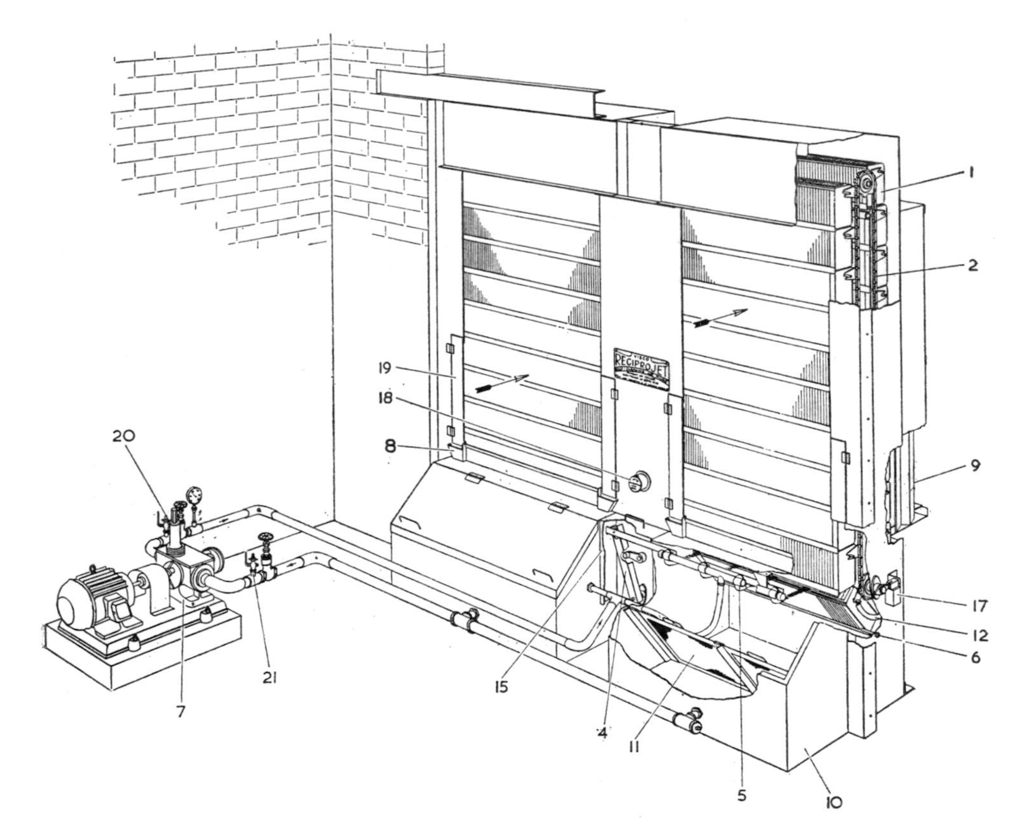

(a)

FIG. 8.28. Continuously cleaned metal viscous filter.

(a) Automatic continuously cleaned filter cells.[888]

1. Filter cells.
2. Cell conveyor chains.
3. Filter drive motor.
4. Reciprocating lever.
5. Reciprocating header with nozzles.
6. Cell tilting pin.
7. Main oil pump.
8. Oil traps.
9. Oil eliminators.
10. Oil tank.
11. Removable strainer screens.
12. Cell guides.
13. Cell conveyor chain tensioning bolts.
14. Worm gear.
15. Removable guard for reciprocating lever.
16. Safety cut-out switch.
17. Limit switch.
18. Hand drive.
19. Loading doors.
20. Hose union for filling oil tanks.
21. Hose union for emptying oil tanks.

tion mechanism, and these filters are only satisfactory when a comparatively low degree of cleaning is required.

When dust burdens are very high, a unit where the filter cells are continuously cleaned can be used. The cells are built on an endless moving belt or curtain arrangement (Figs. 8.28a and 8.28b), where the cells are dipped into an oil bath at regular intervals, normal practice being one complete revolution of the belt every day. The units are built in panels 0·9 m or more wide and between 1·5 m and 4·5 m high, increasing in 100–125 mm steps, depending on cell construction. Face velocities are about 2·8 m s^{-1} and the pressure drop

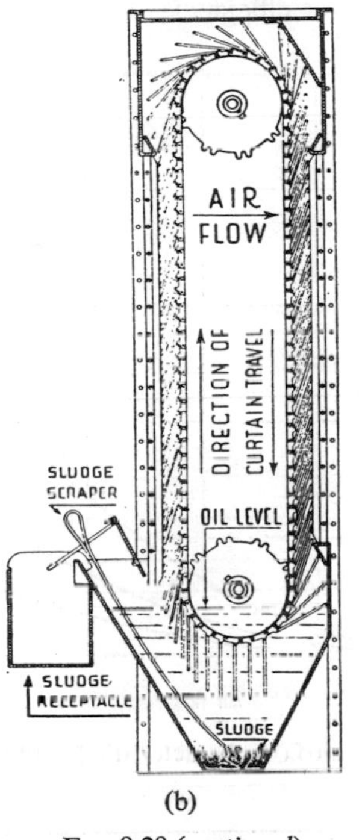

(b)

FIG. 8.28 (*continued*)
(b) Curtain type of filter cells.[18]

varies, depending on cell structure, between 8·7 and 110 Pa. The efficiency of these systems is similar to the fixed wire-mesh panels.

When better cleaning than that provided by the wire mesh filter is required, glass or chemical fibre media are used. In their simplest form, these consist of disposable cells of standard dimensions (0·5 m square and 25, 50 or 100 mm deep), packed loosely with the filter medium, which may be treated to cross-link the fibres, and which is retained between cardboard or metal punched plates. The cross-linkage treatment is a low volatility plastic resin, and is necessary where damp atmospheres cause matting together of the loose fibres. The face velocities recommended for these filters are about 1·5 m s^{-1}, although special cells can be obtained which operate at velocities 50 per cent higher, similar to metal viscous filters. The recommended velocity should not be exceeded or internal breakdown of the filter material occurs as well as re-entrainment, and filtration efficiency is reduced permanently.

The resistance of this type is somewhat greater than the metal viscous filter, and a pressure drop–flow-rate curve is shown in Fig. 8.29.

A development from the rather thick loose fibre packing is the filter mat where synthetic fibres are laid in graduated layers and bonded by spraying with a polymer solution which is then baked on under infra-red heating. This gives a filter medium which no longer needs separate support when enclosed in the approximately 0·6 m square standard frames. The medium is either washable or "throw-away", and recommended operating velocities are 1·8–2·5 m s^{-1}. The highest efficiency achieved is 80 per cent (A.I.F. test) and the filter will hold up 0·38 kg m^{-2}. At lower efficiencies higher dust loadings can be obtained.

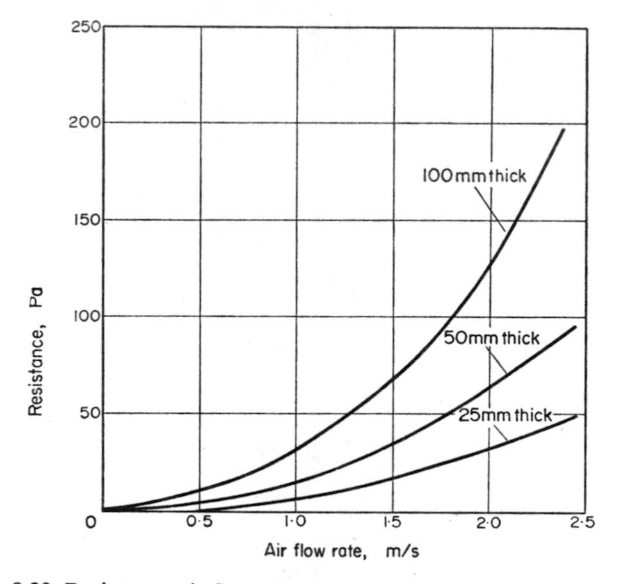

FIG. 8.29. Resistance–air-flow characteristic for fibre-glass filter pack.[889]

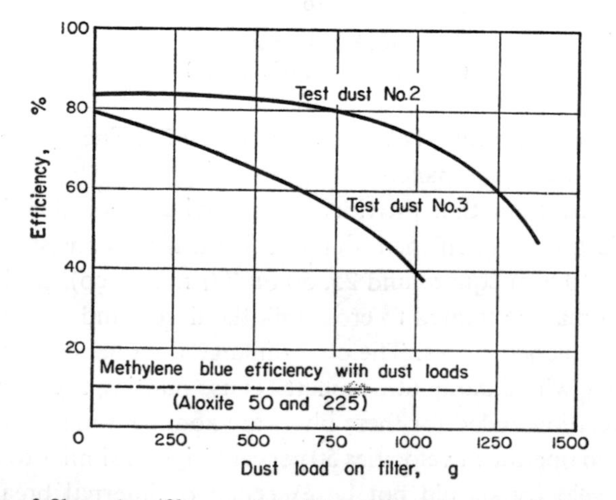

FIG. 8.30. Efficiency of fibre glass filter packs with standard test conditions (B.S. 2831) showing effect of reduced efficiency with high dust loads.[889] (Air flow: 1·5 m s^{-1}.)

The collection efficiency characteristic for a typical fibre-glass filter, tested under British Standard test conditions,[130] is shown in Fig. 8.30. The test dusts used have average particle sizes of 5 μm (No. 2 test dust), 18 μm (No. 3 test dust) and 0·3 μm (Methylene blue fume). The efficiency is high (80–85 per cent) for new filters for test dusts 2 and 3, but falls off with high dust loads, indicating that re-entrainment occurs at this stage. The retention of sub-micron-sized particles is very inefficient, as indicated by the 10 per cent efficiency with methylene blue, and these filters cannot be used for clean rooms for photographic, biological or radiochemical work.

To improve the efficiency of this type of filter, it is important to avoid high dust loads, and an automatic filter medium replacement unit, the "roll away", has been developed (Fig. 8.31). The actual medium is a comparatively close-packed layer of glass or chemical fibre, bonded together, and supported on a loose woven cloth. The clean filter medium is exposed to the dirty gases, and either after a predetermined time or after the pressure

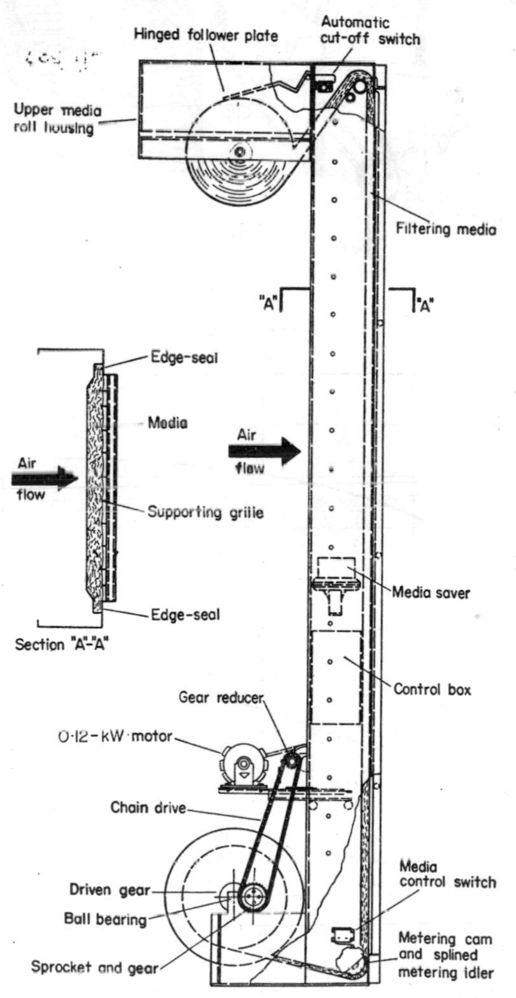

FIG. 8.31. Automatic filter cloth roll system.[171]

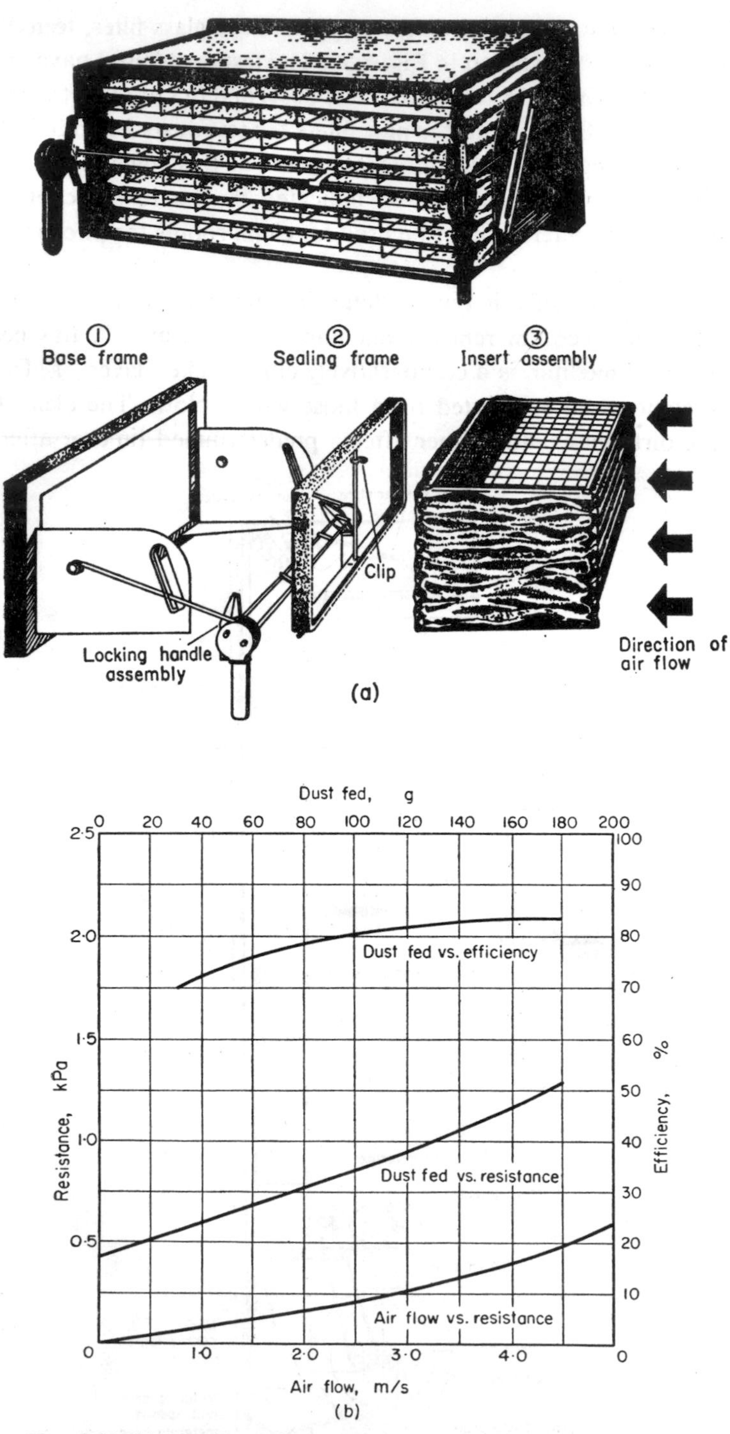

FIG. 8.32. (a) Multilayer filter, showing one method of assembly.[889] (b) Typical operating characteristics for multilayer type filter.

drop has reached an optimum value 87–100 Pa an automatic mechanism exposes a new surface to the gases, rolling up the dirty medium. Face velocities of 2·5 m s^{-1} are recommended, and 95 per cent collection is frequently achieved with dusts comparable with the British Standard dust Nos. 2 and 3, with loadings up to 0·43 kg m^{-2} (which gives filter resistances of the order of 90–100 Pa). For coarser dusts higher fabric loadings up to 1·0 kg m^{-2} can be tolerated without the pressure drop exceeding 100 Pa.

An alternative solution to improving the filter efficiency with high dust loads is to increase the filter surface area, by stretching the filter cloth over a number of frames, as shown in Fig. 8.32. The superficial face velocity of this type of unit is about 1·5 m s^{-1}, which gives a velocity of about 120 mm s^{-1} through the filter fabric, which is sufficiently low to avoid re-entrainment. The resistance of this filter is about 40 Pa for a new filter fabric.

Some manufacturers supply a heavier filter medium with higher efficiencies and superficial face velocities of 1 m s^{-1}. This reduces the velocity through the cloth to 65 mm s^{-1}, with a pressure drop of 80 Pa. The efficiency of this filter has been shown to be 97 per cent for the A.S.H. and V.E. test and about 88 per cent for the methylene blue test.[88]

An interesting development of the deep-bed filter which is useful when very dirty atmospheres are encountered is the "bag"-type unit, in which small bags or shapes are stretched over a framework. The design is such that standard face velocities of 2·3–2·5 m s^{-1} can be used. Average efficiency (B.S. 2831 Specification) for No. 2 dust is from 84 per cent to over 99 per cent depending on the medium used, while methylene blue efficiency is approx. 30 per cent. Dust holding capacity is 4·3–5·4 kg m^{-2} of superficial surface area, 10 times that of the flat dry arrestance filter made of a polymer fibre mat. This unit is lower in cost than the deep-bed filter by about 20 per cent.

If the filter medium is of chemical fibre, the cloth can sometimes be washed in special dry-cleaning tanks and re-used. However, some fibre matting and felting is likely to occur in the cleaning process, and there may be enlargement of pores between the fibres so that these will pass air preferentially without removing particles. Washing the filter medium is therefore not recommended if high efficiencies are to be maintained.

When even higher efficiencies are required than is possible with the multilayer filter, it becomes necessary to construct the filter medium of very fine fibres. Asbestos fibres in particular have proved suitable for this because of their fire resistance as well as very fine fibres. Instead of methylene blue, American workers prefer dioctylphthalate (DOP) smoke which has a mean particle size of 0·3 μm similar to the methylene blue, and so is directly comparable with it.

Asbestos bearing filter papers have DOP efficiencies of 99·85 per cent for new papers, which increase to 99·999 per cent after 2 hr use, with flow rates of 27 mm s^{-1},[698] and pressure drop of 236 Pa. To get practicable flow rates, face velocities of 625 mm s^{-1}, the papers are folded into compact panels, 0·6 m square and 0·2 m deep. The panel is able to handle 850 m^3 h^{-1}, with a velocity through the paper of about 20 mm s^{-1}. These filters, which are in commercial production, are usually called "absolute" filters. They are made of Esparto grass and carded asbestos. Atomic Energy Authorities that require their filters to be fireproof and able to withstand 550°C, use an all glass-fibre paper. These filters are in commercial production and have DOP efficiencies slightly below the experimental types.

For slightly less rigorous duties, and where asbestos fibres are undesirable, an Esparto grass filter can be obtained. The efficiencies of various filters in common use are listed in Table 8.14.

For use in pharmaceutical work and the food industry, the efficiency of capture, or alternatively the penetration of micro-organisms is of interest, and the effectiveness of these filters towards bacteria (1·0 μm) and viruses (0·03 μm) is listed in Table 8.15.

TABLE 8.14. EFFICIENCIES OF FILTER MATERIALS USED IN AIR CONDITIONING FILTERS (D.O.P. OR METHYLENE BLUE TESTS)

	Flow rate (mm s^{-1}) at paper surface	Resistance kPa	Efficiency (%)
Asbestos bearing paper[798]	27	0·20	99·980
at start			
after 205 min	27	0·24	99·993
Commercial "absolute" filters[889]			
Esparto grass-asbestos	20	0·28	99·95
glass fibre-asbestos	20	0·28	99·99
glass fibre (suitable to 550°C)	20	0·28	99·99
cellulose paper-asbestos	20	0·09	90
Esparto grass only	20	0·06	65
(For use in photographic industry where asbestos in undesirable)			
Glass fibre pads			
3 μm fibres loose 12 mm	170	0·01	63
1·3 μm fibres loose 12 mm	150	0·45	91
1·3 μm fibres loose 25 mm	140	0·38	99·4
12 mm 3 μm followed by 12 mm 1·3 μm	150	0·23	94
Vacuum cleaner bag paper[798]	70	0·30	30
Woven glass fibres[798]			
fine weave fabric	26	0·02	48
coarse weave fabric	26	0·005	22
Wool felt (reverse jet system)[798]			
at start	100	0·20	30
after 840 h (constant)	100	0·70	92

TABLE 8.15. PENETRATION OF MICRO-ORGANISMS THROUGH FILTER MEDIA[584]

Type of filter	DOP penetration (%)	Bacteria (1·0 μm) penetration (%)	Virus (0·03 μm) penetration (%)
Glass paper	0·01	0·0001	0·0036
Asbestos—cellulose paper	0·05	0·0002–0·0003	—
Asbestos—cellulose (lower grade)	5·0	0·4	3·5
Cotton-asbestos cannister	0·001	0·005	—

CHAPTER 9

PARTICLE COLLECTION BY LIQUID SCRUBBING

9.1. INTRODUCTION

The effectiveness of rain in removing airborne dust from the atmosphere has been recognized for a very long time. Industry has used the idea of this natural process to develop a variety of liquid scrubbing equipment. The wet collection of particles has a number of advantages compared to dry methods, such as reduced dust explosion risk, as well as a number of disadvantages, largely associated with plant corrosion and effluent liquid disposal. These are summarized in Table 9.1.

The collection mechanisms in scrubbers are the same as in filters: inertial impaction, interception and diffusion, which were discussed quantitatively in Chapter 7. In addition, condensation effects and particle entrainment may play important parts in collection.

Although some electrostatic charge is produced on spray droplets during atomization, this has been shown to be too small to be a major aid in collection,[256] except when the droplets are deliberately charged by an external source.[463] Similarly, thermal precipitation is unlikely to be a major force in attracting particles, because the droplets are volatile, and the temperature difference for effective thermal precipitation is such that the droplets would evaporate. Where spray towers and scrubbers are used for hot gases they perform the multiple task of gas cooling and humidification as well as coarse-particle collection, before passing the gases to a unit for fine particle removal.

The condensation effect is probably of some importance in the venturi scrubber where care is taken to use gases which have been saturated before passing through the reduced pressure zone in the venturi throat where more liquid is added. The subsequent condensation takes place in the diffuser where the pressure rises, as the velocity is reduced.[466] The dust particles tend to act as nuclei for the condensing vapour, increase in size, agglomerate and are more easily separated in the collection chamber. Entrainment of particles can occur in the wake of droplets, and may lead to particle capture when the droplets are collected.

It is not possible to classify scrubbers by the principal collection mechanism which is used in each case, giving one type a particular application. It has in fact been shown that improved scrubber performance, being the ability to collect particles of decreasing sizes, is a function of the energy consumed in the plant.[752] Thus, low resistance scrubbers such as spray towers collect coarse particles, while high pressure loss units of the venturi type are

TABLE 9.1. ADVANTAGES AND DISADVANTAGES OF WET AND DRY COLLECTION[259]

Wet	Dry
Advantages:	
(1) Can collect gases and particles at the same time. (2) Recovers soluble material, and the material can be pumped to other plant for further treatment. (3) High-temperature gases cooled and washed. (4) Corrosive gases and mists can be recovered and neutralized. (5) No fire or explosion hazard if suitable scrubbing liquor used (usually water). (6) Plant generally small in size compared to dry collectors such as bag houses or electrostatic precipitators.	(1) Recovery of dry material may give final product without further treatment. (2) Freedom from corrosion in most cases. (3) Less storage capacity required for product. (4) Combustible filters may be used for radioactive wastes. (5) Particles greater than 0·05 μm may be collected with long equipment life and high collection efficiency.
Disadvantages:	
(1) Soluble materials must be recrystallized. (2) Insoluble materials require settling in filtration plant. (3) Waste liquids require disposal which may be difficult. (4) Mists and vapours may be entrained in effluent gas streams. (5) Washed air will be saturated with liquid vapour, have high humidity and low dew point. (6) Very small particles (sub-micron sizes) are difficult to wet, and so will pass through plant. (7) Corrosion problems. (8) Liquid may freeze in cold weather.	(1) Hygroscopic materials may form solid cake and be difficult to shake off. (2) Maintenance of plant and disposal of dry dust may be dangerous to operatives. (3) High temperatures may limit means of collection. (4) Limitation of use for corrosive mists for some plants (e.g. bag houses). (5) Creation of secondary dust problem during disposal of dust.

very effective with fine fumes. Here scrubbers will be classified primarily by the method of droplet formation, and secondly by the mechanism used to collect the droplets. Thus, in simple spray towers, the droplets are formed by atomizing sprays, and are collected by gravitational attraction, while in centrifugal spray scrubbers, the droplets, still formed by atomizing sprays, are collected by centrifugal forces. In other scrubbers the gas stream is used to break up the liquid and form droplets. Irrigated or wetted wall collectors, where the liquid serves mainly to prevent particle entrainment but not initially to collect the particle, are excluded from the present discussion. These have been included with the principal mechanism used in particle collection. For example irrigated cyclones have been found to have a better performance characteristic in comparison to ordinary cyclones.

9.2. SPRAY TOWERS

The simplest type of scrubber is a spray tower. Liquid droplets are produced by spray nozzles, and are allowed to fall downwards through a rising stream of dirty gases. In order not to be entrained by the gas stream, the droplets must be sufficiently large to have a falling speed greater than the upward velocity of the gas stream, which, in practice, is about 0·6 to 1·2 m s^{-1}. Droplets smaller than 1 mm, in the case of water, tend to be approximately spherical, and their falling speeds can be estimated from the falling speeds of spheres in still air (Table 5.1).

Because the droplets used in spray towers are of the order of 0·1–1 mm diameter, the particles collected by these droplets are comparatively large and the predominant collection mechanisms are inertial impaction and interception. Stairmand[801] has calculated that the optimum collection efficiency by inertial impaction for droplets falling under gravity through still air is independent of the droplet size, and is obtained when the droplets are about 0·8 mm (800 μm) diameter (Fig. 9.1): 0·8 mm diameter droplets have a falling

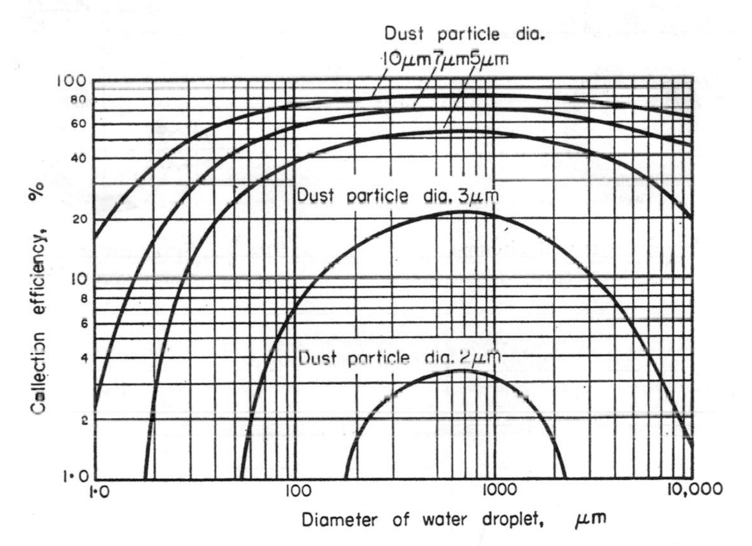

FIG. 9.1. Optimum droplet size for collection by inertial impaction by droplets falling in the earth's gravitational field, i.e. in a simple spray tower.[801]

speed of over 3 m s^{-1} while 0·4 mm drops fall with a relative velocity of 1·58 m s^{-1}, so a nozzle producing a coarse spray with droplets just below 1 mm is the most satisfactory for a simple spray tower. For these an impingement-type nozzle (section 9.4) has been found to be suitable as it is very robust and has little tendency to block up or wear when the scrubbing liquid is recirculated and contains a certain fraction of solid material[344]. In practice about 30–35 per cent of the liquid is recycled in the case of blast furnace gas scrubbers,[911] this leading to a reduced load on the water purification plant. It is also thought, although no evidence will confirm this at present, that the solids content of the scrubbing liquid reduces the surface tension, improves the wetting characteristics and so assists in collection.

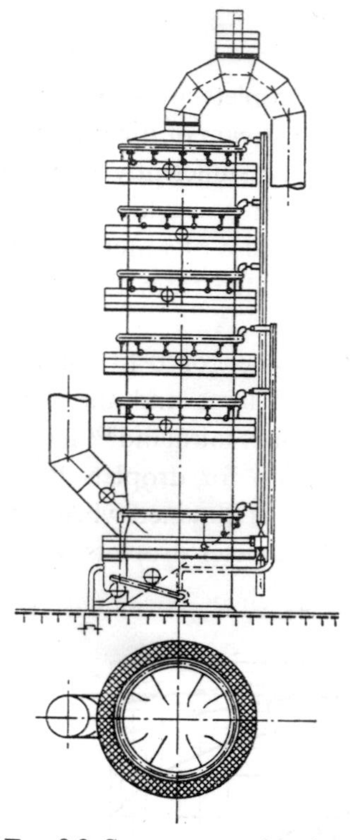

FIG. 9.2. Spray tower with circumferentially placed sprays.[344]

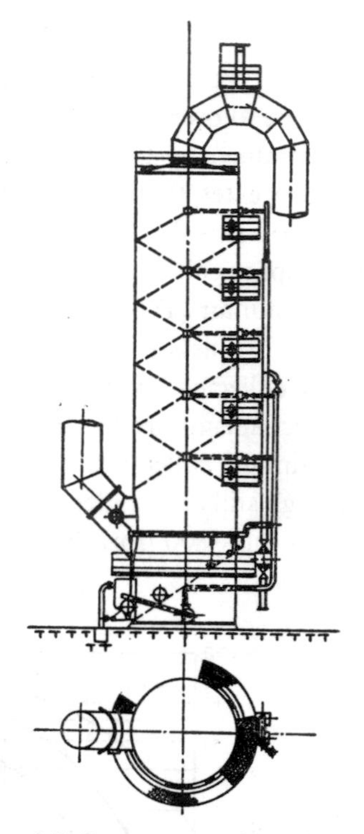

FIG. 9.3. Spray tower with axially placed sprays.[344]

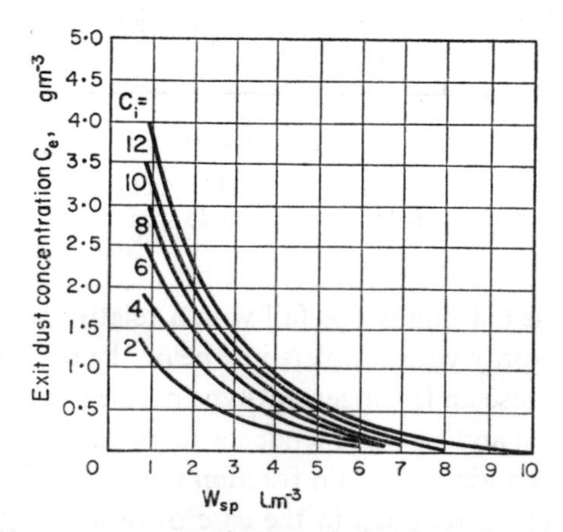

FIG. 9.4. Specific water consumption correlation for a simple spray tower.[344]

C_i = dust concentration at inlet g m^{-3}.
C_e = dust concentration at exit g m^{-3}.
W_{sp} = specific water consumption l. m^{-3}.

A typical large spray tower is shown in Figs. 9.2 and 9.3. The design shown is 5 m diameter, has an effective height of 16 m and in the original pattern had 5 circumferentially arranged rings of sprays with 14–16 spray nozzles per ring (Fig. 9.2). The tower handled 50,000 $m^3 h^{-1}$, and the gas velocity was about 0·75 m s^{-1}. The second arrangement (Fig. 9.3) shows the same tower with nine centrally placed solid cone impact sprays, which reduced the solids concentration in the gas from 2·2 g m^{-3} to 0·22 g m^{-3}, an efficiency of 90 per cent, compared with 73 per cent achieved with the circumferentially placed sprays. The central sprays had a service life of 18 months, and were found to be effective in towers up to 7·5 m diameter handling 200,000 $m^3 h^{-1}$ (gas velocity 1130 mm s^{-1}).

Water consumption for spray towers can be found from the data in Fig. 9.4, which plots specific water consumption (in 1 m^{-3}) for exit dust concentrations using inlet concentrations as parameters, in the case of blast furnace gas dusts.

9.2.1. Scrubbing with cooling (droplet evaporation)

When a counter-current spray tower acts as a gas cooler as well as particle collector, the design of the tower from first principles presents a number of problems, as the droplet size decreases because of the evaporation which takes place. To carry out the calculation it is necessary to divide the tower into a number of sections, each with one ring of spray nozzles, and then find the evaporation which takes place within this zone, based on the gas temperature and the droplet residence time. The first region to be tackled is the one at the top of the tower, where exit gas conditions are specified, and new droplets are introduced. From the evaporation in this zone the conditions at the top of the penultimate zone can be found, but in this zone we have both freshly-introduced droplets and falling partially evaporated ones. A complete calculation has not yet been completed.

However, Alonso and Strauss (unpublished calculations) have attempted to determine analytically the different factors which are involved in a single-stage scrubber. In their analyses they have considered the interaction between particulate material, liquid droplets, and the wall film for a simple cylindrical scrubber with a vertical axis.

In the development of a mathematical model it was assumed that the flow was axisymmetric, the variation in axial velocity along the radials is small, and can be described by an average value, axial dispersion and diffusion due to concentration gradients are accounted for by the appropriate differential terms, radial diffusion and droplet transport are described by transport equations and empirical correlations, while film type transport coefficients describe the transport process.

Material balance equations were written and expressed in non-dimensinal form, as was the droplet-gas mass transfer interaction. These were then solved analytically, The specific simplifying cases which were then considered were

(i) Neglecting droplet evaporation and droplet capture by the walls.
(ii) Droplet evaporation was neglected, but droplet interaction with the walls was considered and assumed to follow the correlation given by Alexander and Coldren.[8]
(iii) Droplet evaporation was accounted for by the methods of Sjenitzer[775] and Frössling,[284] and the droplet interaction with the walls was treated as in (ii).

The solution of the resultant equations also used the size distribution for particles and droplets suggested by Khrgian and Mazin:[872]

$$N(r) = Ar^2 e^{-br}$$

where r = droplet or particle radius, while
A, b = constants.

The resultant equations are extremely complex, and are not given here. On the basis of Sjenitzer's work, however, it can be shown that the heat for evaporation of the droplets comes almost wholly from the sensible heat of the gas, and not radiation from the walls or the heat content of the water or liquid making up the drops. An experimental programme being undertaken by the authors[826] will determine the mass transfer coefficients and lead to empirical correlations in terms of the flow parameters which will give a comparatively simple solution.

The rate of evaporation from pure water droplets travelling at their terminal velocity, which is a reasonable initial assumption, can be estimated from the graphs plotted by

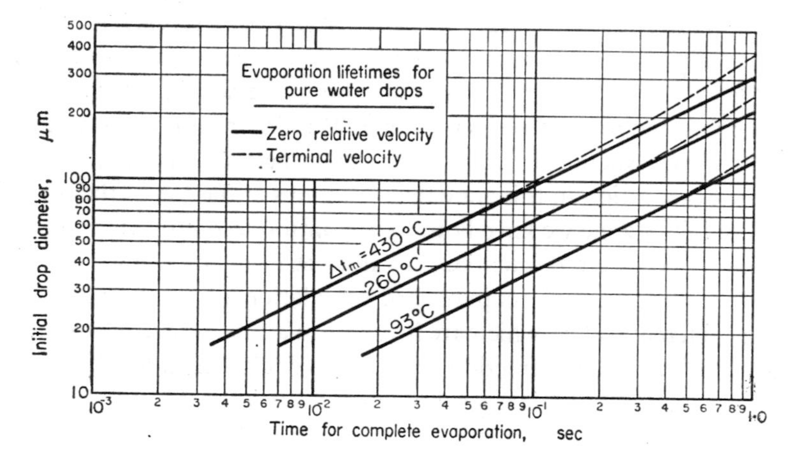

FIG. 9.5. Relation between water drop diameter and time for complete evaporation.[552]

Marshall,[551, 552] one of which is shown in Fig. 9.5. The complex problem of evaporation from drops with some dissolved solids has also been attempted by Marshall as part of his studies on spray drying, and if detailed calculations are to be carried out, the monograph[551] on this subject should be referred to.

9.2.2. Scrubbing with cooling (condensation on droplets)

This is dealt with in Section 9.8.1.

9.3. CENTRIFUGAL SPRAY SCRUBBERS

The efficiency of collection of particles smaller than those recovered in a simple spray tower can be improved by increasing the relative velocity of the droplets and the gas stream. The force on the droplets can be increased by using the centrifugal force of a spinning gas stream rather than the gravitational force in the simple tower. For example, when gases are spinning with a tangential velocity of 17·5 m s^{-1} at a radius of 0·3 m, the centrifugal force is 100 *g*. Johnstone and Roberts[405] calculated the efficiency of collection by inertial impaction of particles by droplets of various sizes under the influence of a force of 100 *g* (Fig. 9.6). The curves show that droplets of about 100 μm are most effective, larger droplets have smaller inertial impaction parameters while smaller droplets are entrained in the gas stream. The curves also show the vastly improved collection efficiency of the centrifugal spray-type scrubber compared to the gravity spray tower, particularly in the 1–10 μm particle-size range. Collection by diffusion is not very effective, except for particles less than 0·01 μm. Commercial centrifugal spray scrubbers have operating efficiencies of 97 per cent or better for particles greater than 1 μm. A selection of performance data is listed in Table 9.2.

Commercially designed centrifugal spray scrubbers are of two patterns. In the first, the spinning motion is imparted to the gas stream by a tangential entry with velocities between 15 and 60 m s^{-1}. The liquid is directed outwards from sprays set in a central pipe[447] (Fig. 9.7a). In the second type (Fig. 9.7b) the spinning motion is given to the gas stream by fixed vanes, and the water spray in the chamber is directed downwards from a centrally placed

TABLE 9.2. PERFORMANCE DATA FOR CYCLONIC SPRAY SCRUBBER[447] (PEASE–ANTHONY TYPE)

Source of gas	Type of dust or mist	Particle size range μm	Dust loadings g m^{-3}		Efficiency (%)
			inlet	outlet	
Boiler flue gas	Fly ash (pulverized coal)	> 2·5	1·1–5·8	0·04–0·1	88–98·8
Blast furnace (iron)	Iron ore coke	0·5–20	6·7–53·5	0·07–0·18	99
Lime kiln (Kraft mud)	Lime	1–25	17·2	0·56	97
Lime kiln (raw stone)	Lime	2–40	20·5	0·18	99
Reverberatory lead furnace	Lead compounds	0·5–2+	1·1–4·5	0·05–0·09	95–98
Rotary dryer	Ammonium nitrate	large, unstable agglomerates	...	...	99+
Superphosphate den and mixer	Fluorine compounds	mist	0·31	0·007	97·8
Air bodying of castor oil	Castor oil	mist	0·006	0·0013	78

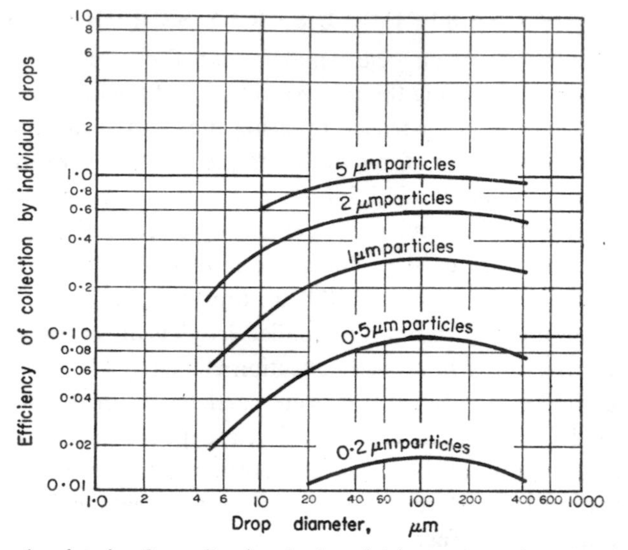

FIG. 9.6. Optimum droplet size for collection by inertial impaction when droplets are moving in field of 100×g: i.e. a centrifugal spray tower.[405]

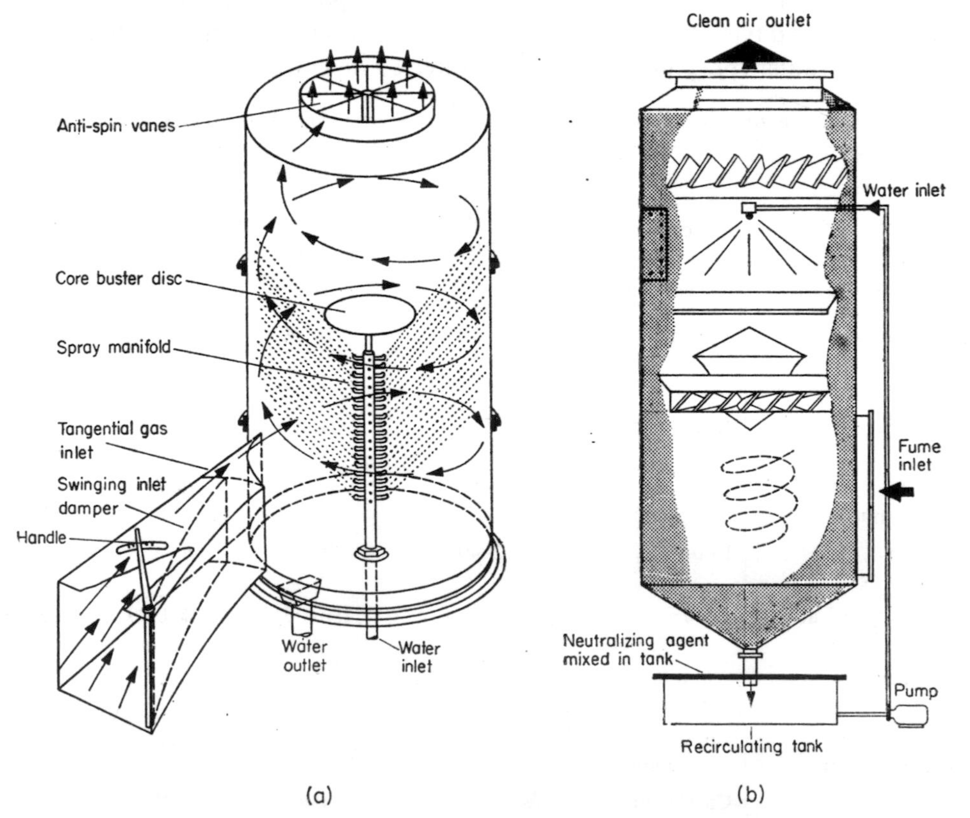

FIG. 9.7. Centrifugal spray scrubber. (a) Tangential entry, centrally placed sprays (Pease–Anthony design).[410] (b) Vane entry.

single nozzle. Both types of plant are built to handle volumes ranging from 850 to 68,000 $m^3\ h^{-1}$.

There are also two horizontal commercial centrifugal spray scrubbers, where some venturi effect is achieved by a narrowing of the duct in which a spinner is arranged, which gives an additional swirl to the gases after the tangential entry. One of these, the Davidson Sirocco V-type collector,[202] is shown in Fig. 9.8, with dimensions for a size to suit 15,000 $m^3\ h^{-1}$.

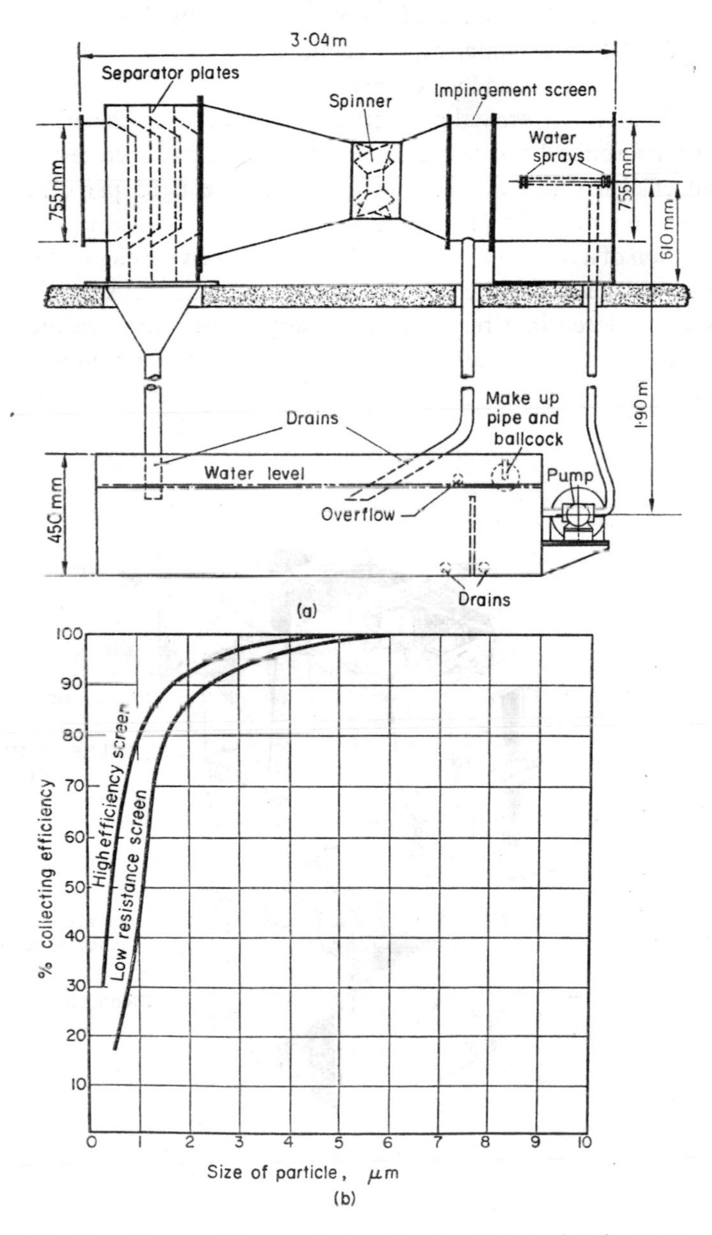

FIG. 9.8. (a) Davidson Sirocco V-type collector (size for 15,000 $m^3\ h^{-1}$). (b) Fractional efficiency curves for this collector.

The pressure loss is 1·9 kPa and the water pressure at the nozzles is 280 kPa, with 0·85 l. s^{-1} water circulation. A typical fractional efficiency curve is also shown in Fig. 9.8. Very similar to this is the Carter–Day Vortex Venturi wet scrubber, where, instead of narrowing the duct, thick, multiple, curved blades give a similar effect, with similar pressure drop and performance characteristics. These units tend, in general, to be larger than the Davidson units.

A third type of centrifugal scrubber, not in commercial production, has been extensively tested by First *et al.*[259] This is essentially a conventional cyclone, but with the gas entry at the base of the cylindrical section, and water sprays directed across the entry. High pressure water jets from a swirl atomizer proved most effective, a collection efficiency of 90 per cent being obtained with micronized talc test dust. A similar unit has been tested elsewhere,[657] and a 95 per cent efficiency was obtained with a coarse dust (85 per cent greater than 10 μm), a loading of 4·6 g m^3 and a pressure loss of 523 Pa.

A commercial version of a multicyclone wet collector has been developed by the American Air Filter Company, which is called the type R-Rotoclone (Fig. 9.9). A number of straight-through cyclones are enclosed in a rectangular housing where the gases are scrubbed for an initial separation, while the fine particles and remaining scrubbing liquors are removed in the small cyclones.

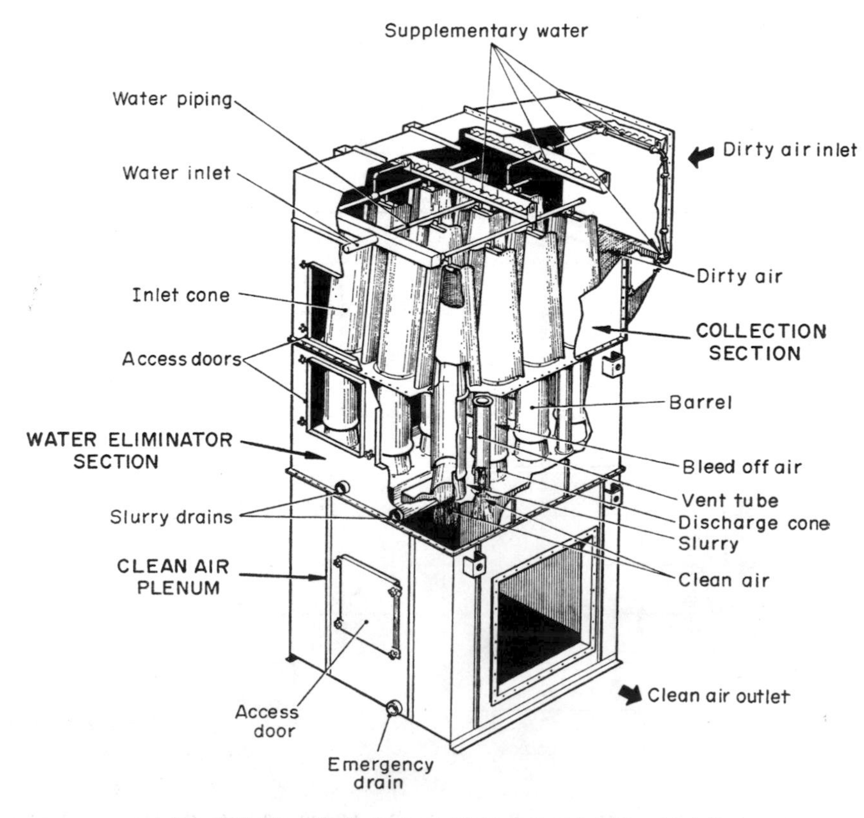

FIG. 9.9. Type R-Rotoclone (American Air Filter Co.).[18]

The liquid droplets in these centrifugal scrubbers are much smaller than in the simple gravitational spray towers, and it is therefore not desirable to have further droplet evaporation taking place in the scrubbers as they would no longer act effectively. It is usual practice to saturate the gases before they enter the scrubber, particularly if they are warmer than the liquid, to reduce droplet evaporation.

9.4. SPRAY NOZZLES

Spray scrubbers require droplets which are fairly closely sized in order to avoid entrainment and act most effectively. As the scrubbing liquid is often recirculated to reduce liquid consumption and the provision of very large separation tanks, the nozzles must be capable of handling liquids with a fairly high solids concentration. The most extensive study of nozzles for this type of duty has been in connection with spray drying[551] and also blast furnace gas cleaning.[344] Six nozzle types are in use:

(i) Impingement type.
(ii) Spiral spray nozzle.
(iii) Spinning-disc type.
(iv) Two liquid jet impingement nozzle.
(v) Pneumatic nozzle.
(vi) Sonic nozzle.

(i) In *impingement nozzles* a high velocity liquid jet is directed at a solid target. The jet is disintegrated by the impact and the liquid attenuation caused by the jet's change of direction. Depending on the type of target, either a directional spray or a hollow cone spray is produced. The two targets used are either a solid wall (Fig. 9.10a) for the directional spray, or the point of a hook, inserted in the axis of the nozzle (Fig. 9.10b) for the hollow

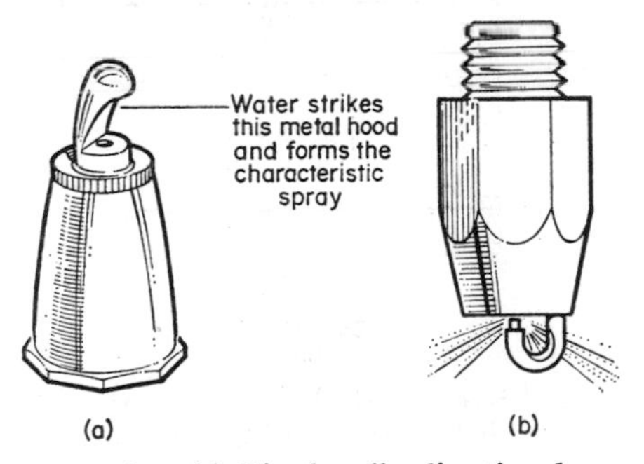

FIG. 9.10. Impingement type nozzles. (a) Fixed wall—directional spray. (b) Hook type—hollow cone spray.[551]

cone spray. The nozzles are cheap, robust and simple in construction and are widely used where coarse sprays are required, as for example in the blast furnace gas spray towers.[344]

(ii) The atomizing action in *swirl-type nozzles* is caused by the spinning motion imparted to the liquid before it leaves the nozzle. This spinning motion is obtained from a spin chamber within the nozzle, the liquid entry to which consists of a series of tangential holes

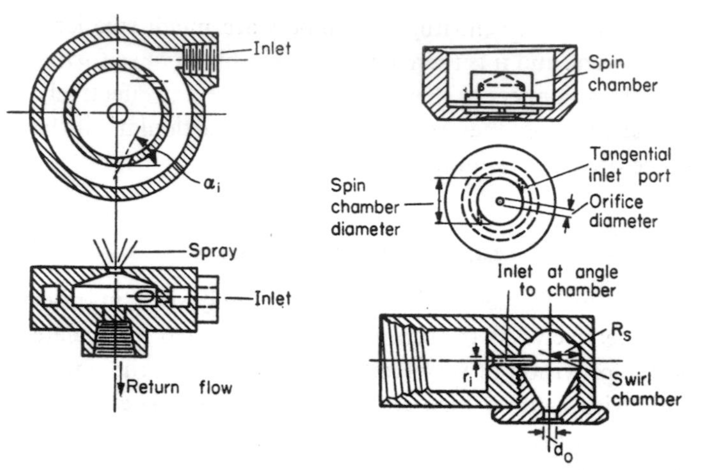

(a)

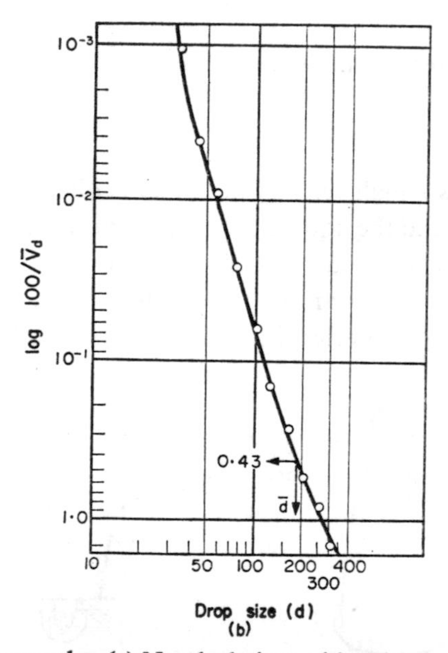

FIG. 9.11. Swirl-type spray nozzles. (a) Nozzle designs with swirl chamber.[551] (b) Droplet size distribution from swirl-type nozzle.

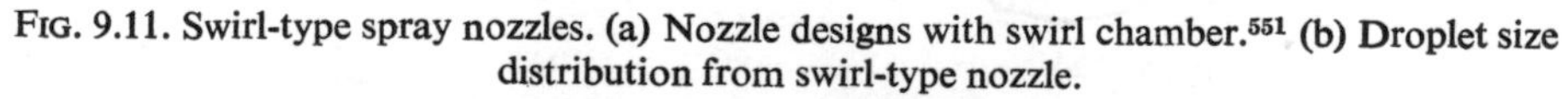

$\bar{V}_d = 100\, e^{(-d/\bar{d})^q}$ where $\bar{V}_d$ = volume per cent oversize. d = droplet diameter. $\bar{d}$ = Rosin-Rammler mean. q = dispersion coefficient (slope of line in Fig. 9.11(b).) (After Fraser and Eisenklam, *Trans. Inst. Chem. Engnrs.* (London), **34,** 294 (1956).)

(Fig. 9.11a). The entry holes have a tendency to block up, and this nozzle is not recommended where a suspension is being recycled. Droplet sizes from this nozzle (with 350 kPa water pressure), vary between 150 and 400 μm, depending on the orifice diameter[648] (Fig. 9.11b). With somewhat higher pressures (2·8 MPa) this type of nozzle has proved most effective for centrifugal spray scrubbers because of the large proportion of droplets about 100 μm which are produced.[400]

(iii) *Spinning disc atomizers* are discs from which liquid droplets are discharged after being accelerated to a high velocity on a rotating disc (Fig. 9.12). The disc can be driven by the liquid or mechanically and the liquid can move radially through tubes or across the flat disc. The droplets produced are uniform in size, and the size can be controlled by changing the disc speed and flow rate, which makes these atomizers most useful in producing the droplets for fundamental studies of their behaviour and performance. Details of their construction can be found elsewhere.[280] At present spinning disc atomizers are not used in commercial spray towers, but may find an application in scrubber developments where closely sized droplets are required.

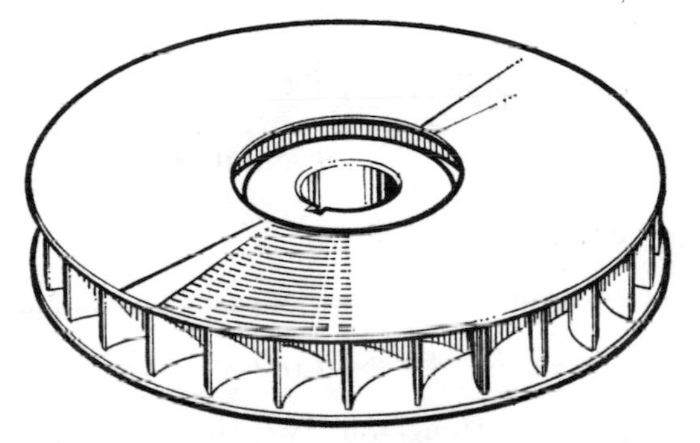

FIG. 9.12. Typical disc in spinning disc atomizers.[551]

(iv) When two liquid jets impinge, a ruffled sheet of liquid forms and then disintegrates intermittently to form groups of drops which appear to originate from the waves formed in the liquid at the point of impingement. The frequency of the waves increases with the jet velocity and with decreasing impingement angle, and rises to 4 kHz, forming large numbers of fine droplets. The viscosity of the liquid affects the spray formation characteristics, although these seemed to be little affected by the surface tension.[551] These atomizers, which are called *liquid jet impingement nozzles*, have the advantage that fine droplet sizes can be produced without the complex mechanical construction of the two foregoing types, or the wearing surfaces of the impingement nozzles. They are not yet used in commercial scrubbers.

(v) In the case of *pneumatic* or *two fluid atomization*, the liquid spray is produced by the impact of a gas stream on a liquid jet rather than of two liquid jets. Pneumatic nozzles produce very fine droplets, which were found in practice to be unsuitable for centrifugal scrubbers.[400] The fact that a high-pressure air or steam supply has to be provided as well as the water or other liquid adds to the installation and maintenance costs, and they are therefore not normally used for scrubbers.

The droplet formation in pneumatic sprays has been studied in considerable detail, largely because of its relevance to the action of a venturi scrubber, where, essentially, a gas stream impinges on a liquid wall. The average drop size D and surface area of the drops per unit volume of gas can be estimated from the empirical equations of Nukiyama and Tanasawa.[610]

$$D = \frac{1{\cdot}85 \times 10^7}{u} \sqrt{\left(\frac{\sigma}{\varrho_L}\right)} + 355\left(\frac{\mu_L}{\sqrt{(\sigma\varrho_L)}}\right)^{0{\cdot}45} \left(\frac{1000 Q_L}{Q}\right)^{1{\cdot}5} \tag{9.1}$$

where D = average droplet diameter (surface diameter) (μm)
u = relative velocity between air and liquid stream (m s^{-1}),
Q_L/Q = ratio of liquid volume flow to gas volume flow at venturi throat,
ϱ_L = liquid density (kg m^{-3}),
σ = liquid surface tension (N m^{-1}),
μ_2 = liquid viscosity (Pa s).

For the air–water system, equation (9.1) simplifies to

$$D = \frac{16\,050}{u} + 9{\cdot}75 \times 10^5 L_1^{1{\cdot}5} \tag{9.2}$$

where L_1 = m^3 water/m^3 air.

The specific surface of the droplets, A (m^2/m^3 of gas), can be obtained from

$$A = 800 L_1/D. \tag{9.3}$$

Experimental studies reported by Houghton[376] were in excellent agreement with the predictions of the above equations for pneumatic nozzles. while fair agreement was obtained for venturi atomizers.[507]

(vi) *Sonic spray nozzles*, recently developed, have a number of advantages over the traditional high pressure liquid or pneumatic types. The construction is shown in Fig. 9.13.

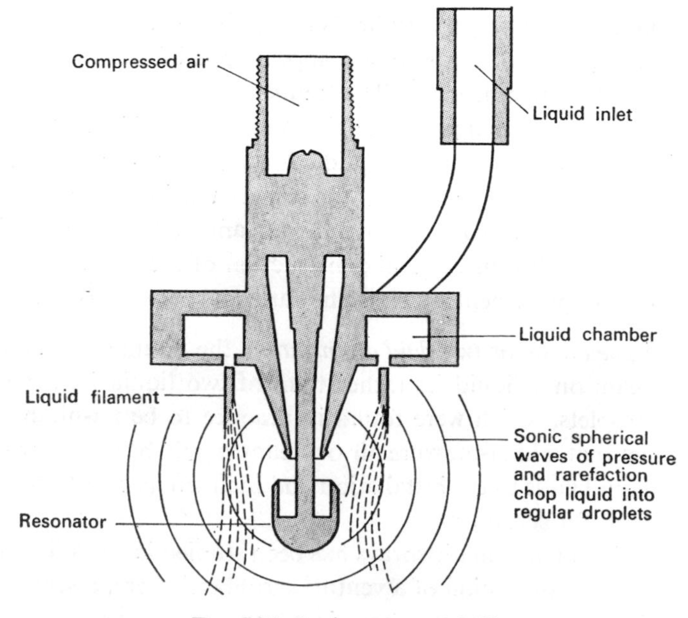

FIG. 9.13. Sonic spray nozzle.[26]

The liquid flows out of a simple annular chamber through holes under low pressures (70 kPa), and the liquid filaments are shattered by the sound waves (at a frequency of 9·4 kHz), producing very uniform droplets. The sound waves are produced by the impingement of a jet of compressed air or steam (100–400 kPa) impinging on the resonator placed centrally between the holes. The liquid throughput of the nozzle is about 45 g s^{-1}.[26]

9.5. SCRUBBERS WITH SELF-INDUCED SPRAYS

In many types of scrubbers the droplets which scrub the gases are formed by the gas stream breaking through a sheet of liquid or impinging on a pool. The simplest unit of this type, which is often applied to cupola furnaces, has water running over a cone-shaped disc,

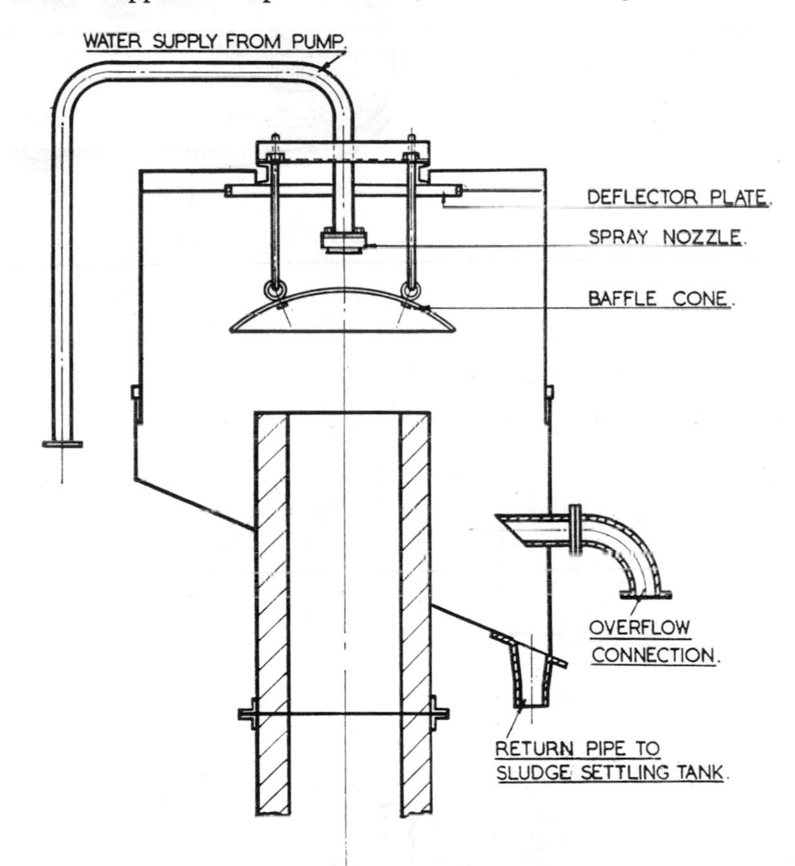

FIG. 9.14. Cupola top scrubber.[657]

forming an annular water sheet, which is broken up by the furnace gases (Fig. 9.14). This system has a low-pressure drop, less than 60 Pa, and can be installed very cheaply, and without induced-draught fans. The efficiency with coarse dusts is claimed to be high.

A more elaborate device is shown in Fig. 9.15, where the gas stream breaks through a pool of liquid, and then creates a liquid curtain because of the specially designed orifices.

FIG. 9.15. Action of a self-induced spray scrubber (Rotoclone N).[18]

FIG. 9.16. Self-induced spray type scrubber (Rotoclone N).[18]

The complete unit is shown in Fig. 9.16, and a similar idea in Fig. 9.17. In the former (Rotoclone type N) the impingement gas velocity is about 15 m s^{-1}, which creates droplets of 300 to 400 μm, while in the latter (Doyle scrubber) the impingement velocity varies between 35 and 55 m s^{-1}.[228]

These plants are extensively used in the metallurgical industry for dusts and sticky materials such as metal buffings, as well as for the collection of explosive dusts. Power require-

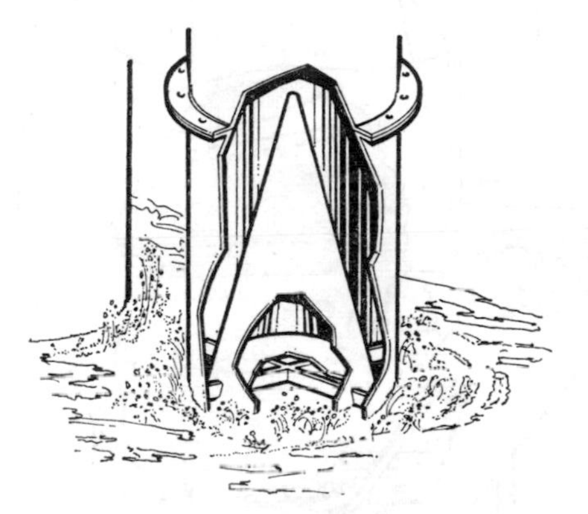

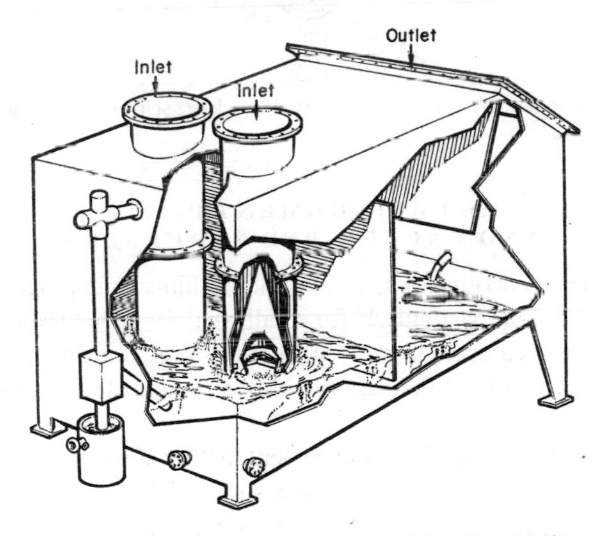

FIG. 9.17. Self-induced spray-type scrubber (Doyle).[228]

ments are between 1·0 and 1·3 kW per 1000 m^3 h^{-1} gas, while water is needed only to make up the losses by evaporation and entrainment. The plants can be rubber or PVC lined for operating with corrosive gases. The collected material settles out and can be removed from the bottom of the unit with a hopper in which a drag-type sludge ejector is placed (Fig. 9.18) or, alternatively, by sluicing the sludge to a central disposal unit. In very small units, or where the dust loading is small, manual methods may be used. Typical operating characteristics for a wide variety of applications are listed in Table 9.3.

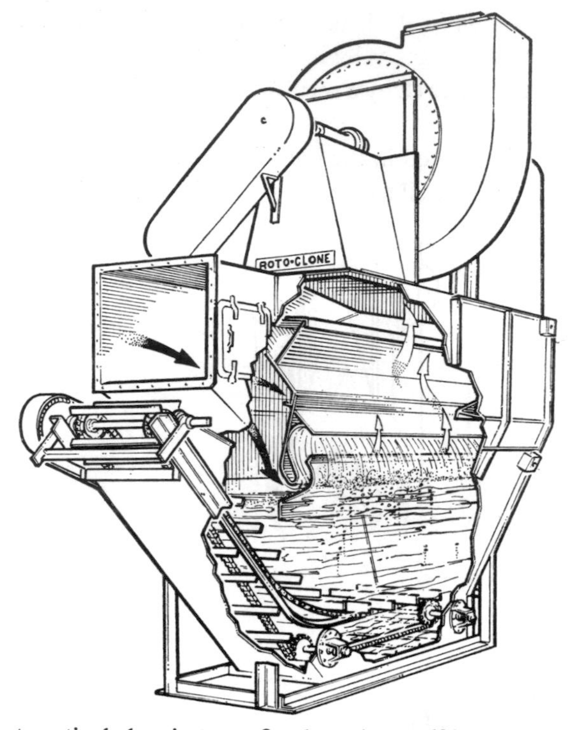

FIG. 9.18. Automatic sludge ejector as fitted to a large self-induced spray scrubber.[18]

9.6. IMPINGEMENT PLATE AND PACKED-BED SCRUBBERS

If the dirty gas stream encounters a series of liquid films or impinges on a succession of pools, then better collection, particularly for small particles, is achieved than for the simple scrubbers with self-induced sprays, but with greater pressure drop. The design of units incorporating these ideas follows the pattern either of modified sieve plates, or of a packed-bed scrubber.

The conventional sieve plate, with impingement baffles directly above the holes in the trays, is the usual modification (Fig. 9.19), and superficial gas velocities 5 times the velocities used for distillation are used for particle collection. Also fewer holes are provided in the trays so that the velocity through the orifices is of the order of 5–7 m s^{-1}. The gas–liquid mixture on the trays is formed into jets which impinge on special baffles, in contrast to bubbles rising through the liquid on the trays of a conventional gas absorption or distillation column. For maximum particle collection efficiency the gas should be saturated before entering the sieve plate section of the column, using water or steam jets.

Similar principles are used in the plant shown in Fig. 9.20 where expanded metal is used instead of sieve plates and baffles. The unit is made to handle gas flows varying between 2000 and 42,000 m^3 h^{-1}. For cleaning the exhaust air from polishing and grinding the unit is filled with a low vapour pressure oil as a collection fluid, while with water as a scrubbing

TABLE 9.3. PERFORMANCE DATA ON SELE-INDUCED SPRAY COLLECTORS

Source of dust	Inlet conc. ($g\ m^{-3}$)	Outlet conc. ($g\ m^{-3}$)	Efficiency (%)	Water consumption ($g\ l^{-1} s^{-1}$)
	Doyle scrubber[228]			
Boiler, coal fired (P.F.)	23·2	0·34	98·4	3·0
Lead sinter crusher	1·9	0·007	99·6	0·6
Lead sinter preparation wet-mix dryer	4·8	0·101	97·9	1·1
Phosphate rock-dryer cyclone in series	17·6	0·47	97·4	1·5
Coal-dryer cyclone in series	4·4	0·064	98·6	1·1
	Type N Rotoclone[89]			
Electric arc furnace	0·6	0·15	75·5	
Fly ash	3·0	0·012	99·4	
Brown coal dust	4·0	0·038	99	
Sinter dust	6·9	0·045	99·3	
Carbon black	0·5	0·005	99	
Asbestos fibres	1·0	0·005	99·5	
Granite dust	10·0	0·045	99·5	
Limestone	10·0	0·39	96	
Ceramic polishing	0·9	0·018	98·8	
Sandblasting	1·4	0·055	96·9	
Metal polishing	0·3	0·029	90	

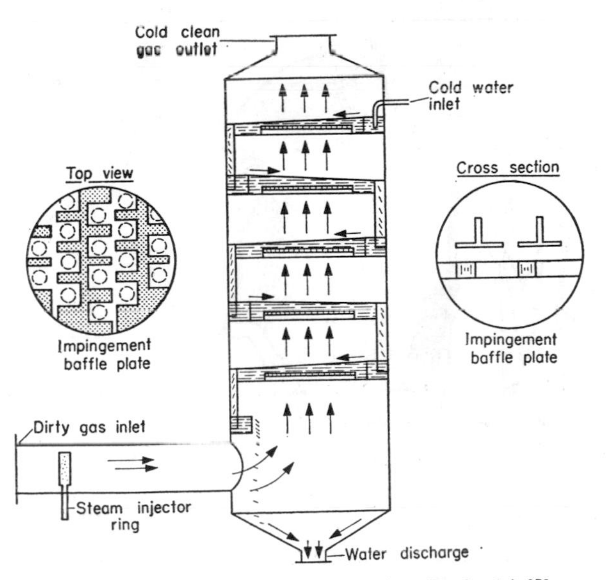

FIG. 9.19. Impingement plate scrubber (Peabody).[673]

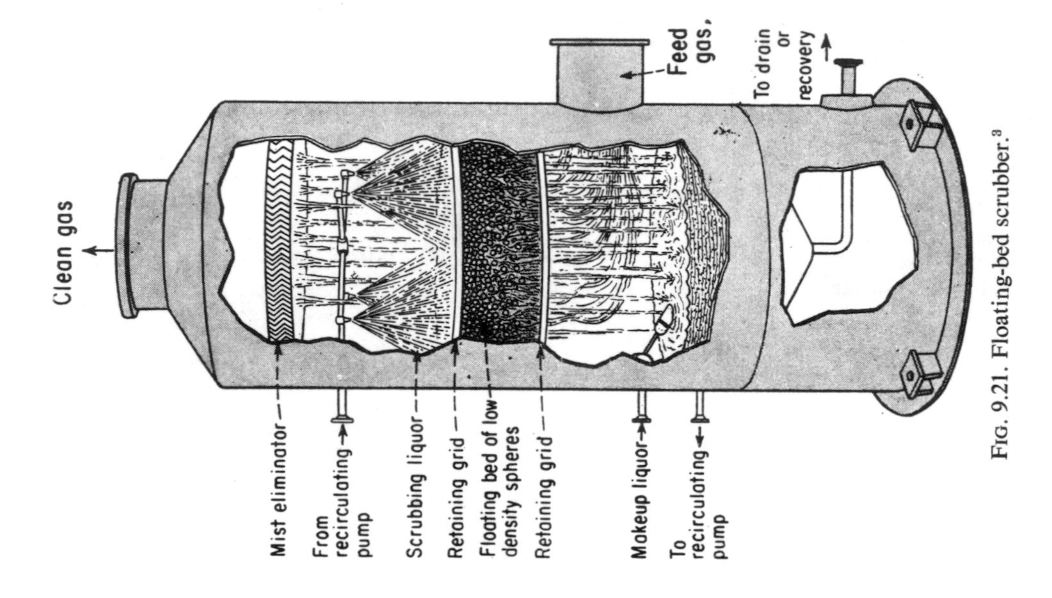

FIG. 9.21. Floating-bed scrubber.[3]

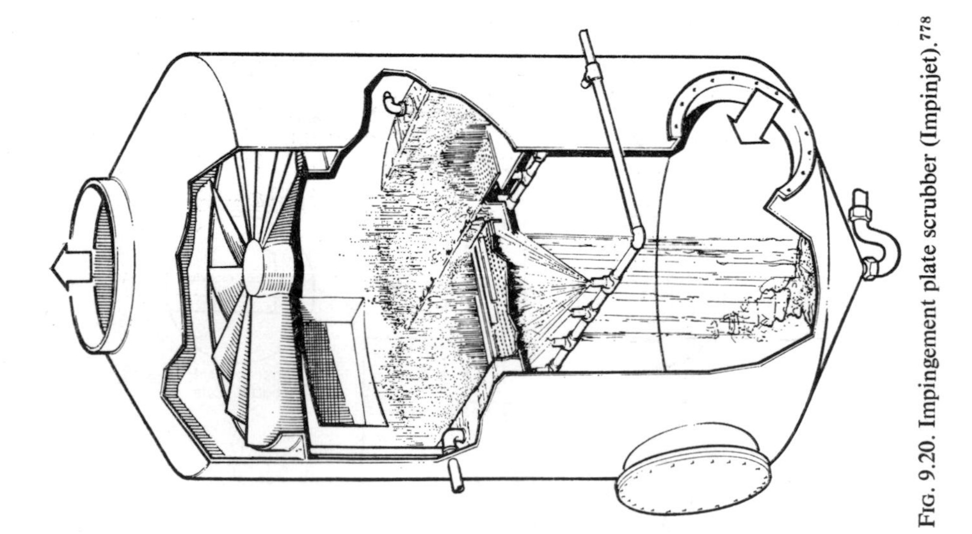

FIG. 9.20. Impingement plate scrubber (Impinjet).[778]

liquid it is used for mist collection. It has been used, after lining with PVC, for sulphuric-acid mists, which were effectively reduced from concentrations of 58 mg m^{-3} to 7 mg m^{-3} (88 per cent).[89]

Because of its high resistance to gas flow, the conventional packed bed is rarely used for particle collection, although sometimes used for mist collection because there is no problem of particle removal from the packing. The packings usually encountered are coke, Raschig rings, saddles or stone. A modified unit has been applied to particle collection. Here low-density spheres are contained between grids (Fig. 9.21), and the bed tends to float on the rising gases. In some ways the action is similar to the sieve plate column because little jets are formed between the spheres and impinge on the sphere above. The fluidized state of the bed prevents plugging, and so avoids the chief problem of operating a sieve plate scrubber for particles and sticky materials. The pressure losses for the floating-bed scrubber average 1·0 kPa, and the efficiencies obtained are similar to the other systems mentioned in this section.

9.7. DISINTEGRATOR SCRUBBERS

The scrubbers mentioned in the previous sections are all used for dealing with particles in the size ranges greater than 1 μm, but for collecting sub-micron-sized particles a scrubber must provide a very finely dispersed liquid so that while the droplets are sufficiently small

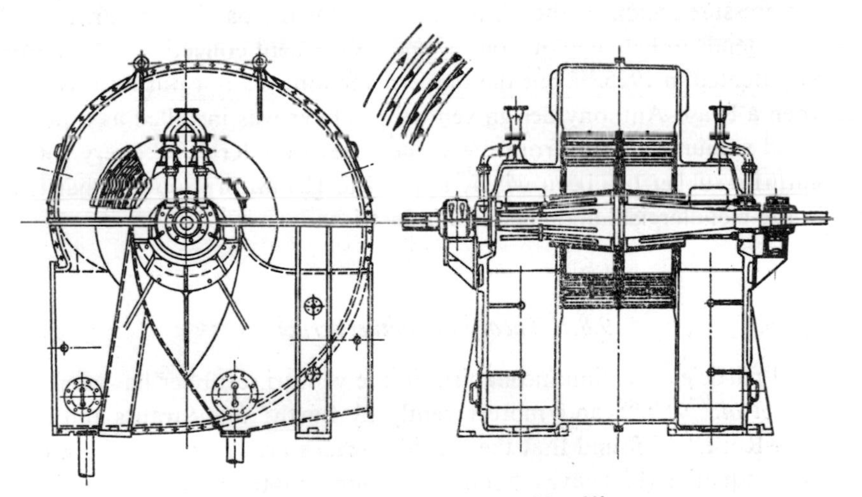

FIG. 9.22. Disintegrator scrubber.[344]

to have a low inertial impaction parameter, together they provide a very large surface area for diffusion to take place. One method of providing these fine liquid droplets is by shearing the liquid between a stator and a rapidly moving rotor and allowing the gas to pass through the apparatus. This is done in the disintegrator scrubber (Fig. 9.22), where water is injected axially and separated into fine droplets by the bars of the rotor and stator while the relative velocity of the gas is maintained at 60–90 m s^{-1} through the system. The particles, particularly those larger than 10 μm, tend to erode the vanes of the scrubber, and it is frequent practice to preclean the gases to a concentration below 2·3 g m^{-3} before passing them to the disintegrator. Power consumption is high: of the order of 5·2–7 kW/m^3 s^{-1} gas treated. The scrubber is very effective, removing 90 per cent of the 1 μm particles present and 70 per cent of the 0·5 μm particles, and has the additional advantage of being very small compared to other types of plant with similar efficiencies handling the same gas volumes. It is therefore often used as a standby unit for electrostatic precipitators on blast-furnace gas treatment.

9.8. VENTURI SCRUBBERS

While in the disintegrator scrubber the atomization of the liquid is caused by the mechanical action of the rotor and stator, in the venturi scrubber the velocity of the gases alone causes the disintegration of the liquid. The energy required for the plant is therefore almost wholly accounted for by the gas-stream pressure drop through the scrubber, apart from the small amount used in the liquid sprays. The other factor which plays a part in the effectiveness of the venturi scrubber is the condensation effect mentioned in the introduction to this chapter. If the gas in the reduced-pressure region in the throat is fully saturated (or preferably supersaturated) condensation will occur on the particles (acting as nuclei) in the higher pressure region of the diffuser. This helps the particles to grow, and the wet particle surface tends to help agglomeration and subsequent collection. The venturi scrubber was first patented in 1925,[345] but the modern version was not put into use for another 20 years, when a Pease–Anthony design venturi scrubber was installed as a pilot plant for the recovery of sodium sulphite from the waste gases of a Kraft recovery furnace. Since then the venturi scrubber has been widely applied to gas absorption and particle-removal problems in the metallurgical and chemical industries.

9.8.1. *Theory of venturi scrubbers*

628.53

The detailed particle collection mechanism in the venturi scrubber has been investigated by Johnstone *et al.*[403, 404, 405] and more recently by Barth[59] and Strauss and Lancaster.[830] Johnstone and Roberts[405] found that the specific surface area of the liquid in the scrubber calculated from equation (9.3) gave a good correlation with the particle collection efficiency, and also the rate of sulphur dioxide absorption or humidification of air (Fig. 9.23). In fact, the number of transfer units, N_t (where $N_t = -\ln(1-\eta)$), equation (9.9), for humidification can be found from the slope of the line in Fig. 9.23. This is greater than the slope for sulphur-dioxide absorption by a factor of 2·2, corresponding to the ratio of the diffusivities of water vapour and sulphur dioxide in air.

It was further found that the predominant mechanism in the venturi scrubber of conventional design is inertial impaction,[404] and the following equation correlates the experimental

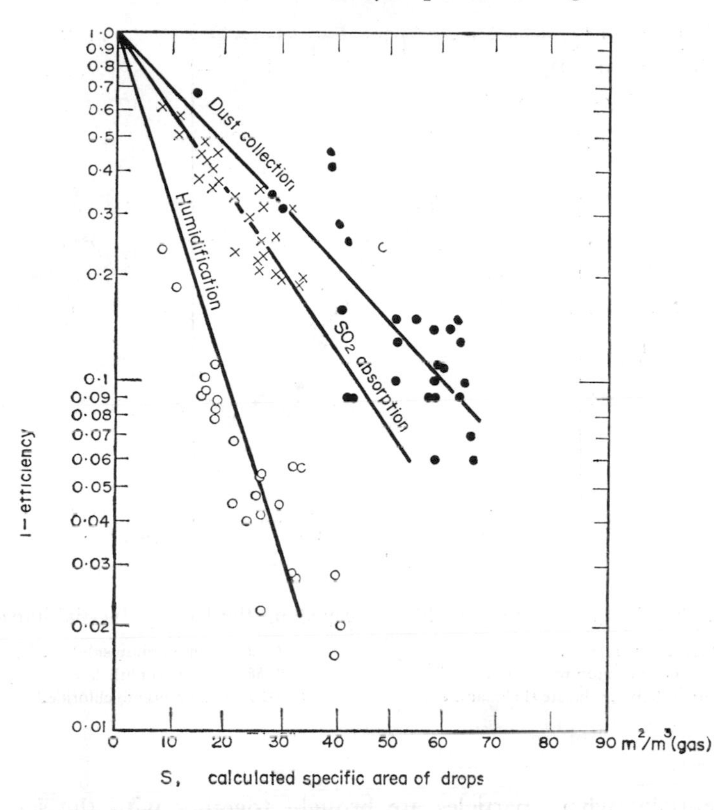

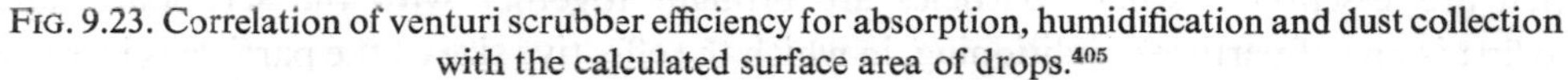
FIG. 9.23. Correlation of venturi scrubber efficiency for absorption, humidification and dust collection with the calculated surface area of drops.[405]

results. The efficiency of collection is given by

$$\eta = 1 - \exp\,(-Kn\sqrt{\psi}) \tag{9.4}$$

where n = concentration of droplets,
K = constant, which is a function of path length and specific surface area of droplets,
ψ = inertial impaction parameter.

A graph of the correlation of the experimental results with equation (9.4) is shown in Fig. 9.24.

Further experiments showed that to collect sub-micron size particles effectively, high throat velocities between 100 and 130 m s^{-1} were needed, but for given efficiencies with larger particles, lower throat velocities would suffice if the liquid consumption was very high.

The basic mechanisms in venturi scrubbers have been investigated both experimentally as well as theoretically by Lancaster and Strauss.[830] Two basic processes occur in scrubbers,

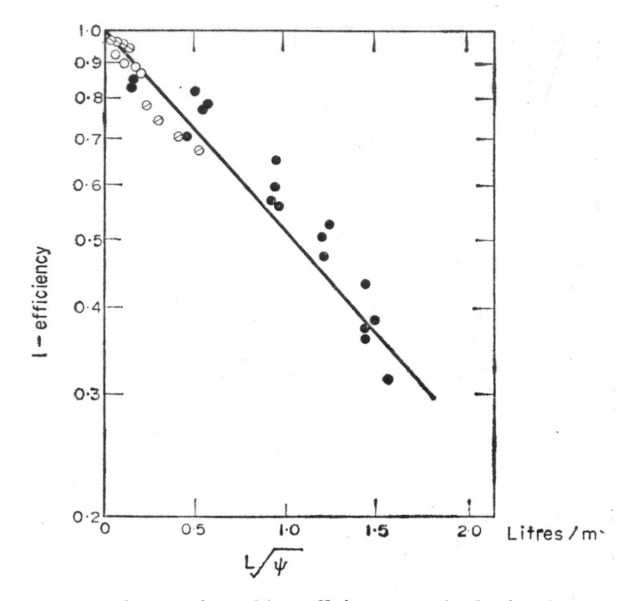

FIG. 9.24. Correlation of venturi scrubber efficiency on the basis of inertial impaction.[404]

L = Liquid flow rate—l/m³.
ψ = inertial impaction parameter.
● 10-μm dibutyl phthalate (Eckman).
⌀ 1·22-μm ammonium sulphite.
○ 0·58-μm dibutyl phthalate.
⦰ 0·27-μm ammonium chloride.

which are essentially where particles are brought together with the scrubbing liquid. The first is one of particle conditioning, in which the effective size of the particle is increased, making it less difficult to capture. The second is that of deposition of the particles on the surface of the scrubbing liquid. The particle conditioning may be either by particle agglomeration or by condensation of vapour on the particles, or a combination of both. These are, in turn, combinations of other fundamental processes. Thus particle conditioning proceeds by a nucleation process, followed by droplet growth and agglomeration, while the particle collection is a combination of inertial impaction and coagulation, thermophoresis, sweep diffusion (also called diffusiophoresis and Stephan flow) and the surface losses from droplets. The relative importance of all these has not yet been fully explored.

(i) *Particle conditioning*. The condensable vapour used in particle conditioning must be evenly distributed among the particles acting as nuclei to obtain the most effective usage of this vapour. This could be achieved by intimate mixing before initiating condensation, possibly by adiabatic expansion of the vapour and gas, which should produce this in a uniform manner. As an alternative, a jet of vapour can be injected into the aerosol at ambient pressure, relying on the quenching of the vapour by the aerosol to promote condensation. In both these processes we have homogeneous nucleation, where clusters of seventy to eighty molecules are produced by the vapour, and heterogeneous "nucleation" where the dust particles, acting as nuclei, adsorb a thin film of liquid on the surface, and this liquid-coated particle acts as a droplet of equivalent size.

The theory of homogeneous nucleation in a jet has been studied by Amelin and Belakov,[17] Higuchi and O'Konski[369] and Levine and Friedlander.[506] The latter produced a theory for the mixing zone in a jet for systems in which the Lewis number (Le) (which is the ratio of the Schmidt and Prandtl numbers; Le = Sc/Pr) is one for the vapour, which approximates the steam–air system. Based on Levine and Friedlander,[506] a zone of supersaturation conditions can be determined in which homogeneous nucleation is possible. Following experiments using a turbulent jet with glycerol vapour, these workers concluded that a very high degree of supersaturation must be obtained in a rapid mixing process for this effect to be detectable. The presence of gas ions increased the mass concentrations of droplets in the jet by several orders of magnitude.

In heterogeneous nucleation, the simplest theory is given by the Thompson–Gibbs relation, but this is not adequate, because it has been shown that a supersaturated vapour will not condense on the plane surface which has adsorbed a thick liquid film. On the other hand, the Volmer theory,[891] which has been verified experimentally by Twomey,[873] argues that the vertical supersaturation rises with increasing contact angle between liquid and solid. Qualitative results show that condensation on a wetted solid begins at the dew point, but for an unwetted solid supercooling of 0·015–0·020°C is necessary, which is equivalent to supersaturation of about 101 per cent.

(ii) *Droplet growth rate.* The growth of droplets, in the absence of Stephan flow and thermophoresis, can be deduced from Ficks Law [equation (3.1)], which, on solution, gives a mass growth rate $dm/d\theta$

$$\frac{dm}{d\theta} = I_0\left(1+\frac{r}{\sqrt{(\pi \mathcal{D}\theta)}}\right) \tag{9.5}$$

where I_0 = quasistationary droplet growth rate,
$\mathcal{D}$ = diffusivity of the vapour,
θ = time, and
r = droplet radius.

(iii) *Particle agglomeration.* This can be calculated from the Smoluchowski theory of Brownian coagulation (Section 11.2). The contribution of turbulent agglomeration is negligible in scrubbers.

(iv) *Particle collection.* The contribution by thermophoresis or thermal precipitation can be calculated by the equations given in section 11.6, while the contribution by inertial impaction can be deduced from the equations for droplet size (9.2) and surface (9.3), and by equation (9.4).

Near the surface on which vapour is condensing there is a hydrodynamic flow of the gaseous medium which is directed to the surface, called *Stephan flow*, *diffusiophoresis*, or *sweep diffusion*. Waldmann and Schmitt and, more recently, Sparks and Pilat,[786] have given a detailed review of the theory of diffusiophoresis.[897] The rate of the flow u_s is given by

$$u_s = \frac{\mathcal{D}}{c}\cdot\frac{dc}{d\varrho} \tag{9.6}$$

where c = vapour concentration,
ϱ = distance from the droplet centre.

For a droplet, radius r

$$u = \frac{\mathcal{D}(c_s - c_\infty) r M_g}{c_\infty \varrho^2 M_v} \tag{9.7}$$

where c_s = vapour concentration at drop surface,
c_∞ = concentration in vapour (i.e. at $\varrho = \infty$), and
M_g, M_v = molecular weights of gas and vapour respectively.

If the gas through which the vapour is diffusing contains small particles, these will move with the flow and at about the same velocity, u_ϱ. This has been considered for small spherical particles ($r < \lambda$) on the basis of the Chapman–Enskog theory by Waldmann[896]

$$u_\varrho = \frac{(M_v)^{\frac{1}{2}}}{\{y_v \sqrt{(M_v)} + y_g \sqrt{(M_g)}\}} \cdot \frac{\mathcal{D}}{y_g} \cdot \frac{dy_v}{dx} \tag{9.8}$$

where y_v and y_g are the concentrations of vapour and gas, respectively. It has been shown experimentally by Goldsmith, Delafield and Cox[308] that the velocities of sub-micron particles in a vapour gradient in air is in excellent agreement with this equation.

In carefully controlled experiments in which the effects of the different mechanisms were separated out by adjusting gas, vapour and liquid temperatures and concentrations, Lancaster and Strauss[478] have been able to show that the dominant mechanisms are particle build up and subsequent impaction collection. Improved particle build up resulting from steam injection had also been reported in the earlier work by Fahnoe, Lindroos and Abelson.[247] The work by Lancaster and Strauss shows that substantially all condensation and build up takes place in the mixing zone of the jet. As a result of this investigation it was concluded that inducing condensation offers a method of obtaining considerable improvement in scrubbing plant. The technique, however, because of the relative inefficiency of steam usage, is not economical, except when low-cost low-pressure waste steam is available. Improvements in jet design and steam mixing could overcome these economic disadvantages.

In another fundamental approach to estimating collection efficiency Barth considers the relative velocity of droplet and gas as the droplet is released in the venturi throat and then accelerated. So far this has also not led to a correlation which is easily applied to the prediction of venturi scrubber efficiency. In general however it appears that effective scrubbing is directly related to the energy expended in the process. The change in kinetic energy, ΔE_K, on liquid–gas impact can be found from

$$\Delta E_K = \frac{1}{2} \frac{m_G m_L}{m_G + m_L} (U_G - U_L) \tag{9.9}$$

where m_G and m_L are the mass of gas and liquid respectively used per unit time, and U_G and U_L are the gas and liquid velocities before impact, along the axis of the venturi.

This relation shows that if the liquid is sprayed across the throat, high energy losses occur because U_L is zero. However, if the liquid is sprayed from a central nozzle facing downstream, with a high velocity U_L, the energy requirements are much lower. In practice the principal difference between venturi scrubbers is the way in which the liquid is introduced and subsequently removed.

It has been suggested by some workers that the wetting ability of the dust being collected plays a major role in the collection efficiency of scrubbers, particularly venturi scrubbers. The addition of quite large quantities of a wetting agent in the collection of carbon black was not found to be effective. The study of the wetting characteristics under standard conditions of a number of dusts which are commonly collected in different types of scrubbers has been carried out by Weber[912] who measured the rate of capillary rise of water in a tube, 0·5 cm^2 in area, containing the dust. His results, shown in Table 9.4, give a great variation in

TABLE 9.4. WETTING RATES OF DIFFERENT DUSTS BY WATER[912] (*Note.* Sample cross-section 5 mm^2, sample height 5 mm)

Dust source	Dust density ($kg\ m^{-3}$)	Capillary rise ($mm\ s^{-1} \times 10^2$)
Soot from boiler	2570	$4·46 \times 10^{-3}$
Hot blast cupola (basic)	2440	$9·48 \times 10^{-2}$
Hot blast cupola (acid)	2680	$2·26 \times 10^{-1}$
Cold blast cupola	2710	5·98
Blast furnace	4680	46·8
Boiler	2300	96·3
Fine sand (quartz)	2510	192·6

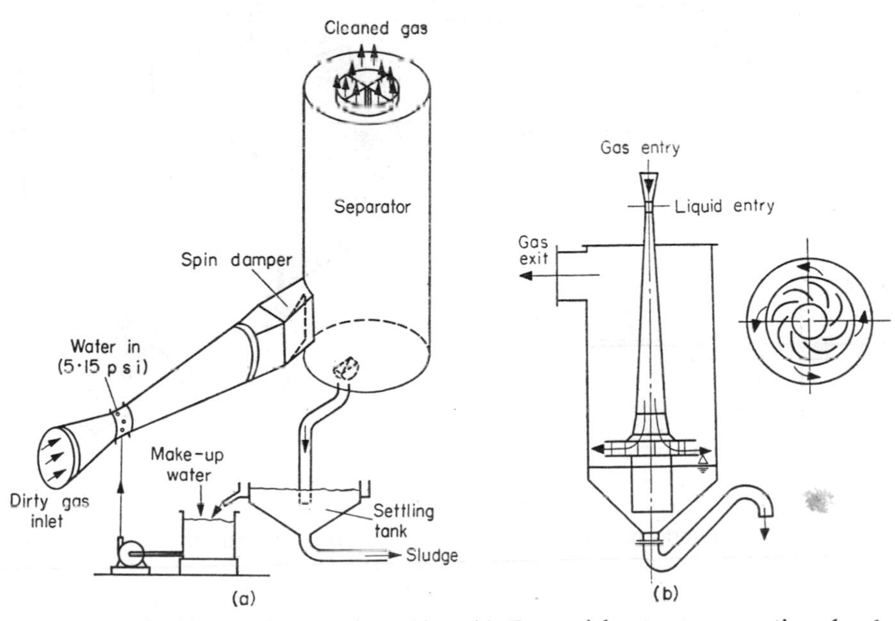

FIG. 9.25. Pease–Anthony type venturi scrubber. (a) Tangential entry to separating chamber.[410] (b) Turbine entry to separating chamber.[326]

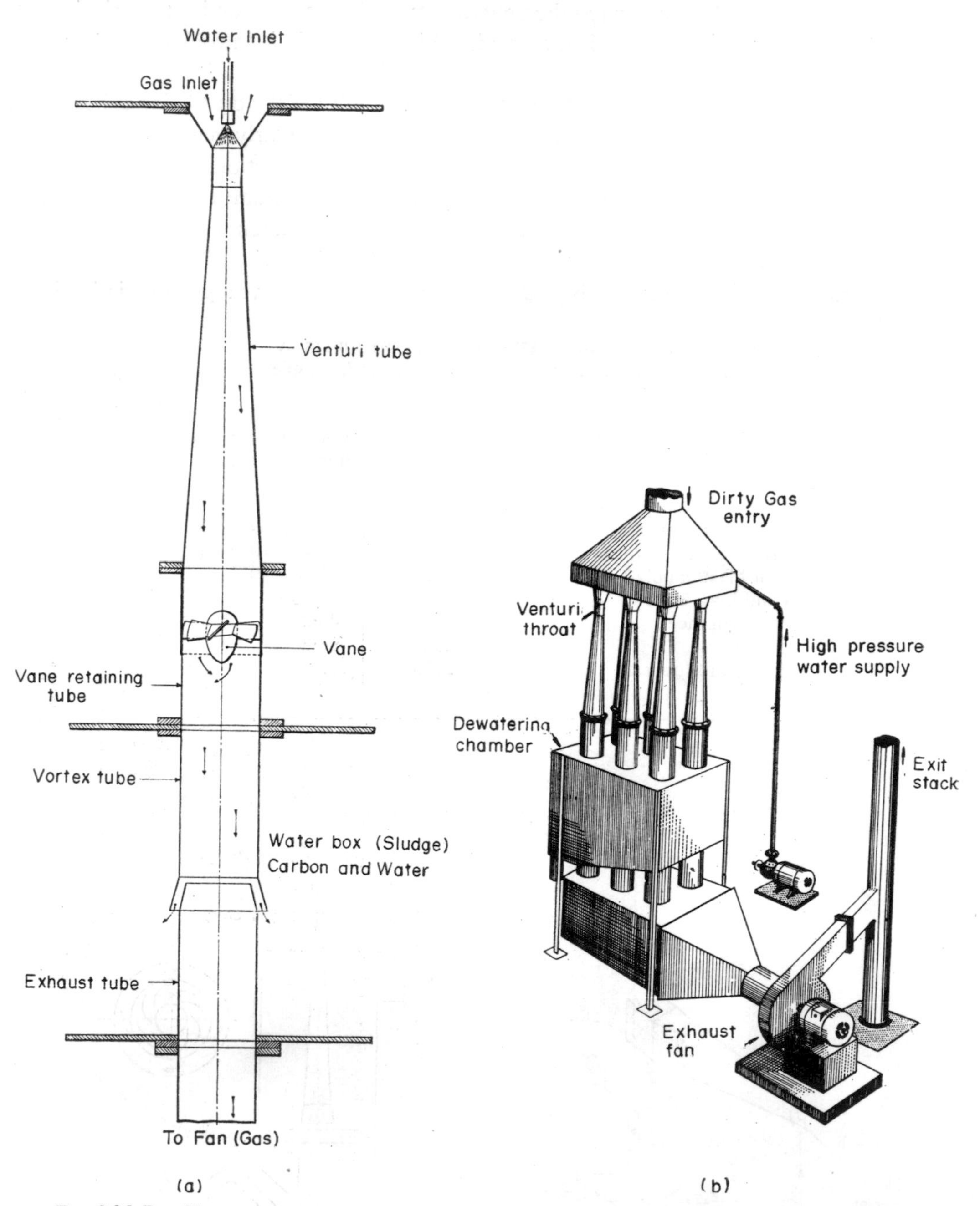

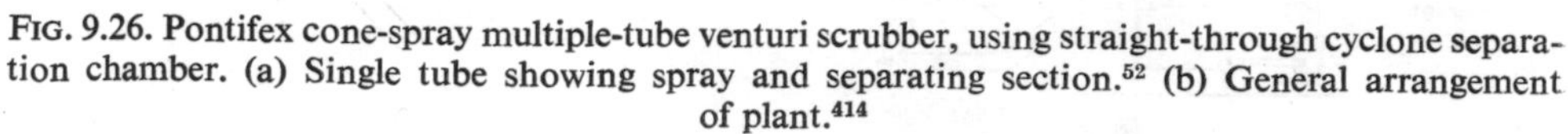
FIG. 9.26. Pontifex cone-spray multiple-tube venturi scrubber, using straight-through cyclone separation chamber. (a) Single tube showing spray and separating section.[52] (b) General arrangement of plant.[414]

the measured rise, which is not observed by a similar difference in collection efficiency. This indirect evidence tends to confirm the earlier observation, that dust wettability or the addition of a wetting agent plays little part in the actual mechanism of collection. However, as the addition of wetting agents can affect the size of droplets produced in sprays, this could play a secondary role.[826]

9.8.2. Practical venturi scrubbers

In the first design developed by Anthony *et al.*[411] and called the Pease–Anthony scrubber, the liquid is sprayed in at right angles to the throat, and is removed by a cyclone separation chamber (Fig. 9.25a), or a turbine entry cyclone settling chamber (Fig. 9.25b). In the second design, the liquid is introduced from a cone spray just ahead of the throat, and is subsequently removed using either a straight-through cyclone separator in the Pontifex pattern[52] (Fig. 9.26) or a fairly deep settling tank in the Waagner–Biro design[620] (Fig. 9.27). The third type uses a jet spray directed through the venturi throat, and by employing a steam or air-pressure nozzle the scrubber can produce its own draft instead of forcing the gases through the throat. This design (Fig. 9.28), developed by Schütte and Korting,[910] is an effective odour and organic vapour collector, but has low efficiency with respect to particles. In the fourth design, by the Svenska Flaktfabriken, called the S.F. venturi (Fig. 9.29) the

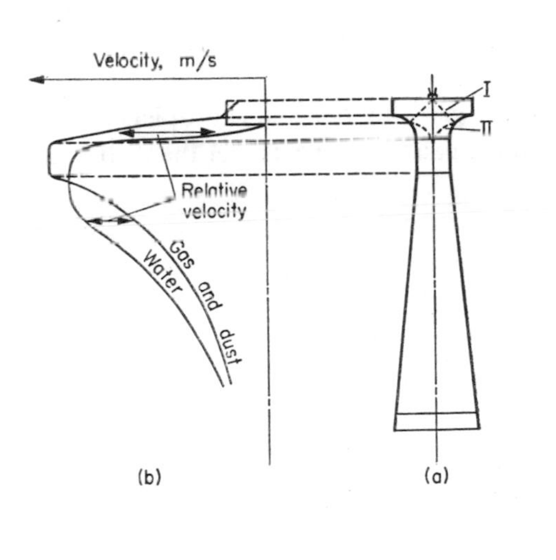

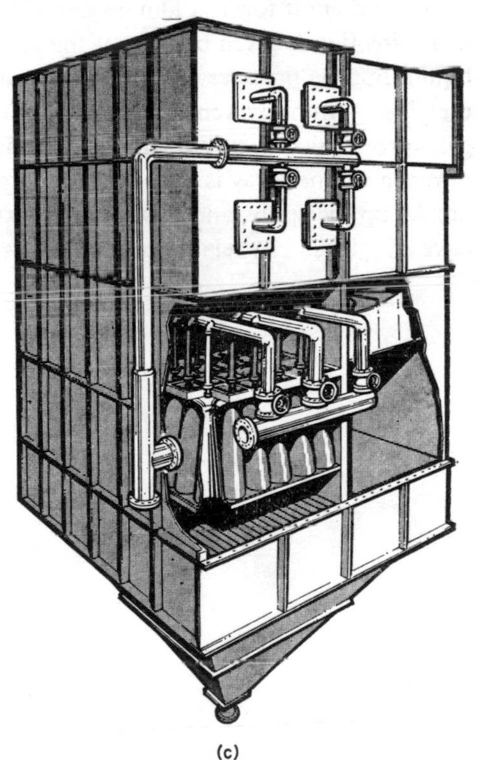

FIG. 9.27. Waagner–Biro cone spray multiple tube scrubber.[620] (a) Cone spray showing double cone. (b) Relative velocities of gas, dust and water. (c) General arrangement of plant.

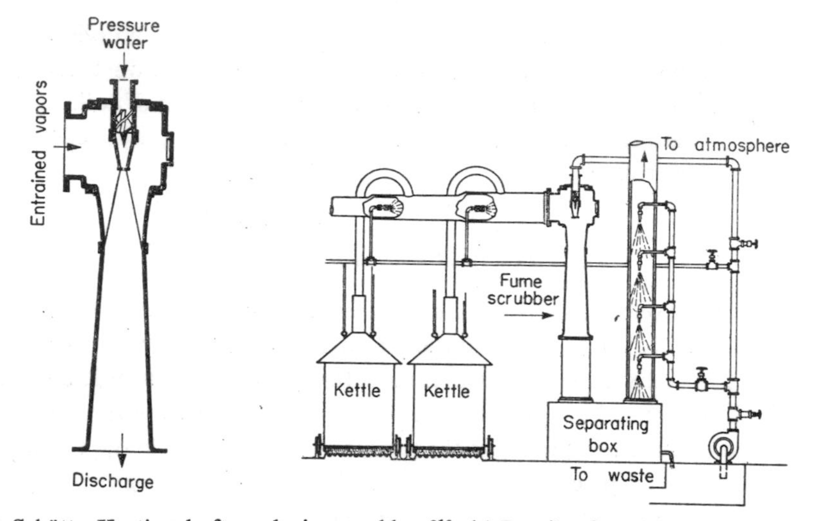

FIG. 9.28. Schütte–Korting draft-producing scrubber.[910] (a) Details of scrubber. (b) General arrangement of plant for varnish kettles fumes. The spray nozzles at each juncture of the collecting lines prevent possible flame propagation between kettles.

water is not directed across the throat by nozzles, but is accelerated through the converging section, where it forms a film on the wall, and then across the throat by the gas itself. The final direction is given by a lip at the edge of the throat. There is no power input into the liquid, but the pressure drop by the gas stream is very high, being of the same order as in the Pease–Anthony venturi scrubber. The S.F. venturi has been widely used for waste gases from calcium carbide furnaces, cement kilns and steel-making processes.

When a cone spray is introduced before the throat it has been shown that the gas stream breaks up the spray, and this is a first scrubbing of the gases by the coarse droplets, which have a high velocity relative to the gas stream. When the coarse droplets enter the throat the

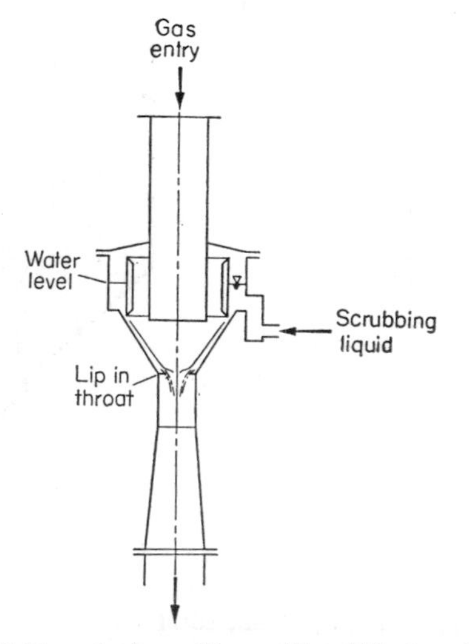

FIG. 9.29. S–F venturi scrubber with self-induced spray.[326]

THE POLYTECHNIC OF WALES
LIBRARY
TREFOREST

high throat velocity disintegrates them, and the fine droplets moving with the gas stream collect particles by a combination of condensation and agglomeration as well as impingement. This pattern uses more liquid than the Pease–Anthony model but this is compensated for by the lower pressure drop in the unit.

Several other venturi scrubber-type devices have been developed commercially in recent years. The first of these is Research Cottrell's *Flooded Disc Scrubber* (Fig. 9.30) where

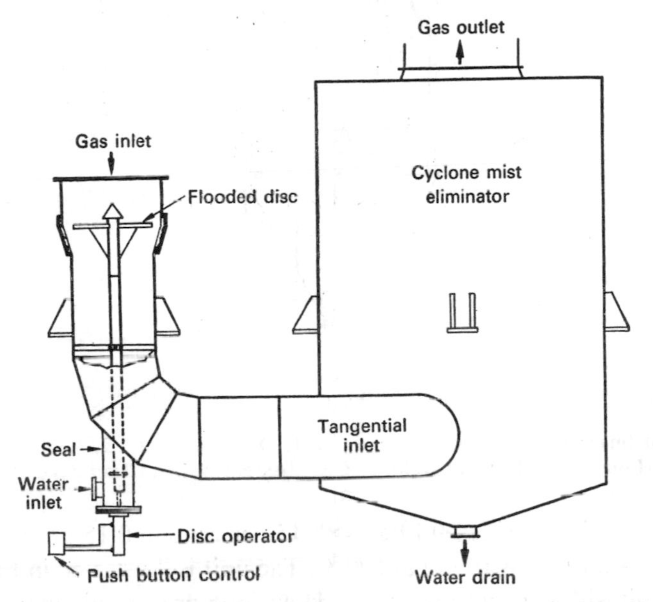

FIG. 9.30. Flooded disc scrubber and demister (Research Cottrell).

a circular disc is mounted coaxially in a conical vertical duct section. The disc is flooded by the liquid which impinges on the distribution cone, while the gas, passing across the disc and across the periphery, shears the liquid into droplets of the order of 50–150 μm. The disc is able to be moved manually or automatically so that the velocity in the annular space can be kept constant with varying gas-volume rate. The size of the commercial unit is from 0·3 to 2 m diameter, with a corresponding height (H) from 2 m to 6·5 m.

Walker and Hall[899] report satisfactory operation of the flooded disc scrubber on a lime kiln, an electric furnace, an open-hearth furnace, a basic oxygen furnace and a blast furnace. Gas volume varied from about 34,000 to 340,000 $m^3\ h^{-1}$, with pressure drops averaging 10 kPa and fume outlet concentrations generally below 0·2 $g\ m^{-3}$, representing efficiencies better than 97 per cent.

Another development is the Kinpactor (American Air Filter Co. Inc.), which consists of a comparatively narrow-slit venturi throat into which jets of water are introduced in either side, just upstream, and protected by deflector plates. Manual or automatic reamers are provided for the nozzles to guard against jet plugging due to solids in recirculated water.

Another commercial venturi scrubber which has been reported by Storch,[821] called the *mirror-image flow* scrubber, has been used in Czechoslovakia for ferromanganese fumes and openhearth furnaces in which oxygen lancing is used. In the latter case an overall efficiency

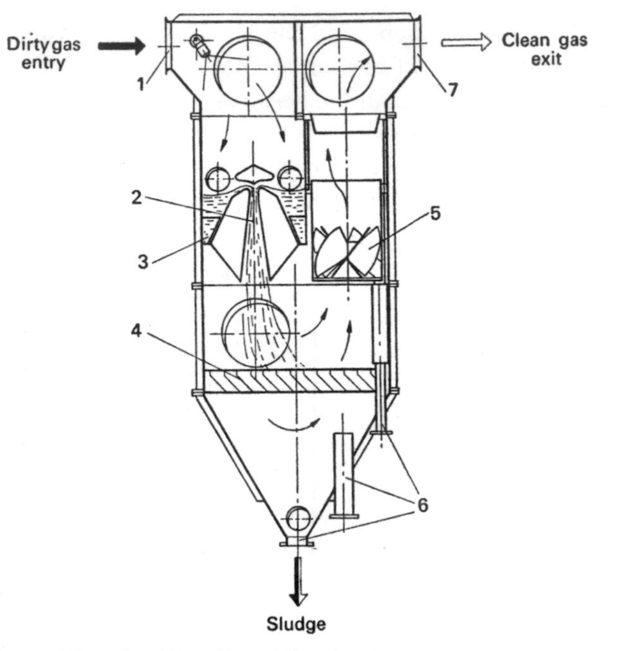

FIG. 9.31. Mirror-Image Flow Scrubber (Storch[821]). 1, Dirty gas entry. 2, Venturi. 3, Water entry. 4, Louvres. 5, Demister cyclone. 6, Sludge removal. 7, Clean gas exit.

of 97 per cent was achieved with 7·0 kPa, with 15 per cent $< 0{\cdot}15\ \mu m$ and 65 per cent $< 0{\cdot}5\ \mu m$, using a gas-entry temperature of 500°C. The unit is illustrated in Fig. 9.31 and consists of a single unit with the venturi throat and cyclone droplet eliminator.

Details of industrial installations can be found in the literature, and a number of references for particular applications are listed in Table 9.5. Table 9.6 gives the efficiency found with

TABLE 9.5. APPLICATIONS OF VENTURI SCRUBBERS IN INDUSTRY

Application	Type of scrubber	Reference	Year
Electric arc furnace	Pease–Anthony	Hohenberger[374]	1961
Oxygen lancing pig iron in ladle	Pease–Anthony	Gledhill, Carnall and Sargent[305]	1957
	Irsid	Septier[755]	1958
L.D. furnaces	Pease–Anthony	Baum[64]	1959
	Pease–Anthony	Harr, Wagner and Willmer[339]	1959
	Pease–Anthony	Krijgsman[465]	1960
	Waagner–Biro	Pallinger[620]	1962
Open hearth	Pease–Anthony	Jones[410]	1949
Carbon black	Pontifex	Bainbridge[52]	1961
Salt cake	Pease–Anthony	Collins, Seaborne and Anthony[166]	1948
Sulphuric acid mist	Pease–Anthony	Jones and Anthony[411]	1952
Phosphoric acid mist	Pease–Anthony	Brink and Contant[124]	1958

TABLE 9.6. PERFORMANCE DATA ON PEASE–ANTHONY VENTURI SCRUBBER[155]

Source of gas	Dust or mist	Particle size (μm)	Loading ($g\ m^{-3}$)		Average efficiency of collection %
			inlet	exit	
Iron and Steel Industry					
Gray iron cupola	Iron, coke, silica dust	0·1 – 10	2·3 – 4·6	0·115 –0·346	95
Oxygen steel converter	Iron oxide	0·5 – 2	18·4 –23	0·115 –0·184	98·5
Steel open-hearth furnace (scrap)	Iron and zinc oxide	0·08– 1·00	1·15– 3·46	0·69 –0·138	95
Steel open-hearth furnace (oxygen lanced)	Iron oxide	0·02– 0·50	2·3 –13·8	0·02 –0·16	99
Blast furnace (iron)	Iron ore and coke dust	0·5 – 20	6·9 –55	0·018 –0·115	99
Electric furnace	Ferro-manganese fume	0·1 – 1	23 –28	0·092 –0·184	99
Electric furnace	Ferro silicon dust	0·1 – 1	2·3 –11·5	0·23 –0·69	92
Rotary kiln—iron reduction	Iron, carbon	0·5 – 50	6·9 –23	0·23 –0·69	99
Crushing and screening	Taconite iron ore dust	0·5 –100	11·5 –58	0·0115–0·023	99·9
Chemical Industry					
Acid—humidified SO_3	Sulphuric acid mist				
(a) scrub with water		—	0·30	0·0016	99·4
(b) scrub with 40% acid		—	0·405	0·00275	99·3
Acid concentrator	Sulphuric acid mist	—	0·133	0·0032	97·5
Copperas roasting kiln	Sulphuric acid mist	—	0·200	0·0021	99
Chlorosulfonic acid plant	Sulphuric acid mist	—	0·745	0·0078	98·9
Phosphoric acid plant	Orthophosphoric acid mist	—	0·193	0·0038	98
Dry ice plant	Amine fog	—	0·025	0·0021	90
Wood distillation plant	Tar and acetic acid	—	1·07	0·0575	95
Titanium chloride plant, titanium dioxide dryer	Titanium dioxide, hydrogen chloride fumes	0·5 – 1	2·3 –11·5	0·115 –0·23	95
Spray dryers	Detergents, fume and odor	—	—	—	
Flash dryer	Furfural dust	0·1 – 1	4·6 –13·8	0·115 –0·345	95
Non-ferrous Metals Industry					
Blast furnace (sec. lead)	Lead compounds	0·1 – 1	4·6 –13·8	0.115 –0·345	99
Reverberatory lead furnace	Lead and tin compounds	0·1 – 0·8	2·3 –4·6	0·276	91
Ajax furnace—magnesium alloy	Aluminium chloride	0·1 – 0·9	6·9 –11·5	0·046 –0·115	95
Zinc sintering	Zinc and lead oxide dusts	0·1 – 1	2·3 –11·5	0·115 –0·23	98

[continued]

TABLE 9.6 *(continued)*

Source of gas	Dust or mist	Particle size (μm)	Loading ($g\ m^{-3}$)		Average efficiency of collection %
			inlet	exit	
Reverberatory brass furnace	Zinc oxide fume	0·05– 0·5	2·3 –18·4	0·23 –1·15	95
Mineral Products Industry					
Lime kiln	Lime dust	1 – 50	11·5 –23	0·115 –0·345	99
Lime kiln	Soda fume	0·3 – 1	0·46–11·5	0·023 –0·115	99
Asphalt stone dryer	Limestone and rock dust	1 – 50	11·5 –34·5	0·115 –0·345	98
Cement kiln	Cement dust	0·5 – 55	2·3 – 4·6	0·115 –0·23	97
Petroleum Industry					
Catalytic reformer	Catalyst dust	0·5 – 50	0·207	0·0115	95
Acid concentrator	Sulphuric acid mist	—	0·136	0·0032	97·5
TCC catalyst regenerator	Oil fumes	—	0·760	0·0080	98
Fertilizer Industry					
Fertilizer dryer	Ammonium chloride fumes	0·05– 1	0·23– 1·15	0·115	85
Superphosphate den and mixer	Fluorine compounds	—	0·308	0·0055	98
Pulp and Paper Industry					
Lime kiln	Lime dust	0·1 – 50	11·5 –23	0·115 –0·345	99
Lime kiln	Soda fume	0·1 – 2	4·6 –11·5	0·023 –0·115	99
Black liquor recovery boiler	Salt cake	—	9·3 –13·8	0·92 –1·38	90
Miscellaneous					
Pickling tanks	Hydrogen chloride fumes	—	0·0253	0·0023	90
Boiler flue gas	Fly ash	0·1 – 3	2·3 – 4·6	0·115 –0·184	98
Sodium disposal incinerator	Sodium oxide fumes	0·1 – 0·3	1·15– 2·3	0·046	98

The efficiencies shown above are average values for a particular plant or group of installations operating under a specific set of conditions.

venturi scrubbers in plant installations. The draft-producing venturi scrubber has been used in the paint industry for varnish-kettle fume scrubbing.[910]

To cover the throat of a large venturi scrubber with liquid presented a number of problems. In an early design several sprays were distributed within the throat, but the high velocities rapidly eroded the nozzles and pipes. Present practice is to use a narrow, rectangular throat, with sprays from both sides. If the gas volumes handled vary, the throat can be constructed in such a way that the cross-section can be altered by moving one side. When a cone spray is used and is to cover the whole orifice, this is difficult for a large tube, and a number of small tubes in parallel are used (Figs. 9.26b and 9.27c). The capacity of the draft-producing

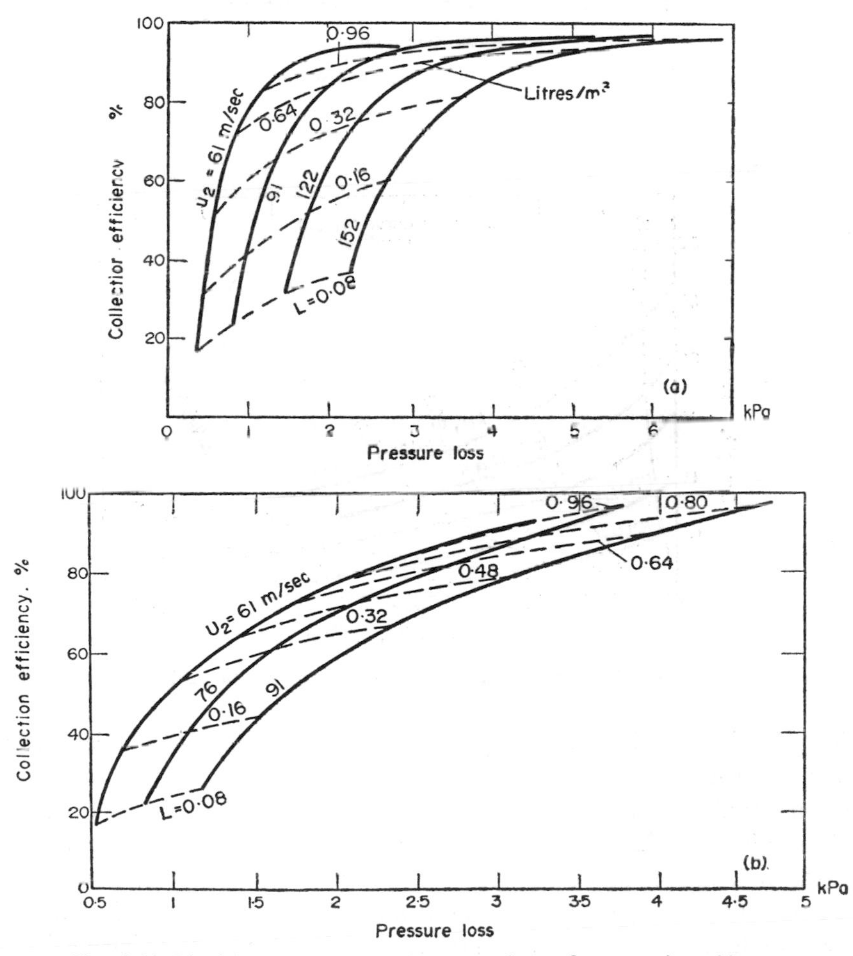

FIG. 9.32. Liquid consumption and pressure losses for venturi scrubbers.
(a) Pease–Anthony ventury scrubber performance for solid particles (salt cake fume from Kraft Mill).[403]
(b) Pease–Anthony venturi scrubber performance for mist droplets (sulphuric acid mist).[403]

FIG. 9.32 *(continued)*

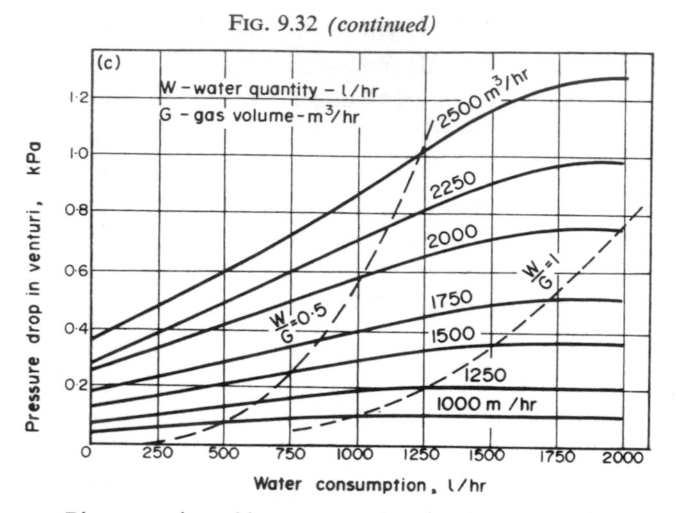

(c) Waagner–Biro venturi scrubber pressure drop/water consumption curves.[620]

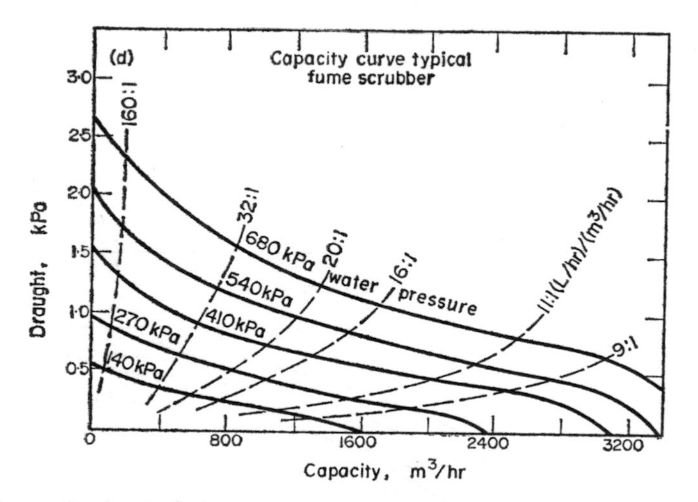

(d) Schutte–Korting draft-producing venturi scrubber performance draft/capacity.[910]

venturi is fairly high, and it is only used on small plants, so one or sometimes two units in parallel are adequate.

Water-consumption and pressure-drop data have been published and are shown in graphical form in Fig. 9.32. Curves (a) and (b) are for Pease–Anthony scrubbers with solid particles (salt cake fume) and mist droplets (sulphuric acid mist), while sections (c) and (d) show the data for the Waagner–Biro and draft-producing venturi respectively.

In an attempt to reduce the high-pressure losses which occur in the high efficiency venturi scrubbers, a design called the Solivore, which emphasizes the condensation on the particles as droplets, has been developed (Fig. 9.33a). The dust enters at the top of the chamber, where it is saturated with a fine spray which also precipitates the coarse particles. The saturated gases then enter the venturi section, where, with the velocity rise, the pressure falls, and more drops evaporate. Further on the gases are slowed down, the pressure increases again, and condensation then takes place on the particles, which agglomerate readily and are deposited by a coarse spray. In a single section the pressure drop is only 250 Pa, while for a four-stage unit (Fig. 9.32b) the pressure drop is less than 1500 Pa.

The efficiency of this unit is high. For example, electric furnace fume concentration was reduced from 5·5 to 0·45 g m^{-3}, representing an efficiency of 99·1 per cent. Although this scrubber has low direct-energy requirements for the gases, efficient collection needs fine atomization of the liquid, and this requires further energy in the form of compressed air or water pumps, so the net energy gain in this unit may not be so high as it appears, as well as the fact that the water requirements are rather high.

9.9 FOAM SCRUBBERS

The collection of fine particles and gases requires a large collecting surface area. As an alternative to spray droplets, foam has been suggested as being a suitable collecting agent.

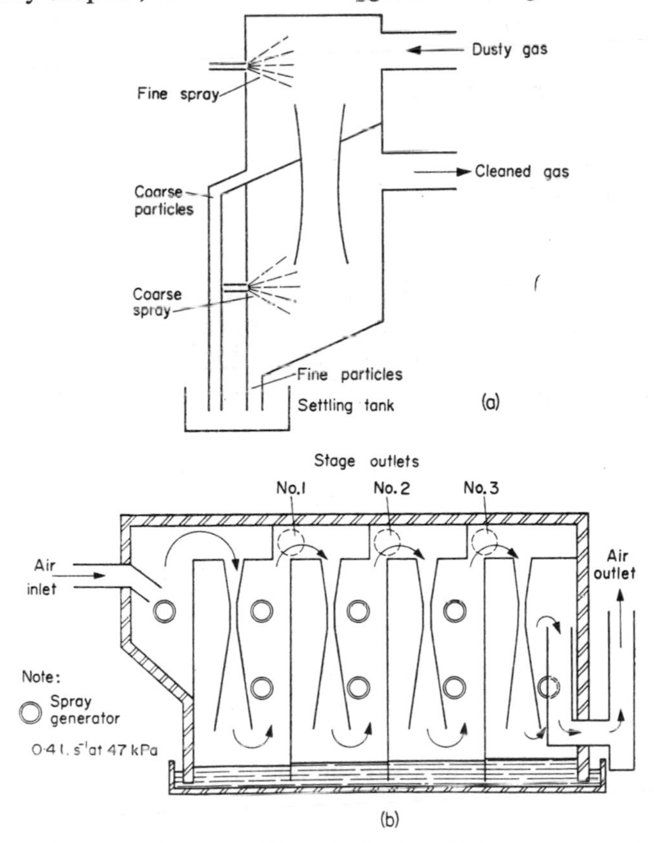

FIG. 9.33. Condensation venturi scrubber:[466] (a) Single stage. (b) Four-stage experimental "Solivore" collector.

The simplest type of equipment, which appears to be widely used in the Soviet Union,[23] is of the sieve-tray pattern.

The development seems to be largely based on the work of Pozin *et al.*[656] who have found that the efficiency of collection depends not only on the physical properties of the dust but also on the operating conditions which determine the height of the mobile foam layers on the plates, and have developed empirical equations by dimensional analysis that can be used in the design of foam scrubbers. The efficiency, η, of this scrubber for dust which is easily wetted can be calculated from

$$\eta = 0{\cdot}89\left(\frac{Ul}{g(h_c - h_b)^2}\right)^{0{\cdot}005}\left(\frac{\varrho_s d^2 U}{g\mu D'}\right)^{0{\cdot}04} \tag{9.10}$$

while if the dust is not easily wetted

$$\eta = 0{\cdot}89\left(\frac{Ul}{g(h_c - h_b)^2}\right)^{0{\cdot}005}\left(\frac{\varrho_s d^2 U}{g\mu D'}\right)^{0{\cdot}235} \tag{9.11}$$

where U = free space velocity of the gases (m s^{-1}),
l = liquid flow rate on the tray,
h_c = height of overflow orifice (m),
h_b = height of baffle (m),
g = gravity acceleration (9·81 m s^{-2}),
ϱ_s = density of dust (kg m^{-3}),
d = diameter of dust (m),
μ = viscosity of liquid (PaS),
D' = diameter of holes in sieve tray (m).

It was also reported by the Russian workers that advantages in collection could be gained by the addition of a surface-active agent,[962] in contrast to the work reported below. The formula holds well for foam heights between 40 and 200 mm, which are common practice. It was found experimentally that apatite and nepheline dust, with an average particle size of 20–25 μm and a concentration of 2 g m^{-3}, was effectively collected with foam heights of 60–100 mm and completely removed by 200 mm of foam.

Following the earlier Russian work reported above, Taheri and Calvert[849] carried out an extensive laboratory study in both a 75 mm square and a 50 mm diameter circular column, with hydrophilic and hydrophobic aerosols, with a range of particle sizes from $< 0{\cdot}6$ μm to 10 μm. The gas-flow rates ranged from 1050 kg h^{-1} m^{-2}, the liquid flow rates from 2500 to 12,500 kg h^{-1} m^{-2}, and the diameter of the sieve-tray perforations from 1·5 mm to 4·5 mm. The experimental results show a minimum collection efficiency of about 5 per cent for 0·6 μm particles, about 30 per cent for 1·4-μm particles, and between 95 and 100 per cent for 10-μm particles. Higher gas-flow rates had a marked effect in improving collection efficiency (i.e. at 3 μm, three times the flow gave an improvement from 60 to 90 per cent) while the effect of liquid flow was much less marked (i.e. at 3 μm, five times the flow, 70 to 80 per cent). Smaller perforations improved collection efficiency, while the hydrophilic aerosols were much more effectively collected than the hydrophobic one. In contrast to the Russian work, the addition of a surface-active agent reduced collection efficiency; it was observed that surface-active agents decreased the foam density and the circulation within the bubble, thus decreasing the effectiveness of the scrubbing. Taheri and Calvert[849] correlated the experimental collection efficiency η for their range of concentrations and particle sizes of hydrophilic aerosols by

$$\eta = 1-\exp(-40F_1^2\psi) \tag{9.12}$$

Here, ψ is an inertial impaction parameter based on the velocity through the perforation u_p, and the perforation diameter D_h, viz.

$$\psi = u_p \varrho_p d^2/9\mu D_h$$

the other terms having their usual meaning. F_1 is the foam density; the ratio of volume of clear liquid over the volume of foam. The total foam height can be found during operation, while the volume of clear liquid on a tray at the end of a run can be measured after quickly topping both the air and water flows. The correlation is shown in Fig. 9.34.

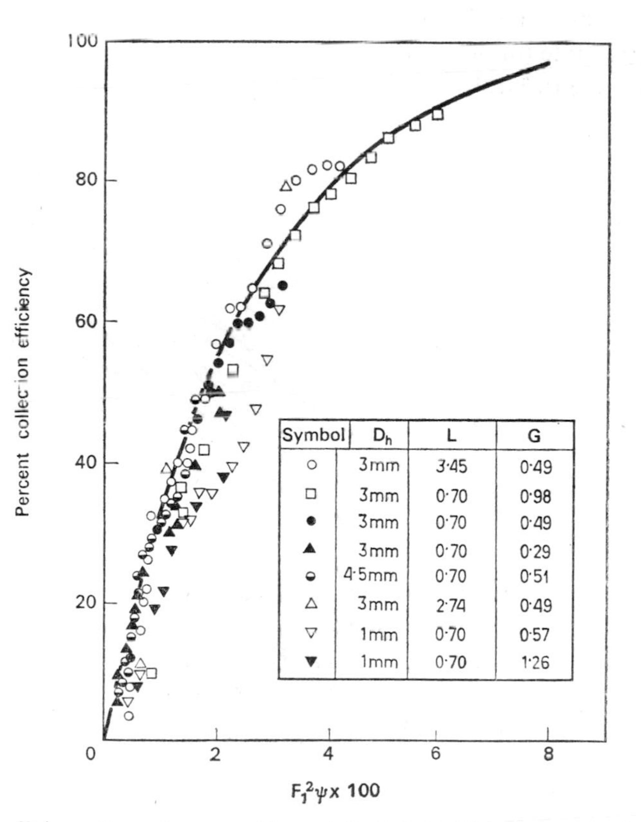

FIG. 9.34. Collection efficiency for a foam scrubber, plotted against the parameter $F_1^2\psi$, where F_1 is the foam density $\left(\text{i.e. } \frac{\text{Volume of clear liquid}}{\text{Volume of foam}}\right)$ and ψ is the inertial impaction parameter, based on diameter of holes (D_h). D_h = orifice diameter (mm), L = liquid flow rate kg m^{-2} s^{-1}, G = air flow rate kg m^{-2} s^{-1}.

9.10. CORRELATION OF SCRUBBER EFFICIENCY

It has been mentioned several times in the preceding sections, particularly with reference to the efficiency of venturi scrubbers, that for the same dusts higher efficiencies require greater power consumption, or for finer dusts, the same efficiency needs more power.

These ideas have been correlated on a quantitative basis, particularly by Semrau.[494, 752, 753, 754] The total pressure loss for the scrubber, P_T, is made up of two parts, the pressure loss of the gas passing through, P_G, and of the spray liquid during atomization, P_L. Semrau has given formulae which give approximate value for P_G and P_L based on power consumption (in kW) per unit volume of gas (1000 m³ h⁻¹). These are

$$P_G = 2{\cdot}724 \times 10^{-4}\,\Delta h \text{ (kWh/1000 m}^3) \tag{9.13}$$

where Δh = pressure loss across the unit in Pa, and

$$P_L = 0{\cdot}28\,\Delta p_{\text{atm}} Q'_L/Q'_G \text{ (kWh/1000 m}^3) \tag{9.14}$$

where Δp_{atm} = pressure in nozzle (100 kPa)
Q'_L = liquid rate (m³ h⁻¹)
Q'_G = gas rate (m³ h⁻¹).

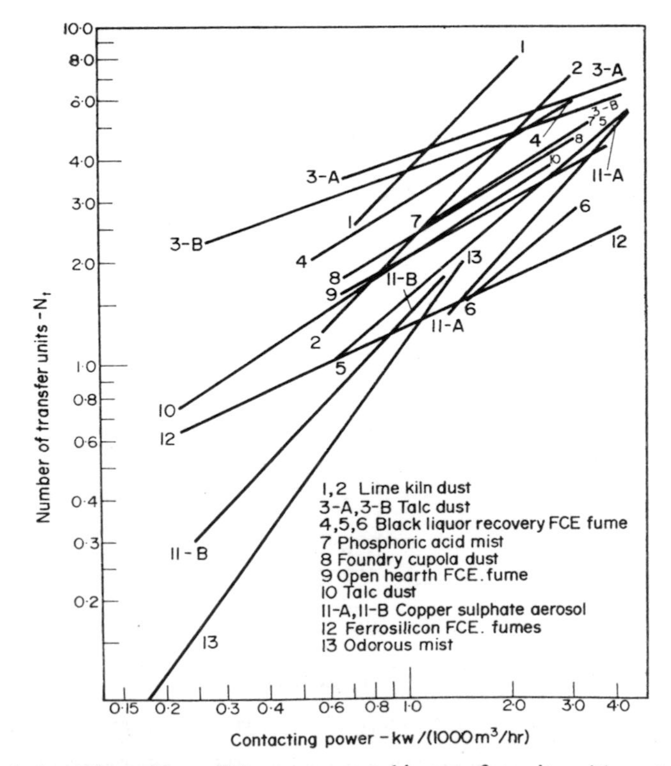

FIG. 9.35. Correlation of scrubber efficiency, measured in transfer units, with contacting power for a series of aerosols, independent of type of scrubber.[752]

The efficiency of a collector is correlated by a dimensionless transfer unit N_t defined in terms of the collection efficiency η by

$$N_t = \ln (1/(1-\eta)) \tag{9.15}$$

which gives comparatively low values of transfer units for efficiencies better than 99 per cent. A table of comparative values is given (Table 9.7).

TABLE 9.7. SCRUBBER TRANSFER UNITS AND PER CENT EFFICIENCIES

Number of transfer units	Collection efficiency (%)
0·5	39·35
1·0	63·21
2·0	86·47
4·0	98·17
6·0	99·752
10·0	99·9955

TABLE 9.8. SCRUBBER EFFICIENCY CORRELATION PARAMETERS[752]

Curve in Fig. 9.35	Aerosol	Scrubber	Correlation parameter	
			α	β
	Lime kiln dust and fume (Kraft mud kiln)			
1.	Raw gas (lime dust and soda fume)	Venturi and cyclonic spray	1·47	1·05
2.	Prewashed gas (soda fume)	Venturi, pipe line and cyclonic spray	0·915	1·05
3.A.	Talc dust	Venturi	2·97	0·362
3.B.		Orifice and pipe line	2·70	0·362
4.	Black liquor recovery furnace fume	Venturi and cyclonic spray	1·75	0·620
	Cold scrubbing water (humid gases)			
5.	Hot fume solution for scrubbing (humid gases)	Venturi, pipeline and cyclonic spray	0·740	0·861
6.	Hot black liquor for scrubbing (dry gases)	Venturi evaporator	0·522	0·861
7.	Phosphoric acid mist	Venturi	1·33	0·647
8.	Foundry cupola dust	Venturi	1·35	0·621
9.	Open hearth steel furnace fume	Venturi	1·26	0·569
10.	Talc dust	Cyclone	1·16	0·655
11.A.	Copper sulphate	Solivore (A) with mechanical spray generator	0·390	1·14
11.B.		(B) with hydraulic nozzles	0·562	1·06
12.	Ferrosilicon furnace fume	Venturi and cyclonic spray	0·870	0·459
13.	Odorous mist	Venturi	0·363	1·41

Semrau has plotted the transfer units against total power consumption P_T for a series of scrubbers and dusts, and a linear relation, independent of the type of scrubber, on a log–log plot has been obtained in each case (Fig. 9.35). It was possible to express this relation as

$$N_t = \alpha P_T^{\beta} \tag{9.16}$$

where α and β are characteristic parameters for the dust being collected. These parameters are given in Table 9.8. The efficiency of collection can thus be expressed in terms of total power used and the characteristics for the dust being collected, and independently of the actual *type* of scrubber being used.

CHAPTER 10

ELECTROSTATIC PRECIPITATION

10.1. INTRODUCTION

In the process commonly called electrostatic precipitation, small droplets and particles are first charged by gas ions produced by the electrical breakdown in the gases surrounding a high tension electrode, and then drift towards earthed collector electrodes. On arrival at the earthed collector, the particles adhere and are discharged to earth potential. When a layer of particles has formed, these are shaken off by "rapping" and fall into a hopper. Because the system is not quite static, as the charges carried by the particles and gas ions produce a small current, many workers prefer to call this type of plant electro-precipitators or electro-filters. In this book, however, the conventional term "electrostatic precipitator" will be used.

Electrostatic precipitators are applied wherever very large volumes of gases have to be cleaned and there is no explosion risk. The plants are invariably used for fly ash collection in modern base-load pulverized-fuel fired power station boilers and for the collection of dusts in the cement industry. Precipitators are also employed for large-scale fume collection systems in the metallurgical industry; for the collection of particles and mist droplets (tar, phosphoric acid, sulphuric acid) in the chemical and allied industries; for dust removal in air-conditioning systems; in fact for fine particle and mist collection problems where large gas volumes are involved.

That small fibres were attracted by a piece of amber after it had been rubbed was known to Thales of Miletus, one of the Greek philosophers, about 600 B.C., and there are a number of later reports, variously credited to Theophrastus (300 B.C.), Pliny (first century) and Solinus (third century) about the attraction of straws, leaves and other light bodies to rubbed amber.[694, 697] Modern work dates from the observation of William Gilbert (1600) who observed that amber, sulphur and other dielectrics, when charged by friction, would "entice smoke sent out by an extinguished light". Boyle made similar observations (1675), while at much the same time Otto von Guericke (1672) constructed an electrostatic generator consisting of a frictionally charged sulphur sphere, and discovered the effectiveness of pointed conductors in attracting charged bodies. Some 30 years later, Francis Hauksbee (1709) reported the discovery of "electric wind" to the Royal Society, and Isaac Newton subsequently discussed corona glow and electric wind (1718), but without reference to particle behaviour.

During the next century there are a number of reports of electrostatic phenomena, notable among these being the work of Benjamin Franklin (1747) on the effect of point conductors in drawing electric currents, and Coulombo (1785) who investigated the loss of charge from an insulated conductor by the connection of charged particles through air. The first demonstration of electrostatic precipitation is attributed to Hohlfeld, who in 1824 showed that a fog was cleared from a glass jar which contained an electrically charged point. Similar demonstrations by other workers were published later in the nineteenth century, an example being the precipitation of tobacco smoke in a glass cylinder 450 mm long and 230 mm diameter by Guitard about 26 years after Hohlfeld's paper.

The successful commercial application of the principle of electrostatic precipitation, however, dates to the first years of the twentieth century and is associated with the names of Lodge in England, Cottrell in the United States and Moeller in Germany. Lodge had been experimenting with electrostatic precipitation since 1880, and indicated the commercial possibilities of the method in a paper published in *Nature* in 1883.[525] Together with two others, Walker and Hutchings, he installed the first commercial electrostatic precipitator at a lead smelter in Bagillt, North Wales. The design is shown in Fig. 10.1, and illustrates

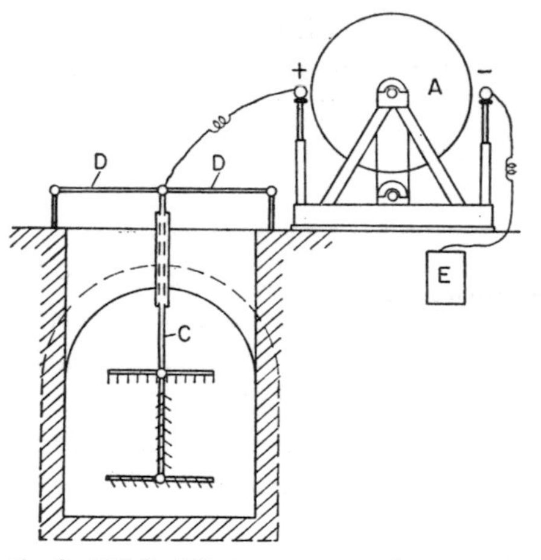

FIG. 10.1. Illustration from the first United States patent on electrostatic precipitation. A. O. Walker, U.S. Pat. No. 342,548 (1886).[930]

the discharge points installed in a duct, as well as the two Wimshurst machines, each 130 mm diameter, which supplied the high voltage, and which were driven by 0·75 kW steam engines. Unfortunately the plant was unsuccessful, for which there were two main reasons;[735, 930] the primitive method of producing the high-voltage power, and the fact that lead smelter fume is very fine and has a high resistivity, making it one of the most difficult fumes to collect. Had the fume consisted of conducting particles or droplets, such as sulphuric-acid mist which was used by Cottrell in his first investigations, then electrostatic precipitation would undoubtedly have come into commercial use 25 years earlier.

The development of alternating current technology and electrical machinery made new sources of high-voltage direct current available from combinations of transformers and synchronous mechanical rectifiers or mercury arc rectifiers. Lodge patented the latter for the purpose of electrostatic precipitation in 1903, while Cottrell experimented with the mechanical rectifier when he found that the discharge from a spark coil proved inadequate for a corona discharge from more than just one or two points in a spark chamber. Cottrell also found that a cotton-covered wire gave a continuous glow, indicating corona formation, over the whole of its surface, and developed the *pubescent* discharge electrode of non-conducting fibrous material covering the conductor (Fig. 10.2). The combination of alternating

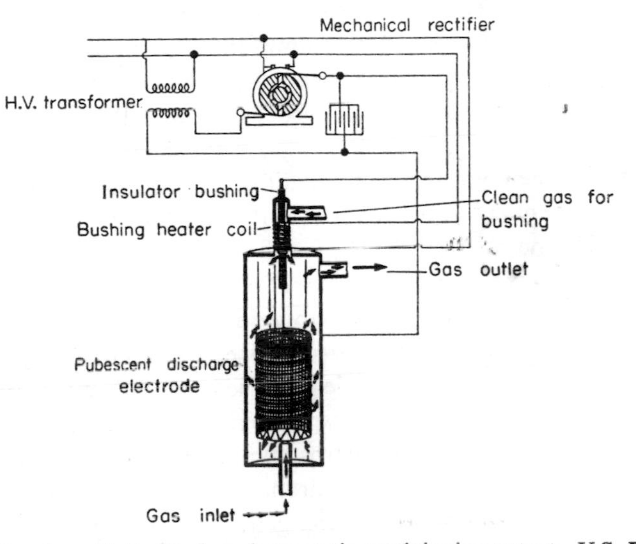

FIG. 10.2. Illustration from Cottrell's first electrostatic precipitation patent—U.S. Pat. No. 895,729 (1908). The construction of the electrode, and the purge gas stream to prevent the insulation shorting is clearly marked.[930]

current transformer, mechanical rectifier and pubescent electrode led to a successful precipitator, able to remove sulphuric acid mist on a laboratory scale with flows of several cubic metres per hour.[181, 182, 183] Cottrell and his colleagues, notably W. A. Schmidt, who was responsible for many of the later improvements, applied the electrostatic precipitator on an industrial scale: first at a powder works at Pinole (near Berkeley) and then at the Selby smelter where there as an acute air pollution problem. At Selby the precipitator operated on the gases from the parting kettles for precious metal recovery, and collected 0·15 l. s^{-1} of sulphuric acid from 140 m^3. Other features of Cottrell's patent which are shown in Fig. 10.2 are the heating and ventilating system provided for keeping the high voltage insulation dry. Cottrell used a negative corona because he found that this could carry higher currents than the positive corona.

The next plant was installed at the lead smelter at Balaclala, where 425,000 m^3 h^{-1} of lead and zinc fume laden gases were to be treated. As Lodge *et al.* had found, this was a difficult problem. However, the newer techniques of high voltage direct current production enabled the plant to operate at efficiencies between 80 and 90 per cent.[554] Many of the features found in modern electrostatic precipitators were developed by W. A. Schmidt when he installed the precipitator at the Riverside Portland Cement Company (Southern California) in 1912.[734] This plant handled 1,700,000 $m^3 h^{-1}$ at 400–500°C. Fine wire discharge electrode operating at 45 kV were used here for the first time, and the plant was still in operation 45 years later.

The development of electrostatic precipitators has been very largely empirical, and over 1000 patents cover all aspects of precipitator construction. The theory of precipitation has lagged far behind its practical application. Thus a theoretical expression for collection efficiency in the absence of turbulence was derived by Deutsch in 1922.[222] A modified theory was presented by Williams and Jackson[945] to allow for turbulent remixing at a rate controlled by the gas-stream turbulence (1962), and similar studies have been carried out by Inyushkin and Averbukh (1962–3),[388-9] Cooperman (1960–6)[172-4] and Robinson (1967)[692] extending this approach.

Equations describing the bombardment and diffusion processes of particle charging were suggested by Pauthenier and Moreau-Hanot[625] in 1932 for the former process and by Arendt and Kallmann[35] (1925) and White[925] (1951) for the latter process. A detailed model of the precipitation process of particles in the presence of others has not yet been attempted. However, the extensive studies of electrostatic precipitation published in the past 15 years enable the calculations based on the fundamental processes to be correlated with experimental behaviour. It is therefore becoming possible in some cases either to design the units from first principles or to extrapolate the sparse experimental data to plants with larger capacity.

The basic processes which occur in electrostatic precipitators are discussed in detail in this chapter. These are the formation of the corona or ionized zone around the high-tension wire, which may be positively or negatively charged; the charging and drift of the particles; particle deposition and discharge and the possibility of particle re-entrainment. This is followed by details of some aspects of the construction of modern electrostatic precipitators.

10.2. CORONA FORMATION

When an electrical potential is established between two parallel plates, the field between the plates is uniform, and its magnitude can be expressed in terms of the voltage gradient between the plates ($V m^{-1}$). When this voltage gradient is increased to a critical value—about 3000 $kV m^{-1}$ in ambient air—electrical breakdown occurs and a spark flashes between the plates. However, if a non-uniform field is produced, for example between a sharply curved surface such as a point or a fine wire, and an enclosing tube or a plate, then electrical breakdown can occur near the curved surface, producing a glow discharge or "corona", without sparkover. The corona is of great importance in other industrial applications beside precipitation of particles, as, for example, it accounts for much of the power losses in high-voltage power transmission lines, and so has been studied extensively. Detailed accounts of electrical breakdown in gases and corona can be found in specialized references such as Cobine's *Gaseous Conductors*[160] or Craggs and Meek's *Electrical Breakdown in Gases*.[187]

Because an a.c. corona produces oscillating motions in charged particles, and a d.c. corona produces a steady force driving particles towards the passive collector electrode, conventional precipitation requires a unipolar discharge. The negative corona is more stable than the positive corona, which tends to be sporadic and cause sparkover at lower voltages than for the negatively charged conductors (Fig. 10.3), and so these are generally used for industrial application. The negative corona, however, does give rise to higher ozone concentrations, and so is not used for air-conditioning plant.

The corona on a positive wire has the appearance of a bluish-white sheath covering the entire wire, while the corona on a negative wire is concentrated in reddish tufts along the

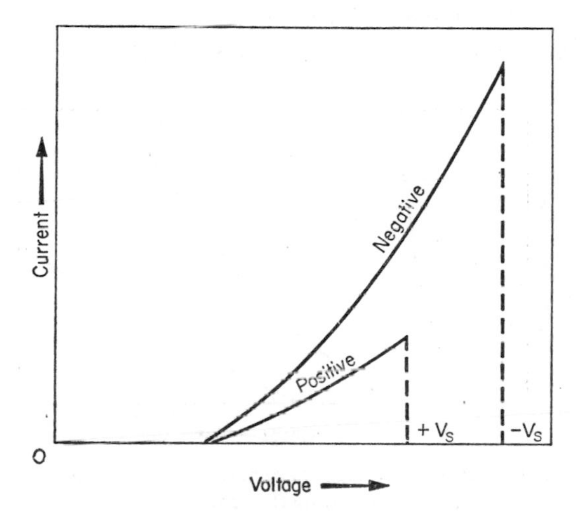

FIG. 10.3. Representative current–voltage characteristics for positive and negative corona in air.[926]

wire. On a polished conductor these tufts are spaced more or less regularly, their number increasing with the current. The uniform spacing of the beads can be explained by the mutual space charge repulsion of the beads. On an unpolished wire, or a wire provided with barbs or other discharge points, the discharge tufts concentrate on imperfections in the wire surface, or on the special points.

The *mechanism of negative corona* formation is as follows: The gases normally, when no current flows, contain some ionized gas molecules, about 1/mm^3, formed by cosmic rays. With current flowing, more ionized molecules are formed from the ultra-violet radiation from the corona glow. The positive gas ions and photons formed in this way are accelerated towards the negative conductor and release electrons from the surface of the conductor. These electrons, moving through the strong field near the conductor, generate new electrons and positive ions by molecular impacts. The electrons then swarm outside this region, slowing to a speed less than that necessary for ionization by collision, and attach themselves to gas molecules, forming gas ions. These gas ions then move towards the passive electrode with a speed proportional to their charge and the intensity of the electric field.

The space outside the corona is therefore filled with a dense cloud of unipolar ions, which in practice is of the order of 5×10^4 ions mm^{-3}, and it is in this zone that the majority of the dust particles acquire their negative charge. The dust particles passing through the corona zone are likely to become positively charged because of the predominance of positive gas ions and high mobility of electrons in this zone, and are then precipitated on the discharge electrode. The ionic mobility of the gas ions and the dust concentration outside the corona zone are the major factors affecting the voltage–current characteristics of the precipitator.

The mobilities of the ions and electrons depend on the gas and on the presence of particles. Pure nitrogen, for example, absorbs only a few of the electrons. The current is mainly transferred by these electrons, which have a high mobility, and so the corona current rises very steeply. On the other hand, electro-negative gases such as oxygen or methyl chloride absorb electrons very easily, and the gas ions, with much lower mobilities, become the current carriers and the current is much lower (Fig. 10.4). Similarly dust particles or mist droplets

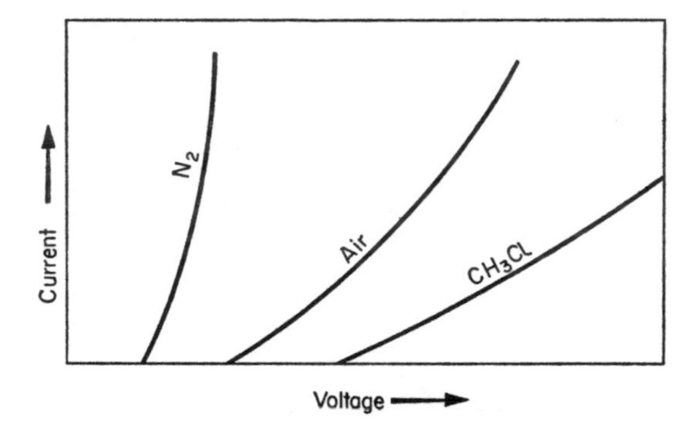

FIG. 10.4. Variations in corona current–voltage characteristics for the negative corona in typical gases (nitrogen, air, methyl chloride).[926]

when present in large concentrations can act as absorbers of electrons and gas ions, and also reduce the corona current. This is called corona quenching or corona suppression. When it occurs (for example with very fine fumes) it is common practice to use precipitators with several stages in series.

The corona current–voltage characteristic for air (without particles) is about half-way between oxygen and nitrogen, and the ionic mobility of the gas ions is of the order of $0{\cdot}18$ mm s^{-1} per V m^{-1}. In a typical gas mixture such as flue gases, the speed of the negative ions is of the order of 30 m s^{-1}.

The *mechanism of positive corona* formation differs markedly from the negative corona, and accounts for the different properties such as lower flash-over voltage, and lower ozone formation. The electrons in the gas move towards the corona region near the discharge wire where electron avalanches form to maintain the corona region. The positive gas ions formed by these electron avalanches move slowly away from the wire, at much lower speeds than the electrons in the negative corona region, and consequently fewer ionizing collisions occur in their movement to the passive electrode. At the low field strengths near this electrode they gain little acceleration so few electrons are emitted by cathode bombardment and most of the current is carried by the positively charged gas ions. Because fewer ionizing collisions occur in the high intensity corona region, the formation of ozone and oxides of nitrogen is much less than in the case of the negative corona.

10.3. FIELD STRENGTH AND CORONA CURRENT

The strength of an electric field E (V m^{-1}) at any point is defined as the potential gradient at that point. If the applied voltage or potential difference between two concentric electrodes, such as a wire and tube system, radius R_1 and R_2 respectively, is $\mathcal{V}$, then:

$$\mathcal{V} = \int_{R_1}^{R_2} E\,\mathrm{d}r. \qquad (10.1)$$

Before the corona has been formed a negligible number of gas ions are present, and in the absence of any ionic current the field strength at radius r is given by the integrated form of equation (10.1):

$$\mathcal{V} = Er \ln R_2/R_1 \qquad (10.2)$$

or

$$E = \frac{C_0}{r} \qquad (10.3)$$

where

$$C_0 = \mathcal{V}/\ln (R_2/R_1) \qquad (10.4)$$

which is a function of the applied voltage and the geometrical dimensions of the system.

It can be seen from equation (10.3) that the field strength has its maximum value for the lowest r; that is at the surface of the wire, at radius R_1. For the corona to start it is necessary for electrical breakdown to occur at the wire, and so this maximum field strength value must be greater found experimentally to require approximately equal electrical field strengths for both positive and negative corona. For air in coaxial cylinders at 25°C the critical field strength, E_c, is

$$E_c = 31\times 10^5 \frac{m\varrho}{\varrho_s}\left(1+\frac{0{\cdot}308}{\sqrt{[(\varrho/\varrho_s)R_1]}}\right) \quad \text{V m}^{-1} \qquad (10.5)$$

while for parallel wires

$$E_c = 31\times 10^5 \frac{m\varrho}{\varrho_s}\left(1+\frac{0{\cdot}301}{\sqrt{[\varrho R_1/\varrho_s]}}\right) \quad \text{V m}^{-1} \qquad (10.6)$$

where R_1 = radius of the wire, m.
ϱ/ϱ_s = relative air density
= 1 at 100 kPa (1 atm) pressure, 25°C
= $(2{\cdot}95P)/T$ (for other conditions),

where P = atmospheric pressure in kPa,
T = absolute temperature, °K

while m is an irregularity factor which is 1 for polished wires, 0·82 for a general corona on stranded wires, or 0·92 for roughened and weathered wires.*

* The constants of "30×10⁵", "31×10⁵" and "0·308" given in equations (10.5) and (10.6) are based on direct experimental evidence, and are widely quoted. The values derived from equation (10.7) should be used for gases other than air.

The equations (10.5) and (10.6) can be written in the more general form

$$E_c = A_g \frac{m\varrho}{\varrho_s}\left[1+\frac{B_g/A_s}{\sqrt{[(\varrho/\varrho_s)R_1]}}\right] \quad \text{V m}^{-1} \tag{10.7}$$

where A_g and B_g are constants, listed by Robinson.[697] The value of B_g is $12{\cdot}3\times 10^4$ V $m^{-\frac{1}{2}}$ and a small selection of values of A_g for some common gases is given in Table 10.1.

TABLE 10.1. VALUES FOR CONSTANT A_g AT 25°C AND 100 kPa (1 atm) FOR COMMON GASES[697]

Gas	A_g, 10^5 V m^{-1}	Gas	A_g, 10^5 V m^{-1}
Air	35·5	CO_2	45·5
H_2	15·5	CO	26·2
He	4·0	C_2H_2	75·2
A	7·2	C_2H_4	21·3
Kr	9·5	C_3H_6	87·2
O_2	29·1	CH_3Br	97·0
N_2	38·0	C_2H_5Br	98·0
Cl_2	85·0*	C_3H_7Br	155·0
NH_3	86·7	CH_3I	75·0
H_2S	52·1	CH_3OH	62·5
SO_2	67·2	C_2H_5OH	97·0

* This value seems too high.[697]

The applied voltage required for starting the corona can be found by substituting equations (10.5 or 6) in (10.2):

$$\mathcal{V}_c = 3100\frac{m\varrho}{\varrho_s}\left(1+\frac{0{\cdot}308}{\sqrt{[\varrho R_1/\varrho_s]}}\right) R_1 \ln R_2/R_1 \tag{10.8}$$

where $\mathcal{V}_c$ = the corona starting voltage (kV).

After the corona has started, the presence of an ionic space charge modifies the field strength in the precipitator, and the dust particles present in the precipitator during operation introduce further complications. Allowances can, however, be made for these.[160, 531, 704, 867, 948]

In the presence of the ionic space charge, the field strength distribution is given by Poisson's equation for coaxial cylinders,[161]

$$\frac{1}{r}\frac{\mathrm{d}}{\mathrm{d}r}(Er) = 4\pi\sigma \tag{10.9}$$

where σ is the space charge, per unit volume (m^3). When a gas, but no particles, are present, this is determined by i, the ionic current per unit length of conductor. In this case, the ionic current is

$$i = 2\pi r \sigma u_i E \text{ (A m}^{-1}\text{)} \tag{10.10}$$

where u_i = ionic mobility in a unit field ($m^2 s^{-1} V^{-1}$)
and $u_i \times E$ = ionic velocity ($m s^{-1}$).

Transforming equation (10.9) and substituting in equation (10.8) gives

$$\frac{dE}{dr} + \frac{E}{r} - \frac{2i}{u_i r E} = 0. \tag{10.11}$$

This can be integrated, using as boundary conditions $r = R_1$, when $E = E_c$ the critical field strength for starting the corona. Then:

$$E = \sqrt{\left[\left(\frac{R_1 E_c}{r}\right)^2 + \frac{2i}{u_i}\left(1 - \left\{\frac{R_1}{r}\right\}^2\right)\right]}. \tag{10.12}$$

From equation (10.3), C_0 can be substituted for $R_1 E_c$, and for zero ionic current, equation (10.11) reduces to equation (10.3). The other important case is for a large current i and when $R_2 \gg R_1$. Then,

$$E = \sqrt{\left(\frac{2i}{u_i}\right)} \tag{10.13}$$

which gives an equation for constant field strength in the region of a precipitator tube some distance from the corona wire, which can be used for approximate precipitator calculations (*wire in tube case*). This equation has been confirmed experimentally by the measurements of Pauthenier and Moreau-Hanot.[626]

Robinson[697] points out that equation (10.12), which is widely used, may seriously underestimate the field at the relatively low discharge currents commonly used in industry. At the higher currents, which are used in laboratory investigations, this equation is reasonably accurate.

In the case of the *wire and plate* type precipitator Troost[867] has shown experimentally that

$$E = \sqrt{\left(\frac{8iL}{u_i W}\right)} \tag{10.14}$$

where L = distance between wire and plate,
W = distance between successive wires.

Where the wires are spaced approximately the same distance apart as the plates ($W = 2L$):

$$E = \sqrt{\left(\frac{4i}{u_i}\right)}. \tag{10.15}$$

The ion mobility can be calculated from the kinetic theory of gases:[948]

$$u_i = \frac{e}{\varrho \bar{u} \pi \sigma_{AB}^2} f(\mathcal{M}) \quad (\text{m}^2\,\text{s}^{-1}\,\text{V}^{-1}), \tag{10.16}$$

e = charge on an ion (Coulombs)
ϱ = gas density = $PM'/\mathbf{k}T$ (ideal gas law),
P = gas pressure,
$\bar{u}$ = mean gas molecular velocity

$$= \sqrt{\left(\frac{8\mathbf{k}T}{\pi M'}\right)} \quad \text{or} \quad \sqrt{\left(\frac{8RT}{\pi M}\right)},$$

M = molecular weight,
M' = weight of a molecule,
$\mathbf{k}$ = Boltzmann's constant,
T = absolute temperature,
σ_{AB} = sum of radii of charge carrier and gas molecule,
$f(\mathcal{M})$ = function of $\mathcal{M}$ where $\mathcal{M} = m/(M'+m)$,
m = weight of a charge carrier (i.e. charged ions).

The values of $f(\mathcal{M})$ in some typical cases are listed in Table 10.2.

TABLE 10.2. VALUES OF $f(\mathcal{M})$ FOR VARIOUS μ IN EQUATION (10.16)[947]

Gas-ion system	$\mathcal{M}$	$f(\mathcal{M})$
Free electron in nitrogen	2×10^{-5}	284·3
Monomolecular ions	0·5	1·38
Normal gas ions	0·9	0·837
Coarse particles	1·0	0·75

By substituting for the gas density and mean molecular velocity, equation (10.16) can be rewritten to demonstrate the dependence of the ionic mobility on the gas type, the pressure and absolute temperature and on whether the charge is being carried by gas ions or electrons:

$$u_i = \frac{ef(\mathcal{M})}{\sigma_{AB}^2 P} \sqrt{\left(\frac{\mathbf{k}T}{8\pi M'}\right)}. \tag{10.17}$$

While the calculated ion mobilities using this method are an excellent guide, experimental mobilities should be used when these are available. A short list is given in Table 10.3. Further values may be found in the paper by Robinson[697] and in the *International Critical Tables*.[387]

TABLE 10.3. MOBILITIES OF SINGLY CHARGED GAS IONS IN THEIR PARENT GASES AT 0°C AT 100 kPa: $\times 10^4$ ($m^2 s^{-1} V^{-1}$)[697]

Gas	−	+	Gas	−	+
Air (dry)	2·1	1·36	C_2H_4	0·83	0·78
N_2	*	1·8	CO	1·14	1·10
O_2	2·6	2·2	CO_2	0·98	0·84
H_2	*	12·3 (H_3^+)	H_2O (100°C)	0·95	1·1
Cl_2	0·74	0·74	HCl	0·62	0·53
SO_2	0·41	0·41	NH_3	0·66	0·56
N_2O	0·90	0·82	H_2S	0·56	0·62

* No electron attachment in pure gas.

Robinson points out that the exact mobility to be used in a particular situation is subject to some uncertainty due to the effect of traces of impurities; this is especially so for pure non-attaching gases, the negative mobility of which is the electron mobility.

So that the values in Table 10.3 can be applied at a temperature T (Kelvin) and a pressure P (kPa), they should be multiplied by the ratio 0·365 T/P.

When particles as well as gas ions are present in the space between the high-tension wire and the earthed electrode, the space charge in equation (10.9) must be modified to allow for these also:

$$\frac{1}{r}\frac{d}{dr}(Er) = \begin{pmatrix}\text{charge on}\\ \text{gas ions}\end{pmatrix} + \begin{pmatrix}\text{charge on}\\ \text{particles}\end{pmatrix}. \tag{10.18a}$$

If the charge on the particles is assumed to be the limiting charge received by bombardment charging (equation 10.31) this equation is

$$\frac{1}{r}\frac{d}{dr}(Er) = \frac{2i}{u_i rE} + 4\pi\left\{\frac{3\epsilon}{\epsilon+2}E\sum\frac{d^2}{4}\right\}. \tag{10.18b}$$

Here d is the diameter of the particles, and the surface area of all the particles in a unit volume, A, is given by $\pi\sum d^2$. Substituting A and integrating equation (10.18) gives

$$E^2 = \frac{K}{r^2}\exp\left(\frac{6\epsilon Ar}{\epsilon+2}\right) - \frac{i}{u_i}\left[\frac{2(\epsilon+2)}{3\epsilon Ar} + \left(\frac{\epsilon+2}{3\epsilon Ar}\right)^2\right] \tag{10.19}$$

where ϵ = dielectric constant of the particles,
K = constant of integration.

This constant has been shown to be[626]

$$K = C_o^2 + \frac{i}{u_i}\left(\frac{\epsilon+2}{3\epsilon A}\right)^2 \tag{10.20}$$

where C_0 has been defined in equation (10.4). The field strength, considering both the dust and the space charge present, is now

$$E = \left[\left\{\frac{C_o^2}{r^2}+\frac{i}{u_i}\left(\frac{\epsilon+2}{3\epsilon A}\right)^2\right\}\exp\left(\frac{6\epsilon Ar}{\epsilon+2}\right)-\frac{i}{u_i}\left\{\frac{2(\epsilon+2)}{3\epsilon Ar}-\left(\frac{\epsilon+2}{3\epsilon Ar}\right)^2\right\}\right]^{\frac{1}{2}}. \quad (10.21)$$

When $3\epsilon A/(\epsilon+2)$ is small (much less than unity) the exponential term in equation (10.21) can be expanded and the equation simplified:

$$E = \left[\frac{2i}{u_i}\left(1+\frac{2\epsilon Ar}{\epsilon+2}\right)+\frac{C_o^2}{r^2}\right]^{\frac{1}{2}}. \quad (10.22)$$

When there is no dust present, this equation reduces to equation (10.12), while for large radii, i.e. small C_o^2/r^2, this term can be neglected, and then equation (10.22) simplifies further to

$$E = \left[\frac{2i}{u_i}\left(1+\frac{2\epsilon Ar}{\epsilon+2}\right)\right]^{\frac{1}{2}}. \quad (10.23)$$

It has been suggested[626] that for the dust burden normally encountered, the term in the parentheses can be expanded as a series, and all terms other than those of the first order neglected. Then, for a wire in tube precipitator, equation (10.22) reduces to

$$E = \left(1+\frac{\epsilon Ar}{\epsilon+2}\right)\cdot\sqrt{\left(\frac{2i}{u_i}\right)}. \quad (10.24a)$$

By analogy a similar equation can be written for a plate type precipitator:

$$E = \left(1+\frac{\epsilon Ar}{\epsilon+2}\right)\cdot\sqrt{\left(\frac{8iL}{u_iW}\right)}. \quad (10.24b)$$

Equation (10.24) shows that the field strength is a function of the ionic current i, the ion mobility, u_i, and is greater when particles are present, being also dependent on their aggregate surface area A.

Lowe and Lucas[531] calculated the strength across a 130 mm radius tubular precipitator with an applied voltage of 40 kV in the following cases:

(i) Zero dust burden and zero ionic current [equation (10.2)].
(ii) Zero dust burden, and an ionic current of 80 $\mu A\ m^{-1}$ of discharge conductor.
(iii) A dust burden, assumed to be equivalent to fly ash with a concentration of 18 $g\ m^{-3}$ and an effective surface area of 8·3 $m^2\ m^{-3}$ gas, and an ionic current of 80 $\mu A\ m^{-1}$ of discharge conductor.

The calculated field strengths are plotted in Fig. 10.5.

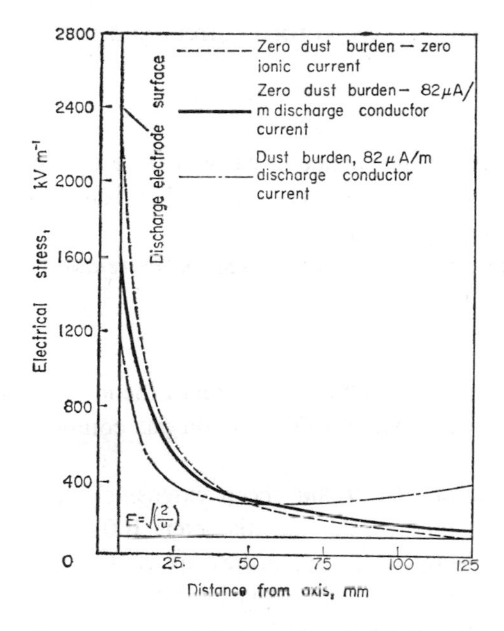

FIG. 10.5. Field strengths across a precipitator tube at different conditions (calculated).[531]

The *corona current* i in a precipitator is a function of the applied voltage $\mathcal{V}$, the ion mobility u_i and the precipitator dimensions. For a wire in tube type unit, with low currents, Townsend[866] and others obtained the following equation by integration of equation (10.12):

$$i = \frac{2u_i}{R_2^2 \ln (R_2/R_1)} \mathcal{V}(\mathcal{V}-\mathcal{V}_c). \tag{10.25}$$

This equation has been modified by Fazel and Parsons[940] to allow for the width of the corona on the discharge electrode, S, the outer edge of which is the critical field strength E_c:

$$i = \frac{2u_i}{R_2^2 \ln (R_2/R_1+S)} \mathcal{V}(\mathcal{V}-\mathcal{V}_c). \tag{10.26}$$

Here it is assumed that the width of the corona annular layer increases with the potential $(\mathcal{V}-\mathcal{V}_c)$. More satisfactory are empirical relations, such as that suggested by Shaffers:[948]

$$i = \frac{u_i}{2}\left[\frac{\mathcal{V}-\mathcal{V}_c}{R_2 \ln (R_2/R_1+S)}\right]^x \tag{10.27}$$

where $x = 2$ when $R_2/R_1 > 1000$
$= \log_{10} (R_2/R_1+S)$ when $R_2/R_1 < 1000$

as long as x remains less than 2, otherwise this should be used as the limiting value. The width of the corona layer is assumed in this equation to be constant at 0·3 mm.

Another empirical relation between corona current and voltage is given by Koller and Fremont:[459]

$$i = \beta \mathcal{V}^{\alpha} \tag{10.28}$$

where α and β are constants for the particular gas which were found by Koller and Fremont to be $\alpha = 4{\cdot}2$ for air and $\alpha = 2{\cdot}8$ for methyl chloride. This is higher than that found by other workers who obtained $\alpha = 1{\cdot}6$ for air.

Generally, the corona current can be expressed as the relation:

$$i = \beta(\mathcal{V} - \mathcal{V}_c)^x \tag{10.29}$$

where β is an empirical constant which is a function of the ion mobility and the precipitator geometry and the type of dust, while x depends on gas composition and has a value between 1 and 2.

This last empirical relation (10.29) has been extensively investigated by Winkel and Schütz[948] using a tube precipitator, 35 mm diameter, 100 mm long, and with a 0·3-mm nichrome-wire discharge electrode. Temperatures up to 600°C were used, and the experimental values obtained for V_c, β and x in equation (10.29), which are listed in Table 10.4. The experimental points are also shown in Fig. 10.6.

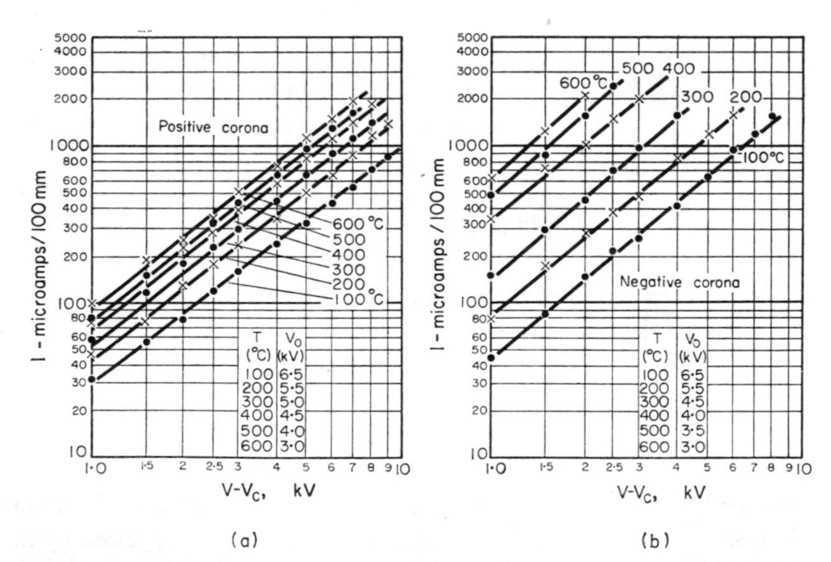

FIG. 10.6. Dependence of ionic current on potential at various temperatures.[948] (a) Positive corona. (b) Negative corona.

If the value of β varied as the square root of the absolute temperature, which is to be expected from equation (10.17) for the ionic mobility u_i, then the ratio $\beta/\sqrt{T}$ would be approximately constant. The upwards trend with temperature of this ratio, shown in the last column of Table 10.4 indicates that the model for ionic mobility is an over-simplified one, and much further work is required for a more comprehensive theory to be established.

TABLE 10.4. VALUES OF x, β AND $\beta/\sqrt{T}$ FOR CORONA CURRENT RELATION AT VARIOUS TEMPERATURES*[948]

Air—negative corona

T°C	$\mathcal{V}_c$(kV)	x	β	$\beta/\sqrt{T}$
100	6·5	1·72	4·5	0·234
200	5·5	1·62	8·5	0·390
300	4·5	1·76	14·5	0·605
500	4·0	1·62	34·0	1·31
400	3·5	1·75	48·0	1·72
600	3·0	1·84	62·0	2·10

Air—positive corona

T°C	$\mathcal{V}_c$(kV)	x	β	$\beta/\sqrt{T}$
100	6·5	1·52	3·0	0·155
200	5·5	1·56	4·3	0·197
300	4·5	1·58	5·3	0·220
400	4·0	1·60	6·8	0·260
500	3·5	1·55	7·8	0·280
600	3·0	1·61	9·0	0·304

Air with water vapour—positive and negative corona, 405 g m^{-3} —Dewpoint 72°C

T°C	$\mathcal{V}_c$	x	β	$\beta/\sqrt{T}$
200	5·0	1·65	2·1	0·097
300	4·5	1·67	3·2	0·133
400	4·0	1·60	5·0	0·192

* Results of other experimental investigations of corona characteristics are given in Table 10.5.

10.4. PARTICLE CHARGING

As particles or droplets move into the field of a precipitator they acquire an electrostatic charge by two mechanisms, bombardment charging and diffusion charging. The gas ions, and also electrons in the case of a negative corona, move normally across the gas stream carrying the particles under the influence of the electric field, and charge the particles with which they collide. This is called *bombardment charging*. In addition the gas ions (and electrons where present) land on the particles because of their thermal motion, and this is *diffusion charging*.

Although the two mechanisms act at the same time, a comprehensive theory combining the two has not yet been developed, and it is usual to consider each mechanism separately. The error due to this is not very great, because the bombardment charging mechanism is of primary importance for particles greater than 1 μm, while diffusion charging is more important for particles smaller than 0·2 μm. Because these small particles normally only

TABLE 10.5. EXPERIMENTAL CORONA CHARACTERISTICS (SPROULL)[791]

Parallel plate type precipitator with plates 100 mm apart and a 2·8 mm wire, 0·3 m long stretched midway

Gas	Temp. (°C)	Polarity: negative threshold voltage (kV)	negative sparkover voltage (kV)	negative current at sparkover (mA)	Polarity: positive threshold voltage (kV)	positive sparkover voltage (kV)	positive current at sparkover (mA)
Air	20	20	54	2·8	20	28	0·2
	225	9	35	3·1	8	22·5	2·1
Air							
+20% steam	225	4	45	3·0	5	31	2·6
+35% steam	225	4	46	2·2	4	48	2·2
Air							
+1% carbon tetrachloride	20	24	62	2·3	26	63	2·2
Carbon dioxide	20	21	49	1·2	29	49	0·7
Methane	20	2·5	29	17·0	25	21	0·2
Water (steam)	225	9	67	5·2	9	> 40	> 1·0

(i) Steady (not pulsating) d.c. was used.
(ii) Data for water (steam) was obtained with less accurate apparatus, and the power supply could not reach sparkover with positive polarity.

represent a small fraction of the dust entering a precipitator, they are neglected in most precipitation theories. When these small particles are a large proportion of the dust and fume load, a number of equations (e.g. 10.18b; 10.39–10.44) have to be modified.

In bombardment charging the magnitude of the electric field, the aggregate surface area of the particles and their dielectric properties play a major part in the process, while in ion diffusion charging the number of ions, their mobility, which is a function of temperature, and the time available for the process are the most important factors. The theoretical calculations of the charge acquired by the particles make the following assumptions:

(i) The particles are spherical.
(ii) Interparticle distances are large compared to the particle diameters.
(iii) The ion concentration and electric field is initially considered to be uniform.
(iv) There is no effect on the charging process on one particle by the fields of other charged particles.
(v) The mean free path of the ions is small compared with the diameter of the particles.

These assumptions do not seriously invalidate the theoretical calculations. For example, Smith and Penney[780a] have shown that in practice, departures from sphericity are not a serious source of error, and the second assumption is valid for the particle concentrations encountered in industrial waste and process gases; the last is valid except for extremely small particles and/or very low pressures, while the others are approached approximately in real situations.[630a, 780a]

Bombardment charging has been studied by Rohmann[699] and Pauthenier and Moreau-Hanot.[625] It is assumed that the gas ions move along the lines of force between the high tension electrode and the passive electrode. Some of the gas ions will be intercepted by the uncharged particles, and their charge will be deposited. The particles are now charged, and the lines of force will distort, and some of the gas ions will be repulsed by the charged particle, thus reducing the rate of charging. After a time the charge on the particle will reach limiting value.

The charge on a particle, q (coulombs), which is a product of the electronic charge e and the number of these charges n, can be found from the equation by Pauthenier:[531]

$$ne = q = \frac{3\epsilon}{\epsilon+2}\,\frac{d^2}{4}\,\frac{Et}{t+\tau}\text{ coulombs} \tag{10.30}$$

where t = time of charging,
τ = time constant = $1/\pi N_i e u_i$,
ϵ = dielectric constant of particle,
N_i = ion concentration per unit volume,
$N_i e$ = space charge density.

The limiting charge, which is also the maximum, can be found by allowing t to approach infinity.

$$q = \lim_{t\to\infty}\frac{3\epsilon}{\epsilon+2}\,\frac{d^2}{4}\,\frac{E}{1+\tau/t}$$

$$= \frac{3\epsilon}{\epsilon+2}\,\frac{d^2}{4}\,E \tag{10.31a}$$

$$\doteqdot \frac{3\epsilon}{\epsilon+2}\,\frac{d^2}{4}\sqrt{\left(\frac{2i}{u_i}\right)}. \tag{10.31b}$$

Lucas and Lowe[531] have pointed out that 91 per cent of the limiting charge is attained in 10τ sec, which in the case of a tubular precipitator 9·25 m diameter, with an 8 mm diameter corona wire, operating at 40 kV and a current of 135 μA m^{-1} is 0·2 sec. If the current is reduced to 34 μA m^{-1}, the 91 per cent value is reached in 1 sec.

The influence of particle size, field strength and, to a lesser extent, the dielectric constant and the charge on a particle is indicated by the terms in equation (10.31). For particles consisting of a material with insulating properties, where $\epsilon = 1$ the ratio $3\epsilon/(\epsilon+2)$ is also 1, while for a good conductor, where ϵ has a high value, the ratio approaches 3. The influence of temperature on the dielectric constant function is negligible.[948] However, the temperature effect on the rate of charging is appreciable, and at constant discharge current, where it is inversely proportional to the ionic mobility, it is also proportional to the inverse of the square root of the absolute temperature [equation (10.17)].

Diffusion charging has been studied in detail by Arendt and Kallmann,[35] and White,[925] and is primarily a function of the thermal movement of the gas ions. The formula developed by Arendt and Kallmann is limited because it applies only to those particles that already possess some charge. Also, because of the particular differential form, it can only be integrated numerically. The equation is

$$\frac{\mathrm{d}n}{\mathrm{d}t}\left(1+\frac{d^2\sqrt{u^2}}{16n\mathbf{k}e}\right)=\frac{\pi}{4}d^2\sqrt{(u^2)}N_i\exp\left(-\frac{2ne^2}{d\mathbf{k}T}\right) \tag{10.32}$$

where $\sqrt{(u^2)}$ = root mean square molecular velocity
= $\sqrt{(3\mathbf{k}T/M')}$.

White's approach to the diffusion charging process is simpler. It is shown that the density of a gas in a potential field is not uniform, but varies according to

$$N = N_i \exp\left(\mathcal{V}/\mathbf{k}T\right). \tag{10.33}$$

where $\mathcal{V}$ is the potential energy.

In the case of gas ions near a suspended particle, with charge, ne, the potential energy of an ion of the same polarity at a distance r from the centre of the particle is $\mathcal{V} = -ne^2/r$. The ion density near the particle is then

$$N = N_i \exp\left(-2ne^2/d\mathbf{k}T\right). \tag{10.34}$$

From the kinetic theory, the number of ions which strike the surface of the particle each unit of time is $\pi d^2 N\sqrt{(u^2)}/4$. If it is assumed that all ions which reach the particle attach themselves, the ion current to the particle (i.e. the rate of charging) will be given by

$$\frac{\mathrm{d}n}{\mathrm{d}t}=\frac{\pi d^2 N_i\sqrt{(u^2)}}{4}\exp\left(-\frac{2ne^2}{d\mathbf{k}T}\right). \tag{10.35}$$

Integrating this equation, using the uncharged particle at zero time as a boundary condition, gives the total charge as

$$q = ne = \frac{d\mathbf{k}T}{2e}\ln\left(1+\frac{\pi d\sqrt{(u^2)}N_i e^2}{2\mathbf{k}T}t\right). \tag{10.36}$$

Equation (10.36) is commonly used to calculate the charge on a particle, but its derivation neglects the contribution of the external field E to the potential energy V. Pauthenier[624] has taken this additional quantity into account for a conducting particle and, as a result, the term associated with t within the parentheses is multiplied by a quotient "sinh $(Ed/2kT)/(Ed/2kT)$". This approaches unity for very fine particles.

Besides Arendt and Kallmann[35] a number of other workers have also used the steady state diffusion equation, using the rate of ion capture as equal to the ionic flux.[118, 226, 325, 595, 624] However, Liu *et al.*[522] point out that the steady state process requires an ion concentration higher than the *maximum* of about 10^{15} ions m^{-3} found in precipitators. These workers used a more rigorous procedure than that by White, but the solution is identical with (10.36).

From equations (10.30) and (10.36) Lowe and Lucas[531] calculated the charges (n) acquired by particles with bombardment or ion diffusion charging. The calculations were based on a tubular precipitator with the following conditions (Table 10.6):

Tube diameter: 0·25 m.
Discharge wire diameter: 7·6 mm
Temperature (T) = 300 K.
Ions per unit volume (N_i) = 5×10^{13} ions m^{-3}.*
Field strength (E) = 200 kV m^{-1}.
Dielectric constant function $(3\epsilon/(\epsilon+2)) = 1{\cdot}8$.
Electronic charge (e) = $4{\cdot}8 \times 10^{-11}$ cC.
Corona current (i) = 135 $\mu A\ m^{-1}$.
Boltzmann's constant (**k**) = $1{\cdot}38 \times 10^{-23}$ J K^{-1} per molecule.
Ion mobility = $1{\cdot}8 \times 10^{-4}\ m^2\ V^{-1}\ s^{-1}$

TABLE 10.6. NUMBERS OF CHARGES ACQUIRED BY PARTICLE[531]

Particle dia. (μm)	Ion bombardment				Ion diffusion			
	period of exposure (sec)				period of exposure (sec)			
	0·01	0·1	1	∞	0·01	0·1	1	10
0·2	0·7	2	2·4	2·5	3	7	11	15
2·0	72	200	244	250	70	110	150	190
20·0	7200	20,000	24,400	25,000	1100	1500	1900	2300

For approximate calculations, Heinrich and Anderson[358] recommend the equation by Ladenburg[472, 473]

$$ne = d \times 10^5 \qquad (10.37)$$

where n = number of elementary charges,
e = elementary charge (esu),
d = particle diameter in mm,

as adequate for practical precipitators.

Combining bombardment and diffusion charging

Both ion-bombardment charging and diffusion charging act simultaneously, but the former is of greater importance for larger particles in the micron region, while the latter is more important for sub-micron particles. Single addition of the charges as calculated by equations (10.30) and (10.36) give reasonable agreement with experiment.[226, 267, 463]

* Hignett[368] points out that industrial precipitators commonly operate at space charge densities as low as 5×10^{12} ions m^{-3}, which is one order of magnitude below this value, which is in line with that quoted for laboratory precipitators.[531, 704, 932]

Cochet[161–2] has derived a combination of the two mechanisms by considering the mechanisms as ion current flow to the particle:

$$q = ne = \left[\left(1+\frac{2\lambda_i}{d}\right)^2 + \frac{1}{\frac{1}{2}+\lambda_i/d}\cdot\frac{\epsilon-1}{\epsilon+2}\right]\pi\epsilon_0 Ed^2\,\frac{t}{t+\tau} \qquad (10.38)$$

where λ_i = mean free path of ions (in air) ($\approx 0{\cdot}1$ μm), ϵ_0 = specific inductive capacity (permittivity) of free space = $8{\cdot}86\times10^{-12}$ F m^{-1}.

However, it has been pointed out by Liu and Yeh[523] that the *ionic* mean free path λ_i is an order of magnitude less than the value used by Cochet, and the irregular thermal motion becomes more significant for the smaller particles, and this cannot be expressed as ordered movement along the lines of force. These workers have, therefore, developed a new theory incorporating random motion as well as that due to an applied field. These relations yield numerical solutions which give good agreement with experiment.

10.5. PARTICLE DRIFT

As soon as the dust particles acquire some charge they will be influenced by the field in the precipitator. Most of the particles will migrate towards the passive collector electrodes away from the discharge electrode with the same polarity as the particles, while a few particles very close or within the corona zone are charged with gas ions of the opposite polarity to that of the corona and collect on the discharge electrode. The overall picture is a very complex one, as the electric field decreases away from the corona and the particles become more fully charged as they move through the precipitator. Near the passive electrode the concentration of the charged particles will be high, and interparticle interferences will occur, as well as the effect of the partially discharged layer of particles on the collector electrode.

The calculation of particle drift velocity which is used to predict precipitator size and efficiency is of necessity based on a much simpler model and the following assumptions are made:

(i) The particle is considered fully charged during the whole of its residence in the precipitating field.

(ii) The flow within the precipitator is assumed to be turbulent, giving a uniform distribution of particles through the precipitator cross-section.

(iii) Particles moving towards the electrode normal to the gas stream encounter fluid resistance in the viscous flow regime and Stokes law can be applied.

(iv) There are no repulsion effects considered between the particles which are charged with the same polarity.

(v) There are no hindered settling effects in the concentrated dust near the wall.

(vi) The effect of the movement of the gas ions, sometimes called the electric wind, is neglected.

(vii) The velocity of the gas stream through the precipitator does not affect the migration velocity of the ions.

(viii) The particle moves at its terminal velocity.

The migration or drift velocity, which is the velocity of the particles normal to the gas stream based on these eight assumptions, can hardly be realistic, but it has been found to give reasonable estimates of the cross-stream drift velocity in many cases while in others it tends to be conservative. This is ascribed to the effects of turbulence in the gas stream, the favourable direction of the electric wind, and the gas resistance which is lower than predicted for viscous flow. On the other hand, hindered settling and the residual charge on the deposited particles tend to resist rapid precipitation.

The force F on a particle, with charge q, in a field of strength E directed towards the passive electrode is

$$F = qE \tag{10.39}$$

and the resistance of the gas, assuming the particle is moving with its terminal velocity, is

$$F' = \frac{C_D}{C} A \frac{1}{2} \varrho \omega^2 \tag{10.40}$$

where ω = migration velocity of the particles.

Combining equations (10.39) and (10.40) gives the migration velocity as

$$\omega = \sqrt{\left(\frac{2qCE}{C_D A \varrho}\right)}. \tag{10.41}$$

The direct solution of this equation is not usually attempted without substituting 24/Re for C_D (i.e. assuming Stokes' Law), assuming the particles are spheres, and assuming that they have been fully charged by bombardment (equation 10.31) in a field of strength E'.

Then

$$F = \frac{3\epsilon}{\epsilon+2} E'E \frac{d^2}{4} \tag{10.42}$$

and

$$F' = 3\pi\mu d\omega/C. \tag{10.40a}$$

These two equations can be combined to give the drift velocity

$$\omega = \frac{\epsilon CE'Ed}{(\epsilon+2)4\pi\mu}. \tag{10.43}$$

Where the precipitator has only a single zone for both charging and precipitating E and E' are the same, and (10.43) simplifies to

$$\omega = \frac{\epsilon CE^2 d}{4\pi(\epsilon+2)\mu}. \tag{10.44}$$

When the particles are of a conducting material with large dielectric constant

$$\omega = 0{\cdot}08 CE^2 d/\mu. \tag{10.45}$$

For a particle with low dielectric constant, where the ratio $3\epsilon/(\epsilon+2)$ is taken as 1·75:

$$\omega = 0{\cdot}046CE^2d/\mu. \tag{10.46}$$

If the particles are small, and Ladenburg's approximate charging by diffusion equation (10.37) is used:

$$\omega \doteqdot \frac{CeE\times 10^5}{\mu}. \tag{10.47}$$

This formula indicates that the terminal migration velocity for very small particles (less than 0·2 μm) is approximately constant and independent of particle size.

In Section 10.8 the migration velocity calculated from the formulae in this section are compared with the effective migration velocities (e.m.v.) ω' obtained from experimental precipitator efficiencies and the specific collecting surface which, in turn, is a function of the precipitator dimensions and gas volume throughput.

10.6. COLLECTION EFFICIENCY OF PRECIPITATORS

If there were no particle re-entrainment in a precipitator, it would be possible, in theory, to construct a precipitator which would collect all the particles entering the unit. The dimensions of this precipitator can be calculated from the drift velocity, ω, which is assumed constant, the average velocity of the gas stream, the diameter of the discharge wire, its potential and current, and the relative diameters of tubes or distances between the plates acting as the passive electrode.

To consider first a tubular precipitator, radius R, with centrally placed discharge electrode. If the flow in the tube is streamline (unlikely in an industrial precipitator, but readily possible

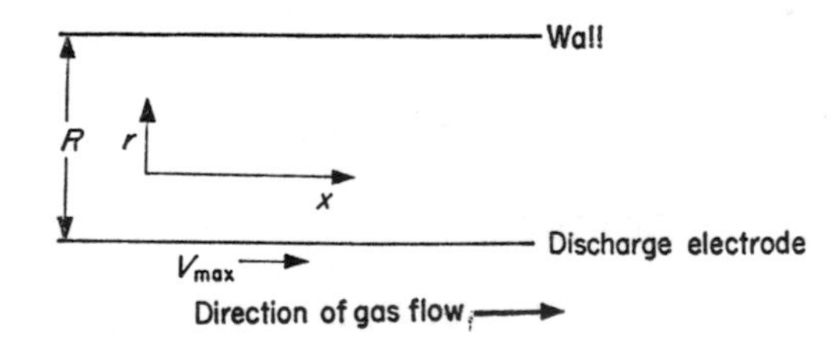

FIG. 10.7. Direction of gas flow in relation to the dimensions of a precipitator.

in special research units used for gas sampling), then the streamline flow pattern will be approximately parabolic, and the average velocity, v_{av}, can be shown to be one-half of the maximum velocity v_{max}. Then, at a time t, the velocity at a radius r (Fig. 10.7) will be

$$\frac{dx}{dt} = v_{max}\left(1-\frac{r^2}{R^2}\right) \tag{10.48a}$$

$$= 2v_{av}\left(1-\frac{r^2}{R^2}\right). \tag{10.48b}$$

The drift velocity at radius r will be ω

$$\text{and} \quad \omega = \frac{dr}{dt}.$$

The equation for the particle path will be

$$\frac{dx}{dr} = \frac{dx}{dt} \cdot \frac{dt}{dr} = \frac{2v_{av}}{\omega}(1-r^2/R^2). \tag{10.49}$$

If the drift velocity is assumed to be approximately constant across the cross-section of the precipitator, the equation (10.9) can be integrated to give:

$$x = \frac{2Rv_{av}}{\omega}\left\{\frac{R}{r} - \frac{1}{3}\left(\frac{r}{R}\right)^3\right\} + \text{constant}. \tag{10.50}$$

The constant of integration is zero, because when $r = 0$, $x = 0$ for a particle starting at the centre of the tube (or between the plates). When a particle starts at the axis, it has to traverse the distance R before collection. So for all particles to be collected the length of the precipitator has to be greater than

$$x = \frac{4}{3} \frac{Rv_{av}}{\omega}. \tag{10.51}$$

If the flow in the tube is turbulent, piston flow is assumed, and the minimum length for 100 per cent collection is equal to the distance travelled by the gas stream while the particle starting from the centre of the tube reaches the wall at constant drift velocity ω. Thus,

$$x = vR/\omega. \tag{10.52}$$

Similar equations for the length of a precipitator of plate and wire or plate and plate type can be calculated,[704] and are listed in Table 10.7.

TABLE 10.7. THEORETICAL LENGTH OF PRECIPITATOR FOR COMPLETE REMOVAL OF DUST, ASSUMING CONSTANT DRIFT VELOCITY

Electrode system	Flow pattern	Precipitator length
Tube type	Streamline Turbulent	$4v_{av}R/3\omega$ $v_{av}R/\omega$
Plate electrodes (discharge and passive)	Streamline Turbulent	$v_{av}L^*/\omega$ $v_{av}L^*/\omega$
Wire (active) and plate (passive) electrodes	Streamline Turbulent	$v_{av}L^\dagger/\omega$ $v_{av}L^\dagger/\omega$

L^* = distance between charged and passive plate.
$L^\dagger$ = distance between charged wire and passive plate.

In practice re-entrainment occurs, and it was found experimentally[20] that the efficiency of a precipitator was an exponential function of the gas stream residence time in the precipitator field. Deutsch[222] derived an equation of this form based on the assumptions that the dust is uniformly distributed at the beginning, that the uncollected dust remains uniformly distributed, and that the drift velocity is effectively constant. Then the efficiency η is found from

$$\eta = 1-\exp\left(\frac{-2\omega x}{R v_{av}}\right) \quad \text{(tube precipitator)} \qquad (10.53a)$$

$$= 1-\exp\left(\frac{-\omega x}{L v_{av}}\right) \quad \text{(plate precipitator)} \qquad (10.53b)$$

$$= 1-\exp\left(\frac{-\omega \mathcal{A}}{Q}\right) \quad \text{(either type)} \qquad (10.53c)$$

where $\mathcal{A}$ = precipitator collecting plate area,
Q = gas flow.

It has been pointed out recently that the unrealistic assumption of uniform particle concentration is not necessary to obtain the efficiency equation with the exponential term as in (10.53c). If the particle concentration near the wall is designated by C_w while $\bar{C}$ is taken as the average particle concentration across the precipitator, and if the ratio $C_w/\bar{C}$ is a constant (χ), independent of the cumulative area of collecting surface downstream of the precipitator entrance, then a modified form of the exponential relation can be derived[697]

$$\eta = 1-\exp\left(-\frac{w\chi\mathcal{A}}{Q}\right). \qquad (10.53d)$$

In the classical Deutsch equation χ is unity.

It should be noted that the Deutsch equation assumes complete charging of particles immediately at the precipitator inlet. In industrial precipitators the errors introduced by this assumption are small, but in high-velocity experimental units and in laboratory sampling precipitators they may be significant.[689] In this case the efficiency can be found from a modified form of equation (10.53):

$$\eta = 1-\left\{\frac{x+vt_0+v\tau_c}{vt_0+v\tau_c}\right\}^{2\omega\tau_c/R} \qquad (10.54)$$

where x = downstream distance from the entrance to the precipitator (wire in cylinder),
v = velocity of gas (average),
R = radius of cylinder,
t_0 = equivalent precharging time, from $q_0 = q_{max}t_0/(t_0+\tau_c)$,
where q_0 = charge on particles at entrance,
q_{max} = maximum charge attainable by particles,
τ_c = particle charging time constant
$= 4\epsilon_0/N_i e u_i$, (10.55)

ϵ_0 = permittivity ($8{\cdot}86 \times 10^{-12}$ F m),
N_i = ion concentration per unit volume,
e = ionic charge,
u_i = ionic mobility (see equation (10.16) and Table 10.3).

These are the most common equations used for predicting the efficiency of a precipitator. When a migration velocity—called the effective migration velocity, ω'—is calculated from a measured precipitator efficiency, equation (10.53c) is expressed as

$$\eta = 1 - \exp(-\omega' \mathcal{S}) \tag{10.56}$$

where $\mathcal{S}$ is the specific collecting surface per unit volume gas flow.

An equation identical with the Deutsch equation has been derived on a more general basis by White[928] who considered the probability ($\mathcal{P}$) of capture of a particle within a precipitator. For a particle to be collected during its residence time Δt, it must be within a distance $\omega\,\Delta t$ of the passive electrode. Thus, for a particle in a tubular electrode, the probability of collection is equal to the ratio of outer annulus, width $\omega\,\Delta t$, and the cross-sectional area of the tube:

$$\mathcal{P} = \frac{\omega\,\Delta t \times 2\pi R}{\pi R^2} = 2\omega\,\Delta t/R. \tag{10.57}$$

The possibility of avoiding capture ($1-\mathcal{P}$) in n sections of a precipitator is given by

$$\begin{aligned}(1-\mathcal{P})^n &= (1-2\omega\,\Delta t/R)^n \\ &= (1-2\omega t/Rn)^n.\end{aligned} \tag{10.58}$$

For large n, the possibility of escaping (which is $(1-\eta)$) is then given by $e^{-2\omega t}$, and expressing this in terms of the average gas velocity v_{av} and the length of the precipitator $x (t = x/v_{av})$, the efficiency is found to be given by an equation identical with (10.53a).

A modification of this equation, which will account for efficiency in capturing particles with a range of diameters, d_1, to d_2, has also been given by White

$$1-\eta = \exp\left(-\frac{\mathcal{A}EE'}{2\pi\mu Q}\right) + \frac{\mathcal{A}EE'}{2\pi\mu Q}\int_{d_1}^{d_2}\left\{f(d)\exp.\left(-\frac{\mathcal{A}EE'd}{4\pi\mu Q}\right)\right\} \mathrm{d}d \tag{10.59}$$

where $f(d)$ is the function expressing the cumulative size distribution of the particles. Because the cumulative size distribution is usually logarithmic, equation (10.59) can be conveniently solved by plotting the experimental particle size distribution on log-probability graph paper combining this with the function $\mathcal{A}EE'/2\pi\mu Q$, and then integrating numerically.

Turbulence, eddy diffusion and re-entrainment. In recent years a number of modifications have been introduced into the Deutsch equation to allow for turbulence, eddy diffusion and re-entrainment. These modifications have been reviewed by Robinson,[691–2, 597] who has also contributed[692] in this. Friedlander, the first to attempt this, developed an equation to account for simultaneous eddy diffusion and movement under an external force field.[276] Friedlander assumes in this case that the particle flux normal to the wall P_f g m^{-2} s^{-1} is given by

$$P_f(x) = \mathcal{D}_t \frac{dc}{dr} + \omega c \tag{10.60}$$

where x = the distance along the precipitator,
$\mathcal{D}_t$ = eddy diffusion coefficient (or diffusivity),
c = particle concentration (g m^{-3}),
r = the distance from the central electrode normal to the collecting surface.

If a wall surface without a laminar boundary layer is assumed, and if the particle concentration and gas velocity are zero near the wall, increasing towards the centre, then it has been shown that the migration velocity ω can be replaced by an effective value ω', the two being related by the equation

$$\omega' = \frac{\omega}{1-\exp(-2\omega/\varkappa v)} \tag{10.61}$$

where v = average gas velocity across the duct, and
$\varkappa$ = D'Arcy resistance coefficient [equation (8.7)].

In turbulent flow, $\varkappa v$ increases with v for a given geometry, and so an increase in gas velocity is accompanied by an increase in effective migration velocity. This is seen in practice, at least until velocities are reached where there is serious re-entrainment. More simply it can be said that turbulent diffusion reinforces electrostatic particle migration, and turbulent diffusion increases with gas velocity.

In another approach to this problem, Williams and Jackson[945] assume that there is remixing of the unprecipitated particles, which is a function of the eddy diffusion in the turbulent core of the gas stream. Their other assumptions are the same as those used by Deutsch. Essentially, the differential equation for diffusion is used (equation 3.1) with additional terms for the superimposed particle drift due to the electrostatic force. The equation is transformed using two dimensionless parameters τ expressing the path length in the precipitator (x) in terms of the distance between wire and plate (L), and φ the drift velocity (ω), in terms of the stream velocity (v_{av}):

$$\tau = \frac{x}{L}\left[4(N+1)^2 \frac{\mathcal{D}_t}{v_{av}L}\right] \doteqdot 7{\cdot}41\frac{x}{L}, \tag{10.62a}$$

$$\varphi = \frac{\omega}{v_{av}}\left[2(N+1)\frac{\mathcal{D}_t}{v_{av}L}\right]^{-1} \doteqdot 5{\cdot}67\frac{\omega}{v_{av}}, \tag{10.62b}$$

where N = no. of points in the precipitator at which calculation is carried out.

At the velocities commonly used for precipitation, the eddy diffusivity is directly proportional to the stream velocity, and so the ratio $\mathcal{D}/v_{av}L$ is 0·0042 for a Fanning friction factor of 0·0035.* The equation was solved numerically at 20 points through the precipitator

* $\mathcal{D} = 0{\cdot}071 v_{av} L \sqrt{(\text{Fanning friction factor})}$.

($N = 20$) using a digital computer, and the solution obtained graphs the precipitator efficiency in terms of the parameters τ and φ (Appendix IV). The constants in equation (10.62) are based on this. Excellent agreement between experimental and predicted efficiencies were obtained for a pilot plant precipitator.

Incorporating turbulence effects in an electrostatic precipitator efficiency equation has also been attempted by Inyushkin and Averbukh[388–9] (reported by Robinson[692, 697]). These workers used a wetted wall precipitator, avoiding re-entrainment and measured efficiencies for Reynolds numbers for the tube ranging from the streamline region up to Re = 20,000, As the flow Reynolds number of the gas increases in turbulence (Re > 2000), the turbulent deposition contribution to the migration velocity ω_t increases. In this case, the effective migration velocity term ω' is the sum of the ordinary, electrostatic migration velocity, and the turbulent and the exponential efficiency relation can be written as

$$\eta = 1-\exp\{-\chi(\omega+\omega_t)\mathcal{A}/Q\} \tag{10.63}$$

where $\chi = c_w/\bar{c}$, the particle distribution term.

Robinson[691–2] has further modified this equation in order to allow for re-entrainment. His modification incorporates the assumptions made by Inyushkin and Averbukh concerning inertial penetration of the boundary layer and turbulent redistribution of particles, but without specifying the exact cross-sectional concentration profiles. This equation is

$$\eta = 1-\left[1-\frac{\beta}{1-\alpha}\right]\exp\{-\chi(1-\alpha)(\omega+\omega_t)\mathcal{A}/Q\}\frac{\beta}{1-\alpha}. \tag{10.64}$$

Here α and β are re-entrainment parameters describing two particle fractions, one having non-zero probability of permanent capture and the other having zero probability.[692] In practical terms, α is the erosion coefficient (dimensionless) representing the mass of dust eroded per unit mass precipitated by inertial impaction (see Section 4.7) while β, the other erosion coefficient (dimensionless), is the mass of "problem" dust eroded per unit mass of total dust precipitated, i.e. with very high dust concentrations. Because the total erosion cannot be greater than the precipitating dust flux, the erosion conditions are limited by

$$0 \leqslant (\alpha+\beta) \leqslant 1.$$

When there is no re-entrainment, $\alpha = \beta = 0$, and the Robinson efficiency equation (10.64) becomes the same as (10.63). On the other hand, when a precipitator functions as an agglomerator, with total re-entrainment (i.e. when operating with carbon black), then $\alpha+\beta=1$ and equation (10.64) gives zero efficiency.

The most extensive approach to the problem of precipitator efficiency incorporating eddy diffusion, electrostatic migration and re-entrainment, has been by Cooperman.[172–4] Both positive and negative turbulent particle transfer is postulated, but with a turbulent boundary layer, the injection of particles through a laminar layer cannot be used to explain the increased deposition with increased Reynolds number. Instead, as indicated by Friedlander, positive diffusion is considered to sweep particles from the region of higher concentration away from the wall into the region of lower concentration, adjacent to the wall, where, under conditions of low entrainment, prescribed concentration gradients exist. With heavy re-entrainment, a dense cloud of particles is formed near the wall and net particle transport by diffusion is negative, i.e. away from the wall. With low diffusivity, gas flow is nearly laminar, and the efficiency is not exponential. For larger diffusivities, the efficiency is exponentially related, but the exponent is different to the Deutsch equation exponent. So far, Cooperman has not provided absolute numerical solutions, because estimates of re-entrainment give unrealistic efficiency, while empirical selection of re-entrainment parameters can give desired collection efficiencies. However, further work is in progress.[697]

A different approach has been used by Hignett,[367] who sums radially the electrostatic and (assumed constant) turbulent forces. Numerical solutions based on this lead him to conclude that the effect of turbulence on the motion of particles in a precipitator is negligible when the particles are greater than 10 μm diameter. Below 10 μm diameter turbulence begins to have some effect on their motion, and consequently on the charge the particles acquire, as these particles may be carried by turbulence towards the discharge wire where the electric field is high. The motion of particles less than 1 μm is turbulence dominated, and precipitation of these particles only occurs if they are projected by turbulence into the laminar boundary layer adjacent to the collector electrode, or if the particle is projected by turbulence into the very high electric field next to the discharge electrode.

Electric wind. The phenomenon of "electric wind", also called "corona wind" or "electric aura", refers to the movement of gas induced by the repulsion of ions from the neighbourhood of the high-voltage discharge electrode. While it was one of the earliest phenomena of gaseous discharge to be investigated, throughout the eighteenth and nineteenth centuries,[690] its importance as a contributing mechanism in electrostatic precipitation has only been considered in quite recent work.[695]. Robinson studied the electric wind in a positive corona model unit using an injected helium tracer, which dispersed, drifting towards the wall, showing a net wire-to-wall gas flow. Robinson[697] shows that the drift velocity contribution, due to electric wind (ω_{EW}) can be calculated from:

$$\omega_{EW} = K_1\left(\frac{i}{\varrho u_i}\right)^{\frac{1}{2}} \qquad (10.65)$$

where K_1 is a system constant which is a function of the geometry, i is the current, u_i the ion mobility and ϱ the gas density, as in earlier equations. This electric wind contribution to the effective migration velocity ω' can be calculated by summing with the Deutsch equation migration velocity[697]

$$\omega' = \omega_{EW}+\omega. \qquad (10.66)$$

The contribution of electric wind is an important second order effect, and should be included in any comprehensive analysis of precipitator behaviour.

For example, in a discussion of ash accumulation, which is observed on precipitator discharge wires, and for which special rapping gear has to be installed, Shale[758] suggests that this dust deposition is largely due to the electric wind caused by gas ions of sign opposite to that of the discharge wire, which are formed as part of the corona pattern.

10.7. SECONDARY FACTORS IN PARTICLE COLLECTION – DUST RESISTIVITY

The previous sections have discussed the removal of particles and droplets from gas streams with electrostatic forces. The practical efficiency of a precipitator, however, depends on a number of secondary factors governed by the behaviour of the dust on arrival at the collection electrodes, and its removal from the electrodes. These factors are a function of the type of dust, its physical properties—particle size and resistivity—and, to a certain extent, of the overall gas velocity in the precipitator. They are allowed for in the effective migration velocity (e.m.v.) which is calculated from the measured efficiency of a precipitator [equation (10.56)] and the specific collecting surface area (projected) per unit volume flow $\mathcal{S}$. The most important of these factors is the particle resistivity, which determines whether or not electrostatic precipitation can be applied to a particular dust-removal problem. When particles or droplets arrive at the collecting electrode, they are partially discharged by the electrode, and adhere to it by a combination of forces—molecular adhesive forces of the London–van der Waals type, surface tension forces from the moisture present and electrostatic forces. The extent of the electrostatic adhesion depends on the rate at which the charge leaks away from the particles to the earthed collector, which depends on the resistivity of the dust, and which, in turn, depends on the conductivity of the dust particles, and of any moisture present.

Liquid droplets and some metallic and carbon particles are very good conductors, while the particles from most industrial smelting plant consist of metallic oxides which are excellent insulators when dry. In industrial type fumes traces of impurities are always present as well as moisture. So, if the temperature is sufficiently low, enough moisture will be there to form a conducting film.

The resistivity of dusts handled by electrostatic precipitators covers a range from 10^{-5} Ωm for carbon black to 10^{12} Ωm for dry lime rock dust at 90°C.[794] For most effective operation the dust resistivity should lie in the range from 10^2 to 5×10^8 Ωm. When particles have low resistivity as in the case of carbon black, they are rapidly discharged on meeting the earthed electrode. Because the molecular and surface tension forces are inadequate to hold the carbon black particles on the collector electrode, they are re-entrained in the gas stream. Carbon particles in flue gas tend to "hop" or "creep" through the precipitator when the electrodes are flat plates and means of preventing re-entrainment are not provided. In the case of air-cleaning precipitators for occupied spaces it is usual to cover the collector plates with a sticky, soluble oil to prevent re-entrainment. The dust quantities collected from the atmosphere are small and barely cover the plates before the oil is washed off and replaced at intervals of 1 to 6 weeks.

When the resistivity of the particles is very high (greater than 5×10^8 Ωm), the rate of

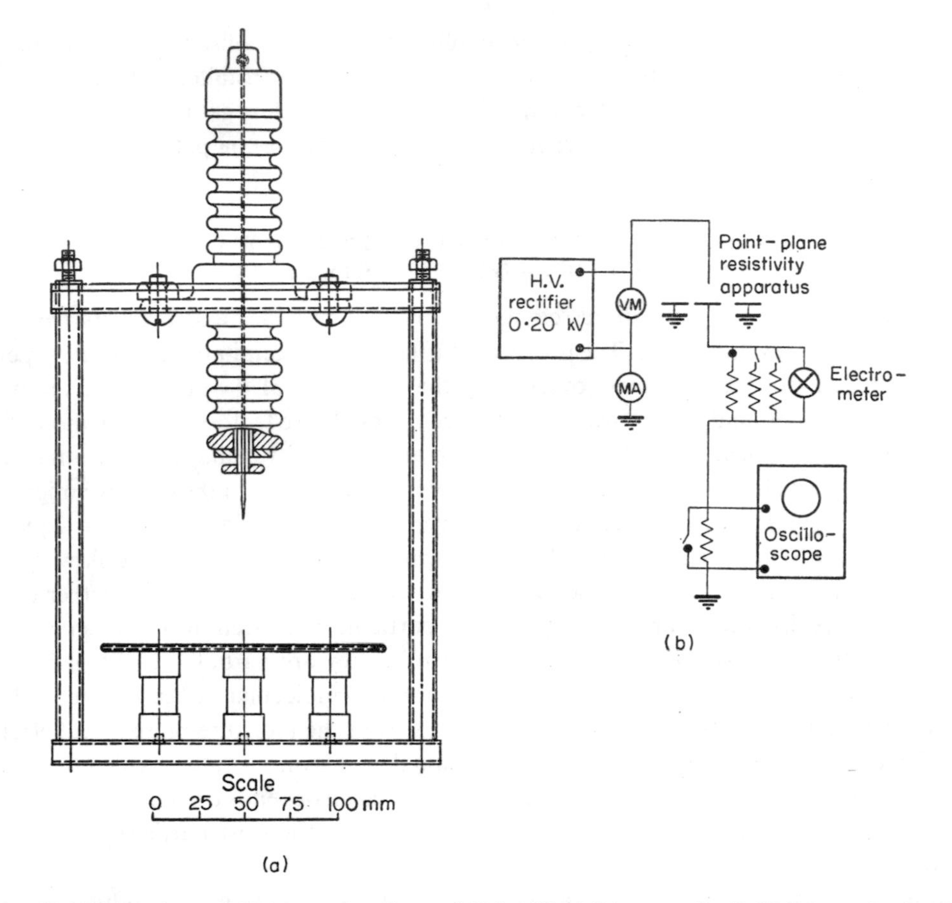

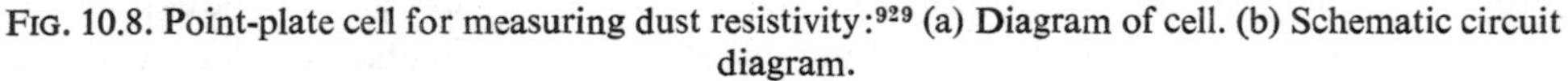

FIG. 10.8. Point-plate cell for measuring dust resistivity:[929] (a) Diagram of cell. (b) Schematic circuit diagram.

One type of laboratory cell is shown. The cell is mounted in an electrically heated and thermostated oven capable of reaching temperatures of 315–370°C. A particle layer several millimeters thick is generally used. The layer may be placed on the plate electrode manually or else precipitated on by means of a d.c. corona from the high-tension point. The high tension disc electrode is lowered on the dust layer for the actual resistivity measurement.

In another type of cell, humidity as well as temperature may be controlled, with humidity values ranging from bone dry up to 30 or 40 per cent by volume. The higher humidities are possible, of course, only for temperatures above the dew point. The schematic electric circuit used with these cells is shown. The electrometer is sensitive to 10^{-12} A, permitting measurement of resistivities as high as $10^{18}\,\Omega$ m. Field measurements of particle resistivities in plant flues may be made by use of a compact, light-weight apparatus similar in principle to that used in the laboratory. The cell may be shielded against direct gas impact where high gas velocities are met. In another arrangement gas may be withdrawn from the flue through a side chamber and resistivity measured under controlled lower gas velocities.

Experiment shows that the resistivity depends somewhat on the compactness of the particle layer and also on the applied voltage. The compactness effect normally may introduce an uncertainty factor of about two in the measured resistivity. Applied voltage is limited by breakdown or sparking through the layer. The breakdown strengths of particle layers usually range from a few thousand volts per metre to 1000–2000 kV m^{-1}, with the latter being more common. Since the current through a particle layer usually increases somewhat faster than the applied voltage, the measured resistivity will be lower at higher voltages; therefore resistivity measurements generally are made at voltages near breakdown or at least at values corresponding to a field strength of a few kilovolts per centimetre.

discharge from the collected particle layer is very small, and a charge builds up on the collected particles until electrical breakdown of the gases occurs, first in the spaces between the particles, and then on the surface of the dust layer. This is called "back ionization" or "back corona". The theory of back corona has been discussed extensively by Robinson[697] and Böhm.[96]

Some ions of opposite polarity to that of the charge on the dust are formed and these migrate back towards the discharge electrode. They reduce the charge in the charging zone and neutralize the charge on the charged particles. To overcome this problem the charge on the dust layer has to be given more time to leak away. This can be done by temporarily reducing the precipitator current, and so, temporarily, the precipitator efficiency.

When the resistivity of the deposited dust is in the range of 10^2 to 10^8 Ωm these problems are avoided. Considerable measurements of the resistivity of deposited dusts have been carried out to see conditions under which their resistances lie within this range.[929, 790, 794] A laboratory cell (point-plate type) of the type shown in Fig. 10.8, mounted in a thermostat oven, is used for these. The electrical circuit includes an electrometer sensitive to 10^{-12} A, which allows the measurement of resistivities as high as 10^{13} Ωm.

Two alternatives to the point plate system described above have been developed. The first is the cyclone collector where the hopper is the measuring cell.[145, 146] In this cell a central rod electrode, with varying potential, is surrounded by an earthed cylinder. The whole system uses insulating polymer insulation (e.g. Teflon) and is retained in an air thermostat. While this system is simple to use, it relies on a cyclone collector, and may not collect a dust sample representative of electrostatic precipitator performance. In a different system the collector is a precipitator where the earthed electrode consist of the meshing prongs of two combs, which do not touch one another. The high-tension electrodes are ribbons fastened to conductors outside the combs. When dust has been collected on the earthed electrodes, it fills the gaps between the prongs and the combs can now be used as the electrodes of the electrometer. The whole system is enclosed in a thermostat and this enables measurements over a range of temperatures and gas compositions to be carried out, using the same or different dust samples.[235]

It is necessary to carry out the resistivity measurements near the breakdown voltage (or at least at field strengths of the order of several kV m^{-1}), because the measured resistivity is lower at these high voltages, and these are conditions similar to those within an actual precipitator. A number of curves showing the experimentally measured dust resistivity for industrial dusts with various moisture contents is shown in Figs. 10.10a–h. It is seen that in all of these, a high moisture content lowers the resistivity, particularly at temperatures below 90°C, when conduction through the moisture enclosing the particles takes place. The curves reach a maximum between 90–180°C, and then fall off again as conduction through the particles themselves, either by electronic or ionic processes, takes over. At about 250–300°C the effect of moisture becomes negligible. In some semi-conductors such as lead sulphide the conduction is electronic, while in others such as lead chloride, it is ionic. In all cases the resistance falls off rapidly with rising temperature.

At temperatures above the commercial electrostatic precipitator range, which was investigated by White,[929] Sproull and Nakada[790], [794] the resistivity of low carbon fly ash and high carbon (30 per cent) residue from synthesis gas manufacture has been investigated by Shale *et al.*[761] The results of this work are shown in Fig. 10.9. The curve for the low carbon ash shows a resistivity of 10^{11} Ωm at 38°C, rises slightly to reach a maximum of 120°C, and then decreases, apparently levelling off at approximately 10^5 Ωm at 815°C. The high carbon ash, on the other hand, has a resistivity of 5×10^3 Ωm at 38°C, and this increases, first gradually to 375°C and then sharply to 480°C. After this, the resisitivity decreases gradually to about 10^5 ohm m at 850°C. The high conductivity below 375°C is explained by the presence of

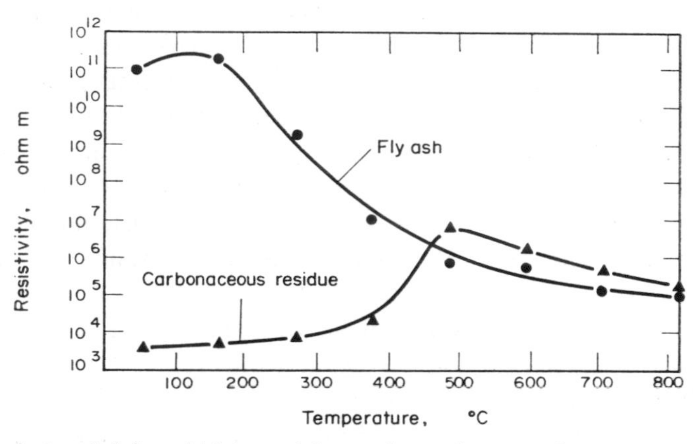

FIG. 10.9. Electrical resistivity of high- and low-carbon ashes as a function of temperature.[761]

graphitic carbon which is a good conductor, with increasing temperature this material is oxidized, and thus there is a marked increase in resistance. At still higher temperatures there is a falling off in resistance as this becomes a function of electronic conduction.

The resistivity decrease has the form:[794]

$$r = a\, e^{b/T} \tag{10.67}$$

where a and b are constants which have to be determined experimentally.

An exception to this behaviour, even at ambient temperatures, is sulphur in the powdered form. Sulphur is hydrophobic and remains a perfect insulator.

The effect of moisture in the gas on resistivity can be estimated from an empirical equation of the type suggested by Masuda[558]

$$r_{moist} = r_{dry} \exp\left(T_{DP}\, a_1\, e^{b_1 T}\right) \tag{10.68}$$

where r_{moist} = resistivity of the "moist" deposit,
r_{dry} = resistivity of the "dry" deposit,
T_{DP} = gas dew point temperature (°C),
T = actual gas temperature (°C),
a_1, b_1 = constants, determined experimentally.

Maartmann[541] has shown that predictions based on this equation give reliable maximum resistivities for fly ash (at about 50°C), but tend to underestimate the resistivity at higher temperatures (200°C).

When certain dusts and fly ash types are very difficult to precipitate, the gases can be conditioned with water, sulphur trioxide, sulphur dioxide or ammonia to improve the effectiveness of the precipitation. In the case of a cement dust, 0·5 per cent sulphur dioxide was used to increase the e.m.v. from 35 to 50 mm s^{-1} at 220°C, and the resistivity was reduced to 10^9 Ωm from a much higher value.[453] When coals are low in sulphur (less than 1·5 per cent) and with high ash content, such as those found in the coal fields of Scotland and burned at Kincardine or in the coal fields of New South Wales (Australia), then chemical conditioning is very effective in improving collection efficiency.

In the case of Kincardine,[145, 199] the coal contains 0·47 per cent sulphur, and continuous dosing with SO_3 at a rate of 7–10 ppm is used (25–35 mg m^{-3}). The efficiency is increased from 94 to 99·3 per cent under test conditions. The SO_3 concentration in the flue gases leaving the precipitator is approximately 1·3 ppm (5 mg m^{-3}), which is of the same order if no conditioning agent had been used. When SO_3 injection at a rate of 10 ppm or less is used, then this is an economical alternative to increasing the size of the precipitator. The SO_3 dosing system used is described in British Patent No. 933,286 and consists of heated storage tanks (to prevent solidification and polymer formation of the SO_3), a flow regulator, a vaporizer, an air entrainment system and disperser.[146]

SO_3 conditioning, at the rate of 20 ppm, was also tried at Tallawarra (S in coal, 0·4 per cent) and improved the efficiency from 80 to 99 per cent [4·8 g m^{-3} to 0·23 g m^{-3}], but as SO_3 is not manufactured commercially in Australia this was abandoned.[909] It was also noted that the fly ash from the southern N.S.W. coalfields was acidic (pH 3·5–4) and this was assumed to account for the large quantities used. Sulphuric-acid injection was also tried, ahead of the economizer, where the temperature is such that dissociation into SO_3 and H_2O was calculated to take place. While this was not successful at Tallawarra, moderate success was experienced with this[909] at Pyrmont power station (S in coal, 0·38 per cent). In the power stations in northern N.S.W. the deposited fly ash has a pH of 11 and SO_3 conditioning is anticipated to be much more effective.

Ammonia injection, at a rate of 15 to 20 ppm, was used at Tallawarra and raised efficiency from 85 to 98 per cent with a collecting area of 60 $m^2/m^3\ s^{-1}$ (135°C). Later work, in another Tallawarra precipitator, showed that injecting 17 ppm ammonia improved efficiency from 80 to 96 per cent. This also resulted in a reduction of back corona, very common with N.S.W. fly ash, but the measured resistivity of the dust remained in the range 10^{11} to 10^{12} Ω m. It is assumed that the ammonia changes the gas characteristics (reducing back corona) rather than that of the deposited fly ash.[909]

Ammonia injection has also been used to improve the performance of precipitators on boilers burning high sulphur coals (2·5–3·5 per cent) with gas exit temperatures of 130°C, which is the gas dew point at the TVA power stations of Colbert and Widows' Creek.[679] In this case the unconditioned gases give the fly ash particles in the gas a film of conducting sulphuric acid, and they no longer retain the static charge needed to let the electrostatic field act effectively. Injection of ammonia neutralizes the acid, but total neutralization requires ammonia at the rate of 60 ppm, which would cost $160,000 p.a. (1968) for two boilers, and would be uneconomical. Using only 15 ppm, however, gave effective precipitator efficiency of 90 per cent and coupled with savings from reduced corrosion in air preheaters and induced draught fans, the cost is now acceptable.[679]

An example of the effectiveness of ammonia injection in improving precipitator efficiency is that of 20 ppm ammonia into the gases entering a precipitator after petroleum cracking carrying powdered aluminium silicate catalyst. The resistivity in this case was reduced from 5×10^9 to 5×10^8 Ωm, with resultant improvement in efficiency from 96 to 99·8 per cent.[929]

Water conditioning has also been used when the coals were low in sulphur and also dry, as at Pyrmont (N.S.W.). The water added was of the order of 18 to 16 g m^{-3} gas, and the precipitator efficiency was improved from 89 to 97 per cent. When this amount of water is introduced as water vapour (steam), then this affects the economics of the boiler; but if it is introduced as water droplets, then the time taken for evaporation requires plant extensions, and water deposition causes corrosion.[145] On the other hand, if a low sulphur coal is associated with large amounts of water (e.g. Yallourn, Australia, brown coal), no problems are experienced with electrostatic precipitation.

Sproull and Nakada[794] suggest an equation for estimating the potential build up across the dust layer, considering this as a condenser, and assuming that an equilibrium is established between the arrival of new charged dust particles and the leaking away of the charge to earth:

$$\mathcal{V}_d = rx\varphi\left(1 - \frac{r\epsilon}{36 \times 10^9 \pi t}\right) \tag{10.69}$$

where $\mathcal{V}_d$ = potential difference across dust (V),
r = resistivity (Ω m),
φ = current density (A m^2),
ϵ = dielectric constant (dimensionless),
x = thickness of dust layer (m).

When r is not very high (less than 10^{11}) equation (10.58) simplifies to

$$\mathcal{V}_d = rx\varphi. \tag{10.70}$$

The actual attractive force F of the particle towards the passive electrode can be found from[531]

$$F = d^2(K\mathcal{V}_d ir - \mathcal{V}_d^2/32) \tag{10.71}$$

where K = constant. The first term in parentheses represents the attractive force based on the charge on the particle, and the second term the induced potential repulsive force. When dust resistivity is low, the repulsive force may be larger than the attractive force. There are still the molecular and surface tension forces holding the particle. If the surface tension forces are negligible, as is the case at high temperatures generally, the molecular forces alone retain the particle, and these vary directly with the diameter d of the particle. Hence large particles may be more likely to leave the electrode than small ones (the repulsive force varying with $(\mathcal{V}_d d)^2$).[531] Strict confirmation of these relations is not available.

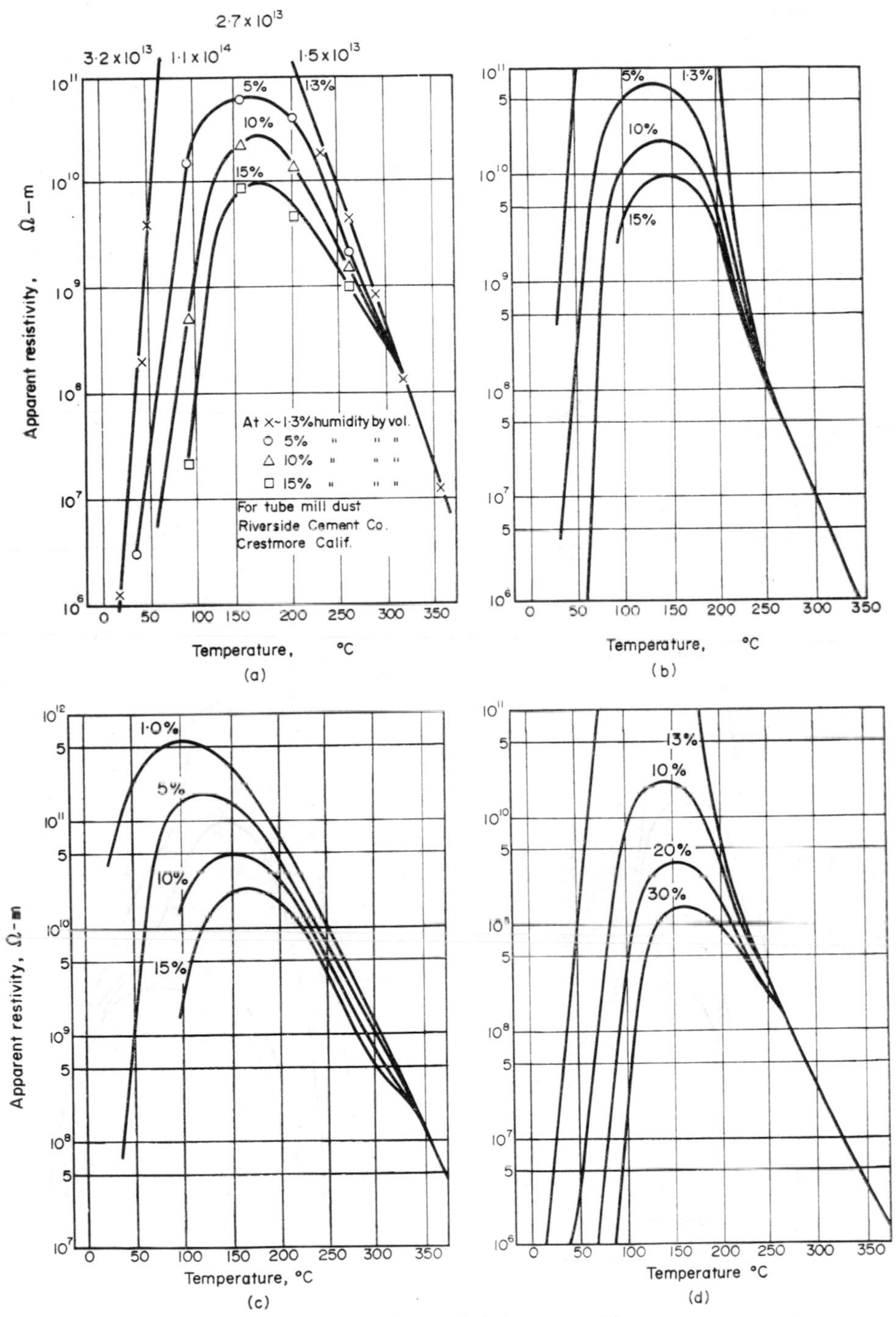

FIG. 10.10. Dust resistivity curves.[794]

(a) Apparent resistivity of powdered lime rock used in making Portland cement (various moisture concentrations: 1·3, 5, 10, 15 per cent).

(b) Apparent resistivity of lead fume from a sintering plant, which contains about 3% zinc, otherwise mainly lead sulphate (various moisture concentrations: 1·3, 5, 10, 15 per cent).

(c) Apparent resistivity of lead fume from a lead blast furnace, which contains about 13% zinc (various moisture concentrations: 1·0, 5, 10, 15 per cent).

(d) Apparent resistivity of lead fume from a slag treatment plant, which contains about 20 per cent zinc, otherwise chiefly lead sulphate (various moisture concentrations: 1·3, 10, 20, 30 per cent).

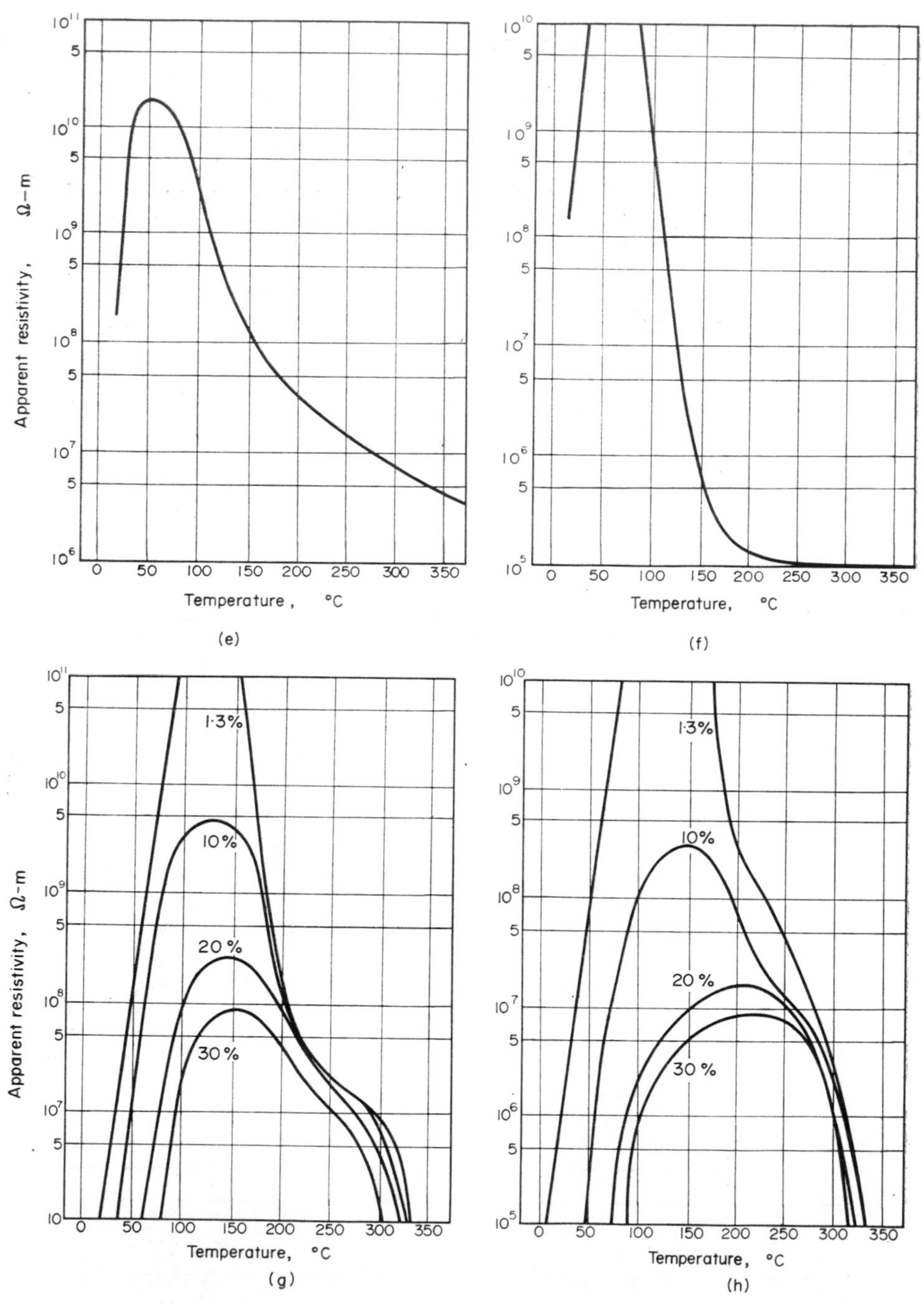

FIG. 10.10 (*continued*)

(e) Apparent resistivity of zinc fume from a slag fuming plant, with about 40% zinc (moisture concentration: 1·3 per cent).

(f) Apparent resistivity of zinc fume from a melting plant, which contains about 50 per cent zinc (moisture concentration: 1·3 per cent).

(g) Apparent resistivity of fume from an open hearth furnace. The furnace handles large quantities of scrap (moisture concentrations: 1·3, 10, 20, 30 per cent).

(h) Apparent resistivity of fume from an open hearth furnace different to (g) (moisture concentrations: 1·3, 10, 20, 30 per cent).

10.8. EFFECTIVE MIGRATION VELOCITY (e. m. v.)

The apparent migration velocity of particles in a precipitator, calculated from the collector surface area of the precipitator, the gas flow, and the measured collection efficiency (equation 10.56) is called the effective migration velocity. It incorporates the effects of such factors as particle resistivity and entrainment losses during rapping.

Early workers, such as Mierdel,[571] found that the e.m.v. values obtained for cement and brown coal dust agreed with the migration velocity calculated from equation (10.44). Similarly the extensive experimental work on fly ash,[867] flue dust, catalyst dust and cement dust,[355] by later authors give reasonable agreement between theoretical and experimental migration velocities, particularly for particles larger than 20 μm. However, these values were

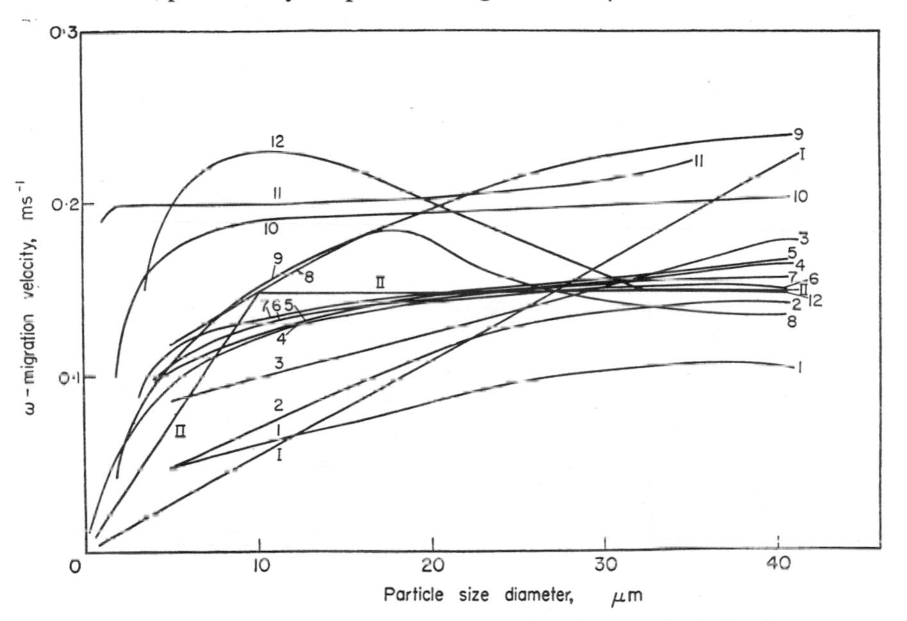

FIG. 10.11. Effective migration velocities as a function of particle size for boiler flue dust, catalyst dust, cement dust and coal dust.[355]

I Theoretical curve based on $E = \sqrt{4i/U_i} = 130$ kV m^{-1}.
II Assumed average curve for average conditions.

1, 2, 3 = boiler flue dust precipitator.
4 = catalyst dust precipitator.
5 = cement kiln dust precipitator.
6 = boiler flue dust precipitator.
7 = boiler flue dust precipitator.
8 = boiler flue dust precipitator (vertical).
9, 10 = boiler flue dust precipitator.
11 = coal dust precipitator.
12 = boiler flue dust precipitator (vertical).

all obtained over a limited range of gas-stream velocities of 1–2 m s^{-1}. Heinrich[355] who studied the effect of particle size on e.m.v. (Fig. 10.11) found that theory gives much lower velocities than those obtained experimentally for particles below 15 μm.

Although the theoretical equations (10.43) and (10.44) indicate that the migration velocity should be independent of the gas-stream velocity, experiments have shown that this has in fact considerable bearing on the e.m.v.[144, 195, 355, 356]

Thus Dalmon[195] found a maximum e.m.v. for gas-stream velocities of 2–2·2 m s^{-1}, while Busby and Darby[144] found a maximum at 5–6 m s^{-1}. Even Kalaschnikow in 1934[420] observed a slight maximum although he attributed this to experimental error. Dalmon investigated the relation between e.m.v. and gas-stream velocity for different electrode types (tulip and

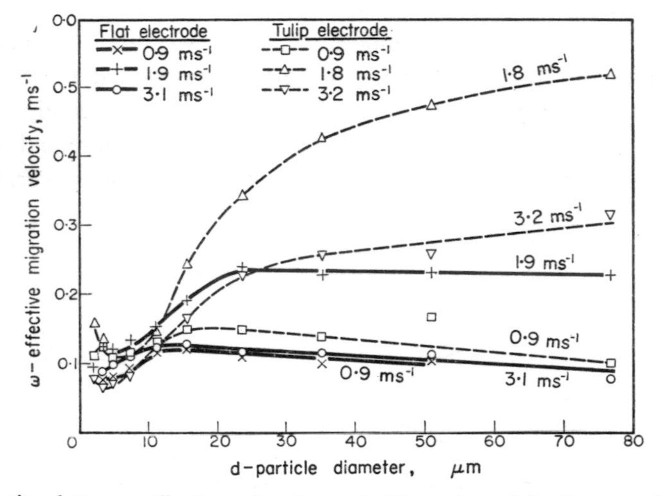

FIG. 10.12. Relation between effective migration velocity and particle diameter for different gas velocities with flat and tulip electrodes in a pilot plant.[195]

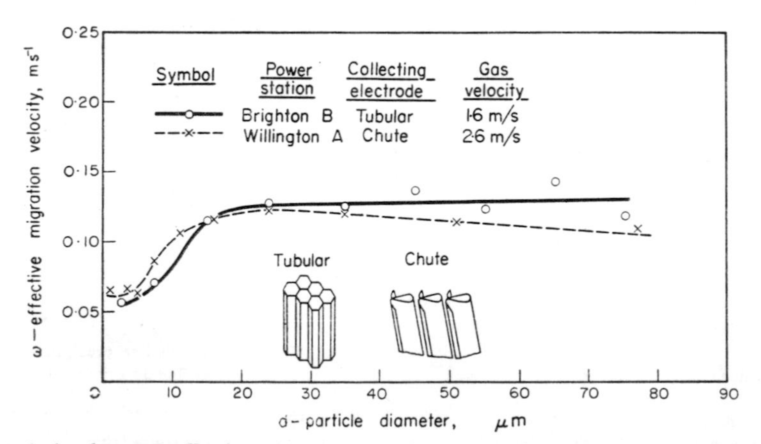

FIG. 10.13. Relation between effective migration velocity and particle diameter with tubular and chute electrodes in commercial sized plant.[195]

flat plate) and the complex relation found is shown in Fig. 10.12 for pilot plants and in Fig. 10.13 for commercial plant (chute and hexagonal tube electrodes).

To overcome the velocity-dependence problem in practical design calculations, Koglin[451] suggested an empirical modification of the e.m.v., ω'' which he successfully applied to pocket plate-type collector electrodes. The modified e.m.v., which is also the maximum,

ω''_{max}, can be calculated from an experimental e.m.v., ω', at a particular gas stream velocity v (m s^{-1})

$$\omega''_{max} = (50\omega' + v^2)/200v^2 \text{ (m s}^{-1}\text{)} \quad (10.72)$$

The efficiency of a precipitator at any velocity, v can then be found from

$$\eta_T = 1 - \exp\left[\frac{-(\sqrt{(200\omega''_{max})} - v)x}{50L}\right] \quad (10.73)$$

where x = length of field (m),
L = distance between wire and plate (m).

Koglin has also investigated the effect of dust concentrations in dry systems[452, 454] and when precipitating dusts and mists in wet electrostatic precipitators.[453] He finds that both in wet and dry systems there is virtually no effect by dust concentrations on the precipitation efficiency in the concentration range of 2·0–34 g m^{-3} for a wide range of dusts (e.g. copper oxide, cinder dust, fly ash from black and brown coals). Koglin obtains the precipitator efficiency in terms of his modified migration velocity ω'' from

$$\eta = 1 - \exp\{(25vL/\omega''x) - 10\} \quad (10.74)$$

where $\omega'' = (2{\cdot}5\omega' + 0{\cdot}1v)^2$,
ω' = experimental effective migration velocity [equation (10.56)].

The most economical gas-stream velocity for operating the precipitator occurs at the maximum e.m.v. The gas stream velocity is then

$$v = \sqrt{(50\omega''_{max})} \text{ m s}^{-1}. \quad (10.75)$$

The effect of gas temperature on the e.m.v. is complex as it consists of the fo l wing aspects:

(i) The temperature dependence of corona and the sparkover characteristics of the system.
(ii) The temperature dependence of gas viscosity.
(iii) The effect of temperature on the resistivity of the deposited dust.

From the discussion of corona and sparkover characteristics it was seen (Section 10.3) that lower field strengths must be used at higher temperatures, and this results in lower migration velocities. The viscosity of all gases increases with temperature, and this also reduces the speed of particle migration. The temperature dependence of dust resistivity is not simple, but at high temperatures it has been shown that this generally decreases exponentially with rising temperature.[794] The combination of these temperature-dependent terms to give a comprehensive method of predicting the effect of temperature on the e.m.v. has not been attempted at this stage, so experiments have to be carried out when attempts are made to increase the temperature in precipitators.[762]

The effect of increased dust resistivity is also not readily predictable. For example, Sproull[790] found that the e.m.v. of a cement test dust in a single-stage precipitator fell from 0·15 to 0·035 m s^{-1} as the dust resistivity increased from 10^8 to 10^9 Ω m (curve AA in Fig. 10.14), while in a two stage precipitator an approximately constant e.m.v. of 0·10 m s^{-1} was maintained (curve BB).

The depth of a layer of high resistivity dust also affects the e.m.v. Brandt[115] found that a 1-mm fly ash layer on the passive electrode resulted in an e.m.v. of 0·098 m s^{-1}, but this decreased to 0·058 m s^{-1} when the layer had increased to 10 mm (Fig. 10.15).

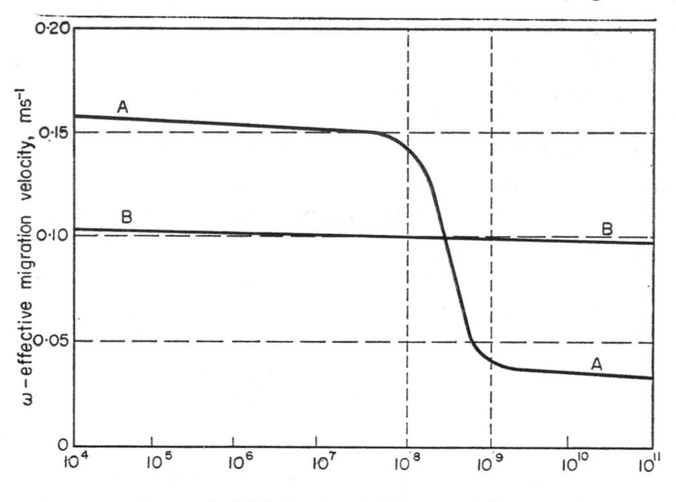

FIG. 10.14. Relation between effective migration velocity and apparent resistivity in a single stage (curve AA) and double stage (curve BB) precipitator collecting lime rock test dust.[790]

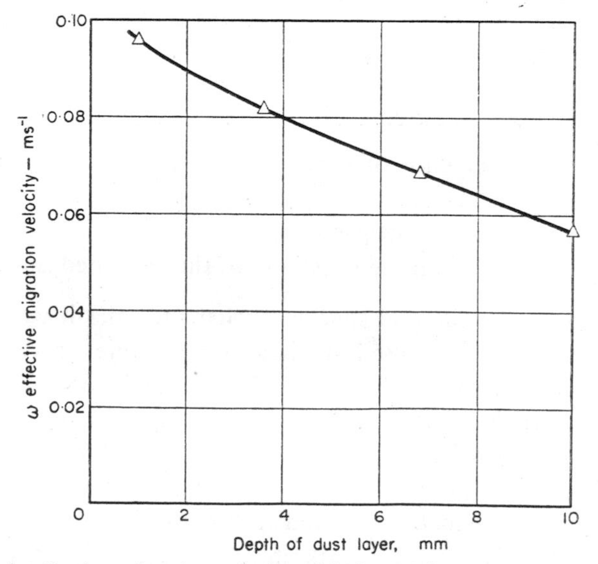

FIG. 10.15. Changes in effective migration velocity with depth of deposited dust layer on flat electrodes (fly ash).[115]

The effect of changes in the specific power input on the e.m.v. is that, for a given *size distribution* (30 per cent less than 10 μm), a 50 per cent rise in e.m.v. results from doubling the specific power input (mA/100 m^2).[355] For individual particle sizes the migration velocity for each increases in direct proportion with the power input. The migration velocity tends to remain constant with constant power input, regardless of whether this is achieved by high current and low potential or high potential with low current.

Results of tests were correlated by Heinrich[355] for a large number of power stations. These had dry bottom pulverized coal-fired boilers and precipitators with 4 mm squared, fluted, or barbed-wire discharge electrode and pocket or chute-type passive electrodes 0·30–0·32 m apart. Good correlations were obtained between e.m.v. and specific current or power input, but very little correlation was found to exist between e.m.v. and the applied high-tension voltage.

10.9. REMOVAL OF DEPOSITED MATTER

When mist droplets are deposited on the passive electrode of a precipitator, these run together forming larger droplets, which in turn run down the plates or tubes and are collected. Often mist precipitators are irrigated with additional liquid streams to increase the rate of droplet removal from the plates. When solid particles are deposited, these can also be washed or scraped off the electrodes, but most frequently they are dislodged from the electrodes by *rapping* with hammers or vibrators.

Vertical flow irrigated electrostatic precipitators (Fig. 10.16) are widely used in the chemical industry for precipitation of acid and tar mists. They have to be constructed from corrosion resistant materials for these duties, which increases their initial cost. Water consumption is usually fairly high with these units. When used for hot gases, they will be cooled by the precipitator and a cold gas plume will be emitted. This cold plume settles in the immediate surroundings of the plant without dispersing, and may cause difficult or harmful conditions if insoluble toxic or obnoxious gases are present in the gases.

A new type of horizontal irrigated precipitator has been described by Parkington and Laurie-Walker[623] which has small plate spacings and continuously irrigated passive electrodes. This results in a more compact design, and greater effective field strength at lower operating voltages (15 kV) than with the conventional design (operating voltages 45–60 kV).

A special case of a precipitator with many of the characteristics of the irrigated precipitator is the two-stage, positive corona, air-cleaning precipitator used for cleaning air for air-conditioning applications. The plates in this precipitator are covered with a soluble oil to which the dust sticks, and which are washed down at intervals between 2 and 8 weeks to remove the accumulated dust, and are then recoated with clean oil.

If a dry precipitator is washed down during shut-down periods, great care must be taken to remove all the water from the plates before passing dirty gases through the plant. The residual liquid tends to form a crust with the particles, and this may be difficult to remove without further shut down and washing.

Removing the accumulated dust by scraping the electrodes during operation has only been applied where the electrodes are of a semi-conducting material such as concrete

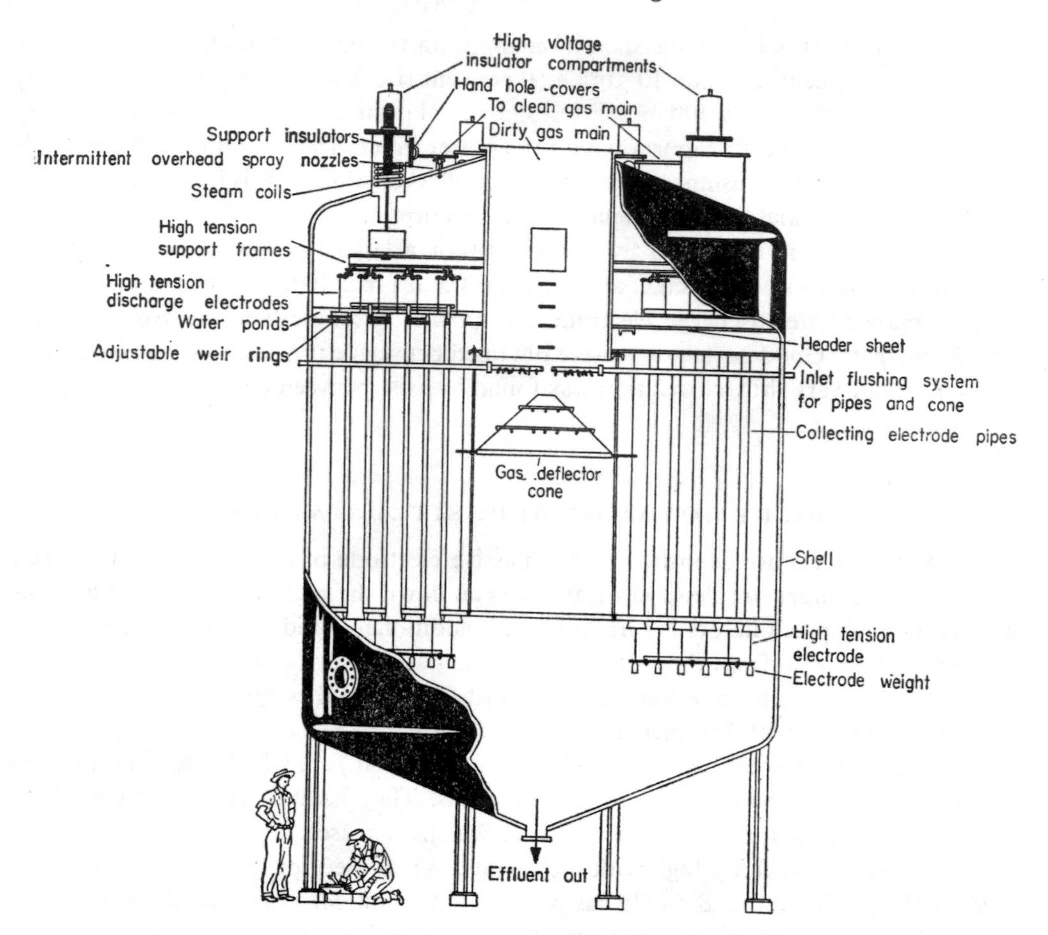

FIG. 10.16. Vertical gas-flow tube-type irrigated precipitator.[920]

reinforced with conducting rods. These electrodes are used where there is a tendency to discharge at a potential below that required for efficient precipitation. The resistance in the electrode tends to suppress the discharging of the dust and stabilize the field. The dust is scraped off by dragging scraper chains across the electrode, usually with the gas stream shut off to avoid-re-entrainment.

Continuous brushing of electrodes has been applied on a very small scale, where precipitators have a continuous belt of electrodes, operating at comparatively low voltages, and where a brush is installed in the base of the unit.[756] In an industrial scale precipitator where there are large passive collecting plate areas, this design has not proved economic, and rapping is used.

The rapping devices are of three types:

(i) Cam-operated drop hammers or spring-loaded hammers.
(ii) Magnetic or pneumatic impulse rappers, controlled by timing switches.
(iii) Electromagnetic vibrators.

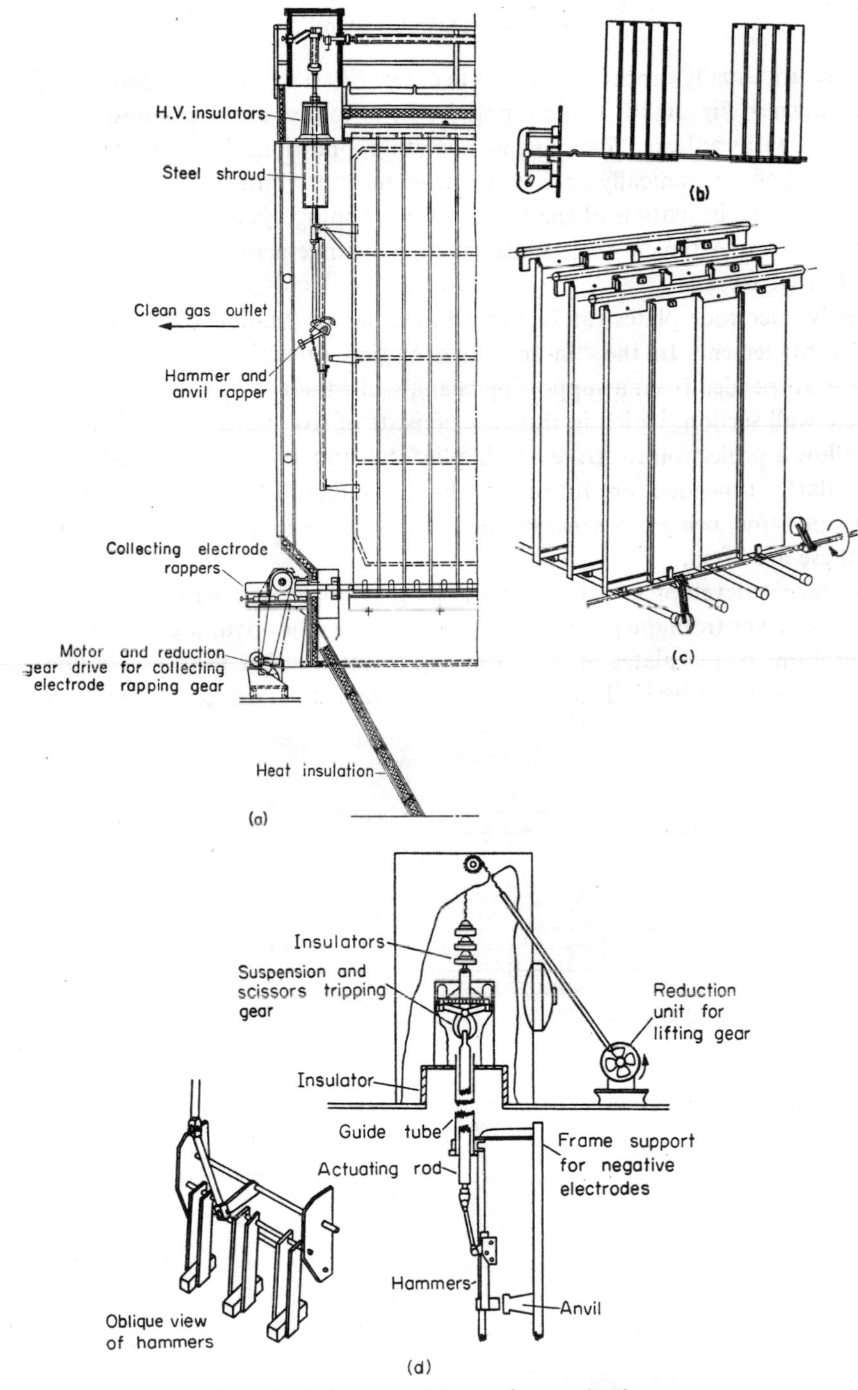

FIG. 10.17. Various plate rapping mechanisms:

(a) Hammer and anvil rappers—in line.[377]
(b) In line impulse rappers (two sections).[520]
(c) Cross rapping, using hammers.[356]
(d) High-tension discharge wire rappers.[520]

The plates are usually rapped by devices in categories (i) or (ii), and diagrams illustrating these are shown in Fig. 10.17. Modern practice tends to favour the impulse rappers because these are more easily adjusted for changing operating conditions. The high-tension electrodes are rapped either mechanically or with electromagnetic vibrators. Care must be taken to ensure the electrical insulation of the high-tension rapping system from the earthed casing of the precipitator. A typical mechanical rapping arrangement for high-tension electrodes is shown in Fig. 10.18.

The passive electrode plates can be rapped in either an "in-line" (Fig. 10.17b) or "cross-rapping" arrangement. In the "in-line" two section precipitator shown, the plates are individually suspended from a supporting beam, while their lower ends hang between rails. A complete wall section, which in this case consists of five plates, is displaced in the same plane to allow a projection to strike a rigid platform. In a common alternative method, the plates, similarly suspended, are hit directly by a hammer. The energy consumed by either method is the same, being governed by the weights of the hammers (or the plates) and the distance these are lifted.

A fairly recent development is a cross-rapping arrangement, which shakes rather than shears the dust layer from the plates. This system also has the advantage that the acceleration of the vibrations of the plates in cross rapping is much larger than in "in-line" rapping, making it more efficient.[115, 356] The plates have a fundamental frequency of vibration depend-

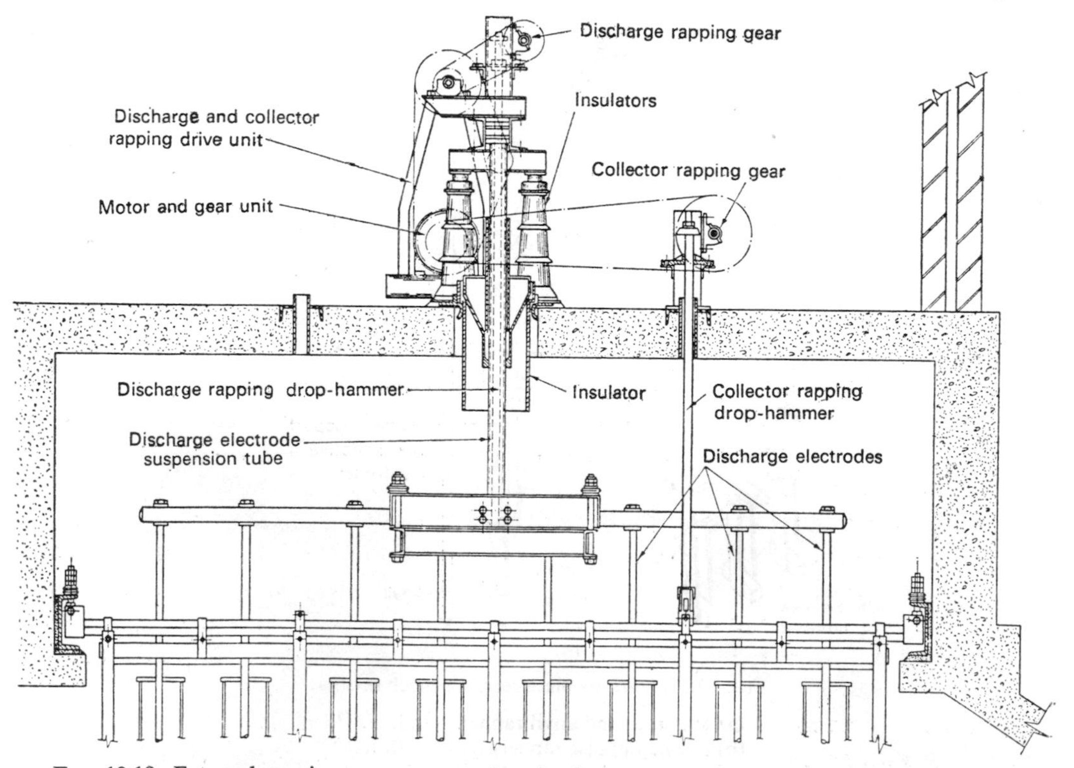

FIG. 10.18. External rapping arrangement for simultaneous rapping of discharge and collector electrodes.[907]

ing on the type and position of the suspension, which may be a pin (Fig. 10.17b) or springs, and the point where the plates are struck by the hammer. The striking of the plates superimposes harmonics on the fundamental frequency which lead to varying accelerations in the plate which shake off the dust. These accelerations have been measured in some cases[115] and these are listed in Table 10.8.

TABLE 10.8. ACCELERATION (IN g) ON RAPPED PLATES[115]

(A) Single plate, 7·5 m high, 0·77 m wide and 1·5 mm thick, suspended by a pin and held between rails.

Exited by:	Hammerblow 2·6 kilo (force)-meter			Vibrator 67 Hz cross vibration
Point of application	Right-angle stroke at base of plate	Right-angle stroke at centre of plate	In line-stroke at base of plate	
Maximum	9300	65,000	550	590
Minimum	540	500	64	101
Mean	2950	4000	215	290

(B) Groups of plates—two of five plates each $3\frac{1}{4}$ m tall, 0·47 m wide, 3 mm thick.

Exited by:	Hammerblow 2·6 kilo (force) meter				Vibrator 67 Hz		
Suspension	*Pin*		*Springs*				
Point of application	Cross base	Cross centre	In-line base	In-line top	Cross base	In-line base	In-line top
Maximum	2300	232	14,000	92	191	97	91
Minimum	28	50	16	6	46	79	26
Mean	133	108	58	37	91	90	63

On the basis of this work it can be concluded that vibration of the plates has not been as successful as striking them with hammers because of the relatively small accelerations obtained when using a vibrator (see column 4, Table 10.8A). In many cases also the even vibrations may tend to compact the dust on the plates instead of shaking it off.

Rapping may be either continuous or intermittent, and the timing of the intermittent rapping is determined by the amount of dust being collected. It is usual practice to allow 6–12 mm of material to collect before rapping. If the quantities collected are too small, the agglomerates will not be sufficiently large to fall into the hopper, but will re-entrain. If the accumulated dust is too thick, re-entrainment may take place, and dust resistivity factors can also cause difficulties.

As different sections of precipitators accumulate dust at different rates, the rappers should be in independently controlled sections, so that the optimum dust layer accumulates in each section before rapping.

It has been found in practice that intermittent rapping tends to give higher precipitator collection efficiencies than continuous rapping.[265, 520] Little[520] found that the maximum efficiency of $95\frac{1}{2}$ per cent for a fly ash precipitator was obtained by rapping at intervals of 90–120 min, isolating the section being rapped during cleaning, compared with 93 per cent with continuous rapping. Longer rapping intervals than 120 min tended to give reduced efficiencies. By grading the intensity of rapping better performance has been obtained.[10] For example, moderate impacts can be used initially during rapping to loosen the dust layer, and then larger blows, increasing in intensity, for shaking off the finer dusts.

Practical rapping cycles for optimum precipitator performance have to be established experimentally in the individual cases, and depend on such factors as electrode types, temperature and humidity. Dalmon and Lowe,[195] for example, found in full scale plant with flat electrodes, maximum efficiency was obtained with 20–30 min intervals between rapping but a similar plant with tulip electrodes (Section 10.10) gave better performances with 45-min rapping intervals. These periods are shorter than those found by Little, but are still considerably longer than the 4–10-min rapping cycles frequently used in industrial plant. When vibrators are used, it has been found that 3–4 sec vibration periods remove nearly all the deposited dust, and no further dust is shaken off after 6 sec.[711]

10.10. ELECTROSTATIC PRECIPITATOR ELECTRODES AND OTHER STRUCTURAL FEATURES

The main structural features of electrostatic precipitators are the high tension electrodes suspended on an insulated framework, the passive electrodes which are earthed to the shell, and the enclosing shell. Provision must be made for dust removal from the plates and the collecting hoppers, for thermal insulation if the gases are hot and the temperature is to be maintained; and for housing for the electrical equipment associated with the installation. Care must be taken to ensure that no corrosion takes place in the precipitator, and that in the case of direct short circuiting the system fails "safe".

The type of precipitator used depends on the duty. In industrial units the air flow is either horizontal or vertically upwards. Downward vertical flow is not used because during rapping the falling dust would re-entrain in the gas stream leaving the precipitator. The vertical-tube type of plant is used generally for smaller gas flows than horizontal flow plant, and in the case of special mist precipitation problems. For dry dusts, a hexagonal tube pattern (Fig. 10.19) is used because this gives the largest collecting surface possible with no space to be blocked off between the tubes. The tubes have to be rapped as a group to dislodge the dust.

Circular tubes can be easily equipped with weirs around the top to give an even liquid flow down the walls (Fig. 10.16). Irrigated tubes do not require rapping, and so can have lower mechanical strength. They can therefore be made of a soft, corrosion-resistant material such as lead, which is an important factor when precipitating dilute sulphuric acid mists and other corrosive mists.

The discharge wires in vertical precipitators can be either fine round wires, wires with small barbs or wires with a square or star-shaped cross-section. Because the discharge wires are often more than 6 m long, round wire which is thin enough to give a stable corona may

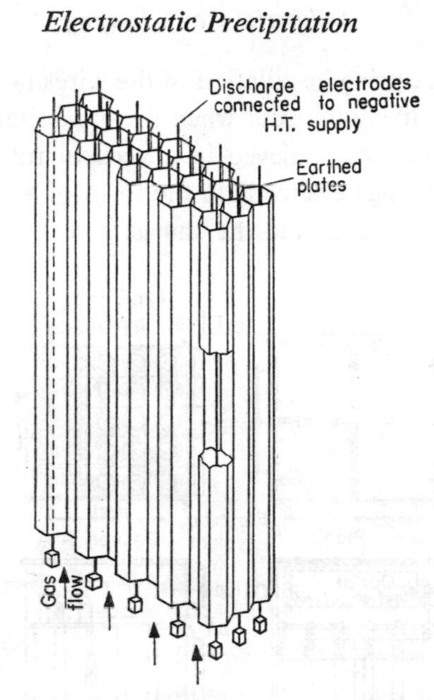

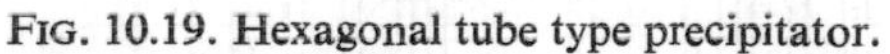

FIG. 10.19. Hexagonal tube type precipitator.

not be sufficiently strong, particularly as it will be subjected to vibrations during rapping. Heavier gauge square or star sectioned wires, with the sharp edges giving stable coronas, are used. Barbed wire has been favoured in some precipitators, a recent application being for the precipitation of iron oxide fume from oxygen steel making.[354]

The wires are generally hung from an insulated frame and are individually weighted. They are loosely retained near the bottom in a guide assembly, thus allowing for individual wire expansion, preventing buckling of the wires when there is an uneven temperature rise. For example, a 6-m stainless-steel discharge wire electrode expands 35 mm when being heated from ambient temperatures (15°C) to an operating temperature of 370°C. The wire shapes, sizes and materials used in some typical industrial applications are listed in Table 10.9.

TABLE 10.9. DISCHARGE WIRES USED IN ELECTROSTATIC PRECIPITATORS[474]

Application	Electrode wire shape	Materials of construction	Wire size (effective) (mm)	Operating tempera-ture (°C)
Acid mist	Star	Lead with monel wire core	9	45
Cement dust	Barbed	Mild steel	19	370
Open heart fume	Barbed	Mild steel	19	290
Fly ash	Square	Mild steel	5	145
Environmental air	Round	Tungsten	0·13	20

Great care must be taken during installation of the wires to avoid damaging them, otherwise frequent breakages are likely to occur when the precipitator is operating.[354]

Two-stage precipitation can be achieved in a vertical tube precipitator by having the precipitating section high-voltage electrode in the form of a heavy wire or a tube, without corona formation at the upper end, and the charging section with a fine wire on which a

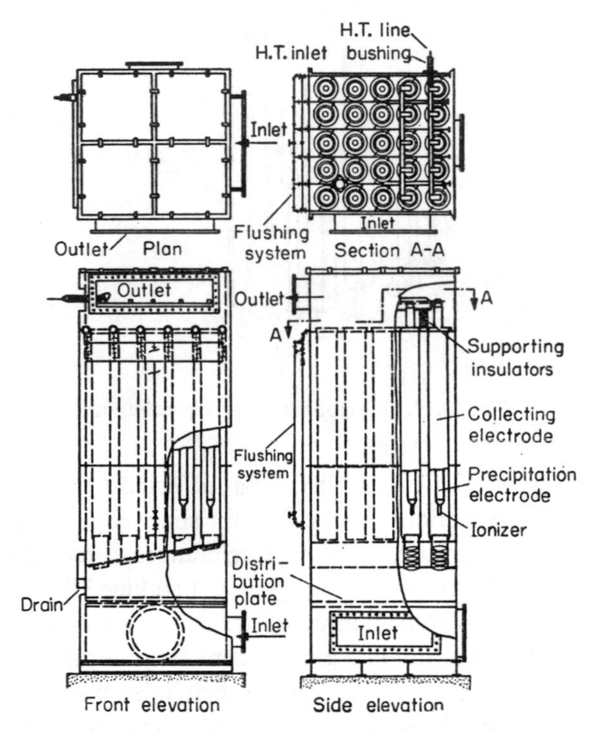

FIG. 10.20. Two-stage discharge electrode vertical flow tube precipitator.[493]

corona forms, attached to the bottom of the precipitating high-voltage electrode (Fig. 10.20).

Vertical-flow plate electrodes can be either simple flat plates, a design used in fuel gas de-tarring plant (Fig. 10.21b), or, when dust loads are very high, "tulip" electrodes.[265] These are made of bent steel trips welded on to two sides of a frame (Fig. 10.21a), so that when rapping occurs the dust falls down the space between the plates, reducing re-entrainment.

Horizontal-flow precipitators (Fig. 10.22) are more frequently applied on a large scale and in cases where high dust resistivity makes collection difficult in a single-stage unit. Thus, in the case of a zinc oxide fume precipitator, where a particularly high fume resistivity is encountered, a four-zone precipitator has been successfully used. In the first zone, where there is high charge density, particle charging and agglomeration takes place, and currents of only 2·5 mA are passed, while in the subsequent zones more even field with different field strengths are used, with the currents increasing to 25 mA in the second zone, 32 mA

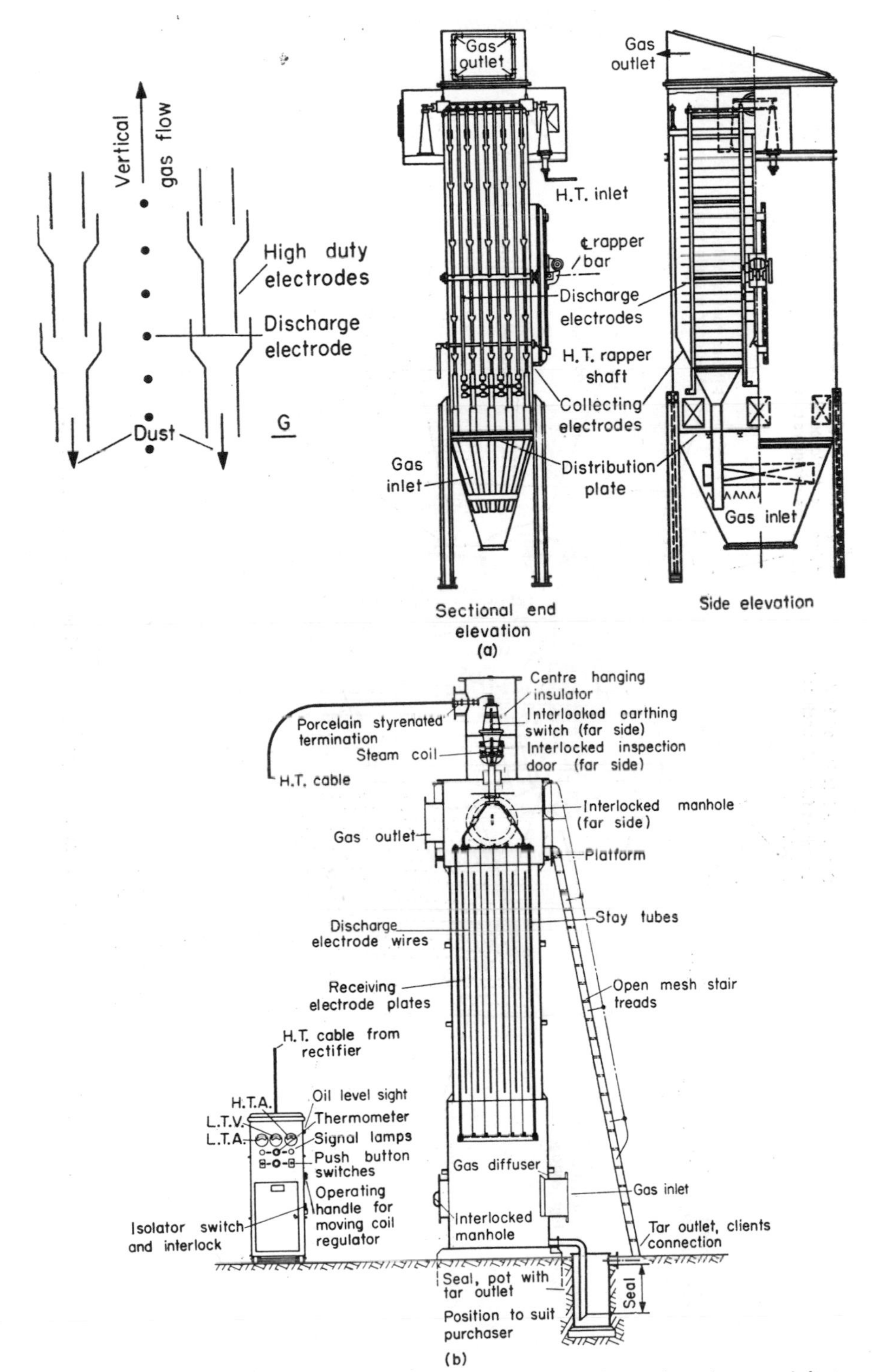

FIG. 10.21. Vertical flow plate type precipitators: (a) Tulip-type heavy-duty dust precipitators.[493] (b) Flat plate de-tarrer for mist collection.[958]

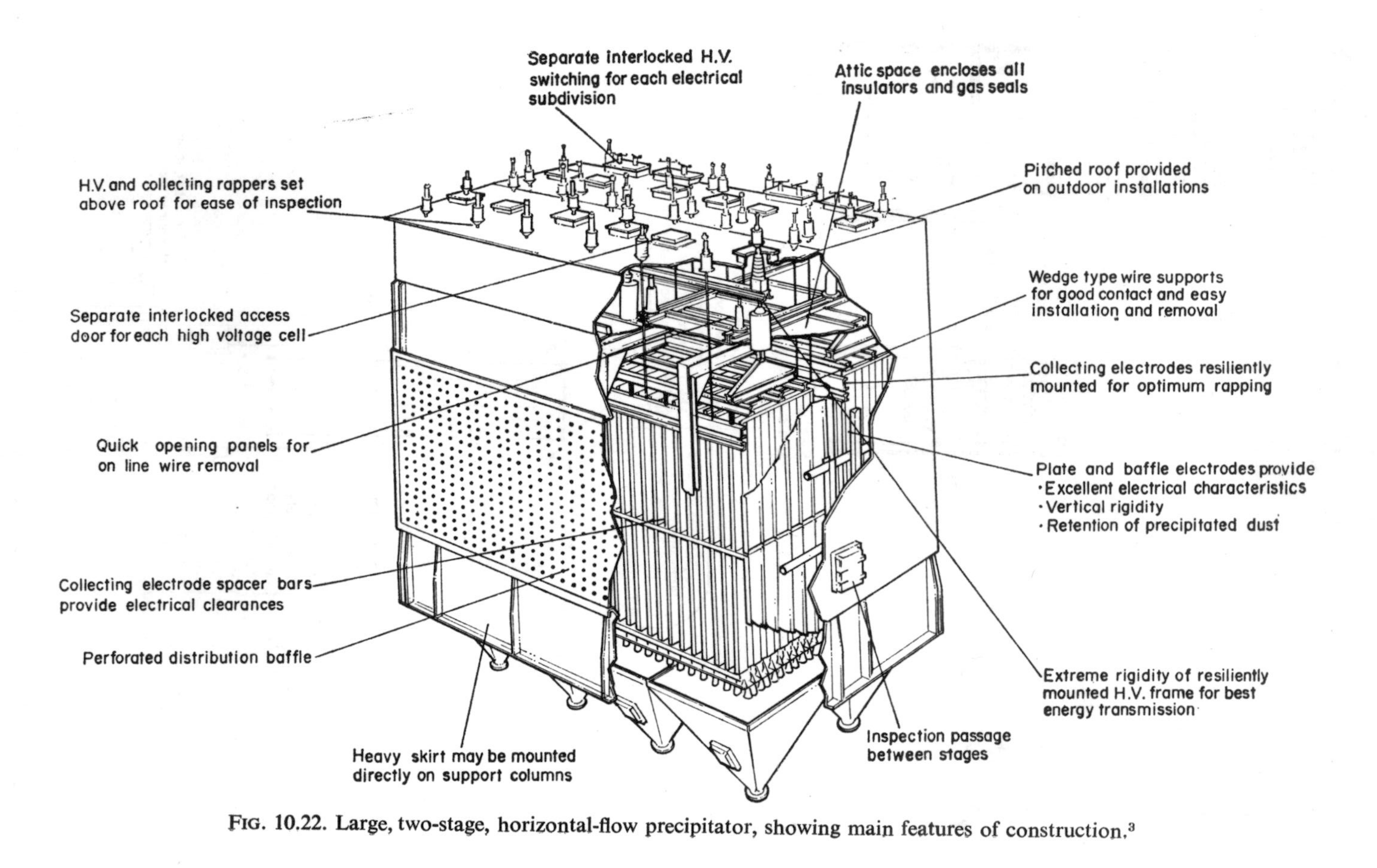

FIG. 10.22. Large, two-stage, horizontal-flow precipitator, showing main features of construction.[3]

in the third zone and 52 mA in the fourth zone. In this last zone 95 per cent of the dust was precipitated, while the overall efficiency was about 99 per cent.[358] The rapping in the different zones must here be timed for different intervals to give most effective deposition.

In the first zone the discharge electrodes are often long steel barbs cut from steel strip, while the electrodes in the precipitating zones consist of star-shaped wires which are more

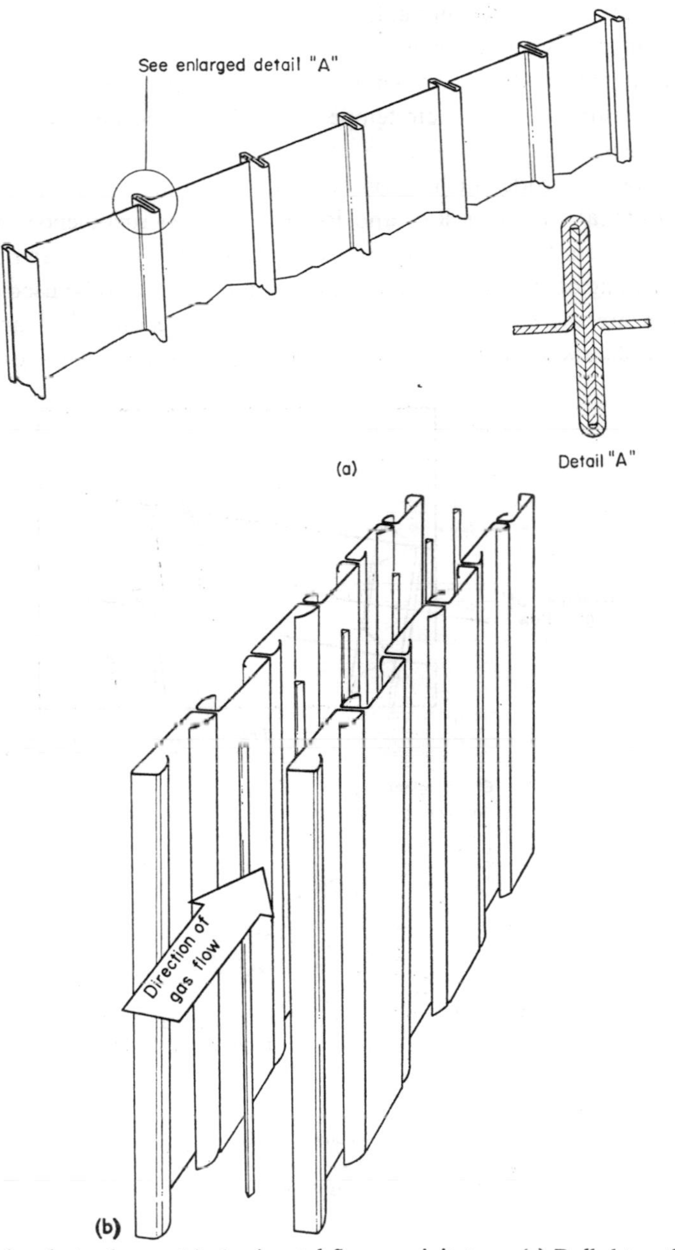

FIG. 10.23. Passive electrodes used in horizontal flow precipitators: (a) Rolled together plates with small projections.[3] (b) *C* electrodes, bolted together.[377]

closely spaced. The discharge electrodes are suspended from a metal framework, which sometimes consists of tubes, and it is this framework which is rapped to remove deposited material from the wires. The framework is, in turn, suspended on insulators, usually of porcelain, and rated for the voltage of the precipitator. Care must be taken to avoid dust and moisture settling on the insulators, causing short circuits. The chief method is to blow clean air over the insulators to prevent fouling. If the inside of the precipitator is at pressures above atmospheric, clean compressed air is used, while for precipitators at sub-atmospheric pressures, clean external air can be drawn in over the insulators. In many cases heating coils surround the insulators to raise their temperature above the dew point and prevent the deposition of moisture.

Shapes of passive electrodes in horizontal flow precipitators are numerous. Flat plates have the best electrical characteristics and induce the least turbulence in the gas stream. However, re-entrainment tends to occur more easily from flat plates than other sections, and also, to avoid buckling of the plates, a heavy gauge material has to be used for construction. Profile electrodes are therefore often found to be more economical and to give better performance. The shapes used are narrow sections rolled together with comparatively small

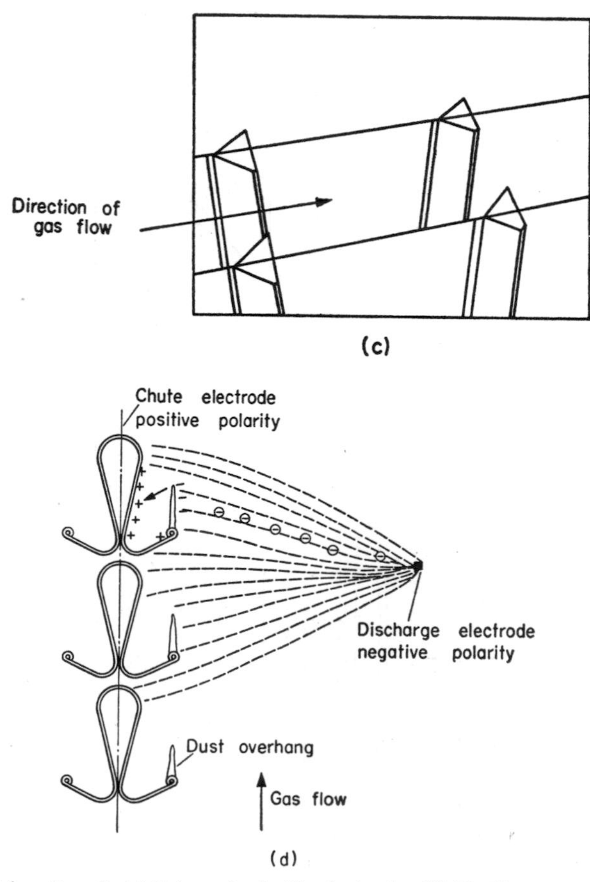

FIG. 10.23 (*continued*). (c) Triangular baffle electrodes.[933] (d) Chute electrodes.[358]

projections (Fig. 10.23a), "C" section channels rolled individually and bolted together[724] (Fig. 10.23b), triangular baffles welded on[933] (Fig. 10.23c), chute electrodes (Fig. 10.23d),[358] or similar sections not shown here. The profiled electrodes provide low-velocity areas at the plate surface downstream from the projections or inside the "C" of a "C" electrode, from which the dust is less readily entrained. Rod curtains and zigzag plates (Figs. 10.23e and g) are also used. Pocket electrodes (Fig. 10.23f), where catch pockets are provided on the outside of the plates and the dust falls down the dead space between the plates (as with "tulip" electrodes), are useful for heavy dust burdens. The suspension of these plates has been discussed in the section on rapping. A recent development[740] is the "hole electrode plate", where the gas stream passes through holes in the passive collector instead of moving parallel to the electrode, which forces the particles closer to the collecting electrode and enhances their chance of collection. These electrodes can be set on a rotating disc and the dust removed by brushes or doctor blades instead of the conventional rapping technique (Fig. 10.24).

Another design using circular electrodes, but with the gas flowing past these, and not through holes, is the MABA-TRON electrofilter, developed by Kaufmann.[748] This has a central dirty-gas plenum from which the gases flow past slowly rotating passive electrodes interspersed with cartwheel-like discharge electrodes. Doctor blades remove collected material continuously, and so this unit proves useful for difficult and sticky dusts, such as

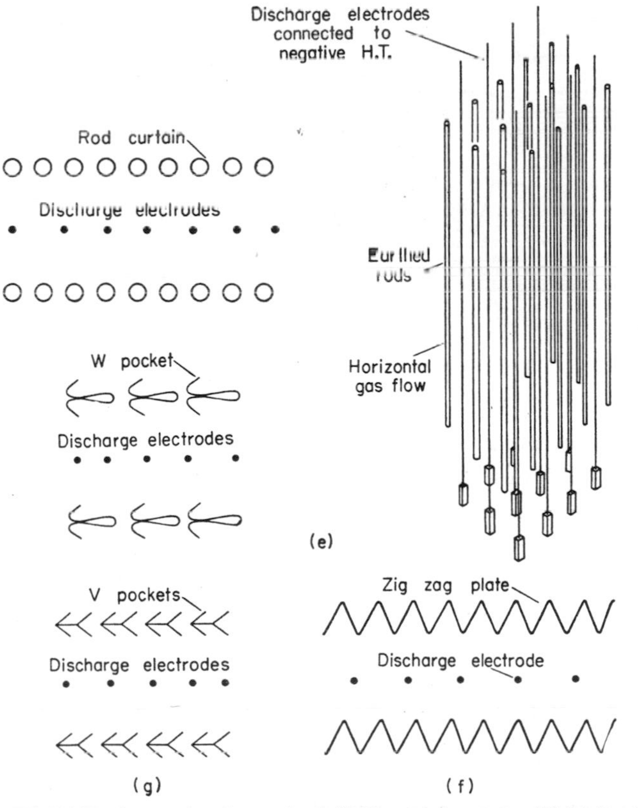

FIG. 10.23 (*continued*). (e) Rod–curtain electrodes.[3] (f) Zigzag electrodes.[926] (g) Pocket electrodes.[920]

those containing tars. Typical units have a gas through-put varying between 12 to 15,000 m h^{-1}, i.e. with discharge electrode potential of 14–18 kV. The efficiency is not as high as for fly ash, but the type of material collected is very difficult. Thus, for carbide dust 94–96 per cent was collected from an initial concentration of 0·400 to 0·435 g m^{-3}, giving 0·015 to 0·025 g m^{-3} in the waste gas (50–105°C); tarry dusts were reduced from 4·27 g m^{-3} to 0·13 g m^{-3} (93 per cent collection) at 70–89°C; and pigment dusts were collected at efficiencies of 90–98 per cent giving residual concentrations generally less that 0·01 g m^{-3}.

The gas flow distribution in precipitators is controlled by flow distributors, dampers and deflector plates. These are required to reduce the gas stream velocity and distribute the dirty

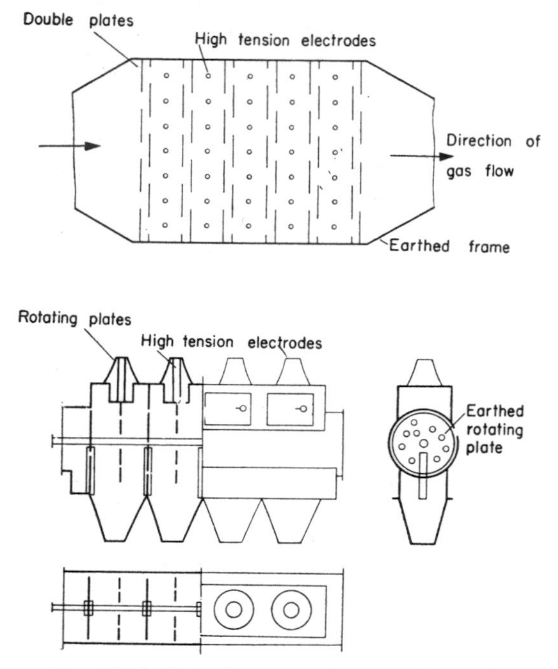

FIG. 10.24. Hole electrode plate precipitator.[740]

gases evenly across the precipitator. Perforated baffles are frequently used for this. The damper system sometimes operates a by-pass for the gases during rapping to reduce losses by entrainment. Air and water models of precipitators are used to determine the shape and positions of baffles and dampers to give the best possible flow distribution.[749, 949]

The construction of the shell for a precipitator depends on whether plates or tubes are used and on operating conditions. If the unit is pressurized, a cylindrical shell is most suitable. Tubular electrodes can be readily arranged in a cylindrical shell, but when plate electrodes are used, a rectangular housing is preferred. Only when operating at particularly high (or low) pressures will plates be housed in a cylindrical shell. Rectangular or cylindrical shells are normally fitted with hoppers for collecting the dust. The hoppers should be large, because fine fumes are not compacted on collection and tend to occupy far more space than is indicated by the bulk density of the material.

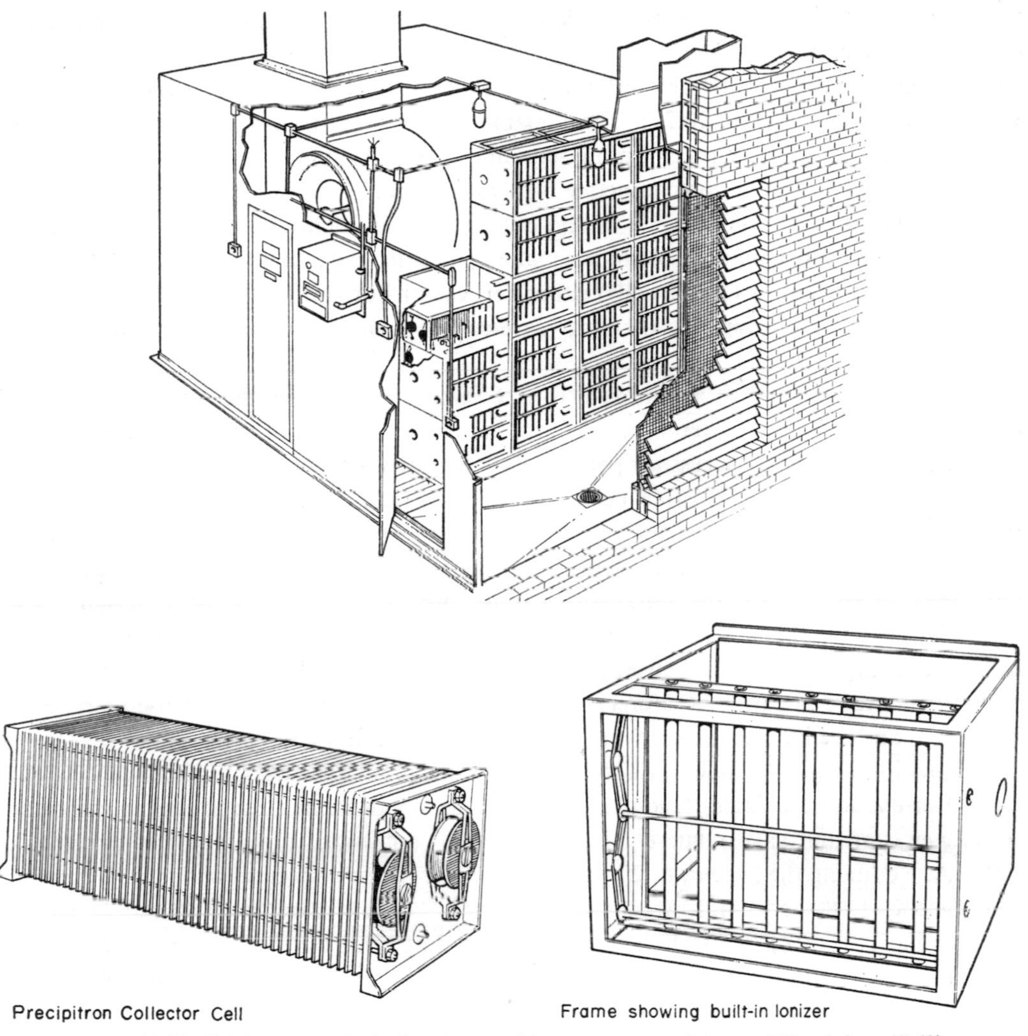

FIG. 10.25. Environmental air-cleaning positive corona precipitator ("Precipitron").[921]

The precipitator shells can be built of brickwork, reinforced concrete or steel, the choice of the material depending largely on the operating conditions (temperature and corrosion) and the local availability of the materials. Steel shells have the advantage that the plant can be prefabricated in the shop, with many of the attachments in position, before removal to the site. Concrete shells are to be avoided when operating temperatures exceed 260°C. For high-temperature operation, where lowering of the temperature may be undesirable, the plant has to be thermally insulated. Metal doors and other metallic conductors have to be arranged so that heat conduction through these is minimized. Construction of experimental precipitators to operate up to 800°C has been discussed by Shale and Moore.[762] (See Section 10.11.)

In all cases access doors and staircases have to be provided. Special locks have to be fitted so that no access can be gained to the plant without switching off the high tension and discharging the electrodes.

The standard electrostatic precipitator for cleaning environmental air is a two-stage unit with positive corona (Fig. 10.25). The ionizing stage has thin (0·13 mm) tungsten wires, charged to 13 kV, and the precipitating section consists of aluminium plates, spaced 6 mm apart, which are in alternation charged to 6 kV or earthed. The gas-flow rate in the precipitator varies between 2·0 and 2·5 m s^{-1}, and efficiencies considerably better than 95 per cent are usually achieved. Complete plants are made up of cells, each handling about 2400 $m^3 h^{-1}$, and for very small units (1 or 2 cells) either vertical or horizontal flow can be used. Large flow rates, frequently over 250,000 $m^3 h^{-1}$, are readily handled in multicell horizontal flow plants.

As the accumulated dust is removed from the plates by washing them down at intervals of upwards of a week to several months, depending on the dust burden, drainage must be provided as well as ready access for personnel. Some sort of protection from coarse particles and insects for the plant is often necessary, as these can cause frequent short circuits if many are present.

Stringent requirements for such applications as nuclear-powered submarines have led to the development of environmental air-cleaning precipitators operating at much higher velocities (10 m s^{-1}) and at efficiencies of the order of 99·8 per cent.[934] The units consisted of a plate charging section followed by horizontal hexagonal tubes with 12 mm cylindrical high-tension tube electrodes as a precipitating section. The charging section used a potential of 38 kV, and 20 kV was used in the precipitation section. A full-scale unit with shorter (1 m) precipitating field tubes used velocities of 8 m s^{-1}. Other operating conditions were tried. For example, a single-stage plant, using velocities of 30 m s^{-1}, reduced the efficiency to 91 per cent and gave relatively high ozone concentrations (1·43 ppm) at currents of 121 mA. Reducing the current to 27 mA, reduced the ozone concentration to 0·32 ppm and the efficiency to 80 per cent. Lower velocities 15 m s^{-1} gave efficiencies of 98 and 96 per cent with similar corona currents. This work indicates that if high ozone concentrations of the order of 1 ppm can be tolerated, high throughput precipitators of much smaller dimensions than those at present in use for environmental air can be constructed.

10.11. ELECTROSTATIC PRECIPITATION AT HIGH TEMPERATURES AND PRESSURES

There are a number of systems[697] in which it is necessary to remove particles from a gas stream at high temperatures, high pressures, or both. Examples are the pressurized fluidized bed combustion system, where the combustion gases are used for a gas turbine[429] (1000 kPa, 900°C), the direct fuel-fired gas turbine (650 kPa, 800°C),[759] waste gases from oxygen steel making (0·1 kPa, 1600°C),[834] the gases circulating in a pressurized gas-cooled high-temperature nuclear reactor (800°C, 5000 kPa),[829] the solid state high temperature fuel cell (100–200 kPa, 850°C)[693] and others.[697]

Large pilot-scale experimental work on high-temperature (moderate pressure) electrostatic precipitation has so far been limited to the project at the U.S. Bureau of Mines (Morgantown, W.Va.) which was carried out in conjunction with Research Cottrell (650 kPa, 820°C). The problem of electrostatic precipitation under extreme conditions has three aspects, which must be elucidated if a workable design is to be developed. First, under what conditions will a stable corona be maintained, and what will be the optimum potential; second, what will be the migration velocity of the particles under these conditions; and third, what special features must be used for safe, effective and continuous operation under these conditions.

10.11.1. High-pressure/high-temperature corona characteristics

The sparkover voltages between parallel plates increase with pressure, but this configuration is not used in electrostatic precipitation with the exception of positive corona ambient air cleaners (see above). The wire in tube pattern (or in the extreme case—wire and plate considered as a tube of infinite diameter) is much more relevant and has been, in part, investigated experimentally.[693, 696]

Robinson has shown[693] that the corona starting voltage for positive corona in air can be determined from equation (10.7) up to relative densities (i.e. ϱ/ϱ_s) of 35 and with wire diameters in the range $0{\cdot}18\text{ mm} < R_1 < 6{\cdot}3\text{ mm}$ as long as the critical density is not exceeded. The restriction is that the tube to wire ratio R_2/R_1 must be greater than 10 or 15,[697] or sparkover will occur without corona, even at ordinary pressures.

The critical pressure (or density) is that value where the sparkover voltage, having previously reached a maximum, declines until it intersects with the corona starting voltage. Beyond this critical value, sparkover will occur without antecedent corona. The positive relative critical density ($(\varrho/\varrho_s)_{CR}$) for wire-pipe electrode systems, in air at room temperatures, can be found from the empirical relation[696]

$$(\varrho/\varrho_s)_{CR} = k_p\left(\frac{1}{R_1}+\mathscr{C}_0\right) \tag{10.76}$$

where $\mathscr{C}_0 = 900\text{ m}^{-1}$, and

$$k_p = 0{\cdot}11(1+0{\cdot}1R_2)$$

over the range $20\text{ mm} < R_2 < 80\text{ mm}$, and possibly for $R_2 > 80\text{ mm}$.

The four curves in Fig. 10.26 show how available pressure range and potential decreases with increasing R_1/R_2. The results show the advantages of a wide stable corona range using fine wires. Robinson explains that the corona phenomena are primarily a function of the relative gas density. The observed corona characteristics are due to two opposing effects: first, in denser gases there are shorter mean free paths, which impede ionization and raise the sparkover potential; second, enhanced photo-ionization and reduced ion diffusion facilitate streamer propagation from the anode across the gap. With increasing pressure, the former first predominates, but then the latter effect takes over, and at the critical pressure, there is spark breakdown.

With increasing gas temperature, the critical pressure is raised, and the greater gas diffusivity quenches the streamers before they bridge the electrode gap. Wider gaps also give greater opportunity for streamer dissipation.

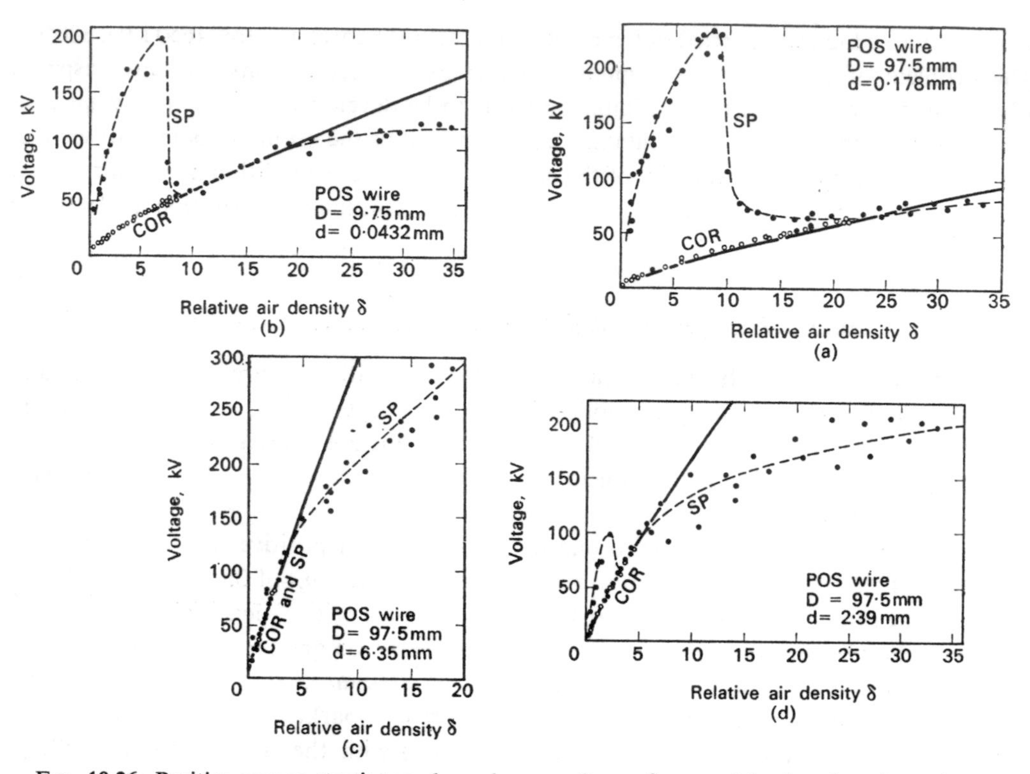

FIG. 10.26. Positive corona-starting and sparkover voltages for coaxial wire-pipe electrodes in air (25°C). D and d are the respective pipe and wire diameters. The voltage is unvarying d.c. The solid lines are in the corona-starting curves according to equations (10.7) and are in good agreement with the data below the critical density; extensions of these curves beyond the critical density are without physical significance.[697]

Experimental *negative corona* starting voltage scatters over a broad band of values well below those calculated by equation (10.7). The unstable nature of this corona has been explained by Robinson in terms of cathode surface impurities and imperfections, which affect electron emission. Negative corona, as can be seen from the curves in Fig. 10.27, give higher maximum sparkover potentials and higher critical pressures (densities). The negative corona critical density in air for wire in pipe systems at ambient temperatures can be estimated from

$$(\varrho/\varrho_s)_{CR} = 0{\cdot}12R_2\left(\frac{1}{R_1}+700\right) \tag{10.77}$$

over the range 23 mm $< R_2 <$ 77 mm.

Furthermore, with negative corona a semi-stable region exists above the critical density, with sporadic, self-quenching sparks. No data are available on precipitator performance in this region.

The effect of high temperatures (820°C) on positive and negative corona has been investigated by Shale and others.[759] The corona starting voltage in air at atmospheric pressure is

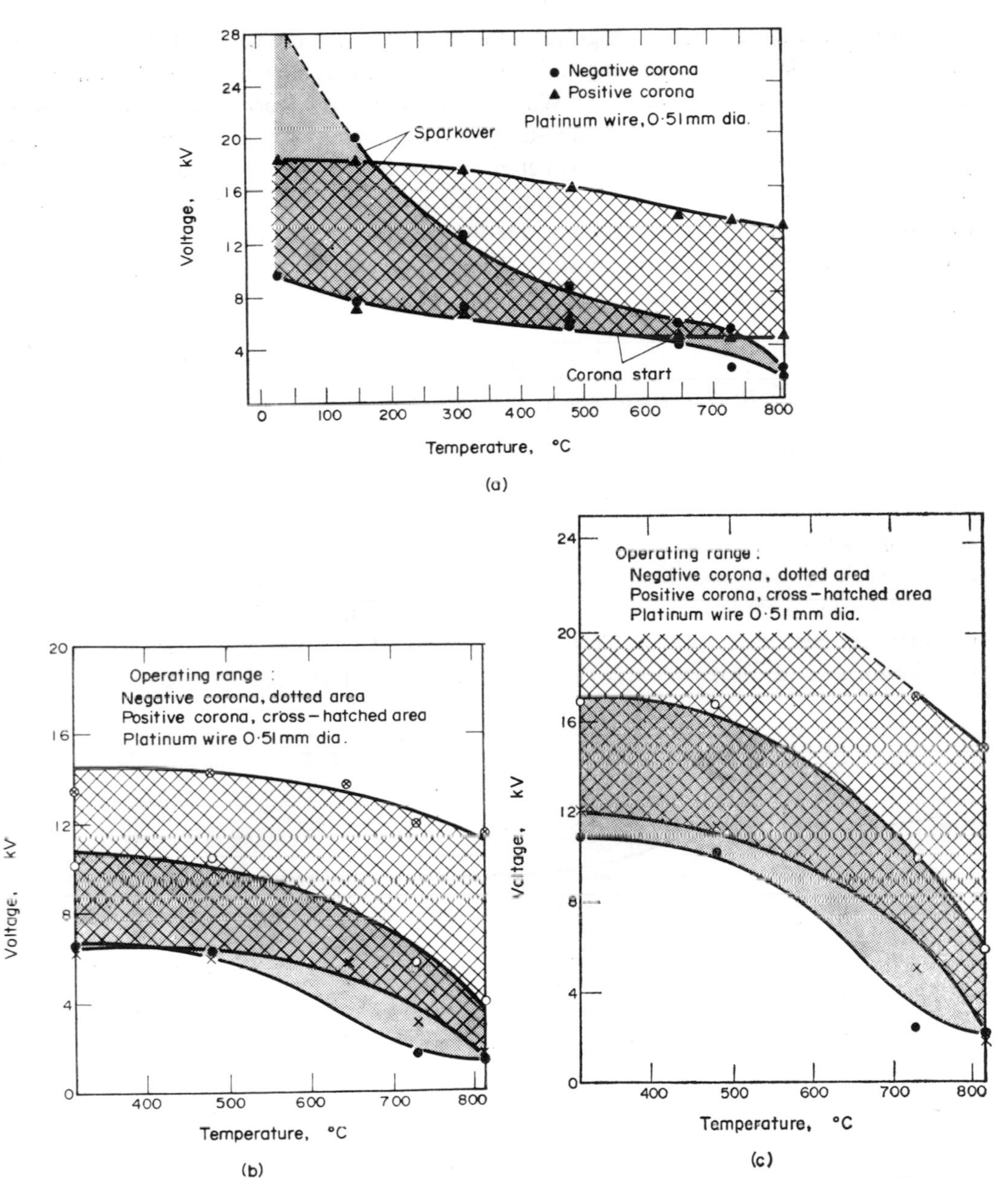

FIG. 10.27. (a) Positive and negative corona characteristics of air at atmospheric pressure. (b) Electrical characteristics of dust-laden air in a precipitator at 135 kPa. (c) Electrical characteristics of dust-laden air in a precipitator at 340 kPa.

the same for both positive and negative corona, to about 650°C. Above this temperature the corona starting voltage drops rather sharply (Fig. 10.27a). The negative corona sparkover voltage drops rapidly and is less than than for positive corona above 190°C. At 650°C there is very little stable corona region with negative potential. However, the *positive* corona sparkover potential remains at a high value (above 14 kV) for the whole of the temperature range, and should give a wide range of stable corona. Similar results were obtained with dust-laden gases (Fig. 10.27 b and c). While this experimental work indicates that positive corona would be more suitable for high-temperature precipitation, Shale[757] reports that in practice the negative corona system is better for precipitation at the conditions of 820°C, 650 kPa, which he used on a pilot plant, which is described later. He attributes this to the much higher negative corona input, for in spite of the fact that the negative corona operating voltage is less than the positive, being limited by sparkover considerations, the negative current is much higher.

10.11.2. Particle migration at extreme conditions

The relative drift velocity of particles at high temperatures and pressures are a function of the parameters which affect this function. These are given in terms of equation (10.43) as the effective potential (discussed in the previous section), the Cunningham Correction factor C [equation (4.30)] and the gas viscosity (equation (4.31) and Appendix Table A.8.1). Other factors, such as the dielectric constant and diameter of the particle, are not significantly changed by temperature and pressure. The effect of temperature in air at atmospheric pressure has been discussed by Thring and Strauss,[834] and the calculated relative drift velocity for a range of particles is shown in Fig. 10.28. The effect of both high pressure

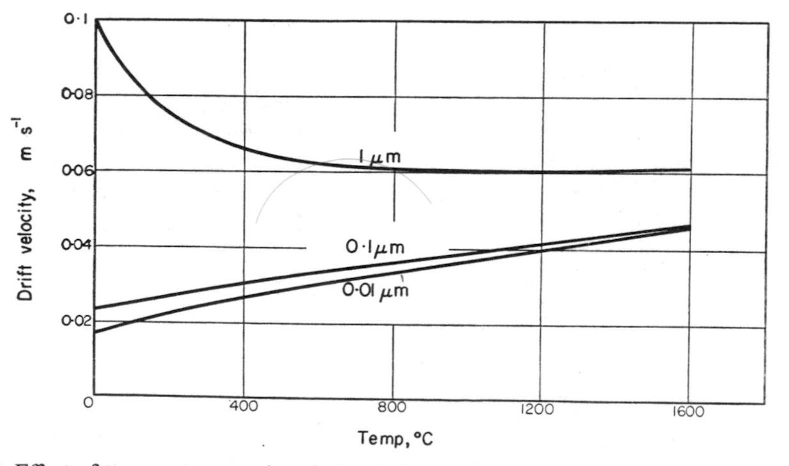

FIG. 10.28. Effect of temperature on the relative drift velocity of 1 μm, 0·1 μm and 0·01 μm particles for an electrostatic precipitator where the drift velocity of a 1-μm particle at ambient conditions is 10 cm s^{-1}.

(or density) and temperature for BeO particles in compressed carbon dioxide has been considered by Lancaster and Strauss,[829] and the results of these calculations are shown in Fig. 10.29. From these it can be seen that high temperatures, without increases in pressure, help the precipitation of very small particles, but constant or increasing density reduces the drift velocity.

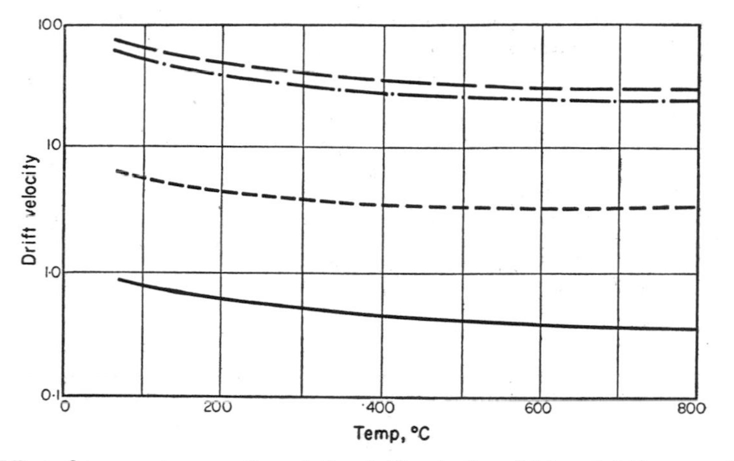

FIG. 10.29. Effect of temperature on the relative drift velocity of 1·0 and 0·01 μm radius beryllia particles in carbon dioxide in an electrostatic field at gas densities of 0·002 and 0·05 g/cm³ (the drift velocity is based on a 1 μm particle at ambient conditions moving at a speed of 100 units/unit time; e.g. 100 cm s in a field of $V_e^* V_p = 1000$)

—— —— ——	Gas density 2 kg m^{-3}	Particle diameter 2·0 μm.
——·——·——	Gas density 50 kg m^{-3}.	Particle diameter 2·0 μm.
- - - - - - - - - - -	Gas density 2 kg m^{-3}.	Particle diameter 0·02 μm.
————————	Gas density 50 kg m^{-3}.	Particle diameter 0·02 μm.

10.11.3. Precipitator design for extreme conditions

Precipitators operating at conditions of high temperature and pressure have to be specially designed to contain these conditions. Thus the rectangular design of plate precipitator is unlikely to be suitable and virtually all units will be of the wire in tube pattern. The expansion of the metal parts has to be allowed for, and the joining of different metals avoided so that distortion from differential expansion is prevented. Problems of electrical insulation and sealing at temperatures above 400–500°C require careful design and selection of materials.

One design which has been built and tested to 820°C, 750 kPa is shown in Fig. 10.30.[757, 760] The passive electrodes consist of sixteen tubes, 1·83 m long, 0·15 m diameter, with 2·1 μm wire discharge electrodes. The precipitator with negative corona was 91–96 per cent effective at these conditions in removing fly ash from a gas stream with a linear velocity of 1·5 m s^{-1}, gas density 1·5, potential 38 kV, current 6·6 mA m^{-1}. With positive corona the efficiency was 75–77 per cent. Serious problems were encountered in rapping the electrodes at this temperature, as this caused severe misalignment resulting in incapacitating the unit. Shale concludes, as a result of his work, that electrostatic precipitation to 700°C with negative corona is possible, that an adequate relative density is retained and the unit is designed structurally rigid.

Modifications to conventional precipitators, which have enabled them to operate up to 450°C, have been described by Liesegang.[511] This worker also found that there were serious

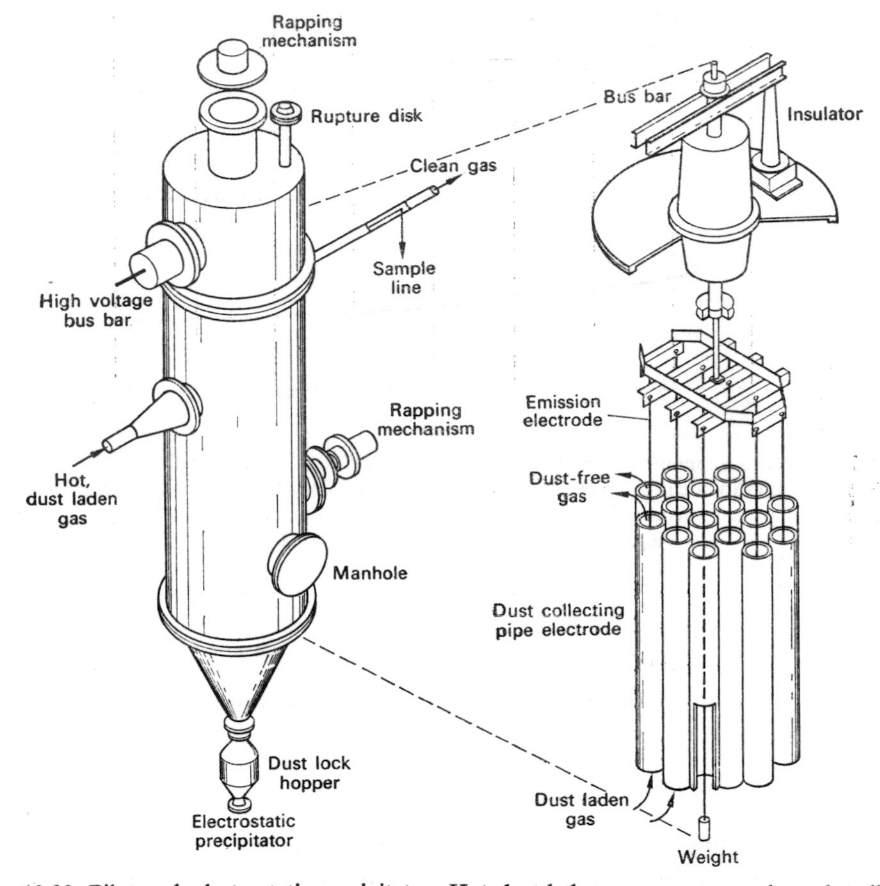

FIG. 10.30. Pilot-scale electrostatic precipitator. Hot dust-laden gas sweeps up through collecting electrodes and is electrified. Dust particles collect inside the tubes and drop to bottom. Rapping mechanism knocks dust from tubes. Clean gas leaves at top of tubes and passes out through exit pipes. Removal efficiency is determined from concentration of dust in inlet and exit streams. Precipitator is about 10 m high. Equipment of this size can give valuable information for scale-up to commercial size equipment.[757, 760]

distortions of the discharge wire frame. Furthermore, these led to breaking up of the insulators, and contact of discharge wires and plates during rapping.

10.12. ELECTRIC POWER SUPPLY FOR PRECIPITATORS

A reliable high-voltage d.c. supply is essential for electrostatic precipitators where negative potentials up to 90 kV are used for industrial plant and positive potentials up to 13 kV for environmental air cleaning. The current supplied to industrial precipitators varies according to the size and duty, between 30 and 500 mA, and so transformers and rectifiers with capacities up to 40 kVA are required. Because the migration velocity is a function of the charging and precipitating field strengths, the largest voltage possible without arcing should be applied. The arcing potential, however, varies with the type of gas—its composition and the humidity and temperature conditions—the dust concentration and the physical dimensions of the precipitator. These are affected by the dust layers deposited on the electrode, and by rapping. Arcing should be avoided not only because it lowers the effective potential in the precipitator, but also because it loosens the deposited dust, assisting re-entrainment, and tends to melt the discharge electrode wire. When arcing does occur, the applied potential must be reduced to zero in the migration velocity and, in turn, the efficiency of the precipitator. Voltage-control systems are therefore usually included in all industrial precipitator installations.

The power supply to a precipitator consists of three sections: the voltage control system, the transformer which steps up the line voltage (usually 415 V) to the precipitator potential, and the rectifier which converts a.c. to d.c. In some small older installations the voltage control may be manual, but automatic voltage control is universal practice with all large modern installations. It has been shown[931] that automatic voltage control gave a 60 per cent increase in corona power in one plant. A low-tension variator may be used for this, as it can be either an automatically stepped or continuously variable transformer. A typical installation has a sixteen step transformer which supplies power from 230–380 V in 10-V steps, and is fed with 50-Hz/415-V single- or two-phase mains power. The automatic control is based on current-sensitive relays which maintain the voltage in the precipitator either just below flashover or at the maximum rating of the variator-step-up transformer set. Alternative methods use transducers to monitor the number of flashovers during a short time interval. White[932] suggests that 0·16–1·6 sparks s^{-1} per high-tension section is the range for optimum efficiency in industrial precipitators.

The transformers are of the usual types, fin cooled for operation at temperatures up to 45°C, and with additional water cooling if the transformer is to operate in warmer conditions. If selenium rectifiers are used, these are included in the units with the transformers at ratings

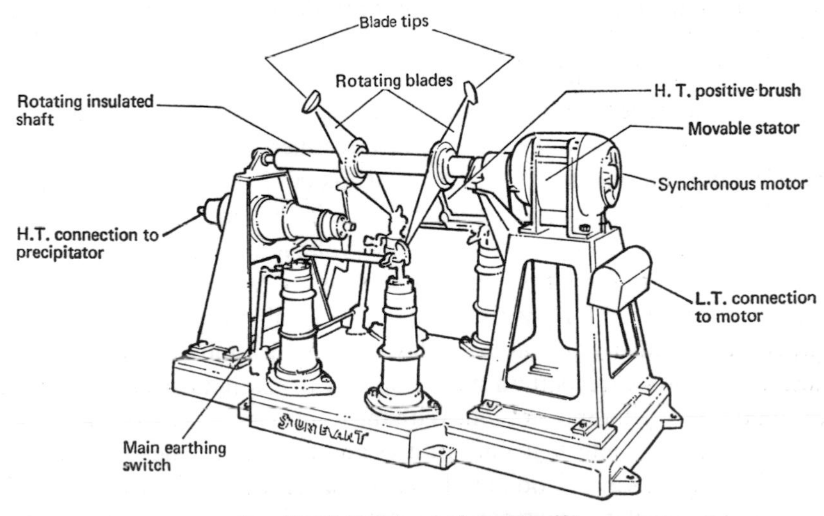

FIG. 10.31. Mechanical rectifier.[839]

up to 40 kVA.[743] The transformer output voltages used are 30, 60, 75 or 90 kV, the two central values, 60 and 75 kV, being the most usual. Three types of rectifiers are used:

(a) Mechanical rectifier.
(b) Solid state rectifier (selenium, copper oxide or more recently silicon).
(c) Electronic valve rectifier.

Mechanical rectifiers (Fig. 10.31) were used in the first commercial precipitators installed by Cottrell and remained popular till 1945 because of their low cost, rugged construction, and the fact that their performance is not affected by fluctuations in process conditions.

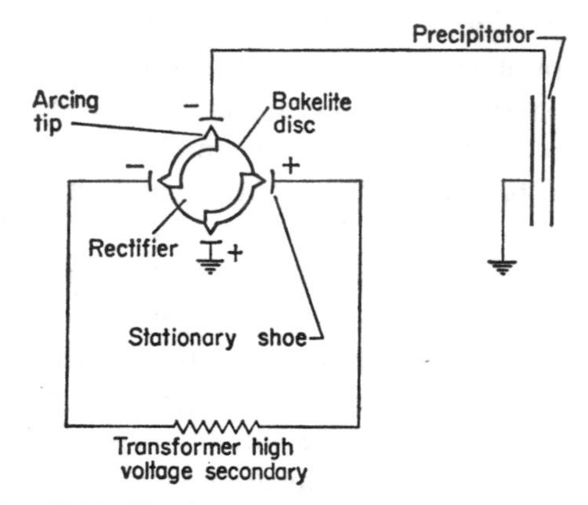

FIG. 10.32. Circuit including mechanical rectifier.[920]

However, oxides of nitrogen are produced at the contacts, so these require frequent maintenance, and a well-ventilated housing has to be provided. Suppressors to reduce radio and television interference must be fitted. A circuit incorporating a mechanical rectifier plant is shown in Fig. 10.32.

The most common rectifiers in use at present are those of the solid state type, generally selenium rectifiers. These are slightly more costly to install than the mechanical type, but they do not require special shielded and ventilated housing as they do not produce radio or UHF receiver interference, or nitric oxide gases. The early solid state rectifiers used copper oxide, and present trends are to the use of silicon rectifiers.

High vacuum hot cathode diodes with a life of 25,000–30,000 hours have been developed in the United States and this has led to their wide use in that country. The circuit (Fig. 10.33) is a special four-tube one which gives a full-wave voltage wave form. In industrial installations it is often possible to mount the tubes on sockets on the transformer bushings. Electron tube rectifiers are also used for positive corona air-cleaning precipitators.

In industrial units either half-wave of full-wave rectification is used from either one or more phases. The half-wave pattern allows sparks to extinguish in less than one cycle, leading to smooth precipitator operation. With the full-wave form there is a greater tendency for the sparks to develop into high current arcs and operation tends to be rougher.

White[926] has developed a method of pulse energization which gives a peak voltage several kilovolts higher and greater corona current than is possible with conventional rectification.

The equipment is similar to that developed for microwave radar, and consists of a line pulser in which energy is accumulated in a high-voltage capacitator, and is then discharged rapidly into the load. The pulse time is of the order of some hundreds of microseconds, repeated with a frequency of several hundred pulses a second. The pulse output may be commutated to several precipitator sections. This mode of operation results in 50–60 per cent more efficient collection, halving the losses from the plant.

Another method of producing a high potential close to the discharge electrodes wires has been achieved by using support electrodes, consisting of charged plates, at discharge potential between, and in line with, the discharge electrodes.[95] If this arrangement is used in

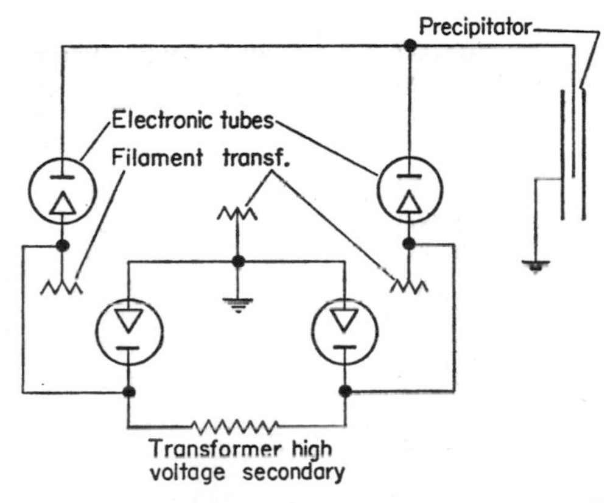

FIG. 10.33. Circuit including diodes as rectifiers.[920]

conjunction with pulse energization, and dusts with resistivity below $10^3\ \Omega$ m, it can lead to more effective precipitation, of the order of 70 per cent better than in conventional electrostatic precipators.

If a precipitator is built in sections, because these have different power requirements, it is best to have a separate power supply for each. It may also be necessary in the case of very large installations where sections in parallel are separately rapped, to have separate supplies to these sections to give the highest efficiency. For a tube precipitator it has been suggested that a bank of 900 tubes is the maximum to run off a single supply.[623]

The difference in elevation between power supply and the insulators on the high tension electrode should be kept to a minimum, not only because of the very high cost of the high-voltage cable, but also because the insulation impregnation compound tends to migrate to the lower level, and leads to insulation failure. The best arrangement is to mount the precipitator voltage control, transformer and rectifier unit on top of the precipitator and connect the high tension electrodes with busbars.

10.13. PRACTICAL PRECIPITATOR OPERATING DATA

Gas velocities. The average gas velocity used in a precipitator should be the one which gives maximum efficiency and also maximum effective migration velocity. For fly ash this appears to be about 2·0–2·3 m s^{-1} just below the fly ash entrainment velocity. For carbon black,

TABLE 10.10. TYPICAL PERFORMANCE DATA FOR ELECTROSTATIC PRECIPITATORS[358]

Type of plant	Gas flow $m^3 h^{-1}$ at temperature	Dust concentrations g/m^3 at operating temp.		Collecting efficiency (%)	Power consumption $W/1000\ m^3 h^{-1}$
		inlet	outlet		
1. *Power Stations:*					
Pulverized fuel fired boilers	275,000	13·16	0·163	98·67	115·6
Pulverized fuel fired boilers	245,000	10·95	0·062	99·43	131·6
Refuse burning boilers	85,000	16·80	0·58	96·6	140·4
Lignite stoker fired boilers	400,000	1·6–2·0	0·017–0·037	98·15	120·4
Lignite pulverized fuel fired boilers (hammer mills)	1,600,000	4·63	0·16	96·5	40·1
2. *Coal Industry:*					
Lignite rotary type steam dryer	29,000	35·24	0·28	99·25	30·0
Lignite rotary type steam dryer	26,000	18·31	0·090	99·40	28·1
Lignite plate type steam dryer	25,000	7·87	0·063	99·20	25·0
Combustion gas lignite dryer	42,000	14·40	0·110	99·50	50·1
Lignite mill dryer	40,000	25·1	0·332	98·67	80·2
Lignite conveying system de-dusting	20,500	54·7	0·278	99·40	20·0
Bituminous coal tube type steam dryer	43,000	16·30	0·086	99·50	70·2
Bituminous coal conveying system de-dusting	11,000	22·4	0·150	99·30	290·9
Bituminous coal-coke grinding plant	4800	13·85	0·056	99·59	300
3. *Coal Gas Industry:*					
Peat gas producer	4500	5·34	0·008	99·85	702
Cracking plant for natural gas	8700	0·224	0·002	99·20	120·4
Producer gas from lignite briquettes	13,000	37·7	0·20	99·47	652
Producer gas from sembitu-minous lignite	48,000	28·7	0·10	99·7	602
Shale-gas cleaning plant	33,980	40·0	0·006	99·9	903
Coke oven town gas cleaning	3100	24·15	0·010	99·9	903
Coke oven town gas cleaning	2300	17·0	0·003	99·9	1605
Coke oven gas cleaning	14,000	28·0	0·078	99·8	752
Oil carburetted water gas cleaning	12,000	4·73	0·039	99·2	1404
Tar carburetted water gas cleaning	4000	10·0	0·050	99·5	1805
4. *Paper Industry:*					
Black liquor burning plant	132,500	2·84	0·13	95·3	265

TABLE 10.10 (*continued*)

Type of plant	Gas flow $m^3 h^{-1}$ at temperature	Dust concentrations g/m^3 at operating temp.		Collecting efficiency (%)	Power consumption W/1000 $m^3 h^{-1}$
		inlet	outlet		
5. *Cement Industry:*					
Rotary kiln dry process 520 ton/day	136,000 (0·5–1 m/s)†	20·7	0·193	99·06	90·3
Lepol rotary kiln dry process 470 ton/day	127,000 (0·8–1 m/s)†	6·33	0·074	98·85	30·0
Rotary kiln wet process 350 ton/day	145,000 (1·2–1·5 m/s)†	21·3	0·067	99·68	85·5
Rotary kiln with calciner, wet process 350 ton/day	145,000 (0·5–0·9 m/s)†	11·0	0·24	98·2	100
Vertical kiln	125,000	1·8	0·048	97·3	90·3
Raw material dryer	36,000	49	0·119	99·75	471
Cement mill	24,000	51·3	0·085	99·8	450·8
Packing machine	15,000	37·7	0·110	99·7	471
6. *Chemical Industry:*					
Pyrites roaster 25 ton/day	9500	3·22	0·047	98·5	683
Pyrites roaster 29 ton/day	11,000	2·10	0·037	98·3	802
Pyrites roaster 35 ton/day	13,000	1·20	0·0036	99·7	962
Pyrites roaster 36 ton/day	14,000	4·11	0·036	93·5	552
Acid mist from sulphur burning furnace 7·5 ton/day	2500	6·88	0·041	99·4	802
Sulphuric acid mist following cooler tower	4300 (1·2–1·5 m/s)*	12·3	0·061	99·5	953
Sulphuric acid mist following cooler tower	5000 (1·2–1·3 m/s)*	7·57	0·070	99·1	729
Blende roaster	15,530	5·13	0·075	98·5	513
Arsenic and sulphuric acid mist removal	14,000	2·71	0·000005	99·99	856
Tail gas for sulphuric acid concentration	23,500	12·37	0·05	99·6	360
Elemental sulphur fume from hydrogen sulphide combustion plant	4300	25·8	0·20	99·2	1700
7. *Mineral Earths and Salts Processing:*					
Bauxite dryer 180 ton/day	20,000	15·5	0·05	99·69	115
Bauxite calcining and processing kiln 220 ton/day	44,000	5·0	0·06	98·8	180
Alumina calciner with multi-cyclone precleaner 45 ton/day	14,500	300	0·03	99·99	465
Potassium chloride dryer	29,000	8·0	0·08	99·0	190
Fuller's earth dryer	30,000	4·32	0·02	99·54	210

TABLE 10.10 (*continued*)

Type of plant	Gas flow $m^3 h^{-1}$ at temperature	Dust concentrations g/m^3 at operating temp.		Collecting efficiency (%)	Power consumption W/1000 $m^3 h^{-1}$
		inlet	outlet		
8. *Non-ferrous Metallurgical Industry:*					
Vertical blast furnace: lead ore	10,000	12·0	0·66	99·5	220
Vertical blast furnace: lead ore	16,000	6·35	0·15	97·5	190
Rotary kiln processing: zinc ores	7000	40	0·44	98·9	200
Rotary kiln processing: zinc ores	12,500	13·2	0·061	99·53	160
Vertical blast furnace: tin ores	9700	5·0	0·034	99·29	180
Vertical blast furnace: tin ores	3600	6·85	0·10	98·70	240
Vertical blast furnace: antimony ores	6200	3·77	0·007	99·8	421
Copper convertors	14,500	4·53	0·136	97·0	70·2
Rotary kiln for nickel bearing iron ores	45,000	27·7	0·065	99·76	270

* Lineal velocity through the precipitator: data from R. L. Cotham, Paper No. 29, Clean Air Conference, Sydney, 1962.

† Lineal velocity through the precipitator: data from G. Funke, *Zement, Kalk, Gips*, **12,** 189 (1959).

the entrainment velocity can be as low as 0·6 m s^{-1}, while for cement dust it is 4 m s^{-1}. Mist re-entrainment velocities are even higher: 5–6 m s. Practical precipitator flow rates are normally kept well below these values to about 1–2 m s^{-1}, although experimental fly ash precipitators have used velocities in the region of 10–13 m s^{-1}.[265]

Power consumption and efficiency. The efficiencies, inlet and exit concentrations of a number of industrial precipitators have been listed by Heinrich and Anderson, and are given in Table 10.10. In some instances in this table gas velocities have been added from other sources. Power consumption in environmental air-cleaning positive corona precipitators is somewhat greater per unit volume of gas treated than in negative corona precipitators.

Pressure loss. The pressure loss in electrostatic precipitation is very low, much lower than for other industrial gas-cleaning methods. It is generally of the order of 125 Pa and rarely exceeds 250 Pa except when experimental high-velocity plants are being tested.

In the case of environmental air-cleaning precipitators, the pressure drop is 7·5 Pa at 2 m s^{-1} and increases to 10 Pa at 2·5 m s^{-1}. This is so low that additional gas resistance has to be added to give uniform flow over the cell bank. Aluminium mesh filters, with a pressure drop of 25 Pa, are normally used *after* the electrostatic precipitator, and this has the additional purpose of preventing water droplets from the washing cycle being carried past the plant.

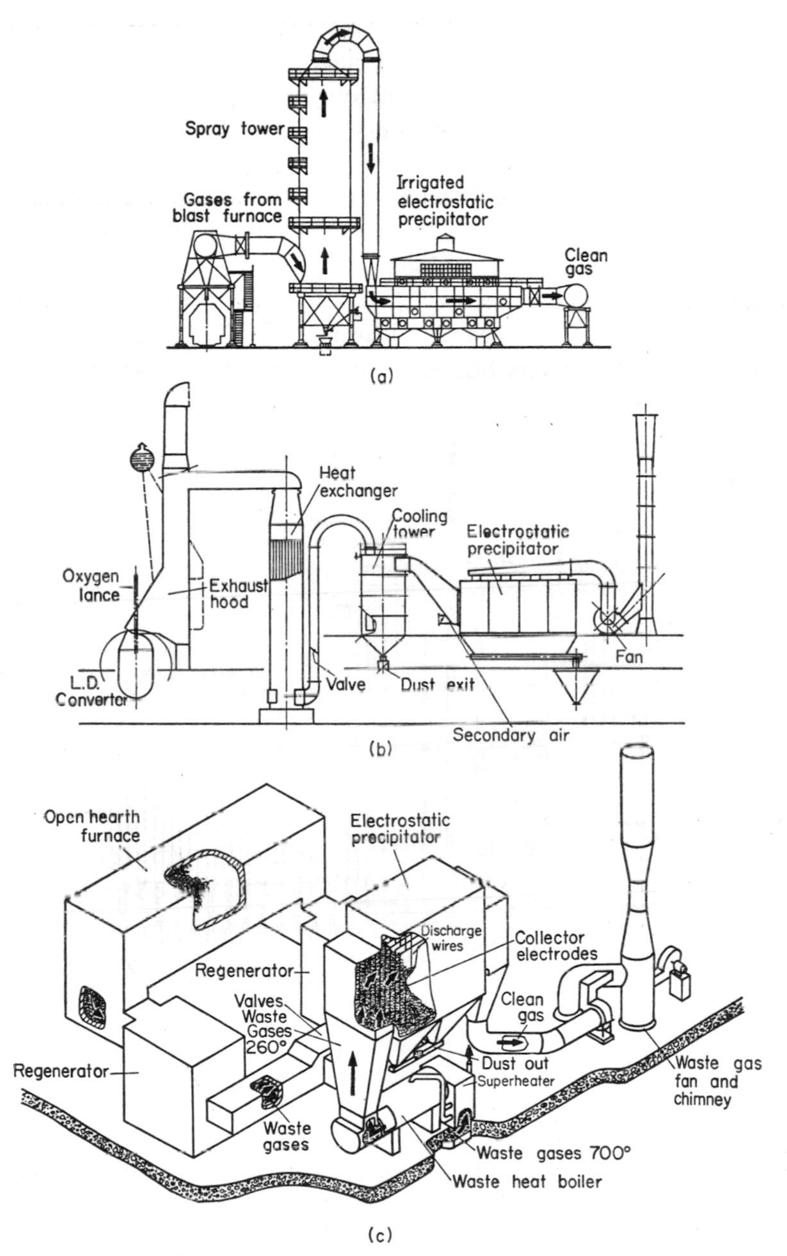

FIG. 10.34. Electrostatic precipitators in the metallurgical industry:[328]

(a) Spray tower and precipitator in series on blast furnace gases.
(b) Waste heat boiler, ring scrubber and precipitator in series on a Linz–Donauwitz convertor plant.
(c) Checkerwork brick regenerator, waste heat boiler and precipitator in series on an open-hearth furnace.

Combination of electrostatic precipitators with other gas cleaners. Electrostatic precipitators are often combined with mechanical or other collectors in cases where gases have exceptionally high dust concentrations or need conditioning, or both. For example, blast-furnace gases are often passed through a spray tower before entering the precipitator (Fig. 10.34a), while low density waste gases are cooled in a waste heat boiler and scrubbed in a ring scrubber (Fig. 10.34b). Open-hearth furnace gases are also passed through waste-heat boilers, having given up some of their heat to the chequerwork brick regenerator, and are then cleaned in electrostatic precipitators having been conditioned by water sprays, before being emitted to the atmosphere (Fig. 10.34c).

The combination of venturi scrubbers of the Imatra type with electrostatic precipitators has been used on a number of European blast furnaces.[137] This system requires a considerable amount of space as well as very high entry to the venturi scrubber. This can be avoided

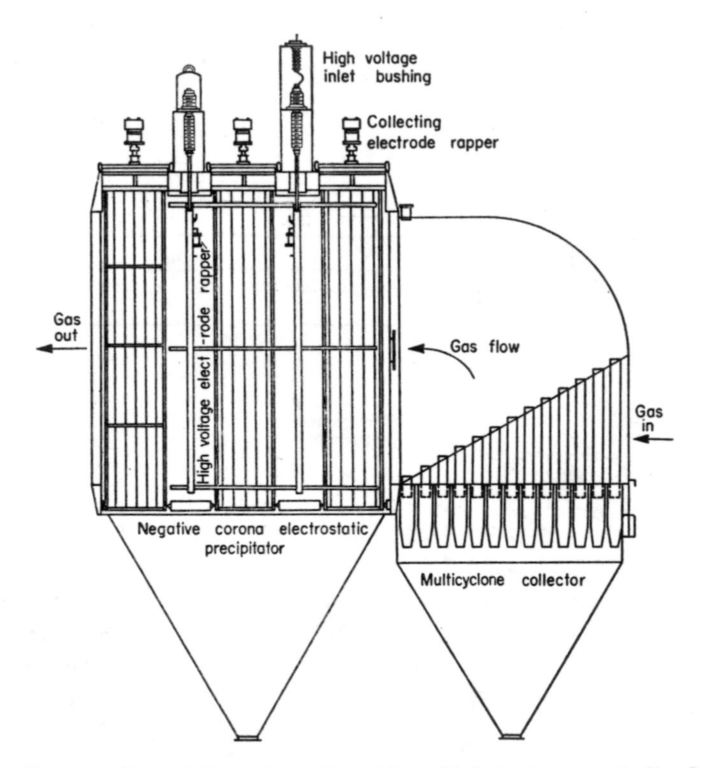

FIG. 10.35. Electrostatic precipitator in series with multiple cyclones on boiler flue gases.[920]

by placing the scrubber horizontally below the electrostatic precipitator (Venturion type) (Fig. 10.36). This gives a compact design coupled with reduced water consumption and lower costs because clarifiers are smaller, foundations for separate gas washers are dispensed with and the pressure in the system reduces the overall size of the plant. The requirements for the clean gas are residual dust contents less than 0·9 mg m^{-3} otherwise gas-burner nozzles become eroded.

For flue gases, power stations frequently use centrifugal collectors of the multicellular cyclone type (straight-through or reverse-flow patterns) (Fig. 10.35) before final cleaning with the electrostatic precipitators. This combination, however, cannot always be used with success. For example, in the collection of fly ash from boilers burning New South Wales coal, which has exceptionally high resistivity, the combined unit gave lower efficiencies than

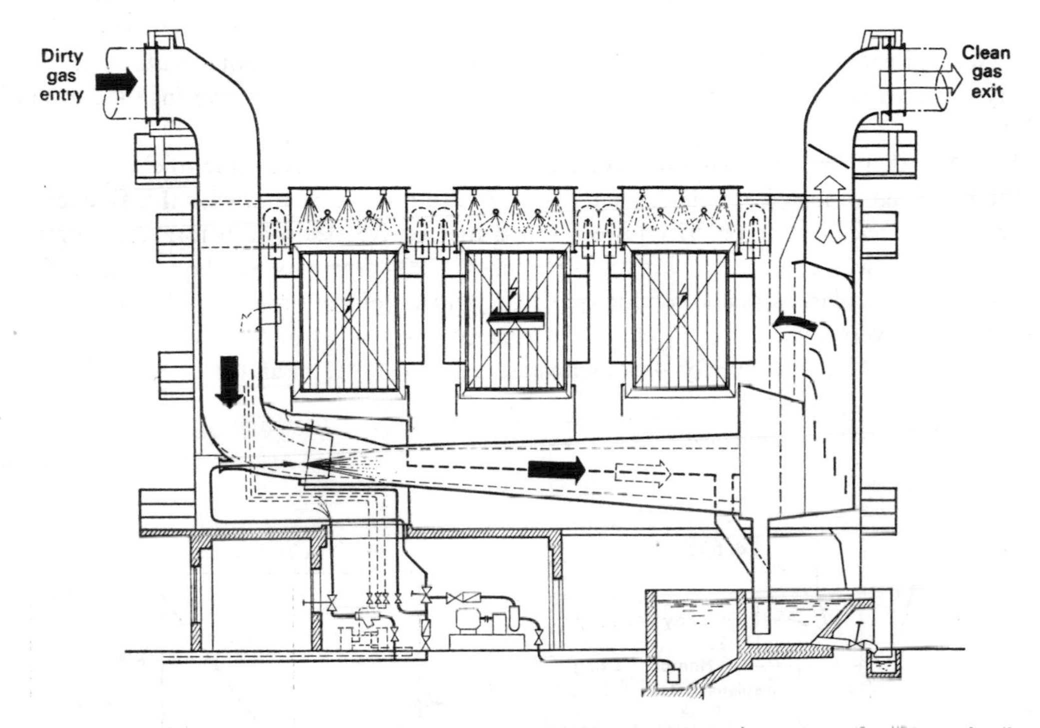

FIG. 10.36. Combination electrostatic precipitator with "Imatra" Venturi scrubber—the "Venturion" precipitator.[137]

the precipitator alone. This is explained by the mechanical collector removing the coarse particles which appear to assist agglomeration in the precipitator. The removal of the coarse fraction also results in a fine dust adhering to the electrodes and being much more difficult to remove by rapping.[440] In the case of carbon black, it is not possible to collect all of the carbon in the precipitator because of its low resistivity, but the precipitator does act as an agglomerator. The precipitator is followed by low-velocity cyclones, and then bag filters or venturi scrubbers are used to achieve the final cleaning.

The enhancement of scrubber efficiencies using electrostatic agglomeration before the scrubber has been shown on a pilot plant treating black liquor boiler fume by Walker.[898] The dominant mechanism in such a situation is that agglomeration in the precipitator occurs by collection and re-entrainment. The design of these agglomerators is on the basis of simple, existing theory, provided the conditions for formation of stable agglomerates are met.

10.14. NOVEL METHODS OF ELECTROSTATIC PRECIPITATION

10.14.1. Alternating current electrostatic precipitation

Krug and his associates, in the high-tension laboratory of Karlsruhe University, observed that the sparkover potential is greater with alternating current in an inhomogeneous field than with direct current.[496] The surface potential was measured in a point-plate system of the type used for determining particle resistivity, with an insulating layer covering the earthed plate.

It was further observed that with 50 Hz a.c. there is ion movement due to electric wind —the repulsion of ions from the corona discharge electrode—and that with this frequency nearly all changes of one polarity have reached the passive electrode before the polarity changes. This was confirmed with high-frequency (2000 frames s^{-1}) photography. The method is particularly suitable for high-resistivity dusts (of the order of $10^{11}\,\Omega$ m) which are normally very difficult to collect. As a general rule, based on laboratory tests, this technique may be applicable for dusts with resistivity $> 10^9\,\Omega$ m and is quite useless for

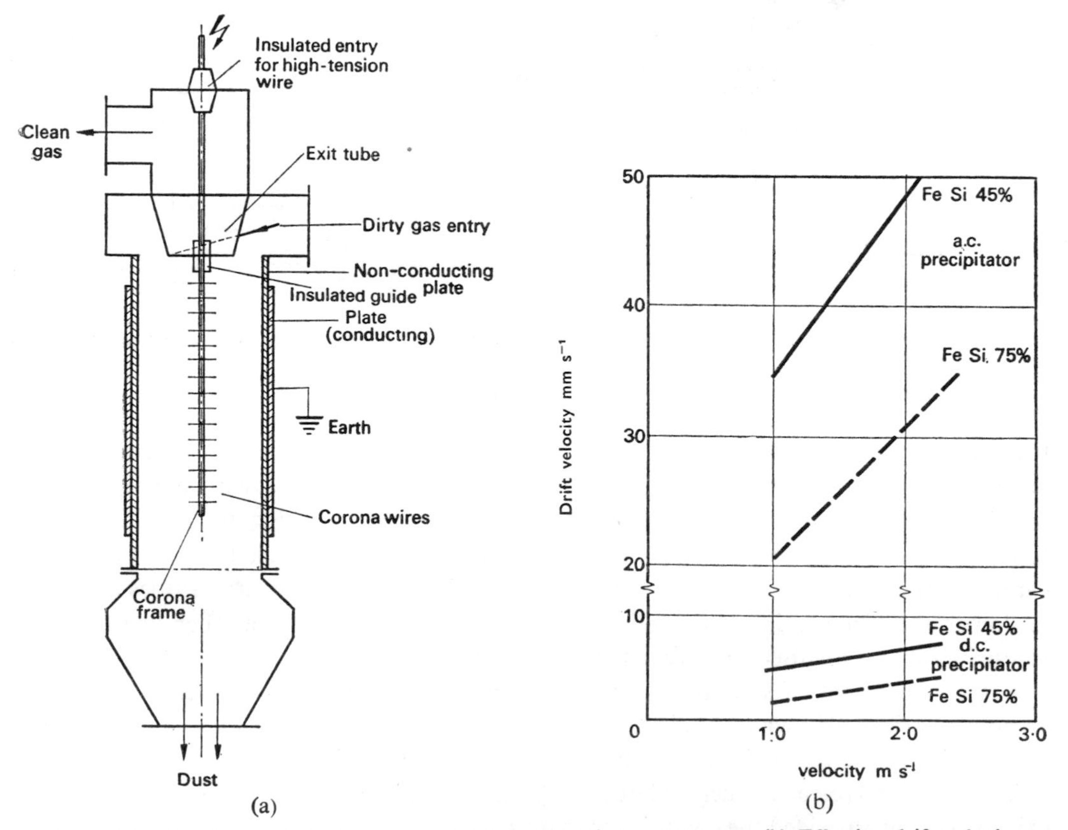

FIG. 10.37. (a) Proposed design of a.c. electrostatic cylcone separator. (b) Effective drift velocity vs. gas velocity for comparable a.c. and d.c. precipitators with two types of ferro-silicon dust (45% $1{\cdot}3\times10^{11}\,\Omega$m, 75% $4{\cdot}3\times10^{11}\,\Omega$m).[496]

those with resistivity < 10^7 Ω m. The a.c. precipitator was developed on a pilot-plant scale, using a cylindrical precipitator with tangential entry, 0·19 m diameter, with effective length 2·2 m and an 8-mm corona "wire" which has points at 30-mm intervals. This unit was tested on ferro-silicon dusts (45 per cent FeSi, concentration 0·6–1·8 g m^{-3}, resistivity $1·3 \times 10^{11}$ Ω m; 75 per cent FeSi, concentration 1·5–3·5 g m^{-3}, resistivity $4·3 \times 10^{11}$ Ω m) with gas temperature of 90°C, dew point 7°C. With the latter dust the sparkover potential was 20–35 kV (d.c.) and 25–50 kV (a.c.). The collected dust with a.c. showed that it was finer grained and had been subject to agglomeration. Furthermore, with a.c. precipitation the dust does not adhere to the collecting plates and falls into the collecting hopper without rapping. A suggested full-scale design is shown in Fig. 10.37.

An economic problem may be that in raising the precipitator temperature to a higher range (100–200°C) a high temperature (and high cost) polymeric insulation has to be used, such as polytetrafluoroethylene or polyphenylene oxide. None the less, the technique appears promising for difficult, high resistivity dusts.

10.14.2. Electrostatic precipitation by space charge in turbulent flow

Conventional electrostatic precipitators, as discussed above, use the field between a discharge wire (or a charged plate) and a tube (or plate) to precipitate the charged particles or droplets. However, it has been shown recently by Hanson, Wilke and others[332–3] that the space charge on the particles and droplets could replace the conventional precipitating section. The physical characteristics of such a system, which would be built directly into the duct carrying particle-laden gases, and may be a multi-stage system, could be described as follows: "Water is sprayed into the gas to give a mist in which the drops are about the same size as the original particles. This mixture passes into a tube bundle made of thin metal tubes, electrically grounded by their contact with each other and the main duct wall. At the entrance to each tube a high-voltage wire stub is used to produce a corona which charges both the particles and drops with the same sign. As the gas flows down the tube the radial field produced in the tube by the space charge of the particles and drops causes both drops and particles to migrate to the grounded tube wall. On contact with the wall, both particles and droplets discharge and coalesce to form a flowing suspension of water and particles. If the duct is sufficiently inclined, the precipitated stream will flow from the tubes and can be easily removed."

Since the precipitating field is produced by the particles and drops, the field will diminish as precipitation proceeds, and precipitation will slow down. If a high degree of precipitation is required, a second stage, with further introduction of charged droplets, can be used to precipitate the remaining particles. A typical system of this type is shown in Fig. 10.38.

The theory of this arrangement was discussed by Faith, Bustany, Hanson and Wilke[332] and the last two[333] analysed this unit in comparison with conventional electrostatic precipitators, with and without the effect of turbulent diffusion. It was found that the estimated wall area requirement was similar to the conventional unit and the power use was somewhat greater. However, the real problem is getting enough waste droplets of small size. If 5-μm droplets were used, a gas flow of 50 m^3 s^{-1} would require 20 kg s^{-1}. If an air-atomizing

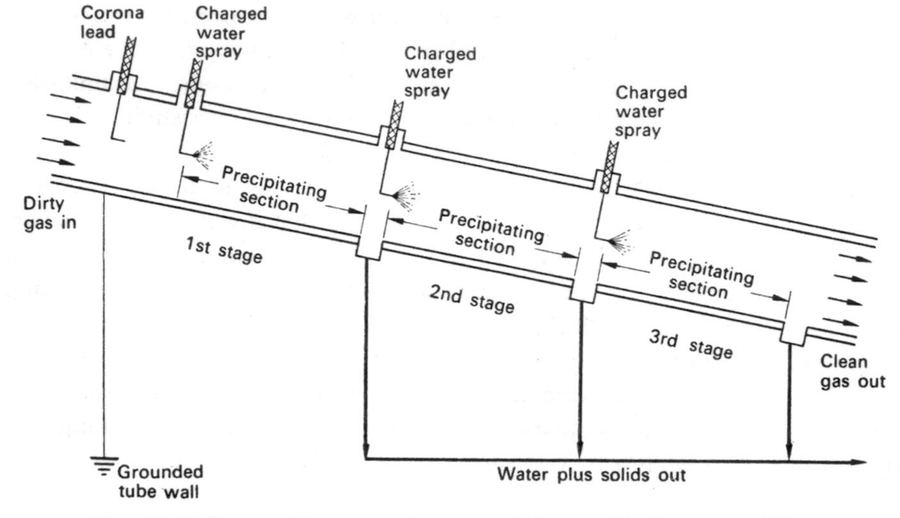

FIG. 10.38. Proposal for space charge precipitator with three stages.[333]

nozzle were used for this, the amount of air needed would be about 14 $m^3 s^{-1}$, which is not feasible. If other means of producing smaller droplets (e.g. sonic nozzles) were to be used this method of precipitation may, in some special cases, be used to advantage.

CHAPTER 11

SUNDRY METHODS AND MECHANISMS IN PARTICLE COLLECTION

11.1. INTRODUCTION

Some properties of particles which may at times be important to their collection have not been discussed to any extent in the previous chapters. They should, however, be included because mechanisms such as agglomeration occur in all clouds of particles, while others, such as thermal precipitation, may be the basis of new developments in gas cleaning. Increasing the rate of agglomeration by sound waves has been used to make agglomerates of very small particles which can then be collected by conventional means. Although this method has been proved technically, it has had only limited commercial success.

Thermal precipitation of particles occurs whenever a cold body is placed in dust laden gases. Although the phenomenon of thermal precipitation was known in the last century, and a satisfactory quantitative theory was developed over thirty years ago, the effect has not yet been used for commercial gas cleaning plant.

This chapter outlines the theories of agglomeration and thermal precipitation, and indicates the extent of their commercial exploitation. Other aspects of particle mechanics, such as the movement of charged particles in magnetic fields, are mentioned briefly because these may lead to future developments in particle-separation techniques.

11.2. AGGLOMERATION OF PARTICLES BY BROWNIAN MOTION

A simple theory of agglomeration of particles has been developed by Smoluchowski,[781] which indicates that the rate of decrease of the number of particles is approximately equal to the square of the number present. This has been generally confirmed by experiment.[868, 939] A comprehensive review has recently been published by Zebel.[961]

A space contains a number of particles, c, randomly distributed, and a sphere of influence, diameter d', is considered to exist within this space. Whenever a particle moves within this sphere of influence it will become part of it, and so a dust free region will exist immediately around the sphere.

Because of the dust free region, a concentration gradient will exist between the sphere and the bulk of the particles in the space, and particles will diffuse to the sphere with a diffusivity

$\mathcal{D}$. If Ficks' law (in spherical coordinates) is applied, then the particles removed will be a function of the surface of the sphere (area $\pi d'^2$) in unit time, the number removed is $2\pi d'\mathcal{D}c$. Now all the particles present can be considered as centres of spheres of influence for removal of particles, so the rate of removal is

$$-\frac{dc}{dt} = \frac{c}{2}.2\pi d'\mathcal{D}c = \pi d'\mathcal{D}c^2. \quad (11.1)$$

The diffusivity of the two colliding particles consist of the sum of their individual diffusivities, $\mathcal{D}_1$ and $\mathcal{D}_2$. $\mathcal{D}$ must be replaced by this sum; and similarly the diameter d' must be replaced by $(d_1'+d_2')/2$ where d_1' and d_2' are the diameters of the spheres of influence of the two particles. Substituting these in equation (11.1)

$$-\frac{dc}{dt} = \frac{\pi}{2}(\mathcal{D}_1+\mathcal{D}_2)(d_1'+d_2')c^2. \quad (11.2)$$

The diffusivity $\mathcal{D}$, was found by Einstein (Section 7.5) to be

$$\mathcal{D} = C\mathbf{k}T/3\pi\mu d \quad (7.23)$$

This equation was derived by assuming Stokes' law and incorporating the Cunningham Correction for slip, C. Substituting equation (7.23) for the diffusivities $\mathcal{D}_1$ and $\mathcal{D}_2$ for particles with diameters d_1 and d_2,

$$-\frac{dc}{dt} = \frac{C\mathbf{k}T}{3\pi\mu}\left(\frac{1}{d_1}+\frac{1}{d_2}\right)\frac{\pi}{2}(d_1'+d_2')c^2$$

$$= \frac{C\mathbf{k}T}{6\mu}\left(\frac{1}{d_1}+\frac{1}{d_2}\right)(d_1'+d_2')c^2. \quad (11.3)$$

If the particles' spheres of influence are such that they join on touching, then d_1' and d_2' are identical with the particle diameters. If, however, other forces exist between the particles, such as electrostatic or thermal forces, then the spheres of influence may be greater (or smaller) than the actual particle dimensions. It is convenient to express the particle's sphere of influence as the product of the actual particle dimension d and an influence factor S, where S is the ratio of the diameter of the influence sphere divided by the actual particle diameter. Equation (11.3) can then be modified

$$-\frac{dc}{dt} = \frac{C\mathbf{k}TS}{6\mu}\frac{(d_1+d_2)^2}{d_1 d_2}c^2. \quad (11.4)$$

If the particles are all the same size, i.e. the aerosol is monodisperse, then $d_1 = d_2 = d_3 = \ldots = d_n$, and equation (11.4) simplifies to

$$-\frac{dc}{dt} = \frac{2C\mathbf{k}TS}{3\mu}c^2. \quad (11.5)$$

So, except for the particle size occurring as a first-order correction factor in the Cunningham Correction C, the rate of agglomeration of a monodisperse aerosol is independent of particle size. Equation (11.5) can be integrated, assuming constant conditions and a constant Cunningham correction factor:

$$\frac{1}{c} = \frac{1}{c_0} + \frac{2}{3} \frac{CkTS}{\mu} t \tag{11.6}$$

where c_0 = concentration of particles at time t_0.

TABLE 11.1. COAGULATION CONSTANTS OF VARIOUS AEROSOLS[299, 316]
(air at ambient conditions—100 kPa (1 atm) Hg, 25°C)

Substance	$\varkappa \times 10^9$ $m^3 s^{-1}$
Ferric oxide	0·66
Magnesium oxide	0·83
Cadmium oxide	0·80
Stearic acid	0·51
Oleic acid	0·51
Resin	0·49
Paraffin oil	0·50
p-Xylene-azo-β-naphthol	0·63
Ammonium chloride (ambient)	0·51
Ammonium chloride (46% relative humidity)	0·43
Zinc oxide formed in arc	1·9
Silica powders in electric field	2·8–3·7
Theoretical value	0·51

The group $2CkTS/3\mu$ is called the coagulation constant $\varkappa$, and this has been experimentally found in a number of cases (Table 11.1). The theoretical value for the coagulation constant under the same conditions (air at 100 kPa, 25°C) and assuming that $S = 2$, is $0{\cdot}51 \times 10^{-9}$ $m^3 s^{-1}$, which is in excellent agreement with the cases of stearic and oleic acid aerosols, which are virtually monodisperse and have no electric charge. Differences in the values for other aerosols, which were always larger than the theoretical one, are explained by the following:

(a) The aerosol was not monodisperse, which can account for small increases in $\varkappa$.
(b) Electrostatic charges were present on the particles which can account for large increases in $\varkappa$.
(c) The presence of humidity tends to decrease the rate of coagulation, until at very high humidities (relative humidity greater than 60 per cent) this trend is reversed.

It is observed in mixtures of small and larger smoke particles, that the fine particles tend to disappear as they agglomerate with the larger ones. This is indicated by the term incorporating particle size in equation (11.4). The changes in rates of agglomeration can be calculated from the ratio $(d_1+d_2)^2/d_1d_2$, which is unity for a monodisperse aerosol. When the particles are in two distinct groups, and the ratio of their diameters is large, e.g. $d_1:d_2::1:50$, then the effect on the rate of coagulation is large (Table 11.2A). When the particles are of a range of sizes, which is the practical case, then the effect is much smaller. Values in Table 11.2B have been calculated on the assumption that the particles occur in groups of equal sizes in ratios from 1 : 1 to 1 : 8 (1–8 times the diameter).

TABLE 11.2. EFFECT OF POLYDISPERSIVITY OF THE COAGULATION CONSTANT[939]

A

Size of particle colliding with unit size particle	$\frac{(d_1+d_2)^2}{4d_1d_2}$
1	1·00
2	1·13
5	1·80
10	3·02
25	6·76
50	12·74
100	25·5

B

Diameter ratio range	$\frac{(d_1+d_2)^2}{4d_1d_2}$
1 : 1	1·0
1 : 1 to 2 : 1	1·06
1 : 1 to 5 : 1	1·19
1 : 1 to 8 : 1	1·27

If the particles are small (below about 0·5 μm) the first-order correction introduced by the Cunningham correction factor becomes appreciable and is written most simply as

$$C = 1+2A\lambda/d \tag{4.21}$$

where λ = mean free path of the gas molecules and

$$A = 1{\cdot}257+0{\cdot}400 \exp\left(-\frac{1{\cdot}1d}{2\lambda}\right)$$

which may be considered approximately constant in a first order correction. This can be introduced in the integrated form of the rate equation:

$$\frac{1}{c} = \frac{1}{c_0}+\frac{4\mathbf{k}T}{3\mu}\left(1+\frac{2\lambda A}{d}\right)t. \tag{11.7}$$

From this it is seen that the coagulation "constant" $\varkappa$ is not a constant, but decreases as the coagulation proceeds and the value of C falls off and approaches unity. Experimentally this has only been found, as would be expected, for very small particles.

A secondary factor which influences the coagulation constant $\varkappa$ and tends to give slightly greater values than are calculated from the classical Smoluchowski equation is the presence of van der Waals' forces between the particles. The correction introduced by these forces, however, is only of the order of a few per cent (at most, 10 per cent),[314] and so can be neglected for most gas cleaning plant calculatins.

The theory of Smoluchowski (which considers the medium as a continuum) and its modification for small particles, for diameters of the order of half the mean free path of the gas, λ, gives excellent agreement for Knudsen numbers ($2\lambda/d$) as low as 5. Modifications of the theory have been attempted by Friedlander,[277] Zebel,[960] Fuchs[285] and Hidy and Brock.[365] Friedlander suggested that these very small particles should diffuse like gas molecules, and a Stokes–Einstein diffusivity should be replaced by a molecular diffusivity for the gas. Zebel,[960] on the other hand, assumed that the medium behaves as a continuum except close to the particle, within a distance λ. As Hidy and Brock[365] pointed out, in this close region, the collision with other gas molecules is neglected, which oversimplifies the situation, and leads to erroneous results. Fuchs proceeded similarly to Zebel, but suggested that in the aerosol boundary layer there is a jump in particle concentration. Fuchs[285] used an empirical approach by using known boundary conditions for high and low Knudsen numbers, and forcing his results to agree with these. Hidy and Brock[365] used the kinetic theory of gases. Figure 11.1 shows the modified classical Smoluchowski theory, the Fuchs theory, and the "free molecule" theory of Hidy and Brock[365] compared with experimental

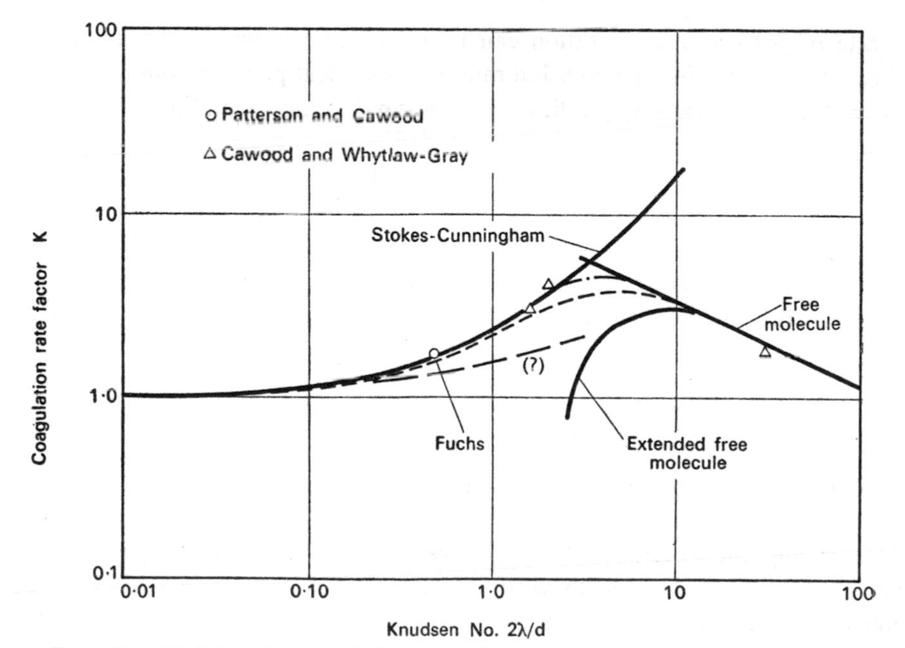

FIG. 11.1. Variation in coagulation rate with Knudsen number for particles.[365]

coagulation rates of Patterson, Cawood and Whytlaw-Gray. It can be seen that the free molecule theory is probably excellent for particles and pressures with Kn = 30, while the classical theory is applicable for Kn < 2. In the intermediate region the Fuchs theory may be the most effective.[365] The calculation of the correction factor C for high Knudsen numbers is difficult, but for approximate calculations, a value based on Fig. 11.1 is adequate.

The effect of charges on the aerosol particles is very complex, and experimental evidence is conflicting. If all particles carry the same charge, this tends to slow coagulation, while bi-polar charges, which were present on the particles in the powerful electric field introduced by Gillespie,[299] speed up agglomeration. If calculations including electric charges on particles have to be considered, they may be found elsewhere.[315] The effect of temperature, pressure and viscosity on the rate of agglomeration can be calculated from the changes in the temperature, viscosity and Cunningham correction term (which is a complex function of pressure, temperature and viscosity effects on the mean free path of gas molecules) on the coagulation constant $\varkappa$, i.e. $(4C\mathbf{k}T/3\mu)$ (where $S = 2$).

Brock and Hidy[134] and Zebel[961] have considered the effect of a range of forces, including electrical, magnetic and centrifugal fields, and non-uniform states of the suspending gas, i.e. temperature and pressure gradients and shear fields. Their results are not, unfortunately, directly applicable in simple calculations, but the former do point out that the directed motion of aerosols resulting from these non-equilibrium effects can strongly influence the coagulation rates, even for particles of small size.

11.3. INCREASING AGGLOMERATION RATES BY STIRRING

The rate of particle agglomeration can be increased by stirring the gas to introduce turbulence, thus increasing the collision rate. For spherical particles, diameter d, in a fluid in laminar flow with a velocity gradient $\mathrm{d}y/\mathrm{d}x$ normal to the streamlines, the rate of agglomeration is

$$-\frac{\mathrm{d}c}{\mathrm{d}t} = \frac{4}{3}d^3c^2\frac{\mathrm{d}y}{\mathrm{d}x}. \tag{11.8}$$

When a cloud of particles is fanned, it has been shown that the losses are due to the combined effect of Brownian coagulation, and losses to the surfaces of the fan and the enclosing vessel,[298]

$$-\frac{\mathrm{d}c}{\mathrm{d}t} = \varkappa c^2+\beta c = -\left(\frac{\mathrm{d}c}{\mathrm{d}t}\right)_c - \left(\frac{\mathrm{d}c}{\mathrm{d}t}\right)_l \tag{11.9}$$

where β is a "loss constant", and the subscripts c and l stand for "coagulation" and "loss" terms.

Experimentally it has been shown that the value of $\varkappa$ varies about 50 per cent with time, but the values of β increase lineally with time. It is therefore necessary to determine these experimentally if it is intended to use the stirring of an aerosol as a method of agglomeration.

In general of course, the increasing rate of stirring not only increases turbulence, but presents new surfaces to the aerosol to which the particles can stick, and once there, form part of the surface. Since aerosol particles are removed each time they contact the blades of a fan, faster fan rotation will achieve a more rapid rate of removal.

A similar argument can be applied to passing a gas through a packed bed where the passages in the bed present a tortuous path to the gas, and so produce increased turbulence in the gas, as well as new surfaces, as the gas winds its way through.

11.4. ACCELERATED FLOCCULATION BY SOUND

Perhaps the most effective way of rapidly agglomerating particles and droplets to larger units, which are then capable of being collected by conventional mechanical collectors, such as cyclones, is by passing the cloud of particles or mist through a column in which the gas is subjected to standing sound waves. When a cloud in a narrow tube has low intensity sound waves passed through it, which give a simple standing wave pattern, the smoke first takes on a banded appearance, as the particles start to migrate to the antinodal regions. Flocculation then becomes visible, and the smoke appears granular. The flocs grow larger and collect on the walls, or become suspended at the antinodal planes, representing wafer-like layers, in some ways similar to the piles of dust formed at the antinode in the classical Kundt tube.[720] A most comprehensive review of the whole field of sonic agglomeration and its application has been published by Mednikov[567] and is available in an English translation.

The mechanism of sonic flocculation is not fully understood, but is probably a combination of three factors:[719]

(i) Co-vibration of the particles with the vibrating gas called orthokinetic coagulation.[114]
(ii) Sonic radiation pressure.[438]
(iii) Hydrodynamic attractive and repulsive forces between neighbouring particles.

Orthokinetic coagulation. When standing sound waves pass through a gas containing a cloud of particles, the particles, depending on their size and the frequency of vibration, may vibrate with the gas if the sound frequency is low, and tend to lag when the frequency is increased. When the frequency is very high, the larger particles remain almost stationary while the smaller ones follow the vibrations. An equation giving the degree of participation of a spherical particle in the gas vibrations was first deduced by König.[460]

$$\frac{u_g}{u_p} = \left[\frac{1+3b+9b^2/2+9b^3/2+9b^4/4}{a^2+3ab+9b^2/2+9b^3/2+9b^4/4}\right]^{\frac{1}{2}} \qquad (11.10)$$

where u_g = velocity amplitude of the gas,
u_p = velocity amplitude of the particle,
$a = \frac{1}{3}(1+2\varrho_p/\varrho)$,
$b = 2/d.\sqrt{(\mu\tau/\varrho\pi)}$,
ϱ_p = particle density, ϱ = gas density,
τ = period of vibration.

This equation is not readily soluble, and a simpler one has been deduced by Brandt, Freund and Hiedemann[114] which has been shown to give virtually the same result.[721] This theory neglects the buoyancy effect of the gas on the particle, and it assumes that Stokes' law can be applied to the relative motion of the particle through the gas. This last assumption is justifiable when the particle Reynolds number is less than 0·2 (Section 4.1). Appropriate particle velocities to which this can be applied in air are listed in Table 11.3.

TABLE 11.3. PARTICLE VELOCITIES AND CUNNINGHAM CORRECTION FACTORS FOR PARTICLE Re = 0·2[114]

Particle diameter μm	Cunningham correction	Velocity in air at Re = 0·2 m s^{-1}
1	1·3	6
2	1·15	3
4	—	1·5
10	—	0·6

Because the velocity of an air segment at the antinode is about 2·5 m s^{-1} at frequencies of 10 kHz, the assumption of Stokes' law is a reasonable one for particles smaller than 3 μm at this or lower frequencies. For higher frequencies the assumption can be justified for smaller particles.

The ratio of the amplitude of particle vibration X_p to the amplitude of vibration of the gas X_g has been found[114] to be with the stated assumptions:

$$\frac{X_p}{X_g} = \frac{1}{\left\{\left[\frac{\pi d^2 \nu \varrho_p}{9\mu C}\right]^2 + 1\right\}^{\frac{1}{2}}} \tag{11.11}$$

where ν = frequency,
C = Cunningham correction.

For constant particle density (ϱ_p) and gas viscosity, $(\pi\varrho_p/9\mu)$ is a constant, and also the Cunningham correction can be considered approximately constant, so that (11.11) simplifies to

$$\frac{X_p}{X_g} \doteqdot \frac{1}{(kd^4\nu^2+1)^{\frac{1}{2}}} \tag{11.12}$$

where $k = \pi\varrho_p/9\mu C$.

Equation (11.12) can also be deduced from König's formula (11.10) when certain restrictions are placed on it.[114]

For particles of unit density vibrating in air, the relative amplitude has been plotted (Fig. 11.2) for frequencies varying from 1–100 kHz. For the highest frequencies (50 and

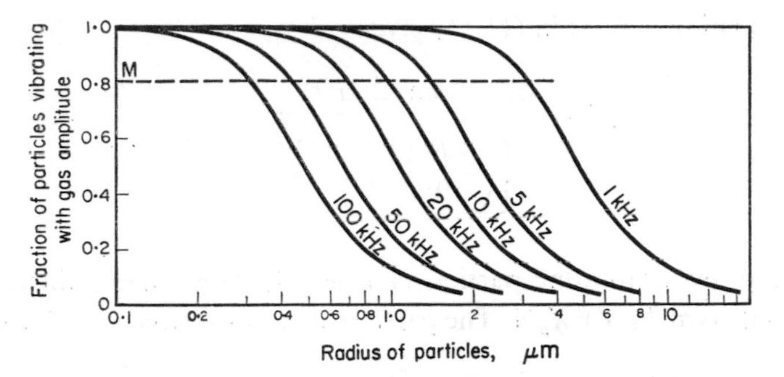

FIG. 11.2. Fraction of particles of unit density of various radii vibrating with gas at sound intensities between 1 and 100 kHz amplitude. The broken line at 0·8 (80%) indicates the size above which nearly all particles are considered to vibrate with the gas.[114]

100 kHz) the curves can only be taken as approximations, as the assumptions regarding the Stokes' law region no longer hold.

Figure 11.2 shows how, until a certain particle size, the aerosol particles swing along with the vibrations in the gas. This size can be called the critical particle size for the particular frequency, while a critical frequency for a particular particle size can be similarly calculated. The critical particle size is indicated by the point where the curves enter the steep gradient after their initial shallow decrease, which occurs at approximately the 80 per cent value for X_p/X_g, shown by the broken line in Fig. 11.1. Substituting 0·8 for X_p/X_g in equation (11.12) gives the critical particle diameter d_c as approximately

$$d_c^2 \nu \doteqdot 4 \times 10^{-8}\ \text{m}^2\ \text{s}^{-1} \tag{11.13}$$

or more generally, from equation (11.11)

$$\left[\frac{d_c^2 \varrho_p \nu}{C\mu}\right]_{\text{critical}} \doteqdot 2{\cdot}16. \tag{11.14}$$

Equations (11.13) and (11.14) indicate that, as a first approximation, a unique relation exists between the critical particle size and the frequency of vibration of the gas, which for particles less than 7 μm occurs in the frequencies greater than 1 kHz and extends into the ultrasonic region. These equations enable the critical frequency or particle sizes to be readily calculated.

From the movement of a particle it becomes possible to calculate the volume in which the particle will collide with other particles and coagulate with them (the aggregation zone). The amplitude of the gas vibration can be calculated from

$$X_g = \mathcal{A} \sin (2\pi a/\lambda) \tag{11.15}$$

where $\mathcal{A}$ = amplitude at the antinode (maximum),
λ = wavelength,
a = distance from the node

and substituting equation (11.15) in (11.11) gives

$$X_p = \frac{\mathcal{A} \sin (2\pi a/\lambda)}{\left\{\left(\frac{\pi d^2 v \varrho_p}{9\mu c}\right)^2 1 + \right\}^{\frac{1}{2}}}. \tag{11.16}$$

The volume swept by a vibrating particle can be calculated, neglecting the ends of the cylinder (Fig. 11.3a), as $\pi(d+d')^2 X_p/4$. The particle also moves towards the antinode owing

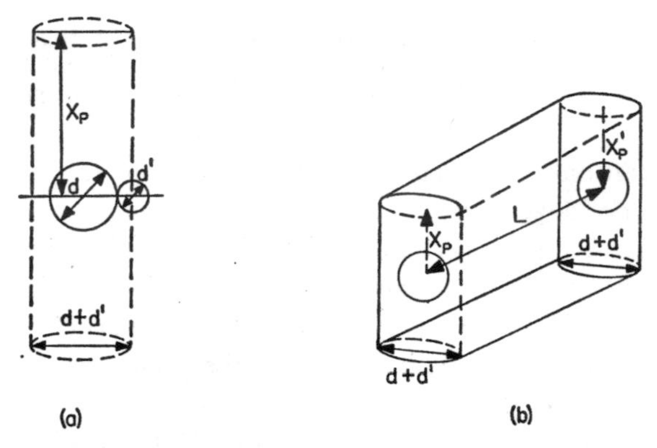

FIG. 11.3. (a) Volume swept by particle vibrating at one point. (b) Total volume swept by vibrating particle as it approaches the antinode.

to radiation pressure, and after time t, having moved distance L, will have a new amplitude X_p'. The swept volume during this period will be (Fig. 11.3b):

$$V_a = (X_p + X_p')(d+d')\left[L + \frac{\pi}{4}(d+d')\right]. \tag{11.17}$$

If the time has been sufficient to reach the antinode, the value of X_p will have reached the maximum possible:

$$X_{p_{\max}} = \frac{\mathcal{A}}{\left\{\left(\frac{\pi d^2 v \varrho_p}{9\mu C}\right)^2 + 1\right\}^{\frac{1}{2}}} \tag{11.18}$$

and the swept volume will be

$$V_{a_{\max}} = (X_p + X'_{p_{\max}})(d+d')\left[\lambda - a + \frac{\pi}{4}(d+d')\right]. \tag{11.19}$$

The efficiency of flocculation can be calculated from the ratio V_a/V where V is the total volume of the chamber enclosing the cloud, basing the values of X_p on the critical particle size for the frequency of vibration.

The speed of the particle in moving towards the antinode can be estimated from a consideration of the *radiation pressure* of the vibrating gas.[438, 721] When the diameter of a sphere is small compared to the wavelength λ, then the force can be found approximately from

$$F_r = \frac{5}{12}\frac{\pi^2 d^3}{\lambda}\bar{E}\sin(2\pi a/\lambda) \tag{11.20}$$

when F_r = radiation pressure (N),
$\bar{E}$ = energy intensity (J m^{-3}).

This will be a maximum when $a = \lambda/4$, i.e. half-way between node and antinode and will be

$$F_{r_{max}} = \frac{5}{12}\frac{\pi d^3}{\lambda}\bar{E}. \tag{11.21}$$

This force will be opposed by the resistance of the gas, which is assumed to be in the Stokes' law region. The velocity of the particle towards the antinode is given by

$$\frac{dl}{dt} = F_{r_{max}}\frac{C\sin(2\pi a/\lambda)}{3\pi\mu d} \tag{11.22}$$

which has been solved to give

$$l = \frac{\lambda C}{2\pi}\,\text{artan}\left(\tan\frac{2\pi a}{\lambda}e^{Bt}\right) \tag{11.23}$$

where

$$B = \frac{5}{9}\frac{\pi^2 d^2 \bar{E}}{\lambda^2 \mu}. \tag{11.23a}$$

At both the node and antinode the velocity due to radiation pressure is zero. An order of magnitude calculation can be done using $2\pi a/\lambda = \pi/6$ as the initial point and $2\pi a/\lambda = \pi/3$ as the final point; then $t = (\ln 3)/B$ and for a 2 μm particle at 20°C in air, with a frequency of 10 kHz, $t = 1{\cdot}07\times 10/\bar{E}$. So if the energy intensity is approximately 1 J m^{-3}, then t is approximately 10 sec. Calculations showing the time taken for a particle to move to within a short distance (particle radius) of the antinode will indicate the desirable residence time for particles in the sonic field.

When spheres are at rest in a vibrating medium, they will attract one another when their line of centres is normal to the direction of the vibrations, and will repel one another if their line of centres is parallel to the vibrations. These *hydrodynamic forces* were used by König[460] to explain the striations observed at antinodes in the Kundt tube.

For two spheres, diameters d and d', distance l apart from one another, the attractive force F_h along the connecting line at right angles to the direction of vibration is given by

$$F_h = \frac{3\pi\varrho}{128}d^3 d'^3\frac{(\Delta v)^2}{l^4} \tag{11.24}$$

where Δv is the relative velocity amplitude of the spheres. When the two particles are the same size ($d = d'$) equation (11.24) reduces to

$$F_h = \frac{3\pi\varrho}{128} \cdot d^6 \frac{(\Delta v)^2}{l^4}. \tag{11.25}$$

If the resistance of the gas opposing the particles coming together follows Stokes' law, the velocity of the spheres towards one another will be

$$\frac{dl}{dt} = \frac{128\mu l^4}{\varrho d^5(\Delta v)^2} \tag{11.26}$$

and the time taken for the particles to come from a distance l to touch one another (d) can be found by integrating (11.26),

$$t = \frac{(\Delta v)^2 \varrho}{384\mu}\left(d^2 - \frac{d^5}{l^3}\right). \tag{11.27}$$

Calculations based on the equations by Brandt *et al.*[114] show that at the frequencies up to about 50 kHz, orthokinetic coagulation is the main mechanism, and hydrodynamic forces between the particles do not contribute to their agglomeration. At ultra-high frequencies — of the order of hundreds of kHz—when orthokinetic coagulation plays a negligible part, hydrodynamic forces become the major agglomerating force.

It has been pointed out[109] that Hiedemann's original theory was based on several simplifying assumptions, which become invalid under certain conditions. For example, Hiedemann used Bernoulli forces to express the attraction between two spheres, and then used the Stokes resistance equation in a Reynolds number range (greater than 0·2) where this is not realistic.

The influence of acoustically generated turbulence in high intensity fields at low turbulence was also ignored by these earlier workers, and has recently been stressed by Matula[564] and Podoshernikov.[651–2] The theoretical importance of hydrodynamic (Oseen) forces has also been reviewed by Pshenai-Severin[664] who concluded that these are an important factor in the agglomeration of particles with diameters in the range of 3 to 30 μm in relatively low-frequency sound fields, as well as orthokinetic coagulation. Furthermore, Timoshenko has studied the interaction between particles in an acoustic field, in Stokes flow, and has derived mathematical expressions which describe the process involved in the agglomeration of two dissimilar particles. These have been used on systems typical for industrial gas-cleaning plant; viz. a pair of 1- and 2-μm-sized, and 3- and 4-μm-sized particles respectively, with densities of either 1 or 2·5 g cm^{-3}, interparticle distances of 100–200 μm, and sound intensities between 70 and 170 mm s^{-1}. The curves show maxima in the mutual particle displacement, which shifts towards lower values of the frequency as the particle radius increases, while with increases in the sound intensity, the mutual displacement increases proportionately. When particle density is increased, the mutual displacement grows and the maximum is shifted towards lower frequencies. The analysis shows further that the mutual displacement per unit time increases abruptly with increasing frequency, reaching a maximum for heavy particles with large diameters at several hundred Hz, and around several kHz for smaller particles. This indicates that high intensity sound at frequencies below 2 kHz may markedly increase the agglomeration rate between particles.

As indicated, the exact interrelation of these mechanisms of sonic agglomeration has not been found, but the above may help to present an order of magnitude as to the effectiveness of sound waves of a certain frequency and intensity in agglomerating a cloud of particles or droplets.

The application of sound waves to aerosol collection depends on a number of factors:[108, 598] the sound frequency and intensity, the aerosol concentration and turbulence, and the exposure time. It has been shown [equations (11.13) and (11.14)] how the vibration of particles depends on the sound frequency. A smoke or mist cloud contains a mixture of particle sizes, so in practice a wide range of frequencies greater than a few kHz is suitable. In industrial plant the sound generators usually operate in the range 1–4 kHz[198] because at higher frequencies it becomes more difficult to produce the required sound intensity. The acoustic power or sound intensity requirements of sonic agglomeration systems are very high. The threshold value for noticeable flocculation is 10–10·8 W m^{-2} while values over 11·5 W m^{-2} are necessary for industrial plants.[598]

The particle concentration should not be less than 1–2 g m^{-3} for particles in the 1–10 μm range, and concentrations of 5 g m^{-3} have been found most suitable. When concentrations become very high (say 200 g m^{-3}) sonic energy will be lost because of increasing attenuation of the sound in the aerosol.[198] In some cases it is helpful to add water mist to the aerosol.[108]* It has also been shown that turbulent flow introduced by the acoustic field enhances agglomeration.

The length of the exposure of the aerosol to the sound waves has been shown by the theory to have a significant effect on the degree of flocculation achieved. The index of agglomeration I, which is the ratio of the final to the initial mean particle diameters, has been found experimentally to be a function of the product of exposure time and field intensity W m^{-2} (Fig. 11.4). Industrial practice uses contact times of about 4 sec, which may be reduced to 2 sec when intensities greater than 13 W m^{-2} are used.[108]

Sound generators. High-powered sound waves can be produced by one of four methods:

(a) The vibration of piezoelectric crystals and ceramics (quartz, tourmaline or Rochelle salt, barium titanate ceramics).
(b) The sound caused by the vibration of a cylinder.
(c) Whistles (static generators).
(d) Sirens (dynamic generators).

The first two types of sound generators are largely for laboratory use. Piezoelectric crystals are used for high-frequency sounds, but will not produce the high sound intensities required for plant-size generators. The sound waves generated by a vibrating metal rod were used in the classical Kundt tube, and these two devices for producing intense high-frequency sound can be useful, particularly on a limited scale.

* See also J. Olaf, *Staub*, **22,** 513 (1962).

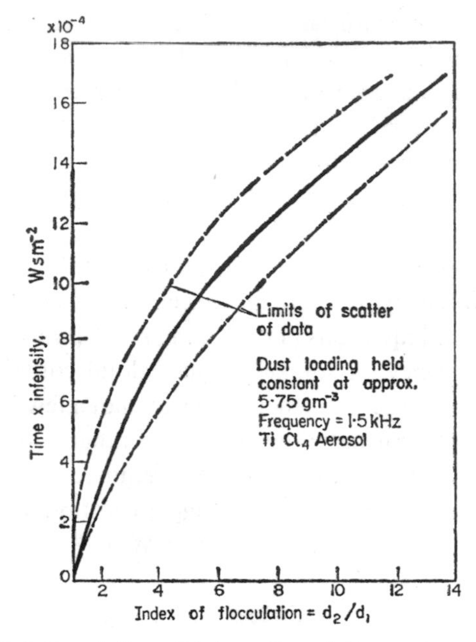

FIG. 11.4. Correlation of data on flocculation and sound intensity.[598]

An electromagnetic sound generator described by St. Clair[718] (Fig. 11.5a) consists of a solid duralium cylinder with a supporting web and a driving ring machined from the same piece. The driving ring projects into the radial gap of a pot magnet, and acts as a one turn secondary of a transformer, of which the primary is the field coil of the magnet. An input of 200 W into the coil from an amplifier produced a high intensity sound of 10–20 kHz.

A second vibrating-rod-type generator is the air-driven stem jet whistle (Fig. 11.5b) which consists of a cylindrical rod in the axis of a nozzle resonator system. This device will produce high-intensity sound levels (12 W m^{-2}) and frequencies between 9 and 15 kHz with low nozzle pressures (210 kPa).

Whistles fall into two two main categories. In the first an air jet impinges on a resonating cavity, while in the second the air is introduced tangentially into a circular tube, creating a vortex and then escaping axially, producing a loud sound.

The original impingement-jet-type whistles were developed by Galton (1883), and later modified by Hartmann, and are usually called Hartmann whistles (Fig. 11.5c). They have been used for frequencies from 10 to 100 kHz, and in their original (Hartmann) form have efficiency of about 4 per cent. The dimensions suggested by Hartmann were to have the same jet diameter A as opening diameter B and depth of resonating cavity b.

The intensity of sound $\mathcal{J}_0$ from a Hartmann whistle can be calculated from[341]

$$\mathcal{J}_0 = 3{\cdot}0B^2\sqrt{(P-0{\cdot}9)}\ \text{W} \tag{11.28}$$

where B = 2–6 mm,
P = gauge pressure = 210–350 kPa,
$A = B = b$ in all cases.

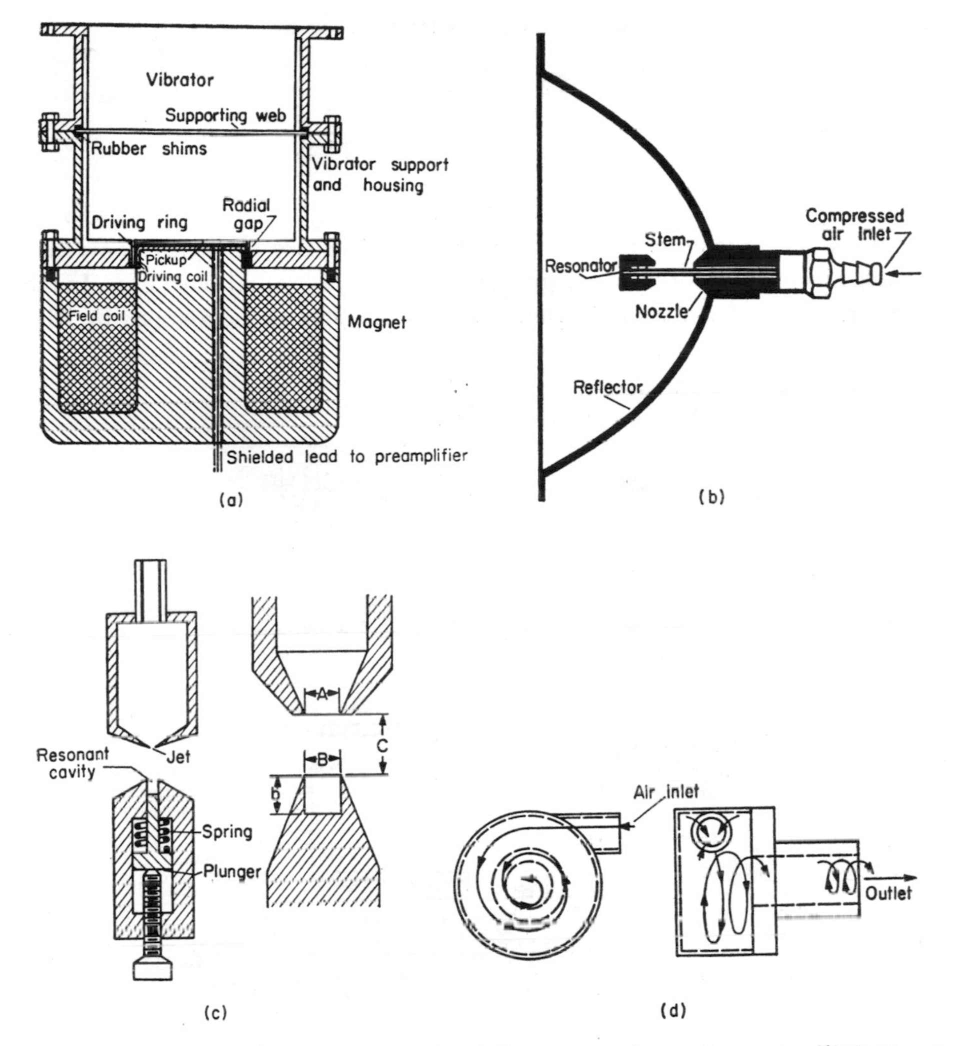

FIG. 11.5. Sound generators (other than sirens). (a) Electromagnetic sound generator.[718] (b) Stem jet whistle.[108] (c) Hartmann whistle.[108] (d) Vortex whistle.[108]

A modified Hartmann whistle[106] with efficiencies twice that of the original pattern has been made by increasing the relative width of the resonating aperture to a diameter more than 30 per cent greater than the jet width ($B/A \geqq 1{\cdot}3$). In addition, placing the nozzle in a secondary resonance chamber has increased the overall efficiency of the whistles up to 20 per cent.

Vortex whistles (Fig. 11.5d) have been studied by Vonnegut[892, 893] who gave the following approximate formula for the frequency of the emitted sound:

$$\nu = \alpha\left(\frac{U_s}{\pi D}\right)\sqrt{\left(\frac{P_1-P_2}{P_2}\right)} \tag{11.29}$$

where α = constant (< 1) which accounts for friction losses,
U_s = velocity of sound,
D = tube diameter,
P_1 = inlet pressure,
P_2 = exhaust pressure.

Frequencies obtained experimentally with the vortex whistle were about 15 kHz.

Dynamic sound generators (sirens) have higher efficiencies of energy conversion into sound energy from other forms of energy than other types of sound generators. In the form used industrially, particularly in the United States, the siren was developed by Allen and

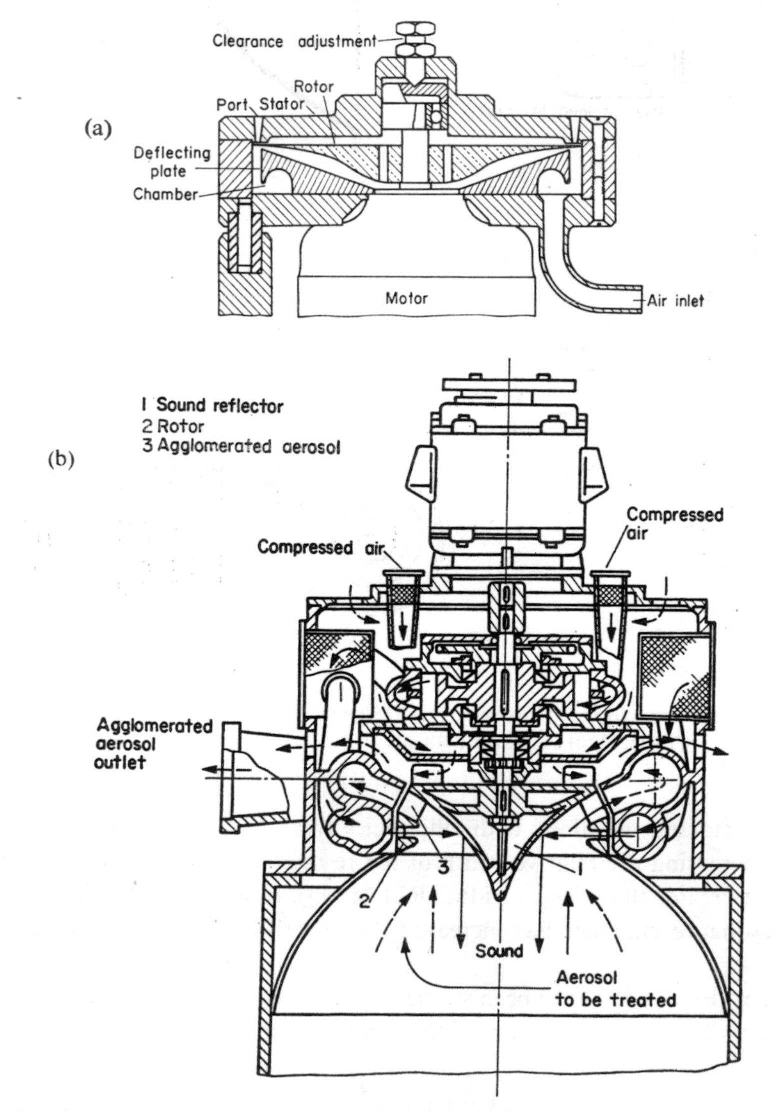

FIG. 11.6. Sound generators (sirens). (a) Allen–Rudnik siren.[12] (b) C.I.O.P. siren.[108]

Rudnik[12] (Fig. 11.6a). It consists essentially of a high-speed rotor, which interrupt the flow of a gas (usually air) through the stator ports.

In Allen and Rudnik's original design, the rotor had 100 ports evenly spaced around a 120 mm disc, and the adjacent rotor and stator surfaces were very closely fitted. The rotor speed was 130–330 Hz. The conversion efficiency in the Allen–Rudnik pattern was 17–34 per cent, with frequencies of 3–19 kHz and power outputs between 80 and 176 W.

More recent designs use rotor speeds up to 50,000 rev/min and efficiencies up to 50 per cent have been reported. If a comparatively pure sound is required, the stator ports should be circular and the rotor ports rectangular.

The major problem with sirens is the maintenance of the high speed rotor. Deposits on the rotor produce poor balance, bearings tend to overheat, and the frequency of the sound cannot be varied except by changing the number of openings, or by changing the rotor speed. Recent developments[543] have led to a modified siren, which has intensities as high as 14 Wm^{-2} in the zone close to the siren, and is coupled directly to a centrifugal extraction fan which removes the agglomerates (Fig. 11.6b).

For sirens the sound intensity $\mathcal{J}_0$ can be found from[543]

$$\mathcal{J}_0 = \frac{\varrho U_s}{2}(2\pi\nu\mathcal{A})^2, \tag{11.30}$$

$\mathcal{A}$ = amplitude of vibrations.

A number of design equations for sonic agglomerating systems have been suggested by Inoue and agree satisfactorily with experiment.[108] The average sound intensity in the agglomeration chamber, $\mathcal{J}$, can be found from

$$\begin{aligned} \mathcal{J} &= \beta\mathcal{J}_0/\text{cross-sectional area of chamber} \\ &= \frac{4\beta\mathcal{J}_0}{\pi D^2} \text{ (for cylindrical chamber with diameter } D) \end{aligned} \tag{11.31}$$

where β is a constant characteristic for the plant so that $0{\cdot}07 < \beta\eta_A < 0{\cdot}16$ where η_A is the acoustic efficiency of the generator. Boucher[108] has suggested that $\eta_A\beta$ is closer to 0·07 than 0·16.

The index of agglomeration I can be found from

$$I = \exp \frac{t}{3}\sqrt{\left(\frac{4\beta\mathcal{J}_0}{\pi D^2}\right)} \tag{11.32}$$

where t is the contact time (sec).

The ideal tower height H (m) for the agglomerator can be estimated from

$$H = 3v \ln I\sqrt{\left(\frac{\pi D^2}{4\pi\mathcal{J}_0}\right)} \text{ m} \tag{11.33}$$

where v = mean linear velocity of aerosol ($m\ s^{-1}$) (usually about 1 $m\ s^{-1}$).

The design of a sonic agglomeration plant is extremely simple. The source of the high intensity sound is placed at one end of an agglomeration chamber through which the gases to be cleaned are passed, and the agglomerated particles or droplets are then collected by a cyclone. Two typical arrangements are shown in Fig. 11.7. In the first arrangement, the gases move towards the source of the sound, while in the second, after passing through a preliminary cyclone, the gases move away from the sound.

An ultrasonic demister[38] has also been tested extensively[110]. This unit is shown in Fig. 11.8. An intense sonic field is directed above the mesh, and the radiation pressure field tends to retard flooding of the mesh and droplet re-entrainment. The sound field also promotes higher

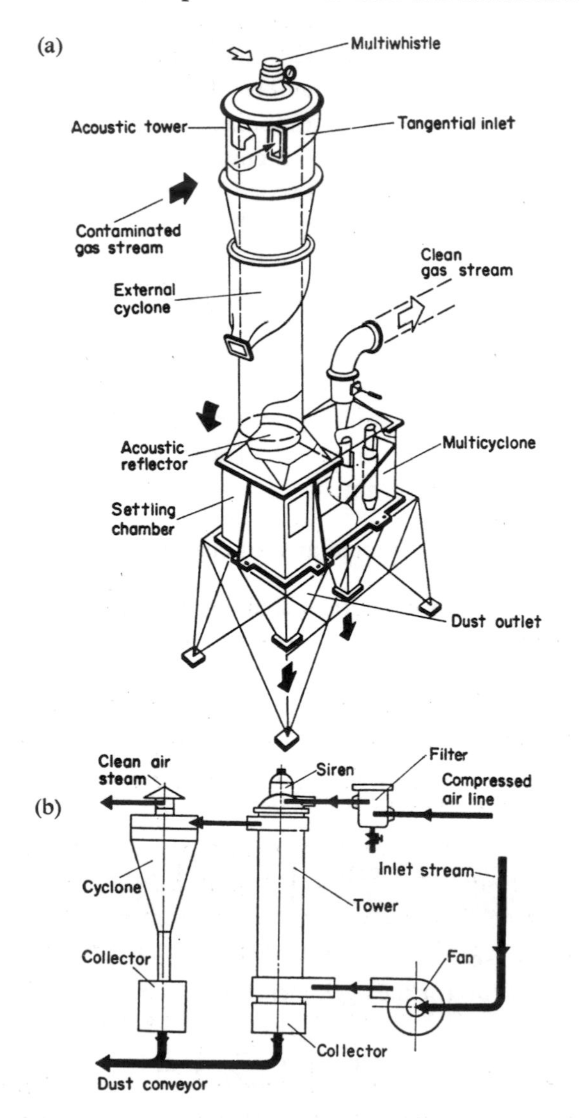

FIG. 11.7. Arrangement of sonic agglomeration–collection systems: (a) Gases move away from source of sound.[108] (b) Gases move towards source of sound.[108]

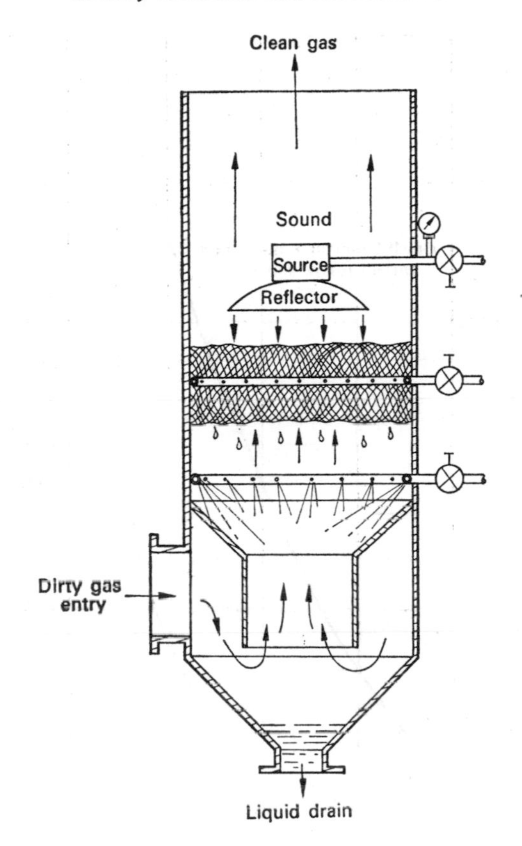

FIG. 11.8. Ultrasonic demister.[38]

impingement rates between the wires and the vibrating droplets. Tests were carried out using a 120 mm mesh and a flow of 5 m s^{-1}. The pressure drop was 0·85 kPa and the aerosol loading was reduced from 0·354 mg m^{-3} to 0·0128 mg m^{-3}, giving an efficiency of 96·5 per cent. An acoustic field of 60–80 W, with a frequency of 9·8 kHz, was obtained from a stem jet whistle. The energy requirement of this was in the range of 4–5 kW/m^3 s^{-1}. A comparison with a two-stage demister, which has more than twice the pressure drop, is shown in Fig. 11.9.

Gas velocities of 1 m s^{-1} have been suggested as suitable, and from this velocity the tower diameter required to handle the gas volume can be found.

It has also been suggested by Mednikov[567] that combining an acoustic agglomerator with an electrostatic precipitator would result in reduced size of electrostatic precipitator, as less space and precipitating area will be needed to collect the fumes which are agglomerated in the sonic unit. Uzhov[882] has reported that on a blast-furnace gas-cleaning installation, the residual particulate concentration was reduced to 0·005 to 0·010 g m^{-3} from 0·020 g m^{-3}, and the capacity of the electrostatic precipitator was increased by a factor of 2.

Sound generators have been suggested by Boucher[108] for improving the efficiency of scrubbers. A sound generator has been tested in a scrubber with self-induced spray, made

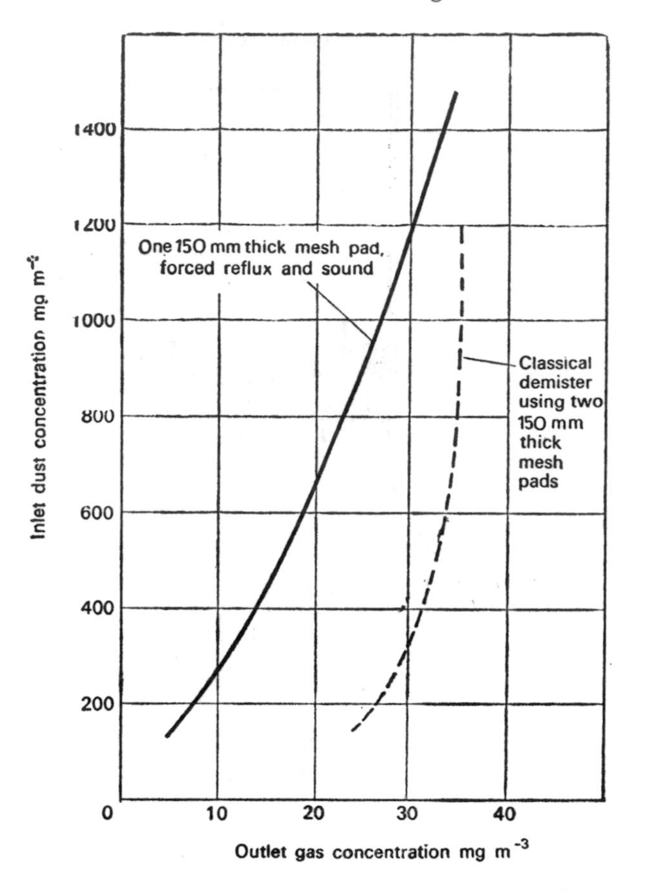

FIG. 11.9. Comparison of ultrasonic demister with one 150-mm pad and conventional two-stage demister.[110] (110 $m^3 h^{-1}$ of gas, velocity 4·4–4·65 $m s^{-1}$, pressure drip: two stage, without sound, 1·07 kPa–1·19 kPa, one stage with sound 0·75–0·87 kPa.)

by Schmieg Industries,[111] which is similar to the Rotoclone type N (see Figs. 9.15 and 9.16). Best results were obtained when the radiation pressure was directed counter-current and a definite increase in collection efficiency was obtained. However, in view of the increased cost and complexity of apparatus, this does not appear economical.

Another suggestion by Boucher[108] involved the installation of a sound generator in an S-F type venturi scrubber. This creates surface cavitation at the atomization point for the liquid and increases the inertial impaction collection efficiency.

As mentioned in Chapter 8, one manufacturer uses sonics to remove the collected cake in high-temperature bag filters.

Sonic agglomeration has been used successfully on a number of difficult collection problems.

Sulphuric acid mist[198] has been agglomerated successfully, and inlet concentrations of 3·5 g m^{-3} have been reduced to exhaust concentrations varying between 0·14 and 0·018 g m^{-3}, i.e. efficiencies of 96–99·5 per cent, depending on the mist residence time between

0·6 and 3·0 s (see Fig. 11.10).[198] The plant consisted of a long column where the gases were exposed to the sound waves, followed by multicyclones.[606]

Carbon black[819] has been successfully agglomerated in a sound field of 13 Wm^{-2} with a frequency of 3–4 kHz. When the concentration of the carbon black was 8–10 g m^{-3}, and the residence time 4·5 s, an efficiency of 82 per cent was achieved. When the concentra-

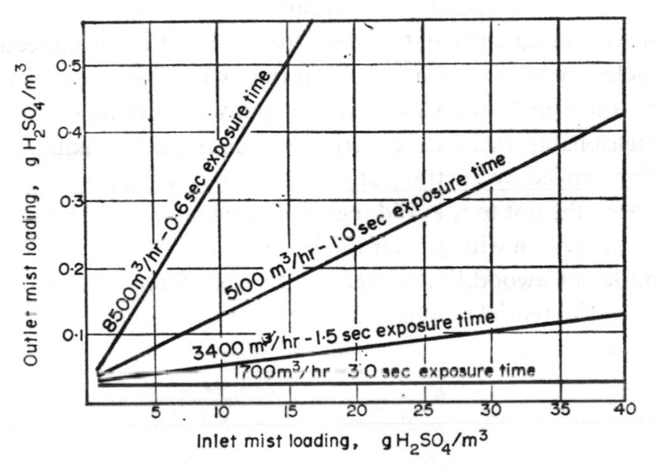

FIG. 11.10. Effect of inlet mist loading and exposure time to sound on the outlet mist loading in a plant collecting sulphuric acid mist.[198]

tion was less than 5 g m^{-3} the efficiency decreased. The introduction of a water mist improved collection at low concentrations.

Sonic agglomeration has been tried on fume collection for several metallurgical operations:[108] ferro-alloy furnaces,[396] ferro-manganese blast furnace gas,[127] zinc oxide fumes from copper recovery furnaces[108] (average efficiency 78 per cent), lead oxide fumes (95–98 per cent efficiency, with 15 kHz siren), open-hearth furnace fumes,[783, 883] carbide furnace fumes,[107] cracking gas condensate[384, 386] and coal tar.[384–6]

The operating and maintenance costs of sonic agglomeration–collection systems are rather high, although installation costs tend to be about 15 per cent lower than for electrostatic precipitators of similar capacity.[739] The efficiency of sonic agglomeration is virtually independent of temperature, and the technique could find application in high-temperature precipitation. When the fumes are corrosive it is easily possible to construct the agglomerator shell from corrosion-resistant materials, while a sound whistle can be very robust. Thus sonic agglomeration is applicable when highly corrosive conditions are likely to occur. Further improvements in sound-generation efficiency may lead to considerable industrial applications of sonic agglomeration techniques.

11.5. THERMAL PRECIPITATION

The force that drives particles away from hotter towards colder regions was observed first by Tyndall[874] and later by Lord Rayleigh[674] as the dust-free region or "dark space" which surrounds a hot body placed in a smoke cloud. Aitken[4] was able to show that this dust-free space extended completely around the hot body and was caused neither by gravity, evaporation from the surface, electrostatic forces nor centrifugal forces, but by a purely thermal force which exists in regions of unequal temperature, driving particles away from hot and towards cold surfaces.

The reason for the thermal force is partially explained by the theories that have been put forward to predict the magnitude of the force, and seems to depend on whether the particle is considerably smaller or larger than the mean free path of the gas molecules.

When the particles are smaller than the mean free path (λ) of the gas molecules, the suggested relation by Einstein[197] and Cawood,[154] which was later modified by Waldmann,[896] appears to agree reasonably well with experimental findings. The equation is based on the thermal force being applied by the translational motion of the gas molecules which are conducting the heat from the hot to the cold region, and which tend to bombard the side of the particle facing the hot region with greater force than the side facing the cold.

Einstein,[197] and later Cawood,[154] deduced the thermal force on the small particle as the resultant force of the differential bombardment by the gas molecules, using the kinetic theory of gases. This gave:

$$F_t = -\frac{1}{2}\lambda P \frac{\pi d^2}{4} \cdot \frac{dT}{dx} \tag{11.34}$$

where P = gas pressure,

λ = mean free path of gas molecules,

$\frac{dT}{dx}$ = thermal gradient in the gas ($\Delta T/\Delta x$),

ΔT = temperature difference through distance Δx

and the negative sign indicates that the force is in the opposite direction to the rise in temperature.

Substituting equations (3.3) and (4.29) for λ in equation (11.34) gives

$$F_t = -\frac{\pi}{8}\frac{P\mu d^2}{\varrho}\sqrt{\left(\frac{\pi M}{2RT}\right)} \cdot \frac{dT}{dx}. \tag{11.35}$$

A more sophisticated calculation with the same basic ideas by Waldmann[896] which introduces the translational part of the thermal conductivity of the gas $\varkappa_{g_{tr}}$ gives

$$F_t = -\frac{4}{15} d^2 \varkappa_{g_{tr}} \sqrt{\left(\frac{\pi M}{2RT}\right)} \cdot \frac{dT}{dx} \tag{11.36}$$

where $\varkappa_{g_{tr}} = 2{\cdot}5 c_v \mu$

$= (\frac{15}{4})(R/M)\mu$ (Eucken's Theory), (11.37)

c_v = specific heat at constant volume.

Substituting (11.37) in (11.36):

$$F_t = -d^2\mu \sqrt{\left(\frac{\pi R}{2MT}\right)} \cdot \frac{dT}{dx}. \tag{11.38}$$

The resistance by the gas to the movement of these small particles can be found from Epstein's equation[243] for fluid resistance

$$F = \frac{4}{3}\pi d^2 P \sqrt{\left(\frac{M}{2\pi RT}\right)} \cdot \left(1+\frac{\pi}{8}a\right)u \tag{11.39}$$

where u = velocity of particle,
a = coefficient of diffuse reflection (Millikan) or accommodation constant (Epstein),
when a = 0, all collisions are perfectly elastic,
when a = 1, all collisions are diffuse.

Experimentally it has been found that "a" can be taken as 0·81.[732]

From equations (11.38) and (11.39) it was deduced by Waldmann[896] that the speed of very small particles in a thermal gradient is

$$u_t = -\frac{1}{5\left(1+\frac{\pi}{8}a\right)} \frac{\varkappa_{gtr}}{P} \cdot \frac{dT}{dx} \tag{11.40}$$

This equation is similar to that by Einstein, with the assumption that the coefficient of diffuse reflection is zero and the constant $\frac{1}{5}$ is substituted by $\frac{1}{4}$. It has been shown[732] that for particles smaller than $\frac{1}{35}$ of the mean free path (i.e. $< \lambda/35$) with the coefficient of diffuse reflection taken as 0·8 to 1, equation (11.40) agrees well with experimental values (Fig. 11.11), while for particles of the size of λ, Epstein's equation is more applicable.

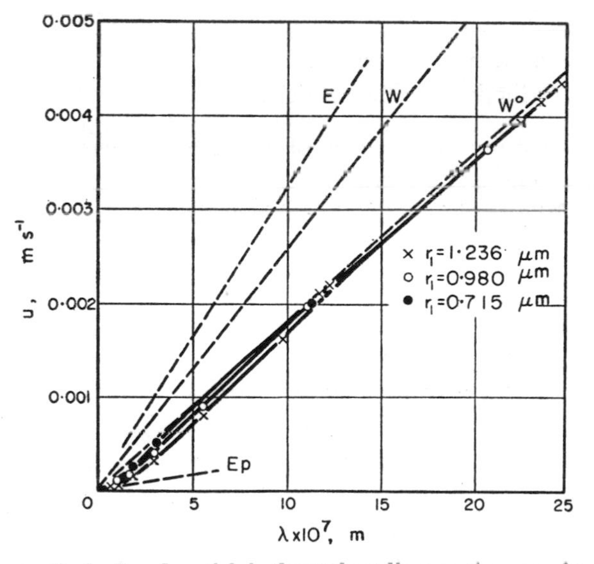

FIG. 11.11. Dependence of velocity of particle in thermal gradient on the mean free path of molecules. The experimental results of Schmitt are compared to the theories of Einstein, Epstein and Waldmann.[732]

×, ○, • } Schmitt's experimental results—particles in argon.

E = Einstein's equation.
E_p = Epstein's equation (eqn. 11.42).
W = Waldmann's equation (a = 0) (eqn. 11.40).
W^0 = Waldmann's equation (a = 1) (eqn. 11.40).

For particles of the size order of the mean free path λ or larger, the *thermal creep* theory of thermal precipitation can be used. This is based on the force set up at the gas–solid interface between a particle and the surrounding gas. When the gas temperature increases along the surface, the molecules leaving the surface will have a greater component of velocity in the direction of the temperature increase than when they arrived at the surface. The net result is a creeping flow of gas from the colder to the warmer regions along the surface of the particle. In turn the surface experiences a force in the cold direction. Epstein[244] assumed the following in his derivation:

(a) The particle was large compared to the mean free path of the gas molecules.
(b) At a great distance from the particle the temperature gradient in the gas is uniform.
(c) Fourrier's equation for heat conduction without convection could be used, although this state did not exist because of the thermal creep.

Epstein then obtained an equation for the thermal force by considering the heat-conduction problem, the equation for thermal creep and the equations of motion of the particle, neglecting the inertia terms, but using the creep velocity as the boundary conditions. The thermal force was then derived as the integral over the surface of the component of stress parallel to the direction of heat flow.

This gave

$$F_t = -\frac{9\pi d\mu^2}{2\varrho T(2+\varkappa_p/\varkappa_g)}\frac{\mathrm{d}T}{\mathrm{d}x} \tag{11.41}$$

where $\varkappa_g$ = thermal conductivity of the gas,
$\varkappa_p$ = thermal conductivity of the particle.

The resistance of the gas to the movement of the particle is given by Stokes' law [equation (4.34)] including the Cunningham correction.

Then the velocity of the particle in the thermal gradient may be found from

$$u_t = -\frac{3C\mu}{2\varrho T(2+\varkappa_p/\varkappa_g)}\cdot\frac{\mathrm{d}T}{\mathrm{d}x}. \tag{11.42}$$

Einstein[133] also deduced an equation for large particles by substituting the product of the circumference of the particle by the mean free path for the area of the small particle. However, it has been shown that this modification gives only one-third of the thermal force predicted by (11.42) (in which $\varkappa_g$ is made much smaller than $\varkappa_p$ to get equivalent equations).[705]

Good agreement was obtained between the particle velocity in a thermal gradient, calculated from Epstein's equation (11.42), and experimental velocities measured by a number of workers for tricresylphosphate,[705] paraffin oil, castor oil,[723] and stearic acid droplets.[727] The width of the dust-free space which was accurately measured by Watson[908] for magnesium oxide smoke around a copper wire has also been calculated with reasonable precision[964] by combining the air velocity due to the convection currents with the thermal force velocity from Epstein's equation.

The particles for which good agreement has been obtained between experiment and theory all had low thermal conductivities (of the order of 10^{-2} W m^{-1} K^{-1}), which is not so much different from that of air. For particles of sodium chloride and particularly iron[727] the agreement was much less, these particles being attracted to the cold surface by a force of 30 and 48 times the value predicted by Epstein's relation.

In view of these apparent discrepancies, Brock[133] reconsidered the problem of thermal precipitation, but with a full set of Maxwell's classical boundary conditions, taking into account the temperature jump and friction slip as well as the thermal creep. The thermophoretic force is then

$$F_t = -\frac{9\pi\mu^2 d}{2\varrho T}\cdot\frac{(\varkappa_g/\varkappa_p+C_t 2\lambda/d)}{(1+3C_m 2\lambda/d)(1+2\varkappa_g/\varkappa_p+4C_t\lambda/d)}\cdot\frac{dT}{dx}. \tag{11.43}$$

The constants C_t and C_m are functions of the thermal and momentum accommodation constants, and can be ascribed values $1{\cdot}875 < C_t < 2{\cdot}48$, and $1{\cdot}00 < C_m < 1{\cdot}27$.[133] The usual values that are used are $C_t \doteqdot 2{\cdot}0$ and $C_m \doteqdot 1{\cdot}25$, although Brock used $C_t = 2{\cdot}5$ and $C_m = 1$.[133] To get the velocity u_t of a very small particle in a thermal gradient the frictional resistance of the gas is found from the Knudsen–Weber equation,[450a] using the Millikan values[572] for the numerical constants:

$$F = 3\pi\mu\, du[1+2{\cdot}5(\lambda/d)+0{\cdot}84(\lambda/d)\exp(-1{\cdot}74\lambda/d)]^{-1}. \tag{11.44}$$

Combining (11.43) and (11.44) gives

$$u_t = -\frac{[1+2{\cdot}5(\lambda/d)+0{\cdot}84(\lambda/d)\exp(-1{\cdot}74\lambda/d)](1+2C_t\lambda\varkappa_p/d\varkappa_g)}{(1+6C_m\lambda/d)[1+2(\varkappa_g/\varkappa_p+4C_t\lambda/d)]}\frac{3\mu}{2\varrho T}\cdot\frac{dT}{dx}. \tag{11.45}$$

Brock showed that these equations not only account for the thermal drift velocity of poor conductors, but also give values for conducting particles to within 25 per cent.[133]

A new approach has been used by Derjaguin and Bakanov,[218] who observed that the gaseous distribution function near a wall gives a thermal slip coefficient which is much lower ($\frac{1}{35}$) then Maxwell's original value ($\frac{1}{5}$), and introduced a heat flux within the gas, which is a function of the local pressure in the gas. The problem was solved by these workers by using a transform based on the Onsager reciprocity principle, and this gave the thermal velocity as

$$u_t = -\frac{\mu}{\varrho T}\left(\frac{4+\varkappa_p/2\varkappa_g}{2+\varkappa_p/\varkappa_g}\right)\cdot\frac{dT}{dx}. \tag{11.46}$$

This is easier to use than Brock's equation, and does not approach a zero velocity for large $\varkappa_p$, which is the difficulty of using Epstein's equation with conducting particles. More recently Derjaguin and Yalamov[220] enlarged the boundary conditions to include the temperature jump. This gives:

$$u_t = -\frac{1}{2}\frac{\mu}{\varrho T}\cdot\frac{[1+8(\varkappa_g/\varkappa_p) - 4C_t\lambda/d]}{[1+2(\varkappa_g/\varkappa_p)+4C_t\lambda/d]}\cdot\frac{dT}{dx}. \tag{11.47}$$

For very small particles ($d \to 0$) this gives

$$u_t = -\frac{1}{2}\frac{\mu}{\varrho T}\frac{dT}{dx}. \tag{11.48}$$

Equation (11.47) gives a thermal force velocity about twice that found from Brock's equation, (11.45), but gave good agreement with experiments by Derjaguin and Rabinovich[219] who used particles of parafin and sodium chloride. On the other hand, Schmitt's measurements[733] with low conductivity oil droplets ($\varkappa_g/\varkappa_p = 1{\cdot}2$), and a Knudsen number of 0·1, using $C_t = 2{\cdot}16$ gives a u_t 1·9 times the Epstein value, while the Derjaguin equations give 2·7 times the Epstein value. Thus, if correct choice is made of the adjustable parameters, quite good estimates of the thermal force velocity can now be obtained for a wide range of particle thermal conductivities.

The practical application of thermal precipitation to gas cleaning plant has only rarely been attempted. Blacktin experimented with the effect of drawing dirty gases through a heated gauze which repelled the particles originally in the gas. When the gauze was heated to 85°C, efficiencies averaging 94 per cent (maximum 98 per cent) on a particle count basis were obtained with gas velocities equivalent to 0·23 $m^3\ m^{-2}\ s^{-1}$ through the gauze surface.[86, 87]

Another patent suggests that the purification of air be carried out by passing it through the narrow channels between the hot and cold fins in a duct, the fins being attached to separate heating and cooling elements. The dust is subjected to a sharp temperature gradient and deposits on the cold fins.[767]

It has also been suggested that thermal precipitation may play a part in the deposition of particles from a hot gas when this is passed through a cold packed bed. The passages in the bed are narrow and so even a temperature difference of 50°C may give rise to a temperature gradient of 10°C mm^{-1} in the passages. Calculations show that this would result in deposition of 98·8 per cent of the particles of 0·1 μm in a 0·2 m deep bed at 500°C.[823]

In such a system the temperature dependence of the thermal force is of considerable importance. This has not yet been investigated experimentally. However, it is possible to calculate this effect by determining the change in the ratio of the terminal velocity and the temperature gradient $u_t/(dT/dx)$ with the temperature, using Epstein's equation (11.42).[833] This shows (Fig. 11.12) that when particles are greater than 1 μm, the thermal precipitation velocity increases with temperature, while for particles smaller than 1 μm, it decreases with increasing temperature, independently of the thermal gradient.

Major variations in pressure may also be present. Thus in a proposed nuclear reactor, gas cooled, and operating at pressures of about 5 MPa, large temperature gradients may occur and result in precipitation of particles on heat exchanger surfaces. The effect on the drift velocity of 2 μm diameter beryllium oxide (BeO) spheres in carbon dioxide over a range of pressures corresponding to gas densities from 2 kg m^{-3} to 50 kg m^{-3} has been calculated, using Epstein's equation[831] (Fig. 11.13). These calculations show that in a thermal precipitator which is 85 per cent effective at atmospheric pressure, this is reduced to 10 per cent at 5 M Pa.

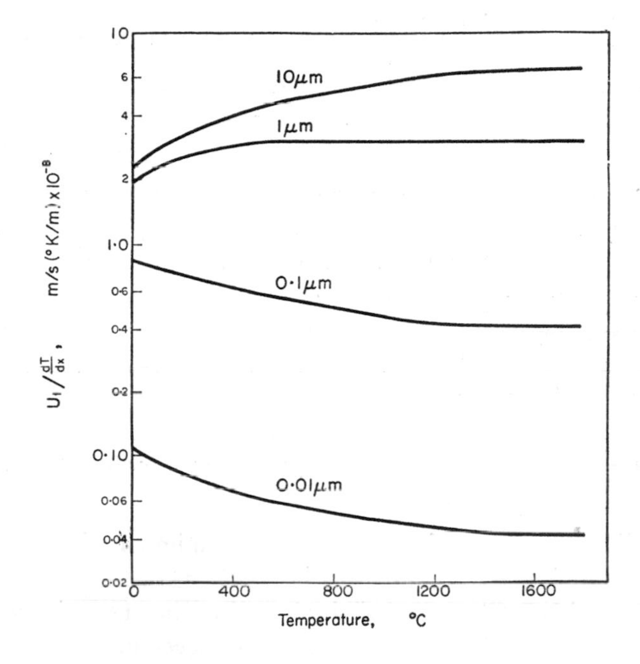

FIG. 11.12. Temperature dependence of thermal forces (calculated).[833] (0·01 to 10 μm particles.)

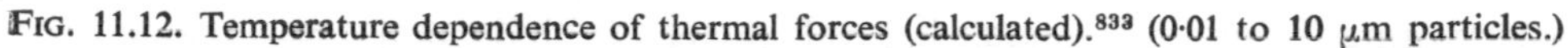

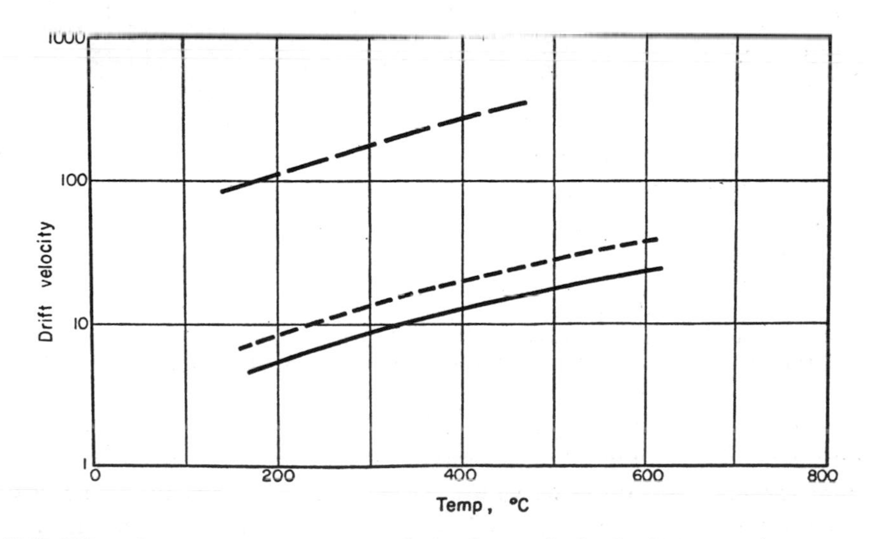

FIG. 11.13. Effect of temperature on the drift velocity (in m $s^{-1} \times 10^{-6}$) of 1·0 μm radius BeO particles in CO_2 at different gas densities in a thermal gradient of 40°C mm^{-1}. ——— Gas density 2 kg m^{-3}. - - - - Gas density 30 kg m^{-3}. ———Gas density 50 kg m^{-3}.

11.6. COLLECTION BY BEDS OF STATIONARY SOLIDS

It has been shown by Strauss and Thring[831] on a 0·3 m diameter pilot plant that a bed of coarse granular material, approximately 6 mm diameter, such as crushed insulating brick, is about 90 per cent effective in collecting open-hearth furnace fume to 455°C, with pressure drops below 1 kPa. Some of the experimental results are given in Table 11.4. In a later

TABLE 11.4. EFFICIENCIES OBTAINED ON A PILOT PLANT GRANULAR BED FILTER (Strauss and Thring[831])

Bed thickness (mm)	25	75	230	270
Typical efficiencies, %	59·3–88·2	65·2–85·3	74–902	87–96
Temperatures, °C	360–455	255–460	303–352	230–290
Pressure drop (kPa)	0·13–0·21	0·18–0·59	0·55–1·2	0·55–1·0

paper[834] it was shown that thermal precipitation plays an important part in fume collection from a hot gas by a cooler bed.

A commercial gravel bed filter, the Lurgi M–B filter, has been found an effective dust collector to temperatures of 350°C.[240] The gravel bed is placed in a basket located on springs. The gases enter from below, and at intervals the bed is vibrated and the collected dust is shaken into the hopper below (Fig. 11.14).

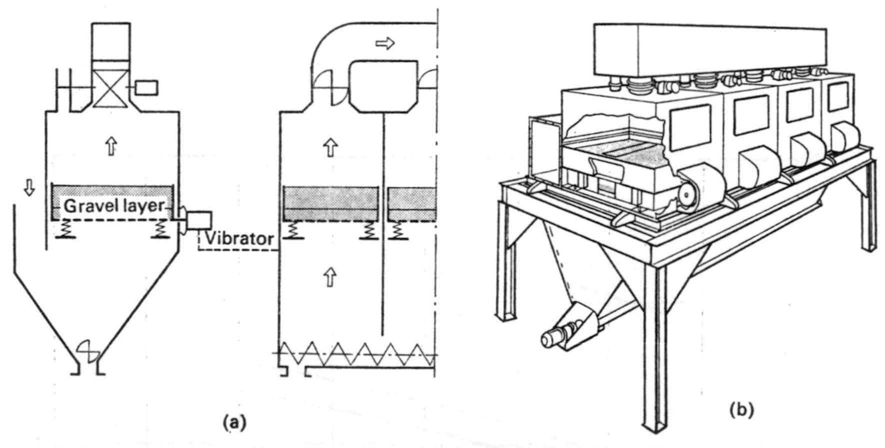

FIG. 11.14. Commercial gravel bed filter (Lurgi M–B filter).[240] (a) Shows the gravel layer on a spring supported tray through which the hot dirty gases are passed. (b) Shows a four section filter with parallel passing of gases.

Typical results of the unit are reported as follows:[307]

(a) A unit for coke dust operated with a 95 per cent efficiency, producing a dust exit concentration range of 0·035 to 0·025 g m^{-3}. The pressure drop in this case rose to 2 kPa, because the effective filtration was only obtained when some of the dust had been deposited as a filter medium.

(b) A unit handling 4000 $m^3 h^{-1}$ was tested with phosphate dust and a 1–3 g m^{-3} concentration was reduced to 0·01 to 0·02 g m^{-3}. The pressure drop was 1·1 to 1·3 kPa.
(c) A filter was used on a carbide furnace producing 70,000 $m^3 h^{-1}$ with a dust concentration of 0·5 to 1·3 g m^{-3}, which was reduced to 0·120 g m^{-3}.

It appears that the pebble or gravel bed filter finds particular application on large-scale operation in inorganic chemicals manufacture.

11.7. COLLECTION BY BEDS OF FLUIDIZED SOLIDS

The use of fluidized beds, where solid particles are held in turbulent suspension in a gas stream, would appear an obvious application for particle and mist collection, because the turbulent movement of the particles in the bed tends to favour agglomeration and impaction of the droplets with the particles in the bed. However, only a limited amount of experimental work has been carried out.[568, 745] This has indicated that efficiencies over 90 per cent would not be obtained and so the equipment would be inadequate for commercial plant.

The experimental studies covered the collection of sulphuric acid mist droplets, 2–14 μm diameter, with fluid beds of glass beads, silica (both non-porous), alumina and silica gel (both porous); dioctyl phthalate droplets, essentially 0·6–1·1 μm diameter with fluid beds of alumina; and ammonium nitrate dust, 0·25–2·5 μm diameter, with fluids beds of glass beads.

It was found in all cases that the collection efficiency was independent of the inlet concentration. As would be expected, for the larger droplets (2 to 14 μm) inertial impaction appeared to be the predominant collection mechanism, efficiency improving with increasing superficial gas velocity. For the smaller dioctyl phthalate droplets, diffusion collection predominated, as was shown by the opposite trend of a decreasing efficiency with increasing superficial gas velocity. The work with ammonium nitrate was very limited and no definite trends were discerned.

For particles in the micron ranges (averaging 8 μm) Meissner and Mickley[568] found that an empirical equation of the form

$$-\ln c_2/c_1 = 0{\cdot}455 u_s \left(\frac{W}{4{\cdot}88}\right)^n \qquad (11.49a)$$

where c_1 = inlet concentration (kg m^{-3})
c_2 = outlet concentration (kg m^{-3}),
u_s = superficial gas velocity (m s^{-1}),
W = weight of bed per unit cross-sectional area (kg m^{-2}),
n = empirical constant (0·16–0·34)

could be fitted to their limited experimental data. Here the exponent n was specific for the bed particle mixture.

For sub-micron sized particles, the limited work indicated that an equation of the form

$$-\ln c_2/c_1 \propto u_s^{0{\cdot}78} \qquad (11.49b)$$

could be fitted to the experiments[745] but no bed weight effect was noticed, and other empirical constants were not suggested, although collection efficiencies close to 90 per cent were found in some runs.

In a later investigation, Black[85] used a fluidized bed to collect sub-micron-sized ammonium chloride and tobacco particles. As expected, the highest efficiencies were found at low gas flows and large bed heights. The efficiency of collection was correlated by

$$\eta = 56{\cdot}5 \frac{H^{0{\cdot}4}}{u_s^{0{\cdot}1}} \quad \text{per cent} \tag{11.50}$$

where H = bed height, and
u_s = superficial velocity.

Black found no effect of either bed age or changes in aerosol concentration. In an analysis of his results, Black concludes that interception and Brownian diffusion are the important mechanisms under consideration, and equation (11.50) can be deduced from (7.51).

Fluidized beds of alumina, 50–300 mm deep, are being used for the removal of gaseous and particulate fluorides from the emissions from primary aluminium smelters. In the process developed by ALCOA (Process A-398),[707] the gaseous and particulate emissions are

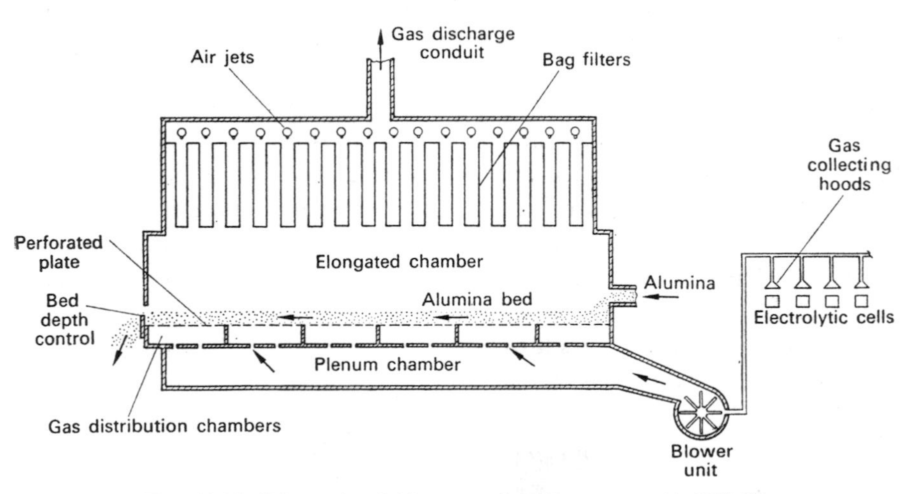

FIG. 11.15. Schematic of Alcoa gas-cleaning process (A-398).[707]

conducted through the fluidized bed composed of finely divided alumina particles, maintained at 65–160°C, and then the gases are passed through a bag-filter system situated directly above the fluidized bed (Fig. 11.15). The filter bag is cleaned intermittently by reversing the gas flow, when the filter cake falls back into the bed. Alumina of a grade normally used in the smelter is fed continuously to the bed, and finally discharged and used in the cells after a residence time between 2 to 14 hr. The ratio of alumina to treated gas ranges between 30 and 150 to 1, and the pressure drop is 0·75–1·5 kPa. The effectiveness of gas absorption is better than 99 per cent, while the removal of particulates is better than 90 per cent (Table 11.5). The plants using this process do not show a visible plume, and emissions are as low

TABLE 11.5. PERFORMANCE OF THE ALCOA A-398 PROCESS[707]

Plant	Total fluoride (mg m^{-3})		Fluoride-removal efficiency (average) (%)		
	Inlet	Outlet	Total	Gaseous	Particulate
1	195	1·6	99·2	99·5	98·3
2	125	1·9	98·5	99·4	96·3
3	100	3·0	96·7	99·2	91·8

as 1·35 g kg^{-1} of aluminium produced. Furthermore, this process has the advantage that it does not produce a polluted liquid waste stream or a solid disposal problem. The process economics are very favourable because the spent alumina from the collector is feed material for the pots, while the adsorbed fluoride maintains the fluoride level in the molten bath, reducing the need for cryolite make-up.

11.8. SEPARATION OF PARTICLES IN A MAGNETIC FIELD

If a particle with no intrinsic magnetic properties is charged to a charge q by bombardment charging [equation (10.30)] or ion diffusion charging (10.36), and is then introduced into the field of a magnet (field strength, **H** oersted), it will be acted on by a force at right angles to both the direction of the field and its direction of motion, and so will be diverted from its original path. The equation for the particle in a vacuum is

$$\mathbf{H}qeu = mu^2/R$$

where m = mass of the particle,
e = electronic charge,
u = particle velocity,
R = radius of circular path.

Consider the case when the particle, moving with the gas at a velocity u_g is given a velocity component at right angles to the gas stream equal to the terminal drift velocity, while passing through the field of a narrow magnet. The terminal drift velocity may be obtained by equating the magnetic force to the resistance of the gas, calculated from the Stokes–Cunningham law:

$$u_t = \frac{C\mathbf{H}qeu_g}{3\pi\mu d}. \quad (11.51)$$

Here u_t is a function of u_g and higher gas velocities would favour higher particle velocities out of the gas stream. To the author's knowledge this phenomenon has not yet been utilized for gas-cleaning plant.

If small magnetic particles were introduced into a magnetic field a different case would exist. As the particles are free to rotate, it may be assumed that they will align themselves in the magnetic field, their opposite ends being attracted to the two poles of the magnet. The net force on the particle in any particular position can be calculated by the algebraic addition of the attractive and repulsive forces. If the particle is midstream between the two poles, the forces will cancel out, and the particle will move straight through. Detailed calculation of the path of a particle and whether or not it will be collected would require a knowledge of the magnetic field distribution, the geometric configuration of the magnet and of the gas-flow pattern.

CHAPTER 12

THE ECONOMICS OF INDUSTRIAL GAS CLEANING*

12.1. INTRODUCTION

In the introductory chapter, the problem of economics of gas cleaning was briefly mentioned, but it should be realized that the extent to which gas cleaning is carried out, and the type of plant selected, is a question of plant economics, process economics, and community economics.

If we consider the simplest case, in which the "waste" gases produced by a process do not cause harm, and are therefore not "pollutants", but contain a valuable ingredient, then the optimum recovery is that which produces the amount of ingredient with the effective minimum cost. However, if the ingredient is harmful even in small quantities, then recovery must be such that the maximum concentration allowable in the exit gases is not exceeded. Even if the ingredient does not do harm, an arbitrary maximum may be set by an outside organization, such as a government body.

At this point in time, these arbitrary levels are set because of lack of adequate knowledge of the factors making up what can be called the "net social cost" (N.S.C.) equation. This, in simplest terms, is:

$$\text{N.S.C.} = \text{Damage cost if no controls} - \begin{pmatrix}\text{Reduction in damage} & - & \text{Cost of control} \\ \text{with control equipment} & & \text{equipment}\end{pmatrix}$$

The reader will realize the tremendous difficulties involved in assessing the damage caused by pollution; while health costs may be estimated in some ways, and damage to property by corrosion and soiling may also have a monetary value, loss of visibility or aesthetic losses are impossible to assess financially. In the same way, the reduced damage which occurs with the installation of controls is not generally assessable. The last term of the N.S.C. equation, the cost of control (which includes any profit from by-product sales) is, however, assessable, and is of greatest interest to those installing gas cleaning or other pollution control equipment.

In fact, it is important to keep this cost at a minimum commensurate with adequate plant

* All costs in this chapter have been converted into $U.S., and are at mid-1970 levels unless otherwise stated.

performance in preventing health, biological or property damage, as well as keeping within terms of any legislative restrictions. The discussion of N.S.C. and Damage Costs is beyond the scope of the present book, but the reader involved in actual plant design and in decisions concerning competing types of plant and equipment will be concerned with the basic methods of plant costing and of equipment as a function of cost. The integration of the recovery of material products or of heat into the plant cost structure will also be considered.

12.2. THE COST OF GAS CLEANING

The cost of cleaning gases is a complex function of capital cost (including interest rates), depreciation, operation and maintenance. In general, equipment which is more complex costs more and is more difficult and costly to operate and maintain. With the exception of some types of small "unit cleaners" installed next to some equipment such as grinders, most industrial gas-cleaning plant such as scrubbers, bag filters or electrostatic precipitators are very large. They are therefore individually engineered for each application, and a considerable part of the total cost is in "on-site" installation.

The unit, in order to work most effectively, must be engineered as part of a total cleaning system, and, if possible, as an integral part of the total process. A typical example of some of the major parts of such a system are illustrated in Fig. 12.1. It should be noted that the

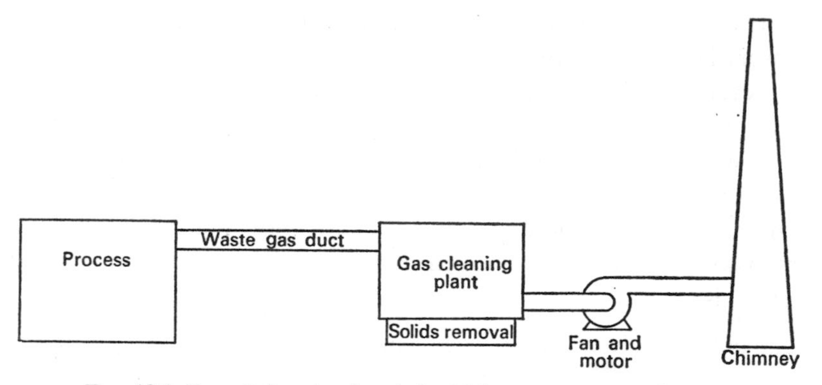

FIG. 12.1. Essential parts of an industrial waste gas control system.

position of the fan relative to the gas-cleaning unit depends on the type of unit. Preferably, if the type of unit allows this, the fan should be on the "clean-gas" side, minimizing erosion and corrosion. The next sections will discuss the capital cost of the different types of plant.

12.2.1. Capital cost of particulate collection

The common types of particulate collection plant are settling chambers, cyclones, bag filters, scrubbers and electrostatic precipitators. The many different types of inertial collectors have costs similar to those of more expensive cyclones. In all cases, the plant essentially consists of sections of metal plate, assembled, welded, and held in a framework. Important additions to this are the cost of filter elements ("bags") for fabric collectors. Electrical equipment, bus bars, insulators and "rappers" have to be added for electrostatic precipitators.

One method that is used for approximate costing is to estimate the total weight of metal involved, and multiply this by a cost per unit weight. Simple, small units are frequently built of 0·9-mm steel, while larger plants are of 1·6-mm or heavier sections. Approximations can be made by using a cost of $200/tonnes for mild steel (in 1970). Structural work is some $25–$45 less per tonne. Stainless-steel work may cost up to $1000/tonne. Dalla Valla[194] has produced graphs showing the approximate weights of different types of plants, correlated with volume throughput. These are shown in Fig. 12.2. The costs obtained using these

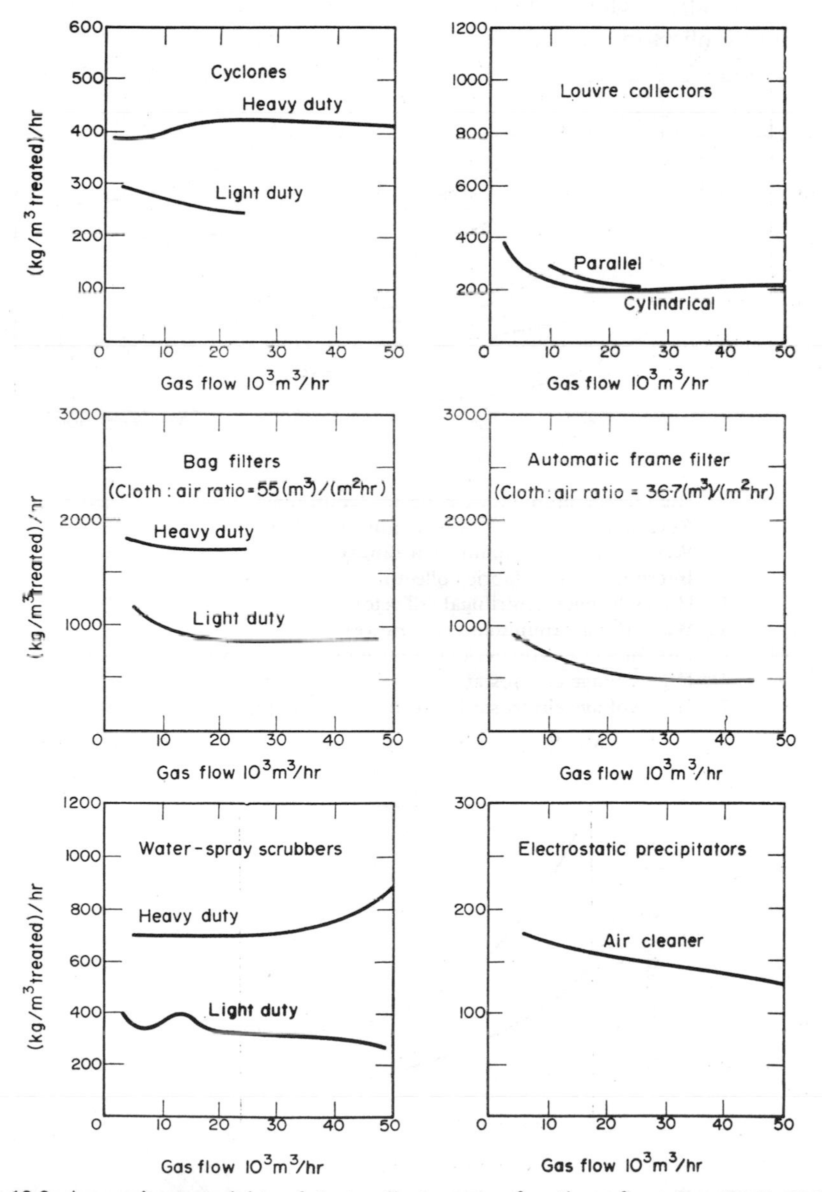

FIG. 12.2. Approximate weights of dust collectors as a function of gas flow (Dalla Valla[194]).

weights are average costs at the point of plant manufacture (F.O.B.). They do not include fans, motors, supporting frame, duct work, or other parts. The costs for the fabric filters (bag and automatic frame) include the shaking gear and actual filter "bags".

Alternative methods that can be used for approximate estimating, which also have a built-in "time factor", are to use the cost of steel involved, assuming that this is one-fifth of the erected cost. If more exotic materials, such as special alloys, stainless steels or non-ferrous metals, are used, then the relative cost of manufacture is less.

A more rigorous approach is to combine the cost of the metal involved with the labour charges. In simple collectors, such as cyclones, it is reasonable to assume that one skilled

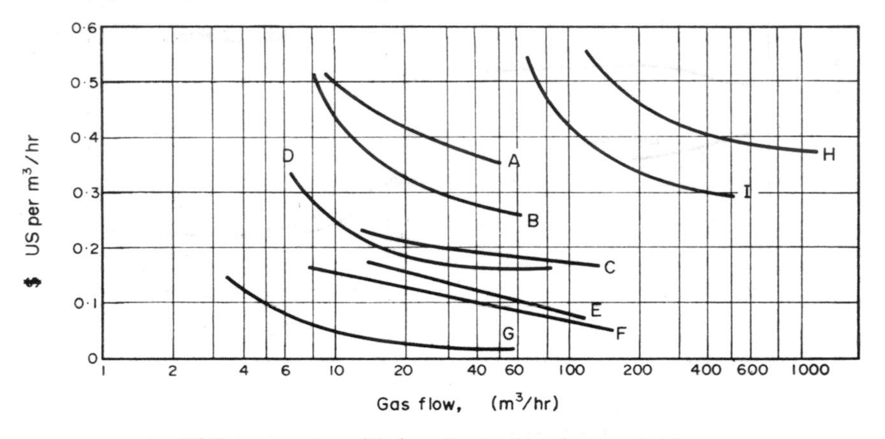

A. High temperature fabric collector (continuous duty)
B. Reverse jet fabric collector (continuous duty)
C. Wet collector (maximum cost range)
D. Intermittent duty fabric collector
E. High efficiency centrifugal collector
F. Wet collector (minimum cost range)
G. Low pressure drop cyclone (maximum cost range)
H. High voltage electrostatic precipitators (for fly ash)
I. High voltage electrostatic precipitators (minimum cost range)

FIG. 12.3. Cost of dry mechanical collectors for dusts \$ per $m^3 h^{-1}$ against airflow. (From Munson.[588])

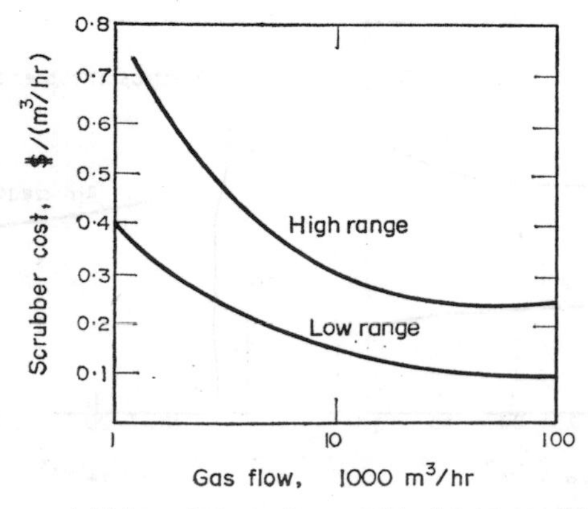

FIG. 12.4. Cost of wet scrubbing of dusts \$ per $m^3 h^{-1}$ against airflow. (From Imperato.[382])

worker can handle about 2·25 m² of steel (1·6 mm) per 8-hour day. However, if this method of estimating is to be used, the manufacturer should be contacted, as the amount of automation in his factory can affect the work output and manpower charges.

Direct costs of dry mechanical collectors can be estimated directly from the data collected by Munson (Fig. 12.3).[588] It will be noted how these costs decrease with gas-flow rate.

These costs do not include fans, motors, drives, pumps, dust-conveying equipment, insulation, weather protection or special materials of construction. A range of costs for wet scrubbers, supplementing the values that can be obtained from Fig. 12.3, is shown in Fig. 12.4. As these costs are not independent of time, they should be modified by using the Chemical Engineering Plant Cost Index, which was 113 > in 1968, and was rising approximately 5 points each year in 1969 and 1970, reaching 126 by July 1970.

The well-known 6/10th factor, which was suggested by Williams[946] for scale up calculations, tends to underestimate the cost of gas cleaning systems. This rule is given by

$$\text{Cost of } a = \text{cost of } b \times \left(\frac{\text{capacity of } a}{\text{capacity of } b}\right)^{0\cdot6} \tag{12.1}$$

Peters and Timmerhaus[636] find that while the 6/10th rule works for carbon steel heat exchangers, in the case of centrifugal blowers the exponent should be 0·96, and for centrifugal compressors as high as 1·22. The use of materials other than mild steel also tends to change the exponent to values greater than one.

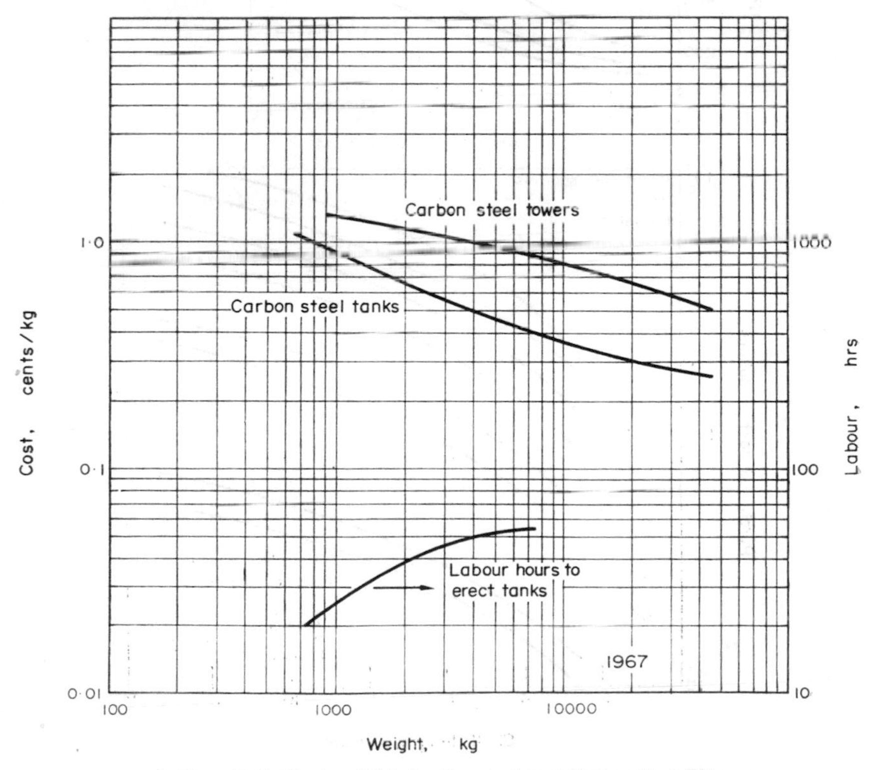

FIG. 12.5. Costs of fabrication and installation time.[636]

12.2.2. Capital cost of scrubbing, absorption and incineration

The central element in a gas scrubber is an absorption column, which may be a simple spray tower, or a column packed with selected packings such as Raschig rings, Berl saddles, Pall rings, etc. Some towers may incorporate trays of different patterns (grid trays, bubble cap trays, etc.).

The cost of the basic tower and of ancillary tanks can be estimated from the weight of metal in the tower or tank (Peters and Timmerhaus[636]), using Fig. 12.5. The cost of ladders, platforms and handrails must be added to this, as is indicated in Table 12.1. The plates or

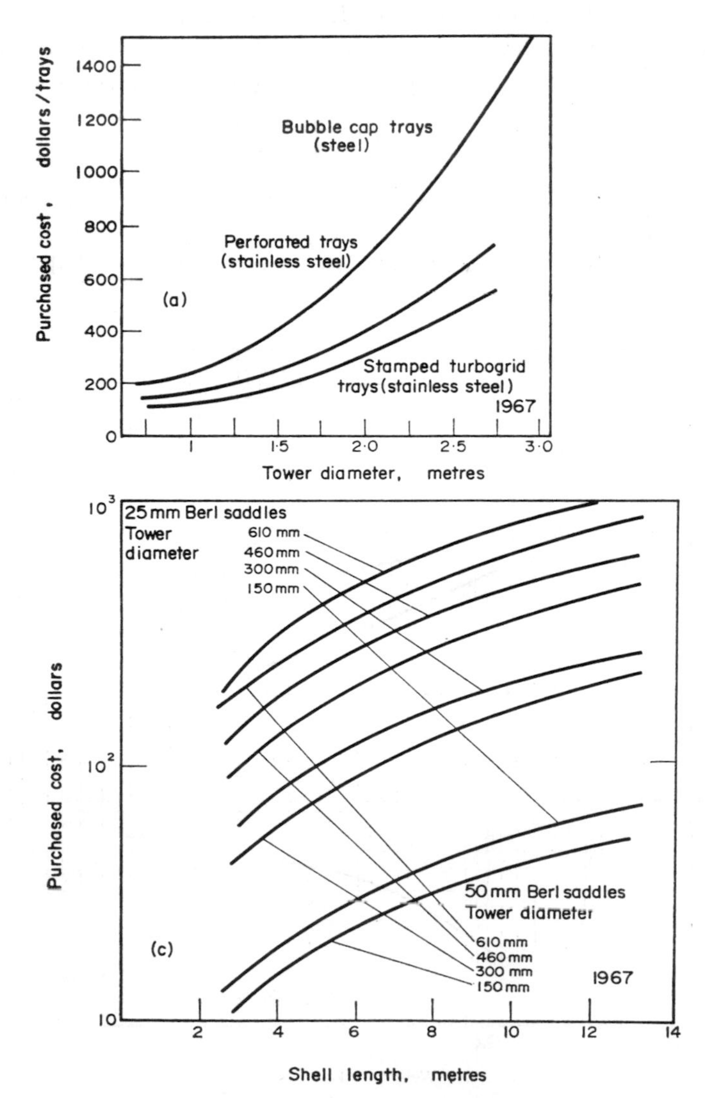

FIG. 12.6. Cost of trays and packings for absorption towers.[636]

packings have also to be added when these are used, and their cost can be estimated from Fig. 12.6.

Total purchased cost of towers, per meter height, for a variety of materials can be estimated from Fig. 12.7.

The cost of adsorption systems, using a solid adsorbent, can be based on the cost of the empty tower (Fig. 12.5), to which the packing is added. Some approximate costs of common packings are shown in Table 12.2.

The most widely used of these adsorbents is activated carbon (see Section 3.8). This comes from many sources, both wood and coal, and is available in many grades. Some of the best

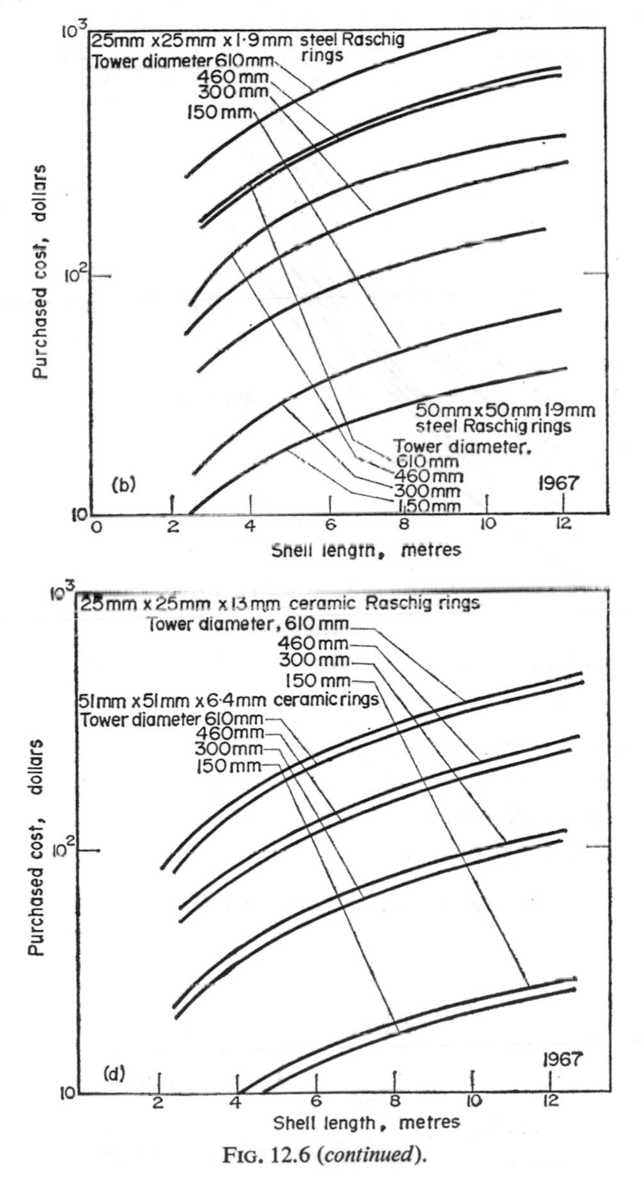

FIG. 12.6 (*continued*).

TABLE 12.1. COST OF TOWER AUXILIARIES
(after Peters and Timmerhaus[636])

Item	Cost, January 1967	Amount for typical tower	
Ladder	\$0·44/kg	42 kg/m height	
Platforms and handrails	\$0·44/kg	Tower diameter: m	Weight: kg
		1·2	770
		1·8	1050
		2·4	1280
		3·0	1550

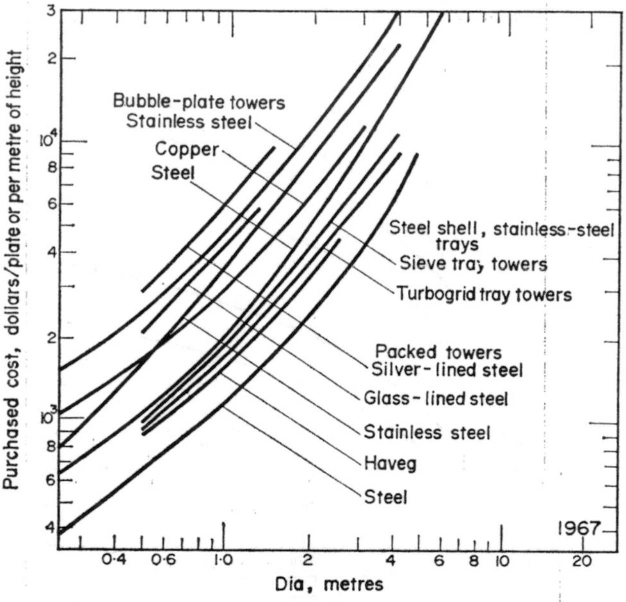

FIG. 12.7. Cost of towers including installation and auxiliaries.[636]

grades, with exceptionally high surface areas, are made from coconut shell. Prices range from 75c/kg to \$1·65/kg for the different grades of carbon, and this can constitute a considerable fraction of the cost of the unit. If the quantities justify this, and it is technically possible, a carbon reactivation plant on site should be considered, because replacing of packing is a major operating expense.

TABLE 12.2. APPROXIMATE COST OF ADSORBENT PACKINGS[636]

	\$U.S. m^{-3}
Activated carbon	430
Alumina	390
Coke	53·5
Crushed limestone	105
Silica gel	890

Fume incineration systems differ in cost depending on the complexity of the system, and as to whether it uses a catalyst and/or heat exchangers to reduce fuel consumption. Typical costs for a number of different aplications have been published, and they are summarized in Table 12.3.

TABLE 12.3. CAPITAL COST OF GAS INCINERATION PLANT, $U.S./m^3 h^{-1}
Units are of different size and for different applications

Type of fume and volume	Thermal incineration			Catalytic incineration		
	Basic unit	with heat exchanger	with heat exchanger and recuperator	Basic unit	with heat exchanger	with heat exchanger and gas return
Gases from metal decorating oven						
210°C (8500 m^3 h^{-1})†	1·50			1·75		
150°C (7700 m^3 h^{-1})†		2·84			3·40	
Dryer exhaust gases (10, 000 m^3 h^{-1})* 180°C	1·56	2·23	—	2·15	3·20	3·00
Solvent exhaust (17,000 m^3 h^{-1})†						
180°C	1·16	1·39	1·93	1·33	1·65	—
290°C	1·13	1·32	1·90	1·33	1·95	—
Phthalic anhydride waste gases (40,000 m^3 h^{-1})† 67°C	—		1·85	—	1·97	—

* From Vollheim and Domin.[890] † From Brewer.[117] ‡ From Cock and Strauss.[828]

This table can be used to give estimates of capital cost of gas incineration plant. However, in a total assessment of capital and operating costs (Section 12.3) the units using heat exchangers, although less attractive in initial cost, have far lower heat requirements, and their total charges are generally well below those of the basic units.

12.2.3. Cost of ancillary plant

The major ancillary items that have to be added in to the capital cost of gas-cleaning plants are for ductwork, foundations and structures, fans and blowers, controls, chimneys, and the equipment needed to remove collected material. Another important allowance which should be included is for the engineering, installation contracts and contingencies. Duct work can be approximately estimated, as has been suggested earlier, either on a weight or an area basis. Much more reliable cost estimates can be obtained by listing all the sections (straight runs, bends, transitions, joints, etc.) that are to be used. Then, from the correlations presented by Alonso,[15] it is possible to find the man-hours that will be needed for the fabrication, as well as the area of metal. Installation of ductwork generally adds approximately 80 per cent to the fabrication cost.

Structural supports and foundations costs can also be obtained from a weight based estimate ($ 155–175/tonne) allowing 140 per cent of the structure cost for erection and foundations.

Simple chimneys, which are merely a single-walled or double-walled steel tube, can be estimated in the same way as absorption columns. Comparatively small chimneys are not difficult to erect, and some suppliers now provide standard sections. Larger chimneys are commonly made of reinforced concrete with a resistant liner, usually of masonry, which will withstand the inner temperatures and corrosive conditions.

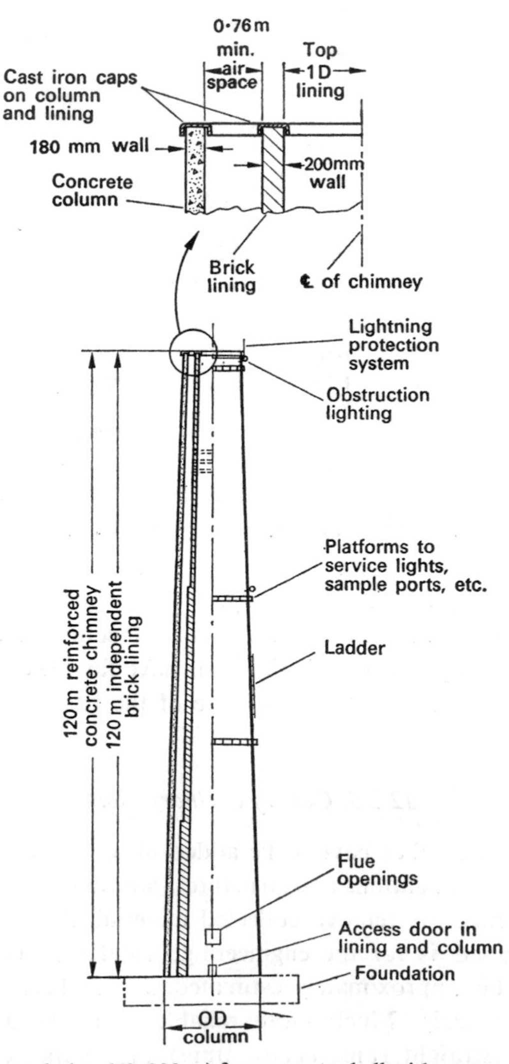

FIG. 12.8. Typical chimney design (60–200 m) for concrete shell with masonry or steel heat resistant and corrosion proof liners (free standing) (Carlton-Jones and Schneider).[148]

A typical design is shown in Fig. 12.8, and approximate costs can be obtained from Fig. 12.9 (Carlton-Jones and Schneider)[148] (1968 values). The costs obtained from this graph do not include foundations, which add 10 per cent in normal areas, and 15 per cent if earthquake resistance is necessary. Chimneys taller than 200 m have been built in recent years for large power stations and non-ferrous smelters. A more economical design than

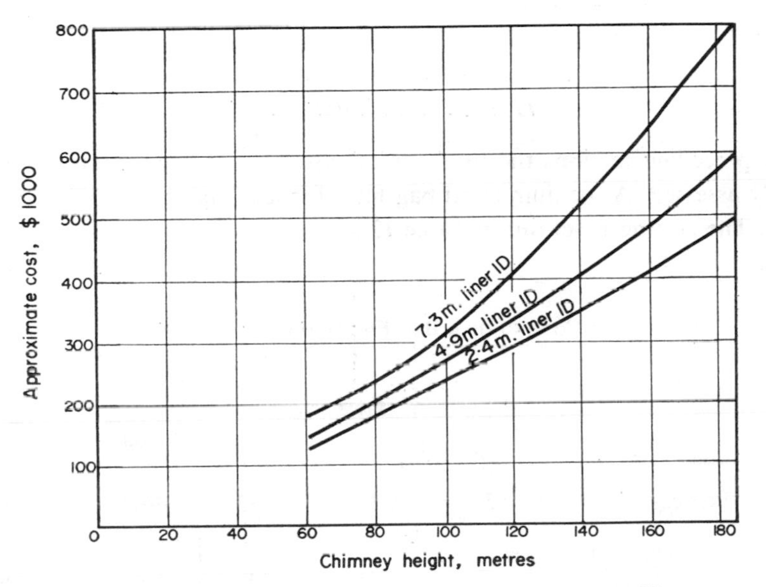

FIG. 12.9. Cost of tall chimneys with resistant liners.[148]

the reinforced-concrete shell has been to use a steel framework in which one or more flues are suspended.

Engineering and contracting charges tend to vary, depending on the nature and location of the plant. Peters and Timmerhaus[636] give a range of 4–21 per cent of the fixed capital for engineering charges (Table 12.4), but this may be even higher if the proposed plant is at a

TABLE 12.4. TYPICAL VARIATION IN CHARGES FOR ENGINEERING AND SERVICES, AS A PER CENT OF FIXED CAPITAL (Peters and Timmerhaus)[636]

Component	Range, %	Median, %
Engineering	1·5–6	2·3
Drafting	2·0–12·0	5·0
Purchasing	0·2–0·5	0·3
Accounting, cost engineering	0·2–1·0	0·3
Reproduction, communication	0·2–0·5	0·3
Travel and living	0·1–1·0	0·3
Total engineering and supervision	4·0–21·0	13·0

remote site. Similarly the contractor's fee of 2–7 per cent may be as high as 10 per cent for remote locations. Contingencies are estimated generally at 8–20 per cent, with 10 per cent as a fair value. These add up to 35 per cent for an average installation, but may be as high as 50 per cent for remote locations. A further establishment charge are "start-up" expenses. These are generally low for simple gas-cleaning plant, but should include the independent acceptance testing of the unit.

12.2.4. *Total capital charges*

From the preceding sections the total capital cost of simple gas-cleaning plants can be fairly readily assessed. A medium sized bag filter for lead oxide dust will be here used as an example. The costing is set out in Table 12.5.

TABLE 12.5. COSTING OF A BAG FILTER HANDLING 17,000 $m^3 h^{-1}$

	Price	Installation		Total
		%	Amount	
Basic filter cost (from Fig. 12.3)	$7500	80	6000	13,500
Support (estimated $4\frac{1}{2}$ tonnes mild steel) at $160/tonne	720	140	1000	1720
Fan (Class 3) (approx. $U.S. 1970)	750	100	750	1500
Electric motor (approx. $U.S. 1970)	200	100	200	400
Starter (approx. $U.S. 1970)	95	100	95	190
Collector and duct	2800	80	2240	4040
				21,350
Engineering and Supervision (13%)				2280
Contracting (6%)				1275
Contingencies (10%)				2135
				27,040
Profit (approx) 25%				6760
				$33,800

The simple example given above has not involved the use of fixed charges, working capital, and other items, as it is assumed that such small plants are installed either as an integral part of a larger system, in which these are included, or as an additional plant being built as a "maintenance" item. Larger and much more costly systems, such as those being suggested for sulphur dioxide recovery from flue gases (Section 3.7.1) require a more complex costing basis.

Katell[425] and Strauss[825] have worked out detailed costing for a number of these processes, and one of these—the Alkalized Alumina Process—will be given as an example (Table 12.6). It should be noted that estimating the individual items in the equipment section involved a long series of engineering estimates.

TABLE 12.6. ESTIMATED CAPITAL AND WORKING CAPITAL FOR THE ALKALIZED ALUMINA PROCESS FOR SO_2 REMOVAL (Katell)[425]

Item	Cost $	%	Estimated operator needs
A. CAPITAL (INVESTMENT)			
Absorber-regenerator	4,171,500	49·0	1
Gas producer (for reactivation)	645,000	7·6	–
Sulphur recovery	1,780,000	20·9	1
Plant facilities	329,800	3·9	$\frac{1}{2}$
Plant utilities	484,800	5·7	$\frac{1}{2}$
Total construction cost	7,411,100	87·1	3
Initial catalyst requirement	341,600	4·0	
Total plant cost insurance and tax basis)	7,752,700	91·1	
Interest during construction	193,800	2·3	
Subtotal for depreciation	7,946,500	93·4	
Working capital (see B)	563,500	6·6	
Total investment	8,510,000	100	
B. WORKING CAPITAL	$	%	
Coal supply for producer, 60 days ($4/tonne)	86,400	15·3	
Absorbent makeup, 60 days ($0·55 kg)	164,000	29·1	
Direct labour, 3 months (direct + maintenance)	62,200	11·0	
18·5% payroll overhead, 3 months	11,500	2·0	
Operating supplies, 3 months	11,700	2·1	
Indirect cost, 4 months	60,800	10·8	
Fixed cost, 0·5% of insurance base	38,800	6·9	
Spare parts	60,000	10·7	
Miscellaneous expense	68,100	12·1	
	563,500	100	

12.2.5. *Operating and maintenance charges*

The most common operating charge is for electric power for fans, blowers, screw conveyors, motorized valves and other equipment. In most units the horsepower requirement for fans and blowers far exceeds that of other ancillaries, with the exception of the electrical power for electrostatic precipitators.

In converting the motor horsepower into electrical units, the motor efficiency can be assumed to vary between about 85 per cent for a 45-kW motor to 93 per cent for 400 kW. The cost of electricity varies widely and tends to depend on the size of the consumer. The cost of electric power in industry varies between about 0·8 to 3 cents per kWh for purchased electric power or 75 per cent of this if the electricity is generated "on site".

If cooling water is used to pre-cool gases either directly, or with a surface condenser, then either the amount of water added, or the amount which is evaporated and has to be replaced, can become a significant item. If river, sea or well water is available, this may cost as little as 0·72 cents/m^3, but city mains water, if needed, usually costs between 11 and 15 cents/m^3.

For incineration with natural gas, a price of 1·2–4·3 cents/m^3 is an appropriate range, with the lower prices applicable in the Gulf States and Oklahoma, and the higher prices for north-eastern United States. Other common fuel costs are as follows:

> (1968 prices) kerosene 3·74–4·35 cents/l.; distillate oil 3·74–4·15 cents/l.; liquified petroleum gas (LPG) 3·4 cents/l. (New York), \$12·5/m^3 (1·6 cents/l.) (Baton Rouge); residual oil \$13·4/m^3 (Gulf region) to \$17·9/m^3 (New York).

If chemicals are used as adsorbent or reactant, then their cost tends to be an appreciable part of operating expenses.

Maintenance costs tend to rise with increasing age of plant. It has been found[24] that maintenance charges average 1 per cent of the product

$$\text{Investment} \times \text{Age of plant (years)}.$$

There may be a "base line" charge in addition to this. For gas-cleaning equipment this "base line" charge can, in general, be neglected. An exception would be fabric filters, where regular filter element replacement is essential, and catalytic incinerators where the catalyst is replaced regularly. The cost of filter elements has been given in Table 8.4.

Another guideline to maintenance costs can be gleaned from the suggestions made by Leonard in 1951, which have been increased by 50 per cent to bring them into line with 1970 estimates. These are:

> Electrostatic precipitators: (17,000–25,000 $m^3\ h^{-1}$): \$7500–12,000
> Packed towers: \$90 p.a/1000 $m^3\ h^{-1}$
> Fabric filters: \$180–220 p.a./1000 $m^3\ h^{-1}$

However, in the present author's opinion, these values are much higher than those predicted by other methods.

Supervisory labour costs associated with most forms of gas-cleaning plant are very low, as these plants tend to operate without attendance, as part of the process. Intermittent inspection is generally adequate, even with large electrostatic precipitators.

An exception to this is specialized plant of the type used for sulphur-dioxide removal from flue gases and similar applications. An example of this will be given in the section on total cost estimation. In such cases specific allowances have to be made for labour and supervision, in addition to the maintenance costs, which are also largely a labour cost.

12.2.6. Fixed charges

Fixed charges associated with plant of this type are interest on capital, depreciation, taxes and insurance. The interest is charged on the capital, less any accrued depreciation allowance (which has been reinvested). This interest charge tends to vary between 7 to 14 per cent, depending on whether the plant is on a public utility or in a process plant, and on the size of the organization, and the type of industry.

Depreciation can either be a direct fraction of the cost of the plant, divided by the estimated life, or a fixed proportion of the depreciated value of the plant, which will amortize it in a selected time. The estimated life of the plant may depend either on the process, which could become obsolete in 5 to 7 years, or on the actual life of the plant which may be twice this, depending on the product. This could be very different from the depreciation permitted by government taxing authorities. Some typical service lives of selected types of operation are given in Table 12.7.

TABLE 12.7. SERVICE LIVES OF PROCESS INDUSTRY PLANT[636]

Type of process	Service life (years)
Pharmaceutical or chemical plant becoming obsolescent	5–7
Chemicals, plastics	11
Paper	12
Non-ferrous metallurgy, rubber	14
Clay products	15
Petroleum refining	16
Ferrous metallurgy	18
Nuclear power generation	20
Steam thermal power generation	28

Other "fixed charges" such as local taxes and insurance are difficult to estimate, but approximately 1·5 per cent of the capital investment should cover these items. Katell (Table 12.6)[425] used only 0·5 per cent for this in the case of a major public utilities plant, but the higher value of 1·5 per cent is more realistic for the more usual types of plant.

12.3. TOTAL OPERATING COSTS

The manufacturer or works engineer in deciding which type of equipment to install must ask the two questions about the item; first, what is the total capital cost (dealt with in Section 12.2.4) showing the money required before and during construction, and second, what is the total operating cost, including appropriate operation, depreciation and maintenance charges. When there is a choice between several different types of equipment with equivalent performance, the total operating costs of the alternatives, placed on a realistic common basis must guide the decision.

In the example used earlier, of the bag filter handling 17,000 $m^3 h^{-1}$ (4·7 $m^3 s^{-1}$), it would be reasonable to assume, with a throughput of a fairly erosive dust, that the bags will have to be replaced each year. If a flow rate of about 0·01 $m^3 m^{-2} s^{-1}$ is used, then a reasonable design pattern would be possible if 150 mm tubes each 4 m long are hung in four compartments, with eighty bags per compartment. This gives a total of 320 tubes. An estimate of total operating cost could be as outlined in Table 12.8.

TABLE 12.8. TOTAL OPERATING COSTS FOR A BAG FILTER HANDLING 17,000 $m^3 h^{-1}$, FOR 8400 HOURS/YEAR (i.e. 50 weeks) ON A CONTINUOUS BASIS

Initial cost of unit $33,800

Cost item	Remarks	Year of operation		
		1	2	3
Interest	12% reducing	$4050	$3720	$3380
Depreciation	Constant over 12 years	2820	2820	2820
Power	12 h.p. total—1·2 cents/unit 8400 hr/year	908	908	908
Replacement bags	$6·67/bag when purchased in in 100s, 320 needed	2130	2130	2130
Maintenance	Capital × life (years) × 0·01	338	675	1050
Fixed charges (insurance, taxes)	$1\frac{1}{2}$% of initial capital	507	507	507
		$10,753	$10,760	$10,795

This uses a constant depreciation for the plant, based on a 12-year service life, and a reducing interest, based on the depreciated capital. The annual charges on this basis remain almost constant, as the reducing interest is balanced by increasing maintenance. Other possibilities, such as calculating depreciation as a constant proportion of the depreciated value of the plant, would give a falling cost. In other examples very high maintenance charges could give marked increases in operation with increasing plant age.

None the less, such an assessment of total operating expenses is most valuable, even if the estimates tend to be optimistic. If the material collected has commercial value, such as a metal oxide which can be either recycled in the plant or sold to an outside firm, then the "break-even" price for the material can be calculated as follows for the case of the bag filter considered above.

If the average fume, i.e. metal oxide concentration, is x g m^{-3}, then 142·8 x tonnes of material will be collected each year. The ratio of \$10,800/142·8 x gives a break-even

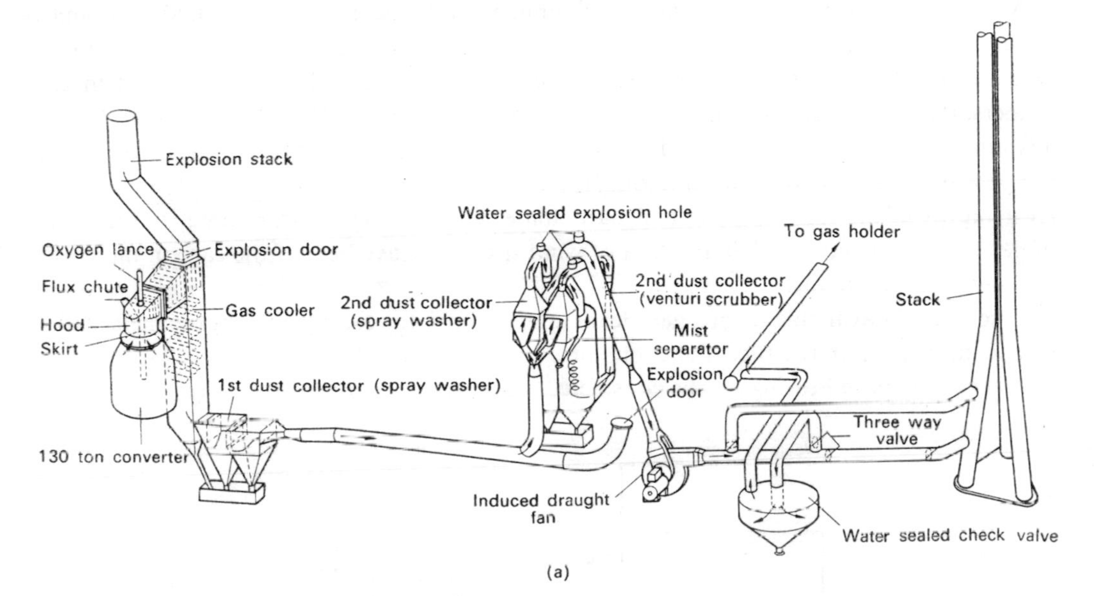

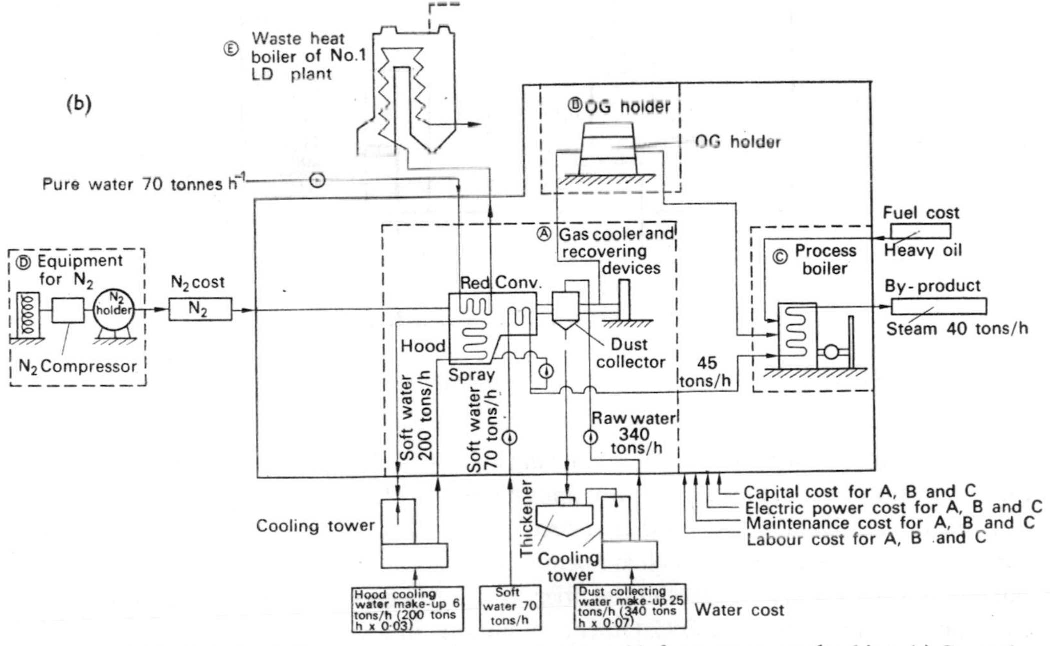

FIG. 12.10. The L.D./O.G. Process, recovering carbon monoxide from oxygen steelmaking. (a) General layout of O.G. plant at the Tobata No. 2 L.D. Converter (Yawata Iron and Steel Co. Ltd, Japan). (b) Flow sheet for calculating the economics of the plant.

price of \$75·6/x for the collected material. For many nonferrous metals, therefore, whose oxides are worth \$300–\$400/tonne, the cost of collection for a plant with a dust burden of 0·5 g m^{-3} or greater, and with the operating costs as calculated, is more than balanced by the cash return on the recovered material.

As well as collecting solids or liquids of commercial value, it has been found economical in the case of the L.D. (Basic Oxygen) steel-making process to collect the carbon monoxide which is produced during 75 per cent of the melting process. This gas is stored in gas-holders, and then used for firing boilers and reheating furnaces (Fig. 12.10). An additional advantage of this process which is called the OG process[581] is a lowering of waste gas temperatures because the carbon monoxide is not burned at the mouth of the converter, reducing problems with cooling and particulate gas cleaning. Furthermore, the quantity of gas to be handled is only a fraction of what it would have been if the carbon monoxide had been burned in air.

When heat is available from a gas-cleaning operation, for example, from the incineration of a combustible fume, then the economics of using the heat in either the process (e.g. a drier) or in space heating can be considered. A typical example of this (Fig. 12.11) shows

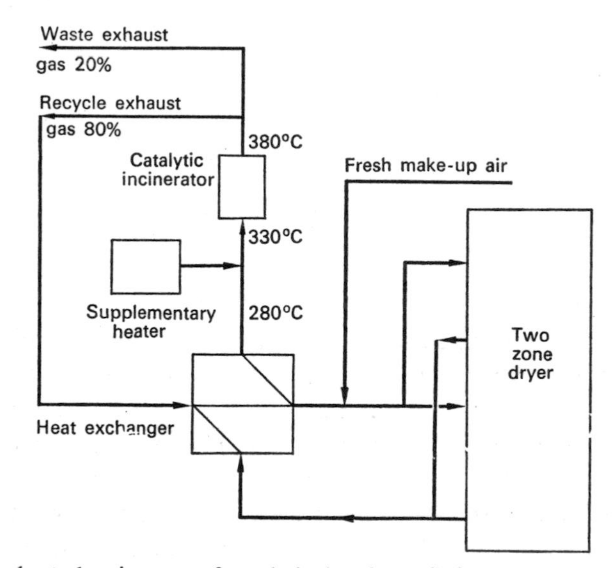

FIG. 12.11. Flow sheet showing use of catalytic (or thermal) fume incinerator providing process heating.[917]

how a catalytic incinerator is used for removing hydrocarbons and then recycles 80 per cent of the hot gases for first preheating its own input gases and then for heating a two-zone drier.[917]

Vollheim and Domin[890] give the cost structure for thermal and catalytic incineration on the waste gases from a drier which are being extracted at a rate of 10,000 m^3/hr. The gases can be used for heating the gases in the drier, which has an inlet temperature of 320°C and an exhaust temperature of 180°C. Other assumptions in this example, which has a pattern similar to Fig. 12.11, are that the catalyst inlet temperature is 250°C and the thermal incineration temperature 750°C. Furthermore, catalyst life is assumed to be 12,000 hr, heat exchanger life 35,000 hr, and other plant items 50,000 hr. This last is a period of 10 years if the plant operates for 5000 hours/year. The interest is assumed to be 4 per cent p.a., and maintenance is not included. The cost analysis, which has been converted to $U.S., is as shown in Table 12.9.

TABLE 12.9. COMPARISON OF OPERATING COSTS FOR THERMAL AND CATALYTIC INCINERATION PLANT (10,000 m^3 h^{-1}) WITH HEAT EXCHANGERS AND HOT GAS RECYCLE AS OPTIONS[890]

	Thermal incineration			Catalytic incineration		
	Basic unit	with heat exchanger	with heat exchanger and recuperation	Basic unit	with heat exchanger	with heat exchanger and recuperation
Capital costs total	15,700	22,500	22,500	21,600	31,600	29,800
$/$m^3$ h^{-1}	1·57	2·25	2·25	2·16	3·16	2·98
Operating costs $/hr						
Depreciation						
Catalyst	—	—	—	0·54	0·54	0·54
Heat exchanger	—	0·26	0·26	—	0·29	0·12
Other items	0·31	0·27	0·27	0·30	0·30	0·38
Fuel oil	6·55	3·57	3·57	1·51	0·30	0·70
Electricity	0·11	0·32	0·32	0·11	0·32	0·32
Interest (4%)	0·12	0·18	0·18	0·17	0·26	0·24
Subtotal	7·09	4·60	4·60	2·63	2·01	2·30
Savings on drier heating			2·50			2·50
Net hourly cost	7·09	4·60	2·10	2·63	2·01	−0·20
$ hr/1000 m^3 h^{-1}	0·79	0·46	0·21	0·26	0·20	−0·02

In the case of a thermal or catalytic incineration system it would be desirable to know how large to make a heat exchanger for a particular application. Vollheim has derived an equation which should give this, but because of the difficulties encountered in applying this equation, it will not be given here. None the less, the basic principle of adding the calculated annual cost of the heat exchanger to the cost of the fuel used over the same period, for a series of heat exhanger sizes should produce a series of totals, which have a minimum at the optimum combination. The heat-exchanger calculation must involve the gas flow, unit cost, capital interest and depreciation, pressure drop, power requirement, and cost as well as other charges. Similarly, the combustion chamber for supplementary heating involves cost items which are also a function of size, as well as the fuel consumption.

The problem of assessing the appropriate heat exchanger for thermal or catalytic incineration is illustrated by the gas flow versus temperature curves shown in Fig. 12.12 for these units.

The gas, going from points 1 to 2, gives up heat to the atmosphere, a drier, space heater, or other application. If there were no heat exchanger, then the gases would have to be heated from the temperature at 3 to the temperature at 4, at which point the catalyst or thermal incinerator would act, burning up the combustible material in the gases and raising

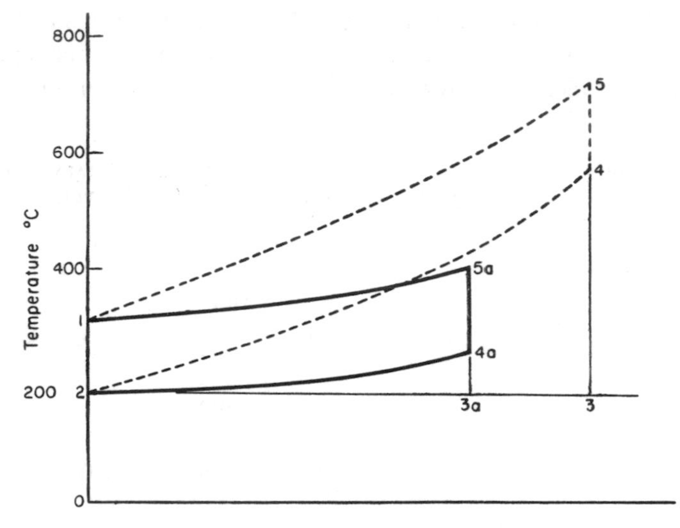

FIG. 12.12. Temperatures and heating surfaces in thermal and catalytic incineration systems.

them to their final highest temperature. If there is a smaller heat exchanger, or there is little combustible material in the gas, and the gases are not raised to the autogenous combustion temperature, then supplementary heating has to be provided in the temperature range of 4 to 5.

It follows that the more effective the heat exchanger is, and this generally implies a greater heating surface, the less supplementary heating has to be provided. Another factor is the temperature of the gases entering the heat exchanger and combustion system. If these are process gases, leaving a hot region or chamber, less supplemetary heat has to be provided. On the other hand, if these gases are cooler, there is a greater temperature difference between these and the gases leaving the incinerator entering the heat exchanger (at 5) and a smaller exchanger gives a greater thermal rise to the gases.

In the case of catalytic incineration the temperature rise required to reach the autogenous combustion temperature on the catalyst (at temperature 4a) is less than for the thermal unit, and also requires a smaller exchanger. In some cases the preheating in the heat exchanger leads to a self-supporting reaction on the catalyst, while in others supplementary heating is required. It can be seen that with a smaller heat exchanger more supplementary heat is required. The greater heat-exchange requirements for high-temperature thermal incineration, in contrast to the lower temperature catalytic process, are also apparent.

In some cases partial preheat is gained by returning part of the hot gases to the incinerator inlet. This modifies the patterns of Fig. 12.12, reducing 4 and 5 in proportion.

When the gas-cleaning plant becomes more complex and much larger, as is the case of flue gas sulphur-dioxide-removal plants, then the total operating costs have to include carefully calculated costing for labour and overheads, which was not necessary in the case of simple, unattended plants. As an example of this, Katell's estimation[425] of the annual operating cost of the Alkalized Alumina Process is shown in Table 12.10. It may be noted from this table how the indirect costs contribute one-third of the total. Furthermore, how much this indirect cost is affected by the "capital charges"; i.e. interest and depreciation, which in this case are at the low level of 14 per cent. For standard industrial applications a value close on 20 per cent would be more appropriate.

TABLE 12.10. ESTIMATED ANNUAL OPERATING COSTS FOR THE ALKALIZED ALUMINA PROCESS FOR SULPHUR DIOXIDE REMOVAL FROM FLUE GASES (800-MW POWER PLANT) (Katell[425])

Direct cost	$	$
Raw materials and utilities		
Absorbent make up	901,700	
Coal: 15 tph × 7920 hr/yr × $4/tonne	475,200	
Power: 1900 kWhr/hr × 7920 hr/yr × × $0.006/kWhr	90,300	
Heat: 23,000,000 W × 7920 hr/yr × × $0.74 GJ	435,500	
Water: 119 × 1000 m³ h⁻¹ × 7920 h/yr $0·028 /1000 m³	26,100	
Credit heat: 23,000,000 W × 7920 hr/yr × × $0·527 GJ	−311,100	1,617,700
Direct labour		
72 man-hr/day: $2.75/man-hr 365-day year	72,300	
Supervision: 15% of labour	10,800	83,100
Plant maintenance		
23 men: $6000/yr	138,000	
Supervision: 20% of maintenance labour	27,600	
Material	69,000	234,600
Payroll overhead: 18·5% of payroll		46,000
Operating supplies: 20% of plant maintenance		46,900
Total direct cost		2,028,300
Indirect cost (administrative and general overhead): 50% of labour, maintenance and supplies		182,300
Total capital charges: 14% of total investment (Table 12.6)		1,191,400
Gross annual operating cost		3,402,000

12.4. THE RELATIVE COST OF DIFFERENT TYPES OF GAS CLEANING PLANT

While the previous sections set out methods of costing individual plants, the question must now be asked as to how different types of plant compare on a cost-efficiency basis. Such a comparison is exceptionally difficult because of the different operating characteristics of the plants, but such a comparison for particle collectors has been carried out by Stairmand.[804, 805]

Stairmand used the fractional efficiencies for a range of gas cleaning plants, and applied these to a standard dust (see page 244). He then costed these plants to clean 100,000 $m^3\ h^{-1}$ at 20°C with a gas-stream particle loading of 11·5 g m^{-3}. The more recent plant capital and operating costs (1965) are given in Tables 12.11 and 12.12.[805] Stairmand plotted these costs against efficiency on a log-log scale, and found that most of the points lay within a narrow band (Fig. 12.13). All costs have been converted to $U.S.

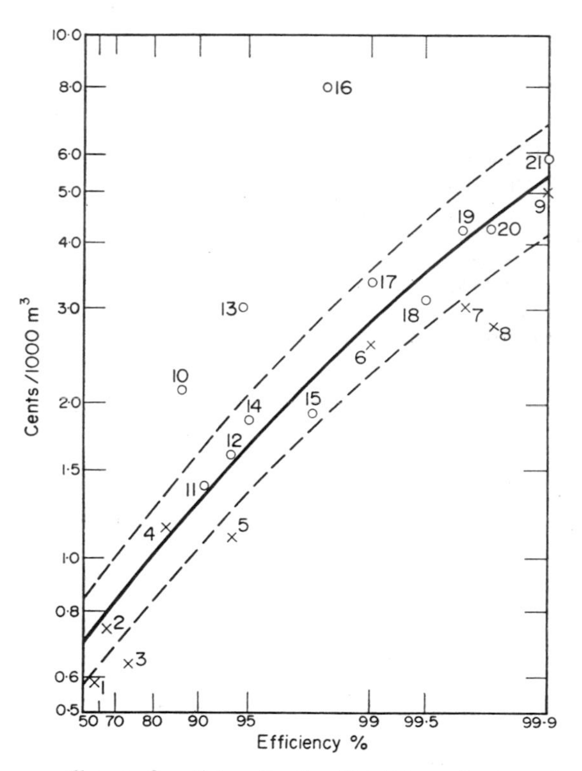

FIG. 12.13. The overall cost of particle collection.[805] × Dry collectors. ○ Wet collectors.

1 Inertial collectors.
2 Medium efficiency cyclones.
3 Cellular cyclones.
4 High efficiency cyclones.
5 Tubular cyclones.
6 Electrostatic precipitators.
7 Shaker type fabric filters.
8 Low velocity bag filters.
9 Reverse jet fabric filters.
10 Jet impingement scrubbers.
11 Irrigated cyclones.
12 Self-induced spray dedusters.
13 Spray towers.
14 Fluidised bed scrubbers.
15 Irrigated target scrubbers.
16 Disintegrators.
17 Irrigated electrostatic precipitators.
18 Annular throat scrubbers—low energy.
19 Venturi scrubbers—medium energy.
20 Annular throat scrubbers—medium energy.
21 Venturi scrubbers—high energy.

A similar correlation for filters for air-borne dusts (for cleaning ambient air) has also been prepared by Stairmand, but with total annual cost for treating 100,000 $m^{-3}\ h^{-1}$. The radioactive aerosol used for this has the particle-size distribution as shown in Table 12.14, with a median mass size of 0·3 μm.[777]

TABLE 12.11. COMPARISON OF DIFFERENT METHODS OF CONTROL OF FUMES FROM ENAMELLING, TIN PLATE BAKING, AND PVC DRYING[828]

Unit	Description
A	A packed column scrubber, using a mineral oil absorbent followed by steam stripping (for fumes from varnish coating)
B	Direct-flame thermal incinerator for fumes from tin-plate lacquer oven, with a heat exchanger (oil fired)
C	Catalytic afterburner on same operation as B—gas fired (gas price: $10.3 per GJ)
D	Direct fuel oil-fired thermal incinerator for PVC fume
E	Direct gas-fired thermal incinerator for PVC fume

	A	B	C	D	E
Volume treated $m^3\ h^{-1}$	20,000	7700	7700	1000	1700
Running time hours per week	144	144	144	144	24
hours per year	7200	7200	7200	7200	1200
Power consumption kWh	112	5·7	n.a.	n.a.	n.a
Steam consumption kg h	1000	—	—	—	—
Fuel or natural gas	—	75 l. h^{-1}	30 $m^3\ h^{-1}$	32 l. h^{-1}	14 $m^3\ h^{-1}$
Capital cost $	70,400	22,000	26,400	1650	1760
Annual charges:					
Depreciation* (10% of capital)	7040	2200	2640	165	176
Fuel	—	14,984	7700	6908	610
Steam	8910	—	—	—	—
Make up absorbent	2622	—	—	—	
Power	8140	871	—	110	110
Maintenance	8800	220	1240	44	44
Total	35,512	18,185	11,580	7227	1040
Total annual cost per 1000 $m^3\ h^{-1}$	1770	2400	1530	7500	620
$/$h^{-1}$/1000 $m^3\ h^{-1}$	0.246	0.334	0.213	1.04	0.52

* A simple 10 per cent p.a. depreciation on initial capital is used as a depreciation allowance, and there is no allowance for interest on capital.

Detailed relative costs of sulphur dioxide removal have been considered by Katell (for the Reinluft, Alkalized Alumina and Cat-Ox Processes).[425] More recently, on a similar common cost basis, Strauss has assessed these processes together with a number of others.[825] These included the Dry Limestone Addition, Sulfacid, Fulham-Simon-Carves and the Molten Carbonate Processes. As this analysis is lengthy, and the principles have been discussed with reference to the Alkalized Alumina Process, they will not be further considered here.

A comparison of different processes for scrubbing or incinerating fumes from metal enamelling or painting, in which organic solvents are evaporated, has been made by Cock and Strauss.[828] It was possible to obtain power, steam, maintenance and other charges and

TABLE 12.12. EFFICIENCY AND COST OF DRY DUST-ARRESTING EQUIPMENT: 100,000 $m^3\ h^{-1}$ (30 $m^3\ s^{-1}$) AT 20°C: DUST LOADING 11·5 $g\ m^{-3}$ (1965)[805]

Equipment	Efficiency on standard dust %	Capital cost		Average pressure drop Pa	Power cost \$/yr	Maintenance \$/yr	Total operating cost \$/yr	Capital charges \$/yr	Total cost including capital charges	
		\$ Total	\$/$m^3\ h^{-1}$ capacity						\$/yr	\$/m^{-1}
Inertial collectors	58·6	26,200	0·262	420	1850	240	2120	2620	4730	0·047
Medium efficiency cyclones	65·3	18,000	0·180	920	4050	144	4200	1900	6100	0·061
Low-resistance cellular cyclones	74·2	32,400	0·324	350	1580	144	1730	3240	4960	0·049
High-efficiency cyclones	84·2	36,800	0·368	1230	5420	144	5550	3680	9250	0·093
Tubular cyclones	93·8	39,200	0·392	1080	4750	144	4900	3980	8900	0·089
Electrostatic precipitators	99·0	177,000	1·77	230	2400	960	3360	17,700	21,100	0·211
Shaker-type fabric filters	99·7	125,000	1·25	625	4490	7700*	12,100	12,500	24,600	0·246
Low velocity fabric filters	99·8	115,000	1·15	500	4050	7200†	11,300	11,500	22,800	0·228
Reverse-jet fabric filters	99·9	175,000	1·75	750	9500	14,400*	24,000	17,500	41,500	0·415

* Bags changed twice each year.
† Bags changed once each year.

Table 12.13. Efficiency and Cost of Wet Dust-Arresting Equipment: 100,000 m³ h⁻¹ (30 m³ s⁻¹) at 20°C (1965)[806]

Equipment	Efficiency on standard dust %	Capital cost		Average pressure drop k Pa	Power cost \$/yr	Water usage kg/1000 m³	Water cost \$/yr	Main-tenance \$/yr	Total operating cost \$/yr	Capital charges \$/yr	Total cost including capital charges	
		\$ total	\$/m³ h⁻¹ capacity								\$/yr	\$/m³
Jet impingement scrubber	87·9	72,000	0·72	2·0	9000	160	336	770	10,100	7200	17,300	0·173
Irrigated cyclone	91·0	45,000	0·45	0·98	4870	800	1580	384	6850	4500	11,300	0·113
Self-induced spray deduster	93·6	50,400	0·50	1·53	6750	120	264	504	7540	5040	12,600	0·126
Spray tower	94·5	105,000	1·05	0·35	5720	3600	7930	770	14,400	10,600	25,000	0·250
Fluidized bed scrubber	95·0	40,000	0·40	0·60	3720	2900	6120	1200	11,000	4000	15,000	0·150
Irrigated target scrubber	97·9	62,400	0·62	1·53	6950	600	1320	768	9000	6240	15,300	0·153
Disintegrator	98·5	103,000	1·03	—	54,300	1000	2040	504	57,000	10,300	67,200	0·672
Irrigated electrostatic precipitator	99·0	228,000	2·28	2·15	2660	500	1080	1010	4750	22,800	27,600	0·276
Low-energy annular throat venturi scrubber	99·5	77,000	0·17	3·1	14,100	1400	2900	770	17,700	7700	25,500	0·255
Medium energy venturi scrubber	99·7	82,000	0·82	5·0	22,600	1400	2900	770	26,300	8150	34,400	0·344
Medium energy annular throat scrubber	99·8	82,000	0·82	5·1	23,200	1400	2900	770	27,000	8150	35,000	0·350
High-energy venturi scrubber	99·9	89,000	0·89	7·9	35,800	1400	2900	770	39,900	8900	50,400	0·504

TABLE 12.14. PARTICLE SIZE DISTRIBUTION OF RADIOACTIVE AEROSOL USED FOR TESTING AIR FILTER (Skrebowski and Sutton)[777]

Particle size range (μm)	% by weight	Particle size range (μm)	% by weight
0–0·10	13	0·35–0·37	10
0·10–0·20	17	0·37–0·43	10
0·20–0·25	10	0·43–0·55	10
0·25–0·30	10	0·55–1·0	9
0·30–0·35	10	1·0 –1·2	1
		> 1·2	0

place them on a common basis. The analysis, converted to $U.S. and modified to allow for the lower cost of natural gas instead of towns (manufactured) gas, is shown in Table 12.11.

It is interesting to note how the apparently expensive oil scrubber is in fact comparable to the other low cost processes, and would be more so with reduced maintenance.

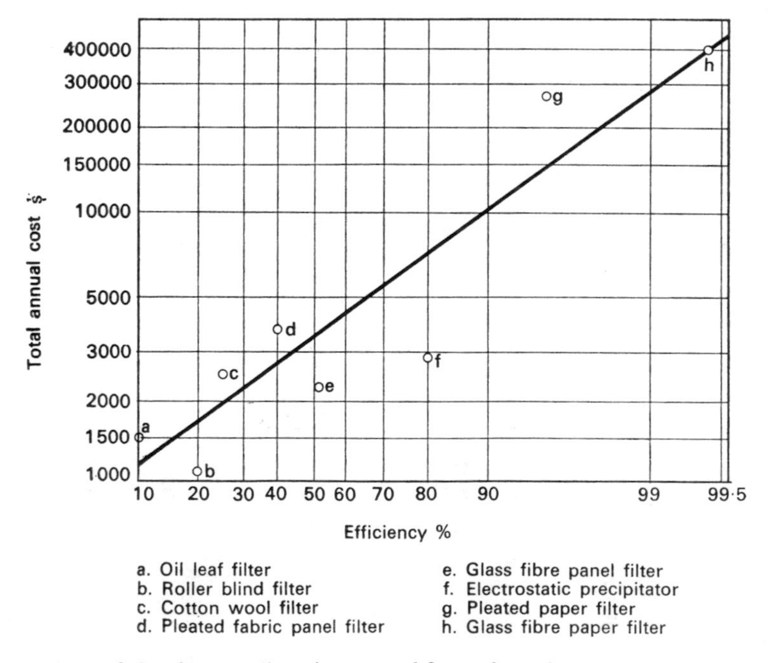

FIG. 12.14. Cost of cleaning a radioactive aerosol from air at the rate of 100,000 m^3 h^{-1}.[777]

APPENDIXES

I. EXAMPLE OF CALCULATION OF GAS-FLOW RATE FROM FUEL CONSUMPTION AND COMPOSITION

A boiler is to produce 100,000 kg h^{-1} steam at 1151 kPa abs. and 480°C from water at 20°C. An efficiency of around 83 per cent can be expected. The fuel is coal of a gross calorific value of 30·4 MJ kg^{-1} and an ultimate analysis of 74 per cent carbon, 4·5 per cent hydrogen, 6·5 per cent oxygen, negligible nitrogen and sulphur, 7 per cent ash and 8 per cent moisture. The air supply will have an average temperature of 21°C and an average relative humidity of 65 per cent. An excess air consumption of 50 per cent may be assumed. The flue gases will leave at 220°C. What is the flow rate of flue gases to the gas cleaning device?

From steam tables, we get the following enthalpies,

Feed water at 20°C	0·084 MJ kg^{-1}
Superheated steam at 1151 kPa, 480°C	3·320 MJ kg^{-1}
∴ Enthalpy added	3·236 MJ kg^{-1}

Assuming efficiency is based upon gross calorific value of fuel,

∴ Fuel consumption,

$$= 100{,}000 \times 3{\cdot}236 \times \frac{100}{83} \times \frac{1}{30{\cdot}4}$$

$$= 12{,}800 \text{ kg h}^{-1} = 3{\cdot}58 \text{ kg s}^{-1}$$

Assume complete combustion of the coal,

Basis: 100 kg of coal as fired.

$$\text{Carbon in} = 74 \text{ kg} = \frac{74}{12} = 6{\cdot}17 \text{ kg moles}$$

$$\text{Hydrogen in} = 4{\cdot}5 \text{ kg} = \frac{4{\cdot}5}{2} = 2{\cdot}25 \text{ kg moles}$$

$$\text{Oxygen in} = 6{\cdot}5 \text{ kg} = \frac{6{\cdot}5}{32} = 0{\cdot}20 \text{ kg moles}$$

$\therefore$ Theoretical oxygen $= 6{\cdot}17+\frac{1}{2}(2{\cdot}25)-0{\cdot}20 = 7{\cdot}1$ kg moles

$\therefore$ Actual oxygen $= 7{\cdot}1\times\frac{150}{100} = 10{\cdot}65$ kg moles

$\therefore$ Nitrogen in with air,

$$= 10{\cdot}65\times\tfrac{79}{21} = 40{\cdot}06 \text{ kg moles}$$

$\therefore$ Flue gas will consist of,

(i) Carbon dioxide = 6·17 kg moles
(ii) Oxygen = 3·55 kg moles
(iii) Nitrogen = 40·06 kg moles
(iv) Water

(a) from moisture in coal

$$= 8 \text{ kg} = \tfrac{8}{18} = 0{\cdot}44 \text{ kg moles}$$

(b) from combustion of hydrogen = 2·25 kg moles
(c) in with combustion air.
From a psychrometric chart, absolute humidity of air

$= 0{\cdot}01$ kg Water/kg dry air.
$= 0{\cdot}01\times\frac{29}{18} = 0{\cdot}016$ moles Water/mole dry air

$\therefore$ Water in $= (0{\cdot}016)\,(10{\cdot}65+40{\cdot}06) = 0{\cdot}81$ kg moles
$\therefore$ Total water in flue gases $= 3{\cdot}5$ kg moles
$\therefore$ Total flue gas,

$$= 6{\cdot}17+3{\cdot}55+40{\cdot}06+3{\cdot}5$$
$$= 53{\cdot}28 \text{ kg moles/100 kg of coal}$$
$$= \frac{53{\cdot}28}{100}\times 3{\cdot}58 = 1{\cdot}91 \text{ kg moles s}^{-1}$$

Assuming that gases are at 101 kPa pressure, Volume flow rate,

$$= 1{\cdot}91\times 22{\cdot}4\times\frac{493}{273}$$
$$= 77{\cdot}2 \text{ m}^3\text{ s}^{-1} \text{ or } 250{,}000 \text{ m}^3\text{ h}^{-1}$$

II. EXAMPLE OF SUCTION PYROMETER CALCULATION

Find the gas temperature using a suction pyrometer with a double-shield finned head. The refractory material of the head has a thermal conductivity of 6·3 $\text{Nm}^{-1}\,°\text{C}^{-1}$, a wall thickness of 3 mm and an emissivity of about 0·4 at 1600°C. The gases, when aspirated past the couple with a velocity of 90 m s^{-1}, give a temperature reading of 1635°C, at zero velocity, of 1453°C and at 25 m s^{-1} of 1562°C.

Method 1

The ratio $w/\varkappa = 0{\cdot}30/6{\cdot}3 = 0{\cdot}48$.
From Table 2.6, at about 1600°C,

$$f = 0{\cdot}7.$$

From Table 2.7, with Emissivity = 0·4, $f = 0{\cdot}7$.
Effective number of simple metallic shields per refractory shield (without fin allowance)

$= 2{\cdot}1$

Fin allowance $= 1{\cdot}4$

$\therefore$ Equivalent total number $= 2{\cdot}1 \times 1{\cdot}4 \times 2$

$= 5{\cdot}9.$

The aspiration velocity of 90 ms^{-1}, gives, from Table 2·3, an equivalent number of shields at 150 m s^{-1}.
No. of equivalent shields (150 ms^{-1}) = 0·82
$\therefore$ Total equivalent (150 m s^{-1}) $= 5{\cdot}9 \times 0{\cdot}82 = 4{\cdot}8$
From Table 2.2, efficiency of pyrometer = 81 per cent
$\therefore$ Actual gas temperature $= 1453 + (1635 - 1453)/0{\cdot}81$
$= 1678$°C.

Method 2

An approximate gas temperature can be found by first calculating the shape factor:

$$\text{Shape factor} = \frac{T_{max} - T_0}{T_{max} - T_{\frac{1}{4}}} = \frac{1635 - 1453}{1635 - 1562} = 2{\cdot}5$$

From Table 2.8, efficiency = 80 per cent.
Actual gas temperature $= 1453 + (1635 - 1453)/0{\cdot}80$
$= 1681$°C.

III. METHODS OF EXPRESSING THE EFFICIENCY OF COLLECTION

1. *Gravimetric efficiency*

This is the efficiency of collection based on weight collected compared to the total passing through the system. It is usually expressed as a percentage.

$$\eta_T = \frac{\text{wt. of material collected}}{\text{total weight entering system}} \times 100 \text{ per cent.}$$

This is the most common method used, and it is the easiest determined in practice.

2. *Fractional efficiency*

The variation of efficiency with particle size of a particle collection system is best expressed in the form of a fractional efficiency curve (sometimes called a grade efficiency curve). The abscissa is used for particle size and the ordinate for the efficiency of collection. A collector, under specific conditions of gas-flow rate, temperature, particle composition and distribution will have a specific efficiency at each size.

A fractional efficiency curve is set up by considering the efficiency in a series of narrow ranges of the particle-size spectrum. This can be obtained by analysing for particle-size distribution the original dust and the dust collected. From this data the efficiency in each size range (say 0–5, 5–10 μm, etc.) can be found from

$$\eta_i = \frac{\text{fraction in range “}i\text{” collected}}{\text{quantity in range “}i\text{” entering system}} \times 100 \text{ per cent}$$

if m_i = quantity in range "i" entering the collection system,
M = total quantity of dust entering system.

The gravimetric efficiency for the system can then be deduced from the fractional efficiency data[495]

$$\eta_T = \frac{\sum_{i=1}^{n} m_i \eta_i}{M} \quad \text{per cent}$$

where n = number of size fractions in the sample.

The construction of a fractional efficiency curve is shown in Fig. A.3.1.

Fractional efficiency curves are generally related to particles with a specific density. If it is necessary to obtain a fractional efficiency curve for particles with a different density, points on the curve have to be multiplied by the ratio $\sqrt{(\varrho_n/\varrho_0)}$, where ϱ_n is the new density, ϱ_0

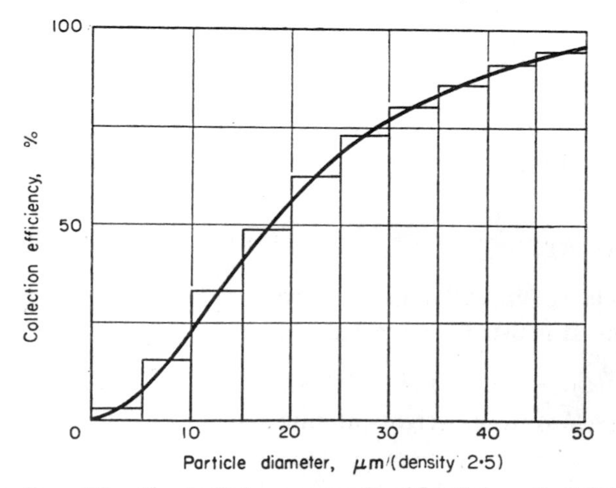

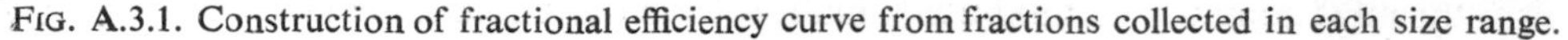
FIG. A.3.1. Construction of fractional efficiency curve from fractions collected in each size range.

the original density, in the case of cyclones, fibre filters, scrubbers or other devices where inertial impaction is the predominant collection mechanism; or ϱ_n/ϱ_0 for settling chambers and electrostatic precipitators.

The abscissa of a fractional efficiency curve is sometimes shown in terms of falling speeds of particles of stated density instead of particle size. This is very useful as it can be obtained directly from a particle-size analysis based on aero- or hydrodynamic properties of the particles, and avoids the problems connected with non-spherical particles, where equivalent diameters have to be used. (See Chapter 4.)

3. *Number fraction efficiency*

Efficiency based on numbers of particles η_N

$$\eta_N = \frac{\text{no. of particles collected}}{\begin{array}{c}\text{total number of particles}\\ \text{entering collector}\end{array}} \times 100 \text{ per cent.}$$

4. *Surface area efficiency*[704]

Based on the surface area of particles. This is of great importance when obscuration is used as a measure of collection efficiency.

$$\eta_A = \frac{\text{surface area of particles collected}}{\begin{array}{c}\text{total surface area of particles}\\ \text{entering collector.}\end{array}} .$$

5. *Penetration* P

This is the inverse of the efficiency of collection η_T. The penetration, in per cent, is given by

$$\mathbf{P} = 100 - \eta_T .$$

6. *Decontamination factor* D.F.

This is a logarithmic scale of measuring efficiency

$$\text{D.F.} = [1 - \eta_T/100]^{-1}.$$

The logarithm to the base 10 of the D.F. is known as the *decontamination index.*

IV. CALCULATION OF EFFICIENCY OF ELECTROSTATIC PRECIPITATORS WITH EDDY DIFFUSION ALLOWANCE

The method suggested by Williams and Jackson[945] (discussed on page 434) can be used as follows. Two dimensionless parameters, representing a distance function (τ) and a velocity function (φ) are calculated, and the efficiency can then be obtained from Fig. A.4.1.

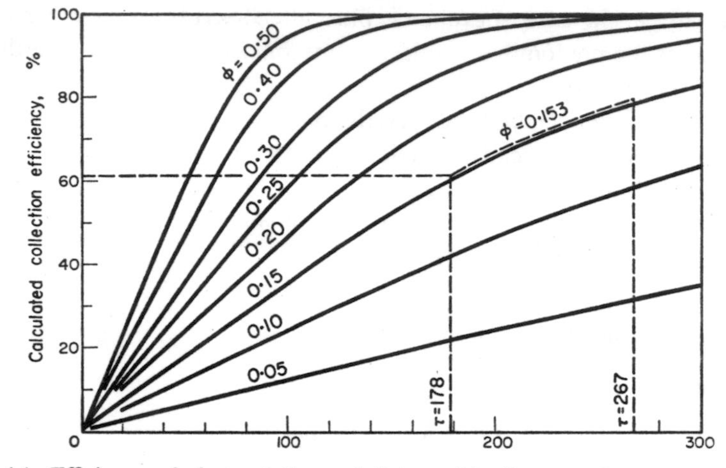

FIG. A.4.1. Efficiency of electrostatic precipitator with allowance for eddy diffusion.[945]

Equation (10.62a) $\tau = \left[4(N+1)^2 \frac{D}{v_{av}L}\right] \frac{x}{L} \doteqdot 7{\cdot}41 \frac{x}{L}$.

Equation (10.62b) $\varphi = \left[2(N+1) \frac{D}{v_{av}L}\right] \frac{w'}{v_{av}} \doteqdot 5{\cdot}67 \frac{w'}{v_{av}}$.

($D/v_{av}L = 0{\cdot}0035$ and ($N = 20$)).

The functions can be found from:

$$\tau = 7{\cdot}41 \frac{x}{L}$$

where x = path length in precipitator,
L = distance between wire and plate (plate-type precipitator) or wire and tube (tube-type precipitator),

$$\varphi = 5{\cdot}67 \frac{\omega'}{v_{av}}$$

where ω' = apparent drift velocity,
v_{av} = stream velocity.

The apparent drift velocity is best determined from test results for a precipitator with similar dimensions, and stream velocity. If this is not available, then the former can be estimated from measured efficiencies at various gas velocities.

The following problem given by Williams and Jackson illustrates this.

Problem. A parallel-plate precipitator with gas passages 3·6 m long, having a wire to plate distance of 150 mm, is found to have collection efficiencies of 61, 49, 40 and 35 per cent for gas stream velocities of 0·9, 1·2, 1·5 and 2 m s^{-1}, respectively. What would be the effect of increasing the length of the gas passages to 5 m and reducing the wire to plate distance to 125 mm, keeping the other operating conditions constant?

Solution. The parameter τ for the 3·6 m long precipitator ($x = 3{\cdot}6$) with wire to plate distance 150 mm ($L = 0{\cdot}15$) is:

$$\tau = \frac{7{\cdot}41 \times 3{\cdot}6}{0{\cdot}15} = 178.$$

From a vertical line drawn at $\tau = 178$ the values of φ for the given collection efficiencies are found to be 0·153, 0·118, 0·0970, 0·0825. If the precipitator dimensions are altered to $x = 5$ m and $L = 0{\cdot}127$ m, then $\tau = 267$, and the new efficiencies, for the same values of φ, are then 80, 67, 57 and 50 per cent. This is summarized in the following table:

Stream velocity m s^{-1}	Efficiency %	φ	Efficiency % $\tau = 267$
	$\tau = 178$		
0·9	61	0·153	80
1·2	49	0·118	67
1·5	40	0·0970	57
2·0	35	0·0825	50

V. DESIGN OF HOODS AND PIPELINES

For containing and extracting gases, fumes and dusts from processes, hoods of various types are required. The design of the hood and accompanying ductwork is a function of the type of process, the size of the plant and its situation in relation to the gas cleaning plant. Details of design of hoods and ducts are discussed elsewhere, and only a brief outline of major features is given here.

The hood should enclose the process as fully as possible compatible with plant operation, reducing the amount of air that is drawn in to a minimum, and in turn, minimizing the quantity of gas to be treated subsequently in the cleaning plant. In the case of an electric furnace, total enclosure is often feasible with a hood of the type shown in Fig. A.5.1, which permits tilting of the furnace without breaking the gas ducts. An alternative arrangement for enclosing a furnace is shown in Fig. A.5.2, although in this case the exhaust duct, which is water cooled, swings away from the furnace when this is being tilted for tapping. In steel making, where fumes during tapping tend to be at a minimum,[832] this would not cause any difficulties. When considerable access has to be gained to a process, as in electrolytic plating, it is usual to have hoods which are open at one or more sides. Some typical designs are shown in Fig. A.5.3. When a hood is required for a grinding operation it may only be possible to draw air and dust away to one side, and much higher air velocities are necessary to ensure that the dust does not leak past the entrance to the duct.

Recommended practice as to the type of hood and the velocities of air required over the open face for a number of common processes is given in Table A.5.1. Gas velocities in ducts are generally in the range of 12·5 to 25 m s^{-1}, which is well above the "pick-up" velocities listed in Chapter 5.

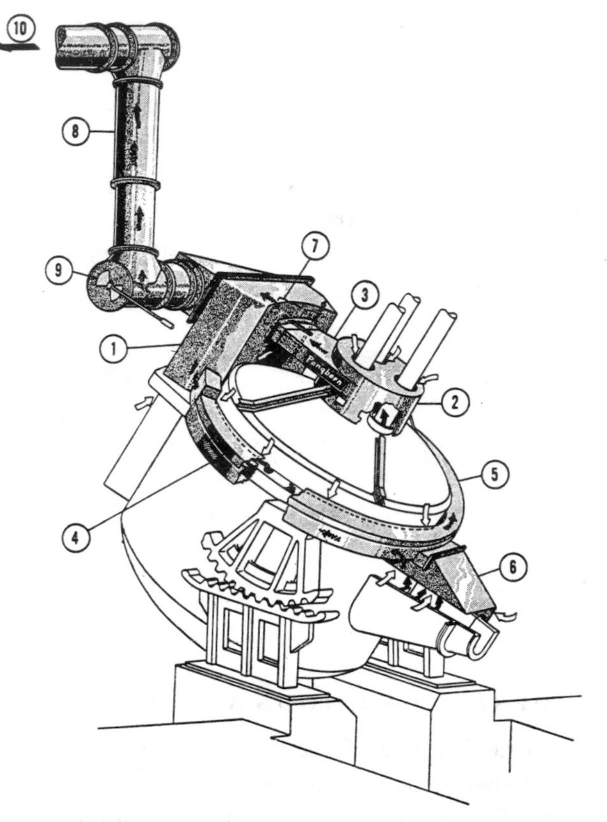

FIG. A.5.1. Total enclosure hood for arc furnace (frequent tilting).[621]

1. Distribution section and slag door hood.
2. Electrode hood and support arms (fastens to roof frame and moves with roof).
3. Connecting duct.
4. Manifold ring.
5. Manifold ring baffle.
6. Pouring spout hood.
7. Adjustable spring loaded damper.
8. Telescoping and swivel connection to exhaust system for continuous ventilation during furnace tilt.
9. Tempering damper.
10. Duct to dust collector.

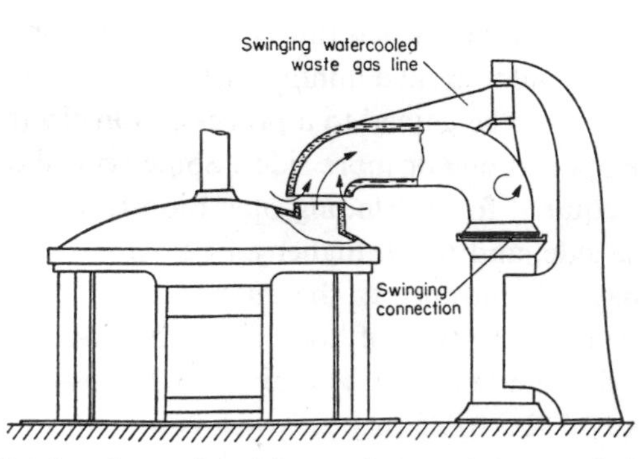

FIG. A.5.2. Total enclosure hood for arc furnace (water-cooled gas line).[328]

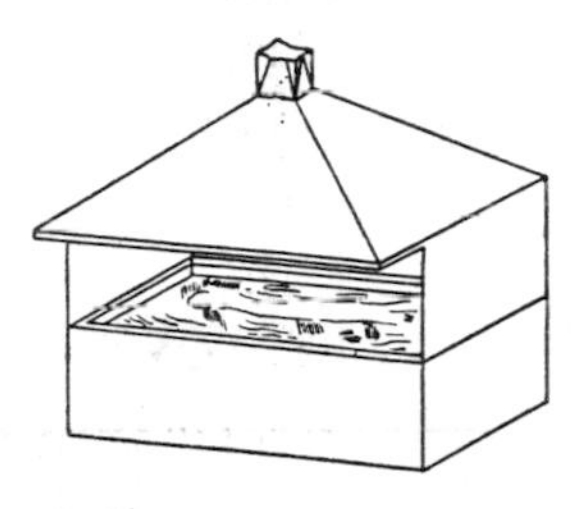

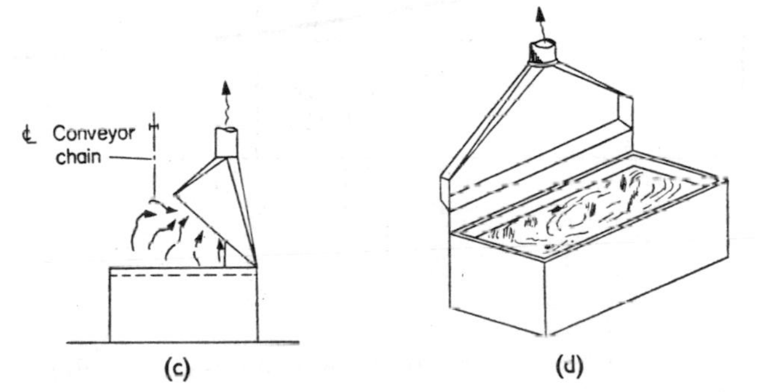

FIG. A.5.3. Various hoods:[7] (a) Double hood. (b) Canopy hood with sides. (c) Semi-canopy hood. (d) Low rear hood with slit entrance.

TABLE A.5.1

Process	Usual type of hood	Face velocity ($m\ s^{-1}$)
Pharmaceuticals coating pan	Narrow side hood	1 –2
Pickling tanks	Canopy hood	1·2 –1·5
Soldering	Enclosed—open at one side	0·15–1
Varnish kettle	Canopy type	1·2 –1·5
	Slot type—50 mm slot	10
Dust from grain flour, wood	Canopy	2·5 –3
	Slot (50–100 mm slot)	10
Aluminium furnaces	Enclosed hood—open one side	0·75–1
	Canopy hood	1 –1·2
Brass furnaces	Enclosed hood—open one side	1 –1·2
	Canopy hood	1·2 –1·5

The pressure loss in the ducts and pipelines can be estimated by the following:

1. In cylindrical pipes, the pressure drop per metre length, in kPa is:

$$\Delta p = 9{\cdot}45 \times 10^{-4} f \frac{u}{D} \qquad \text{(A.5.1)}$$

where Δp = pressure drop per m in kPa,
u = average pipe velocity (m s^{-1}),
D = pipe diameter (mm),
f = friction factor, which is a function of the Reynolds number

$$(\text{Re} = uD\varrho/\mu)$$

and can be obtained from the curve in Fig. A.5.4.

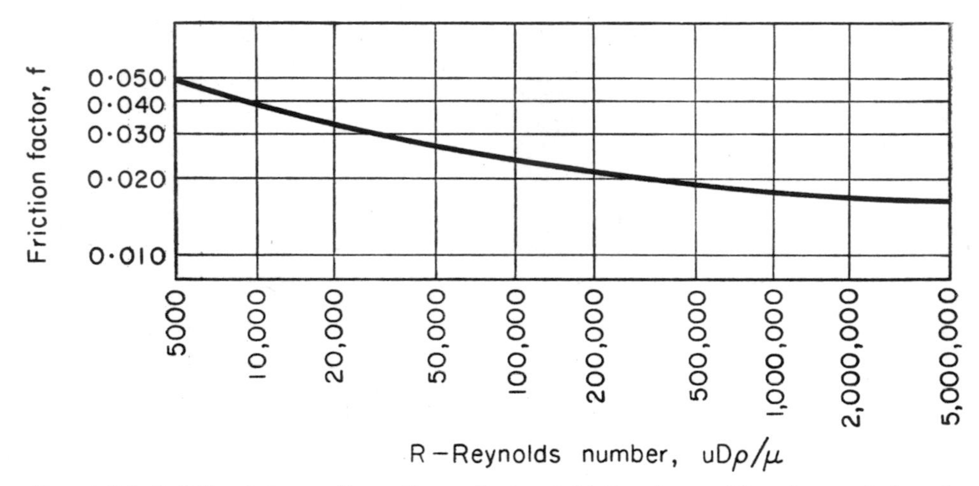

FIG. A.5.4. Relation between Reynolds number and friction factor f for pipes and ducts.[7]

2. Rectangular ducts, cross section $a \times b$; the diameter in equation (A.5.1) is replaced by the equivalent hydraulic diameter D_h where

$$D_h = \frac{2ab}{a+b}.$$

The pressure loss at a bend or fitting can be expressed in terms of a pipe length equivalent to a number of pipe diameters which are a function of the type of bend or fitting. Typical examples are given in Table A.5.2.

TABLE A.5.2. FRICTION LOSS OF FITTINGS, EXPRESSED IN TERMS OF EQUIVALENT PIPE DIAMETERS

90° elbow—standard radius	32
90° elbow—long sweep radius	20
180° close return bend	75
Tee—used as elbow entering run	60
Tee—used as elbow entering branch	90
Gate valve (open)	7
Globe valve (open)	300

VI. DESIGN OF HOPPERS

The hoppers fitted below the collection section of gas-cleaning equipment (cyclones, settling chambers, bag filters or electrostatic precipitators) are usually the shape of either an inverted cone or pyramid. The design of a hopper for a particular application is not simple, as it depends on the rheological behaviour of the bulk dust or powder which has been collected. With good design, smooth flow of the collected material will occur under gravity without arching or partial retention. These problems have been studied in detail by many workers and are reviewed elsewhere,[221, 252, 633, 869] particularly by Richards.[682] Collected powders behave essentially as a cohesive solid, which, on application of a limiting stress, "fails" and proceeds to flow as a plastic. It is important to design hopper walls and openings so that arching does not occur even when damp conditions improve the internal cohesive properties of the powder.

An estimate of the minimum opening can be approximately obtained from one of a number of equations listed by Richards[682] for the distance between walls for supporting stable arches. One of these (by Gardner[290]) is

$$L = \frac{2\mathscr{C}}{\varrho_b} \cdot \frac{\cos\eta(1+\sin\eta)}{\cos(\alpha+\eta)} \tag{A.6.1}$$

where $\mathscr{C}$ = the cohesion of the material,
ϱ_b = bulk density,
α = angle of friction on the wall (radians),
L = radius of a conical opening or the wall to wall distance of a pyramidal opening,
η = angle of internal friction of powder (degrees).

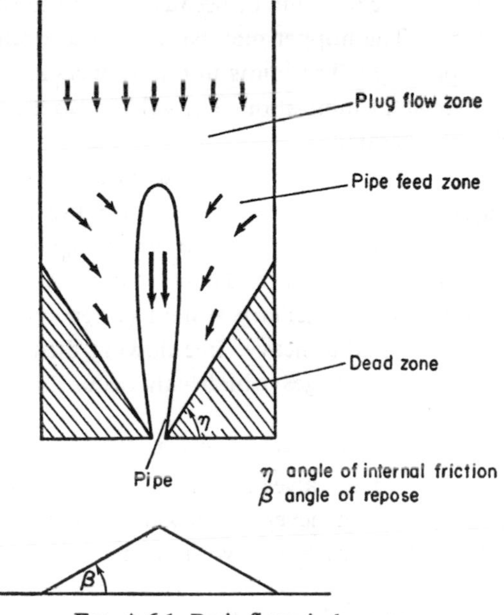

FIG. A.6.1. Basic flows in hoppers.

The cohesion, $\mathscr{C}$, is defined in terms of the normal and shear stresses, σ and τ, and the angle of internal friction of the powder η:

$$\tau = \mathscr{C} + \sigma \tan \eta. \tag{A.6.2}$$

The simplest design of hopper is a cylinder with a flat base with an orifice. The powder will collect in this and flow through the orifice, leaving a "dead" zone of powder. The angle between the horizontal and the dead zone boundary (the angle of internal friction) is approximately twice the angle of repose for the material. The angle of sloping hopper walls is generally approximately of this order. The angle may be decreased by using very smooth walls, so that a lower hopper can be used. It may be found, however, that a significant amount of "core flow" occurs in this case, with some of the powder remaining in the hopper for long periods, except when it is completely drained, in which case the whole of the powder flows from the hopper. In other cases, the shape of the exit, and the possible insertion of a double cone with steep sides may be used to obtain good flow of the collected powder. In some cases it may also be useful to line the hopper walls with a non-corrosive, smooth, hard metal or ceramic liner to obtain a lower angle of slide. Vibrators or rappers may be installed for some cohesive powders. The lowering of the slope angle of the hopper walls enables the gas-cleaning equipment to be installed closer to the base level, and may reduce costs.

In practice the hopper sides are frequently sloped at an angle of 60° to the horizontal. Steeper sides may be used if the collected material does not flow easily, or a vibrator can be installed to help to empty the hopper. The size of the hopper should be such that only occasional opening of the hopper valve is required, because when the valve is open, the cyclone will not operate satisfactorily. The hopper may be used as a product storage, and there is no upper limitation on hopper size. The joints in the hoppers should be carefully made to ensure complete sealing against air infiltration. The valves also must seal easily, and may be either hand or mechanically operated.

On small cyclones the hand-operated valves may be of the gate-valve pattern, with a push–pull opening (Fig. A.6.2a), or of the spring-loaded mushroom valve type. Another possibility is to have a counter-weighted flap valve (Fig. A.6.2b) which opens automatically when there is a sufficient head of dust in the hopper. These cannot be used if the pressure in the cyclone is positive compared to the ancillary system (when the fan is located before the cyclone), and are generally restricted to heavy, free-flowing dusts. Their application should also be limited because of the air leakages which can occur and remain undetected when the valve is not closed properly.

Two mechanical systems frequently used are the double-flap valve (Fig. A.6.2c) and the rotary airlock (Fig. A.6.2d). The flap valves are alternately opened by a motor-driven cam arrangement so that outside air can never be sucked through the cyclone hopper. This arrangement is recommended for heavy-duty work, such as the handling of abrasive materials.

The rotary air lock is ideal for continuous discharge of material and consists of a slowly

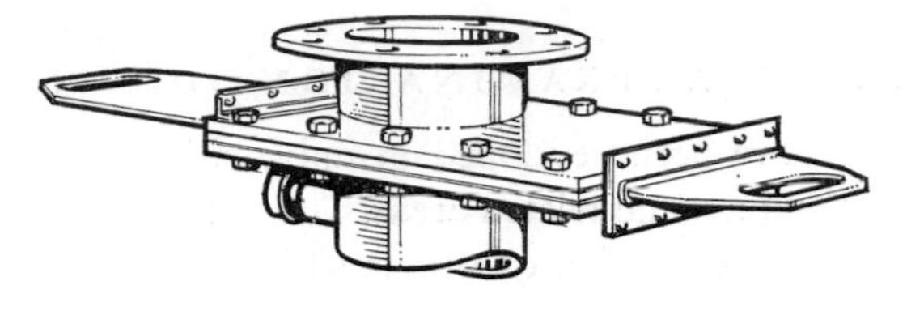

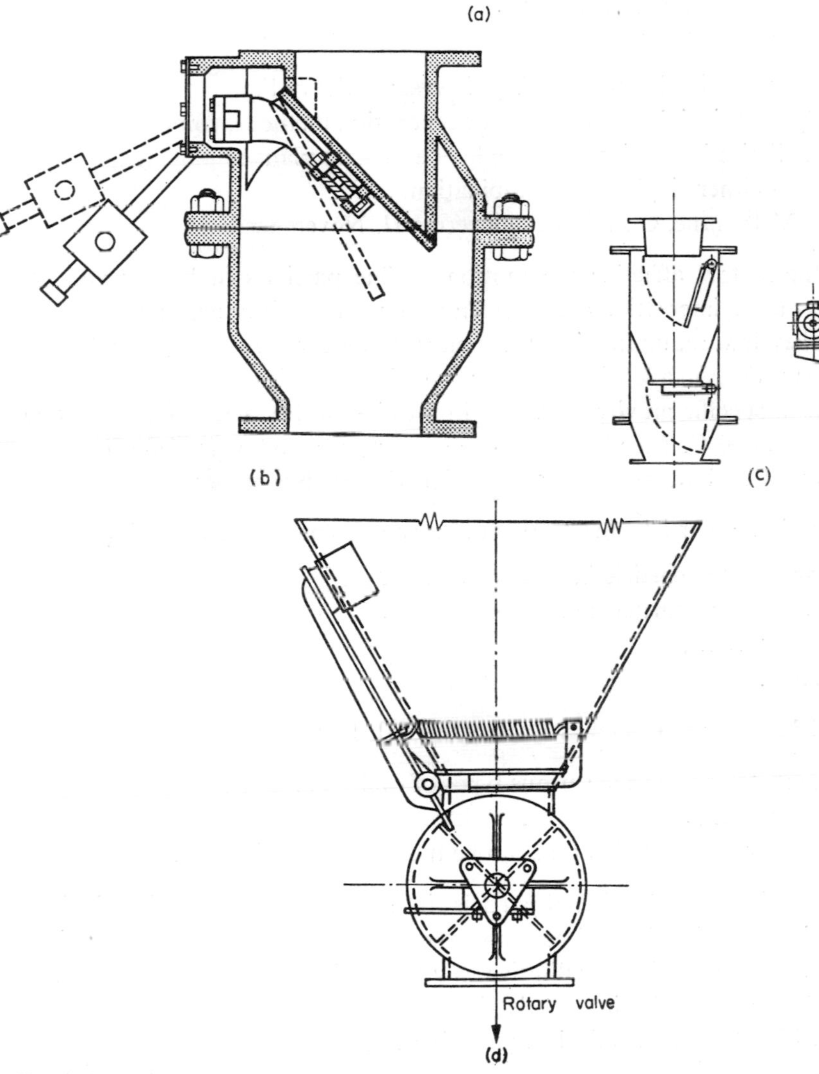

FIG. A.6.2. Valve types fitted into base of hoppers. (a) Push–pull valves (manual). (b) Counterweighted flap valve (self-actuating). (c) Double-flap valve (mechanical). (d) Rotary valve (mechanical).

moving four- or six-bladed rotor with dust-tight glands and sealed-in shaft, driven by a geared-down motor. This unit requires less headroom than the double-flap arrangement.

The valve can discharge the product into bags, bins, dust conveyors, or some other means of disposing of the collected material.

VII. TYPICAL EXAMINATION QUESTIONS

A number of courses have been developed in universities in recent years dealing with gas cleaning, particle mechanics, and related areas. The author has been involved in several of these, and some typical examination questions which were given may be of assistance in the formulation of others. The answers to the questions were generally to take one-half hour.

1. What methods are available for removing dispersed solids or liquids from gas streams? Indicate briefly the principles of the various separation methods, commenting on their advantages and limitations. Give an example of a process plant employing one or more of these methods either singly or in combination.
(University of Melbourne, Chemical Engineering II, November 1961.)

2. (i) Discuss the factors affecting the motion of fine particles in liquids and gases. Comment on the interaction effects when groups of particles are present.
(ii) Using Stokes' law, set up fractional efficiency curves for the collection of flyash and iron oxide particles in a settling chamber 6 m wide, 5 m high and 9 m long.
Flyash has a specific gravity of 2, and iron oxide of 4·5. The gas volume flow is 23·5 $m^3 s^{-1}$ at 400°C. The viscosity of the carrier gas at 400°C is 32·8 μ Pa s.
(University of Melbourne, Chemical Engineering II, November 1962.)

3. Discuss the motion of particles in a static fluid with reference to the following:
(i) drag on an isolated particle in an open system,
(ii) drag on an isolated particle in a closed system,
(iii) particle interactions,
(iv) shape factors.

(University of Melbourne, Chemical Engineering III, November 1963.)

4. Discuss the characteristics of the various types of cyclones for removing particles from gas streams. Comment particularly on the methods that are used for assessing cyclone performance and derive a relation that can be used for predicting this.

5. The basic mechanisms of collection of particles in filtration by fibres are inertial impaction, interception and diffusion. Discuss how these mechanisms operate and their relative importance. Comment in a qualitative way on the interaction of the mechanisms, and how this knowledge helps in the selection of filter materials.
(University of Melbourne, Chemical Engineering III, November 1964.)

7. Discuss the movement of a fully submerged particle and the resistance to its motion. Consider the following aspects:
(a) The different flow regimes in a continuous medium.
(b) Particle size and the effect of a non continuous medium.
(c) The effects of boundaries and other particles.
(d) Particle shape.

How do these particle properties affect the design of an elutriator for separation of different types of dust?

8. A particle-collection system is required for the removal of flyash (density 2) from furnace waste gases flowing at the rate of 15·5 $m^3 s^{-1}$ at 350°C. It has been decided that the initial collection of coarse grit (greater than 100 μm) shall be in a settling chamber followed by some other device.
Discuss the systems that could be used if

   (a) collection efficiency of 99 per cent for all particles is required,
   (b) collection efficiency of about 90 per cent for those particles 5 μm and larger is adequate.

   Estimate the dimensions of a settling chamber and sketch the proposed unit. Dynamic viscosity of air at 350°C: 30·9 μ Pas.

9. Write notes on *three* of the following:

   (a) The basic collection mechanisms of particle collection in fibrous filters.
   (b) The problems of spray scrubbers for particles.
   (c) The main processes in electrostatic precipitation.
   (d) The flow patterns assumed for cyclones for calculating cyclone efficiencies.

   (University of Melbourne, Chemical Engineering III, November 1965.)

10. Discuss the different theories that have been used to calculate the critical particle-separation size or other, similar parameter for cyclones. Develop this theory for a particular theoretical model.
Calculate the critical particle size for a gas cyclone handling 1·4 $m^3 s^{-1}$ of air at 20°C, with particles density 2·5 g cm^{-3}, with a wide range of sizes, dispersed in air.
The cyclone dimensions are:

    diameter 0·9 m, height 3·6 m, outlet duct diameter 0·45 m inlet duct, 0·45 m high and 0·21 m deep, air density = 0·0013 g cm^{-3}, air viscosity = 18·1 μPas.

    (University of Melbourne, Chemical Engineering III, November 1966.)

11. "Filtration of particles from gases is a multistage process requiring a series of mechanisms to be effective".
Discuss this statement and indicate how it can be used to determine suitable properties for a gas filtration medium.
(University of Melbourne, Chemical Engineering III, August 1968.)

12. Discuss the following:

    (a) The behaviour of non-spherical and irregular particles during free fall in a fluid.
    (b) The different equivalent diameters which can be used for calculation of the fluid resistance to non-spherical particles moving through a homogeneous fluid.

(c) Shape factors and their use:
Calculate *one* equivalent diameter and *one* shape factor for both a cube and a rod (length is twice rod diameter).
(University of Melbourne, Chemical Engineering III, August 1968.)

13. Discuss *three* of the following:

(i) Particle agglomeration.
(ii) Thermal precipitation.
(iii) Fluid resistance to non-spherical particles.
(iv) Resistance to an accelerating particle.

14. Outline some of the advantages and disadvantages in using a liquid scrubbing system for control of air pollutants.
Write notes on the following:

(i) spray nozzles;
(ii) optimum droplet sizes in spray towers and centrifugal spray scrubbers;
(iii) venturi scrubbers;
(iv) wire-mesh mist eliminators and their principles of operation.

15. The equation commonly quoted for the drift velocity of particles in an electrostatic precipitator is

$$\omega = \frac{\epsilon}{\epsilon+2} \cdot \frac{CEE'd}{4\pi\mu}.$$

(i) What is the meaning of these terms and what were some of the vital stages and major assumptions made in the derivation?
(ii) Indicate how the drift velocity can be used in an expression for the efficiency of an electrostatic precipitator.
(iii) What is the effective drift velocity, and how does it differ from the one indicated in (ii)? How is it obtained?
(iv) Discuss briefly the following aspects of electrostatic precipitation: dust resistivity; rapping; positive and negative corona.

(University of Melbourne, Chemical Engineering III, August 1969.)

16. Calculate the following shape factors:

(a) surface diameter;
(b) volume diameter;
(c) area diameter;
(d) sphericity;
(e) circularity.

The shapes are:

(1) cube, in terms of one side;

(2) a rectangular particles where the end is square and the sides are twice as long as one side of the square end;
(3) a rod twice the length of the cross-sectional diameter of the end.

17. Calculate the drag diameter for the following in the viscous flow region:

(a) an ellipsoid of revolution, where the minor axis is half the length of the major axis, falling lengthways and at right angles to this direction;
(b) a flat disc plate, horizontal and at right angles to this.

18. A new stainless-steel fibre has been developed, with an average diameter of 5 μm which can be made into a felt, with a permeability of 0·019 $m^3\ s^{-1}$ at N.T.P.
Discuss the possibility of using this "felt" to filter fumes from an electric arc steel making furnace.

(i) The gases contain 12 per cent CO_2, 20 per cent H_2, 58 per cent N_2 and 10 per cent O_2. (Assume linear interpolation for behaviour of mixture characteristics.)
(ii) The gases are at 500°C.
(iii) The particles of fume are largely iron oxide (Fe_3O_4). Particle density approx. 4·5 g cm^{-3} and with a mean particle size of 0·8 μm 95 per cent of the particles are in the range of 0·08 to 5 μm while the next 4 per cent are approximately 0·01 to 15 μm. Fume density is 6·7 g m^{-3}.
(iv) The pad of fibres gives depth of 8 mm, and has an equivalent of forty fibres in one linear path of the particles and gases.

It is suggested that 0·05 m s^{-1} is a suitable linear velocity through the bed. (*Note.* Permeability is arbitrarily defined as the volume that passes through a square metre of filter surface with 0·12 kPa pressure drop in 6 seconds.)
Suggest a design of a filter chamber that could be used for filtering 2·4 $m^3\ s^{-1}$ of gases (N.T.P.) at the temperature.

19. (a) (i) For a simple gravity settling chamber, show that, if Stokes law is assumed, the collection efficiency is given by

$$\eta = \frac{L_s \, . \, B_s d^2(\varrho_p - \varrho)g}{18Q\mu} \qquad (\text{for } \eta \leqq 1)$$

where L_s = length of chamber,
B_s = breadth of chamber,
Q = volumetric flow rate,
μ = gas viscosity
ϱ_p and ϱ = densities of particles and gas, respectively,
d = particle diameter.

Hence find an expression for the smallest particle size which can be completely collected in the chamber.
What other assumption is necessary?

(ii) The gases from a rotary cooler cooling Portland cement emerge at 95°C, and were expected to contain particles as small as 200 μm in diameter which are to be separated in a simple settling chamber. The mass flow of air is 89,000 kg h^{-1}. To avoid re-entrainment of the particles, the velocity of air in the chamber should not exceed 3 m s^{-1}. For space reasons, the length of the chamber is limited to 9 m. Find the approximate dimensions of the chamber required. Take the density of cement as 1·3 g cm^{-3}.

(b) After operation was commenced, it was found that particles as small as 50 μm were passing through the chamber. It was decided to convert the chamber to the Howard type by installing a series of horizontal shelves. How many shelves are needed to ensure complete collection?

20. Sketch the design of a bag house with cylindrical sleeves, 0·2 m diameter, for filtering 25,000 m^3 h^{-1} gases (at 150°C). What are the significant parts? The gases are relatively non-corrosive, and the dust burden is approximately 10 g m^{-3}.

21. A simple scrubbing tower is to be designed for treating 7 m^3 s^{-1} at ambient conditions. Suggest a suitable design to avoid entrainment of 90 per cent of the droplets, if 4 per cent of droplets are less than 0·2 mm. Estimate the efficiency of collection for a dust (see below) in a 6-m-high tower, if the liquid-flow rate is assumed to be 1 dm^3 m^{-2} of tower cross-section per second, and an average droplet size of 0·8 mm is assumed.

Size Distribution of Standard Fine Sand (Stairmand)

Particle size (μm)	Cumulative per cent by weight less than stated size
150	100
104	97
76	90
60	80
40	65
30	55
20	45
10	30
5	20

VIII. VISCOSITY AND DENSITY DATA FOR GASES

Table A.8.1. Dynamic Viscosity of Gases in S.I. units

(Modified data from *Technical Data on Fuels*, Ed. Spiers)[789]

Temperature, °C	Pa s × 10^5									
	Air	N_2	O_2	CO_2	CO	H_2	CH_4	C_2H_4	C_2H_6	Water vapour
—160	78·1	77·1	87·9		74·4	46·0	44·9			
—140	92·0	90·0	103		87·4	51·5	52·8			
—120	105	103	117		99·9	56·8	60·5			
—100	117	114	131		112	61·8	68·1	62		
— 80	129	126	144		124	66·5	75·1	68		
— 60	140	136	157	108	135	71·1	82·5	74		
— 40	151	147	169	118	145	75·6	89·3	81		
— 20	162	157	181	127	156	79·9	95·9	88		
0	172	166	192	137	166	84·1	102	94	86	
20	181	175	203	146	175	88·2	109	101	92	95·6
40	191	184	213	156	185	92·2	115	108	98	103
60	200	193	223	165	194	96·1	121	115	104	110
80	209	201	233	173	202	100·0	127	122	109	117
100	217	209	243	182	211	104	133	128	115	125
150	238	229	266	203	231	113	147	142	129	143
200	257	247	288	222	251	122	160	156	142	161
250	275	265	309	241	269	130	173	169	155	179
300	293	282	329	259	287	139	184	181	166	197
350	309	298	348	276	304	147	196	192	176	215
400	325	313	367	293	320	154	207	202	184	233
450	340	328	385	309	336	162	217	212	193	251
500	355	342	402	324	352	169	227	222	204	269
550	369	355	419	339	367	177	237			287
600	383	368	436	354	382	184	246		230*	306
650	396	381	452	368	396	191	256			327
700	409	393	468	382	410	198	265		249*	348
800	433	417	500	408	437	211	283		269*	387
900	457	440	530	434	464	223	300		283*	424
1000	479	461	559	459	490	235	316		299*	456
Sutherland* Equation factors										
$K = 10^{-7} \times$	150	143	176	158	135	66	—	—	—	—
C	124	110	131	240	102	72	—	—	—	—

* These values were taken from the 5th edition of *Technical Data on Fuels* and were calculated from Sutherland's formula:

$$\text{gas viscosity} = \frac{KT^{\frac{3}{2}}}{T+C},$$

where T = absolute temperature and K, C = constants.

The constants given at the foot of the table can be used for estimating gas viscosities when temperatures exceed 1000°C. The values calculated in this way can only be used as rough approximations.

TABLE A.8.2. DENSITIES OF SOME COMMON GASES

(at 0 or 20°C, 100 kPa (1 atm))

	$kg\ m^{-3}$	
Air	1·2928	(20°C)
Blast furnace gas	1·121	(20°C)
Coal gas		
Horizontal retort	0·48	(20°C)
Continuous vertical retort (no steaming)	0·52	(20°C)
Continuous vertical retort (steaming)	0·58	(20°C)
Producer gas		
Coal (mechanical producer)	1·05	(20°C)
Coke (mechanical producer)	1·08	(20°C)
Carburetted water gas	0·82	(20°C)
Oxygen	1·428	(0°C)
Nitrogen	1·257	(0°C)
(Air)	1·293	(0°C)
Carbon dioxide	1·963	(0°C)
Carbon monoxide	1·250	(0°C)
Hydrogen	0·090	(0°C)
Methane	0·715	(0°C)
Ethane	1·037	(0°C)
Propane	2·020	(0°C)
Ethylene	1·251	(0°C)
Hydrogen sulphide	1·521	(0°C)
Sulphur dioxide	2·922	(0°C)
Sulphuric acid vapour	4·380	(0°C)
Ammonia	0·771	(0°C)
Hydrochloric acid gas	1·639	(0°C)
Fluorine	1·696	(0°C)
Chlorine	3·214	(0°C)
Argon	1·783	(0°C)

TABLE A.8.3. CONVERSION FACTORS. MEASURES OF CONCENTRATION

(M = molecular weight)

To convert from	to	multiply by
g/m^3	g/ft^3	0·02832
	$lb/1000\ ft^3$	0·06243
	$gr./ft^3$	0·43701
$lb/1000\ ft^3$	mg/m^3	$16{\cdot}018 \times 10^3$
	g/ft^3	0·35314
	$gr./ft^3$	7
$gr./ft^3$	g/m^3	2·288
$gr./1000\ ft^3$	mg/m^3	2·288
ppm by volume	ppm by weight	$M/28{\cdot}8$
	lb/ft^3	$M/3{\cdot}851 \times 10^8$

MEASURES OF PRESSURE

To convert from	to	multiply by
mm Hg	Atmospheres	0·001316
	Millibars	1·333
in. W.G. (water gauge) (60°F)	mm Hg (0°C)	1·8663
	Pascals	249·1

MEASURES OF VELOCITY

To convert from	to	multiply by
m/sec	ft/sec	3·281
	m.p.h.	2·237
	km/hr	3·6

IX. S.I. UNITS

During the 1970s the metric system of weights and measures is being introduced progressively. As a result engineering and science students at Universities and Colleges of Advanced Education are being taught in these units, textbooks and tables increasingly use metric units, and more and more of our products and materials are being measured and sold in units based on a metric system.

The metric system being used is known as the S.I. System, because of its French title "Système International d'Unités", and it is simpler and more restrictive than the metric "c.g.s." (centimetre, gramme, second) or "m.k.s." (metre, kilogramme, second) systems that many of us have learned; it uses neither the calorie as a unit of heat, nor the poise as a unit of viscosity. The confusions of the "e.s.u." system of electrical units are avoided, and no longer will one have to worry whether it is an Imperial gallon or a U.S. barrel.

Basic S.I. units

The basic units in the S.I. system are the kilogramme (mass), metre (length), second (time), Kelvin (temperature), ampere (electric current), candela (the unit of luminous intensity) and radian (angular measure). All but the last are commonly used by the air-pollution engineer. The Celsius scale of temperature (0°C = 273·15°K) is commonly used instead of the absolute or Kelvin scale.

Derived S.I. units

The important derived units are the newton (S.I. unit of force), the joule (S.I. unit of heat and work), the watt (S.I. unit of power), the pascal (S.I. unit of pressure), the hertz (unit of frequency); as well there are a number of electrical units, viz. coulomb (charge), farad (capacity), henry (inductance), volt (potential) and weber (magnetic flux).

Metric Multipliers

One of the major advantages of the metric system is that larger and smaller units are given in powers of ten. In the S.I. system a further simplification is introduced, by using only those units with multiples of 10^3. Thus for lengths, in engineering, the micrometre (previously micron), millimetre, and kilometre will be used, and the centimetre will be abandoned. A further simplification is that the decimal point will be substituted by a comma (as in France, Germany and South Africa), while the other numbers, before and after the comma, will be separated by spaces between groups of three, i.e. one million dollars will be "$1 000 000,00." Other countries, such as the United Kingdom, will be retaining the decimal point, and using the comma as before; i.e. $1,000,000.00.

TABLE A.9.1. PREFIXES, SYMBOLS AND MULTIPLIERS FOR THE S.I. SYSTEM

Prefix	Symbol	Multiplier
Tera	T	10^{12}
Giga	G	10^{9}
Mega	M	10^{6}
kilo	k	10^{3}
—	—	—
milli	m	10^{-3}
micro	μ	10^{-6}
nano	n	10^{-9}
pico	p	10^{-12}

Table A.9.1 gives the prefixes and multipliers for the S.I. system. It should be noted that the units of time have not been decimalized, and the minute, hour and day remain with their traditional multipliers of 60 and 24. There are 86,400 sec in a day. Table A.9.2 gives the basic and derived units commonly used in the S.I. system, and some of the most important conversions to traditional units. Table A.9.3 lists some of the secondary measures commonly used, and Table A.9.4 the special combinations of units in frequent use by air-pollution-control engineers and scientists. Some conversions from traditional to S.I. units are given in Table A.9.5.

All of the industrial nations in the world have either been metric for some generations, have commenced metrication, or are about to do so. Even the United States, which for a time was the only major industrial country contemplating staying with traditional units, has now decided to go over to S.I. units.

It is interesting to note that a decimal system, based on the earth as a fundamental unit, was first suggested in 1670 by Gabriel Mouton, a vicar of Lyons. The French Revolution was a period of change, during which the metric system of weights and measures was introduced. This was detailed in the French legislation of 1795, and confirmed by decree on the 19th Frimaine, in the eighth year, otherwise the 10th December 1799, under Napoleon.

TABLE A.9.2. BASIC AND DERIVED UNITS

Quantity	Name	Symbol	Definition	Conversion
Length	metre	m	—	3·281 ft 39·370 in
	millimetre	mm	10^{-3} m	0·03937 in
	micrometre	μm	10^{-6} m	$3·937 \times 10^{-5}$ in
	kilometre	km	10^{-3} m	5/8 mile
Mass	kilogram	kg	—	2·2046 lb
	gram	g	10^{-3} kg	0·03527 oz 15·432 grains
	milligram	mg	10^{-6} kg	0·01542 grains
Electric current	ampere	A	—	—
Time	second	s	—	
	minute	min	60 s	—
	hour	hr	3600 s	—
	day	day	86400 s	—
Frequency	hertz	Hz	S^{-2}	—
Energy	joule	J	kg m^2 s^{-2}	0·239 cal. $9·48 \times 10$ B.t.u.
Pressure	pascal	Pa	kg m^{-1} s^{-1}	Nm^{-2} $4·146 \times 10^{-3}$ in W.G. (4°C) 0·10197 mm W.G. (4 °C) $2·953 \times 10^{-4}$ in mercury (0°C) $7·50006 \times 10^{-3}$ mm (0°C) (torr) $9·869 \times 10^{-6}$ atm.
Force	newton	N	kg m s^{-1}	J m^{-1} 10^5 dyne 0·2248 lb (force)
Power	watt	W	kg m^2 s^{-3}	J s^{-1} 3·414 B.t.u./hr $1·341 \times 10^{-3}$ H.P.
Electrical charge	coulomb	C	A s	—
Potential difference	volt	V	kg m^2 s^{-3} A^{-1}	JA^{-1} s^{-1}
Resistance	ohm	Ω	kg m^2 s^{-3} A^{-2}	VA^{-1}
Conductance	siemens	S	s^3 A^2 kg^{-1} m^{-2}	AV^{-1} Ω^{-1}

TABLE A.9.3. SECONDARY UNITS

Unit of		Abbreviation	Conversion into	Multiply by
Area	square metre	m^2	cm^2	10^4
			mm^2	10^6
			ft^2	10·764
			yd^2	1·196
Volume	cubic metre	m^3	1	10^3
			ft^3	35·31
			U.S. barrels (oil)	6·29
			in.^{3}	$6{\cdot}102 \times 10^4$
			galls (Imp.)	220
			galls (U.S.)	264
Density	kilogram per cubic metre	kg m^{-3}	g cm^{-3}	10^{-3}
			lb ft^{-3}	0·06243
			lb in^{-3}	$3{\cdot}613 \times 10^{-5}$
Dynamic Viscosity	pascal second	Pa s (or N s m^{-2})	Poise	10
			kg f.s. m^{-2}	0·1020
			lb f.s ft^{-2}	0·02088
Kinematic Viscosity	square metre per second	m^2 s^{-1}	stoke($m^2 s^{-1}$)	10^4
			ft^2 hr^{-1}	$3{\cdot}875 \times 10^4$

The basic unit of the metric system was the length of 1 metre, representing, as suggested by the vicar of Lyon, one ten-millionth of the quadrant of the circumference of the earth. The length, or rather a section of the arc, was based on a special survey by two French surveyors, Delambre and Mechain, who surveyed the meridian running through Barcelona in the south to Dunkirk in the north in the years following 1791.

There was, however, strong prejudice against the new metric system in France, and another decree under Napoleon re-established the old scheme of French weights and measures in 1812. This was withdrawn only 25 years later. Other European countries were slow to follow the road to metrication. For example, the German states only abandoned their old systems, which varied from state to state, seven years after the formation of the Empire, by decree of the Federal parliament in 1877. This decree followed the Metric Convention Treaty of 1875 to which eighteen countries were signatories.

In England, uniform weights and measures were decided upon as part (Chapter 35) of the Magna Carta in 1215. It is this uniformity which probably relieved the pressure towards metrication in the nineteenth century, although the use of metric weights and measures in

TABLE A.9.4. SPECIAL UNITS

Unit of	Abbreviation	Conversion to	Multiply by
Concentration	kg/m^3	lb/ft^3	0·06243
	g/m^3	$grains/ft^3$	0·437
	mg/m^3	$grains/ft^3$	0·000437
		ppm	22·41/M*
Volume Flow	$m^3\ s^{-1}$	ft^3/s	35·3
	m^3/min	ft^3/min	35·3
	m^3/hr	ft^3/hr	35·3
		ft^3/min	0·5883
Velocity (flows through filters)	$m\ s^{-1}$	ft/s	3·2808
		ft/min	196·85
	m/min	ft/min	3·2808
		ft/sec	0·05468
	m/hr	ft/min	0·05468
Deposition (based on 30 day month)	$mg\ m^{-2}\ (day)^{-1}$	$tons/mile^2$ month (30 day month)	0·0765

* M = molecular weight

contracts became legal in 1864, and the metric system as a whole was legalized by act of parliament in 1897.

The United States, following the advice of Thomas Jefferson, toyed with a decimal system developed by Jefferson in 1791. Thirty years later, another future president, John Quincy Adams, recommended metrication after the French system to Congress. However, although the United States has permitted metric measures since 1864, and was a member of the Metric Convention of 1875, it has not yet taken the final steps towards metrication.

Many experienced engineers will find the conversion difficult, as for a time they will lose the "feel" for reasonable quantities, whether they be temperatures, gas flow rates or pollutant concentrations. Nonetheless the effort at complete conversion will have to be made soon, or we will not be able to follow designs and developments made throughout the world.

TABLE A.9.5. CONVERSIONS OF CONVENTIONAL TO S.I. UNITS

Quantity	Multiply by	to obtain S.I. unit
ft	0·3048	metres (m)
ins	$2{\cdot}54\times10^{-2}$	metres (m)
miles	$1{\cdot}609\times10^{-3}$	metres (m)
ft/sec	0·3048	metres/sec (m s^{-1})
ft/min	$5{\cdot}08\times10^{-3}$	metres/sec (m s^{-1})
ft^3/sec	$28{\cdot}32\times10^{-3}$	m^3 s^{-1}
ft^3/min	$0{\cdot}472\times10^{-3}$	m^3 s^{-1}
Imp gall/min	$75{\cdot}77\times10^{-6}$	m^3 s^{-1}
pounds	0·4536	kilogrammes (kg)
ounces	$28{\cdot}35\times10^{-3}$	kilogrammes (kg)
grains	$64{\cdot}8\times10^{-6}$	kilogrammes (kg)
lb/ft^3	16·02	kg m^{-3}
grains/ft^3	2·29	g m^{-3}
lb/sec	0·4536	kg s^{-1}
lb/min	$7{\cdot}56\times10^{-3}$	kg s^{-1}
lb/hr	$126{\cdot}0\times10^{-6}$	kg s^{-1}
ft^2	$92{\cdot}9\times10^{-3}$	m^2
yd^2	0·836	m^2
ft^3	$28{\cdot}317\times10^{-3}$	m^3
Imp. galls	$4{\cdot}546\times10^{-3}$	m^3
lb (f)/in^2	$6{\cdot}895\times10^3$	Pa, (Nm^{-2})
atm.	$101{\cdot}3\times10^3$	Pa, (Nm^{-2})
in. W.G.	249·1	Pa, (Nm^{-2})
in. mercury	$3{\cdot}386\times10^3$	Pa, (Nm^{-2})
mm mercury	133·3	Pa, (Nm^{-2})
Poise	10^{-1}	Pa, (Nsm^{-2})
lb f.sec/ft^2	47.88	Pa, (Nsm^{-2})
kg f.sec/m^2	9·807	Pa, (Nsm^{-2})
Stoke (cm/s)	10^{-4}	m^2 s^{-1}
ft^2/hr	$25{\cdot}81\times10^{-6}$	m^2 s^{-1}
ft^2/sec	$92{\cdot}90\times10^{-3}$	m^2 s^{-1}
B.t.u.	$1{\cdot}055\times10^3$	J
therm	$105{\cdot}5\times10^6$	J
k Wh	$3{\cdot}60\times10^6$	J
calorie	4·1868	J
ft lb (f)/sec	1·356	W
horse power	745·7	W
B.t.u./lb.	$2{\cdot}326\times10^3$	J kg^{-1}
B.t.u./ft^3	37.21×10^3	J m^{-3}
B.t.u./lb °F	$4{\cdot}187\times10^3$	J kg^{-1} K^{-1}
B.t.u./ft^2 hr	3·155	W m^{-2}
tons/$mile^2$ month	13·077	mg m^{-2} $(day)^{-1}$

BIBLIOGRAPHY

1. Abel, E. and Barth, H., *Arch. Eisenhüttenwes.* **29,** 683 (1958).
2. Adams, D. F., Hendrix, J. W. and Applegate, H. G., *J. Agric. Fd. Chem.* **5** (2), 108–16 (1957).
3. Aerotec Industries Inc., South Norwalk, Conn., U.S.
4. Aitken, J., *Trans. Roy. Soc.* (Edinburgh), **32,** 239 (1887).
5. Albany Felt Co., United States Pat. No. 2,896,263, July (1959).
6. Albrecht, F., *Physik. Z.* **32,** 48 (1931).
7. Alden, J. L., *Design of Industrial Exhaust Systems*, 3rd ed., The Industrial Press, New York (1959).
8. Alexander, L. G. and Coldren, C. L., *Ind. Engng Chem.* **43,** 1325 (1951).
9. Alexander, R. McK., *Proc. Australas. Inst. Min. Metall.* **152,** 203 (1949).
10. Alford, H. B., *Combustion*, **32,** 45 (1960).
11. Allcut, A. E., *Proc. Instn Mech. Engrs* **140,** 308 (1938).
12. Allen, C. H. and Rudnik, I., *J. Acoust. Soc. Am.* **19,** 857 (1947).
13. Allen, H. S., *Phil. Mag.* **50,** 323, 519 (1900).
14. Allen, S. and Hamm, J. R., *Trans. Am. Soc. Mech. Engrs*, pp. 851–8 (1950).
15. Alonso, J. R. F., *Chem. Engng* **78** (27) 86–96 (1971).
16. Amelin, A. G., *Kolloid. Zh.* **10,** 169 (1948).
17. Amelin, A. G. and Belakov, M. I., *Kolloid. Zh.* **17,** 10 (1951).
18. American Air Filter Co. Inc., 215 Central Ave., Louisville, 8, Ky., U.S.
19. Anderson, H. C. and Green, W. J., *Ind. Engng Chem.* **53,** 645 (1961).
20. Anderson, E. from Perry, J. H. (Ed.) *Chemical Engineer's Handbook*, 1st ed., p. 1548, McGraw-Hill, New York (1935).
21. Anderson, R. B., McCartney, J. T., Hall, W. K. and Hofer, L. J. E., *Ind. Engng Chem.* **39,** 1618 (1947).
22. Andreasen, A. H. M., *Staub*, **35,** 11 (1954).
23. Anon: *Times Review of Industry*, p. 80 (July 1954).
24. Anon: *Cost Engineering in the Process Industries*, Ed. Chilton, C. H., p. 331, McGraw-Hill, New York (1960). Republished from *Chem. Engng*, **66,** 13th July (1959).
25. Anon: *Engineering*, **189,** 806 (1960).
26. Anon: *Chem. Engng*, **68,** No. 18, p. 84, 4th Sept. (1961).
27. Anon: *Chem. Process*, 10th Feb. (1964) (reprint).
28. Anon: *Chem. Engng*, **71,** 128, 20th July (1964).
29. Anon: *A Compilation of Ambient Air Quality Standards and Objectives*, prepared by Technical Assistance Branch, Divn. of Air Pollution, Public Health Service, Dept. of H.E.W., Taft Sanitary Engng Center, Cincinnati, Ohio, 11th May (1965).
30. Anon: *Environmental sci. Tech.* **1,** 282–6 (1967).
31. Anon: *A Digest of State Air Pollution Laws* (1967 edition), U.S. Dept. of H.E.W., Public Health Service, National Center for Air Pollution Control, Washington D.C. 20201, pp. 556 (1968).
32. Anon: *Electrical World*, **29,** 9th Oct. (1967).
33. Applebey, M. P., *Trans. Soc. Chem. Ind.* **56,** 139 (1937).
34. Arawaki, M., *Mitsubishi Heavy Industries Ltd. Technical Review*, **5,** 208 (1968).
35. Arendt, P. and Kallmann, H., *Z. Physik*, **35,** 421 (1925).
36. Arnold, H. D., *Phil. Mag.* **22,** 755 (1911).
37. Ashman, R., *Proc. Instn Mech. Engrs*, **1B,** 157 (1952).
38. Asklöf, S. H. V., U.S. Pat. No. 3,026,966, 27th Mar. (1962).

39. *Am. Soc. Mech. Engrs, Standard No. APS-***1**, *Recommended Guide for the Control of Dust Emissions—Combustion for Indirect Heat Exchangers*, Am. Soc. Mech. Engrs, New York (1966).
40. ATSUKAWA, M., MISHIMOTO, Y. and MATSUMOTO, K., *Mitsubishi Heavy Industries Ltd. Technical Review*, **2,** 134 (1965).
41. ATSUKAWA, M., MISHIMOTO, Y. and MATSUMOTO, K., *Mitsubishi Heavy Industries Ltd. Technical Review*, **4,** 33 (1967).
42. ATSUKAWA, M., MISHIMOTO, Y. and TAKAHASHI, N., *Mitsubishi Heavy Industries Ltd. Technical Review*, **5,** 129 (1968).
43. AVY, A. P., *Staub*, **37,** 372 (1954).
44. AVY, A. P., in *Problems and Control of Air Pollution*, Ed. MALLETTE, F. S., p. 264, Reinhold (1955).
45. Babcock & Wilcox Ltd., Farringdon St., London E.C. 4, Bull. No. 1467/1.
46. BACHMAN, P. W. and TAYLOR, G. B., *J. Phys. Chem.* **33,** 447 (1929).
47. BACHMANN, D., *Dechema Monographs*, **31** (1959).
48. BADGER, W. L. and BANCHERO, J. T., *Introduction to Chemical Engineering*, McGraw-Hill, New York, p. 425 (1955).
49. BADZIOCH, S., *J. Inst. Fuel*, **33,** 106 (1960).
50. BAGG, J., The formation and control of oxides of nitrogen in air pollution, in *Air Pollution Control*, Ed. STRAUSS, W., pp. 35–94, Wiley–Interscience, New York (1971).
51. BAINBRIDGE, C. A., private communication (1958).
52. BAINBRIDGE, C. A., *Chem. Process.*, Nov. (1961). Brit. Pat. No. 851,555.
53. BALIFF, J., GREENBURG, L. and STERN, A. C., *Am. Ind. Hyg. Ass. Quart.* **9,** 85 (1948).
54. BARBER, R., JACKSON, R., LAND, T. and THURLOW, G. G., *J. Inst. Fuel*, **27,** 408 (1954).
55. BARKER, D. H., Ph.D. Thesis, Univ. of Utah (1951).
56. BARNEBEY, H. L., *Heat. Pip. Air Condit.*, p. 155 (Mar. 1958).
57. BARTH, W., *Ver. Deut. Ing. Tagungsheft*, **3,** 11 (1954).
58. BARTH, W., *Brennst.-Wärme-Kraft*, **8,** 1 (1956).
59. BARTH, W., *Staub*, **19,** 175 (1959).
60. BARTH, W., *Staub*, **21,** 382 (1961).
61. BARTH, W. and LEINWEBER, L., *Staub*, **24** (2), 41–55 (1964).
62. BASSA, G. and BEER, J., *Silikattechnik*, **5,** 529 (1954).
63. BAUM, F. and RIESS, F., *Staub*, **24,** 369 (1964).
64. BAUM, K., *Berg.- u. hüttenm. Mh.* **104,** No. 2, 41 (1959).
65. BECKER, H., RECHMANN, H. and TILLMANN, P., *Kolloid. Z.* **169,** 34 (1960).
66. BECKER, H. A., *Can. J. Chem. Engng*, **37,** 85 (1959).
67. BENARIE, M., *Staub*, **22,** 118 (1962).
68. BENFORADO, D. M., PAULETTA, C. E. and HAZZARD, N. D., *Air Engineering* (Mar. 1967) (reprint).
69. BENFORADO, D. M. and WAITKUS, J., Fume control in wire enamelling by direct flame incineration, *Air Pollut. Control Ass. Annual Conference*, Cleveland, June (1967).
70. BERG, G. J. VAN DEN, *Trans. Instn. Chem. Engrs*, London, **35,** 409 (1957).
71. BERG, R. H., *Am. Soc. Test Mater.*, Special Technical Publication No. 234, p. 245 (1958).
71a. BERGMANN, L., *Wasser, Luft und Betrieb*, **5** (10) Oct. (1968) (reprint).
72. BERNER, A., *Staub Reinhalt. Luft*, **26,** 167 (1966).
73. BERNER, A. and PREINING, O., *Staub*, **24,** 292 (1964).
75. BIENSTOCK, D., BRUNN, L. W., MURPHY, E. M. and BENSON, H. E., Sulphur dioxide, its chemistry and removal from flue gases, *U.S. Bur. Mines Information Circular 7836* (1958).
76. BIENSTOCK, D., FIELD, J. H. and MYERS, J. G., *J. Eng. Pwr. Am. Soc. Mech. Engrs*, Ser. **A, 86** (3), 353 (1964).
77. BIENSTOCK, D., FIELD, J. H., KATELL, S. and PLANTS, K. D., *J. Air Pollut. Control Ass.* **15,** 459 (1965).
78. BILLINGS, C. E., *Effects of Particle Accumulation in Aerosol Filtration*, report from the W. M. Keck Laboratory of Environmental Health Engng, California Institute of Technology, Pasadena, Sept. (1966).
79. BILLINGS, C. E. and SILVERMAN, L., *Int. J. Air Wat. Pollut.* **6,** 455 (1962).
80. BILLINGS, C. E., SILVERMAN, L. and KURKER, C., *J. Air Pollut. Control Ass.* **8** (3) 185 (1958).
81. BILLINGS, C. E., SILVERMAN, L. and KURKER, C., *Industrial Wastes* (1959) (reprint).
82. BILLINGS, C. E., SILVERMAN, L., LEVENBAUM, L. H., KURKER, C. and HICKEY, E. C., *J. Air Pollut. Control Ass.* **8** (1), 53 (1958).

83. Billings, C. E., Silverman, L., Denis, R. and Levenbaum, L. H., *J. Air Pollut. Control Ass.* **10** (4), 318 (1956).
84. Billings, C. E., Silverman, L. and Small, W. D., *J. Air Pollut. Control Ass.* **5** (3), 159 (1955).
85. Black, C. H., Effectiveness of a fluidized bed in filtration of airborne particulates of submicron size, Ph.D. Thesis, Oregon State Univ. (1967).
86. Blacktin, S. C., *Trans. J. Soc. Chem. Ind.* **58,** 334 (1939).
87. Blacktin, S. C., *Trans. J. Soc. Chem. Ind.* **59,** 153 (1940).
88. Blankenburg, R., *Staub,* **21,** 321 (1961).
89. Blankenburg, R., *Staub,* **21,** 426 (1961).
90. Blasewitz, A. G. and Judson, B. F., *Chem. Engng. Prog.* **51,** 6 (1955).
91. Blasewitz, A. G. and Schmidt, W. C., *2nd Int. Conf. on Peaceful Uses of Atomic Energy,* **18,** 184 (1958).
92. Blatz, H., *Introduction to Radiological Health,* chap. 4, McGraw-Hill, New York (1964).
93. Boehme, G., Kling, W., Krupp, H., Lange, H. and Sandstede, G., *Z. angew. Phys.* **16,** 486 (1964).
94. Boehme, G., Krupp, H., Rabenhorst, H. and Sandstede, G., *Trans. Instn Chem. Engrs,* London, **40,** 252 (1962).
95. Höhlen, B., Lüthi, J. and Guyer, A., *Chemie-Ing-Tech.* **39,** 910–13 (1967).
96. Böhm, J., *Staub Reinhalt. Luft,* **30,** 99–106 (1970).
97. Hohne, H., *Allg. Forstzeitschr.* **18** (7), 109–11 (1963).
98. Bol, J., Gebhart, J., Heinze, W., Petersen, W. D. and Wurzbacher, G., *Staub Reinhalt. Luft,* **30,** 475 (1970).
99. Booras, S. G. and Zimmer, C. E., Comparison of conductivity and West-Gaeke analyses for SO_2, Paper No. 67-109, presented at 60th Annual Meeting, *Air Pollut. Control Ass.,* 11th June (1967).
100. Borgwardt, R. H., Harrington, R. E. and Spaite, P. W., *J. Air Pollut. Control Ass.* **18** (6), 387–90 (1968).
101. Bosanquet, C. H., *Trans. Instn Chem. Engrs* (London), **28,** 130 (1950). Appendix to paper by Stairmand, C. J.
102. Bosanquet, C. H., Carey, W. F. and Halton, E. M., *Proc. Instn Mech. Engrs,* **162,** 355 (1950).
103. Bosanquet, C. H. and Pearson, J. L., *Trans. Faraday Soc.* **32,** 1249 (1936).
104. Bostock, W., *J. Sci. Inst.* **29,** 209 (1952).
105. Bostrom, C. E. and Brosat, C., *Atmospheric Environment,* **3,** 407 (1969).
106. Boucher, R. M. G., *Z. Aerosol Forsch.* **6,** 26 (1957).
107. Boucher, R. M. G., *Genie Chim.* **77** (6), 163 (1957); *ibid.* **78** (1), 14 (1957).
108. Boucher, R. M. G., *Chem. Engng,* **68,** 83, 2nd Oct. (1961).
109. Boucher, R. M. G., private communication (1968).
110. Boucher, R. M. G. and Koehler, G. R., *Ultrasonic Demister Project.* Preliminary Report (undated). Macrosonics Corpn, Cartoret, N.J.
111. Boucher, R. M. G. and Weiner, A. L., *Ultrasonic News,* Summer (1962) (reprint).
112. Bouvier, R. M., *Proc. Am. Pwr Conf.* **24,** 138 (1964).
113. Bradley, D. and Pulling, D. J., *Trans. Instn Chem. Engrs* (London), **37,** 34 (1959).
114. Brandt, O., Freund, H. and Hiedemann, E., *Kolloid. Z.* **77,** 103 (1936).
115. Brandt, W., *Staub,* **21,** 392 (1961).
116. Brennan, E., Leone, I. A. and Daines, R. H., *Int. J. Air Wat. Pollut.* **9,** 791–7 (1965).
117. Brewer, G. L., *Chem. Engng,* **75** (22), 160, 14th Oct. (1968).
118. Bricard, J., in *Problems of Atmospheric and Space Electricity,* Ed. Coroniti, S. C., Elsevier, Amsterdam (1965).
119. Brink, J. A., Mist removal—P_2O_5, in *Gas Purification Processes,* ed. G. Nonhebel, Geo. Newnes, London (1964).
120. Brink, J. A., Air pollution control with fibre mist eliminators, Paper presented to the 1962 *Annual Chem. Engng Conf. of the Chem. Inst. of Can.,* Sarnia, Ontario, Oct. (1962) (preprint).
121. Brink, J. A., Burggrabe, W. F. and Greenwell, L. E., *Chem. Engng Prog.* **62** (4), 60–65 (1966).
122. Brink, J. A., Burggrabe, W. F. and Greenwell, L. E., *Chem. Engng Prog.* **64** (11), 82–86 (1968).
123. Brink, J. A., Burggrabe, W. F. and Rauscher, J. A., *Chem. Engng Prog.* **60** (11), 68–73 (1964).
124. Brink, J. A. and Contant, C. E., *Ind. Engng Chem.* **50,** 1157 (1958).
125. Brinkman, H. C., *Appl. Sci. Res.* **A1,** 27 (1949).
126. Brinkman, H. C., *J. Phys. Chem.* **20,** 571 (1952).
127. Brisse, A. H., *Am. Soc. Mech. Engrs,* Apr. (1960) (preprint 50PR1-16).
128. British Standard, Flow measurement, *B.S. No. 1042* (1943–57), British Standards Inst., London.

129. British Standard, Code for the sampling and analysis of flue gases, *B.S. No. 1756* (1952), British Standards Inst., London.
130. British Standard, Methods of test for air filters used in air conditioning and general ventilation, *B.S. No. 2831* (1957), British Standards Inst., London.
131. British Standard, Code for the continuous sampling and automatic analysis of flue gases. Indicators and recorders, *B.S. No. 3048* (1958), British Standards Inst., London.
132. British Standard, Methods for the sampling and analysis of fuel gases, *B.S. No. 3156* (1959), British Standards Inst., London.
133. Brock, J. R., *J. Colloid Sci.* **17,** 768 (1962).
134. Brock, J. R. and Hidy, G. M., *J. Appl. Phys.* **36,** 1857 (1965).
135. Bromley, L. A. and Read, S. M., Removal of sulphur dioxide from stack gases by sea water, *Research Report S-15, Project Clean Air*, University of California, Berkeley, 1st Sept. (1970).
136. Brooks, S. H. and Calvert, W. J., *Iron Steel Inst. Special Report No. 61*, p. 5 (1958).
137. Bruederle, E., Scheidel, C. and Werner, H., *Blast Furnace and Steel Plant* (reprint) Oct. (1960).
138. Brunauer, S., *Adsorption of Gases and Vapours*, Princeton Univ. Press, Princeton (1943).
139. Buck, M. and Giess, H., *Staub Reinhalt. Luft*, **26,** 379 (1966).
140. Buell-van Tongeren System, Buell (1952) Ltd., 3 St. James Sq., London, S.W. 1.
141. Bureau, A. V. and Olden, J. F., *The Chemical Engineer*, **206,** CE55, Mar. (1967).
142. Burgers, J. M., *Proc. K. Ned. Akad. Wetensch.* **44,** 1045, 1177 (1941); **45,** 9, 126 (1942).
143. Burke, E., *Chemistry and Industry*, 1312 (1955).
144. Busby, H. G. T. and Darby, K., *Colloques Internationaux du Centre National de la Recherche Scientifique*, Grenoble, 27th Sept. (1960). (Editions du Centre National de la Recherche Scientifique, Paris, 1961.)
145. Busby, H. G. T. and Darby, K., *J. Inst. Fuel*, **36,** 184–97 (1963).
146. Busby, H. G. T., Whitehead, C. and Darby, K., High efficiency precipitator performance on modern power stations firing fuel oil and low sulphur coals, Paper EN 34 H, *Second Int. Clean Air Congress*, 6th–11th Dec. (1970), Washington D.C.
147. Cadle, R. D., Wilder, A. G. and Schadt, C. F., *Science*, **118,** 490 (1953).
148. Carlton-Jones, D. and Schneider, H. B., *Chem. Engng*, **75** (22), 166, 14th Oct. (1968).
149. Carman, P. C., *Trans. Instn Chem. Engrs* (London), **15,** 150 (1937).
150. Carman, P. C., *Flow of Gases Through Porous Media*, Butterworths, London (1956).
151. Carpenter, S. B., Leavitt, J. M., Thomas, F. W., Frizzola, J. A. and Smith, M. E., *J. Air Pollut. Control Ass.* **18** (7), 458–65 (1968).
152. Casella, C. F. & Co., London. Leaflet 778, Settlement dust counter.
153. Casella, C. F. & Co., London. Leaflet 804, Thermal precipitator.
154. Cawood, W., *Trans. Faraday Soc.* **32,** 1068 (1936).
155. Chemical Construction Corpn, 525 West 43rd St., New York 36, N.Y.
156. Chen, C. Y., *Chem. Revs.* **55,** 595 (1955).
157. Chen, N. H., *Chem. Engng*, **69,** p. 109, 5th Feb. (1962).
158. Chowdhury, K. C. R. and Fritz, W., *Chem. Engng Sci.* **11,** 92 (1959).
159. Christian, J. G. and Johnson, J. E., *Ind. Engng Chem. Prod. Res. Dev.* **2,** 235–7 (1963).
160. Cobine, J. D., *Gaseous Conducteurs*, Dover, New York (1958).
161. Cochet, R., *Compt. Rend.* **243,** 243 (1956).
162. Cochet, R., *Colloques int. Cent. natn. Rech. Scient.*, Paris, **102,** 331 (1961).
163. Cock, W. H. and Ferris, G. K., Fabric filters—selection and application problems, Paper 2.10, *Clean Air Conf.*, Sydney (1969).
164. Coke, J. R., Ph.D. Thesis, Sheffield Univ. (1963).
165. Cole, A. F. W. and Katz, M., *J. Air Pollut. Control Ass.* **16,** 201–6 (1966).
166. Collins, T. T., Seaborne, C. R. and Anthony, A. W., *Paper Trade J.* **26** (3), 55 (1948).
167. Combustifume Premix Burners, Maxon Premix Burner Co. Inc., Muncie, Ind., U.S.A.
168. Committee on *Permissible Doses for Internal Radiation*, Values recommended in the Report of: (1958 revision).
169. Compton, O. C. and Remmert, L. R. (1950), quoted by Thomas, M. D. and Hendricks, R. H. in Section 9, *Air Pollution Handbook*, Ed. Magill, P. L., Holden, F. R. and Ackley, C., McGraw-Hill, New York (1956).
170. Connor, P., Hardwick, H. W. and Laundy, B. J., *J. Appl. Chem.* **9,** 529 (1959).
171. Continental Air Filter Inc., Louisville, Ky., U.S.

172. COOPERMAN, P., Turbulent gas flow and electrostatic precipitation, *Am. Instn Elect. Engrs*, Winter General Meeting, N.Y. (1960).
173. COOPERMAN, P., Eddy diffusion and particle erosion in electrostatic precipitation, Paper 65-132, *Air Pollut. Control Ass., 58th Gen. Meeting*, Toronto (1965).
174. COOPERMAN, P., Boundary layer effects in electrostatic precipitation, Paper 66-124, *Air Pollut. Control Ass.*, **59***th Ann. Gen. Meeting*, San Francisco (1966).
175. CORBETT, P. F. and CRANE, W. M., *B.C.U.R.A. Bull.* **16,** 1 (1952).
176. CORN, M., *J. Air Pollut. Control Ass.* **11,** 566 (1961).
177. CORN, M., Adhesion of particles, in *Aerosol Science*, Ed. DAVIES, C. N., chap. 11, Academic Press, London (1966).
178. CORN, M. and SILVERMAN, L., *Am. Ind. Hyg. Ass. J.* **22,** 337 (1961).
179. CORN, M. and STEIN, F., *Am. Ind. Hyg. Ass. J.* **26,** 325 (1965).
180. CORTELYOU, C. G., Sulphur dioxide removal from flue gas, Reprint 54A, presented at the Symposium on Industrial Research and Development of Sulphur Dioxide Control Processes, *61st Ann. Meeting, A.I.Ch.E.*, Los Angeles, 1st–5th (1968).
181. COTTRELL, F. G., U.S. Pat. No. 895,729 (1908) (also U.S. Pat. Nos. 866,843 and 945,717).
182. COTTRELL, F. G., German Pat. No. 230,570, 17th Mar. (1908).
183. COTTRELL, F. G., *J. Ind. Engng Chem.* **3,** 542 (1911).
184. COULL, J. C., BISHOP, H. and GAYLORD, W. M., *Chem. Engng Prog.* **45,** 525 (1949).
185. COXON, W. F., *Flow Measurement and Control*, Heywood, London (1959).
186. COYKENDALL, J. W., SPENCER, E. F. and YORK, O. H., *J. Air Pollut. Control Ass.* **18** (5), 315–18 (1968).
187. CRAGGS, J. M. and MEEK, J. D., *Electrical Breakdown in Gases*, The Clarendon Press, Oxford (1953).
188. CULLIS, C. F., HENSON, R. M. and TRIM, D. L., *Proc. Roy. Soc.* (London), Series A, **295,** 72 (1966).
189. CUMMINGS, W. G. and REDFERN, M. W., *J. Inst. Fuel*, **30,** 628 (1957).
190. CUNNINGHAM, E., *Proc. Roy. Soc.* (London), A, **83,** 357 (1910).
191. CZAJA, A. T., *Staub*, **22,** 228 (1962).
192. DAESCHNER, H. W., SEIBERT, E. E. and PETERS, E. P., *Am. Soc. Test Mater.*, Special Pub. No. 234 (1958).
193. DALLA VALLA, J. M., *United States Technical Conference on Air Pollution*, Ed. MCCABE, L. C., p. 341, McGraw-Hill, New York (1952).
194. DALLA VALLA, J. M., in *Cost Engineering in the Process Industries*, Ed. CHILTON, C. H., p. 177, McGraw-Hill, New York (1970). (First appeared in *Chem. Engng*, Nov. 1953.)
195. DALMON, J. and LOWE, H. J., *Colloques Internationaux du Centre National de la Recherche Scientifique*, Grenoble, 27th Sept. (1960). (Editions du Centre National de la Recherche Scientifique, Paris, 1961.)
196. DANCKWERTS, P. V., *Ind. Engng Chem.* **43,** 1460 (1951).
197. DANIELS, T. C., *The Engineer*, **203,** 358 (1957). (See also DAVIDSON, I. M., *Proc. Instn Mech. Engrs*, **160,** 243, 1949.)
198. DANSER, H. W., *Chem. Engng*, **57,** 158, May (1950).
199. DARBY, K. and HEINRICH, D. O., *Staub Reinhalt. Luft*, **26,** 464–8 (1966).
200. DAS, P. K., *Indian J. Met. Geophys.* **1,** 137 (1950).
201. Davidson & Co. Ltd., Belfast, Publication Ref. No. 387/61.
202. Davidson & Co. Ltd., Belfast, Brit. Pat. Appl. 25,440/65.
203. DAVIES, C. N., *Proc. Phys. Soc.* **57,** 259 (1945).
204. DAVIES, C. N., Symposium on particle size analysis, *Trans. Instn Chem. Engrs* and *Soc. Chem. Ind.*, p. 25, Feb. (1947).
205. DAVIES, C. N., *Staub Reinhalt. Luft*, **28,** 219 (1968).
206. DAVIES, C. N., *Proc. Phys. Soc.* **63B,** 288 (1950).
207. DAVIES, C. N., *Proc. Instn Mech. Engrs*, **1B,** 185 (1952).
208. DAVIES, C. N., *Dust is Dangerous*, p. 21, Faber, London (1954).
209. DAVIES, C. N., *Proc. Roy. Soc.* (London), A **289** (1417), 235–46 (1966).
210. DAVIES, C. N., *Proc. Roy. Soc.* (London), A **290** (1418), 557–62 (1966).
211. DAVIES, C. N. and PEETZ, C. V., *Proc. Roy. Soc.* (London), A **234,** 269 (1956).
212. DAVIS, R. F. and HOLTZ, J. C., Exhaust gases from diesel engines, in MALLETTE, F. S., Ed., Problems and control of air pollution, *Proc. of First Int. Congress on Air Pollution, New York, 1955*, chap. 9, Reinhold, New York (1955).
213. DAWES, J. G. and SLACK, A., Safety in mines research establishment, *U.K. Min. of Power, Report No.* **105** (1954).

214. DECKER, L. D., Incineration techniques for controlling emissions of nitrogen oxides, *Air Pollution Cont. Ass. Ann. Conf., Cleveland*, Paper 67-148 (1967).
215. DECKER, W. A., SNOEK, E. and KRAMERS, H., *Chem. Engng Sci.* **11**, 61 (1959).
216. DENNIS, R., SAMPLES, W. R., ANDERSON, D. M. and SILVERMAN, L., *Ind. Engng Chem.* **49**, 294 (1957)
217. DENNIS, R. *et al.*, *Air Cleaning Studies Progress Report*, Oct. 1956, p. 13, NYO-4611. Air Cleaning Laboratory, School of Public Health, Harvard Univ., Boston, Mass.
218. DERJAGUIN, B. V. and BAKANOV, S. P., *Kolloid. Zh.* **21**, 377 (1959). *Dokl. Akad. Nauk SSSR (Phys. Chem.)*, **141**, 384 (1961).
219. DERJAGUIN, B. V. and RABINOVICH, Y. I., *Dokl. Akad. Nauk SSSR (Phys. Chem.)*, **157**, 154 (1964).
220. DERJAGUIN, B. V. and YALAMOV, Y. I., *Dokl. Akad. Nauk SSSR (Phys. Chem.)*, **155**, 886 (1964).
221. DEUTSCH, G. P. and CLYDE, D. H., *J. Eng. Mech. Divn., Proc. Am. Soc. Civ. Engrs*, **93**, No. EM6, Proc. Paper 5660, 103–25, Dec. (1967).
222. DEUTSCH, W., *Ann. Phys.* **68**, 335 (1922).
223. DONOGHUE, J. K., *Trans. Instn Chem. Engrs* (London), **33**, 72 (1955).
224. DONOVAN, J. R. and STUBER, P. J., *J. Metals*, **19**, 45, Nov. (1967).
225. DONOVAN, J. R. and STUBER, P. J., Technology and economics of interpass absoprtion sulphuric acid plants, *A.I.Ch.E. Annual Meeting*, Los Angeles, 1st–5th Dec. (1968).
226. DÖTSCH, E. and FRIEDRICHS, H. A., *Staub Reinhalt. Luft*, **30**, 156–9 (1970).
227. DOWIS, E., *The Refrigeration and Air Conditioning Business*, July 1959 (reprint).
228. DOYLE, H. and BROOKS, A. F., *Ind. Engng Chem.* **49 (12)**, 57A (1957).
229. DRATWA, H. and JÜNGTEN, H., *Staub Reinhalt. Luft*, **27**, 301 (1967).
230. EBBENHORST-TENGBERGEN, H. J. VAN, *Problems and Control of Air Pollution*, Ed. MALLETTE, p. 255, Reinhold (1955).
231. EBBENHORST-TENGBERGEN, H. J. VAN, *Staub*, **25**, 486–90 (1965).
232. EINSTEIN, A., *Ann. Phys.* **19**, 289 (1906); *ibid.* **34**, 591 (1911).
233. EINSTEIN, A., Investigations on the theory of Brownian movement, Dover (1956), p. 75, from *Z. Electrochemie*, **14**, 235 (1908).
234. EINSTEIN, A., *Z. Physik*, **27**, 1 (1924).
235. EISHOLD, H. G., *Staub Reinhalt. Luft*, **26**, 12–14 (1966).
236. ELKINS, H. B., *The Chemistry of Industrial Toxicology*, 2nd ed., John Wiley & Sons, New York (1959).
237. EMMETT, P. H., *Advances in Catalysis*, **1**, 65 (1948).
238. ENDRES, H. A. and VAN ORMAN, W. T., *Heat. Pip. Air Condit.*, p. 157, Jan. (1952).
239. ENDRES, H. A. and VAN ORMAN, W. T., *Soc. Plastics Engrs J.* **9**, 26 (1953). Quoted by A. WINKEL, *Staub*, **41**, 469 (1955).
240. ENGELBRECHT, H. L., *J. Air Pollut. Control Ass.* **15** (2), 43 (1965).
241. Engelhard Industries, Baker Platinum Division Data Sheet.
242. ENGELS, L. H., *Staub*, **23**, 98 (1963).
243. EPSTEIN, P. S., *Phys. Rev.* **23**, 710 (1924).
244. EPSTEIN, P. S., *Z. Physik*, **54**, 537 (1929).
245. ESSENHIGH, R. H., Safety in Mines Research Establishment, Ministry of Fuel and Power, *Report No.* **120** (1955).
246. ETTRE, L. S., *J. Air Pollut. Control Ass.* **11**, 34 (1961).
247. FAHNOE, F., LINDROOS, A. E. and ABELSON, R. J., *Ind. Engng Chem.* **43**, 1336 (1951).
248. FAIRS, G. L., *Chem. and Ind.* **62**, 374 (1943).
249. FAIRS, G. L., *Trans. Instn Chem. Engrs* (London), **22**, 110 (1944).
250. FAIRS, G. L., *Trans. Instn Chem. Engrs* (London), **36**, 475 (1958).
251. FALTERMAYER, E. K., *Fortune*, p. 158, Nov. (1965).
252. FARLEY, R. and VALENTIN, F. H. H., *Trans. Instn Chem. Engrs* **43**, 193–8 (1965).
253. FAXÉN, H., *Ann. Phys.* **68**, 89 (1922). This is the corrected version of this equation as given by Faxén (1964) and quoted by Happel and Brenner (ref. 337, p. 327, eqn. 7–4.27).
254. FAXÉN, H., quoted by HAWKSLEY, P. G. W., *B.C.U.R.A. Bull.* **15**, 105 (1951).
255. FEIFEL, E., *Ver. Deut. Ing. Forschungshefte*, **9**, 68 (1938); *ibid.* **10**, 212 (1939).
256. FEILD, R. B., Collection of aerosols particles by atomised sprays, *Univ. Ill. Expt. Stat. Report No. 5* (1951).
257. FIELD, J. H., BIENSTOCK, D. and MYERS, J. G. (a) Process development in removing sulphur dioxide from hot flue gases, Pt. I—Bench scale experiments, *U.S. Bur. Mines, Report of Investigation 5735*

(1961), (b) Pt. II (with KURTZROCK, R. C.) *U.S. Bur. Mines, Report of Investigation 6037* (1963), (c) Pt. III, *U.S. Bur. Mines, Report of Investigation 7021* (1967).
258. FIRST, M. W., *Am. Soc. Mech. Engrs*, Preprint 49-A-27. Paper read at Am. Soc. Mech. Engrs Conference, Nov. (1949).
259. FIRST, M. W., JOHNSON, G. A., DENNIS, R., FRIEDLANDER, S. and SILVERMAN, L., Performance characteristics of wet collectors, *Harvard Air Cleaning Laboratory Report NYO 1587*, 22nd Apr. (1953).
260. FIRST, M. W., MOSCHELLA, R., SILVERMAN, L. and BERLY, E., *Ind. Engng Chem.* **43**, 1363 (1951).
261. FLEMING, E. P. and FITT, T. C., *Ind. Engng Chem.* **42**, 2252 (1950).
262. FOHL, T., Optimization of flow for forcing stack wastes to high altitudes, *Air Pollut. Control Ass.*, 60th Ann. Meeting, Cleveland, paper 67-66, 11th–16th June (1967).
263. FONDA, A. and HERNE, H., in preparation (quoted by H. Herne).[362]
264. FONTEIN, F. J. in *Cyclones in Industry*, Ed. RIETEMA, K. and VERVER, C. G., p. 118, Elsevier, Amsterdam (1961).
265. FORREST, J. S. and LOWE, H. J., Mechanical engineers' contribution to clean air, *Instn Mech. Engrs*, p. 42 (1957).
266. FORSTER, R. H. B., *Proc. Instn Mech. Engrs*, **160**, 246 (1949).
267. FOSTER, W. W., *Brit. J. Appl. Phys.* **10**, 206 (1959).
268. FOWLER, J. L. and HERTEL, K. L., *J. Appl. Phys.* **11**, 496 (1940).
269. FRANCIS, A. W., *Physics*, **4**, 403 (1933).
270. FRANCIS, W. and LEPPER, G. H., *Engineering*, **172**, 36 (1951).
271. FRASER, J. M. and DANIELS, F., *J. Phys. Chem.* **20**, 22 (1952).
272. FRASER, R. P. and EISENKLAM, P., *Trans. Instn Chem. Engrs*, **34**, 294 (1956).
273. FREDERICK, E. R., *Chem. Engng*, **68**, p. 107, 26th June (1961).
274. FRIEDLANDER, S. K., Univ. Illinois Engng Expt. Station, *Tech. Report No.* **13** (1954).
275. FRIEDLANDER, S. K., *A.I.Ch.E. J.* **3**, 43 (1957); *Ind. Engng Chem.* **50**, 1161 (1958).
276. FRIEDLANDER, S. K., Principles of gas–solid separation in dry systems, *Chem. Engng Prog. Symp. Ser.* **55**, 139–49 (1959).
277. FRIEDLANDER, S. K., *J. Meteorol.* **18**, 753 (1961).
278. FRIEDLANDER, S. K., *J. Colloid Interf. Sci.* **23**, 157 (1967).
279. FRIEDLANDER, S. K. and PASCERI, R. E., *Can. J. Chem. Engng*, **38**, 212 (1960).
280. FRIEDMAN, S. J., GLUCKERT, F. A. and MARSHALL, W. R., *Chem. Engng Prog.* **48**, 181 (1952).
281. FRIEDRICH, W., *Staub*, **19**, 281 (1959).
282. FRIEDRICHS, K. H., *Staub Reinhalt. Luft*, **26**, 240 (1966).
283. FRIEDRICHS, K. H., *Staub Reinhalt. Luft*, **28**, 193 (1968).
284. FRÖSSLING, N., *Gerlands Beiträge Geophysik*, **52**, 170 (1938).
285. FUCHS, N. A., *Mechanics of Aerosols*, Revised Ed., trans. Ed. DAVIES, C. N., Pergamon Press (1964).
286. FUCHS, N. A. and STECHKINA, I. B., *Ann. Occup. Health*, **6**, 27 (1963).
287. FURMIDGE, C. G. L., *Brit. J. Appl. Phys.* **12**, 268 (1961).
288. GANS, R., *Ann. Phys.* **86**, 628 (1928).
289. GANZ, S. N., *J. Appl. Chem. U.S.S.R.* **28**, 145 (1955).
290. GARDNER, G. C., *Chem. Engng Sci.* **18**, 35 (1962).
291. GARNER, J. F. and OFFORD, R. S., *The Law on the Pollution of the Air*, Shaw & Sons, London (1957).
292. GARNETT, A., in *Air Pollution*, Ed. THRING, M. W., p. 73, Butterworths, London (1957).
293. GASIEROWSKI, K., *Mitt. Ver. Grosskesselbesitzer*, **83**, 83 (1963).
294. GEBHART, J., BOL, J., HEINZE, W. and LETSCHERT, W., *Staub Reinhalt. Luft*, **30**, 238 (1970).
295. Gelman Instrument Company, P.O. Box 1448, Ann-Arbor, Mich. 48106, U.S.A.
296. GIAMMARCO, G., Italian Pat. Nos 537,564 (1955), 560,161 (1956) and 565,320 (1957).
297. GILLESPIE, T., *J. Colloid Sci.* **10**, 299 (1955).
298. GILLESPIE, T. and LANGSTROTH, G. O., *Can. J. Res.* B **25**, 455 (1947).
299. GILLESPIE, T. and LANGSTROTH, G. O., *Can. J. Chem.* **30**, 1003 (1952).
300. GILLESPIE, T. and RIDEAL, E., *J. Colloid Sci.* **10**, 281 (1955).
301. GILLILAND, E. R., *Ind. Engng Chem.* **26**, 681 (1934).
302. GILLILAND, E. R. and SHERWOOD, T. K., *Ind. Engng Chem.* **26**, 516 (1934).
303. GLASTONBURY, J. R., *Sydney Clean Air Conf.*, Tech. Paper No. 18 (1962).
304. GLAUERT, M., *Aeronautical Research Committee Report No. 2025* (London), H.M.S.O.
305. GLEDHILL, P. K., CARNALL, P. J. and SARGENT, K. H., *J. Iron Steel Inst.* **186**, 198 (1957).
306. GOKSOYR, H. and ROSS, K., *J. Inst. Fuel*, **35**, 177 (1962). See also *B.W.K.* **14**, 60 (1962).

307. GOLDMANN, L., *Staub*, **24**, 449 (1964).
308. GOLDSMITH, P., DELAFIELD, H. S. and COX, L. C., *Quart. J. Roy. Meteorol. Soc.* **89**, 43 (1963).
309. GOLDSTEIN, S., *Modern Developments in Fluid Dynamics*, Vol. II, p. 340, Clarendon Press, Oxford (1938).
310. GOLLMAR, H. A., in *Chemistry of Coal Utilization*, Ed. LAWRY, H. H., Vol. 2, p. 947, Wiley, New York (1945).
311. GOSLINE, C. A., FALK, L. L. and HELMERS, E. N., in *Air Pollution Handbook*, Ed. MAGILL *et al.*, chap. 5, p. 20, McGraw-Hill (1956).
312. GRANVILLE, R. A. and JEFFREY, W. G., *Engineering*, **187**, 285 (1959).
313. GRAUE, G., GRADTKE, W. and NAGEL, H., *Staub*, **25**, 525 (1965).
314. GREEN, H. L. and LANE, W. R., *Particulate Clouds, Dusts, Smokes and Mists*, p. 134, Spon, London (1957).
315. GREEN, H. L. and LANE, W. R., *Particulate Clouds, Dusts, Smokes and Mists*, p. 149, Spon, London (1957).
316. GREEN, H. L. and LANE, W. R., *Particulate Clouds, Dusts, Smokes and Mists*, p. 202, Spon, London (1957).
317. GREEN, H. L. and LANE, W. R., *Particulate Clouds, Dusts, Smokes and Mists*, p. 212, Spon, London (1957).
318. GREEN, H. L. and WOOTTEN, N. W. (unpublished work quoted by GREEN, H. L. and LANE, W. R., *Particulate Clouds, Dusts, Smokes and Mists*, p. 66, Spon, London, 1957).
319. GREEN, T. E. and HINSHELWOOD, *J. Chem. Soc. 1926*, 1709.
320. GREENBURG, L. and SMITH, C. W., *U.S. Bur. Mines Report of Investigation 2392* (1922).
321. GREGG, S. J., Symposium on particle size analysis, *Instn. Chem. Engrs*, and *Soc. Chem. Ind.* (London), p. 27, Feb. (1947).
322. GUCKER, F. T., *Chem. Revs.* **44,** 245 (1949).
323. GUCKER, F. T., O'KONSKI, C. T., PICKARD, H. B. and PITTS, J. M., *J. Am. Chem. Soc.* **69,** 2422 (1947).
324. GUCKER, F. T. and O'KONSKI, C. T., *Chem. Revs.* **44,** 389 (1949); *J. Colloid Sci.* **4,** 441 (1949).
325. GUNN, R., *J. Météorol.* **11,** 339 (1954).
326. GÜNTHEROTH, H., *Staub*, **21,** 430 (1961).
327. GUSTAVSSON, K. A., *Tekn. Tidskv.* **78,** 667 (1948).
328. GUTHMANN, K., *Stahl und Eisen*, **75,** 1571 (1955); *Staub*, **21,** 398 (1961).
329. HAMANN, S. D., *Physico Chemical Effects of Pressure*, p. 78, Butterworths, London (1957).
330. HANSEN, K., *Fifth World Power Conference*, Vienna, **16,** 5829 (1956).
331. HANSEN, N. L. (1930), quoted by ROSSANO and SILVERMAN (1954) (*op. cit.*) and by DAVIES (1952) (*op. cit.*).
332. HANSON, D. N., WILKE, C. R., FAITH, L. E. and BUSTANY, S. N., *Ind. Engng Chem. Fundamentals*, **6,** 519–26 (1967).
333. HANSON, D. N. and WILKE, C. R., Electrostatic precipitator analysis, Draft Paper, private communication from HANSON, D. N. (1968).
334. HÄNTZSCH, S. and PRESCHER, K. E., *Staub Reinhalt. Luft*, **26,** 332 (1966).
335. HAPPEL, J. and BRENNER, H., *Low Reynolds Number Hydrodynamics*, pp. 327, 355, Prentice-Hall Inc., Englewood Cliffs, N.J. (1965).
336. HAPPEL, J., *A.I.Ch.E. J.* **5,** 174 (1959).
337. HAPPEL, J. and BRENNER, H., *Low Reynolds Number Hydrodynamics*, p. 395, Prentice-Hall Inc., Englewood Cliffs, N.J. (1965).
338. HARDWICK, B. A., THISTLETHWAYTE, D. K. B. and FOWLER, R. T., *Atmospheric Environment*, **4,** 379 (1970).
339. HARR, R., WAGNER, K. and WILLMER, T. K., *Berg.-u. hüttenm. Mh.* **104,** No. 2, 50 (1959).
340. HARRIS, W. B. and MASON, M. G., *Ind. Engng Chem.* **47,** 2423 (1955).
341. HARTMANN, J. and LAZARUS, F., *Phil. Mag.* **29,** 140 (1940).
342. HASENCLEVER, D., *Staub Reinhalt. Luft*, **26** (7), 288 (1966).
343. HASENCLEVER, D., *Staub*, **19,** 42 (1959).
344. HAUSBERG, G., *Staub*, **21,** 418 (1961).
345. HAUSBERG, G., quoting German Pat. No. 521,697, 15th Nov. (1925) in *Staub*, **21,** 418 (1961).
346. HAUT, H. VAN, *Staub*, **21,** 52 (1961).
346a. HAUT, H. VAN and GUDERIAN, R., *Staub Reinhalt. Luft*, **30,** 17–26 (1970).

346b. HAUT, H. VAN and STRATMANN, H., *Schriftenreihe d. Landesanst. für Immissions und Bodennutzungsschutz*, Nordrhein-Westf, 7 (1967), quoted by *Staub Reinhalt. Luft*, **27**, 508–9 (1967).
347. HAWKSLEY, P. G. W., in Some aspects of fluid flow, *Inst. Physics Conf.*, p. 114 (1950) (Edward Arnold, London).
348. HAWKSLEY, P. G. W., *B.C.U.R.A. Bull.* **15**, 105 (1951).
349. HAWKSLEY, P. G. W., Physics of particle size analysis, *Brit. J. Appl. Phys.*, Suppl. 3, S. 1 (1954).
350. HAWKSLEY, P. G. W., BADZIOCH, S. and BLACKETT, J. H., *J. Inst. Fuel*, **31**, 147 (1958).
351. HAWKSLEY, P. G. W., BADZIOCH, S. and BLACKETT, J. H., *Measurement of Solids in Flue Gases*, British Coal Utilization Research Association, Leatherhead (1961).
352. HEDLEY, A., A kinetic study of SO_3 formation in a pilot scale furnace, in *Mechanisms of Corrosion by Fuel Impurities*, Ed. JOHNSON, H. R. and LITTLER, D. J., pp. 204–15, Butterworths (1964).
353. HEIMANN, H., *Air Pollution*, p. 159, World Health Organization, Geneva (1961).
354. HEINRICH, D. O., *J. Iron Steel Inst.* **33**, 452 (1960).
355. HEINRICH, D. O., *Trans. Instn Chem. Engrs* (London), **39**, 145 (1961).
356. HEINRICH, D. O., *Staub*, **22**, 360 (1962).
357. HEINRICH, D. O., *Staub*, **23**, 83 (1963).
358. HEINRICH, R. F. and ANDERSON, J. R., *Chemical Engineering Practice*, **3**, 464, Ed. CREMER, H. W., Butterworths, London (1957).
359. HEISS, J. F. and COULL, J., *Chem. Engng Prog.* **48**, 133 (1952).
360. HEITMANN, H. G. and SIETH, J., *Mitt. Ver. Grosskesselbesitzer*, **83**, 82 (1963). German Pat. No. 1,217,535, 26th May (1962).
361. HELWIG, H. C. and GORDON, C. L., *Anal. Chem.* **30**, 1810 (1958).
362. HERNE, H., *Int. J. Air Pollut.* **3**, 26 (1960).
362a. HERSEY, H. J., *Ind. Chem.* **31**, 138 (1955).
363. HERRICK, R. A., OLSEN, J. W. and RAY, F. A., *J. Air Pollut. Control Soc.* **16** (1), 7–11 (1966).
364. HESS, K. and STICKEL, R., *Chemie-Ingr-Tech.* **39**, 334–40 (1967).
365. HIDY, G. M. and BROCK, J. R., *J. Colloid Sci.* **20**, 477 (1965).
366. HIGBIE, R., *Trans. Am. Instn Chem. Engrs* **31**, 365 (1935).
367. HIGNETT, E. T., Particle charging in electrostatic precipitation, *Inst. Elec. Eng.* (London) *Colloq. on Electrostatic Precipitators*, 19th Feb. (1965).
368. HIGNETT, E. T., *Proc. Instn Elect. Engrs*, **114**, 1325 (1967).
369. HIGUCHI, W. I. and O'KONSKI, C. T., *J. Colloid Sci.* **15**, 14 (1960).
369a. HINDAWI, I. J., Injury by sulphur dioxide, hydrogen fluoride and chlorine as they were observed and reflected on vegetation in the field, A.P.C.A. Paper 67-159, *60th Ann. Meeting, Air Pollut. Control Ass.*, Cleveland, 11th–16th June (1967).
370. HIRSCHFELDER, J. O., CURTIS, C. F. and BIRD, R. B., *Molecular Theory of Gases and Liquids*, p. 539, Wiley, New York (1954).
371. HOCKING, L. M., *Quart. J. Roy. Met. Soc.* **85**, 44 (1959).
372. HOCKING, L. M., *Int. J. Air Pollut.* **3**, 154 (1960).
373. HODGESON, J., STEVENS, R. K. and KROST, K. J., Chemiluminescent ozone sensor, *ACS Meeting*, Atlantic City (1968).
374. HOHENBERGER, A., *Stahl Eisen*, **81**, 1001 (1961).
375. HOUDRY, E., German Pat. No. 1,003,192, 2nd July (1954).
376. HOUGHTON, H. G. in PERRY (Ed.), *Chemical Engineer's Handbook*, 3rd ed., p. 840, McGraw-Hill, New York (1950).
377. HOWDEN, James & Co. Ltd., 195 Scotland St., Glasgow, C. 5.
378. HUBBARD, E. H., *J. Inst. Fuel*, **30**, 564 (1957).
379. HUMPHREY, A. E. and GADEN, E. L., *Ind. Engng Chem.* **47**, 924 (1955).
380. IBERALL, A. S., *J. Nat. Bur. St.* **45**, 398 (1950).
381. IHLEFELDT, H., *Staub*, **21**, 448 (1961).
382. IMPERATO, N. F., *Chem. Engng*, **75** (22), 152, 14th Oct. (1968).
383. INGLES, O. G., *Aust. J. Appl. Sci.* **9**, 120 (1958).
384. INOUE, I., Modern sonic gas purification, *Kagaku Koge*, **6**, 225 (1955).
385. INOUE, I., OYAMA, Y. and SAWAHATA, Y., *J. Sci. Res. Inst.* (Tokyo), **49** (1376), 39 (1955).
386. INOUE, I., OYAMA, Y., SAWAHATA, Y. and OKADA, M., *J. Sci. Res. Inst.* (Tokyo), **48** (1369), 260 (1954).
387. *International Critical Tables*, Ed. WASHBURN, E. W., McGraw-Hill, New York (1926).
388. INYUSHKIN, N. V. and AVERBUKH, YA. D., *Soviet J. Non-ferrous Metals* (Engl. trans.), **35**, 35 (1962).

389. INYUSHKIN, N. V. and AVERBUKH, YA. D., *Khim. i. Khim. Tekhnol.* **6,** 1031 (1963).
390. JACKSON, R., *B.C.U.R.A. Bull.* **23,** 349 (1959).
391. JACKSON, R., *B.C.U.R.A. Bull* **24,** 221 (1962).
392. JACKSON, R., THURLOW, G. G. and GOODRIDGE, A. M., *J. Sci. Inst.* **35,** 81 (1958).
393. JACKSON, R., THURLOW, G. G. and HOLLAND, R. E., *J. Inst. Fuel,* **33,** 180 (1960).
394. JACOBS, M. B., *The Chemical Analysis of Air Pollutants,* Interscience, New York (1960).
395. JACOBSON, J. S. and HILL, A. C., *Recognition of Air Pollution Injury to Vegetation: A Pictorial Atlas,* Informative Report No. 1, T.R. 7, Agricultural Committee, Air Pollut. Control Assoc., Pittsburgh (1970).
396. JAHN, R., *Radex Rundschau,* **7,** 625–31 (1955).
397. JARMAN, R. T., *J. Agric. Engng Res.* **4,** 139 (1959).
398. JEANS, J. H., *The Dynamical Theory of Gases,* p. 316, Dover, New York (1954).
399. JENNINGS, R. F., *J. Iron Steel Inst.* **164,** 305 (1950).
400. JOHNSON, G. A., FRIEDLANDER, S. K., DENNIS, R., FIRST, M. W. and SILVERMAN, L., *Chem. Engng Prog.* **51,** 176 (1955).
401. JOHNSON, J. C. and GOODWIN, G. C., Conference on the Mechanical Engineers Contribution to Clean Air, *Instn Mech. Engrs* (London), 20th Feb. (1957) (Preprint).
402. JOHNSTONE, H. F., *Univ. Ill. Engng Exptl Station Circular No.* **20** (1929).
403. JOHNSTONE, H. F. and ECKMAN, F. O., *Ind. Engng Chem.* **43,** 1358 (1951).
404. JOHNSTONE, H. F., FEILD, R. B. and TASSLER, M. C., *Ind. Engng Chem.* **46,** 1601 (1954).
405. JOHNSTONE, H. F. and ROBERTS, M. H., *Ind. Engng Chem.* **41,** 2417 (1949).
406. JOHNSTONE, H. F. and SILOX, H. E., *Ind. Engng Chem.* **39,** 808 (1947).
407. JOHNSTONE, H. F. and SINGH, A. D., *Univ. Ill. Engng Exptl Station Bull. No.* **324,** 31st Dec. (1940).
408. JOHSWICH, F., *Brennst.-Wärme-Kraft,* **14,** 105 (1962).
409. JONES, N. E., Ph.D. Thesis, Sheffield Univ. (1960).
410. JONES, W. P., *Ind. Engng Chem.* **41,** 2424 (1949).
411. JONES, W. P. and ANTHONY, A. W., *United States Technical Conference on Air Pollution,* Ed. MCCABLE, L. C., p. 318, McGraw-Hill, New York (1952).
412. JOOS, E., *Staub,* **35,** 18 (1954).
413. Joy Manufacturing Co., Henry W. Oliver Bld., Pittsburgh 22, Pa., U.S.A.
414. Jukes, Fredrk. Ltd., Garth Road, Surrey, England.
415. JÜNGTEN, H., *Erdöl Kohle,* **16,** 119 (1963).
416. JÜNGTEN, H. and KARWEIL, J., *Erdöl Kohle,* **15,** 898, 985 (1962).
417. JÜNGTEN, H., KNOBLAUCH, K. and KRUEL, M., *Chemie-Ingr-Tech.* **42,** 77 (1970).
418. JÜNGTEN, H. and PETERS, W., *Staub Reinhalt. Luft,* **28,** 89 (1968).
419. KAFKA, F. L. and FERRARI, L. M., The performance of a core oven afterburner, *Clean Air Conf.,* vol. 2, No. 2, *Sydney* (1965).
420. KALASCHNIKOW, S., *Z. Tech. Phys.* **9,** 267 (1934).
421. KAMAK, H. W., *Anal. Chem.* **23,** 844 (1950).
422. KANE, L. J., CHIDESTER, G. E. and SHALE, C. C., *U.S. Bur. Mines,* Report of Investigation No. 5672 (1960).
423. KANGRO, C., *Staub,* **21,** 275 (1961).
424. KASUI, FUJI, Japan Pat. No. 6151 (1954), reported by Atsukawa *et al., op. cit.*
425. KATELL, S., *Chem. Engng Prog.* **62** (10), 67 (1966).
426. KATZ, M., Analysis of inorganic gaseous pollutants, in STERN, A. C. (Ed.), *Air Pollution,* 2nd ed., **2,** Academic Press (1968), also Measurement of air pollutants, a guide to the selection of methods, *World Health Organisation,* Geneva (1969).
427. KATZ, M. and COLE, R. J., *Ind. Engng Chem.* **42,** 2258 (1950).
428. KAY, K., *Anal. Chem.* **29,** 589 (1957); *ibid.* **31,** 633 (1959).
429. KEAIRNS, D. L. and ARCHER, D. H., Fluidized bed boilers—concepts and comparisons, *Second Int. Conf. on Fluidized Bed Combustion,* Hueston Woods, Ohio, 4th–7th Oct. (1970).
430. KEILHOLTZ, G. W. and BATTLE, G. C., Treatment of radioactive airborne wastes from reactors, in STRAUSS, W. (Ed.), *Air Pollution Control,* **2,** Wiley–Interscience, New York (1971).
431. KELSALL, D. F., *Trans. Instn Chem. Engrs* (London), **30,** 87 (1952).
432. KELSALL, D. F. and MCADAM, J. C. H., *Trans. Instn Chem. Engrs* (London), **41,** 84 (1963).
433. KELLY, F. H. C., *Proc. Australas Inst. Min. Metall.,* New Series Nos. **152–3,** 17 (1949).
434. KEMPNER, S. K., SEILER, E. N. and BOWMAN, D. H., *J. Air Pollut. Control Ass.* **20,** 139 (1970).

435. KENNAWAY, T., *Iron and Steel Institute Special Report No. 61*, p. 139, Air and water pollution in the iron and steel industry (1958), *J. Air Pollut. Control Ass.* **7**, 266 (1958).
436. Keram Chemie, reported by ATSUKAWA *et al.* (*op. cit.*).
437. KIELBACK, A. W., *Chem. Engng Prog. Symp. Ser.*, No. 35, **57** (1961), Pollution and environmental health, p. 51.
438. KING, L. V., *Proc. Roy. Soc.* (London), A **147**, 233 (1934).
439. KING, R. A., *Ind. Engng Chem.* **42**, 2241 (1950).
440. KIRKWOOD, J. B., Paper No. 14, *Clean Air Conf., Sydney* (1962), Univ. N.S.W. Press.
441. KIRSH, A. A. and FUCHS, N. A., *Ann. Occup. Hyg.* **10**, 23 (1967).
442. KIRSH, A. A. and FUCHS, N. A., *J. Phys. Soc.* (Japan), **22**, 1251 (1967).
443. KIYOURA, R., *J. Air Pollut. Control Ass.* **16**, 488 (1966); *Staub Reinhalt. Luft*, **26**, 524 (1966).
444. KIYOURA, R., Studies on the removal of SO_2 from hot flue gases, II, *60th Ann. Meeting Air Pollut. Control Ass.* (1967).
445. KLEIN, H., *Staub*, **23** (11), 501–8 (1963).
446. KLEIN, H., *Energie und Technik*, **18** (6), 228–35 (1966).
447. KLEINSCHMIDT, R. V. and ANTHONY, A. W., *United States Technical Conference on Air Pollution*, Ed. MCCABE, L. C., p. 310, McGraw-Hill, New York (1952).
448. KLIMICEK, R., SKRIVANEK, J. and BETTELHEIM, J., *Staub Reinhalt. Luft*, **26**, 235 (1966).
448a. KLYACHKO, L., *Otopl. i Ventil.* (4) (1934). Quoted by FUCHS, N. A., *Mechanics of Aerosols*, Ed. DAVIES, C. N., p. 33, revised edition, Pergamon Press, Oxford (1964).
449. KNEEN, T. and STRAUSS, W., *Atmospheric Environment*, **3**, 55–67 (1969).
450. Knit-Mesh Ltd., 36 Victoria St., London, S.W.1.
450a. KNUDSEN, M. and WEBER, S., *Ann. Phys.* **36**, 982 (1911).
451. KOGLIN, W., *Staub*, **22**, 189 (1962).
452. KOGLIN, W., *Staub Reinhalt. Luft*, **28**, 398–402 (1968).
453. KOGLIN, W., *Staub Reinhalt. Luft*, **30**, 151–2 (1970).
454. KOGLIN, W. and VETTER, H., *Staub*, **23**, 300–4 (1963).
455. KOHL, A. L. and RIESENFELD, F. C., *Gas Purification*, McGraw-Hill, New York (1960).
456. KOLK, H. VAN DER, *Ver. Deut. Ing. Tagungsheft*, **3**, 23 (1954).
457. KOLK, H. VAN DER, *Ver. Deut. Ing. Berichte*, **7**, 25 (1955).
458. KOLK, H. VAN DER, *Cyclones in Industry*, chap. 6, p. 77, Ed. RIETEMA, K., and VERVER, C. G., Elsevier (1961).
459. KOLLER, L. R. and FREMONT, H. A., *J. Appl. Phys.* **21**, 741 (1950).
460. KÖNIG, W. *Ann. Phys.* (Lpzg), **42**, 353 (1891).
461. KORDECKI, M. C. and ORR, C., *Arch. Env. Health*, **1**, 1 (1960).
462. KOTTLER, W., KRUPP, H. and RABENHORST, H., *Z. Angew. Phys.* **24**, 219 (1968).
463. KRAEMER, H. F. and JOHNSTONE, H. F., *Ind. Engng Chem.* **47**, 2426 (1955).
464. KRIJGSMAN, M., De Quantitatieve Bepaling van Stoff in Hoogovengas, p. 6, *K.N.H.S. Centraal Laboratorium Report*, 28th Mar. (1955).
465. KRIJGSMAN, M., *Stahl Eisen*, **80**, 621 (1960).
466. KRISTAL, E., DENNIS, R. and SILVERMAN, L., *T.I.D.*, 7313, p. 203 (1956). U.S. Atomic Energy Commission, Technical Information Service Extension, Oak Ridge, Tenn.
467. KRUPP, H. and SPERLING, G., *Z. Angew. Phys.* **19**, 259 (1965).
468. KRUPP, H., *Advanc. Colloid Interface Sci.* **1**, 111 (1967).
469. KUDLICH, R. (revised by BURDICK, L. R.), Ringelmann smoke chart, U.S. Bur. Mines, Information Circular 7718 (1955).
470. KUWABARA, S., *J. Phys. Soc.* (Japan), **14**, 527 (1959).
471. LACHMAN, J. C., *Inst. and Control Systems*, **32**, 1030 (1959).
472. LADENBURG, R., *Ann. Phys.* **4**, 863 (1930).
473. LADENBURG, R., *Der Chemie. Ingenieur*, **I** (iv), 31 (1934).
474. LAGARIAS, J. S., *J. Air Pollut. Control Ass.* **10** (iv), 271 (1960).
475. LAMB, H., *Hydrodynamics*, p. 601, Dover, New York (1945).
476. LAMB, H., *Hydrodynamics*, p. 612, Dover, New York (1945).
477. LANCASTER, B. W. and STRAUSS, W., *J. Sci. Inst.* **43**, 395 (1966).
478. LANCASTER, B. W. and STRAUSS, W., *Ind. Engng Chem. Fundamentals* **10**, 362-9 (1971).
479. LAND, T. and BARBER, R., *Trans. Soc. Inst. Tech.* **6**, 112 (1954).

480. LAND, T. and BARBER, R., *J. Iron Steel Inst.* **184,** 269 (1956).
481. Land Pyrometers Ltd., Sheffield, Technical Information, *Land Suction Pyrometer*, Type 4.
482. Land Pyrometers Ltd., Sheffield, *The Land Venturi Pneumatic Pyrometer—a First Report*, Technical Note No. 72, 15th May (1962).
483. LANDAHL, H. D. and HERRMANN, R. G., *J. Colloid Sci.* **4,** 103 (1949).
484. LANDAU, R. and ROSEN, R., *Ind. Engng Chem.* **39,** 281 (1947).
485. LANDAU, R. and ROSEN, R., *Ind. Engng Chem.* **40,** 1389 (1948).
486. LANDOLT-BÖRNSTEIN, *Physikalisch-Chemische Tabellen*, J. Springer Verlag, Berlin, 1–6th ed. (1923–62).
487. LANDT, E., *Gesundheits Ing.* **77,** 139 (1956).
488. LANDT, E., *Staub*, **48,** 9 (1957).
489. LANGMUIR, I., O.S.R.D. Report No. 865 (1942).
490. LANGMUIR, I. and BLODGETT, K. B., General Electric Research Laboratory, Schenectady N.Y. Report R.L. 225 (1944–5).
491. LANGMUIR, I. and BLODGETT, K.B., *Am. Air Force Tech. Report*, 5418 (1946).
492. LANTERI, ANNEMAREE, Australian clean air legislation, its history and guidelines for future development, *Clean Air Conf., Sydney* (1969).
493. LAPPLE, C. E., in PERRY, J. H. (Ed.) *Chemical Engineer's Handbook*, p. 1021, 3rd ed., McGraw-Hill, New York (1950).
494. LAPPLE, C. E. and KAMACK, H. J., *Chem. Engng Prog.* **51,** 110 (1955).
495. LARSEN, R. J., *Am. Ind. Hyg. Ass. J.* **19,** 265 (1958).
496. LAU, H., *Staub Reinhalt. Luft*, **29,** 311–14 (1969).
497. LAUER, O., *Staub*, **20,** 69 (1960).
498. LAXTON, J. W. and JACKSON, P. J., *J. Inst. Fuel*, **37,** 12 (1964).
499. LEA, F. M. and NURSE, R. W., Symposium on particle size analysis, *Inst. Chem. Engng* and *Soc. Chem. Ind.*, p. 47, Feb. (1947).
500. LEDERC, E., *Air Pollution*, p. 279, World Health Organisation, Geneva (1961).
501. LEINWEBER, L., *Aufbereitungs Techn.* **7,** 249–56 (1966).
502. LEINWEBER, L., *Staub Reinhalt. Luft*, **27,** 123–9 (1967).
503. LEITHE, W., *The Analysis of Air Pollutants*, trans. KONDOR, R., Ann-Arbor Humphrey Science Publishers, Ann Arbor (1970).
504. LEONARD, J. B., *Cost Engineering in the Process Industries*, Ed. CHILTON, C. H., p. 324, McGraw-Hill, New York (1960).
505. LEONE, I. A., BRENNAN, E. and DAINES, R. H., *J. Air Pollut. Control Ass.* **16** (4), 191–6 (1966).
506. LEVINE, D. C. and FRIEDLANDER, S. K., *Chem. Engng Sci.* **113,** 49 (1960).
507. LEWIS, H. C., EDWARDS, D. G., GOGLIA, M. J., RICE, R. I. and SMITH, L. W., *Ind. Engng Chem.* **40,** 67 (1948).
508. LEWIS, W. K., GILLILAND, E. R. and BAUER, W. C., *Ind. Engng Chem.* **41,** 1104 (1949).
509. LEWIS, W. K. and WHITMAN, W. G., *Ind. Engng Chem.* **16,** 1215 (1924).
510. LIEBSTER, H., *Ann. Phys.* **82,** 541 (1927).
511. LIESEGANG, D., *Staub Reinhalt. Luft*, **28,** 403–5 (1968).
512. LIFSHITZ, E. M., *Soviet Physics J.E.P.T.* **2,** 73 (1956).
513. LIN, C. S., MOULTON, R. W. and PUTNAM, G. L., *Ind. Engng Chem.* **45,** 636 (1953).
514. TER LINDEN, A. J., *Engineering*, **167,** 167 (1949).
515. TER LINDEN, A. J., *Proc. Instn Mech. Engrs* **160,** 233 (1949).
516. TER LINDEN, A. J., *Tonindustrie-Zeitung*, **22(iii),** 49 (1953).
517. TER LINDEN, A. J., in *Problems and Control of Air Pollution*, Ed. MALLETTE, F. S., p. 236, Reinhold (1955).
518. TER LINDEN, A. J. and VAN DONGEN, J. R. J., *Trans. Am. Soc. Mech. Engrs*, **80,** 245 (1958).
519. LINTON, F. L. and BRINK, J. A., *Chem. Engng Prog.* **63** (2), 83–86 (1967).
520. LITTLE, A., *Trans. Instn Chem. Engrs*, (London), **34,** 259 (1956).
521. LITTLE, ARTHUR D., Report for U.S. Public Health Service, *Robert A. Taft Sanitary Engineering Centre Technical Report*, No. A61-34.
522. LIU, B. H. Y., WHITBY, K. T. and YU, H. H. S., *J. Colloid Sci.* **23,** 367 (1967) and *J. Appl. Phys.* **38,** 1592 (1967).
523. LIU, B. H. Y. and YEH, H. C., *J. Appl. Phys.* **39,** 1396 (1968).
524. LOBO, W. E., FRIEND, L., HASHMALL, F. and ZENZ, F. A., *Trans. Am. Instn Chem. Engrs*, **41,** 693 (1945).

525. LODGE, O., *Nature* (London), **28,** 297 (1883).
526. LÖFFLER, F., Untersuchung der Haftkräfte Zwischen Feststoffteilchen und Filterfaseroberflächen, Dr. Ing Thesis, Technische Hochschule Karlsruhe (1965).
527. LÖFFLER, F., *Staub Reinhalt. Luft*, **26,** 274 (1966).
528. LÖFFLER, F., *Staub Reinhalt. Luft*, **28,** 456 (1968).
529. LÖFFLER, F., Collection of particles by fibre filters, in *Air Pollution Control*, Ed. STRAUSS, W., chap. 6, vol. 1, Wiley–Interscience, New York (1971).
530. LORENZ, H., *Abh. th. Physik*, **1,** 23 (1906).
531. LOWE, H. J. and LUCAS, D. H., *Brit. J. Appl. Phys.* **4,** S40 (1953).
532. LUCAS, D. H., MOORE, D. J. and SPURR, G., *Int. J. Air Wat. Pollut.* **7,** 473–500 (1963).
533. LUDWIG, S., *Chem. Engng* **75,** 70, 29th Jan. (1968).
534. LUIZ, A. M., *Chem. Engng Sci.* **22,** 1083–90 (1969).
535. LUIZ, A. M., *Chem. Engng Sci.* **24,** 119–23 (1969).
536. LUNDE, K. E., *Ind. Engng Chem.* **50,** 293 (1958).
537. LUNDGREN, D. A. and WHITBY, K. T., *Ind. Engng Chem. Proc. Des. Dev.* **4,** 345 (1965).
538. LUNNON, R. G., *Proc. Roy. Soc.* A **110,** 319 (1926).
539. LUNNON, R. G., *Proc. Roy. Soc.* A **118,** 680 (1928).
540. Lurgi Apperatebau G.m.b.H., *Sulfacid Process*, pp. 1–9.
541. MAARTMANN, S., The effect of gas temperature and dew point on dust resistivity, *2nd Int. Clean Air Congress*, Washington, D.C., Paper EN-34F, 6th–11th Dec. (1970).
542. MACEY, H. H., The mathematics of chimney heights, paper delivered to the *ANZAAS Symp. on Pollution*, Jan. (1968).
543. MACZEWSKI-ROWINSKI, B., quoted by BOUCHER, R. M. G. and by MACZEWSKI-ROWINSKI, B., *Int. Clean Air Conf., London*, p. 160 (1959), National Society for Clean Air, London (1960).
544. MCCABE, L. C., *Ind. Engng Chem.* **44** (11), 123A, Nov. (1952).
545. MCNOWN, J. S., LEE, H. M., MCPHERSON, M. B. and ENGEZ, S. M., *7th Int. Cong. App. Mech. Proc.* **7 11 (i),** 17 (1948).
546. MCNOWN, J. S. and MALAIKA, J., *Trans. Am. Geophys. Union*, **31,** 74 (1950).
547. MCWILLIAMS, J. A., PRATT, H. R. C., DELL, F. R. and JONES, D. A., *Trans. Instn Chem. Engrs*, **34,** 17 (1956).
548. MAGILL, P. L., HOLDEN, F. R. and ACKLEY, C., *Air Pollution Handbook*, pp. 13–96, table 13–22, McGraw-Hill, New York (1956).
549. MAHLER, E. A. J., Standards of emission under the Alkali Act, *Proc. Int. Clean Air Congress*, London, Oct. (1966). (Reprinted *103rd Annual Report on Alkali Works* (1966), pp. 51–58.)
550. MANTELL, C. L., *Adsorption*, McGraw-Hill, New York (1945). Also in PERRY (Ed.) *Chemical Engineer's Handbook*, section 14, p. 885 (1950).
551. MARSHALL, W. R., Atomization and spray drying, *Chem. Engng Prog. Monogr. Ser.*, No. 2, vol. 50, chaps. I to IX (1954).
552. MARSHALL, W. R. and SELTZER, E., *Chem. Engng Prog.* **46,** 501 (1950).
553. MARTIN, A. E., REID, A. M. and SMART, J., *Control*, **2** (18), 108 (1959); *ibid.* **3** (19), 91 (1960).
554. MARTIN, A. H., *Mining Sci.* **63,** 337 (1910).
555. MARZOCCHI, A., LACHUT, F. and WILLIS, W. H., *J. Air Pollut. Control Ass.* **12** (1), 38 (1962).
556. MASSEY, O. D., *Chem. Engng*, **66** (No. 14), p. 143, 13th July (1959).
557. MASSON, H., *Staub*, **21,** 459 (1961).
558. MASUDA, S., *Electrotech. J. of Japan*, **7,** 108 (1963).
559. MATIJEVIC, E., KERKER, M., KITANI, S., ESPENCHILD, W. F., and FARONE, W. A., *J. Colloid Sci.* **19,** 213 (1964).
560. MATIJEVIC, E., KITANI, S. and KERKER, M., *J. Colloid Sci.* **19,** 223 (1964).
561. MATIJEVIC, E., KERKER, M., ESPENCHILD, W. F. and WILLIS, E., *J. Colloid Sci.* **20,** 501 (1965).
562. MATSUMOTO, K. and SHIRAISHI, Y., *Mitsubishi Heavy Industries Ltd. Technical Review*, **5,** 59 (1968).
563. MATTERN, C. F. T., BRACKETT, F. S. and OLSON, B. J., *J. Appl. Physiol.* **10,** 56 (1957).
564. MATULA, B., *Prace Komis. Mat. Przyrodn, Poznan Towarz, Przyjacio Nauk*, **8,** 21 (1957).
565. MAY, K. R., *J. Sci. Inst.* **22,** 187 (1945).
566. MEDLOCK, R. S., *Instrument Engineer*, **1** (1), 3 (1952).
567. MEDNIKOV, E. P., *Acoustic Coagulation and Precipitation of Aerosols*, U.S.S.R. Akad. Sci., Moscow (1963). Translated by C. V. Larrick Consultants Bureau (1965).
568. MEISSNER, H. P. and MICKLEY, H. S., *Ind. Engng Chem.* **41,** 1238 (1949).

569. MELLOR, J. F., *J. Air Pollut. Control Ass.* **10** (6), 456–7 (1960).
570. Menardi & Co., 1220 East Grand Ave., El. Segundo, Calif., U.S.
570a. MIDDLETON, J. T. and DARLEY, E. F., *Bull. World Health Organisation*, **34,** 477–80 (1966).
571. MIERDEL, G., *Z. Tech. Phys.* **8,** 564 (1932).
572. MILLIKAN, R. A., *Phys. Rev.* **22,** 1 (1923).
573. Millipore Filter Corpn, Bedford, Mass.
574. MILLIKAN, R. C. and KASKAN, W. E., Spectroscopic studies of flame gases, *8th Int. Symp. on Combustion*, Pasadena, Cal. (1960), Proc. Paper 23, p. 262 (1962).
575. MITCHELL, R. I. and PILCHER, J. M., *5th A.E.C. Air Cleaning Conf.*, T.I.D. 7551, p. 67 (1957).
576. MOELLER, W. and WINKLER, K., *J. Air Pollut. Control Ass.* **18,** 324 (1968).
577. MÖLLER, W., *Phys. Z.* **39,** 57 (1938).
578. MORGAN, B. B., *Research*, **10,** 271 (1957).
579. MORGAN, B. B. and MEYER, E. W., *J. Sci. Inst.* **36,** 492 (1959).
580. MORGAN, G. B., OZOLINS, G. and TABOR, E. C., *Science*, **170,** 289 (1970).
581. MORITA, S., Operation and economy of the oxygen converter gas recovery process (OG Process), *I.S.I. Special Report* **83,** p. 109 (1964).
582. MORRIS, G. A. and JACKSON, J., *Absorption Towers*, Butterworths, London (1953).
583. MOSES, H. and CARSON, J. E., Stack design parameters influencing plume rise, Paper No. 67–84, *60th Ann. Meeting Air Pollut. Control Ass.*, Cleveland, June (1967). *J. Air Pollut. Control Ass.* **18** (7), 454–7 (1968).
584. MULCASTER, K. D., *Staub*, **21,** 302 (1961).
585. MULDER, H., reported by TER LINDEN, A. J., *Proc. Instn Mech. Engrs*, **160,** 233 (1949).
586. MÜLLER, P., *Chemie-Ingr-Tech.* **31,** 345 (1959).
586a. MÜLLER, R. H., *Analyt. Chem.* **40,** 109A, May (1968).
587. MUNROE, A. J. E. and MASDIN, E. G., *Brit. Chem. Engng*, **12,** 369 (1967).
588. MUNSON, J. S., *Chem. Engng*, **75,** (22), 147, 14th Oct. (1968).
589. MUSCHELKNAUTZ, E., *Chemie-Ingr-Tech.* **39,** 306–10 (1967).
590. MUSCHELKNAUTZ, E. and BRUNNER, K., *Chemie-Ingr-Tech.* **39,** 531–8 (1967).
591. NACOVSKY, W., *Combustion*, p. 35, Jan. (1967).
592. NAGEL, R., *Ver. Deut. Ing. Tagungsheft*, **3,** 25 (1954).
593. NATANSON, G. L., *Dokl. Akad. Nauk, SSSR* (Physical Chemistry Section), **112,** 100 (1957).
594. NATANSON, G. L., *Dokl. Akad. Nauk, SSSR* (Physical Chemistry Section), **112,** 696 (1957).
595. NATANSON, G. L., *Soviet Phys. Tech. Phys.* (Engl. trans.), **5,** 538 (1960).
596. NATANSON, G. L., *Kolloid Zh.* **24,** 52 (1962).
597. NATANSON, G. L. and USHAKOVA, E. N., *J. Phys. Chem.*, U.S.S.R. (Engl. trans.), **35,** 224 (1961).
598. NEUMANN, E. P. and NORTON, J. L., Ultrasonics—two symposia, *Chem. Engng Prog. Symp. Ser.*, No. 1, vol. 47, p. 4 (1951).
599. NICKEL, W., *Staub*, **23** (11), 508–12 (1963).
600. NICKLIN, T. and HOLLAND, B. H., Cleaning of coke oven gas by the Stretford Process, *Coke Oven Mgrs Ass.*, Cardiff, pp. 1–8, Jan. (1963).
601. NICKLIN, T. and HOLLAND, B. H., Removal of hydrogen sulphide from coke oven gas, *Symposium—Cleaning Coke Oven Gas*, Saarbrücken, pp. 1–20, Mar. (1963).
602. NIETRUCH, F. and PRESCHER, K. E., *Zeit. analyt. Chemie*, **226,** 259 (1967).
603. NONHEBEL, G., *Air Pollution*, Ed. THRING, M. W., chap. 1, Butterworths (1957).
604. NONHEBEL, G. and HAWKINS, J. E., *J. Inst. Fuel*, **28,** 530 (1955).
605. NORD, M., *Chem. Engng*, **62** (1), 236 (1955).
606. NORD, M., *Chem. Engng*, **57,** 116, Oct. (1950).
607. Nordac Ltd., Middlesex, England, private communication (1962).
608. NORMAN, W. S., *Absorption, Distillation and Cooling Towers*, Longmans, London (1961).
609. NOSOV, V. A., *Soviet Progress in Applied Ultrasonics* (1963), Consultants Bureau, New York (1965).
610. NUKIYAMA, S. and TANASAWA, Y., *Trans. Soc. Mech. Engrs* (Japan), **4** (14), 86 (1938). Quoted by LEWIS, H. C. *et al.*, *Ind. Engng Chem.* **40,** 67 (1948).
611. OAKES, B., *Int. J. Air Pollut.* **3,** 179 (1960).
612. OELSCHLÄGER, W., *Staub*, **25,** 528 (1965).
613. O'GARA, P. J., *Ind. Engng Chem.* **14,** 744 (1922).
614. O'KONSKI, C. T. and DOYLE, G. J., *Analyt. Chem.* **27,** 694 (1955).

615. Oldenkamp, R. D. and McKenzie, D. E., The molten carbonate process for control of sulphur emissions, *Air Pollut. Control Ass. Meeting*, St. Paul, June (1968).
616. Oldenkamp, R. D. and Margolin, E. D., *Chem. Engng Prog.* **65** (11), 73 (1969).
617. Owen, P. R., *Int. J. Air Pollut.* **3,** 1 (1961).
618. Oxy-Catalyst Inc. (Berwyn, Penn.), *Basic Engineering Principles of the Oxycat*, p. 18.
619. Pall, D. B., *Ind. Engng Chem.* **45,** 1197–1202 (1953).
620. Pallinger, J., *Staub*, **22,** 270 (1962).
621. Pangborn Corporation, Hagerstown, Maryland, U.S.A.
622. Parker, A., Air pollution legislation, in *Air Pollution*, p. 365, World Health Organisation, Geneva (1961).
623. Parkington, J. W. and Laurie Walker, S., *Colloque Internationaux du Centre National de la Recherche Scientifique*, Grenoble, 27th Sept. (1960). (Editions du Centre National de la Recherche Scientifique, Paris, 1961.)
624. Pauthenier, M. M., *Compt. Rend.* **240,** 1610 (1955).
625. Pauthenier, M. M. and Moreau-Hanot, M., *J. Phys. et Rad.*, Series 7, **3,** 590 (1932).
626. Pauthenier, M. M. and Moreau-Hanot, M., *J. Phys. et Rad.*, Series 7, **6,** 257 (1935).
627. Pearcey, T. and Hill, G. W., *Quart. J. Roy. Met. Soc.* **83,** 77 (1957).
628. Pearson, J. L., Nonhebel, G. and Ulander, P. H. N., *J. Inst. Fuel*, **8,** 119 (1935).
629. Pemberton, C. S., *Int. J. Air Pollut.* **3,** 168 (1960).
630. Pengelly, A. E. S., *J. Inst. Fuel*, **35,** 210 (1962).
630a. Penney, G. W., *J. Air Pollut. Control Ass.* **19,** 596–600 (1969).
631. Pereles, E. G., Safety in Mines Research Establishment, *U.K. Min. of Power*, Report No. 144 (1958).
632. Perry, J. H. (Ed.) *Chemical Engineer's Handbook*, 3rd ed., McGraw-Hill (1950).
633. Perry, M. G. and Handley, M. F., *Trans. Instn Chem. Engrs*, **45,** 367–71 (1967).
634. Persson, G. A., *Int. J. Air Wat. Pollut.* **10,** 845 (1966).
635. Peters, M. S., *Chem. Engng*, **62** (5), 197 (1955).
636. Peters, M. S. and Timmerhaus, K. D., *Plant Design and Economics for Chemical Engineers*, 2nd ed., McGraw-Hill, New York (1968).
637. Petersen, H., *Achema–Jahrbuch*, vol. II, Technical developments in chemical engineering (1965–1967) (reprint).
638. Petroll, J., *Freiberger Forschungshefte*, A **204,** 94 (1961).
639. Petroll, J. and Langhammer, K., *Freiberger Forschungshefte*, A **250,** 175 (1962).
640. Petroll, J., Quitter, V., Schade, G. and Zimmermann, L., *Staub Reinhalt. Luft*, **27,** 115–23 (1967).
641. Pettyjohn, E. S. and Christiansen, E. B., *Chem. Engng Prog.* **44,** 157 (1948).
642. Pich, J., *Staub*, **25,** 186 (1965).
643. Pich, J., Theory of aerosol filtration, in *Aerosol Science*, Ed. Davies, C. N., chap. 9, Academic Press, New York (1966).
644. Pich, J., *Staub Reinhalt. Luft*, **29,** 407 (1969).
645. Picknett, R. G., *Int. J. Air Pollut.* **3,** 160 (1960).
646. Phillips, P. H., The effects of air pollutants on farm animals, 8–8, in *Air Pollution Handbook*, Ed. Magill, P. L., Holden, F. R. and Ackley, C., McGraw-Hill, New York (1956).
647. Pigford, R. L. and Colburn, A. P., in Perry, J. H. (Ed.) *Chemical Engineer's Handbook*, 3rd ed., p. 675, McGraw-Hill, New York (1950).
648. Pigford, R. L. and Pyle, C., *Ind. Engng Chem.* **43,** 1649 (1951).
649. Pitelina, N. P., Chemical purification of effluent gases, Paper VII-119, pp. 119–23, *W.H.O. Int. Regional Seminar*, Moscow–Volgograd, 31st Aug.–20th Sept. (1967).
650. Plumley, A. L., Whiddon, O. D., Shutko, F. W. and Jonakin, J., Removal of SO_2 and dust from stack gases, *Am. Pwr Conf.*, Chicago, 25th–27th Apr. (1967).
651. Podoshernikov, B. F., *Zhur. Prikl. Khimi.* **34,** 2664 (1961).
652. Podoshernikov, B. F. and Tartakovskii, B. D., *Zhur. Prikl. Khimi.* **34,** 2573 (1961).
653. Pollock, W. A., Tomany, J. P. and Frieling, G., Paper 66 WA/CD-4, *Am. Soc. Mech. Engrs Winter Ann. Meeting*, New York (1966).
654. Potter, A. E., Harrington, R. E. and Spaite, P. W., Limestone dolomite process for flue gas desulphurization, *Am. Chem. Soc. Meeting*, Chicago, 11th Sept. (1967).
655. Powell, A. R. in *Chemistry of Coal Utilization*, Ed. Lowry, H. H., vol. 2, p. 921, Wiley, New York (1945).
656. Pozin, M. E., Muhklenov, I. P. and Tarat, E. Ya., *J. Appl. Chem., U.S.S.R.* **30,** 293 (1957).

657. Prat-Daniel (Stroud) Ltd., private communication.
658. PREINING, O., BAUMANN, R. and MOSER, F., *Staub Reinhalt. Luft*, **29,** 443 (1969).
659. PRESSLER, A. F., Ph.D. Thesis, Iowa State College, Iowa (1956).
660. PRIESTLEY, C. H. B., private communication (1959), based on *Quart. J. Roy. Met. Soc.* **82,** 165 (1956).
661. PRIESTLEY, C. H. B., private communication (1964).
662. PRING, R. T., *Air Repair*, **4,** No. 1, May (1954); *ibid.* **4,** No. 3, Nov. (1954) (reprinted).
663. PROCKAT, F., *Glasers Ann.* **106,** 73 (1930); *ibid.* **107,** 43, 47 (1930).
664. PSHENAI-SEVERIN, S. V., *Dokl. Akad. Nauk, SSSR*, **125,** 775 (1959).
665. Pulverizing Machinery Ltd., Parnall Rd., Fishponds, Bristol.
666. PYATT, F. B., *Enviromental Pollution*, **1,** 45 (1970).
667. RADUSHKEVICH, L. V., *J. Phys. Chem. U.S.S.R.* **32,** 282 (1958).
668. RAMMLER, E. and BREITLING, K., *Freiberger Forschungshefte*, A **56,** 5 (1957).
669. RAMSKILL, E. A. and ANDERSON, W. L., *J. Colloid Sci.* **6,** 416 (1951).
670. RANZ, W. E., *Tech. Report No.* **8,** 1st Jan. (1953), Univ. Ill. Engng Exptl Station.
671. RANZ, W. E. and JOHNSTONE, H. F., *J. Appl. Phys.* **26,** 244 (1956).
672. RANZ, W. E. and WONG, J. B., *Ind. Engng Chem.* **44,** 1371 (1952).
673. RAY, A. B., *United States Technical Conference on Air Pollution*, Ed. MCCABE, L. C., p. 355, McGraw-Hill, New York (1952).
674. RAYLEIGH, E. LORD, *Proc. Roy. Soc.* (London), **34,** 414 (1882–3).
675. RAYLEIGH, E. LORD, *Phil. Mag.* (5) B **4,** 59 (1892).
676. REED, L. E., TROTT, P. R. and SUTTON, S., *Removal of Sulphur Oxides from Flue Gas: The Reinluft Pilot Plant*, Report No. LR. 15 (AP), Ministry of Technology, Warren Spring Laboratory, Sept. (1965).
677. REES, R. LL., *J. Inst. Fuel*, **25,** 350 (1953).
678. REES, R. LL., *Instn Mech. Engrs Conf.*, on The Mechanical Engineers Contribution to Clean Air, p. 34 (1957).
679. REESE, J. T. and GRECO, J., *Mechanical Engineering*, **90** (10), 34–37 (1968).
680. REEVE, L., *J. Inst. Fuel*, **31,** 319 (1958). See also GREGORY, S. A., *Brit. Chem. Engng*, **5,** 340 (1960).
681. REGENER, H. V., *J. Geophys. Res.* **69,** 3795 (1964).
682. RICHARDS, J. C. in *Storage and Recovery of Particulate Solids*, Ed. RICHARDS, J. C., *Instn Chem. Engrs* (London) (1966).
683. RICHARDSON, J. F. and WOODING, E. R., *Chem. Engng Sci.* **7,** 51 (1957).
684. RICHARDSON, J. F. and ZAKI, W. N., *Trans. Instn Chem. Engrs*, **32,** 35 (1954).
685. RICHARDSON, J. F. and ZAKI, W. N., *Chem. Engng Sci.* **3,** 65 (1954).
686. RIETEMA, K., *Cyclones in Industry*, Ed. RIETEMA, K. and VERVER, C. G., pp. 46 and 85, Elsevier, Amsterdam.
687. RILEY, H. L., *Iron Steel Inst.*, Special Report No. 61, p. 129 (1958).
687a. RJAZANOV, V. A., *W.H.O. Inter-regional Seminar*, Moscow, Volgograd, 31st Aug.–20th Sept., pp. 53–56 (1967).
688. ROBINSON, E. and ROBBINS, E. C., Sources, abundance and fate of gaseous atmospheric pollutants, Chap. I, *Air Pollution Control*, vol. II, Ed. STRAUSS, W., Wiley–Interscience, New York (1971).
689. ROBINSON, M., *Anal. Chem.* **33,** 109–13 (1961).
690. ROBINSON, M., *Am. J. Physics*, **30,** 366–72 (1962).
691. ROBINSON, M., The role of turbulence in electrostatic precipitation, Paper No. 67-34, *60th Ann. Meeting, Air Pollut. Control Ass.*, Cleveland, 11th–16th June (1967).
692. ROBINSON, M., *Atmospheric Environment*, **1,** 193–204 (1967).
693. ROBINSON, M., *I.E.E.E. Trans.* (Power Apparatus and Systems), **86** PAS, 185–9 (1967).
694. ROBINSON M., *J. Electrochem Soc.* **115,** 131C, May (1968).
695. ROBINSON, M., *J. Air Pollut. Control Ass.* **18,** 235–9 (1968).
696. ROBINSON, M., *J. Appl. Phys.* **40,** 5107–12 (1969).
697. ROBINSON, M., Electrostatic precipitation, in *Air Pollution Control*, Ed. STRAUSS, W., pt. I, chap. 5, Wiley–Interscience, New York (1971).
698. ROGERS, S. M. and EDELMAN, S., *A Digest of State Air Pollution Laws*, U.S. Dept. of H.E.W., Washington 25 D.C. (1962).
699. ROHMANN, H., *Z. Physik*, **17,** 253 (1923).
700. ROSCOE, R., *Brit. J. Appl. Phys.* **3,** 267 (1952).

701. ROSE, A. H., STEPHAN, D. G. and STENBURG, R. L., in *Air Pollution*, p. 307, World Health Organisation, Geneva (1961).
702. ROSE, H. E., *J. Appl. Chem.* **2,** 217 (1952).
703. ROSE, H. E. and SULLIVAN, R. M. E., *Nature*, **184,** 47 (1959).
704. ROSE, H. E. and WOOD, A. J., *An Introduction to Electrostatic Precipitation in Theory and Practice*, Constable, London, 1st ed. (1956), 2nd ed. (1963).
705. ROSENBLATT, P. and LA MER, V. K., *Phys. Rev.* **70,** 385 (1946).
706. ROSIN, P., RAMMLER, E. and INTELMANN, W., *Zeit Ver. Deut. Ing.* **76,** 433 (1932).
707. ROSSANO, A. T. and PILAT, M. J., Recent developments in the control of air pollution from primary aluminium smelters in the United States, Paper En-16-F, *2nd Int. Congress of the Int. Union of Air Pollut. Prevention Assns*, 6th-11th Dec., Washington D.C. (1970).
708. ROSSANO, A. J. and SILVERMAN, L., *Heating and Ventilating*, p. 102, May (1945).
709. ROUSSEL, A. A. and STEPHANY, H., Continental Europe Report, Paper II/6, pp. 29–34, *Int. Clean Air Cong.*, Proc. Pt. I, London, Oct. (1966).
710. ROZOVSKY, H., *Can. Mining and Met. Bull.* **48,** 486 (1955).
711. RUCKELSHAUSEN, K., Über die Beseitigung von Staubansatzen auf technisch glatten Oberflächen durch Klopfen oder Vibrieren, Dr. Ing. Thesis, Technische Hochschule, Stuttgart (1957).
712. RUDORFER, H. and KRAMSER, K., Austrian Pat. No. 184,992 (1956).
713. RUFF, R. J., *Chem. Engng Prog.* **53,** 377 (1957); *Industrial Gas*, Oct. (1955) (reprint).
714. RUFF, R. J. and SUTER, H. R., U.S. Pat. No. 2,658,748 (1950).
715. RUFF, R. J. in *Air Pollution*, Ed. STERN, A. C., vol. 2, p. 356, Academic Press, New York (1962).
716. RUMPF, H., BORHO, K. and REICHERT, H., *Staub Reinhalt. Luft*, **29,** 270–2 (1969).
717. RYDER, C. and SMITH, A. V., *Inst. Gas Engng Comm.* 624, **1** (1962).
718. ST. CLAIR, H. W., *Rev. Sci. Inst.* **12,** 250 (1941).
719. ST. CLAIR, H. W., *Ind. Engng Chem.* **41,** 2434 (1949).
720. ST. CLAIR, H. W., in *United States Technical Conference on Air Pollution*, Ed. MCCABE, p. 382, McGraw-Hill, New York (1952).
721. ST. CLAIR, H. W., SPENDLOVE, M. J. and POTTER, E. V., *U.S. Bur. Mines*, Report of Investigation No. 4218, Mar. (1948).
722. Sartorius Membranfilter G.m.b.H., Göttingen, Germany.
723. SAXTON, R. L. and RANZ, N. E., *J. Appl. Phys.* **23,** 917 (1952).
724. SAYERS, J. F., *Howden Quarterly*, No. 45, p. 21 (1961).
725. SCARINGELLI, F. P., SALTZMAN, B. E. and FREY, S. A., (i) Spectrophotometric determination of atmospheric sulphur dioxide. (ii) Effects of various parameters on the spectrophotometric determination of sulphur dioxide with pararosaniline, *Div. of Water, Air and Waste Chemistry ACS*, Atlantic City, 13th Sept. (1965).
726. SCARINGELLI, F. P., SALTZMAN, B. E. and FREY, S. A., Spectrophotometric determination of SO_2 in the atmosphere with pararosaniline, *Div. of Water, Air and Waste Chemistry ACS*, Pittsburgh, 23rd Mar. (1966).
727. SCHADT, C. F. and CADLE, R. D., *J. Colloid Sci.* **12,** 356 (1957).
728. SCHAUFLER, E. and ZENNECK, Deutsches Bundes Pat. No. 1,092,281, 24th Jan. (1953).
729. SCHIELE, O., *Ver. Deut. Ing. Tagungsheft*, **3,** 20 (1954).
730. SCHILLER, L. and NAUMANN, A., *Z. Ver. Deut. Ing.* **77,** 318 (1933).
731. SCHMIDT, K. R., *Staub*, **23** (11), 491–501 (1963).
732. SCHMITT, K. H., *Z. Naturforsch*, **14a**, 870 (1959).
733. SCHMITT, K. H., *Z. Naturforsch*, **14a**, 870 (1959).
734. SCHMIDT, W. A., *J. Ind. Engng Chem.* **4,** 719 (1912).
735. SCHMIDT, W. A. and ANDERSON, E., *Elect. Engng*, Aug. (1938) (reprint).
736. SCHMIEDEL, J., *Physik Z.* **29,** 593 (1928).
737. SCHNABEL, W., *Staub Reinhalt. Luft*, **28** (11), 449 (1968).
738. SCHNEIDER, E., *Mitt. Ver. Grosskesselbesitzer*, **80,** 354 (1962).
739. SCHNITZLER, H., *Arch. Eisenhüttenwes.* **24,** 199 (1953).
740. SCHNITZLER, H., *Staub*, **23,** 78 (1963).
741. SCHÖNBECK, R., Austrian Pat. No. 188,723 (1957).
742. SCHUSTER, H., *Deutsche Luft- und Raumfart Forschungsbericht*, 67–35 (1967), quoted by LÖFFLER, F.
743. SCHWARTZ, E. and WEPPLER, R., *Siemens Zeitchrift*, **31,** 607 (1957).

744. Schwarz, K., Die Staubemissionen kohlegefeurter Dampfkessel Grossanlagen in der Bundesrepublik Deutschland. Paper V/8, Proc. Pt.I, *Int. Clean Air Cong., London*, pp. 136–41, Oct. (1966).
745. Scott, D. S. and Guthrie, D. A., *Can J. Chem. Engng*, **37**, 200 (1959).
746. Scurfield, G., *Forestry Abstracts*, **21** (3), 1–20 (1960).
747. Seamens, F. L., *Air Pollution Handbook*, Ed. Magill *et al.*, McGraw-Hill, New York (1956).
748. Seidel, W. and Kaufmann, W., *Staub*, **24**, 405–7 (1964).
749. Seidman, E. B., *Anal. Chem.* **30**, 1680 (1958).
750. Sell, W., *Ver. Deut. Ing. Forschungsheft*, 347 (1931).
751. Semrau, K. T., Emission of fluorides from industrial processes, *130th Ann. Meeting, Am. Chem. Soc.*, Atlantic City, Sept. (1956).
752. Semrau, K. T., *J. Air Pollut. Control Ass.* **10** (3), 200 (1960).
753. Semrau, K. T., *Staub*, **22**, 184 (1962).
754. Semrau, K. T., Marynowski, C. W., Lunde, K. E. and Lapple, C. E., *Ind. Engng Chem.* **50**, 1615 (1958).
755. Septier, L. G., *Iron Steel Inst.*, Special Report No. 61, p. 74 (1958).
756. Sfindex, S. A. (Sarnen, Switzerland), Brit. Pat. No. 697,918 (1950).
757. Shale, C. C., *private communication* (1968).
758. Shale, C. C., Ash accumulation on precipitator discharge wires, Paper 68–102, *Air Pollut. Control Ass. Ann. Conf., St. Paul, Minn., 23rd–27th June* (1968).
759. Shale, C. C., Bowie, W. S., Holden, J. H. and Strimbeck, G. R., Characteristics of positive corona for electrical precipitation at high temperatures and pressures, *U.S. Bur. Mines, Report of Investigation No. 6397* (1964).
760. Shale, C. C. and Fasching, G. E., Operating characteristics of a high temperature electrostatic precipitator, *U.S. Bur. Mines, Report of Investigation No. 7276* (1969).
761. Shale, C. C., Holden, J. H. and Fasching, G. E., Electrical resistivity of fly ash at temperatures to 1500°F, *U.S. Bur. Mines, Report of Investigation No. 7041* (1968).
762. Shale, C. C. and Moore, A. S., *Combustion*, p. 42, Dec. (1960).
763. Shale, C. C., Simpson, D. G. and Lewis, P. S., Removal of sulphur and nitrogen oxides from stack gases by ammonia, *A.I.Ch.E. Ann. Meeting, Washington, D.C., 16th Nov.* (1969).
764. Shaw, F. M., *Fuel Econ. Rev.* **37**, 41 (1959).
765. Shepherd, G. B. and Lapple, E. C., *Ind. Engng Chem.* **31**, 972 (1939).
766. Sherlock, R. H. and Stalker, E. A., *Univ. Mich. Engng Research Bull.*, No. 29 (1941).
767. Sherwood, T. K., U.S. Pat. No. 2,833,370 (1958).
768. Sherwood, T. K. and Pigford, R. L., *Absorption and Extraction*, see particularly pp. 270, 282–5, 347, 368, McGraw-Hill, New York (1952).
769. Silverman, L., *Eighth A.E.C. Air Cleaning Conf.*, Oak Ridge, TID 7677, p. 177 (1963).
770. Silverman, L., Conners, E. W. and Anderson, D. M., *Ind. Engng Chem.* **47**, 962 (1955).
771. Silverman, L., Connors, E. W. and Anderson, D. M., Electrostatic mechanisms in aerosol filtration by mechanically charged fabric media, *Harvard School of Public Health, Air Cleaning Laboratory, Report* (1956).
772. Sinclair, D., *O.S.R.D. Report No. 865* (1942) Section VA, *J. Air Pollut. Control Ass.* **17**, 105 (1967). See also *Handbook on Aerosols*, Washington, D.C., U.S. A.E.C. (1950).
773. Sinclair, D. and La Mer, V. K., *Chem. Revs.* **44**, 2 and 5 (1949).
774. Sisk, F. J., private communication (1970).
775. Sjenitzer, F., *Chem. Engng Sci.* **1**, 101 (1952).
776. Skinner, D. G. and Boas-Traube, S., Symposium on particle size analysis, *Instn Chem. Engng* and *Soc. Chem. Ind.*, p. 44, Feb. (1947).
777. Skrebowski, J. K. and Sutton, B. W., *Brit. Chem. Engng*, **6**, 12 (1961).
778. Sly, W. W. Mfg. Co., Cleveland 1, Ohio, Bull. 104.
779. Smith, Sir E., *Proceeding of the Conference on the Mechanical Engineers Contribution to Clean Air*, p.1, London, Feb. (1957).
780. Smith, M. E. *et al.*, Recommended guide for the prediction of the dispersion of airborne effluents, *Am. Soc. Mech. Engrs*, New York, May (1968).
780a. Smith, P. L. and Penney, G. W., *Trans. Am. Instn Elect. Engrs* (Comm. Elect.), **80** I, 340 (1961).
781. Smoluchowski, M. von, *Physik Z.* **557**, 585 (1916); *Z. Physik. Chemie*, **92**, 129 (1918).
782. Societa Industriale Cataversa per Azioni, French Pat. (cl. C. 01 c), Oct. (1966).
783. Soderberg, C. R., *Iron Steel Engng*, **29** (2), 87 (1952).

784. Somers, E. V., Dispersion of pollutants emitted into the atmosphere, in *Air Pollution Control*, vol. I, chap. 1, Ed. Strauss, W., Wiley–Interscience, New York (1971).
785. Spaite, P. W., Stephan, D. G. and Rose, A. H., *J. Air Pollut. Control Ass.* **11** (5), May (1961).
786. Sparks, L. E. and Pilat, M. J., *Atmospheric Environment*, **4**, 651–60 (1970).
787. Sperling, G., Eine Theorie der Haftung von Feststoffteilchen an festen Körpern, Dr. Ing. Thesis, Technische Hochschule, Karlsruhe (1964).
788. Spielman, L. and Goren, S., *Env. Sci. Tech.* **2**, 279 (1968).
789. Spiers, H. M. (Ed.) *Technical Data on Fuels*, 6th ed., British National Committee of the World Power Conf., London (1961).
790. Sproull, W. T., *Ind. Engng Chem.* **47**, 940 (1955).
791. Sproull, W. T., *A.I.E.E. Preprint Winter General Meeting*, Paper No. CP57-46, 21st–25th Jan. (1957).
792. Sproull, W. T., *J. Air Pollut. Control Ass.* **10**, 307 (1960).
793. Sproull, W. T., *J. Air Pollut. Control Ass.* **16** (8), 439–41 (1966).
794. Sproull, W. T. and Nakada, Y., *Ind. Engng Chem.* **43**, 1350 (1951).
795. Spurny, K. and Lodge, J. P., *Staub Reinhalt. Luft*, **28**, 179, 503 (1968).
796. Squires, A. M., *Chem. Engng* **74**, 133, 20th Nov. (1967).
797. Squires, A. M., Cyclic use of calcined dolomite, in *Advances in Chemistry*, **69**, pp. 205–9, Fuel Gasification ACS (1967).
798. Stafford, E. and Smith, W. J., *Ind. Engng Chem.* **43**, 1346 (1951).
799. Stairmand, C. J., Symposium on particle size analysis, *Instn Chem. Engrs* and *Soc. Chem. Ind.*, p. 77, 4th Feb. (1947).
800. Stairmand, C. J., *Engineering*, **168**, 409 (1949).
801. Stairmand, C. J., *Trans. Instn Chem. Engrs*, **28**, 130 (1950).
802. Stairmand, C. J., *Engineering*, **171**, 585 (1951).
803. Stairmand, C. J. *Trans. Instn Chem. Engrs*, **29**, 356 (1951).
804. Stairmand, C. J., *J. Inst. Fuel*, **29**, 58 (1956).
805. Stairmand, C. J., *The Chemical Engineer*, CE 310 (1965).
806. Stairmand, C. J. and Kelsey, R. N., *Chemistry and Industry*, p. 1324 (1955).
807. Standard Filterbau G.m.b.H., 44 Münster (Westf.), Germany.
808. Stanton, T. E. and Panell, J. R., *Phil. Trans.* A **214**, 199–224 (1914).
809. Stechkina, I. B., *Dokl. Akad. Nauk, SSSR*, **167**, 1327 (1965).
810. Stefan, J., *Sitzb. Wien Akad. Wiss.* **77**, 371 (1878); **99**, 161 (1879).
811. Stephan, D. G. and Walsh, G. W., *Ind. Engng Chem.* **52**, 999 (1960).
812. Stephens, E. J. and Morris, G. A., *Chem. Engng Prog.* **47**, 232 (1951).
813. Sterling, P. H. and Ho, H., *Ind. Engng Chem.* **53** (6), 52A (1961).
814. Stern, A. C., Air pollution standards, in *Air Pollution*, Ed. Stern, A. C., vol. 2, p. 451, Academic Press (1962).
815. Stern, A. C., Analysis, monitoring and surveying, in *Air Pollution*, vol. II, 2nd ed., Academic Press, New York (1968).
816. Stern, S. C., Zeller, H. W. and Schekman, A. I., *J. Colloid Sci.* **15**, 546 (1960).
817. Stevens, R. K. and O'Keeffe, A. E., *Anal. Chem.* **42** (2), 143A (1970).
818. Stites, J. G., Horlacker, W. R., Bachofer, J. L. C. and Bartman, J. S., The catalytic-oxydation system for removing SO_2 from flue gas, *Monsanto Report*, St. Louis (1969).
819. Stokes, C. A., *Chem. Engng Prog.* **46**, 423 (1950).
820. Stokes, G. G., *Trans. Camb. Phil. Soc.* **9**(ii), 8 (1850).
821. Storch, O., *Staub Reinhalt. Luft*, **26**, 479 (1966).
822. Strauss, W., unpublished calculations.
823. Strauss, W., Ph.D. Thesis, Univ. of Sheffield (1959).
824. Strauss, W., Aust. Pat. Application 41,506/68, 1st Aug. (1968).
825. Strauss, W., The control of sulphur emissions from combustion processes, in *Air Pollution Control*, Ed. Strauss, W., vol. I, chap. 3, p. 95, Wiley–Interscience, New York (1971).
826. Strauss, W., unpublished experiments.
827. Strauss, W., Cyclone separator, Aust. Pat. No. 403,736, 22nd June (1970).
828. Strauss, W. and Cock, W. H., Paper No. 1 : 1, *2nd Sydney Clean Air Conf.* (1965).
829. Strauss, W. and Lancaster, B. W., *Atmospheric Environment*, **2**, 135 (1968).
830. Strauss, W. and Lancaster, B. W., Condensation effects in scrubbers, chap. 7 in *Air Pollution Control*, vol. I, Ed. Strauss, W., Wiley–Interscience, New York (1970).

831. Strauss, W. and Thring, M. W., *J. Iron Steel Inst.* **196,** 62 (1960).
832. Strauss, W. and Thring, M. W., *J. Iron Steel Inst.* **193,** 216 (1959).
833. Strauss, W. and Thring, M. W., *J. Iron Steel Inst.* **196,** 62 (1960).
834. Strauss, W. and Thring, M. W., *Trans. Instn. Chem. Engrs,* **41,** 248 (1963).
835. Strauss, W. and Woodhouse, G., *Brit. Chem. Engng,* **3,** 620 (1958).
836. Streight, H. R. L., *Can. J. Chem. Engng,* **36,** 3 (1958).
837. Strohm, G. H., Atmospheric dispersion of stack effluents, in *Air Pollution,* 2nd ed., Ed. Stern, A. C., vol. I, p. 227, Academic Press, New York and London (1968).
838. Stuckmeier, H., *Staub Reinhalt. Luft,* **28,** 200–2 (1968).
839. Sturtevant Engineering Co. Ltd., Southern House, Cannon St., London, E.C.4, "Ludgate Collector".
840. Sullivan, J. L., Kafka, F. L. and Pottinger, J., Paper 12, *Sydney Clean Air Conf.* (1965).
841. Sullivan, R. R. and Hertel, K. L., *J. Appl. Phys.* **11,** 761 (1940).
842. Sundelöf, L. O., *Staub Reinhalt. Luft,* **27,** 358 (1967).
843. Sutton, O. G., *Quart. J. Roy. Meteorol. Soc.* **73,** 426 (1947).
844. Sutton, O. G., *J. Met.* **7,** 307 (1950).
845. Svanda, J., *Int. Chem. Engng,* **7** (2), 238–45 (1967).
846. Swift, James & Son Ltd., London. Leaflet on Graticules.
847. Taga, Tahakide, Continental Asia: Report Paper II/4, *Proceedings,* vol. I, pp. 22–24, *Int. Clean Air Congress,* Pt. I, London, Oct. (1956).
848. Taggart, A. F., *Handbook of Mineral Dressing,* Chapman & Hall, p. 9-07 (1947).
849. Taheri, M. and Calvert, S., *J. Air Pollut. Control Ass.* **18,** 240 (1968).
850. Taigel, P. G., Safety in Mines Research Establishment (Sheffield), *Report No. 48* (1952).
851. Taylor, J. R., Hasegawa, A. and Chambers, L. A., *Air Pollution,* p. 293, World Health Organisation, Geneva (1961).
852. Teller, A. J., *Chem. Engng Prog.* **63** (3), 75 (1967).
853. Terabe, M. and Oomichi, S., Relationships between sulphur dioxide concentrations determined by West-Gaeke and conductivity methods, Paper No. 67–108, presented at *60th Air Pollut. Control Ass. Meeting, Cleveland, 11th June* (1967).
854. Theodorsen, T. and Regier A., *N.A.C.A. Report* 793 (1944).
855. Thom, A., *Proc. Roy. Soc.,* London, **41A,** 651 (1933).
856. Thomas, D. G. and Lapple, C. E., *A.I.Ch.E.J.* **7,** 203 (1961).
857. Thomas, J. W. and Yoder, R. E., *A.M.A. Arch. Ind. Health,* **13,** 545, 550 (1956).
858. Thomas, M. D., *Air Pollution,* p. 233, World Health Organisation, Geneva (1961).
859. Thomas, M. D. and Hill, G. R., *Plant Physiol,* **10,** 291 (1935).
860. Thompson, B. W. and Strauss, W., *Chem. Engng Sci.* **26,** 35–45 (1971).
861. Thring, M. W., *J. Inst. Fuel,* **12,** 558 (1938).
862. Thring, M. W., *Science of Flames and Furnaces,* 2nd ed., Chapman & Hall, London (1962).
863. Timoshenko, V. I., *Akust. Zhur.* **11** (1) (1965).
864. Torgeson, W. L., quoted by Stern, A. C., Zeller, H. N. and Schekman, A. I., *J. Colloid Sci.* **15,** 546 (1960).
865. Torobin, L. B. and Gauvin, W. H., *Can. J. Chem. Engng,* **37,** 129, 167, 224 (1959); **38,** 142, 189 (1960); **39,** 113 (1961).
866. Townsend, J. S., *Electricity in Gases,* p. 376, Oxford (1915).
867. Troost, N., *Proc. I.E.E.* **101,** 369 (1954).
868. Tuorila, P., *Kolloidchem. Beih.* **24,** 1 (1927).
869. Turitzin, A. M., *J. Structural Div. A.S.C.E.* **89** (SC2), Proc. Paper 3479, 49–73 (1963).
870. Turk, A. in *Air Pollution,* Ed. Stern, A. C., vol. 2, p. 384, Academic Press, New York (1962).
871. Turnbull, S. G., Benning, A. F., Feldmann, G. W., Lynch, A. L., McHarness, R. C. and Richards, M. K., *Ind. Engng Chem.* **39,** 286 (1947).
872. Tverskoi, P. N., *Physics of the Atmosphere; a Course in Meteorology,* Israel Programme for Scientific Translations, pp. 315–27 (1965).
873. Twomey, S., *J. Phys. Chem.* **30,** 941 (1959).
874. Tyndall, J., *Proc. Roy. Inst.* **6,** 3 (1870).
875. U.K. Ministry of Housing and Local Government Clean Air Act (1956), *Memorandum on Chimney Heights,* H.M.S.O., London (1963); 2nd ed. (1967).
876. Ulrich, H., *Mitt. Ver. Grosskesselbesitzer,* **81,** 413 (1962).

877. UMNEY, L. E. R., National Gas Turbine Est., Pyestock. *Report No. R33* (1948) (quoted by DANIELS, T. C.).
878. Union Carbide Corpn, Linde Div., Catalogue F-1026-B, *Linde Molecular Sieves.*
879. Union Carbide Corpn, National Carbon Division Catalogue.
880. United States of America, Public Law 90–148, 90th Congress, S. 780, 21st Nov. (1967).
881. U.S.S.R. Air Quality Standards (1967), *W.H.O. Inter-Regional Seminar*, Moscow, Volgograd, pp. 165–71, 31st Aug. to 20th Sept. (1967).
882. UZHOV, V. N., *Purification of Industrial Exhaust Gases*, Goskhimizdat, Moscow (1959).
883. VAJDA, S., *Iron Steel Engnr*, **29** (7) 111 (1952).
884. VALENTIN, F. H. H., *33rd Int. Cong. Ind. Chemistry*, Toulouse. Quoted by *Brit. Chem. Engng.*
885. VAN DYKE, M., *Perturbation Methods in Fluid Mechanics*, chap. V, Academic Press, New York (1964).
886. VELICHKO, M. V., and RADUSHKEVICH, L. V., *Dokl. Akad. Nauk, SSSR*, **154**, 415 (1964).
887. VERMEULEN, T. in *Advances in Chemical Engineering*, vol. 2, p. 148, Wiley, New York (1958).
888. Visco Engineering Co. Ltd., Stafford Road, Croydon, Surrey.
889. Vokes Ltd., Henley Park, Guildford, Surrey, England.
890. VOLLHEIM, G. and DOMIN, G., *Catalytic After-Burning*, Schilde Schriftenreihe, Band 10, Mar. (1966).
891. VOLMER, M., *Kinetik der Phasenbildung*, Diedrich Steinkopff Verlag, Leipzig (1939).
892. VONNEGUT, B., *J. Acoust. Soc. Am.* **26**, 18 (1954).
893. VONNEGUT, B., *J. Acoust. Soc. Am.* **27**, 430 (1955).
894. WADELL, H., *J. Frank. Inst.* **217**, 459 (1934).
894a. WAHNSCHAFFE, E., *Mitt. Ver. Grosskesselbesitzer*, **83**, 72 (1963).
895. WAID, D. E., Effective methods of direct flame incineration of organic materials, *Air Pollut. Control. Ass.*, East Central Section, Youngstown, Sept. (1968).
896. WALDMANN, L., *Z. Naturforsch.* **14A**, 589 (1959).
897. WALDMANN, L. and SCHMITT, K. H., Thermophoresis and diffusiopIoresis of aerosols, chap. 6 in *Aerosol Science*, Ed. DAVIES, C. N., Academic Press, London (1966).
898. WALKER, A. B., Enhanced scrubbing of black liquor boiler fume by electrostatic pre-agglomeration, *56th Ann. Meeting of Air Pollut. Control Ass.*, Detroit, Mich., 9th June (1963).
899. WALKER, A. B. and HALL, R. M., *J. Air Pollut. Control Ass.* **18**, 319 (1968).
900. WALLIS, E., *Brit. Chem. Engng*, 7, **833** (1962).
901. WALTER, E., *Staub*, **42**, 678 (1955).
902. WALTER, E., *Staub*, **48**, 14 (1957).
903. WALTER, E., *Staub*, **53**, 88 (1957).
904. WALTON, W. H., Symposium on particle size analysis, *Trans. Instn Chem. Engrs.* and *Soc. Chem. Ind.*, p. 136, Feb. (1947).
905. WALTON, W. H. and WOODCOCK, A., *Int. J. Air Pollut.* **3**, 129 (1960).
906. WANDT, C. J. and DAILEY, L. W., *Hydrocarbon Process*, **46** (10), 155 (1967).
907. WATKINS, E. R. and DARBY, K., The application of electrostatic precipitation to the control of fume in the steel industry, *I.S.I. Special Report* No. 85, pp. 24–35 (1964).
909. WATSON, K. S. and BLECHER, K. J., Investigation of electrostatic precipitators for large P.F. fired boilers, *Clean Air Conf., Sydney*, Paper 2.10 (1965).
910. WEBB, R. L., *Paint Industry Magazine*, Jan. (1958); *ibid.* Oct. (1958).
911. WEBER, E., *Staub*, **20**, 338 (1960).
912. WEBER, E., *Staub Reinhalt. Luft*, **28**, 462 (1968).
913. WEBER, K., *Z. Naturforsch.* **1**, 217 (1946).
914. WEIDNER, G., *Ver. Deut. Ing. Tagungsheft*, **3**, 16 (1954).
915. WELLMAN, F., *Feuerungstechnik*, **26**, 137 (1938).
916. WELLS, A. C. and CHAMBERLAIN, A. C., *Brit. J. Appl. Phys.* **18**, 1793–9 (1967).
917. WERNER, K. D., *Chem. Engng* **75** (25) 179–84, 4th Nov. (1968).
918. WEST, P. W. and GAEKE, G. C., *Anal. Chem.* **28**, 1816 (1966). WEST, P. W., GAEKE, G. C., NAUMAN, R. V. and TRON, F., *ibid.* **32**, 1307 (1960).
919. WESTERBOER, I., *Staub*, **21**, 466 (1961).
920. WESTERN Precipitation Corpn Design, Los Angeles 54, Calif., U.S., *Methods for the Determination of Velocity, Volume, Dust and Mist Content of Gases*, Bull. WP-50, 6th ed. (1958).
921. WESTINGHOUSE Electric Corpn, Pittsburgh, Pa. 15222.
922. WHEELER, A., *Advances in Catalysis*, Ed. Frankenburg, W. G. *et al.*, vol. 3, p. 250, Academic Press, New York (1951).

923. WHITBY, K. T., *Am. Soc. Heat. Refrig. Air Cond. Engrs J.* **7** (9), 56–65 (1965).
924. WHITE, C. M., *Proc. Roy. Soc.* (London), A **186**, 472 (1946).
925. WHITE, H. J., *Trans. A.I.E.E.* **70, II,** 1186 (1951).
926. WHITE, H. J., *Trans. A.I.E.E.* **71, I,** 326 (1952).
927. WHITE, H. J., *Air Repair*, **3** (2), 79 (1953).
928. WHITE, H. J., *Ind. Engng Chem.* **47,** 932 (1955).
929. WHITE, H. J., *Chem. Engng Prog.* **52,** 244 (1956).
930. WHITE, H. J., *Air Pollut. Control Ass. Silver Jubilee Meeting*, June (1957), Paper 57–35 (Reprint).
931. WHITE, H. J., *Colloques Internationaux du Centre National de la Recherche Scientifique*, Grenoble, 27th Sept. (1960). (Editions du Centre National de la Recherche Scientifique, Paris, 1961.)
932. WHITE, H. J., *Industrial Electrostatic Precipitation*, Addison-Wesley, Reading, Mass. (1963).
933. WHITE, H. J. and BAXTER, W. A., *Am. Soc. Mech. Engrs*, Paper No. 59-A-279 (1959).
934. WHITE, H. J. and COLE, W. H., *52nd Air Pollut. Control Ass. Meeting*, Paper 59-48, June (1959).
935. WHITE, P. A. F. and SMITH, S. E. (Eds.) *High Efficiency Air Filtration*, Butterworths, London (1964).
936. WHITELEY, A. B. and REED, L. E., *J. Inst. Fuel*, **32,** 316 (1959).
937. WHITMAN, W. G., *Chem. Met. Eng.* **29,** 146 (1923).
938. WHYNES, A. L., *Trans. Instn Chem. Engrs* (London), **34,** 118 (1956).
939. WHYTLAW-GRAY, R. and PATTERSON, H. S., *Smoke*, E. Arnold, London (1932).
940. WICKERT, K., *Mitt. Ver. Grosskesselbesitzer*, **83,** 74 (1963).
941. WIESELSBERGER, C., *Phys. Z.* **23,** 219 (1922).
942. WIGGINS, E. J., CAMPBELL, W. B. and MAASS, O., *Can. J. Res.* **17,** 318 (1939).
943. WILKE, C. R. and HOUGEN, O. A., *Trans. Am. Instn, Chem. Engrs*, **41,** 445 (1945).
944. WILLIAMS, D. J., *Coal Res. C.S.I.R.O.* **23,** 7, July (1964). (Quoting average equilibrium constants based on Bodenstein and Pohl, Kapustinsky and Shamovsky, and Evans and Wagman.)
945. WILLIAMS, J. L. and JACKSON, R., *Third Cong. of the European Federation of Chemical Engineering —Section*, "The Interaction Between Fluids and Particles", P.C. 51, 22nd June (1962).
946. WILLIAMS, R., *Chem. Engng*, **54** (12), 124 (1947).
947. WILSON, B. W., *Aust. J. Appl. Sci.* **4,** 47 (1953).
948. WINKEL, A. and SCHÜTZ, A., *Staub*, **22,** 343 (1962).
949. WOLF, E. F., VON HOHENLEITEN, H. L. and GORDON, M. B., *Int. Clean Air Conf., London*, p. 239 (1959).
950. WONG, J. B. and JOHNSTONE, H. F., *Univ. Ill. Eng. Exp. Station Report* No. 11 (1953).
951. WOOD, C. W., *Trans. Instn Chem. Engrs*, London, **38,** 54 (1960).
952. WOOLLAM, J. P. V. and JACKSON, A., *Chem. Engng Group of Soc. Chem. Ind.* **27,** 43 (1945).
953. WORLD HEALTH ORGANIZATION, *Inter-Regional Seminar on Air Quality and Methods of Measurement*, Report WHO/AP/23 (1963).
954. WORLD HEALTH ORGANIZATION, *Expert Committee on Atmospheric Pollutants*, W.H.O. Tech. Rep. Ser. 271 (1964).
955. WORLD HEALTH ORGANIZATION, *Air Pollution—A Survey of Existing Legislation*, W.H.O., Geneva, p. 20 (1963).
956. YAVORSKY, P. M., MAZZOCCO, N. J., RUTLEDGE, G. D. and GORIN, E., *Env. Sci. Tech.* **4,** 757 (1970).
957. YOCOM, J. E., *Chemical Engineering*, p. 103, 23rd July (1962).
958. ZAHN, R., *Staub*, **21,** 56 (1961).
959. ZAHN, R., *Staub*, **23,** 343–52 (1963).
960. ZEBEL, G., *Kolloid Zh.* **157,** 37 (1958).
961. ZEBEL, G., Coagulation of aerosols, in *Aerosol Science*, chap. 2, Ed. DAVIES, C. N., Academic Press, London (1966).
962. ZEMSHOV, I. F., STEPANOV, A. S. and MELKIKH, A. V., *J. Appl. Chem. U.S.S.R.* **35,** 11 (1962) (reported by Taheri and Calvert[849]).
963. ZENTGRAF, K. M., *Staub Reinhalt. Luft*, **28,** 94 (1968).
964. ZERNICK, W., *Brit. J. Appl. Phys.* **8,** 117 (1957).
965. ZYSK, E. D. and TOENSHOFF, D. A., Calibration of refractory metal thermocouples, ISA Paper 12.11-4-66, *ISA 21st Ann. Conf.*, New York (1966). See also *Inst. Tech.* **14,** 49 (1967) *and Engelhard Ind. Tech., Bull.* **7,** 137 (1967).

AUTHOR INDEX

SUBJECT INDEX

Bold figures *denote major treatment of topic*

THE POLYTECHNIC OF WALES
LIBRARY
TREFOREST